Smart Innovation, Systems and Technologies

Volume 35

Series editors

Robert J. Howlett, KES International, Shoreham-by-sea, UK
e-mail: rjhowlett@kesinternational.org

Lakhmi C. Jain, University of Canberra, Canberra, Australia
e-mail: Lakhmi.jain@unisa.edu.au

About this Series

The Smart Innovation, Systems and Technologies book series encompasses the topics of knowledge, intelligence, innovation and sustainability. The aim of the series is to make available a platform for the publication of books on all aspects of single and multi-disciplinary research on these themes in order to make the latest results available in a readily-accessible form. Volumes on interdisciplinary research combining two or more of these areas is particularly sought.

The series covers systems and paradigms that employ knowledge and intelligence in a broad sense. Its scope is systems having embedded knowledge and intelligence, which may be applied to the solution of world problems in industry, the environment and the community. It also focusses on the knowledge-transfer methodologies and innovation strategies employed to make this happen effectively. The combination of intelligent systems tools and a broad range of applications introduces a need for a synergy of disciplines from science, technology, business and the humanities. The series will include conference proceedings, edited collections, monographs, handbooks, reference books, and other relevant types of book in areas of science and technology where smart systems and technologies can offer innovative solutions.

High quality content is an essential feature for all book proposals accepted for the series. It is expected that editors of all accepted volumes will ensure that contributions are subjected to an appropriate level of reviewing process and adhere to KES quality principles.

More information about this series at http://www.springer.com/series/8767

Amaresh Chakrabarti
Editor

ICoRD'15 – Research into Design Across Boundaries Volume 2

Creativity, Sustainability, DfX, Enabling Technologies, Management and Applications

Editor
Amaresh Chakrabarti
Centre for Product Design
and Manufacturing
Indian Institute of Science
Bangalore
India

ISSN 2190-3018 ISSN 2190-3026 (electronic)
Smart Innovation, Systems and Technologies
ISBN 978-81-322-2228-6 ISBN 978-81-322-2229-3 (eBook)
DOI 10.1007/978-81-322-2229-3

Library of Congress Control Number: 2014957705

Springer New Delhi Heidelberg New York Dordrecht London

Printed on acid-free paper

Springer (India) Pvt. Ltd. is part of Springer Science+Business Media (www.springer.com)

Preface

Design is ubiquitous; it pervades all spheres of life, and has been around ever since life has been engaged in purposefully changing the world around it. While some designs have transcended time, designs are always in the process of being evolved. Research into design and the emergence of a research community in this area have been relatively new, its development influenced by the multiple facets of design (human, artefact, process, organisation, ecology, micro- and macro-economy by which design is shaped and which it shapes in turn) and the associated diversification of the community into those focusing on various aspects of these facets, in various applications. Design is complex, balancing the needs of multiple stakeholders, and requiring a multitude of areas of knowledge to be utilised, with resources spread across space and time.

The collection of papers in these two book volumes constitutes the Proceedings of the Fifth International Conference on Research into Design (ICoRD'15) held at the Indian Institute of Science, Bangalore, India during 7–9 January 2015. ICoRD'15 is the fifth in a series of biennial conferences held in India to bring together the international community from diverse areas of design practice, teaching and research. The goals are to share cutting edge research about design among its stakeholders; aid the ongoing process of developing a collective vision through emerging research challenges and questions; and provide a platform for interaction, collaboration and development of the community in order for it to address the global and local challenges by forming and realising the collective vision. The conference is intended for all stakeholders of design, and in particular for its practitioners, researchers, teachers and students.

Of the 265 abstracts submitted to ICoRD'15, 196 were selected for full paper submission. 151 full papers were submitted, which were reviewed by experts from the ICoRD'15 International Programme Committee comprising 180 members from over 131 institutions or organisations from 34 countries spanning six continents. Finally, 118 full papers, authored by over 275 (275 unique authors, actually 335 author entries in 118 papers) researchers from 86 institutions and organisations from 24 countries spanning six continents, were selected for presentation at the conference and for publication as chapters in this book. ICoRD has steadily grown over the last four editions, from a humble beginning in 2006 with 30 papers and 60 participants, through 75 papers and 100 participants in ICoRD'09, 100 papers and 150 participants in ICoRD'11, to 114 papers and 170 participants in ICoRD'13.

ICoRD'15 had 14 sessions with 70 podium papers, and 48 papers with brief podium presentations followed by poster display and discussion. It had keynotes from prominent researchers and practitioners from around the world such as: Vincent Floderer from CRIMP, France, Kristin Wood from Singapore University of Technology and Design, Singapore, John Gero from George Mason University, USA, Richard Gardner from The Boeing Company, USA, Sudhakar Nadkarni from Welingkar Institute of Management, India, and Rishikesha Krishnan from Indian Institute of Management Indore, India. It had two panel discussions on "Publishing Research Papers" and "Practice of Design", and five workshops, on Design Cognition, Sustainability, Emotional Engineering, Paper Crumpling, and Design Innovation Centre (DIC) Hubs in India. From 2015, ICoRD started giving **ICON^3** awards (acronym for **I**CoRD **O**utstanding **C**ontribution to desig**N** scie**N**ce and educatio**N**) to outstanding contributors to design education and research. Professor Sudhakar Nadkarni and Prof. John Gero were selected as **ICON^3** awardees for 2015, respectively for their outstanding contributions to design education and design research.

The chapters in this book together cover all three major areas of products and processes: functionality, form and human factors. The spectrum of topics range from those focusing on early stages such as creativity and synthesis, through those that are primarily considered in specific stages of the product life cycle, such as safety, reliability or manufacturability, to those that are relevant across the whole product life cycle, such as collaboration, communication, design management, knowledge management, cost, environment and product life cycle management. Issues of delivery of research into design, in terms of its two major arms: design education and practice, are both highlighted in the chapters in this book. Foundational topics such as the nature of design theory and research methodology are also major areas of focus. It is particularly encouraging to see in the chapters the variety of areas of application of research into design—aerospace, healthcare, automotive and white goods are but a few of the sectors explored. The theme of this year's conference and of this book is "Design Across Boundaries", where boundaries are manifold and span many dimensions—economy, culture, age, gender, religion, caste, class, education, family, digitalisation, geography (rural/urban) and so on.

The book has two volumes. Volume I focuses on Design Theory, Research Methodology, Aesthetics, Human Factors and Education. Volume II focuses on Design Creativity, Sustainability, Design for X, Enabling Technologies, Design Management and Applications in Practice. Volume I broadly focuses on form and human factors, while Volume II focuses on functionality.

On behalf of the Steering Committee, Advisory Committee, Local Organising Committee and Co-Chairs, I thank all the authors, delegates, institutions and organisations that participated in the conference and the International Programme Committee for their support in organising ICoRD'15 and putting this book together.

I am thankful to the Design Society and Design Research Society for their kind endorsement of ICoRD'15. I thank Indian Institute of Science (IISc), Bangalore and its Centre for Product Design and Manufacturing, for their support of this event. I also wish to place on record and acknowledge the enormous support provided by Mr. Ranjan B.S.C., Ms. Kumari M.C. and Ms. Nishath Salma of IISc in managing the review process, and in preparation of the conference programme and this book, and the large and dedicated group of student-volunteers of Indian Institute of Science, Bangalore in the organisation of the conference. Finally, I thank Springer, especially its Editor Ms. Swati Meherishi, for the wonderful support extended in the publication of this book and for sponsoring books and book coupons for Distinguished Paper Awards and **ICON^3** Awards respectively.

Amaresh Chakrabarti

Conference Organisation

Steering Committee

B. Gurumoorthy	Indian Institute of Science, India
John Gero	George Mason University, USA
Kota Harinarayana	National Aeronautical Laboratories, India (Chair)
Uday Athavankar	Indian Institute of Technology Bombay, India
Udo Lindemann	Technical University of Munich, Germany

Advisory Committee

Ananda Kumar	LPerspective, India
Anjan Das	Confederation of Indian Industry, India
Aparajita Ojha	IIIT D&M Jabalpur, India
Arabinda Mitra	International Division, Department of Science and Technology, India
V.S. Arunachalam	Center for Study of Science, Technology and Policy, India
Arun Jaura	Traktion, India
Ashok Jhunjhunwala	Indian Institute of Technology Madras, Chennai, India
Bala Bharadvaj	Boeing Research and Technology Center, India
R. Gnanamoorthy	IIIT D&M, Kancheepuram, India
Imre Horvath	Delft University of Technology, The Netherlands
Larry Leifer	Stanford University, USA
P.J. Mohanram	Indian Machine Tool Manufacturers Association, India
T.S. Mruthyunjaya (Chairman Emeritus, CPDM)	Indian Institute of Science, India

K. Radhakrishnan	Indian Space Research Organisation, India
Raman Saxena	USID Foundation, India
S. Sainath	LPerspective, India
Sam Pitroda	National Innovation Council, India
Denis Bertin	Airbus India
P.S. Subrahmanyam	Aeronautical Development Agency, India
Lt. Gen. V.J. Sundaram	National Design and Research Forum, India
Udayant Malhoutra	CII National Committee on Design, India

Programme Chair

Amaresh Chakrabarti Indian Institute of Science, India

Co-chairs

Lucienne Blessing	University of Luxembourg, Luxembourg
Gaur Ray	Indian Institute of Technology Bombay, India
Raghu Prakash	Indian Institute of Technology Madras, Chennai, India
Steve Culley	University of Bath, UK
Tim McAloone	Technical University of Denmark, Denmark
Toshiharu Taura	Kobe University, Japan
Umberto Cugini	Politecnico di Milano, Italy

Programme Committee

A.V. Gokula Vijaykumar	University of Strathclyde, UK
Ahmed Kovacevic	City University London, UK
Ajith Kumar	T. A. Pai Management Institute, India
Akin Kazakci	Ecole de Mines, France
Alex Duffy	University of Strathclyde, UK
Alison Mckay	University of Leeds, UK
Amitabha Mukerjee	Indian Institute of Technology Kanpur, India
K. Ananthasuresh	Indian Institute of Science, India
Anindya Deb	Indian Institute of Science, India
Anja Maier	Technical University of Denmark, Denmark
Anshuman Tripathy	Indian Institute of Management Bangalore, India
Ashitava Ghosal	Indian Institute of Science, India
Ashok Goel	Georgia Institute of Technology, USA
Ashok Iyer	Manipal University Dubai Campus, UAE

Ashutosh Tiwari	Cranfield University, UK
Axel Thallemer	Universität für künstlerische und industrielle Gestaltung, Austria
Aylmer Johnson	University of Cambridge, UK
Aziz Bouras	Qatar University, Qatar
Ben Hicks	University of Bath, UK
Benoit Eynard	Universite de Technologie de Compiegne, France
Bernard Yannou	École Centrale Paris, France
Biren Prasad	CERA Institute, USA
Bishakh Bhattacharya	Indian Institute of Technology Kanpur, India
C. Amarnath	Indian Institute of Technology Bombay, India
Cees de Bont	Hong Kong Polytechnic University, Hong Kong
Chalapathi Rao Nori	Indian Institute of Science, India
Christian Redlinghuys	University of Cape Town, South Africa
Christian Weber	Technische Universität Ilmenau, Germany
Christopher Magee	Massachusetts Institute of Technology, USA
Claudia Eckart	Open University, UK
Clement Fortin	École Polytechnique de Montreal, Canada
Craig Vogel	University of Cincinnati, USA
Dan Braha	University of Massachusetts, USA
Dan McAdams	Texas A&M University, USA
Darlie O. Koshy	Apparel Export Promotion Council, India
Dave Brown	Worcester Polytechnic Institute, USA
David Rosen	Georgia Institute of Technology, USA
Debkumar Chakrabarti	Indian Institute of Technology Guwahati, India
Denis Cavallucci	INSA Strasbourg, France
Dibakar Sen	Indian Institute of Science, India
Edwin Koh	National University of Singapore, Singapore
Ehud Kroll	Technion, Israel
Elena Mulet	Universitat Jaume I, Spain
Emmanuel Caillaud	Universite de Strasbourg, France
Eric Blanco	Institut Polytechnique de Grenoble, France
Eswaran Subrahmanian	Carnegie Mellon University, USA
Filippo Salustri	Ryerson University, Canada
Francesco Ferrise	Politecnico di Milano, Italy
Gabriela Goldschmidt	Technion, Israel
Gaetano Cascini	Politecnico di Milano, Italy
Gaur Ray	Indian Institute of Technology Bombay, India
Georges Fadel	Clemson University, USA
Glen Mullineux	University of Bath, UK
Guenther Seliger	Technical University of Berlin, Germany
Gul Kremer	Pennsylvania State University, USA
B. Gurumoorthy	Indian Institute of Science, India
Helen Petrie	University of York, UK
Henry Ming	Shanghai Jiao Tong University, China

Herbert Birkhofer	TU Darmstadt, Germany
I.S. Jawahir	University of Kentucky, USA
Indira Thouvenin	Universite de Technologie Compiegne, France
J.E. Diwakar	Indian Institute of Science, India
Jami Shah	Arizona State University, USA
Janet Allen	University of Oklahoma, USA
Jaywant Arakeri	Indian Institute of Science, India
Joaquim Macia	Polytechnic University of Catalunya, Spain
Johan Malmqvist	Chalmers University of Technology, Sweden
Jonathan Borg	University of Malta, Malta
Joost Duflou	Katholieke Universiteit Leuven, Belgium
Jozef Duhovnik	University of Ljubljana, Slovenia
Jozsef Vancza	MTA SZTAKI, Hungary
K. Sudhakar	Indian Institute of Technology Bombay, India
K. Munshi	Indian Institute of Technology Bombay, India
Kalyanmoy Deb	Michigan State University, USA
Kees Dorst	University of Technology Sydney, Australia
Kemper Lewis	University of Buffalo, The State University of New York, USA
Kikuo Fujita	Osaka University, Japan
Kristina Shea	Technische Universität München, Germany
Lauri Koskela	University of Salford, UK
Li Shu	University of Toronto, Canada
LIN Rungati	National Taiwan University of Arts, Taiwan
Linda Schmidt	University of Maryland, USA
L.S. Ganesh	Indian Institute of Technology Madras, India
Lucienne Blessing	University of Luxembourg, Luxembourg
M. Manivannan	Indian Institute of Technology Madras, India
Maik Maurer	Technische Universität München, Germany
Mandeep Singh	School of Planning and Architecture New Delhi, India
Marco Aurisicchio	Imperial College London, UK
Marco Cantamessa	Politecnico di Torino, Italy
Maria Yang	Massachusetts Institute of Technology, USA
Mario Storga	University of Zagreb, Croatia
Mario Fargnoli	University of Rome, La Sapienza, Italy
Martin Grimhelden	Royal Institute of Technology, Sweden
Martin Steinart	Norwegian University of Science and Technology, Norway
Mary Matthew	Indian Institute of Science, India
Mary Thompson	Technical University of Denmark, Denmark
Ming xi Tang	Hong Kong Polytechnic University, Hong Kong
Mitchell Tseng	The Hong Kong University of Science and Technology, Hong Kong
Monica Bordegoni	Politecnico di Milano, Italy

Monto Mani	Indian Institute of Science, India
Moreno Muffatto	Universita di Padova, Italy
Nilesh Vasa	Indian Institute of Technology Madras, India
Ozgur Eris	Delft University of Technology, The Netherlands
P. Radhakrishnan	PSG Institute of Advanced Studies, India
P. Rodgers	Northumbria University, UK
P.V.M. Rao	Indian Institute of Technology Delhi, India
Panos Papalambros	University of Michigan, USA
Peter Torlind	Lulea University of Technology, Sweden
Peter Childs	Imperial College London, UK
Petra Badke-Schaub	Delft University of Technology, The Netherlands
Phillippe Girard	University of Bordeaux, France
Pieter Vermass	Delft University of Technology, Netherlands
Prabir Sarkar	Indian Institute of Technology Ropar, India
Pradeep Yammiyavar	Indian Institute of Technology Guwahati, India
Puneet Tandon	IIIT D&M Jabalpur, India
Rachuri Sudarsan	National Institute of Standards and Technology, USA
Raghu Echempati	Kettering University, USA
Raghu Prakash	Indian Institute of Technology Madras, India
Rajkumar Roy	Cranfield University, UK
Ralph Bruder	Technical University of Darmstadt, Germany
Ravi Punekar	Indian Institute of Technology Guwahati, India
Ravi Poovaiah	Indian Institute of Technology Bombay, India
Reiner Anderl	Technical University of Darmstadt, Germany
Ricardo Sosa	Singapore University of Technology and Design, Singapore
Richard Liu	Chang Gung University, Taiwan
Rikard Soderberg	Chalmers University of Technology, Sweden
Rina Maiti	Indian Institute of Science, India
Roman Zavbi	University of Ljubljana, Slovenia
Rosario Vidal	Universitat Jaume I, Spain
S. Krishnan	Center for Study of Science, Technology and Policy, India
S. Vinodh	National Institute of Technology, Tiruchirappalli, India
Sangarappillai Sivaloganathan	United Arab Emirates University, UAE
Santosh Jagtap	Lund University, Sweden
Satish Kailas	Indian Institute of Science, India
Sean Hanna	University College London, UK
Serge Rohmer	Université de Technologie de Troyes, France
Shayne Gooch	University of Canterbury, New Zealand
Shilpa Ranade	Indian Institute of Technology Bombay, India
Simon Bolton	Birmingham Institute of Art and Design, UK
Somwrita Sarkar	University of Sydney, Australia

Srinivas Kota	Birla Institute of Technology and Science Pilani, India
Srinivasan Venkataraman	Technische Universität München, Germany
Stanislav Hosnedl	University of West Bohemia, Czech Republic
Stephen C.-Y. Lu	University of Southern California Los Angeles, USA
Steve Culley	University of Bath, UK
Steve Evans	University of Cambridge, UK
Steven MacGregor	University of Girona, Spain
Subir Saha	Indian Institute of Technology Delhi, India
Susy Verghese	Indian Institute of Technology Madras, India
Tamotsu Murakami	The University of Tokyo, Japan
Tim McAloone	Technical University of Denmark, Denmark
Tjamme Wiegers	Delft University of Technology, The Netherlands
Tom Vaneker	University of Twente, The Netherlands
Torben Lenau	Technical University of Denmark, Denmark
Tracy Bhamra	Loughborough University, UK
Umberto Cugini	Politecnico di Milano, Italy
Vesna Popovic	Queensland University of Technology, Australia
Vijay Srinivasan	National Institute of Standards and Technology, USA
Vinod Vidwans	FLAME School of Communication, India
Vishal Singh	Aalto University, Finland
William Ion	Strathclyde University, UK
Winifred Ijomah	University of Strathclyde, UK
Wolfgang Wimmer	Vienna University of Technology, Austria
Yasushi Umeda	University of Tokyo, Japan
Ying-Chieh Liu	Chang Gung University, Taiwan
Yong Chen	Jiao Tong University, China
Yong Se Kim	Sungkyunkwan University, Korea
Yong Zeng	Concordia University, Canada
Yoram Reich	Tel Aviv University, Israel
Yosef Oehmann	Technical University of Denmark, Denmark
Yoshiki Shimomura	Tokyo Metropolitan University, Japan
Yoshiyuki Matsuoka	Keio University, Japan
Yrjo Sotamaa	Aalto University, Finland
Yukari Nagai	Japan Advanced Institute of Science and Technology, Japan

Local Organising Committee

G.K. Ananthasuresh	Indian Institute of Science, India
Jaywant Arakeri	Indian Institute of Science, India

Anindya Deb	Indian Institute of Science, India
J.E. Diwakar	Indian Institute of Science, India
Ashitava Ghosal	Indian Institute of Science, India
Satish Vasu Kailas	Indian Institute of Science, India
Rina Maiti	Indian Institute of Science, India
Monto Mani	Indian Institute of Science, India
Mary Matthew	Indian Institute of Science, India
N.V.C. Rao	Indian Institute of Science, India
N.D. Shivakumar	Indian Institute of Science, India
Dibakar Sen	Indian Institute of Science, India
Anumeha Rai	Indian Institute of Science, India
Ashish Verma	Indian Institute of Science, India
Bismita Nayak	Indian Institute of Science, India
Kaushik Choudhury	Indian Institute of Science, India
Dharam Deo Prasad	Indian Institute of Science, India
E. Govindaprasath	Indian Institute of Science, India
Govinda Sharma	Indian Institute of Science, India
Anirudha Bhattacharjee	Indian Institute of Science, India
Murali Krishna	Indian Institute of Science, India
P. Krupakar	Indian Institute of Science, India
Nipun Patil	Indian Institute of Science, India
Salil Sapre	Indian Institute of Science, India
Samrat Sankhya	Indian Institute of Science, India
Shivam Raina	Indian Institute of Science, India
Unnati Jain	Indian Institute of Science, India
Minu Pradeep	Indian Institute of Science, India
Mangaldas Budho Gaonkar	Indian Institute of Science, India
C.S. Susmith	Indian Institute of Science, India
P.S. Suvin	Indian Institute of Science, India
S. Tushar Pawar	Indian Institute of Science, India
A. Vignesh Kumar	Indian Institute of Science, India
S. Chandra Mouli	Indian Institute of Science, India
Chandana Venkatayogi	Indian Institute of Science, India
Dawn Varghese	Indian Institute of Science, India
J.R. Hari Prakash	Indian Institute of Science, India
Hari Narayanan	Indian Institute of Science, India
V.S. Krishna Prasad	Indian Institute of Science, India
M.C. Kumari	Indian Institute of Science, India
Pavan Sridharan	Indian Institute of Science, India
Nilakantha Singh Deo	Indian Institute of Science, India
Biplab Sarkar	Indian Institute of Science, India
B.S.C. Ranjan	Indian Institute of Science, India
B. Santhi	Indian Institute of Science, India
B. Damayanthi Jesudas	Indian Institute of Science, India
Divyanshu Joshi	Indian Institute of Science, India

Gajanan Kulkarni	Indian Institute of Science, India
S. Harivardhini	Indian Institute of Science, India
Kiran Ghadge	Indian Institute of Science, India
N. Madhusudanan	Indian Institute of Science, India
Manoj Kumar Mahala	Indian Institute of Science, India
Nitesh Batia	Indian Institute of Science, India
Praveen T. Uchil	Indian Institute of Science, India
Suman Devadula	Indian Institute of Science, India
Sonal Keshwani	Indian Institute of Science, India
Shakuntala Acharya	Indian Institute of Science, India
V. Prajwal	Indian Institute of Science, India
K. Eazhil Selvan	Indian Institute of Science, India
Vineeth Muralidharan	Indian Institute of Science, India
Anshul Mittal	Indian Institute of Science, India
Rahul Kanyal	Indian Institute of Science, India
G. Ranga Srinivas	Indian Institute of Science, India
Nishath Salma	Indian Institute of Science, India
R. Ravindra	Indian Institute of Science, India
Rohit John Varghese	Indian Institute of Science, India

Contents

Part I Design Creativity, Synthesis, Evaluation and Optimisation

Evolution of Design Intuition and Synthesis Using Simulation Enriched Qualitative Cognitive Models 3
Satish Chandra

Analytical Estimation and Experimental Validation of Acceleration at Spacecraft Solar Array Latch-up Considering Differential Latching 15
B. Lakshmi Narayana, Gaurav Sharma, G. Nagesh, C.D. Sridhara and R. Ranganath

Assessment of the Biomimetic Toolset—Design Spiral Methodology Analysis 27
Pierre-Emmanuel Fayemi, Nicolas Maranzana, Améziane Aoussat and Giacomo Bersano

Decision Uncertainties in the Planning of Product-Service System Portfolios 39
Daniel Kammerl, Oliver Malaschewski, Sebastian A. Schenkl and Markus Mörtl

The DSM Value Bucket Tool 49
Bernard Yannou, Romain Farel and François Cluzel

Development of Creativity Through Heightening of Sensory Awareness 63
N. Divya

Enhancing Creativity by Using Idea-Wheel and Its Validation 75
Avinash Shende and Amarendra Kumar Das

Part II Eco-Design, Sustainable Manufacturing, Design for Sustainability

Designing Technology, Services and Systems for Social Impact in the Developing World: Strong Sustainability Required 89
Gavin Melles, Blair Kuys, Ajay Kapoor, James Rajanayagam, Joseph Thomas and Aswhin Mahalingam

Understanding Consumers' Perceptions of Sustainable Products in India 99
Prabir Sarkar, Srinivas Kota and Bijendra Kumar

The Use of Sugarcane Bagasse-Based Green Materials for Sustainable Packaging Design 113
L. Pereira, R. Mafalda, J.M. Marconcini and G.L. Mantovani

Does the Ecomark Label Promote Environmentally Improved Products in India and What Experiences Can Be Drawn from the Nordic Ecolabel? 125
Jakob Thomsen and Tim C. McAloone

***Taki*, the Community (Sustainable) Sensory Garden** 137
Phebe Valencia and Martin L. Katoppo

Empowerment for Chhattisgarh Craft Clusters 151
Parth Shukla and Satyaki Roy

Sustainable Supply Chain in Product Development 159
Srinivas Kota and Kirthi Bandi

Use of MFF and Concurrent Engineering to Develop a Sustainable Product—Radical Redesign of Flushing System 171
Nikola Vukašinović, Žiga Zadnik and Jože Duhovnik

Biogenic Domestic Waste—Exploring Select Dimensions of Socio Technical Innovation Using Design Probe 181
Amit Kundal, Jayanta Chatterjee and Shatarupa Thakurta Roy

Design for the BOP and the TOP: Requirements Handling Behaviour of Designers 191
Santosh Jagtap, Andreas Larsson, Anders Warell, Deepak Santhanakrishnan and Sachin Jagtap

A Sustainable Design Method Acting as an Innovation Tool 201
Jeremy Faludi

Green Is the New Colour for Menstruation. *Uger* Fabric Pads Show the Way Through a Sustainable Perspective 213
Lakshmi Murthy

Sustainable Machining Approach by Integrating the Environmental Assessment Within the CAD/CAM/CNC Chain 227
Hery Andriankaja, Julien Le Duigou and Benoît Eynard

Idea-to-Market: Product Innovation Sustainability in Challenging Health-Care Context 237
Dipanka Boruah and Amarendra Kumar Das

An Interface Between Life Cycle Assessment and Design 251
Praveen Uchil and Amaresh Chakrabarti

Part III Design for X (Safety, Manufacture and Assembly, Cost, Reliability etc.)

Three Dimensional Form Giving of *Kundan* Jewellery—A Parametric, Cluster Based Approach to Jewellery Design and Prototyping 263
Parag K. Vyas

Design and Implementation of a Line Balance Visualization and Editing Tool 275
Rahul Sharan Renu and Gregory Mocko

Identifying and Utilizing Technological Synergies—A Methodological Framework 291
Michael Roth, Matthias Gürtler and Udo Lindemann

Examples of Poor Design from an Engineering Expert Witness and Consultant 303
Swaminathan Balachandran

Design for Method Study—Work Measurement: Do We Need It? 313
Sangarappillai Sivaloganathan and Rana Yanis

Identification of Distinct Events in an Assembly by Automatically Tracking Body Postures 327
B. Santhi, Amaresh Chakrabarti, B. Gurumoorthy and Dibakar Sen

Validation of Methodology and Tool for Design for Adaptability in Accomplishment of Project Objectives 339
V. Srinivasan, Phillip Schrieverhoff, Cristina Carro Saavedra, Matthias Gürtler and Udo Lindemann

Part IV Enabling Technologies and Tools (Computer Aided Conceptual Design, Virtual Reality, Haptics, etc)

Effectiveness of Tangible and Tablet Devices as Learning Mediums for Primary School Children in India 353
Anmol Srivastava and Pradeep Yammiyavar

Analysing the Innovation Growth of Robotic Pets Through Patent Data Mining 365
Teo Kiah Hwee, Mohan Rajesh Elara, Ricardo Sosa and Ning Tan

Studies in Application of Augmented Reality in E-Learning Courses 375
Mannu Amrit, Himanshu Bansal and Pradeep Yammiyavar

Mobilising Design for Development: An Analysis of a Human-Centered Design Process Used for a South African mHealth Student Project 385
Fatima Cassim and Nina Honiball

A Methodology for the Analysis of the Influence of Odours on the Users' Evaluation of Industrial Products 397
Marina Carulli, Monica Bordegoni, Umberto Cugini and Ding Weibin

Skweezee-Mote: A Case-Study of a Gesture-Based Tangible Product Design for a Television Remote Control 409
Mehul Agrawal, Vero Vanden Abeele, Karen Vanderloock and Luc Geurts

Extraction of Gestures for Presentation: A Human Centered Approach 421
Surbhit Varma, Keyur Sorathia and Abdul Sameer Ashraf

Implementation of an Algorithm to Classify Discourse Segments from Documents for Knowledge Acquisition 433
N. Madhusudanan, Amaresh Chakrabarti and B. Gurumoorthy

Part V Design Management, Knowledge Management and Product Life Cycle Management

From Concept to Specification Maintaining Early Design Intent 445
Ricardo Sosa, Jun Bum Lee, Diana Albarran and Kevin Otto

A Spatio-Temporal Network Representation for Manufacturing 459
Kumari Moothedath Chandran, Monto Mani and Amaresh Chakrabarti

Requirements Evolution: Understanding the Type of Changes in the Requirement Document of Novice Designers 471
Shraddha Joshi and Joshua D. Summers

Integration of Strategic Flexibility into the Platform Development Process 483
Fatos Elezi, Rita Tschaut, Wolfgang Bauer, Nepomuk Chucholowski and Maik Maurer

Research on Development of Liquid Composite Molding Parts: Situation and Framework 495
Paul Bockelmann, Klaus Drechsler and Amaresh Chakrabarti

Knowledge Sharing in Design Based on Product Lifecycle Management System 507
Pierre-Emmanuel Arduin, Julien Le Duigou, Marie-Hélène Abel and Benoît Eynard

Part VI Applications in Practice (Automotive, Aerospace, Biomedical Devices, MEMS, etc.)

Development of a Cellular Lightweight Cement (CLC) Block Cutting Machine for Small Scale Industries 521
Vikas Patil, Aniket Patil, Bhagyesh Deshmukh and Roohshad Mistry

Novel Design Features for Matched Die Moulding for Bio-composites 533
D. Saravana Bavan, P. Kamalbabu and G.C. Mohan Kumar

Design Methodology and Dynamic Simulation of Fixed Displacement Swash Plate Compressor 545
Raushan Kumar Jha, Selvaraji Muthu and S. Sivakumar

Compliance Monitored Clubfoot Brace "PADMAPADA"© 559
Pradyumna K. Kammardi, Kaushik D. Sondur, N.S. Dinesh, Chalapathi Rao Nori, Vrisha Madhuri and Sanjay Chilbule

Lumped Parameter Modeling of Vibratory Soil Compactor 569
Gomathinayagam Arumugam, G.S. Narayana, S. Babu and S. Mohamed Ebrahim

Automated Puppetry—Robo-Puppet© 579
M.A. Aravind, N.S. Dinesh, Nori Chalapathi Rao and P. Ram Charan

Math Based Model for Quick Estimation of Heat Generated in an Automotive Li–Ion Battery Pack at Various Operating Conditions 591
Vikrant Singh, Chiru Venkat Reddy, Justin R. McDade and Rashed S. Rabaa

Ideal Workstation Design of Driver's Cab for New EMU Rake of Mumbai Local 601
Amar Kundu, Kiran Gangadharan, Nishant Sharma and Gaur G. Ray

Characteristics of Jewellery Design: An Initial Review 613
Noor Adila Mohd Rajili, Elin Olander and Anders Warell

QFD Based Methodology to Decide upon Contextually Appropriate Solution Category with Specific Reference to Pineapple Peeling Equipment Design 621
Prakash Kumar and Debkumar Chakrabarti

Role of Colour and Form in Product Choice and Variation of Preferences Across Product Categories: A Review 631
Swathi Matta Reddy, Anirban Chowdhury, Debkumar Charkrabarti and Sougata Karmakar

What Do You See? Research on Visual Communication Design to Promote Positive Change for Unorganized Workers in Karnataka, India 641
Sabina von Kessel

A Study on DSM Partitioning Through Case Study Approach 653
Purva Mujumdar, Soma Bhattacharya and J. Uma Maheswari

A Paradigm-Shift towards User-Centred Empirical Methodology in Interactive Multimedia Communication 663
Manoj Majhi and Debkumar Chakrabarti

Handheld Isobaric Aesthesiometer for Measuring Two-Point Discrimination 675
M. Manivannan, R. Periyasamy and Devasahayam Suresh

An Application of Particle Swarm Optimization Technique for Optimization of Surface Roughness in Centerless Grinding Operation 687
Subhas Chandra Mondal and Prosun Mandal

Author Index 699

About the Conference

Design is ubiquitous; it pervades almost all spheres of life, and has been around as long as life has taken up the task of purposefully changing the world around it. Research in design and the emergence of a research community in this area has been relatively new, its development influenced by the multiple facets of design (human, artefact, process, organization, and the micro- and macro economy by which design is shaped) and the associated diversification of the community into those focusing on various aspects of these individual facets, or various applications. Design is complex, balancing the needs from multiple stakeholders, and requiring a multitude of areas of knowledge to be utilized, from resources spread across space and time.

ICoRD'15 is the fifth in a series of conferences intended to be held every 2 years in India to bring together the international community from diverse areas of design practice, teaching and research, to: showcase cutting edge research on design to the stakeholders; aid the ongoing process of developing and extending the collective vision through emerging research challenges and questions; and provide a platform for interaction, collaboration and development of the community in order for it to take up the challenges to realize the vision. The conference is intended for all stakeholders of design, and in particular for its practitioners, researchers, teachers and students.

The theme of ICoRD'15 is "Design Across Boundaries". Human society is ridden with boundaries: economic, political, religious, educational, cultural, racial, caste and class-based, age-based, ability-based, environmental, oriental-occidental, rural–urban, traditional–modern and so on. Yet designs must transcend boundaries to create value for people from all walks of the global society. Specific forms of design across boundaries are already being encapsulated in for instance "inclusive design" which aims to help melt age- and ability-based boundaries, or "design for the bottom-of-the-pyramid" which extends design to those with extreme economic challenges. The theme of ICoRD'15 is a celebration of the power of design to be pervasive. It is also a call to the design and its research fraternity to acknowledge boundaries, if only to transcend them and reach the benefits of design to everyone.

The conference has:

- Invited presentations from eminent international experts and practitioners;
- Presentations of refereed papers as podium, poster, panel or theme presentations;
- Industrial sessions to present perspectives from industry and studies in practice.

ICoRD'15 is co-located with the 3rd International Conference on Design Creativity (3rd ICDC) held in Bangalore, India, 12–14 January, 2015.

About the Editor

Amaresh Chakrabarti is professor of Engineering Design at Centre for Product Design and Manufacturing, Indian Institute of Science (IISc), Bangalore. He holds a BE in Mechanical Engineering from University of Calcutta (now IIEST), India, an ME in Mechanical Design from IISc, and a Ph.D. in Engineering Design from University of Cambridge, UK. After Ph.D., he led for 10 years the Design Synthesis team at the EPSRC Centre for Excellence Engineering Design Centre at University of Cambridge. His interests are in design synthesis and creativity, eco-design and sustainability, and product informatics. He has authored/edited 10 books, over 250 peer-reviewed articles, and has six patents granted/pending. He co-authored DRM, a methodology used widely as a framework for doing engineering design research. He is an Associate Editor, AI EDAM, Area Editor, Research in Engineering Design (Springer), Regional Editor, Journal for Remanufacturing (Springer), and Advisory Editor for seven international journals including Journal of Engineering Design, Clean Technologies and Environmental Policy (Springer), and International Journal of Design Creativity and Innovation. Professor Chakrabarti has been on the Advisory Board of Design Society, UK, where he is currently a member of the Board of Management. He is a member of the CII National Committee on Design, India, and member of the Jury for India Design Mark, India Design Council. He founded IDeaSLab—the first laboratory in India for research into design creativity, sustainability and innovation. He is Chair for International Conferences on Research into Design (ICoRD), 22nd CIRP Design Conference (CIRP Design 2012), 3rd International Conference on Design Creativity (3rd ICDC 2015), and vice-Chair for AI in Design (AID) and Design Computing and Cognition (DCC) Conferences. He is an Honorary Fellow of the Institution of Engineering Designers, the peer society under the UK Royal Charter in engineering design. Seven of his papers won top paper awards in various international conferences.

Part I
Design Creativity, Synthesis, Evaluation and Optimisation

Evolution of Design Intuition and Synthesis Using Simulation Enriched Qualitative Cognitive Models

Satish Chandra

Abstract Design of engineering systems requires understanding of physical behavior. Previously, design intuition was considered to be developed based on actual experience. However, use of simulations to understand physical behavior is now widespread, and can play a role in developing design intuition rapidly. It can be said that design intuition is essentially a cognitive model in a qualitative form and is descriptive. Research indicates that these cognitive models have a structure, are decomposable and runnable and combine with memories in the mind. In this paper, it is argued simulations help the evolution of the cognitive models, using a predict, test and validate (PTV) approach aided by visualization and imagery. Using case studies, it is seen that while some naïve intuitive models are consistent with physical laws, in other cases, the faulty models are replaced with models that have basis in physical laws when using the PTV approach. Thus, simulations appear to enrich cognitive models of physical behavior, aid design synthesis within a design space. It is seen that simulation enriched design intuition evolves rapidly and can be used effectively in training.

Keywords Cognitive models · Simulation · Design intuition · Synthesis · Qualitative reasoning

1 Introduction

The use of simulations to design products and systems is widespread in engineering. In the past, direct assimilation of quantitative data from simulation provided new experiences of physical behavior. With the advent of visualization and imagery using computer graphics, demands on simulation have been to provide a realistic replication of actual physical behavior and in this lies the sense of

S. Chandra (✉)
Structural Technologies Division, National Aerospace Laboratories, Bangalore, India
e-mail: schandra@nal.res.in

A. Chakrabarti (ed.), *ICoRD'15 – Research into Design Across Boundaries Volume 2*, Smart Innovation, Systems and Technologies 35, DOI 10.1007/978-81-322-2229-3_1

"experience", where conceptual relationships and memories are developed. There are specific reasons to understand the development of design intuition of the engineers who run these simulations, as it would have long term consequences in design synthesis and innovation. Over the years, design intuition was regarded to have been developed by experience of working on new designs, testing the designs and observing their performance in the real world. While today, simulations are in widespread use, there are questions with regard to the design intuition of engineers, who may not have either the full knowledge of the physical laws or mathematical models embedded in software used for simulations, or the real world experience.

In this paper, we focus attention on the development of design intuition by such design engineers. This paper describes studies on the intuitive and counter intuitive notions of designers who use simulations repeatedly in a bounded design space. The use of predict-test-validate (PTV) approach routinely used in engineering design, with simulations used for predictions, tests used to validate and verify is discussed. While, it is generally felt that visualization leads to development of the design intuition, it is argued here that it is the descriptive qualitative models that are generated after visualization and imagery, that can be regarded as structured design intuition or a cognitive model of physical behaviour. The emphasis in this paper is on the development of the qualitative structure of these models, and not on visualization and imagery.

Design intuition can be described in the form of cognitive or mental models. The process of the development and refinement of cognitive models of physical behavior from lay intuition to expert intuition is fascinating. It has been seen that the cognitive models can be organized in terms of objects/agents which operate out of a framework and these replaced as evidence is gathered against it as the model is continuously refined.

There appears to be considerable work regarding the development of cognitive models of physical behavior e.g., Johnson-Laird [1], Gentner and Stevens [2], DiSessa [3], McCloskey [4]. In widely cited work edited by Gentner and Stevens [2], it is noted that analogies and metaphors are structured to form these models. Now, this is interesting in the light of the argument that models are really in a sense relationships that are dynamic representations. How these models are formed in individuals, whether one operates from first principles or just from intuition can be largely explained by structures formed by relationships. In Williams et al. [5], the concept is of an autonomous object, with an explicit representation of state and topological connections to other objects. These objects have a set of internal parameters and rules, which modify its parameters and propagate information via the rules and topology. Running a cognitive model is akin to modifying parameters of the model by propagating information, by objects interacting with a limited number of connected objects.

Staggers and Norcio [6] argued that these models were related to perceptual entities. In [7] it is said that humans construct cognitive models of both concrete and abstract objects in a relational structure which emulates an actual situation. Humans in that sense run models in their minds for description and causal hypothesis testing as models are constructed and modified.

In parallel, there has been considerable work on qualitative reasoning [8–11], driven by interest in how human reason about physical behavior. de Kleer and Brown [8] believe humans could use a qualitative calculus to manipulate knowledge of the physical world. Forbus and Gentner [10], Forbus [11] argues that there is evidence that qualitative representation of spatial and temporal information represents the intuitive positions on physical behaviour. Forbus used qualitative reasoning to model intuitive phenomena such as motion, liquids, and heat. Interestingly, there is exploration of how scientists and engineers reason to enable them to navigate and extract valuable insights about complex systems purely from visual imagery. This has considerable implications for enriching cognitive models using simulation driven imagery. In Chandra [12], it is argued that simulations can also be used to produce descriptions in a qualitative way, as assimilation of behavior in humans is essentially qualitative. While, we could further explore visualization and imagery as basis for learning in a separate paper, it is assumed here that visualization provides the same scope for assimilation as real experience, as the general view about visualization is that all attempts are being made to create a real experience as much as possible. Tenenbaum and co-workers [13–15] in a series of papers see simulation as an engine for understanding physical behaviour, be it video games, graphics etc., describe how intuitive physics is formed in the mind, noting that inferences are made beyond the data and use probability based inferences on representations that are evolving and flexible. Trafton et al. [16] noted that it is possible to generate information from complex visualizations indicating structured cognitive models can be built using visualisation and imagery.

Sanborn et al. [17] shows in using the case of colliding objects, intuitions deviate from classical laws of mechanics arguing for the use of heuristics to make judgement on collisions, only that mass is intuitive, while velocities and trajectories are really something where there could be uncertainty. The argument is that Bayesian inference and memory (recall) could enable consistent intuitions with respect to classical mechanics.

Forbus and Gentner [10] relook at the work in the 1980s and argue that cognitive models are high resolution simulations, where reasoning is like watching a movie of a physical system behaviour in the mind and is essentially qualitative. Thus, previous work indicates that cognitive models of physical behavior have a framework, in which objects are embedded in a topological structure which can be run as a mental simulation. There appears to evidence that early (lay) intuition cognitive models of physical behavior involving motion and trajectory generally can be difficult and has associated uncertainty.

Using case studies, it is shown that engineers without the deeper knowledge of the mathematical model embedded into the simulation can rework and refine their cognitive models by running simulations and visualizations and assimilating the physical behavior represented by these simulations. It is argued that the intuitive cognitive models based on simulations can be changed either completely or in parts (in terms of connections or objects or internal rules of the objects), and replaced with new models. Interestingly, the PTV approach also aids in refining design intuition, and in the process altering the original cognitive models, as data from tests are generally used in replacing faulty design intuition with what is proven. Much of the work done earlier in the 1980s and 1990s were probably aimed at developing artificial intelligence concepts based on the understanding of intuitive models for physical behavior. In this paper, the emphasis is to use the mapping of quantitative models into qualitative descriptions from simulations, from which design synthesis and intuition can be developed rapidly.

2 Development of Design Intuition During the Design Process in Engineering

Typically the design process in engineering can be characterized in the form as shown in Fig. 1. Concept models are developed using the well developed intuition of experienced design engineers. A conceptual model is created by sketches, description and notes, these product concepts are then converted to a computer aided design model. The computer aided design (CAD) model is used to simulate physical behavior. Simulations are performed using software embedded with mathematical models representing the appropriate physics.

Simulations produce visualizations that are assimilated by human beings. Results are compared with experiments or tests and validated, which is known as the Predict-Test-Validate (PTV) approach. This enables certification of the product, which is then manufactured and operated. Development of the various models: conceptual, CAD and the simulation model involve experience, learning and memory, driven by cognitive models that have evolved from the very naïve stage till it can be acknowledged as engineering expertise. Design procedures which are accepted based on certification and regulatory processes are also developed based on cognitive models that have evolved over a number of experiences on real engineering projects. Simulations during a design process could be regarded as

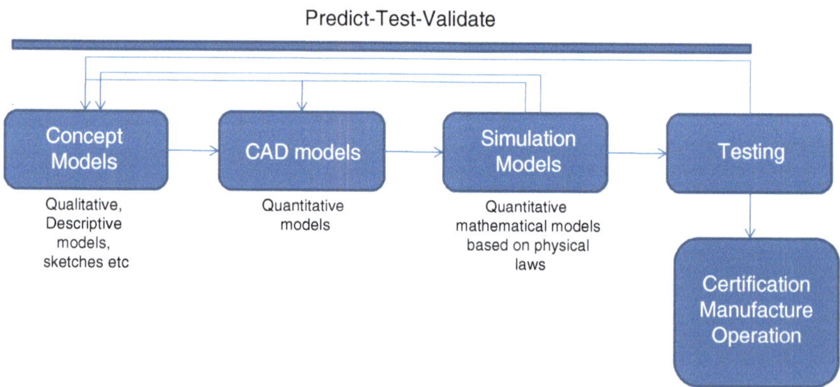

Fig. 1 Design process in engineering

repeated experiments on a computer, where each simulation result is analysed and absorbed as an experience. Typically, any prediction via simulation is tested, results from tests measured and correlated, the test-simulation correlation regarded as validation. Thus, under the PTV approach, a cognitive model of behavior is refined and accepted after validation by tests or in many cases actual experience of the design in real operation. In many ways, design synthesis occurs in individuals through a highly developed cognitive model of the engineered system and sometimes referred to as tacit knowledge, which can in turn be regarded as causal.

2.1 Case Studies in Mechanics Simulations

A series of experiments were conducted regarding the learning from simulations among engineers, especially if the underlying mathematical models were either not fully understood, or was not recoverable from memory. The anecdotal experience was further discussed. Initially, the discussion was to be restricted to understanding the rather confusing issue of energy absorption under impact. However, more basic mechanics behavior was explored as well. Interestingly, these "thought experiments" showed how the intuitive position of strength and safety pervaded the discussion, and generates a causal relationship between strength and safety.

From a qualitative reasoning perspective, the varieties of behaviours as imagined that influence design using mechanics were explored. Fundamentally, the process of reasoning qualitatively is set in descriptions of behavior. Typical behaviours commonly used in mechanics simulation, in a steady state sense, are as follows (see Fig. 2). While we will not deal with visualization and imagery in detail here, it is sufficient to note that whether this imagery is through a real world experience, test or simulation, it will lead to assimilation in a qualitative form lending to description.

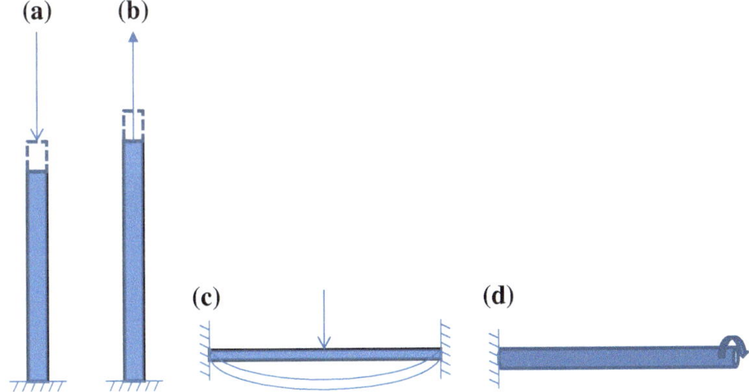

Fig. 2 **a** Compression, **b** tension, **c** bending, **d** torsion

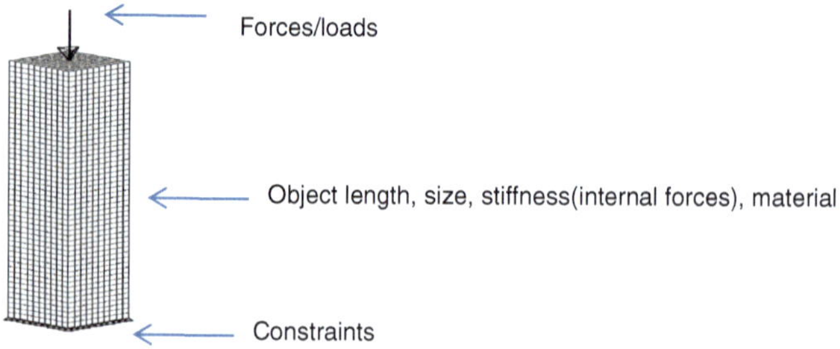

Fig. 3 Simulation model for compression

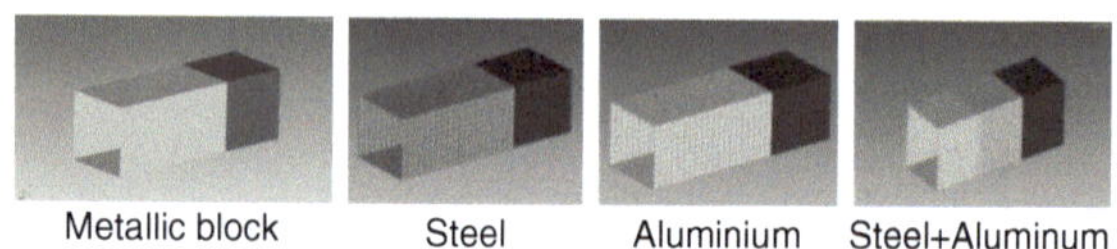

Fig. 4 Metallic block (with steel/aluminium crush zone)

Figure 3 shows a typical compression model in simulation. The model has a certain number of features, which can be assimilated using imagery and represented through qualitative description. These include the loads (forces), constraints for movements, length, size, shape etc. Firstly, we use Newtonian mechanics and mathematical models, to arrive at similar conclusions with equilibrium relationships governing force versus reaction relationships as seen below:

$P = Kx$, where P is load applied, K the stiffness and x the deformation of a block shown in Fig. 4. In a qualitative way, compression can be described in terms of External force, Internal Force, deformation and behavior represented based causal description in the form: *Low deformation if stiffness is high and/or force is low.*

Exploring this further with different materials, consider a simple metallic block undergoing compression: Here we can have this as steel, aluminum and a combination of aluminum and steel.

Results from the simulation done in a static (steady state simulation) shows deformations to be lower for steel than aluminum or a hybrid steel and aluminum crush zone. Here, there is consistency between the lay-intuitive cognitive model and simulations, as steel appears stronger than aluminum.

In more complex states, e.g., buckling, qualitative reasoning based on intuitive physics is difficult. The intuitive notions in dynamic situations were also explored. The following mechanics phenomena can be classified as dynamic: Vibration, Flutter, Impact etc. In each of these cases, the intuitive notions were not easy to articulate during discussions, unless a foundation of physical laws and mathematical models was available. In dynamic situations, there appears to be issues of naïve physics failing as discussed earlier by other researchers as well [13]. So, it was found necessary for the development of cognitive models of the dynamic phenomena to be based on exposure to the mathematical models and observations of tests (PTV). We now explore energy absorption further as there were very clearly different interpretations between lay intuition and the actual physical behavior. Concepts of work done, energy, velocity, acceleration, inertia etc., all part of Newtonian physics are generally not lay-intuitive unless they are taught and learning occurs.

Further exploration revealed that the use of non-colloquial language terminology, basically scientific terminology linked to physical laws was required, for dynamic situations, while these were not required for the simpler model associated with compression etc., though there have been some consistency between intuitive and Newtonian models in steady state.

Typically, using explicit dynamics software, also known as a crash analysis code, e.g., PAMCRASH, a dynamic simulation was carried out (Fig. 5). It was found a metallic block with the steel crush zone deformed the least, but transmitted the highest accelerations, while the block with aluminum crush zone deformed the most, but had the least accelerations. The hybrid aluminum and steel metal block crushed, but the steel block effectively remained intact, representing an automobile with a crush zone, but provided an intact chamber for an occupant.

A first level reasoning of dynamic behavior of impact was carried out with engineers with no exposure to dynamics, especially impact engineering. They felt that if the structure was "strong", it was safe. These engineers were shown simulations and videos of car crashes with cars with did not deform versus cars which had a crush/crumble zone. The first intuitive reaction was one of confusion and skepticism. In fact, it was shown that images of cars which remained fairly intact were not necessarily safe for the occupant, as forces were transmitted to the occupant, resulting in severe injury and fatalities.

A simple experiment, as in the development of the first level model was undertaken. It was shown that the conventional, intuitive view that strong materials and designs were safer was not appropriate in an impact situation, as counter-intuitive physics was at play. The introduction to the above concepts and more importantly, the engineers were led through a predict, test and validate (PTV) process. Repeated exposure to simulations in a parametric way, with number of

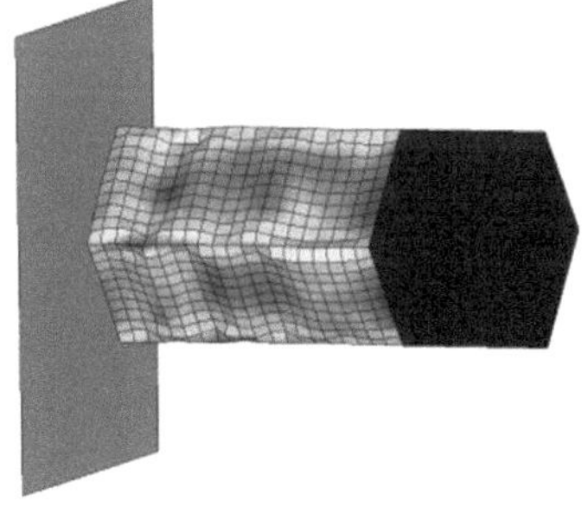

Fig. 5 Visualisation of crushing of the metallic block under impact

case studies demonstrated using the PTV approach led to more consistent qualitative descriptions of dynamic behavior.

In this dynamic situation, the runnable model is function of energy, inertia, damping, stiffness, Force etc. We can relate to the Newtonian laws through concepts of energy, inertia, reaction etc. Other motion attributes in terms of (deformation, velocity or speed, acceleration) will need to be assimilated. This can also be expressed mathematically in the form:

$$\mathrm{M}\ddot{x} + \mathrm{c}\dot{x} + \mathrm{K}x = \mathrm{F}$$

where again the $\mathrm{M}\ddot{x}$ represents Inertial Forces, $\mathrm{c}\dot{x}$ represents damping forces and $\mathrm{K}x$ Restoring forces, which are internal, while an external force F is applied.

Expressing accelerations in the form: $\ddot{x} = (\mathrm{F} - \mathrm{c}\ddot{x} - \mathrm{k}x)/\mathrm{M}$, shows that if the displacements are higher, the accelerations will be lower. This can only occur, if the crushing is higher. Thus, based on this PTV approach, where simulation results pointed to physical behavior that could be validated, new runnable objects/entities that can be used in a qualitative modeling structure were found in terms of energy, work done, inertia, mass etc. During these studies, it was found that a parametric approach to using simulations, which are repeatedly run for a number of scenarios helped refine the cognitive models of typical physical behaviours.

3 Structure of the Simulation Driven Cognitive Model

From the case studies, it appears while a number of intuitive models were fairly representative of simpler physical behaviours. In more complex behaviours, e.g., dynamics, there was a need to enable learning and reasoning using physical laws and mathematical models. In some cases, e.g., impact behavior, the need to revise the intuitive models was observed. However, repeated use of simulation in a PTV environment showed that the cognitive models could undergo refinement and improvement. This led to more realistic qualitative description of behavior. However, this is with the knowledge that simulation models are limited by the capability of the embedded mathematical model. There is a whole body of research which looks at qualitative reasoning [11]. In Chandra [12], there is an argument that spatial reasoning occurs in engineers in a qualitative way. A mapping of quantitative data to qualitative attributes occurs in human beings with regard to qualitative representation of quantities and embedded in the cognitive model. Now this is illustrated in the following form: deformation $\langle 0, 1.0, 10.0, 100.0 \rangle$, all in mm being represented as deformation $\langle$very low, low, medium, high, very high$\rangle$ based on repeated experience on the simulations and the PTV approach.

Developing a qualitative behavior model could be illustrated for two typical cases: Compression and Energy absorption. The qualitative model can be developed using a simple qualitative spatial reference in terms of qualitative descriptions of direction (up, down, left, right), which is again mapped using the reference co-ordinate system of (x, y, z). A typical qualitative model can be represented as follows using a qualitative spatial reference (QSR) Dir {(up(u), down(d), left(l), right(r))} and direction represented as (l-r, r-l) etc. All quantities are mapped using an exceedance limit approach, and resolved into intervals in terms of qualitative attributes, represented as [q]. These could be in the form long, short etc. for length, low, medium, high for stiffness. For example, compression phenomena can be represented for any representative object (A) as:

(i) For steady state events: Compression

$\forall$Object(A)
Attributes {(Length[q]), (Size[q]), Material[q]);
Internal_Force {(Stiffness[q])}
}
Influence: External_Force[q], dir);
Reaction:
Internal_Force $\propto$ Length[q] $\oplus$ Size[q] $\oplus$ Material[q];
Result:
Deformation [q]
{
Deformation $\propto$ Stretch: if Org length < Final length
Deformation $\propto$ Compressed: if Org length > Final length
}
Behaviour Model:
Deformation(Stretch, High)) $\propto$ (Internal_Force(Low) ⍰ External_Force(High, up))
Deformation(Compression, High) $\propto$ (Internal_force(Low) ⍰ External_Force (High, down))

(ii) For dynamic physical behavior events: acceleration

Object (A)
Motion Attributes {Displacement([q], dir)), Velocity [q], dir), Acceleration ([q], dir)}
Mass ([q])
Influence:
External_Force ([q], dir);
If $\nexists$ Internal_force [q]: Acceleration(dir) $\propto$ (Mass ⍰ External Force([q], dir))
If $\exists$ Internal Force [q], dir {
Acceleration $\propto$ (Mass, External Force, Internal Force))
}
Behaviour model
Acceleration (High, l-r) if (Mass, Low, External_Force (High, l-r), Internal Force(Low, l-r))

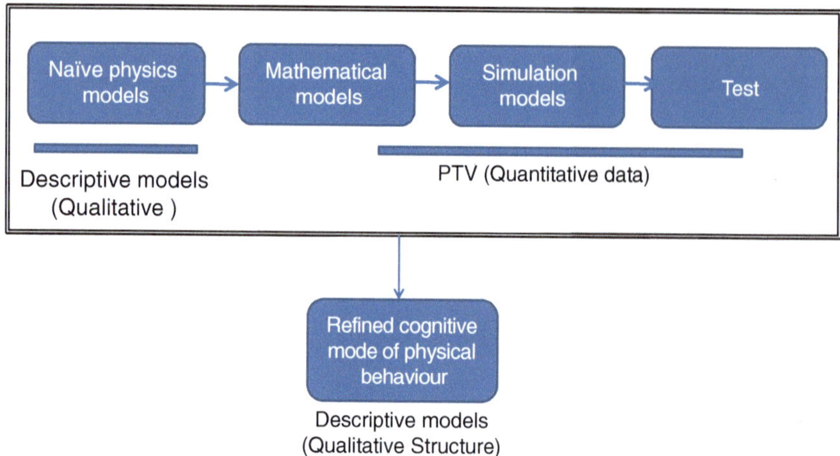

Fig. 6 Flow of cognitive model development

Using a typical structure as noted above, a cognitive model of physical behavior can be elicited from individuals. This model gets continuously refined based on simulations. Case studies indicate that the attributes in fact are constantly refined such that the qualitative representation of the attributes can be used in behavior descriptions. Repeated use of parametric (bounded) simulations helped in refining the qualitative models. Thus, the flow of cognitive model development can be shown as in Fig. 6. During interactions during case studies, most descriptions of behavior were qualitative, even while the underlying training in physical laws and mathematical models was embedded in these descriptions. It can be said that extensive use of simulations produce a qualitative and descriptive model in the mind enabling design synthesis.

Evidence from the case studies also suggests that reasoning occurs at various levels and design synthesis occurs in a constrained requirement envelope. As a result, not all engineers may require to be exposed to the underlying physical models that drive a simulation. This is akin to a driver in a simulation game of a race car or a pilot who uses a simulator to enhance training. It seems that refining cognitive models, (which is in a sense is also design synthesis) about dynamic physical behavior can be optimally done using simulations, as actual tests and experiments are very expensive. It is to be noted that efficient design synthesis will require extensive understanding of dynamic behavior.

There appear to be possibilities of innovation in engineering new products based on the rapid learning from the simulations, using the PTV framework, as experimenting on the simulation model is considerably less expensive compared to testing prototypes. Further detailed studies on the structure of the cognitive models associated with physical behaviour can aid in creating more efficient training procedures for fresh engineers.

4 Conclusion

In this paper, the role of simulation as a source of enriching cognitive models of physical behavior as used in engineering design is examined. As is well known, real experience has traditionally been associated with engineers who have worked on a number of products. As seen here, simulations can rapidly enhance the experience about physical behavior. Intuitive cognitive models can be replaced by cognitive models that are developed based on repeated use of simulations, which are driven by mathematical models embedded with physical laws. It is seen that especially when we consider complex physical behavior, e.g., dynamic behavior, simulation models help in rapidly training fresh engineers. It is argued that cognitive models are runnable, evolve, are descriptive and qualitative. These models can be made to reflect reality through the predict-test-validate (PTV) approach. While, humans are known to use qualitative reasoning, assimilation from a simulation via visualisation and imagery, where spatial reasoning and motion information is absorbed to enable formation or refinement of cognitive models. In summary, it can be said that simulation driven experience can in fact enrich design intuition by synthesis and enable more efficient training processes, especially with regard to understanding of dynamic physical behavior.

References

1. Johnson-Laird, P.N.: Mental Models. Towards a Cognitive Science of Language, Inference, and Consciousness. Cambridge University Press, Cambridge (1983)
2. Gentner, D., Stevens, A.L.: Mental Models. Lawrence Erlbaum Associates, Hillsdale (1983)
3. DiSessa, A.A.: Phenomenology and the evolution of intuition. In: Gentner, D., Stevens, A.L. (eds.) Mental Models, pp. 15–34. Lawrence Erlbaum Associates, Hillsdale (1983)
4. McCloskey, M.: Intuitive physics. Sci. Am. **248**(4), 122–130 (1983)
5. Williams, M.D., Hollan, J.D., Stevens, A.L.: Human reasoning about a simple physical system. In: Gentner, D., Stevens, A.L. (eds.) Mental Models, pp. 131–154. Lawrence Erlbaum Associates, Hillsdale (1983)
6. Staggers, N., Norcio, A.F.: Mental models: concepts for human computer interaction research. Int. J. Man Mach. Stud. **38**, 587–605 (1993)
7. Chandra, S., Blockley, D.I.: Cognitive and computer models of physical systems. Int. J. Hum. Comput. Stud. **43**, 539–559 (1995)
8. de Kleer, J., Brown, J.S.: A framework for qualitative physics. In: Proceedings of the Sixth Annual Conference of the Cognitive Science Society, pp. 11–18 (1984)
9. Kupiers, B.: Commonsense reasoning about causality: deriving behaviour, from structure. Artif. Intell. **24**, 169–203 (1984)
10. Forbus, K., Gentner, D.: Qualitative mental models: simulations or memories. In: Eleventh International Workshop on Qualitative Reasoning, Cortona, Italy (1997)
11. Forbus, K.: Qualitative modeling. WIRES: Cogn. Sci. **2**(4), 374–391 (2011)
12. Chandra, S.: The structure of a physical behaviour description facility. Artif. Intell. Eng. **13**, 91–103 (1999)

13. Hamrick, J.B., Battaglia, P.W., Tenenbaum, J.B.: Internal physics models guide probabilistic judgments about object dynamics. In: Carlson, L., Holscher, C., Shipley, T. (eds.) Proceedings of the 33rd Annual Conference of the Cognitive Science Society, Austin, TX (2011)
14. Tenenbaum, J.B., Kemp, C., Griffiths, T.L., Goodman, N.D.: How to grow a mind: statistics, structure, and abstraction. Science **331**(6022), 1279–1285 (2011)
15. Battaglia, P.W., Hamrick, J.B., Tenenbaum, J.B.: Simulation as an engine of physical scene understanding. Proc. Natl. Acad. Sci. **110**(45), 18327–18332 (2013)
16. Trafton, J.G., Kirschenbaum, S.S., Tsui, T.L., Miyamoto, R.T., Ballas, J.A., Raymond, P.D.: Turning pictures into numbers: extracting and generating information from complex visualizations. Int. J. Hum. Comput. Stud. **53**(5), 827–850 (2000)
17. Sanborn, A.N, Mansinghka, V.K, Griffiths, T.L.: Reconciling intuitive physics and Newtonian mechanics for colliding objects. Psychol. Rev. **120**(2), 411–437 (2013)

Analytical Estimation and Experimental Validation of Acceleration at Spacecraft Solar Array Latch-up Considering Differential Latching

B. Lakshmi Narayana, Gaurav Sharma, G. Nagesh, C.D. Sridhara and R. Ranganath

Abstract Solar array deployment on-board a spacecraft is a mission critical activity. The torque margin over friction torque in the deployment hinge mechanism is an important parameter that ensures positive deployment. Higher torque margin results in higher latch-up moment that affects the design of panel substrates and hinges. Latch-up moment estimated earlier assumes simultaneous latching of hinges. However, in reality, hinges latch at different instants of time due to the presence of closed control loops which has been confirmed by on-orbit observations. The latch-up sequence of the panels influences the distribution of latch-up moments induced at the hinges. In this paper, a mathematical model for a solar array with a yoke two panel configuration is developed using ADAMS software which considers differential latching of hinges. The mathematical model includes the influence of panel flexibility, close control loops, harness, snubbers, ejectors and air drag. The acceleration and moment at latch-up are estimated as a function of time. Deployment time and acceleration at latch-up dictate the magnitude of latch-up moment at hinges and have been compared with experimental value obtained from accelerometer mounted on the outermost panel, during ground deployment test. The novelty in the present work is the formulation of a test validated methodology to analyse the deployment dynamics of a multi-panel solar array considering differential latching which has the potential to simulate the latching of solar array more accurately compared to previously adopted methods. It will be useful in carrying out deployment dynamics of futuristic solar arrays with larger number of panels.

Keywords Solar array · Dynamics · Differential latching · Acceleration

---

B. Lakshmi Narayana (✉) · G. Sharma · G. Nagesh · C.D. Sridhara · R. Ranganath
ISRO Satellite Centre, Old Airport Road, Vimanapura, Bangalore, India
e-mail: narayana@isac.gov.in

A. Chakrabarti (ed.), *ICoRD'15 – Research into Design Across Boundaries Volume 2*, Smart Innovation, Systems and Technologies 35, DOI 10.1007/978-81-322-2229-3_2

1 Introduction

A typical solar array used in satellites consists of a yoke and one or more number of panels. It is kept stowed during launch and deployed in the orbit. The yoke and panels are interconnected by hinges. The deployment of solar array is extensively tested and verified on ground before deployment in the orbit. The ground deployment testing of a solar array consisting of a yoke and two panels is shown in Fig. 1, in stowed, partially deployed and fully deployed configurations.

The ground deployment of solar array is carried out by suspending the yoke and panels on overhead zero-g test set-up. The force in the overhead zero-‘g’ spring is adjusted to balance the weight of the yoke/panels so that the hinges do not experience load. The energy for the deployment is provided by preloaded torsion springs mounted at the hinges along the hinge lines. The Close Control Loops (CCL) provided between Yoke-Panel 1 and Panel 1-Panel 2 enable coordinated deployment ensuring that each panel opens up by a similar angle and latches near

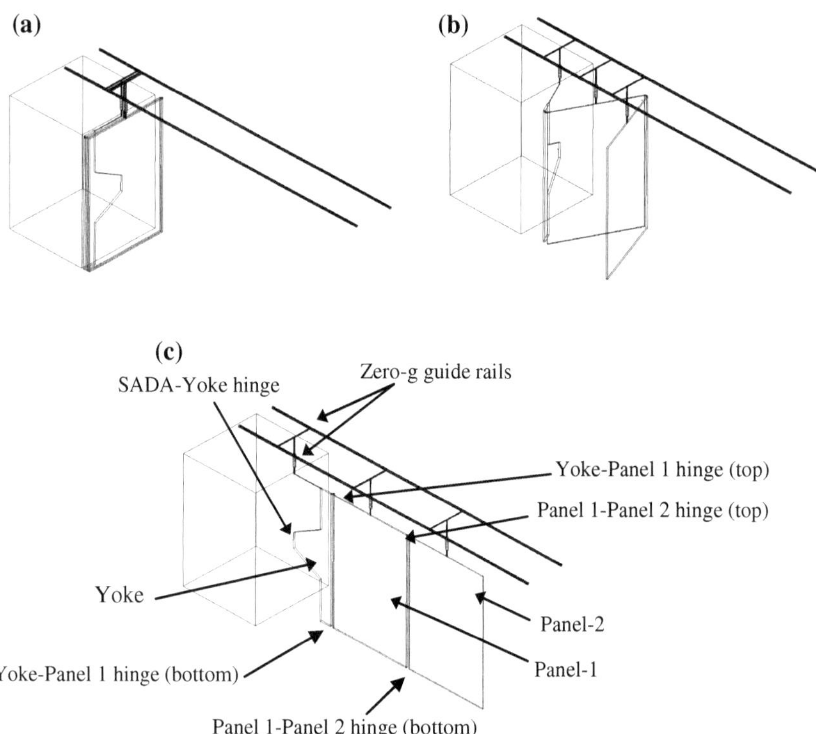

Fig. 1 Schematic of ground testing of solar array deployment. **a** Solar array (stowed). **b** Solar array (partially deployed). **c** Solar array (fully deployed)

simultaneously with respect to other panels (Fig. 6). This minimizes the inter panel latch up shock because the momentum gets countered between the successive joints due to change in the direction of rotation.

The latching of hinges produces latch-up shock at the hinges which needs to be estimated as it is an input for design of hinge mechanism and yoke/solar panel substrates. The dynamics of deploying solar array has been studied by many researchers in the past. Nataraju and Vidyasagar [1] formulated the deployment dynamics of a yoke and three panel configuration using Lagrange's method assuming the panels to be rigid. Subsequently, the latch up shock was estimated for the single solar panel by energy method [2]. Nagaraj et al. [3] studied the deployment dynamics of two flexible links with revolute joints undergoing locking. The flexibility was modeled by finite element method and locking was modeled by momentum balance method. The theoretical results were validated by experiments. The study on yoke and two panel configuration of solar array was extended for 'n' panels by Balaji et al. [4] using matrix approach. The authors also considered the effect of damper and air drag in the model. The energy transfer for each stage of locking for flexible links was addressed by Nagaraj et al. [5]. de Faria et al. [6] evaluated the latch-up force through transient dynamics for the Brazilian solar array. The link flexibility of multi-link flexible structures was modeled by Na and Kim [7] using Timoshenko beam theory and compared with the experimental results of two flexible links undergoing locking, as carried out by B.P. Nagaraj et al.

As observed from above survey, the latch-up moments at the hinges of a deploying appendage on spacecraft were either estimated by energy method or by transient dynamic analysis both of which assume simultaneous latching of hinges. The energy method assumes that the latch up kinetic energy is absorbed by the deployed solar array in the form of strain energy. The loading distribution is assumed triangular and the strain energy of the deployed array is calculated for unit angular acceleration which is then scaled up to obtain equivalent angular acceleration for the latch-up energy. The magnitude of loading is then the product of equivalent angular acceleration, mass per unit length of yoke/panel and distance along the yoke/panel. In transient dynamics, the velocity distribution of the solar array at latch-up is assumed triangular and is obtained from the rigid body dynamics. It is then taken as an initial condition for finite element model of deployed array to estimate the latch up moments at hinges. In actual scenario, latching takes place at different intervals of time due to flexibility of CCL wire ropes connecting yoke to first panel and first panel to second panel. The latch-up moment experienced by the hinges depends on the sequence of latching of the panels. Therefore, a model that predicts the latch-up moment at hinges considering differential latching would be more accurate compared to earlier methods.

In the present work, the deployment dynamics is carried out for solar array having a yoke-two panel configuration in a single software platform ADAMS considering differential latching of hinges. Mathematical modelling includes the influence of panel flexibility, close control loops, harness, snubbers, ejectors and air drag.

Simulations are carried out for ground and on-orbit conditions and acceleration/moment at latch-up are estimated as a function of time. The time taken for array deployment and the acceleration at latch-up are indicative parameters that dictate the magnitude of latch-up moment. These have been measured in the test and used to validate the correctness of analytical estimates. The paper is organized as follows. Section 2 provides the details of modeling, Sect. 3 presents the results of analysis and test and Sect. 4 summarises the essential findings of the study.

2 Details of Modelling

2.1 Modelling of Yoke and Panels Flexibility

Based on the FFT of on orbit body rates at solar array latch-up of an earlier satellite, first bending natural frequency was observed to be 0.73 Hz. The geometry and flexibility parameters of yoke and panels were modeled to achieve 0.73 Hz using PATRAN finite element software. The modal neutral files generated for yoke and panels were imported to ADAMS/View. A revolute joint is created between yoke and ground. A torsion spring of known pre- rotational angle and stiffness is created along the axis of this joint. Panel-1 is positioned keeping the required gap between it and the yoke. Two rigid links are created between the Yoke and Panel-1 facilitating the creation of a revolute joint where the torsion spring is modeled. The procedure is repeated for other inter-panel hinges.

2.2 Modelling of Closed Control Loop (CCL)

A typical CCL used in solar array consists of a pre-tensioned wire rope passing over two pulleys mounted between the successive joints as shown in Fig. 2. The generalized coordinates are shown in Fig. 3.

In the absence of tension springs in CCLs for a given angle α of the yoke, the panel should rotate by angle α_P as given in Eq. 1 but due to tension spring the first

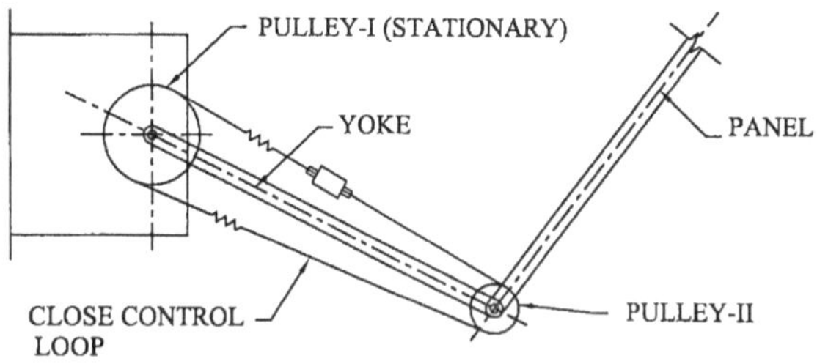

Fig. 2 Schematic of CCL connecting yoke and first panel

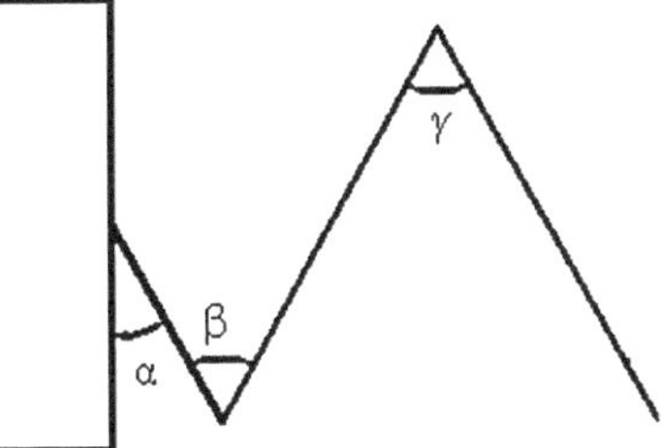

Fig. 3 Generalised coordinates

panel rotates by angle β which changes the length of CCL springs affecting the energy in the tension springs through which the torque between the joints gets adjusted.

$$\alpha_P = \frac{R_{PY}}{R_{P1}}\alpha \tag{1}$$

where, R_{PY} is the radius of yoke pulley and R_{P1} is the radius of panel-1 pulley. The feedback energy of the tension springs is calculated as ½ (difference in the angle of rotation)^2 (radius of pulley)^2 (spring stiffness). It is calculated as:

$$U_{C1} = \frac{1}{2}\left[\frac{R_{PY}}{R_{P1}}\alpha + \beta\right]^2 R_{P1}^2(k_{C11} + k_{C12}) \tag{2a}$$

$$U_{C2} = \frac{1}{2}\left[\frac{R_{P2}}{R_{P3}}\alpha + \gamma\right]^2 R_{P3}^2(k_{C21} + k_{C22}) \tag{2b}$$

where, R_{P1}–R_{P3} are the radii of pulleys 1–3, k_{Cij} are the equivalent stiffness values of CCL loop. Summation of potential energy U_{C1} and U_{C2} gives the total energy of tension springs. Taking $R_{PY}/R_{P1} = 2$ and R_{P1}–R_{P3} as R, the generalized force at various coordinates that have been modeled as external forces in ADAMS/View are given by:

$$\frac{-\partial U}{\partial \alpha} = -2(k_{C11} + k_{C12})R^2(2\alpha + \beta) \tag{3a}$$

$$\frac{-\partial U}{\partial \beta} = -(k_{C11} + k_{C12})R^2(2\alpha + \beta) - (k_{C21} + k_{C22})R^2(\beta + \gamma) \tag{3b}$$

$$\frac{-\partial U}{\partial \gamma} = -(k_{C21} + k_{C22})R^2(\beta + \gamma) \tag{3c}$$

2.3 Modelling of Harness and Snubbers/Ejectors

The power generated by solar cells bonded on to the solar panels is transmitted through power cables that extend from one panel to another and from innermost panel to yoke where they are connected to a solar array drive assembly. Measured torque characteristics of the harness are modeled using spline, as shown in Fig. 4. Preloaded snubbers are provided between yoke and spacecraft deck and between panels to enhance the stowed natural frequency of the stack and limit the amplitude of vibration during launch. They also provide initial push force to the deploying panel acting for a very short duration. These are also provided at hold down interfaces to provide initial push force for deployment of panels. This is represented in the mathematical model by a linearly varying force with maximum value at the start of deployment and reducing to zero (once the snubbers or ejectors loose contact). A typical snubber/ejector force variation with time is shown in Fig. 5.

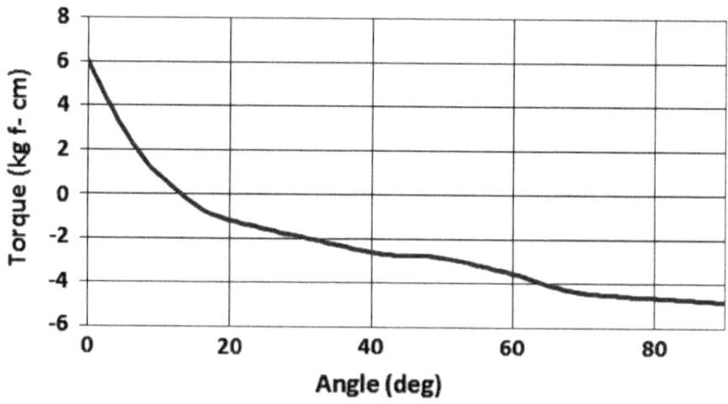

Fig. 4 Typical variation of harness torque

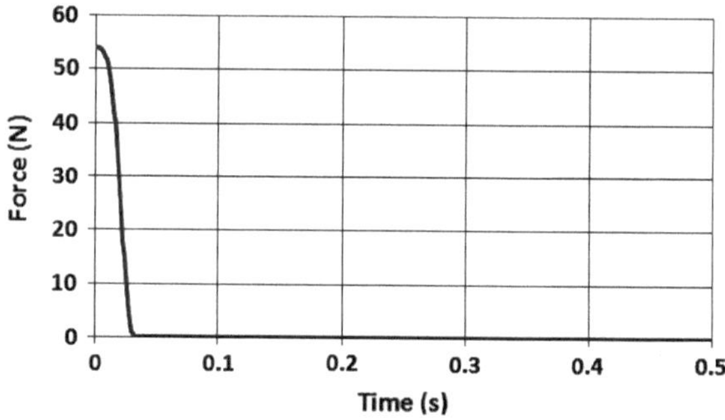

Fig. 5 Typical variation of snubber force

2.4 Modelling of Air Drag

Solar array deployment is carried out on ground to demonstrate the functional requirement of the mechanism. The presence of air drag in the laboratory offers resistance to the deploying panels. Mathematically, the drag force is given by:

$$F_D = \frac{1}{2} C_D \rho A V^2 \tag{4}$$

where, C_D = drag coefficient, ρ = density of air, A = projected area of each strip, V = linear velocity. The panels are divided into ten equal sized strips along the length and the linear velocity of each strip is estimated at each instant of time and the drag force is applied.

2.5 Modelling of Latching

Latching of hinges is modelled using BISTOP [8] function in ADAMS. BISTOP is a two sided impact function that provides resistive torque when the hinge rotates by a predefined angle (90° for SADA-yoke and 180° for inter panel hinges) as a function of angular velocity. This prevents further rotation of the hinges and each latching reduces the degrees of freedom by one. The BISTOP [8] function is defined by following parameters; x: Real variable that specifies the angular displacement to be used for force computation, $\dot{x}$: Real variable that communicates the angular velocity to BISTOP function, x1: Upper limit of x, x2: Lower limit of x, k: Non negative real variable that specifies the boundary surface interaction. In the present model it is hinge stiffness. Experimentally measured hinge stiffness for SADA-Yoke, Yoke-Panel 1 and Panel 1-Panel 2 are used as inputs, e: Positive variable that specifies the exponent of the force deformation characteristics, cmax: Non negative real variable that specifies the maximum damping coefficient, d: Positive variable that specifies penetration at which full damping is applied.

3 Results and Discussion

The simulation is carried out for ground and on orbit conditions. During simulation the torque provided by the BISTOP function are activated once the hinge moves by the predefined angle (90° for SADA-Yoke hinge and 180° for inter panel hinges). As the locking of hinges is not simultaneous, simulation is continued till all the hinges get locked. Locking torques are activated using simulation scripts. The results are presented in following sections.

3.1 Ground Simulation

Simulation provides the deployment time, latch-up velocity of panels, acceleration near Panel 1-Panel 2 hinge, and latch-up moments, during deployment and subsequent to latching. The latching time and peak moments are summarised in Table 1. The variation of angle of opening of yoke and panel joints with time is shown in Fig. 6 and the corresponding latch up moment plots for the hinges are shown in Fig. 7.

It is observed from analysis that SADA-Yoke latches first at 11.88 s, followed by latching of Panel 1-Panel 2 hinges at 12.0 s and final latching of Yoke-Panel 1 hinge at 12.01 s. This matches with the observed deployment time of 12.0 s on ground measurement. Peak moment of 95.0 N m was observed at SADA-Yoke hinge due to first latching whereas the latch-up moments at other two hinges are observed to be nearly equal as they latch nearly at the same time.

The linear acceleration at a point near Panel 1-Panel 2 hinge is measured, as shown in Fig. 8. The predicted and measured acceleration at this location is shown in Figs. 9 and 10 respectively.

It can be observed that the predicted peak acceleration is 0.68 g and matches well with measured acceleration of 0.7 g.

Table 1 Estimated latch-up moment at hinges for ground conditions

	SADA-Yoke	Yoke-Panel 1	Panel 1-Panel 2
Latching time (s)	11.88	12.01	12.0
Latch-up moment (N m)	95.0	30.0	27.0

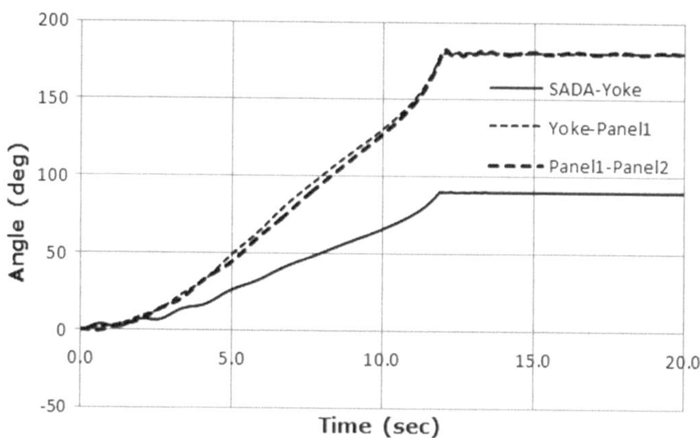

Fig. 6 Angle of opening of yoke and panels with time for ground conditions

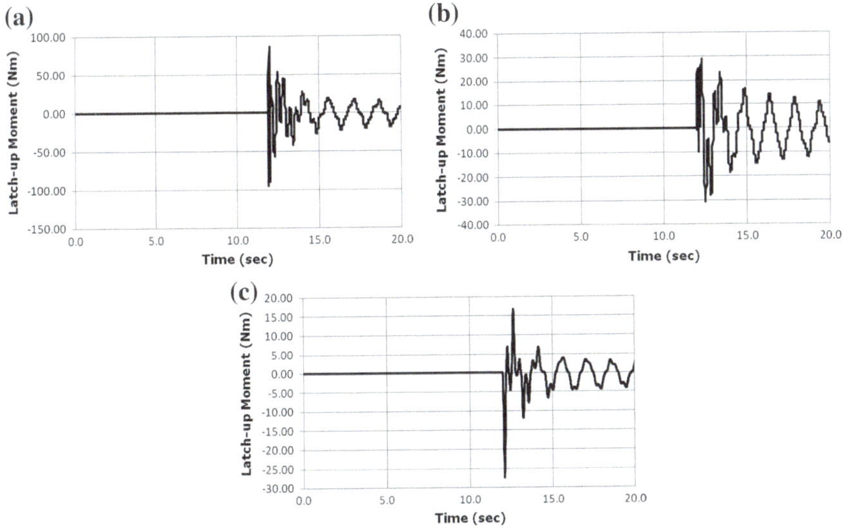

Fig. 7 Variation of latch-up moment at different hinges for ground conditions. **a** Latch-up moment at SADA-Yoke. **b** Latch-up moment at Yoke-Panel 1. **c** Latch-up moment at Panel 1-Panel 2

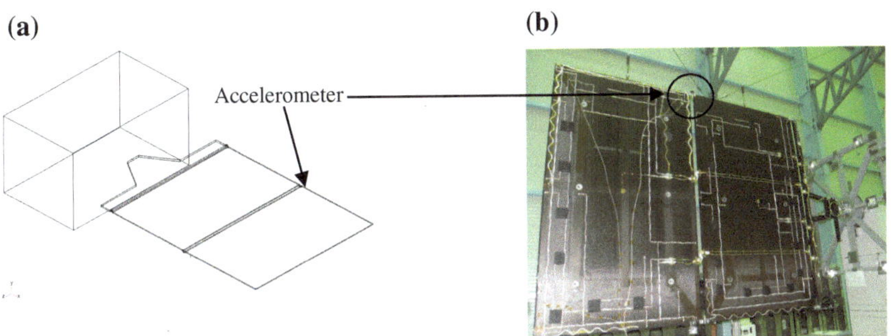

Fig. 8 **a** Accelerometer location. **b** Deployed solar array

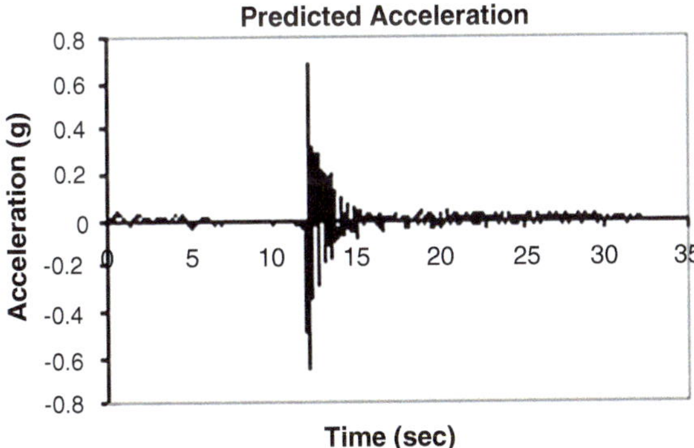

Fig. 9 Predicted acceleration versus time

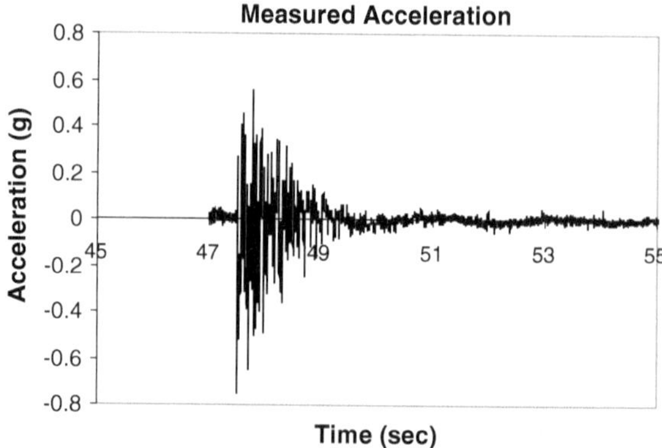

Fig. 10 Measured acceleration versus time

3.2 On Orbit Simulation

Simulation studies are carried out for on-orbit conditions for estimating the deployment time and latch-up moment. The variation of angle of opening of yoke and panel joints with time is shown in Fig. 11. Latching time and peak latch up moments are presented in Table 2. Lesser co-efficient of friction compared to ground conditions, absence of air drag and effects of zero-g simulation hardware, results in faster on-orbit deployment and higher latch-up moments compared to ground conditions. It may be observed that the sequence of latching on-orbit is

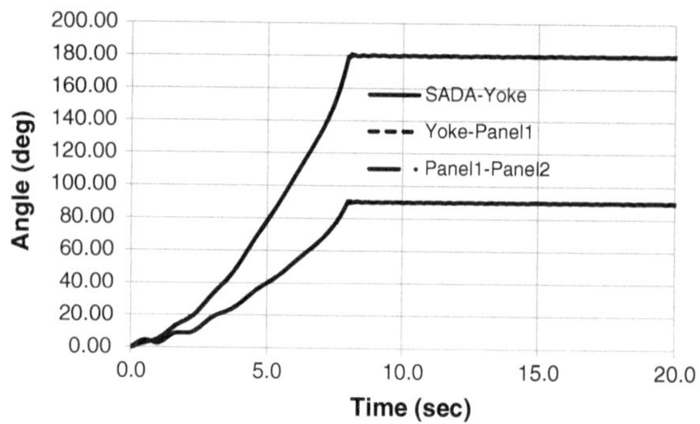

Fig. 11 Angle of opening of yoke and panels versus time for on-orbit conditions

Table 2 Estimated latch-up moment at hinges for on-orbit conditions

	SADA-Poke	Yoke-Panel 1	Panel 1-Panel 2
Latching time (s)	7.9	8.0	8.1
Latch-up moment (N m)	124	52.0	27.0

SADA-Yoke, Yoke-Panel 1 and Panel 1-Panel 2. The peak linear acceleration at a point near Panel 1-Panel 2 hinge is estimated as 1.48 g.

4 Conclusions

The dynamics of deploying solar array consisting of a Yoke and two Panels has been successfully carried out considering the influence of yoke and panel flexibility, close control loops, harness torque variation, snubber force, ejector force and air drag. During deployment, the array is a multibody dynamical system and after latch up, it behaves as a structure. Both these phases are modeled and analysed on a single software platform ADAMS seamlessly considering differential latching. Simulation results have been obtained in terms of deployment time, angle of opening, acceleration and latch-up moment, for ground and on-orbit conditions. The fidelity of simulations is validated on ground by comparing the measured deployment time and acceleration at latch-up with predicted values. The methodology developed provides an effective means to model the dynamics of deploying solar array more accurately compared to previous methods. This will be of immense use in the study of deployment dynamics of future spacecrafts where large number of panels would need to be deployed.

Acknowledgments The authors would like to thank Group Director, Spacecraft Mechanisms Group, Deputy Director, Mechanical Systems Area and Director, ISAC for their support and encouragement. The authors wish to thank colleagues from Spacecraft Mechanisms Group, Structures Group and Environmental Test Facility, for providing guidance and support for carrying out the experiment.

References

1. Nataraju, B.S., Vidyasagar, A.: Deployment dynamics of accordian type solar array considering flexibility of closed control loops. In: 38th Congress of International Astronautical Federation, Brighton, UK (1987)
2. Nataraju, B.S., Chinnasamy, R., Krishnamurthy, T.S., Bonde, D.H.: Modelling of deployment mechanisms for latchup shock. ESA J **13**, 393–400 (1989)
3. Nagaraj, B.P., Nataraju, B.S., Ghosal, A.: Dynamics of a two link flexible system under going locking: mathematical modeling and comparison with experiments. J. Sound Vib. **207**(4), 567–589 (1997)
4. Balaji, K., Nataraju, B.S., Suresha kumar, H.N.: Matrix approach to deployment dynamics of an n panel solar array. Proc. Inst. Mech. Eng. K J. Multi-body Dyn. **217**, 15–27 (2003)
5. Nagaraj, B.P., Nataraju, B.S., Ghosal, A.: Energy transfer during locking of deployable flexible links. J. Spacecraft Technol. **13**(2), 19–32 (2003)
6. de Faria, A.R., Cardozo, L., Fonseca, I.M.: Brazilian multimission platform solar array generator deployment. In: 55th International Astronautical Congress 2004, Vancouver, Canada
7. Na, K.S., Kim, J.H.: Deployment of multi link flexible structures. J. Sound Vib. **294**(1–2), 298–313 (2006)
8. ADAMS Users' Guide, Version 10 (2010)

Assessment of the Biomimetic Toolset—Design Spiral Methodology Analysis

Pierre-Emmanuel Fayemi, Nicolas Maranzana, Améziane Aoussat and Giacomo Bersano

Abstract The potential of bio-inspired design has been illustrated many times over human history. Nevertheless its process is complex, as biologists and engineers have to work together without necessarily sharing the same vision or language. Methods and tools have therefore been developed in order to support these kind of approaches. Following the outlined problem-driven biomimetic process, the article seeks to provide a global view of the existing biomimetic toolset. To do so, tools are classified according to their methodological purpose. For each identified type of tools an evaluation chart has been generated. By its experiment, the article addresses more specifically the assessment of tools originating from the Design Spirals methodology. The article provides thereby an identification of advantages, high accessibility and convenience, and limitations, need to be coupled with other tools, of the assessed tools. This identification leads to the proposal of an axis of research to develop new biomimetics tools.

Keywords Biomimetics · Methods · Biomimicry · Bio-inspired design

1 Introduction

Hornby defined inspiration as 'the process of being mentally stimulated to do or feel something, especially to do something creative' [1]. Designers constantly use sources of inspiration, also called triggers or stimuli, to generate new ideas and to solve problems. Active search for inspiration, but also inspirational process based on unconsciousness or serendipity, can be a source of inspiration in order to work in a systematic way [2].

P.-E. Fayemi (✉) · N. Maranzana · A. Aoussat
Product Design and Innovation Laboratory, Arts et Métiers Paristech, Paris, France
e-mail: Pierre-Emmanuel.Fayemi@Ensam.eu

P.-E. Fayemi · G. Bersano
AIM-Innovation, Palaiseau, France

A. Chakrabarti (ed.), *ICoRD'15 – Research into Design Across Boundaries Volume 2*, Smart Innovation, Systems and Technologies 35, DOI 10.1007/978-81-322-2229-3_3

Biological knowledge has been demonstrated throughout ages as an effective source of inspiration [3]. From mimicking animals' behaviors to hunt and survive, to Leonardo da Vinci mimicking birds' flight to design his flying machine, biological knowledge has been a constantly investigated in order to provide new materials, processes or strategies. In our closest times, inspiration from aquatic organisms has led to at least eleven Nobel Prize rewarded medical breakthroughs [4]: Metchnikoff laid the foundation for the non-organ-specific autoimmune defense through the observation of the universality of a larvae of sea stars mechanism he will called *immunity*; Richet discovered the anaphylactic shock (a heightened immune response) while studying the venom of box jellyfish; O von Warburg identified the mechanism preventing two spermatozoa to penetrate the same ovum by studying sea urchins; Hodgkin and Huxley explained the transmission of nerve impulses by experimenting on squid; Kandel discovered the key elements of long term memory from Aplysia; Hunt et al. identified Cyclin B, essential in cell cycles, from purified sea stars oocytes; Shimomura isolated the GFP, protein used in a wide variety of areas due to its fluorescence, from the jellyfish *Aequorea victoria*. Blackburn, Greider and Szvostak have highly impacted ageing and cancer science thanks to the explanation of chromosome reproduction through from their study of a ciliate protozoa called Tetrahymena.

Regardless of its empirically proven effectiveness, bio-inspiration refers to several concepts/sub-concepts. Biomimicry, bionics and biomimetics are the most common terms that formalize these concepts. Having been used interchangeably since their introduction by the vast majority of people, bio-inspiration related terms have the tendency, in conjunction with the increasing of their current exposure, to drift apart from each other over time.

This article, and its related outcomes, will tackle, with a specific focus, biomimetics. Biomimetics can be defined as an interdisciplinary creative process between biology and technology, aiming at solving antrophospheric problems through abstraction, transfer and application of knowledge from biological model. Biomimetics refer thus to the methodological aspects of bio-inspiration [5].

Consequently, the article will firstly introduce the definition of the problem driven biomimetic process and its needs. Existing tools and methods which may address these identified needs will then be presented. Lastly, one methodology and its forming tools, the Design Spiral methodology, will be assessed. This assessment has been made possible thanks to the formalization and the use of biomimetics criteria of evaluation.

2 Biomimetics

2.1 Biomimetic Process

Biomimetics can be used in two separate ways, solution-driven or problem-driven. The solution driven method assumes a biological system that performs a function which the engineer wants to emulate as a starting point. The process is focused on abstracting the biological system so that the designer can then use the functional

Table 1 Problem driven biomimetic process and its specific steps

	Bio-Inspiration specific						
Assess the situation/identify a problem	Abstract the technical problem	Translate into a biological challenge	Identify potential biological models	Select the biological model of interest	Abstract biological strategie(s)	Translate into a technological challenge	Implement in initial situation
Evaluate	Abstract	Transfer	Applicate	Applicate	Abstract	Transfer	Applicate

model to inspire an engineering design concept. The problem-driven method assumes that there is a specific behavior/function which the designer wishes to fulfill. The process is focused on determining the biological systems which needs to be considered for inspiration. The rest of the article will focus on the problem driven (PD) method of biomimetics.

A generic problem driven biomimetic process, in accordance with the definitions stated in Sect. 1 and work of the ISO TC 266 committee, has been outlined [5]. This process involves 9 sequential steps: define the human needs/challenge; abstract the technical problem; translate it into a biological challenge; identify potential biological models; select the biological model(s) of interest; abstract the biological strategy(ies); translate it into a technological challenge; implementation to the initial situation; evaluate.

As the last step, "evaluate", may trigger another cycle, the generic problem driven biomimetic process can be simplified in a non-linear 8-steps process.

This generic process involves tools from 4 different types, as mentioned in Table 1: Assessment/Problem identification; Abstraction; Transfer; Application.

This classification is essential to assess existing tools, as comparing one tool with another from a different type does not appear as relevant.

2.2 *Methods and Tools Supporting Biomimetic Design*

Methods and tools are the element which enables designers to implement biomimetic design processes. It is therefore essential to tackle them and their methodological basis and purpose in order to understand how they could address the needs of the different outlined steps of the biomimetic process.

2.2.1 Design Spirals Methodology

Biomimicry 3.8 Institute co-founded by Janine Benyus promotes the emergence of new sustainable technologies through the study and imitation of nature. They have developed the Design Spirals Methodology which emphasizes values of biomimicry practice and its related benefits. The methodology is articulated around several tools such as the Biomimicry Taxonomy, Ask Nature [6], or the Living Principles.

2.2.2 BioTRIZ

Based on the "theory of inventive problem solving" of Altshuller [7], the working group upon bio-inspiration of the University of Bath, created in 2001, has developed, in 2008, a new methodology called BioTRIZ. A TRIZ analysis of solutions provided by Nature has shown a substantial difference between "Nature design" and "engineering design": technology tends to solve the problem thanks to the use of energy while Nature is more about using information and structure of matter [8]. Therefore, the main thrust of the approach is to allow to benefit from Nature's strategies thanks to the power of abstraction of TRIZ.

2.2.3 Natural Language Approaches

An alternative methodology is based on lexical analysis and has been proposed by Vakali et al. [9]. The goal is to associate semantically and statically some desired engineering functions or processes to keywords in the easily accessible biological literature in search algorithms. Emphasizing this work, Chiu and Shu incorporated Wordnet to the tool to overcome the high amount of irrelevant matches revealed previously. The lexical database Wordnet provides the ability to distinguish homonyms and to handle troponym trees. Further implementations based on this work have made possible the retrieval of meaningful keywords with sometimes more useful matches than the corresponding engineering keywords alone [10, 11]. However, such approaches need, even with a semi-automatic search, a manual filtering of the results, which could be challenging and time consuming.

2.2.4 Functional Modelling

Biological modelling. Tinsley et al. [12] reported that an effective way to convert biological knowledge and especially design solutions into valuable input for engineering problems is to model it with the Functional Basis language. Based on this statement, Nagel et al. [13] proposed guidelines which offer a functional representation of a biological system with incorporation of its mimicry category and scale. Biological functional models can therefore be implemented in a database. Thus, designers could search and identify analogous biological models for their engineering system.

Databases and modelling approaches. Such approaches focus on bringing to light potential biological models of interest which designers with no or limited background in biology would not be able to identify by their own. Chakrabarti et al. have developed IDEA-INSPIRE [14], a software which uses SAPPhIRE constructs to identify different levels of abstraction to represent biological information. With a similar approach, Vattam et al. have developed a knowledge-based CAD system called Design by Analogy to Nature Engine (DANE) [15, 16]. Both tools are designed around custom-built databases. Thus, they are dependent on the addition of new engineering and biologically representations and they require time and effort. Indeed, the needed level of detail to analyze both engineering and biological systems is time consuming. The wealth of information on biological organisms and furthermore the biological information to be processed, due to current unknown species, suggests that these databases would only contain a minor portion of the existing knowledge.

The identified biomimetic toolset has been synthesized in Table 2.

Regarding the apparent large number of bio-inspired specific tools, the choice has been made for this article only to tackle one method. As Design Spirals would supposedly differ from the other methods, being the only toolbox originating from Life Science—designed by biologists, it will thus constitute the first methodology to be further investigated.

3 Assessment of the Design Spirals Tools

Several tools addressing the same theoretical steps of the biomimetic process exist. It is therefore important to try to understand what their different purpose and provided features are. This will allow biomimetic designers to be able to choose tools according to the specificity and constraints of their bio-inspired design processes. Consequently, there is a need for assessment criteria in order to compare tools. This assessment can be extended to methods with the outcome of allowing the assessment of the global biomimetic toolset maturity.

3.1 Presentation of Tools

The Design Spirals methodology is made of three major design process steps (ethos, reconnect and emulate). Several tools are used among its process [8].

3.1.1 Life's Principles

Life's Principles is a benchmark tool based on Nature's strategies to create conditions conducive to life. The tool allows designers to identify new axes of sustainable improvement regarding their system.

3.1.2 5-WHYs

5-WHYs is an iterative process tool consisting in asking "why?" several times in order to determine what the real challenge is.

Table 2 Identified biomimetic methods and tools

	Assess the situation	Abstract technical problem	Translate into biological challenge	Identify potential biological models	Select biological model(s) of interest	Abstract biological strategie(s)	Translate into technological challenge	Implement in initial situation
BioTRIZ	IFR	Su-Field MSD	Resources	Brainstorming		Technical contradictions	Inventive principles	
Design spirals	Living principles	5-WHYs	Taxonomy	Ask Nature				
Functional modelling		FBS		DANE IDEA-INSPIRE		Biological modelling		
Natural language				BIDlab databases				

3.1.3 Taxonomy

Taxonomy consists in a transferring tool that allows designers to translate a technical problem into a biological one. Taxonomy is a gateway to Ask Nature, corresponding to its functional ontology.

3.1.4 Ask Nature

Ask Nature, known for being the largest database related to bio-inspiration, aims at creating a link between a natural phenomenon, a living organism presenting such phenomenon and a potential expert of this organism. To do so, the database is articulated around its own taxonomy of functional keywords. Three modus operandi co-exist: identifying a verb corresponding to the challenge to solve, defining more global concepts than the challenge to solve and going against the initial request.

3.2 Criteria of Assessment

With these different classes of tools identified, defining their objectives to picture an ideal tool is now possible. In order to do so a list of criteria has been established.

Evaluation tools aim at assessing an initial situation and identifying problem(s). Thus an ideal evaluation tool should: be able to analyze exhaustively the initial situation; be able to analyze precisely and in details the initial situation; provide an ideal situation to aim at; prioritize outlined issues in order for designers to know which are the key sub-problems to solve first.

Abstracting tools, aim at modelling a problem. To fulfill this objective, from the theoretical contribution point of view an ideal abstracting tool should: be able to model complex problems in order to fit as much cases as possible; strongly integrate different systemic levels to allow designers to model their problems precisely; effectively filter information regarding its significance for the problem solving process, to avoid overflowing designers with information they do not need; establish a very strong access to the problem in a generic way in order to allow its translation into a biological challenge; completely maintain specific constraints with respect to the generated generic problem by avoiding an over-generification of the problem which could lead to identification of biological models that do not solve the original technical problem.

Transfer tools, which are involved in translating a technical problem into a biological challenge and vice versa, imply idea generation. To fulfill this objective, an ideal transferring tool should: point to an ideal unique solution only; be able to strongly enlarge the designer(s) knowledge if necessary; allow the designer to completely sub-modularize generated solution(s) to enhance versatility of the generated concept; generate solution(s) with high level of inventiveness.

A practical/operational point of view, in addition to the previous mentioned criteria, was added to the experiment. Thus, an ideal tool: should be able to be implemented in a very short time; could be used intuitively, without the need for any training; should be as efficient as a stand-alone tool as well as in combination with other tools; should be able to used in any scientific or industrial domain without the need for adjustment; should have the same efficiency when used by a single designer than by a group of designers; should facilitate the use of subsequent tools by offering an up-stream support for their completion. These practical criteria remain the same for every type of tools.

3.3 Experimental Protocol

The experiment has taken place in an industrial context. Bubble tree is a small French company which commercializes temporary accommodation through premium eco-tourism solutions. Its "bubbles" are plastic film spherical structures held by an air flow originating from a compressor. In their wish to provide more resilient products, in order for them to stay longer at the same location, the bubbles needed to integrate themselves in a more transparent way in their environment. Thus, a biomimetic problem solving workshop has been set up. The workshop problematic has been defined as: "How to dynamically manage fluxes of energy".

The attending group consisted of equal numbers of problem-solving/technology transfer experts and bio-inspiration experts.

Each use of tools was followed by an evaluation phase where its advantages and drawbacks were assessed thanks to questionnaires. These questionnaires incorporate the theoretical and practical criteria mentioned in Sect. 3.2. Each question corresponds to a single criterion. Replies are formalized in a 4 point Likert-type scale. This evaluation phase was consensus-building oriented.

3.4 Experiment Results

In the ideal case the use of one single tool should be sufficient for the completion of one process step.

Life's Principles is presented as a tool that scores average on theoretical criteria. Participants highlight the fact that they were taken aback by the criteria specificity differences. Some assessment criteria, such as "chemistry in water" are very specific, preventing them from being relevant in every case. On the opposite side, criteria such as "replicate strategies that work" are very generic. Generic criteria assessment implies putting them in perspective while specific criteria can be directly assessed, requiring a certain amount of mental gymnastics to go from one to another (Fig. 1).

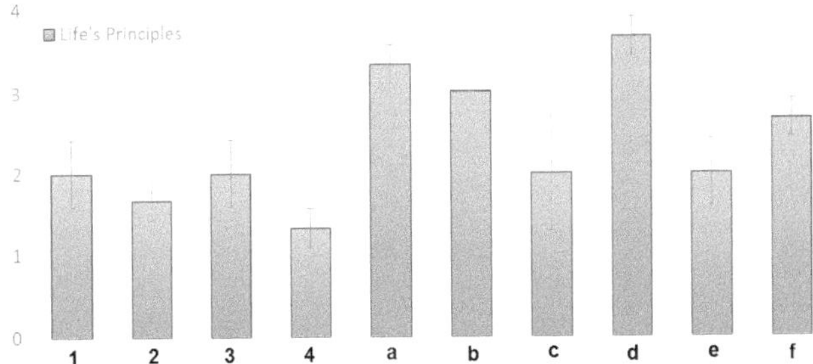

Fig. 1 Life's principles' (assessment tool) results. *1* Completeness of the analysis. *2* Measurement accuracy. *3* Idealized vision. *4* Capacity to prioritize. *a* Implementation time. *b* Ease of use. *c* Stand-alone capacity. *d* Adaptability. *e* Capacity to help single solver. *f* Subsequenttool facilitation

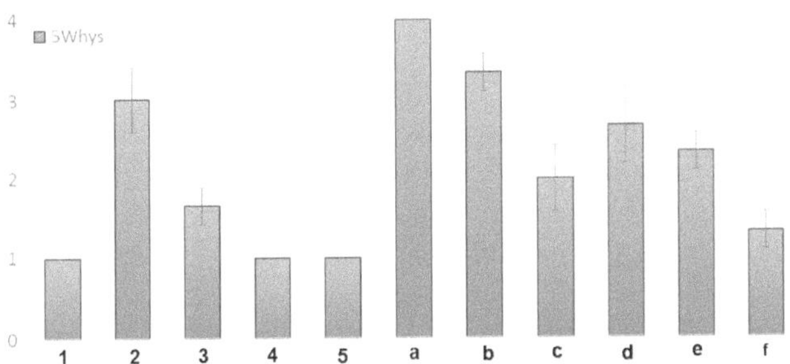

Fig. 2 5-WHYs's (abstraction tool) results. *1* Modelisationcapacity. *2* Sub/supersystems integration. *3* Information filtering capacity. *4* Capacity togeneralize. *5* Capacity to maintain constraints. *a* Implementation time. *b* Ease of use. *c* Stand-alone capacity. *d* Adaptability. *e* Capacity to help single solver. *f* Subsequent tool facilitation

Lastly, participants put forward that despite providing an idealized vision for every criteria, even if Life's Principles include examples both in the biological and in human design field, a global idealized vision is provided by the tool. This allow Life's Principles to formalize a goal to reach, specific to the tackled challenge, even if they are generic.

Despite its subsystem/super systems integration, the 5 why's tools did not convince workshop's participant on the theoretical side. Its lack of capacity to abstract the problem induced the integration of another abstraction tool, System Operator Tool, originating from TRIZ, in order for the experiment to continue (Fig. 2).

While all participants mentioned the fact that they appreciated the use of Ask Nature coupled with its Taxonomy, evaluation results do not seem to reflect it. Raised comments and criticism involve its lack of descriptive text, the limited completeness the solutions displayed and the lack of a ranking of the identified solutions considering their relevance (Fig. 3).

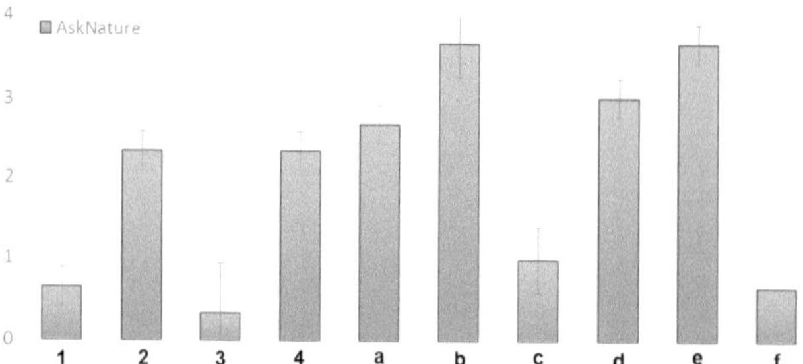

Fig. 3 Ask Nature (transfer tool) results. *1* Capacity to point at unique solution. *2* Capacity to enlarge designer(s) knowledge. *3* Capacity to modularize solution(s). *4* Level of inventiveness of solution generated. *a* Implementation time. *b* Ease of use. *c* Standalone capacity. *d* Adaptability. *e* Capacity to help single solver. *f* Subsequenttool facilitation

4 Conclusion

The article investigates how existing bio-inspired tool might address the need for a problem-driven biomimetic process. From a designer point of view, Design Spiral offers means to fulfill biomimetic problem solving step needs. Criteria displayed in this article can be used as a benchmarking tool in order to compare tools originating from this methodology with others that could have already been integrated by a designer in his bio-inspired process.

With a respective average score of 1.75, 1.5 and 1.4, Living Principles, 5-WHy's and Ask Nature present a low theoretical rating. Looking at these evaluation results, the Design Spiral methodology does not seem capable of serving as the core of a bio-inspired process. During the experiment, options for improvement have been identified. For example, the use of a scoring system instead of checkboxes could serve two purposes. First, the measurement accuracy of the tool could be significantly improved. Second, a designer would not have to ask himself if he should check a box when a criterion is not fully addressed by the object of study. This removal of certain ambiguities would lead to faster usage.

Nonetheless, tools addressed by the article are not devoid of interest or irrelevant to bio-inspired design. The swiftness of the methodology and its ease of use, coupled with its relative adaptability ensure that Design Spiral methodology could be an asset for developing a bio-inspired design process in an industrial context.

Being the only available tools of the biomimetic toolset originating from the Life Science field, and considering the need of bio-inspired approaches for interdisciplinarity, it thus seems important not to overlook the Design Spiral methodology tools. Combining the assessed tools with tools from other methodology in order to compensate their weaknesses, would allow designers to integrate both technical engineers and biologists' vision and language in a single approach. On the other hand, methodologies coming from the engineering field, such as TRIZ which is described as a complex and rigid design method [17], could benefit from the accessibility and convenience of tools from the Design Spiral methodology.

From a more general point of view, the article, and more specifically Table 2 coupled with the experimentation shed light on two facts. The first one is that biomimetic designers are facing a current lack of means to identify potential biological models. Several tools using various methods exist, but do not seem to be publicly available. The second statement is that a methodological gap exists. None of the identified tools focuses on selecting the biological model of interest. These two axes constitute a major challenge for the biomimetic approach democratization, especially in an industrial context, as they could lead to a major improvement of the overall biomimetic process.

Criteria displayed can, in a longer term, be used to assess the global biomimetic toolset. To do so, other tools and methodologies presented in Sect. 2.2 will be assessed with the same protocol in future work. Criteria would thus become a benchmarking tool which allow the biomimetic designers to choose the right design tools to fit their projects.

References

1. Hornby, A.: Oxford advanced learner's dictionary of current English. In: Wehmeier, S. (ed.). Oxford University Press, Oxford (2000)
2. Goldschmidt, G., Sever, A.L.: Inspiring design ideas with texts. Des. Stud. **32**(2), 139–155 (2011)
3. Simon, H.A.: Search and reasoning in problem solving. Artif. Intell. **21**, 7–29 (1983)
4. Boeuf, G.: Océan et recherche biomédicale. Journal de la Société de Biologie **201**(1), 5–12 (2007)
5. Fayemi, P.-E., Maranzana, N., Aoussat, A., Bersano, G.: Bio-inspired design characterisation and its links with problem solving tools. In: Design (Design14) (2014)
6. Baumeister, D.: Biomimicry Resources Handbook. Biomimicry 3.8 (2013)
7. Altshuller, G.: And suddenly the inventor appeared: TRIZ, the theory of inventive problem solving. Technical Innovation Center, Inc. (1996)
8. Vincent, J.F., Bogatyreva, O.A., Bogatyrev, N.R., Bowyer, A., Pahl, A.K.: Biomimetics: its practice and theory. J. R. Soc. Interface **3**(9), 471–482 (2006)
9. Vakili, V., Chiu, I., Shu, L.H., McAdams, D., Stone, R.: Including functional models of biological phenomena as design stimuli. In: Proceedings of ASME 2007 IDETC/CIE (2007)
10. Chiu, I., Shu, L.H.: Biomimetic design through natural-language analysis to facilitate cross-domain information retrieval. Artif. Intell. Eng. Des. Anal. Manuf. **21**(1), 45–59 (2007)
11. Cheong, H., Chiu, I., Shu, L.H., Stone, R.B., McAdams, D.A.: Biologically meaningful keywords for functional terms of the functional basis. J. Mech. Des. **133**(2), 021007 (2011)
12. Tinsley, A., Midha, P.A., Nagel, R.L., McAdams, D.A., Stone, R.B.: Exploring the use of functional models as a foundation of biomimetic conceptual design. In: Proceedings of DETC2007-35604, Las Vegas, NV (2007)
13. Nagel, J.K.S., Nagel, R.L., Stone, R.B., McAdams, D.A.: Function based, biologically inspired concept generation. Artif. Intell. Eng. Des. Anal. Manuf. **24**(4), 521–535 (2010)

14. Chakrabarti, A., Sarkar, P., Leelavathamma, B., Nataraju, B.S.: A functional representation for aiding biomimetic and artificial inspiration of new ideas. Artif. Intell. Eng. Des. Anal. Manuf. **19**(2), 113–132 (2005)
15. Vattam, S., Wiltgen, B., Helms, M., Goel, A.K., Yen, J.: DANE: fostering creativity in and through biologically inspired design. In: Design Creativity 2010, pp. 115–122. Springer, London (2011)
16. Vattam, S.S., Helms, M.E., Goel, A.K.: A content account of creative analogies in biologically inspired design. Artif. Intell. Eng. Des. Anal. Manuf. **24**(4), 467–481 (2010)
17. Glier M.W., McAdams D.A., Linsey J.S.: Concepts in biomimetic design: methods and tools to incorporate into a biomimetic design course. In: IDETC 2011, Washington DC (2011)

Decision Uncertainties in the Planning of Product-Service System Portfolios

Daniel Kammerl, Oliver Malaschewski, Sebastian A. Schenkl and Markus Mörtl

Abstract During the early phase of the product development process of future Product-Service System (PSS) portfolios, the planning phase, solutions ideas are generated and within a decision process the best ones are selected. In this phase, the designer faces the challenge of making decisions under great uncertainty, because many of these decisions have a great impact on the rest of the product development but still with small knowledge of the product. In this paper we present a method for depicting and evaluating the possible uncertainties as a decision support for companies. By analyzing the interconnections between impact factors, decision criteria and PSS-elements and their stability, the uncertainty of each decision criterion related to a PSS-element can be calculated. This way, the designer can get a general idea of the existing uncertainties and their relevance. As a basis for the method, the important criteria regarding uncertainties in the early phase of the product planning were analyzed by means of a literature research. This was followed by a preliminary definition of the approach for evaluating decision uncertainties occurring throughout the planning phase of Product-Service Systems. Finally, the approach was evaluated by the application of the method in two case studies.

Keywords Product-service system · Decision uncertainty · Planning phase · Case study

1 Introduction

Consumer behavior in industrialized countries changes rapidly: A customer no longer buys a product itself, but wants to buy functionality, like mobility. Many enterprises react to this development by changing their portfolio. They develop or expand services to differ from their competitors and to enhance their products.

D. Kammerl (✉) · O. Malaschewski · S.A. Schenkl · M. Mörtl
Institute of Product Development, Technische Universität München, Boltzmannstraße 15, 85748 Garching, Germany
e-mail: daniel.kammerl@pe.mw.tum.de

A. Chakrabarti (ed.), *ICoRD'15 – Research into Design Across Boundaries Volume 2*, Smart Innovation, Systems and Technologies 35, DOI 10.1007/978-81-322-2229-3_4

For example car manufacturers offer the possibility to rent a car from a fleet which is distributed within a municipal area. This so called car sharing offers the possibility to rent and return a car on short notice and more location-independent in contrary to using a car hire. In this case the manufacturer not only has to take care of the product itself but also of the additional infrastructure and the service which is needed to provide the offer. Concepts of integrated product and service offerings are known as Product-Service Systems [1]. According to [2] Product-Service System (PSS) can be thought of as a market proposition that extends the traditional functionality of a product by incorporating additional services. When developing a PSS, in the early phase of a product development process, each product planner faces the problem that many decisions regarding the future product direction have to be made. These decisions address for example the kind of elements which are marketed as product and which as service. The challenge in making decisions in this phase is that the product knowledge is insufficient, so there is a great uncertainty. This is aggravated by the fact that changes in product properties get more expensive the later they are made [3], so the decisions have to be made sophisticated. All decisions that are not depicted or evaluated lead to a suboptimal development. Decisions against the customer's interests could be made because there is a great variety of relevant information that cannot be gathered sufficiently.

In this paper a method for depicting and evaluating the possible uncertainties in the planning phase of a PSS as a decision support for companies is presented. At the beginning, the important criteria regarding uncertainties in the early phase of the product planning were analyzed by means of a literature research. This was followed by a preliminary definition of an approach evaluating decision uncertainties occurring throughout the planning phase of Product-Service Systems. The approach was evaluated by the application in two case studies. The first use case was a student project in which an electrical bike sharing system was developed, and the students had to document all the decisions they made. The second use case was based on data of a PSS-provider offering pay-per wash.

2 State of the Art

In this section the elements of a decision process, the information necessary for a decision as well as uncertainties and finally the challenges when planning a PSS are described. So this section is the basis for the approach described in the next section.

While planning a PSS it is essential to make right, defensible and verified decisions. Especially when decisions are made across organizational border, for example in interdisciplinary PSS-development teams, they are very critical [4]. Possible consequences resulting from a wrong decision can cause immense damage. The information which is necessary or has to be set for the actual decision covers according to [5] and [6]:

- Assessment criteria
- Relative weight of the criteria
- Calculated value of each alternative relative to each criterion

Given a certain amount of alternatives, a decision process aims to select those alternatives which are most compliant to the desired objectives. While planning a PSS it is essential to make right, defensible and verified decisions. Consequences resulting from a wrong decision can have great impact on the product and even the company. As a support, the designer needs decision criteria to enable the decision. They can be differentiated into internal and external criteria [7]. Internal criteria come from the company, external ones come from the customer. In this paper decision criteria are understood as factors leading a decider to a decision without additional data.

Each designer phases the problem that decisions often have to be made with great uncertainty. The lack of knowledge regarding the future behavior of a product, the evolution of need, market and customer requirements as well as the change in economic and political constraints and further factors pose challenges to the designers. According to [8] uncertainties occur if the behavior in a process cannot be determined. Some of the uncertainties can be described and quantified by means of methods of risk analysis. But there are other 'weak' uncertainties which are not measurable directly and in consequence cannot be quantified. Figure 1 shows the model for depicting uncertainties which was defined by Sprenger and Anderl [8].

In this model each uncertainty causes an effect on a property or the behavior of a human or a machine. This effect can be classified as known, not known or assumed. If the latter is the case, the uncertainty cannot be analyzed or interpreted. Consequently, the knowledge about uncertainties is very important to prevent the generation of unwanted or dangerous properties. If the effect is known, the probability of the occurrence can be calculated by means of stochastic or further calculation methods.

The uncertainties in the planning phase according to [5] can be seen in Fig. 2. According to this model there are three layers which allow for identifying and

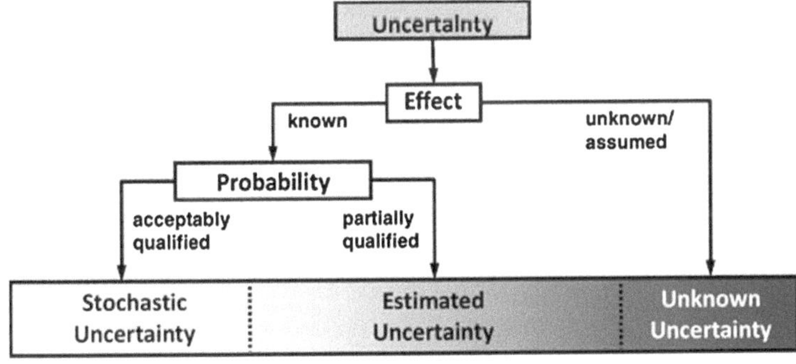

Fig. 1 Model for depicting uncertainties. *Source* [8]

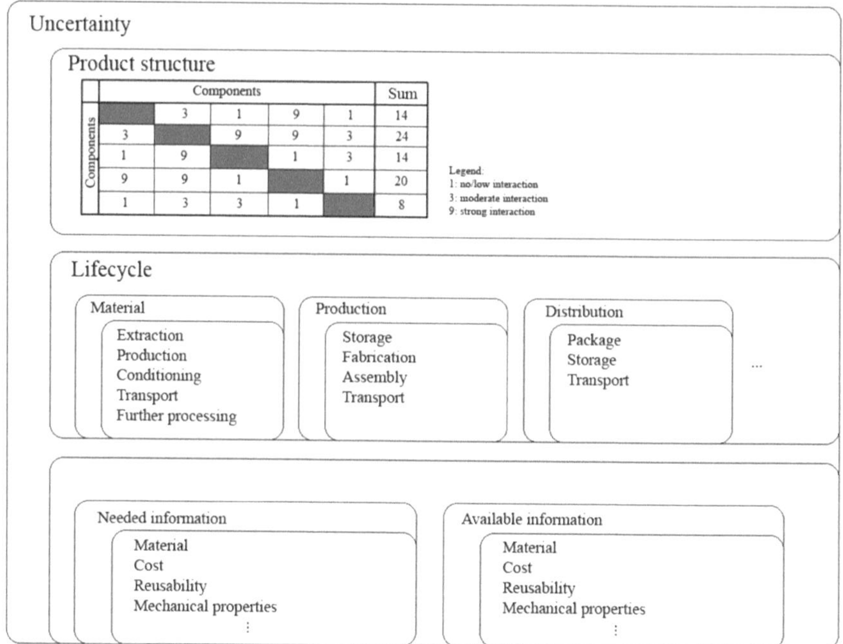

Fig. 2 Uncertain impact factors on the planning phase. *Source* [5]

analyzing uncertainties. The first layer is the product structure layer. It contains information about the individual subsystems or the components within a product and their interconnections. In this layer the designer can see which components influence each other, the certainty of this influence and whether there are interconnections which have not been taken into account so far. For calculating the strengths of the interconnections within the components a matrix can be used. Here the active and passive sum can be depicted as well the uncertainty of a component and the influence on or the behavior of components while interacting with each other.

The second layer depicts properties of the lifecycle, e.g. material, production, sales, use and recycling. Here the environment, e.g. suppliers, customers, …, plays an important role. Reference [5] advice to use Life Cycle Assessment for considering the environment. The last layer is the information layer. Here the available and the needed information for a holistic understanding of the system are in the focus. Based on the result of the analysis of considered and available information the uncertainty of a particular property within a product can be evaluated.

Signposting is a modeling framework which can support design process improvement activities ranging from process description to simulation, automation and use [9]. The approach considers design processes to be composed of discrete tasks which are coupled by shared interactions with input and output parameters.

A distinction is made between interactions that drive task selection and those which inform task execution [10].

The analysis of the state of the art showed that there is no holistic modeling approach for decision uncertainties to systematically address all aspects need.

3 Approach

In this section the description of an approach for evaluating decision uncertainties occurring throughout the planning phase of Product-Service Systems is described. It is a methodic approach for estimating uncertainties in the planning phase of a PSS by depicting all PSS-elements, decision criteria and decision uncertainties. So, PSS-elements with small uncertainties and regarding the decision criteria most important for the enterprise can be detected and selected.

The approach consists of five main steps:

- Identification of data relevant for the decision
- Evaluation of the decision criteria
- Calculation of the uncertainty sources
- Calculation of the uncertainties
- Depiction of the uncertainties

In the first step the information and data relevant for the decision have to be identified. As a basis a detailed market analysis as well as a profound knowledge of the strategic orientation of the enterprise is necessary. For determining the influences on a decision, the designer has a checklist containing a collection of uncertain impact factors which can influence a product development. The checklist is provided within this approach and contains items from the fields of technology, economy, competition, … (e.g. amendments or innovations) based on [11]. Following, the decision criteria have to be determined. There are two peculiarities of decision criteria, external and internal. Internal decision criteria are defined by the business culture (e.g. frequency or probability of changes), whereas external ones have to be identified by customer polls or interviews (e.g. market demand). The same way relevant product and service elements can be gathered.

The identified uncertain factors are evaluated according to their stability in step two. In this context stability is an indicator for the coherency between the probability of change of an uncertain impact factor and the company's possibility to influence the factor. The stability is rated according to a risk evaluation in the style of a FMEA with 1, 3 and 9. A 9 stands for a high stability and with it a low change probability or a high possibility for the company to actively influence or change it. After that two Multiple-Domain Matrices are created where the uncertain impact factors are opposed to the decision criteria and the PSS-elements. The influence of the uncertain impact factors on the other domains is also evaluated with 1, 3 and 9. The rating with 1, 3 and 9 offers the possibility to have a good distinction between low, middle and high impact.

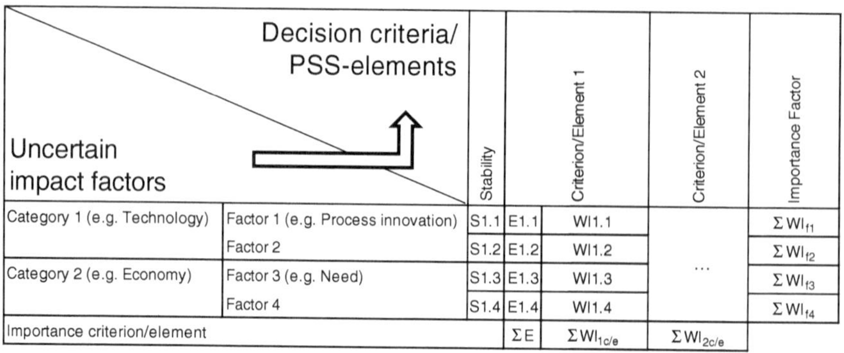

Fig. 3 Uncertainty factors—decision criteria and uncertainty factors—PSS-elements matrices

In the third step the sources for uncertainty, the impact of the uncertain impact factors on the decision criteria and the PSS-elements can be calculated. For this purpose the weighted influence (WI) of each criterion is calculated by means of the following formula:

$$\mathrm{WI} = \mathrm{E} * (10 - \mathrm{S}) \tag{1}$$

WI = weighted influence
E = evaluation where $\mathrm{E} \in \{1, 3, 9\}$
S = stability where $\mathrm{S} \in \{1, 3, 9\}$

The value for the uncertainty is calculated by summing up all the weighted influences of this element.

$$\mathrm{SU} = \Sigma\, \mathrm{WI} \tag{2}$$

SU = Evaluation of the source of uncertainty

Figure 3 shows a scheme of the resulting matrices. With these matrices elements with high uncertainty can be identified. Furthermore the matrices allow for a qualitative sorting of the elements regarding their uncertainty.

In the fourth step the uncertainties resulting from linking the decision criteria to PSS elements can be deduced from the prior calculations. Besides the value of uncertainty is normed to a scale from 0 to 100 with following formula:

$$\mathrm{U} = [\Sigma \mathrm{SU}(\mathrm{e_i}) * \Sigma \mathrm{SU}(\mathrm{c_i}) - \min(\mathrm{Mat}(\mathrm{e},\mathrm{c}))]/[\max(\mathrm{Mat}(\mathrm{e},\mathrm{c})) - \min(\mathrm{Mat}(\mathrm{e},\mathrm{c}))] * 100 \tag{3}$$

U = Uncertainty
SU(e_i) = Source of uncertainty PSS-element
SU(n_i) = Source of uncertainty decision criterion

A scheme of the resulting matrix can be seen in Fig. 4.

In the last step the matrix is converted into a three-dimensional diagram, see Fig. 5. The length of the bars symbolizes the qualitative peculiarity of the uncertainties. The corresponding axes show the combination of decision criterion and PSS-element leading to the uncertainty. Thus, the combination of PSS-element 1 and criterion 3 leads to a high uncertainty (long bar), whereas the combination of PSS-element 3 and criterion 1 leads to a low uncertainty (short bar).

The approach was applied on two case studies, the development of an electrical bike sharing concept within a student's project and existing data of the development of pay-per-wash concept. The application proved that PSS-elements with high uncertainty as well as decision criteria affected by great uncertainty can be identified. The approach also allows for finding PSS-elements with a low uncertainty which are influenced by a small amount of uncertain impact factors. These elements can be integrated into the future product without great risk. Besides, it became apparent that the approach generates the highest value in the early phase of a

Fig. 4 Decision criteria—PSS-element matrix

PSS-elements / Decision criteria		Element 1	Element 2
		SU_{1e}	SU_{2e}
Criterion 1 (e.g. Technology)	SU_{1c}	U1.1	U1.2
Criterion 2	SU_{2c}	U2.1	U2.2

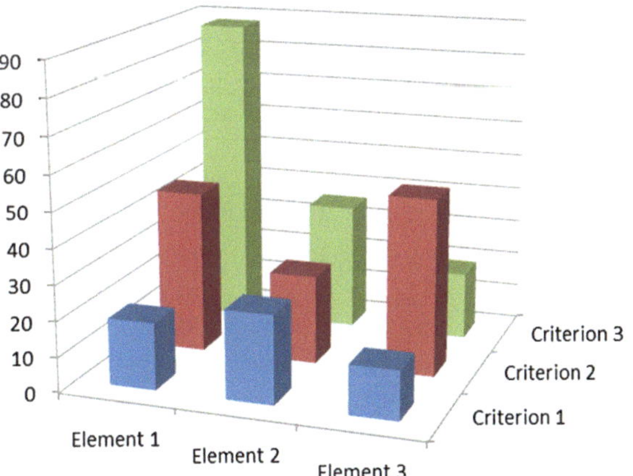

Fig. 5 Peculiarity of uncertainties

development process. Finally, we state that this approach should not provide a firm value, which characterizes the point at which the uncertainty in the decision process is too great. It just gives transparency of the existing uncertainties. For this task, the approach could be applied successfully. But in the end each company has to decide if the calculated uncertainty is acceptable or potential for improvement is present.

4 Discussion

In conclusion, it is never possible to depict all uncertainties comprehensively, so the right selection of the elements in the focus is very important. Thus, the designer has to choose the system boundary very carefully. Furthermore, not all influences can be manipulated or foreseen by the company, because they are random. Nevertheless, this approach offers assistance for depicting a plurality of eventualities and allows the decider to depict possible uncertainties qualitatively.

The approach offers a great degree of freedom and thus it is applicable on manifold subject areas. The method is not based on statistic calculations, so each company can decide which influences and decision criteria are rated as critical or notable. The given list of impact factors does not claim to be exhaustive. It is only an example for possible influences. The decider can and must adjust it to the current situation.

It should be noted that the evaluation is subjective, because humans are involved, so it is only a qualitative comparison that can provide different results depending on the decider. This situation is aggravated by the fact, that it can be difficult to relate PSS–elements and decision criteria with impact factors. The effect on each other not always is well-known. To minimize the subjectivity, an assiduous and extensive analysis of the relevant data has to be performed in advance.

Still, this approach can deliver an overview and a comparison of the emerging or remaining uncertainties and in this way support the decision process. In addition, the decider can be made aware if the decision has to be reviewed and in this case where to start searching for further information.

5 Conclusion and Outlook

In this paper an approach for depicting decision uncertainties occurring throughout the planning phase of Product-Service Systems was presented. Furthermore, a framework of impact factors, decision criteria and PSS-elements relevant for decision processes was explained. The results of a literature review were combined systematically for developing an approach for depicting uncertainties in a decision process. The first step of the approach is the determination of all relevant data. They include results of market analyses, the definition of company specific criteria which are relevant for decisions and the evaluation of the stability of the uncertain impact factors. In the second step two multiple-domain matrices are generated. Thereby the uncertain impact factors are opposed to decision criteria and PSS-elements and are evaluated with a progressive point scale. In the following the weighted influences for each factor can be calculated by combining them with the stability. By summing up the weighted influences for each decision criterion and PSS-element over all uncertain impact factors the sources of uncertainty can be calculated. They can be translated into a further matrix and the characteristic numbers can be calculated by multiplying each PSS-element which each decision criterion. In the last step for better comparability the results are converted into a three-dimensional graphic. The relevance and validity of the approach was evaluated by applying it to data of a student project and an industry partner.

For minimizing the risk for companies offering PSS, it is necessary to examine uncertainties occurring throughout the planning process more explicitly. A more detailed listing of uncertainties in a decision process can help minimizing the risk. Especially a model offering recommendations for selecting the criteria which are taken into account when making decision would be helpful. This is necessary for putting the maximum focus on the most important criteria and neglecting the less important ones. The depiction of uncertainties occurring only in certain target markets could be depicted and more time could be spent on the actual product development. Furthermore uncertainties could be reduced if there was a possibility to determine the sources for these uncertainties, so arrangements for reducing these uncertainties should be developed. A first step to minimize the influencing factors which yet has not been taken into account is human emotion. Questions which have to be investigated are: How decisions change if emotions come into play? How can these be depicted or how can their influence be minimized? Finally, the goal is to eliminate as much influences as possible.

Acknowledgments We thank the German Research Foundation (Deutsche Forschungsgemeinschaft—DFG) for funding this project as part of the collaborative research centre "Sonderforschungsbereich 768—Managing cycles in innovation processes—Integrated development of product-service-systems based on technical products".

References

1. Schenkl, S.A., et al.: Managing cycles of innovation processes of product-service systems. In: 2013 IEEE International Conference on Systems, Man, and Cybernetics (SMC). IEEE (2013)
2. Baines, T., et al.: State-of-the-art in product-service systems. Proc. Inst. Mech. Eng. Part B J. Eng. Manuf. **221**(10), 1543–1552 (2007)
3. Ehrlenspiel, K.: Integrierte produktentwicklung: Denkabläufe, methodeneinsatz, zusammenarbeit. Hanser Verlag (2009)
4. Jupp, J.R., Eckert C., Clarkson J.: Dimensions of decision situations in complex product development. In: DS 58-3 Proceedings of ICED 09, the 17th International Conference on Engineering Design, vol. 3, Design Organization and Management, Palo Alto, CA, USA, 24–27 Aug 2009

5. Kota, S., Chakrabarti,A.: A Method for Comparative Evaluation of Product Life Cycle Alternatives under Uncertainty. 2007
6. Roozenburg, N.F., Eekels, J.: Product design: fundamentals and methods, vol. 2. Wiley, Chichester (1995)
7. Schenkl, S.A., et al.: Deploying decision criteria in a cyclical decision process for the product planning phase. In: Lindemann U., et al., (ed.) 19th International Conference on Engineering Design 2013 (ICED13), Seoul, Korea (2013)
8. Sprenger, A., Anderl, R.: Product life cycle oriented representation of uncertainty. In: Product Lifecycle Management. Towards Knowledge-Rich Enterprises, pp. 277–286. Springer, Berlin (2012)
9. Clarkson, P.J., Hamilton, J.R.: 'Signposting', a parameter-driven task-based model of the design process. Res. Eng. Des. **12**(1), 18–38 (2000)
10. Wynn, D.C., Eckert C.M., Clarkson P.J.: Applied signposting: a modeling framework to support design process improvement. In: ASME 2006 International Design Engineering Technical Conferences and Computers and Information in Engineering Conference. American Society of Mechanical Engineers (2006)
11. Langer, S., Lindemann U.: Managing cycles in development processes-analysis and classification of external context factors. In: 17th International Conference on Engineering Design, Design Society Stanford University, California, USA (2009)

The DSM Value Bucket Tool

Bernard Yannou, Romain Farel and François Cluzel

Abstract The Dependency Structure Modelling Value Bucket (DSM-VB) tool is integrated to Radical Innovation Design (RID) methodology for exploring the front end of innovation in need seeker mode. The determination of value buckets has been automated by matrix representations of dependencies between problems or pain points, usage situations and existing solutions. Three matrices are built along the problem setting stage of a RID process. The first matrix expresses which problems occur during usage scenarios, the second how much existing solutions cover problems and the third how much existing solutions are useful in usage situations. Combining these three matrices results in a matrix of value buckets as being the combinations of important problems occurring during characteristic usage situations and for which few existing solutions are useful or efficient. This outcome allows to perform focused creativity workshops and to result in "blue ocean" innovations with high likelihood to be successful on the market.

Keywords Radical innovation design · RID methodology · Front end of innovation · Need seeker innovation · Value bucket · Dependency structure modelling

1 Introduction

The *Dependency Structure Modelling Value Bucket* (DSM-VB) is a tool integrated in the *Radical Innovation Design* (RID) methodology. RID is a structured process for exploring the front end of innovation in need seeker mode. Indeed, the problem setting stage starts with re-expressing the ideal need to set the issue playground—for usefully thinking in the box—in which two worlds are addressed: the world of

B. Yannou (✉) · F. Cluzel
Laboratoire Genie Industriel, Ecole Centrale Paris, Chatenay-Malabry, France
e-mail: bernard.yannou@ecp.fr

R. Farel
Paris Saclay Efficacité Énergétique (PS2E), Saclay, France

A. Chakrabarti (ed.), *ICoRD'15 – Research into Design Across Boundaries Volume 2*,
Smart Innovation, Systems and Technologies 35, DOI 10.1007/978-81-322-2229-3_5

problems or pain points and the world of situations or usage scenarios. The two spaces—problems and usage scenarios—are populated with real world situations. For this purpose, some modeling techniques as causal graph representations and persona method are used. Then, a first "ideal performances matrix" of the DSM-value-bucket tool allows to cross problems with usage scenarios to express in which usage situations people are subject to pains. Next, existing design solutions—commercial solutions or patents—are identified and their coverage of the two spaces is modeled. Here the DSM-value-bucket tool proposes to represent the coverage effectiveness and efficiency of both problems and usage scenarios by two appropriate matrices: the Solution-Problem matrix and the Usage-Solution matrix. Multiplying both matrices allow to come up with the likelihood for the existing solutions to satisfactorily answer to one problem arising during one usage scenario. Subtracting this matrix with the "ideal performances matrix" results in a final "value buckets" matrix highlighting which problem is worthy to be addressed in an innovation project. The last step of the RID problem setting stage is to select a subset of opportunistic value buckets to further address in the problem solving stage, so as to ensure to perform radical innovation on "blue ocean"—i.e. not yet explored—usage and problem situations. The DSM value bucket tool has been successfully applied on more than 20 company innovation projects for 5 years. In this paper, the DSM value bucket tool is illustrated on the search for radical innovations for a handitennis wheelchair of a champion. Two important value buckets are detected as determining to improve the likelihood to win of the champion. It is then showed that creativity workshops starting from these two value buckets have led to several convincing innovations. The DSM value bucket tool opens the way of automating the radical usage-driven innovations along with a systematic investigation and representation of problems or pain points and usage scenarios.

2 Exploring the Front End of Innovation

Boston Consulting Group has stipulated that firms follow at least one of three innovation strategies: Need Seeker, Market Reader, Technology Driver, depending on the focus put on the customer, the market or the technology. Booz and Company (see [1]) defines them as follow:

- Need Seekers, such as Apple (US), Dyson (UK) and Oxylane (France), make a point of engaging customers directly to generate new ideas. They develop new products and services based on superior end-user understanding.
- Market Readers, such as Hyundai, Caterpillar and Loréal, use a variety of means to generate ideas by closely monitoring their markets, customers, and competitors, focusing largely on creating value through incremental innovations.
- Technology Drivers, such as Google and Bosch, depend heavily on their internal technological capabilities to develop new products and services.

After a recent Booz and Company study (see [1]), following a Need Seekers strategy offers the greatest potential for superior performance in the long term. These companies are effective at both the ideation and conversion stages of innovation and they consistently outperform financially.

Being predominantly Need Seeker is not easy; it can be made by two ways:

- Using lead-users (see von Hippel [2]), their insightful refreshing ideas and dreams and their testimonies on usage and pain points. This is the case of Oxylane company in France—sport equipment and outdoor.
- Having a visionary leader like Steve Jobs (Apple) or James Dyson (Dyson), the company growth and the number of product references being limited by the imagination and control power of a single brain.

There is thus a need for a methodology investigating growth territories or strategic value niches for generating disruptive innovations beyond current customer expectations and in a cooperative and multidisciplinary manner and a secure way. After Motte et al. [3], it can be done thanks to an adapted organization and special methodologies and processes. In terms of organization, Millier [4] insists on the necessity to manage antagonism and balance between exploration and exploitation of new idea territories. Christensen [5, 6] say with different words that for succeeding disruptive innovations, companies must not put too much emphasis on customers' current needs, and work on how to adopt new technologies or business models that will meet customers' unstated or future needs. In terms of methodology, Christensen [5, 6] proposes the jobs-to-be-done concept and defines it as "*a framework which is a tool for evaluating the circumstances that arise in customers' lives. Customers rarely make buying decisions around what the "average" customer in their category may do—but they often buy things because they find themselves with a problem that they need to solve. With an understanding of the "job" for which customers find themselves "hiring" a product or service, companies can more accurately develop and market products well-tailored to what customers are already trying to do*". For this and other works on innovations, Clayton Christensen has been designated as the most influential management thinker in the world (see The Washington Post paper [7]). Ulwick [8] has extended it in a principle of design-outcomes segmentation instead of a conventional a priori customer segmentation.

Inspired by these ideas, Yannou et al. [9] and He et al. [10] have adapted this user-centered perspective to model the market demand model in a design engineering platform through the representation of usage contexts. It has been called the Usage Context Based Design (UCBD). Next, Yannou et al. [11] have proposed the *Design by Usage Coverage Simulation* principle for evaluating with coverage indicators how much a new product or product family [12] may cover in a dominant way a number of usage scenarios characteristics of the targeted user/consumer group. Proceeding that way, they show that innovative designs may be proved to be dominant—i.e. ranked first because performing better—on a subspace of usage situations; these designs are then naturally in a "blue ocean" (after Kim and

Mauborgne [13]) which is almost a guaranty of success when launching an innovative offer.

Beyond design analysis, these principles have been applied to be the core principle of an innovation methodology in need seeker mode, namely Radical Innovation Design® (RID).

3 The RID Methodology and Process

Albert Einstein said "*If I had an hour to solve a problem and my life depended on the solution, I would spend the first 55 min determining the proper question to ask, for once I know the proper question, I could solve the problem in less than five minutes*". Following that maxim, Yannou et al. [14] structured the RID process in two macro stages of *problem setting* (see Fig. 1) and *problem solving* (see Fig. 2). Radical Innovation Design® is a methodology because it is based on (1) structuring principles, (2) a stage-and-gate process (see also [15–17]) very detailed in the early problem setting like Cooper suggested in [18], (3) a list of 9 expected templated deliverables along the process, (4) two computerized tools such as the *DSM-Value-Bucket* tool described in the present paper and the *UIPC-monitor* tool [19], (5) already several successes in company contexts since after RID innovation projects with 20 companies, several innovations are being to be launched on the market.

The goal of RID methodology is to maximize the potential value creation inside a legitimate design perimeter called *ideal need*. RID is a systematic exploration/exploitation process of value creation opportunities through a series of stages making the inventory of usage situations (or *scenarios*) and pain points (or *problems*) users may live. RID uses at the same time 3 perspectives:

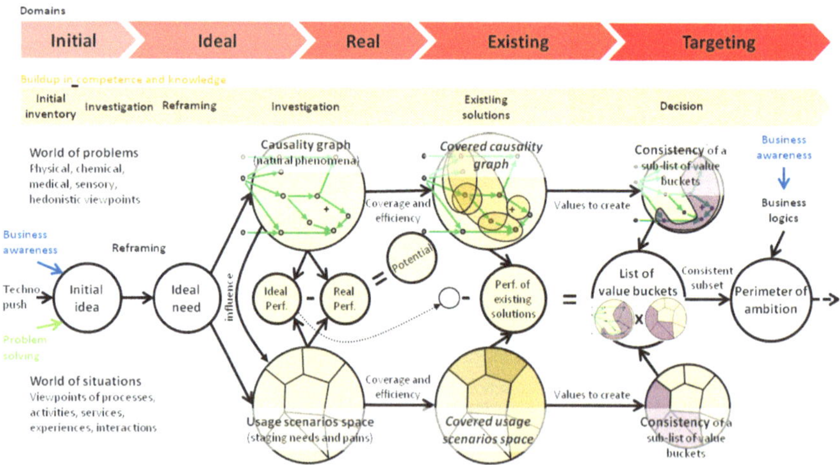

Fig. 1 The *problem setting* macro-stage of Radical Innovation Design® methodology

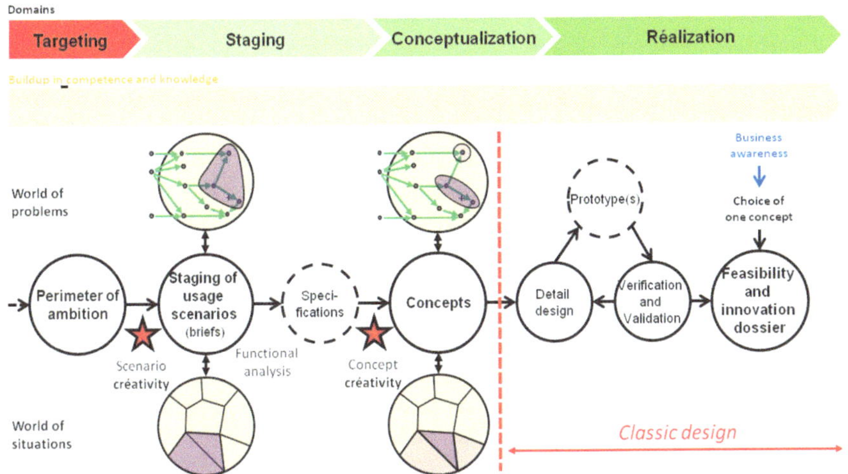

Fig. 2 The *problem solving* macro-stage of Radical Innovation Design® methodology

- The perspective of an economist: design is considered as a probabilistic theory of value creation,
- The perspective of an industrial designer: design starts with the know-how for observing users—their usages, pain points, needs…—and inventing new usages,
- The perspective of a design engineer: knowing how to measure utilities to create, gather evidences and bring serious proofs of concept using the most adequate technologies.

Yannou et al. showed in [14] that the more the design team completes the successive RID deliverables, especially in problem setting, the most likely the innovation outcome is to be successfully launched on the market. To that aim, they use a monitoring with four proofs to consolidate along the design process: Utility, Innovation, Profitability and Concept, this is the UIPC model described in [20].

The problem setting starts with the reframing of the *initial idea* submitted by the innovation project initiator into an *ideal need*. Let us start with the example of need seeker innovation on the wheelchair of a handitennis champion—example free of confidentiality rights. It has been the actual innovation project initiated by a 22 year old handicapped female student who is nearly ranked 30th in the world ranking and who wants to win in Rio-2016 Paralympic Games. She came with the initial idea of "*to lighten at most her handitennis wheelchair*". Such a goal would have led to a carbon fiber high tech wheelchair. Making lighter the wheelchair is not an objective in itself; it has been reframed into the following ideal need: "*to be performing on every tennis point in every game situation*". This ideal need is a "box perimeter" inside which investigation must be pursued at its extreme limit. Contrarily to most of people about creativity, the authors do not believe that "*thinking outside the box*" is the must, but it is more efficient to "*thinking inside the box, providing the box is large enough and well defined*".

Continuing with RID process, two worlds are investigated concurrently within the *ideal need* perimeter (see Fig. 1):

- The world of problems. It consists in inventorying, quantifying and causally ordering the miscellaneous pain points, counter-performances, dissatisfactions, needs, that users may experiment.
- The world of situations. It consists in inventorying, qualifying and sizing the usage situations that users live and in which problems occur with more or less intensity.

Defining the *real* world consists in building a *causal graph* (of problems) and a *usage scenarios space* of characteristic usage situations (see Fig. 1). Next, as *existing* solutions may partly cover problems in usage situations, a *covered causality graph* and a *covered usage scenario spaces* are derived from the careful analysis of the conditions (usage scenarios) and effectiveness/efficiency of service delivery (problems/pain points more or less relieved). Next, in the final *targeting* stage of problem setting, a list of weighed value buckets are derived as being the combinations of important problems occurring during very characteristic (frequent) usage situations and for which few existing solutions exist or are really effective/efficient. From this list of value buckets, a perimeter of ambition is defined by the project team, including (a) a subset of relevant value buckets, (b) other (problems × usage-situations) currently covered by existing situations but that consumers consider as "must have", (c) these previous choices being compatible between them and with the present offer portfolio and customer segmentation of the company (represented by "business logics" in Fig. 1).

4 Building and Covering the Causal Graph of Problems and the Usage Scenarios Space for the Handitennis Wheelchair

The determination of value buckets has been partly automated by a matrix representation of dependencies between *problems*, *usage situations* and *existing solutions* and by a computational mechanism leading to the *DSM-Value-Bucket* tool. This approach and tool may be affiliated to *Dependency Structure Modelling* approaches [21].

In Fig. 3, the causal graph is represented as causal paths leading to *point loss* problem and it is further graphically covered by four existing solutions. Here, some modeling techniques of causal graph representations are borrowed from the system dynamics practice (see for instance [22]). For simplicity, we only retain 4 problems out of 16, namely: *time loss* (moving), *injury of the racquet hand*, *loss of ball power* and extended *tiredness* during the match.

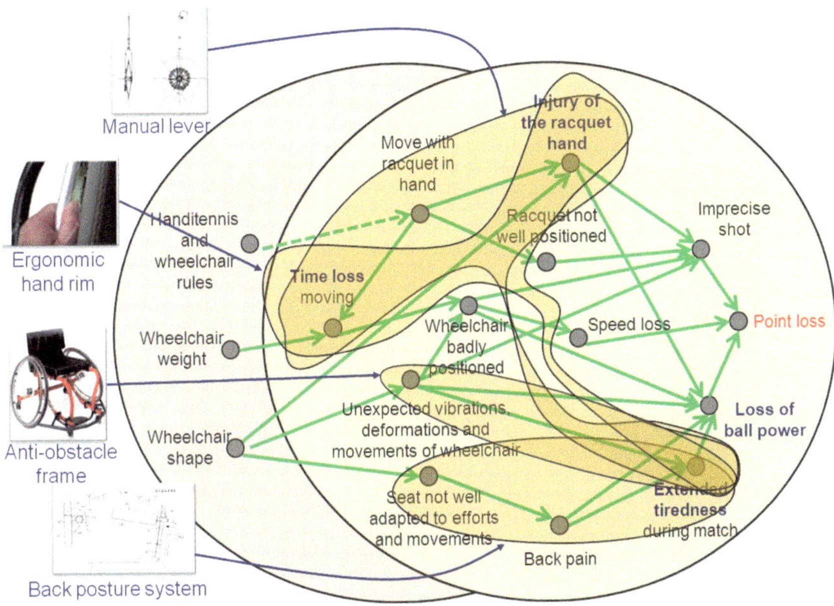

Fig. 3 The *covered causality graph* for the handitennis wheelchair issue (see Fig. 1)

In Fig. 4, a graphical tessellation of typical usage situations during a match is represented. Proximity of two usage situations means a high probability of time precedence (or in other cases, proximity of user types). For simplicity, we only retain 4 usage scenarios out of 8, namely: *serve*, *shot in move*, *ball receiving* and *start moving* to hit ball.

Practically, a pre-screening of problems is made and a first version of the causal graph of problems sketched. Next, the list of typical usage situations is established and for each usage situation an observation protocol is designed and followed to get a deep understanding of the pains/problems possibly occurring in this usage situation, for measuring them (frequency, repeatability, importance, consequences) and carrying out a *root cause analysis*. It goes far beyond the classical *personas method* storyboarding usage situations with weak rationale of the situation representativity and no measurements of pain points. For instance, here, the serve situation has been carefully studied: gestures have been recorded and analyzed, ball speed has been measured as well as serve accuracy, ability to serve aces, double faults rate. In addition, it has been observed that a back and forth translation as well as a rotational twist of the wheelchair occurred during the serve. It is obviously due to the translational freedom of the four wheels and the rotational freedom of the two caster wheels. An additional investigation in root causes led to experiment the gains to block the four wheels during serve (+30 % in ball speed) or to only block the rotation of caster wheels (+20 % in ball speed).

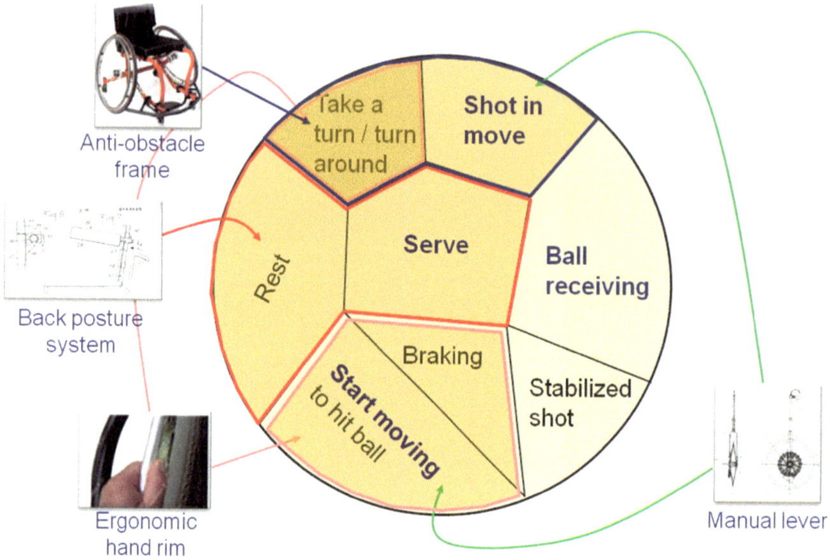

Fig. 4 The *covered usage scenarios space* for the handitennis wheelchair issue (see Fig. 1)

5 The DSM Value Bucket Mechanics

The determination of value buckets has been automated by matrix representations of dependencies between problems, usage scenarios and existing solutions. Three matrices A, B and C are built along the problem setting stage of a RID process described in previous sections. The first matrix A (see Fig. 5) expresses which problems occur during usage scenarios, the second matrix B how much existing solutions cover problems and the third matrix C how much existing solutions are useful in usage situations. Combining these three matrices results in a matrix E of value buckets as being the combinations of important problems occurring during characteristic usage situations and for which few existing solutions are useful or efficient.

Matrix A is named the "Ideal performances matrix" and links problems (columns) and usage scenarios (rows) with an intensity scale from 0 to 5 for expressing how much (or often) a problem occurs in a usage scenario. The meaning of the intensity scale is {0 = null; 1 = weak; 2 = moderate; 3 = average; 4 = important; 5 = very important}. For instance (see Fig. 5, matrix A):

- The *racket hand injury* mainly occurs when the player *starts moving*, pushing with her hand to propel the wheelchair, grasping the racket and the hand rim at the same time.
- There is an important *power loss* during *serve* due to an uncontrolled twist of the wheelchair.

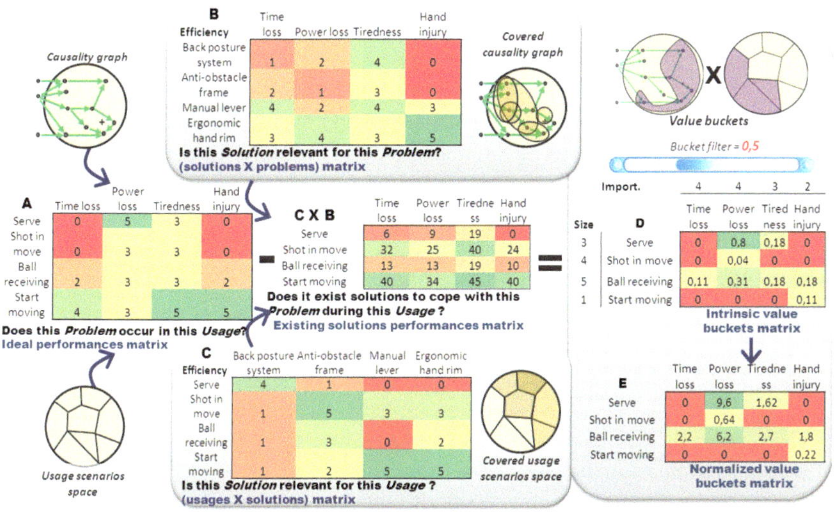

Fig. 5 The *DSM Value Bucket* data streaming and computation mechanics (refer to Fig. 1)

Matrix B is the "(solutions X problems) matrix" and expresses the relevance of an existing solution for a given problem with the same qualitative scale from *null* (0) to *very important* (5). For instance (see Fig. 5, matrix B):

- The *ergonomic hand rim* is very relevant for avoiding *racket hand injury*.
- The *ergonomic hand rim* also partly avoids *time loss*.
- Both the *back posture system* and the *manual lever* are good for relieving the *generalized tiredness*.

Matrix C is the "(usages X solutions" matrix" and expresses the relevance of an existing solution in a given usage scenario with the same qualitative scale from *null* (0) to *very important* (5). For instance (see Fig. 5, matrix C):

- The *manual lever* is very efficient during *start moving* situation and moderately during *shot in move*.
- The *back posture system* is efficient during the *serve* situation.

At this stage, an "Intrinsic Value Buckets matrix" D is computed as the subtraction between the "Ideal performances matrix" A expressing importance of problems to solve in usage situations and the matrix multiplication C × B expressing the average relevance of existing solutions in (usage, problem) cases. Of course, this difference is normalized to get each number at both sides of the subtraction comprised between 0 and 1. Moreover, one introduces a "bucket filter" BF, a real number comprised between 0 and 1 and being 0.5 by default, to eliminate the least important (usage, problem) cases, following formula (1).

$$\mathrm{IVB_{ij}} = \mathrm{Max}\left(0, \frac{\mathrm{A_{ij}}}{\mathrm{Max_{kl}(A_{kl})}} - 2 \times \mathrm{BF} \times \frac{\mathrm{CB_{ij}}}{\mathrm{Max_{kl}(CB_{kl})}}\right) \tag{1}$$

Finally, the *importance* of problems (relatively to the ideal need) and the *size* of usage scenarios are assessed, again through the 0 to 5 intensity scale (see Fig. 5, *size* and *importance* introduced in the surroundings of matrix D). The rationale for weighing problem importance and usage size must be captured. The RID framework encourages keeping the traceability of exploration/exploitation and decision making. For instance, the logic for justifying the problem importance may be:

- Ball power loss and time loss moving should be importantly improved in the champion play.
- Tiredness and hand injury are second order issues for the champion play.

The rationale for justifying the size of usage scenarios may be the scenario frequency (comparing the number of times serving and shooting in move) and of its importance for winning a point (80 % of serves in handitennis are winning points).

A last "Normalized value buckets matrix" E is computed to augment intrinsic value buckets with importance of problems and size of usage scenarios, following formula (2).

$$\mathrm{NVB_{ij}} = \mathrm{IVB_{ij}} \times \mathrm{size_i} \times \mathrm{importance_j} \tag{2}$$

Two important value buckets are revealed for the handitennis wheelchair project; their matrix coordinates are (1,2) and (3,2) (see Fig. 5). The designer team is asked to verbally interpret them and they come up with these natural justifications:

- Value bucket #1 (1,2): The *loss of power* during *serve* is partly due to the (observed) wheelchair twist.
- Value bucket #2 (3,2): The champion player is late on the position for *receiving the ball*, and consequently she returns the ball with *power loss*; this is due to her right hand grasping the tennis racket and at the same time moving the wheel.

6 Validation and Discussion

For briefly illustrating the relevance of such fine interpretations of revealed value buckets, we invite the readers at looking at the two-stage ideation process starting from the value buckets in Fig. 6 and at the fruitful outcomes of this ideation process in Fig. 7 (comments of these outcomes are out of the scope of this paper).

RID methodology may be somewhat compared to known innovative design methodologies like TRIZ, QFD or axiomatic design and design thinking. Compared to TRIZ, RID uses a causal graph for representing the problem structure whereas comparable substance-fields representations in TRIZ are used for representing

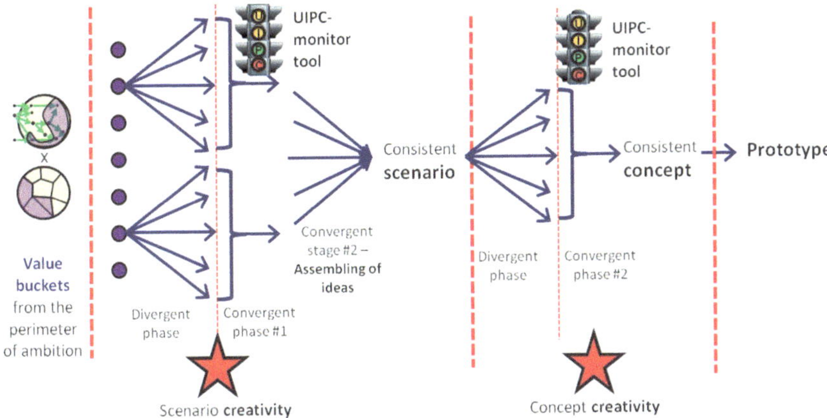

Fig. 6 The two-stage ideation process starting from the value buckets included in the perimeter of ambition (refer to Fig. 2)

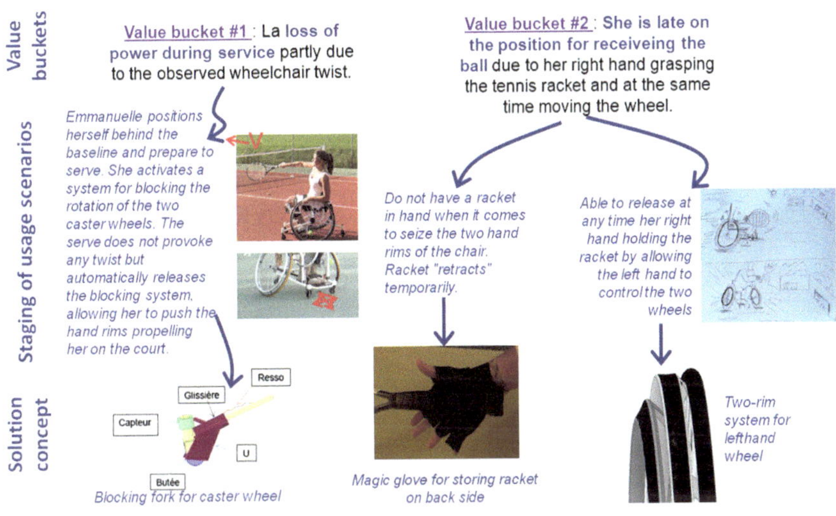

Fig. 7 Illustration of the two-stage ideation process (scenario creativity and concept creativity) starting from the two value buckets identified for the handitennis wheelchair

imperfect solutions. In the same manner QFD and axiomatic design may be used to represent the propagation of the voice of the customer into the product components and design parameters, but little is done to characterize the problem opportunities especially in the light of what the other existing solutions use to efficiently perform or "cover". Finally, RID demonstrates that it exists other ways than the design thinking prototype-and-learn experimental loop, being a more rational manner to investigate need-seeker problems.

7 Conclusions

The authors have proposed a method for structuring and automating the discovery of value buckets during the front end of need seeker innovations. The interest of need seeker innovations have been revealed by people like Christensen [5, 6] and Ulwick [8] but no one before had implemented these ideas in a design engineering process. The DSM Value Bucket tool was designed two years ago and has already been applied successfully to 14 innovation projects with 10 private companies, plus the “handitennis wheelchair project” presented in this paper (for the reason the authors have no non-disclosure agreement). The DSM-VB is a master-piece of the Radical Innovation Design® methodology. Future works will validate the relevance of the most rated resulting value buckets to the light of launching disruptive products after opinion of lead users.

References

1. Ohtonen J.: Is your company a need seeker? http://thenextten.org/is-your-company-a-need-seeker/ (2014)
2. von Hippel, E., Ogawa, S., De Jong, J.P.J.: The age of the consumer-innovator. MIT Sloan Manage. Rev. **51**(1), 27–35 (2011)
3. Motte D., Yannou B., Bjärnemo R.: The specificities of radical innovation. In: Proceedings of ICoRD, Bangalore, India, 10–12 January 2011
4. Millier, P.: Marketing the Unknown: Developing Market Strategies for Technical Innovations. Wiley, NewYork (1999)
5. Christensen, C.: The Innovator’s Solution: Creating and Sustaining Successful Growth. Harvard Business School Press, Boston (2003)
6. Christensen, C.: The Innovator’s Dilemma: The Revolutionary Book That Will Change the Way You Do Business. HarperBusiness, NewYork (2011)
7. McGregor J.: The world’s most influential management thinker? http://www.washingtonpost.com/blogs/on-leadership/wp/2013/11/12/the-worlds-most-influential-management-thinker/, The Washington Post (2013)
8. Ulwick, A.: What Customers Want: Using Outcome-Driven Innovation to Create Breakthrough Products and Services. McGraw-Hill, NewYork (2005)
9. Yannou B., Chen W., Wang J., Hoyle C., Drayer M., Rianantsoa N., Alizon F., Mathieu J.-P.: Usage Coverage Model For Choice Modeling: Principles, IDETC, San Diego, CA, Aug 30–Sep 02, 2009
10. He, L., Chen, W., Hoyle, C., Yannou, B.: Choice modeling for usage context-based design. J. Mech. Des. **134**, 3 (2012)
11. Yannou, B., Yvars, P.-A., Hoyle, C., Chen, W.: Set-based design by simulation of usage scenario coverage. J. Eng. Des. **24**(8), 575–603 (2013)
12. Wang, J., Yannou, B., Alizon, F., Yvars, P.-A.: A usage coverage-based approach for assessing product family design. Eng. Comput. **29**(4), 449–465 (2013)
13. Kim, C.W., Mauborgne, R.: Blue Ocean Strategy—How to Create Uncontested Market Space and Make the Competition Irrelevant. Harvard Business School press, Boston (2005)
14. Yannou, B., Jankovic, M., Leroy, Y., Okudan Kremer, G.E.: Observations from radical innovation projects considering the company context. J. Mech. Des. **135**(2), 021005 (2013)

15. Cooper, R.G.: A process model for industrial new product development. IEEE Trans. Eng. Manage. **30**, 2–11 (1983)
16. Cooper, R.G.: Stage-gate system: a new tool for managing new products. Bus. Horiz. **33**, 44–54 (1990)
17. Cooper, R.G.: Winning at New Products: Accelerating the Process From Idea to Launch, 3rd edn. Basic Books, Cambridge (2001)
18. Cooper, R.G.: Predevelopment activities determine new product success. Ind. Mark. Manage. **17**, 237–248 (1988)
19. Yannou, B., Farel, R.: The UIPC monitor tool. In: Proceedings of to be Submitted to ICDC. Bangalore, India, 12–14 Jan 2015
20. Yannou, B., Zimmer, B., Farel, R., Jankovic, M., Stal Le Cardinal, J.: Proofs of utility, innovation, profitability and concept for innovation selection. In: ICED, Seoul, Korea (2013)
21. Eppinger, S.D., Browning, T.R.: Design Structure Matrix Methods and Applications. MIT Press, Cambridge (2012)
22. Schaffernicht, M.: Causality and diagrams for system dynamics. In: 50th International Conference of the System Dynamics Society, Boston (2007)

Development of Creativity Through Heightening of Sensory Awareness

N. Divya

Abstract In this era of innovation, it is pertinent that our future designers be taught to think and act creatively. However, our design graduates who are highly knowledgeable in the content area do disappoint when it comes to complex creative thinking. The remedy begins with the creation of a pedagogical model that emphasizes on the inherent development of creativity as opposed to teaching it as an application skill. This paper focuses on an experimental study conducted following one of Leonardo DaVinci's seven principles—Sensazione (heightened awareness of senses), as a method of enhancing creativity. An evaluation of the creativity of randomly selected graduate level design students using the Test of Creativity-Drawing Production (TCT-DP) before and after exposure to a series of sensorial exercises illustrate how heightened awareness of senses increases creativity while establishing the scoring dimensions of the TCT-DP as a holistic method for assessing creativity.

Keywords Creativity · Design education · Sensorial activities · Creative thinking

1 Introduction

'Creativity in education' is the buzzword trending in education circles today where more and more educators are looking for ways to teach creatively and teach creativity. Ironically, in an industry like design, that thrives on innovation; teaching and learning how to think creatively has been pushed behind, to make room for core content and practical skill development. As a result, Indian design graduates are not fully equipped to cater to the market like their international counterparts and are often

N. Divya (✉)
National Institute of Fashion Technology, Chennai, India
e-mail: divdes@gmail.com

A. Chakrabarti (ed.), *ICoRD'15 – Research into Design Across Boundaries Volume 2*, Smart Innovation, Systems and Technologies 35, DOI 10.1007/978-81-322-2229-3_6

forced to go abroad for this exposure to creative thinking and living. Hence filling up this gap with creativity focused high value education is the need of the hour.

In the past, creativity was often confused with inborn genius, was considered to be the prerogative of a fortunate few. It was envisaged to be a natural trait that did not require training. Creativity, in truth, is the synthesis of two harmonious processes—Creative thinking (generation of a multitude of ideas) and critical thinking (evaluation of ideas and implementation of those that are considered most suitable to solve the problem at hand) and includes divergent production and insight [11]. Thus in this the fast changing world which functions on cross pollination of ideas and multi domain management, it is imperative that we teach creativity as a vital life skill.

The concept of teaching creativity is not new and has been around since E. Paul Torrance, who advocated that it was skill-specific, requiring intentional instruction [14]. Models to teach creative thinking, like Osborn-Parnes program do exist, however, these models center on teaching creative thinking as an application skill (like mathematics) providing a structure or a formula as opposed to treating it like a core skill (such as language) that has to acquired. As a solution, this paper recommends a human centered model that relies on heightening the awareness of sense–response reaction to stimuli to enhance creativity. The significance of this paper is that it also points to an effective assessment method that can be utilized by the design educators to record, study and analyze student creativity and its response to sensory exercises.

1.1 Sensory Awareness and Creativity

Sensory awareness is the ability to receive and differentiate sensory stimuli that can be perceived by the auditory, tactile, visual, gustatory, proprioceptive, and vestibular systems in our body. Though the function of sense organs cannot be controlled beyond an extent, they can be trained to detect certain productive stimuli that might otherwise go unnoticed. Our pattern of response to the stimuli can be modified to bring about desired response through awareness and conditioning. This experience of exploring, freeing, and deepening one innate potentials can, if followed through, can have far-reaching creativity inspiring consequences in all spheres of life [1, 6].

Sensazione or heightened awareness of senses is the process of continual refinement of the senses, especially sight, as the means to enliven experience. Almost all our indulgences from sporting clothes in various colors and materials to trying out a new cuisine; from wearing perfume to listening to music, revolve around this need to not just satisfy, but also heighten our senses [5].

Great Artist, poet, sculptor, musician, inventor and engineer Leonardo Da Vinci lived by this principle of Sensazione. His life's motto was *saper vedere* (knowing how to see) upon which he built his work. He believed that, refining sensory awareness was the key to enriching life experience. He painted exquisitely by observing the minutest of details, played music and listened to silence, kept fresh flowers in his room, used perfume, introduced crafted meals in the west and proved time and again how this heightening of senses increased his creativity. Yet Leonardo reflected sadly that the average human 'looks without seeing, listens without hearing, touches without feeling, eats without tasting, moves without physical awareness, inhales without awareness of odour, and talks without thinking'. His assessment, centuries later, reads as an invitation to improve our minds and experiences to be a more creative, more intuitive and a more developed person [5, 10].

Sensazione can be pursued by handling different materials and textures, purposeful listening to sounds of varying intensity while feeling the mood or emotions evoked by them, tasting a new cuisine or flavour and learning to describe a smell or a taste. Writing or drawing a description of everyday scenes, practice of voluntary synesthesia and being surrounded by beautiful sense stimulating objects also helps in increasing one's sensitivity to sensory stimulations. According to Da Vinci, the stimulations themselves do not matter as much as ones awareness of them [5, 10].

The premise of this paper is not that this tool would enable students to become a genius like Da Vinci, but the fundamentals of his approach to learning and the cultivation of intelligence can be abstracted and applied to inspire and guide young minds toward the realization of their full creative potential.

1.2 Teaching and Assessing Creativity

The Da Vincian Principles summarized by Michael Gelb state that it is possible to relearn creativity and curiosity as adults, when driven by desire to know, to learn, and to grow. Unlearning myths that confine creativity and curiosity to the 'gifted' domain and opening up to new experiences can increase creative thinking and learning. Being more aware of one's senses while actively processing sense—response reactions can open up new ways of thinking, of processing found knowledge and taking holistic yet creative decisions [1, 5].

While we may have the tools to teach and assess content, creativity is another matter. Educators sometimes shy from assessing creative thought for fear of inhibiting students, or assess it very strongly based on their own perception of the subject at hand, distinguishing the output as right and wrong or good and bad. Thus developing a curriculum tool that enables teachers to enhance the creativity of their students and asses it with regard to relative quality and not necessarily quantity becomes vital. Such a tool should provide the required flexibility to adapt to various classroom situations, individual differences and should lay out insightful, solid, and elaborate understanding of the student's creativity ability, fluency and originality [12].

The Limitations of existing tools prompted Urban and Jellen in 1986 to develop their instrument, the "Test zum schöpferischen Denken—Zeichnerisch (TSD-Z)" or "Test for Creative Thinking-Drawing Production (TCT-DP)", a simple culture fair drawing test that can be used to test the creativity of any age group, independent of their educational qualifications and nurture. It applies a holistic and gestalt-oriented approach to creativity analysis by pursing a 14 point scoring dimension that comes to a total of 72 points. The dimensions are Continuation, Completion, New elements, Connections between fragments, Gestalt—Connections made to produce a theme, Boundary breaking that is both fragment dependent and independent, Perspective, Humor and Speed. It focuses on Unconventionality, as the sum of four processes—(a) manipulation of the material, (b)inclusion of surrealistic, fictional and/or abstract elements, (c) usage of symbols or signs and (d) Unconventional use of given fragments with negative marking when fragments are used to draw conventional images [16, 17].

This test and its scoring dimensions have been around for more than two decades; nonetheless it is used only during entrance exams and is almost nonexistent in classroom evaluations. This research uses TCT-DP as it enables a holistic understanding of student creativity and its development without any ambiguity.

2 Literature Review

The India report (1958) by Charles Eames and Ray Eames, defined the underlying spirit of NID and beginning of design education in India, by recommending an education of creative ideas rather than training [4]. Ruggiero, in the "Becoming a Critical Thinker" [11] discussed how inculcating creativity in students is one of prime pedagogical goals of a 21st century educator. Study of works [13, 15] list Creativity to be the foremost transferable skill that is expected out of Design and arts students in the industry and those of [14, Garett (2011)] affirmed faith in the fact that creativity could be taught and learnt and improved upon with practice [12–14]. The study, Barriers to Creativity in Education (2012–2013) found that 70 % of educators believe they can do more to teach creativity provided they are more equipped with tools and techniques [2]. Upon researching for probable creativity enhancing methods, the work of Micheal Gelb on "How to think like Leonardo Da Vinci" [5] that spoke in depth about Da Vinci's seven principles for creative genius, made an impact. Dr.Shade's effort in converting the first principle curiosita into a standard teaching tool—the "Curiosita teaching program—CTP" [12] established that exercises derived from the other principles too, could also be used to teach creativity. On close examination of each of the seven principles, sensazione emerged as the frontrunner owning to its simplicity in conception and application.

A review of Dissanayake's "What is art?" showed how arousal or intense emotion through aesthetic experience is pleasurable or satisfying as these emotions make us feel alive and engaged with the world, increasing not only our chances of survival but also our quality of life [3]. Gregory [6] in his Eye and the Brain has offered a remarkable amount of thought on how sensations (qualia) affect creativity. Explaining how keen development of senses play an important role in the increase of creative and complex thinking Gregory, illustrated that exposure to various stimuli, notably sensory, the performance of the average human brain improves by making increasingly complex new connections [6]. The study of perceptions by Helmhotlz indicated how our perceptions of sensory stimulation affect cognitive thinking. In the premise of Gestalt theory, problem solving by insight becomes possible (Koffka, Wertheimer, and Köhler), where perception is the product of complex interactions among various stimuli [7]. Montessori [8] believed that sensorial experiences began at birth and the child needs to be a lifelong sensorial explorer whose own experiences contribute to the increase in creative aptitude [8]. Despite the popularity of the Montessori Method of education in India, such sensorial exercises do not receive their due in higher education and students from regular school systems stay unaware of their potential. Sadly, even curriculum inclusions like material handling, field studies or art appreciation (included for the very purpose of sense stimulation) often get viewed as mere accompaniments to core content. Gelb's immense research on the principles of Leonardo Da vinci to enhance productivity of professionals at the workplace forms the backbone of this study and is based on his recommendation to study the effects of sensory stimulations to increase creativity in educational setup.

Designing research methodology for this kind of activity and population called for a Quantitative oriented experimental study [9]. In terms of statistical tests, it was found that experts have not advised complex procedures like structural equations or hierarchical liner modeling for an investigation at this stage, but a paired T test or ANACOVA. A simple yet culture fair test instrument with specific scoring points was found in Urban and Jellen's TCT-DP, a test with high validity and interrater relaibility which has been successfully

adapted and carried out in different Asian cultural contexts [16, 17] thereby making it suitable to study Indian students too.
Based on the necessity to examine processes of sensory stimulation that leads to a change in the creativity of design students and the lack of studies with this specific type of intervention, this paper may present valuable insight and urge additional investigation to reinstate these experiences as the vanguard of design education.

3 Research

The Objective was to discover a significant link between heighted awareness of senses and creativity and whether heightened awareness of senses increases potential to gather new knowledge, organize thoughts in unconventional and unique ways, provide access to out of the box, boundary breaking imagination and the ability to make holistic thematic connections thereby enhancing total creativity.

3.1 Research Hypotheses

1. There is a significant positive link between heighted awareness of senses and creativity. Heightening of sensory awareness increases creativity.
2. Heightening of senses significantly broadens access to new information and new knowledge.
3. Exercises to heighten the awareness of senses increases the ability to make holistic thematic connections.
4. Exercises to heighten the awareness of senses increases the potential for out of the box, boundary breaking imagination.
5. Heightening of senses offers unconventional ways of organising thought and perceiving the surrounding world.

3.2 Research Scope

A quantitative approach of experimental design with a pre-test/post-test control group design was selected. Study participants consisted of 144 undergraduate students from two design institutes in Chennai—National School of Design and INIFD (Fashion design and Interior design departments) in the age group of 18–24. Of these 44 were randomly chosen and divided into two equal groups of 22 each which formed the treatment and control groups as 21 participants per group for a one tailed hypothesis is sufficient detecting moderate effect sizes with 0.80 power at the 5 % level of significance in an experimental study [9]. It was verified that these students were not exposed to this sort of a creativity test or doodling exercises prior to this test being conducted to avoid effects of testing. The control group was kept unaware of the treatment used in the experimental group and the experimental group was not informed that the posttest will be similar to the pretest in advance to avoid effects of testing and experimental effects.

3.3 Method

For a significant effect, the Test instrument—TCT-DP: test of Creative Thinking Drawing Production was chosen.

3.3.1 Test Instrument Design

The TCT-DP requires the respondent to complete two drawings as in Fig. 1 based on six figural fragments that were designed in accordance with the following rules. The fragments are (1) different in design, (2) geometric and non-geometric, (3) round and straight, (4) singular and compositional, (5) broken and unbroken, (6) within and outside a square frame, (7) placed irregularly on the space provided, and (8) incomplete. The big square frame together with the small open square outside the large (boundary) serves to provide information on risk taking, which was seen by the authors as "boundary breaking in a twofold manner." The given figural fragments were kept simple and unique to give rise to a multitude of creative responses along with enough suggestibility for stereotypical responses from less creative students [3]. In the 2 forms—Forms A and B, Form B is designed to be slightly harder than Form A to avoid effects of testing.

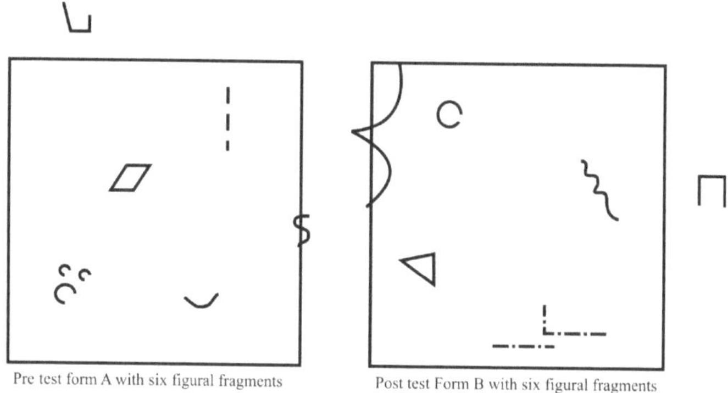

Fig. 1 TCTC-DP test forms. *Source* self designed

3.3.2 Treatment

This treatment focuses on sensory stimulations pertaining to vision, hearing, touch, taste and smell using four exercises.

Exercise# 1—Vision: Participants were exposed to everyday natural occurrences like sun rise, sunset, flow of water, blooming of flowers, movement of birds via audio visuals of the same. This was followed by a presentation exposing them to works of art and design by various masters like Michelangelo, Da Vinci, Van Gogh, Picasso, Ravi Verma, Antonio Gaudi, Raymond Loewy and more. The participants observed how basic elements were manipulated to create more advanced forms.
Exercise# 2—Hearing: An assortment of everyday sounds like water draining, bird song, bell sounds, train whistles were played to sensitize participants to sounds that are generally dismissed as background noise. This was followed by playing of western music (Vivaldi, Beethoven, and Stravinsky) and Indian classical music which are considered to be conducive for complex thinking. The participants were asked to associate these sounds with suitable images in their mind. The music was also supported by visuals of random splashes of color to emphasize color psychology.
Exercise# 3—Touch: Participants were given specially prepared boxes with different tactile textures on each side. The concept for these boxes was adapted from the Montessori's Stereognostic Sense Exercises. They were passed around first with the lights off and the participants were asked to feel and differentiate between the textures for buildup of muscle memory and then with lights on for textural recall.
Exercise# 4—Taste and smell: Tasting of new flavors, different than what one is accustomed to by taking the time to smell, eat and savour develops not only ones taste buds but also sharpens conceptual thinking. As wine or food tasting is difficult to organise in a classroom atmosphere the participants were offered dark chocolate pieces which they were expected to smell and taste slowly to allow full appreciation.

3.4 Data Collection

The pre-test of TCT-DP (form A) was administered to all subjects in both groups ensuring that they experienced the same conditions except that in addition the experimental group experienced the treatment. Four Exercises to stimulate the participant's senses (of vision, hearing, touch, smell and taste) were introduced to the treatment group. The post-test of TCT-DP (form B) was administered to all subjects of both groups. Time taken was noted on all forms for marking of Speed component. The test was evaluated by design faculties following the rules of scoring of TCT-DP, listed in Sect. 1.2 of this paper and the average scores were recorded. All values were tabulated and the highest and lowest scoring sheets were identified and also analysed visually.

Table 1 Total creativity score of treatment group versus control group

	Pretest mean	Posttest mean	t stat	SD
Treatment group	32.409	39.455	4.885	6.765
Control group	29.091	29.727	0.604	4.9403

Source TCT-DP test scores

4 Analysis and Discussion

The amount of change on the value of the dependent variable from the pre-test to the post-test for each group was assessed separately for Total creativity, new elements, Unconventionality, thematic connection and Boundary breaking imagination aspects of TCT-DP using the Paired t-test method at 5 % level of significance as a one-tailed test.

4.1 Heightened Awareness of Senses Increases Creativity

The assessment of the total creativity score, Pretest and posttest of both groups as in Table 1 led to the rejection of the null hypothesis and acceptance of the alternate hypothesis that there is a significant positive link between heightening of sensory awareness and creativity. Heightened awareness of senses increases creativity.

4.2 Heightening of Senses Significantly Broadens Access to Information and New Knowledge

The Investigation of the "New Elements" (Ne) component which is the addition of any new figure, symbol or element drawn by the participant within the box that is not a continuation of the 6 given figural fragments and points to the ability to think of new ideas based on existing concepts resulted in the rejection of the null hypothesis (based on values seen in Table 2) and acceptance of the alternate hypothesis that Heightening of senses significantly broadens access to new information and new knowledge.

Table 2 New elements score of treatment group versus control group

	Pretest mean	Posttest mean	t stat	SD
Treatment group	3.318	3.955	2.134	1.399
Control group	1.545	1.364	(−) 0.699	1.220

Source TCT-DP test scores

Table 3 Thematic connections (gestalt) score of treatment group versus control group

	Pretest mean	Posttest mean	t stat	SD
Treatment group	4.227	5.273	2.596	1.900
Control group	3.636	3.636	0	1.690

Source TCT-DP test scores

Table 4 boundary breaking imagination score of treatment group versus control group

	Pretest mean	Posttest mean	t stat	SD
Treatment group	2.182	3.500	2.851	2.169
Control group	2.227	2.500	0.843	1.520

Source TCT-DP test scores

4.3 Exercises to Heighten the Awareness of Senses Increases the Ability to Make Thematic Connections

The evaluation of the "Thematic Connections" (Cth) component which is based on Gestalt Theory as indicated in Table 3 led to the rejection of the null hypothesis and acceptance of the alternate hypothesis that Exercises to heighten the awareness of senses increases the ability to make holistic thematic connections.

4.4 Exercises to Heighten the Awareness of Senses Increases the Potential for Out of the Box Boundary Breaking Imagination

The assessment of the summation of the two scores of Bfd and Bfi, i.e., Boundary Breaking imagination that is fragment dependent and independent, as seen in Table 4 referred to as 'B' points to the rejection of null hypothesis and acceptance of the alternate hypothesis that Exercises to heighten the awareness of senses increases the potential for out of the box boundary breaking imagination.

4.5 Heightening of Senses Offers Unconventional Ways of Organising Thought and Perceiving the Surrounding World

Unconventionality Score or 'U' as per the scoring system is the summation of the four sub components of UCa, UCb, UCc and UCd, when statistically tested provided values as seen in Table 5 resulting in the rejection of the null hypothesis and acceptance of the alternate hypothesis that Exercises to heighten the awareness of senses offers unconventional ways of organising thought and perceiving the surrounding world.

Table 5 Total unconventional score of treatment group versus control group

	Pretest mean	Posttest mean	t stat	SD
Treatment group	3.3640	4.3640	2.128	2.200
Control group	3.682	2.909	(-)1.557	2.330

Source TCT-DP test scores

5 Conclusion

The results of this study demonstrate that creativity can indeed be taught, learnt and assessed provided the right tools are available. The statistical analysis of the creativity total and selected components reflect favorably towards the research hypotheses. Analyses of the control group scores proved that pretest, history, maturation, or any other external event did not influence the posttest score and that the treatment given was the sole influence of the rise in posttest scores. The values indicate that heightened sensory awareness is the natural way to impact positively upon one's creativity. Heightening of one's senses increases the ability to acquire new knowledge, to form holistic gestalt based compositions that tell a story, ability to achieve out of the box boundary breaking imagination and unconventional thinking. To state simply there exists indeed a significant positive link between heighted awareness of senses and creativity and heightened awareness of senses increases creativity in design students.

Taking a closer look at the statistics, an interesting fact comes to the fore front. Though there was significant difference to prove all five research hypotheses, maximum difference exists in the total creativity score. This coincides with Urban and Jellen's recommendation of how creativity should be judged—as a sum of all components and not simply on the merit of any one of them. This is a direct indication of how Creativity should be assed compared to how it is assessed today. Thus one of the other significant findings to emerge from this study is the existence of a creativity assessment method that is neither too rigid nor too vague—a method that is not ambiguous. The fourteen scoring dimensions of TCT-DP comes forward as valuable tool to assess creativity in classrooms as it is flexible, elaborate and allows fluency and originality which form the four elements of creativity. Although the findings in this study cannot be generalized to be reflect the creative growth of all college students or even all university level design education students in India and abroad, transferability of this study can be counted upon for students of design institutes similar to those studied.

6 Recommendations

Going by the results of the study and the feedback received from participants, it is recommended that design colleges and other creative institutes take to teaching creativity and educators must provide inputs on thinking creatively. The sensazione exercises outlined in this paper along with the scoring dimensions of TCT-DP can be developed into an effective pedagogical method that can be utilized by the teachers of design colleges to record, study and analyze the creative potential of their students. The process is not time consuming or financially exacting and educators can create their own treatment exercises inspired by the ones mentioned in this study, following Da Vinci's principles depending on the age group, background of the students and the time available. It is recommended that further research be carried out, on a larger scale to determine the effectiveness of the scoring dimensions suggested in real time classroom situations.

Acknowledgments I would like to express my deep gratitude to Mr. Raees Ahmad (Director, National School and College of Design) and Ms. Fareeda Khan (Director, INIFD, Chennai) for their help in offering me the resources to conduct this research.

References

1. Ackerman, D.: A *Natural History of the Senses*. Vintage Books, New York (1991)
2. Adobe Systems Inc.: Barriers to Creativity in Education: Educators and Parents Grade the System. Study (2013)
3. Dissanayake, E.: What is Art For? University of Washington press, Seattle (1988)
4. Eames, C., Eames, R.: The India Report (1958). National Institute of Design, Ahmedabad (1997)
5. Gelb, M.J.: How to Think Like Leonardo-Seven Steps to Genius. Bantam Dell, New York (1998)
6. Gregory, R.L.: Eye and Brain: The Psychology of Seeing, 4th edn. Princeton University Press, New York (1990)
7. Koffka, K.: Principles of Gestalt Psychology. Harcourt Brace, New York (1935)
8. Montessori, M.: The Discovery of the Child. Clio Press, Oxford (1997)
9. Onwuegbuzie, A.J., Collins, K.M.T.: A Typology of Mixed Methods Sampling Designs in Social Science Research (2010)
10. Richter, J.P.: The Notebooks of Leonardo da Vinci. Dover Publications, New York (1970)
11. Ruggiero, V.R.: Becoming a Critical Thinker. Cengage Learning, Boston (2011)
12. Shade, A., Garrett, P.: Curiosita Teaching—Integrating Creative Thinking into Your 21st Century Classroom. Pieces of Learning, Saline (2011)
13. Stephenson, J., Weil, S.: Quality in Learning—A Capability Approach in Higher Education. Kogan Page, London (1992)
14. Torrance, E.R., Robert, E.M.: What Next? Futuristic Scenarios for Creative Problem Solving. Hawker Brownlow Education, England (1996)
15. Twigg, C.: Need for National Learning Infrastructure. Educom, Washington DC (1995)
16. Urban, K.K.: Assessing creativity: the test for creative thinking—drawing production (TCT-DP) the concept, application, evaluation and international studies. Psychol. Sci. **46**, 387–397 (2004)
17. Urban, K.K., Jellen, H.G.: Test for Creative Thinking—Drawing Production (TCT-DP). Swets & Zeitlinger, Lisse (1998)

Enhancing Creativity by Using Idea-Wheel and Its Validation

Avinash Shende and Amarendra Kumar Das

Abstract In IIT Guwahati, the students who join the Master of Design program come from four different backgrounds, these are Technical background, Architecture Background, Fine Arts Background and Design and Fashion Design Background. The students of such diverse background react/think differently while solving a design problem. The students try to find a solution to a problem which is generally ill-defined, cannot be resolved by them and this leads them to follow a problem-focusing strategy. Empirical data on such design processes were obtained from a set of protocol studies of 27 Design Students with heterogeneous background, whose designs were evaluated on the basis of overall quality and taking into consideration variety of aspects including creativity. From the generated data, we identify aspects of creativity in design related to the formulation of the design problem and to the concept of originality. We have also introduced a tool "Idea Wheel" to the same set of students, faced with the same design problem, left for them to solve. We then reassess the creativity aspect. It is then observed as to how an ill-defined problem can be resolved and how a problem-focusing strategy changes to a solution-focusing strategy among the students. In addition to this, the paper will highlight the validation part by conducting the identical experiment on another set of students with heterogeneous background.

Keywords Assessing creativity · Heterogeneous background · Design process · Problem solving

1 Background

Design schools are increasing in steady numbers in India in the recent times. There are more than 20 design schools which include both government and private funded. The selection of students for Master of Design at Indian Institute of Technology in

A. Shende (✉) · A.K. Das
Department of Design, Indian Institute of Technology, Guwahati, India
e-mail: savinash@iitg.ernet.in

A. Chakrabarti (ed.), *ICoRD'15 – Research into Design Across Boundaries Volume 2*, Smart Innovation, Systems and Technologies 35, DOI 10.1007/978-81-322-2229-3_7

India is based on their performance in an entrance examination "Common Entrance Examination for Design (CEED)". The eligible students appearing for this examination, could be artists, fashion designers, architects and engineers [18].

In general, the selected students belong to diverse background generally comprising four domains and hence, there is hardly an identical response towards lateral thinking, which is necessary for ideation and post ideation process of designing product. The students with technical background go through rigorous training to obtain technical knowledge for four years or more; similarly the students with architectural, art and fashion design background undergo through architecture, art and fashion design studies for four years. These students are accustomed to showcase a typical approach to a design product or solve design problems. An essential quality in problem solving that is desired in design students is creativity. The Creativity factor amongst the students differs, as their mind are already conditioned during their undergraduate training; students with technical background are more prone to think technically and students from architecture, art and fashion design background shows inclination towards aesthetics. However students from architecture background appear to be more creative, because they are already trained to think that way. Various attempts have been made to negate the assumption that this mental activity, creativity occurs inside the heads of some special people [6]. It is systematic rather than an individual phenomenon. Affective way of increasing creativity were tried out by altering the environment and by introduction of tools like speed-storming [1], mood-board, scamper etc., which indicates that creativity can be enhanced by certain techniques/tools/methods [13].

2 Methods of the Study (Methodology)

The above stated fact has led to a number of questions: What factors lead heterogeneous qualified students to arrive at a certain way of thinking? Whether the students from creative background are more creative than the students from technical background? How can they improve creativity exclusively while problem solving? More specifically, how can Heterogeneous background students generate good and creative ideas? These general questions have been the preliminary point for the development of this research and have led to the refinement of the following research question and null hypotheses.

Question 1: How do Heterogeneous background students respond through the use of tools/techniques/method compared to equivalent background students without using tools/techniques/method in terms of their creative performance with respect to the quality of ideas produced?

Null Hypothesis 1a: The level of creativity among the students with heterogeneous background does not significantly vary with regards to their background.

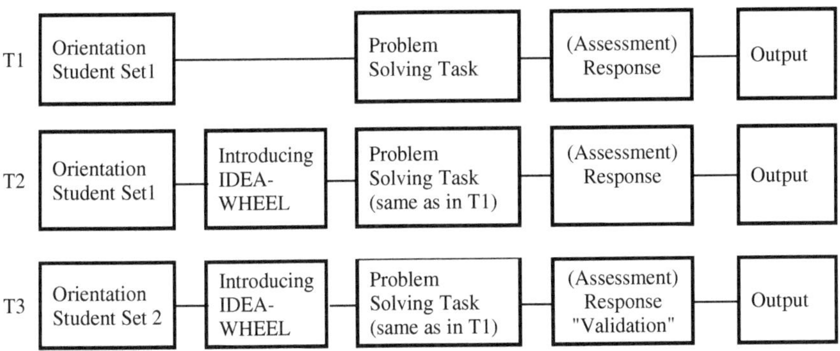

Fig. 1 The formation of tests with or without tool (idea wheel)

Null Hypothesis 1b: There is no significant difference on the level of creativity among students with heterogeneous background with and without proposed method/tool/techniques.

To answer this question and test the hypotheses, a mixed method approach was used by which empirical data were collected [8]. Two Tests between-students for two conditions were employed i.e. T1 and T2, to compare the quantitative differences in idea-generation between the students with heterogeneous background with and without introducing tools/techniques/method. Test T3 employed to validate the result of T2 (Fig. 1).

Test T1 and T2 was conducted among equivalent set of students (27 students who recently joined M.Des. program at Department of Design, IIT Guwahati. Out of 27 students 12 belong to Engineering Background, 4 from Architecture Background, 5 from design background and 6 from Fine arts Background). Tests starts with *orientation* to *problem solving task* performed by the students leading to *assessment* of creativity and concludes with a *result*. The difference between tests T1 and T2 is the introduction of a creativity enhancement tool “IDEA-WHEEL” during test T2 between orientation and problem solving task. Problem solving task given was the same in both tests T1 and T2, whereas the time gap between T1 and T2 is more than 32 weeks. The time gap is decided upon the fact that accelerated long term forgetting phenomenon will occur, and by 8 weeks, students retain about zero percent of the task performed [9], however we extended time gap up to 32 weeks. In addition, during this gap the students were not made to discuss about the task considering the fact that complete disremembering occurs when memories are not re-activated and so they do not benefit from the formation of multiple traces in the Medial Temporal Lobe [19]. Test T3 conducted among another set of students (applicant came for Entrance Exam for M.Des. at Department of Design, IIT Guwahati, by keeping same ratio of background as in T1 and T2).

3 Problem Solving Task

The students were asked to generate ideas in a stipulated period of 1 h for the problem solving task given below.

TASK: "After joining M.Des, in the hostel you are washing your clothes with your hand, using detergent powder, soap and a bucket. Design a Tool, which will assist/help you in this task".

The task given was same in test T1,T2 and T3, the ideas derived from the tests were then evaluated by condition-blind evaluators on counts of non-redundant ideas and three measures of idea quality: non-obviousness [12], utility or effectiveness, and feasibility. This gave room for appropriate comparisons among students with heterogeneous background performance through use and non- use method/tool/ techniques' in answering the first research question and testing the two associated null hypotheses. The notes from observations and the written protocol, along with the responses, were reviewed and coded, which helped in assessment without contradiction.

3.1 *Setting*

The test took place in a midsize class room located at IIT Guwahati, India. The room was spacious with moveable furniture that allowed for adaptations of the space and its organization. The tables and chairs were set up at a minimum distance of 1.5 m. The goal was to keep the room setting as sterile as possible, to avoid any confounds from the students' surroundings. It also took into account that there were no objects or materials lying on shelves or hanging on walls that could potentially influence the participants' thoughts and actions during the course of study.

3.2 *Materials*

The materials used for this study included pencil, ball-point, sketch pen and blank paper for writing down or sketching out ideas, as well as a brief script informing the participants about the purpose of the study, the relevant topics addressed, providing facts and examples for ideational inspiration. All students received the same materials in all treatment conditions. The topic for tests T1, T2 and T3 was chosen from a bank of topics that were developed in preparation for the test activities. Each student had an equal opportunity to contribute knowledge on the given task. Each topic was reviewed and discussed with the researcher's supervisor zeroing down on the ones most suitable for this study. The "Tools for Washing Clothes" topic was developed in conjunction with the students who live in the campus and wash their own clothes all by themselves. The topic was familiar to the students already experiencing the situation. Creativity enhancement tool 'Idea-Wheel' was introduced to students prior to test T2 and T3 clearly through the assignment sheet and ensured that each student understood it clearly and ways to use it.

3.3 Data Analysis

There were two measurement instruments used for this study: one for the *quantitative* data and the other for the *qualitative* data. The quantitative measures were the generated ideas that were hand-drawn on paper by the students. The qualitative measures were written protocol for which a space was provided on the right hand side of drawing sheet (A3 size), and the vacant sections s10, s11, s12 provided on Idea Wheel to fill (Fig. 2). The quantitative and qualitative data were analyzed through the use of the Consensual Assessment Technique [2–4, 11] whereby a number of evaluators independently assessed the generated ideas on measures of quantity and quality. There were three evaluators for this process: two evaluators were design faculty from design schools in India and Milan and the third was an industrial designer and graduate student from the Masters of Design program, also at Milan. First, evaluator independently evaluated the generated ideas on measures of quantity and quality at an individual level. They reviewed each student's idea-sheets and filtered out all redundant ideas for every individual student. They then began rating the quality of the non-redundant ideas based on their non-obviousness, their utility and effectiveness, and their feasibility using 7-point scales for each

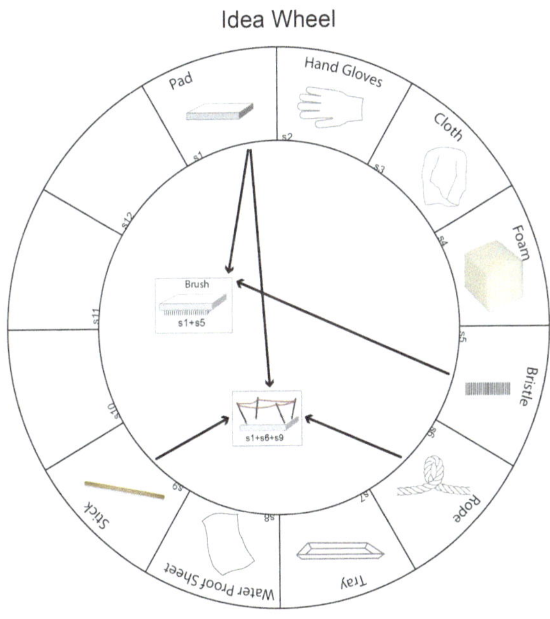

Fig. 2 Idea wheel, introduced to the students during test T2 and T3

factor, where 1 represented very low and 7 represented very high. The inter-judges reliability was determined by computing the alpha-coefficient for the agreement between the judges. The alpha-coefficient for the end measure in this study, the total judgment, was a very reasonable 0.71 [7].

4 Test Conducted Without and with Creative Enhancement Tool

4.1 Test T1: Conducted Without Any Support of Tool

Empirical data on design processes were obtained from a set of protocol studies of 27 design students with heterogeneous background, whose designs were evaluated on the basis of overall quality and taking into consideration a variety of aspects including creativity. One way ANOVA was conducted for a significance level of 0.05 and we found the value of $p(0.034) < 0.05$ and thus the null hypothesis 1a was rejected and concluded that there is a significant difference of level of creativity among the students from heterogeneous background. This led us to design a creativity enhancement tool "Idea-Wheel" that was introduced in the Test T2.

4.2 Test T2: Conducted by Introducing a Tool Called "Idea Wheel"

Idea Wheel was designed on the basis of findings of Literature Review, which are presented below.

- Breaking out of an established pattern either perceptual or cognitive, allowing for the development of a broader range of alternative, perhaps creative solutions [16].
- The initial problem was vague or ill-defined [2].
- Mode of problem solving is 'Solution Focused' [4].
- Creative idea as a combination of thoughts [19]. Creativity is the ability to see and establish new relationships or connections. These juxtapositions set up new patterns and relationships, and created fresh and astonishing way of looking at things.
- Techniques to think creatively—Making Novel combination: The combining and recombining ideas, images thoughts keywords into different combinations in the conscious or subconscious minds converts to the novel ideas by connecting the unconnected: There are things around us that can be connected with each other by coercing relationships [1]. Creative ideas involve "bisociation" or the connecting of disparate thoughts to form a new and relevant idea [17].

Description about Idea Wheel: The idea wheel has 12 sections-s1, s2, s3, s4 … s12 (see Fig. 2). Every section has randomly selected familiar objects, except s10, s11 and s12. One may combine any section "sn" for any number of time to create a new tool. For example (s1 + s5) gives a brush for washing clothes, and (s1 + s6 + s9) has led to an absolutely new tool. Additionally, Idea wheel encourages to add any familiar object in s10, s11, s12 by the user, this may turned out to be an useful contribution in idea wheel to generated new idea.

5 Results

In order to answer the research question and to test the proposed hypotheses, the mean ideational performances of students with heterogeneous background with and without Idea Wheel were compared. A SPSS paired sample t-test with unequal variances was conducted for each measured variable to assess the significance of any differences that were found between them. This was done by taking average of the data produced by judges. The following sections present the results from this comparison for the measures of idea quality.

Table 1 presents the mean total number of non-redundant ideas produced in both the without Idea Wheel and with Idea Wheel conditions. These means are presented for one topic i.e. "designing tool for washing clothes", where, the two tailed significance value p is 0.00, therefore the null hypothesis 1b is rejected and concluded that there is significant differences in terms of Creativity among the heterogeneous background students with and without using the tool "Idea Wheel".

Students with heterogeneous background using Idea Wheel were able to generate quality Idea (M = 6.132, SD = 0.641) in comparison to the equivalent student without using Idea Wheel (M = 3.518, SD = 1.672); The quality of Idea generated using Idea Wheel is enhanced nearly 1.75 times with standard deviation (SD = 0.641), which is much more lower in comparison to the quality of idea generated without using Idea Wheel whose standard deviation is (SD = 1.672). Figure 3 Scatter gram shows very few student's ideas were creative in test T1 (*without idea wheel condition*) and Fig. 4 Scatter gram shows more than 80 % students ideas are reaching towards the peak of creativity in test T2 (*with idea wheel condition*).

Table 1 Paired sample t-Test, showing mean quantity of ideas produced

Topic	Condition	N	M	Sd	T	Df	P
Designing tool for washing clothes	Without idea wheel	27	3.518	1.672	−9		0.00
	With idea wheel	27	6.132	0.641			

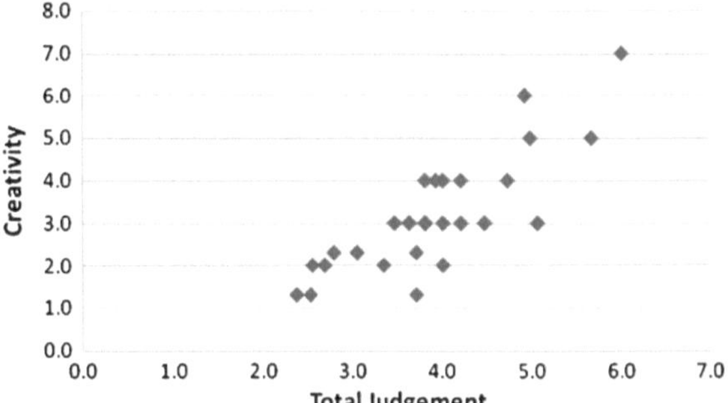

Fig. 3 Scatter gram for the means of 'total judgment' and 'creativity' for Test T1 (*without idea wheel condition*)

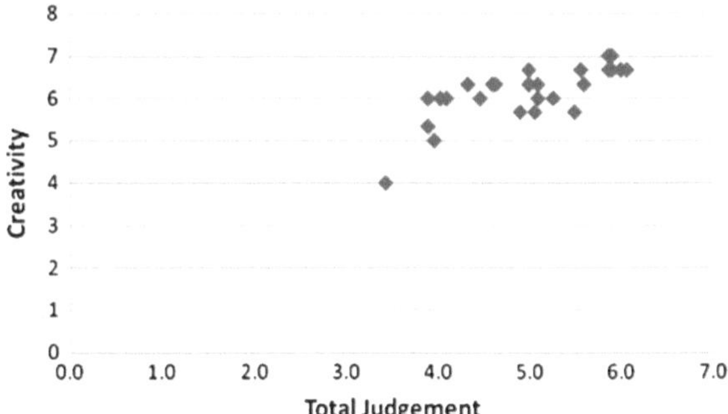

Fig. 4 Scatter gram for the means of 'total judgment' and 'creativity' for Test T2 (*with idea wheel condition*)

6 Validation

In order to validate the result of test T2, *Test T3: was conducted with another set of students (applicants who came for Entrance Exam for M.Des. at Department of Design, IIT Guwahati), by introducing same tool called "Idea-Wheel".*

Empirical data on design processes were obtained, and were evaluated on the basis of overall quality and taking into consideration a variety of aspects including creativity. Later, a SPSS paired sample t-test was conducted for each variable measured to assess the significance of the differences that were found between the two different set of students under same condition i.e. "with Idea-Wheel".

Table 2 presents the mean total number of non-redundant ideas produced by the two different set of students under same *Idea Wheel* conditions. These means are presented for the topic "designing tool for washing clothes" for. As can be seen, the two tailed significance value p is 0.032 > 0.00 therefore can be summarised as" there is no significant difference on level of creativity among the students with heterogeneous background students with introduction of Tool/method/technique (Idea Wheel)". Two different set of students S1 and S2 with heterogeneous background using Idea Wheel were able to generate quality Ideas like for S1-(M = 6.1326, SD = 0.641) and for S2-(M = 5.759, SD = 0.594); Whose mean values are nearly same and with very less difference on Standard deviation. Figure 5 Scatter gram and Fig. 6 Scatter gram are nearly matching to each other.

Table 2 Paired sample t-Test statistic mean quantity of Ideas produced on test T2 and T3

Topic	Condition	Students	N	M	Sd	T	Df	P
Designing tool for washing clothes	With idea wheel	S1-students equivalent T2	27	6.132	0.641	2.26	26	0.032
	With idea wheel	S2-students entrance exam for M.Des.	27	5.759	0.594			

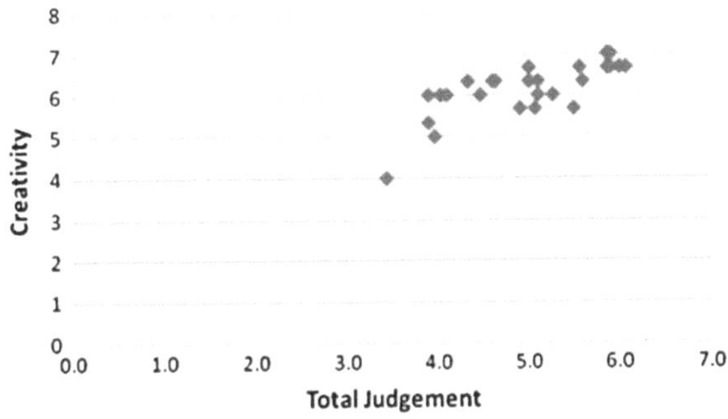

Fig. 5 Scatter gram for the means of 'total judgment' and 'creativity' for test T2 (*for Student S1—with idea-wheel condition*)

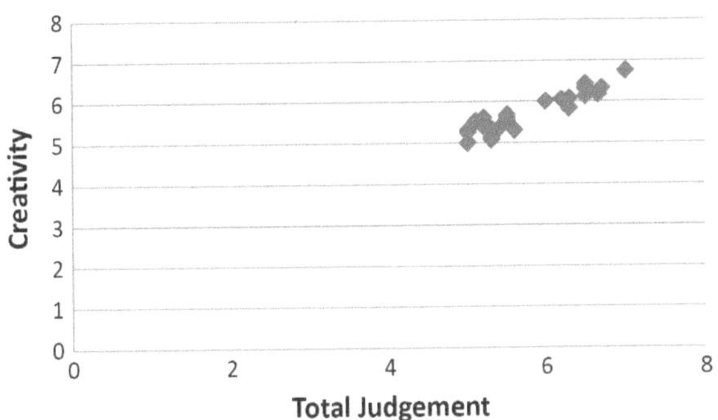

Fig. 6 Scatter gram for the means of 'total judgment' and 'creativity' for test T3 (*for Student S2—with idea-wheel condition*)

7 Observation and Discussion

For Test T1, students' responses were all disjointed and not focused on the actual problem; this may be due to the fact that during test T1 the problem was not defined properly, and students employed Problem-focus approach [4] (Fig. 7). However, for test T2 and T3, students altered their approach completely. They were applying solution-focus approach by using Idea Wheel effectively. Students tried to fill objects of their own in the vacant section of idea wheel (Fig. 8) and used it to design a new tool which was not obvious and feasible initially. It was observed that around fifty percent of students ideas generated were same during the test T2 and T3

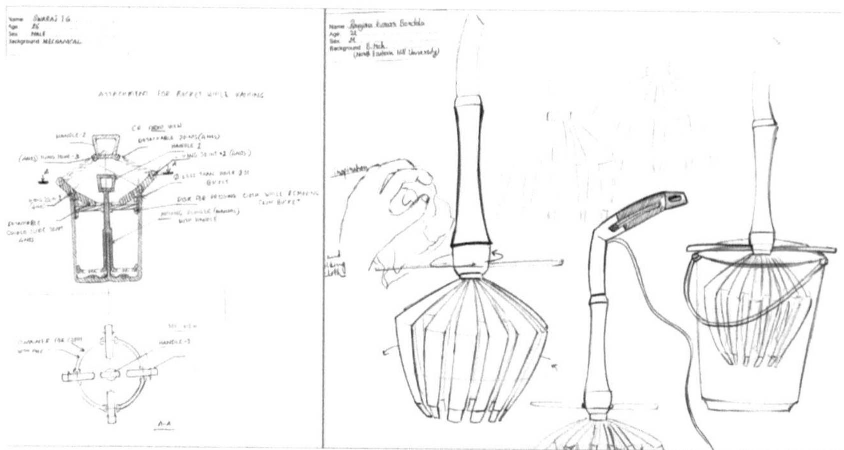

Fig. 7 Students' responses during test T1

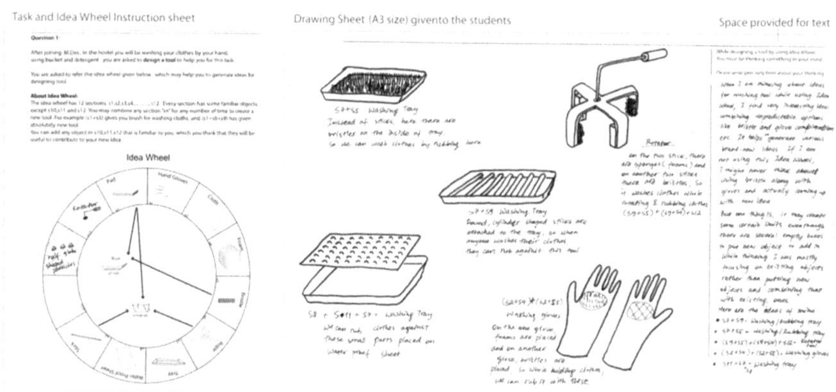

Fig. 8 Student response during test T2

irrespective of their background, which means there is a significant change on the approach of problem solving among the students using Idea wheel. In Figs. 3, 4, 5 and 6 of scatter grams, there is a significant difference in quality of ideas generated without and with 'idea wheel'; also indicates that the student's responses despite belonging to different sets are identical, faced with similar condition of using 'Idea Wheel'. Overall, it shows that a significant number of ideas generated is creative, even if some of them might not perform well at Total Judgment.

8 Conclusion

Table 1 shows that the mean value of the responses with Idea wheel condition is significantly higher than the mean value of the responses without Idea Wheel condition. Hence, it can therefore be safely concluded that there is a significant difference in terms of creativity among students with heterogeneous background with and without Idea-Wheel conditions. Table 2 shows that the mean value of responses of two different set of students with same Idea-Wheel condition are nearly identical, hence it can be concluded that the responses of students with heterogeneous background are more creative while using the Idea wheel. The two different set of student' performance using Idea-wheel shows significant enhancement on Creativity and also on the total judgment collectively, and hence, creativity enhancement tool "Idea-Wheel" was suggestively effective in generating quality ideas among the students with heterogeneous background. It can also be perceived that students' approach on solving problem changed from problem-focused to solution-focused, when the Idea-wheel began to act like a catalyst. In test T1, the results of students among themselves were diverse to a great extent compared to test T2 and T3, which shows that in test T2 and T3 the problem was defined properly by introducing Idea-Wheel. Further study is planned to be carried out by conducting more experiments with students with heterogeneous background of similar institutes and also to investigate on the results that could be arrived at, if there is a change of "Idea wheel".

Acknowledgments Authors acknowledge M.Des students of 2012 batch at Department of Design, IIT Guwahati and applicants who appeared for the departmental entrance examination for M.Des program at IIT Guwahati in 2013. Authors also acknowledge Dr. Pratul Kalita for his input in data analysis, that forms an essential part of this research.

References

1. Adams, J.L.: Conceptual Blockbusting: a guide to better ideas. W.W. Norton and Company, New York (1979)
2. Amabile, T.M.: Social psychology of creativity: a consensual assessment technique. J. Pers. Soc. Psychol. **43**, 997–1013 (1982)
3. Baer, J., McKool, S.: Assessing Creativity Using The Consensual Assessment. Handbook of Assessment Technologies, Methods, and Applications in Higher Education. IGI Global, Hershey, Pennsylvania (2009)
4. Cross, N.: Designerly Ways of Knowing. Springer, Berlin (2006)
5. Csikszentmihalyi, M.: Creativity: Flow and the Psychology of Discovery and Invention. HarperCollins, New york (1996)
6. De Bono, E.: Lateral Thinking. Academic Press, Waltham (1968)
7. Dorst, K., Cross, N.: Creativity in the design process: co-evolution of problem–solution. Des. Stud. **54**, 425–437 (2001)
8. Elam, J., Mead, M.: Can Software Influence Creativity? The Institute of Management and science, Howard University, Boston (1990)

9. Elliott, G., Isaac, C.L., Muhlert, N.: Measuring forgetting: a critical review of accelerated long-term forgetting studies. Cortex **54**, 16–32 (2014)
10. Haats, T.: Creative synergy in interdisciplinary design collaboration: using speed storming to generate ideas. M.Des. thesis, The Faculty of Graduate and Postdoctoral Affairs, Carleton University (2012)
11. Hennessey, K., Kim, G.: A Multi-cultural application of the consensual assessment technique. Int. J. Creativity Probl. Solving **18**, 87–100 (2008)
12. Lopez-Mesa, B., Vidal, R.: Novelty metrics in engineering design Experiments. International design conference-Design, vol. 1, pp. 557–564. (2006)
13. Mayer, R.E.: Guiding students' cognitive processing of scientific information in text. In: Pressley, M., Harris, M., Guthrie, J.T. (eds.) Promoting Academic Competency and Literacy in School. Academic Press, Waltham (1992)
14. Michalko, M.: Cracking Creativity: The Secrets Of Creative Genius. Ten Speed Press, Berkeley (2001)
15. Nadel, L., Moscovitch, M.: Memory consolidation, retrograde amnesia and the hippocampal complex. Curr. Opin. Neurobiol. **7**, 217–227 (1997)
16. Osborn, A.: Applied Imagination: Principles and Procedures of Creative Thinking. Scribner's, New york (1963)
17. Runco, M.A., Pritzker, S.R.: Encyclopedia of creativity. Academic Press, Waltham (1999)
18. Shende, A., Das, A.K.: Challenge of teaching product design in master course with heterogeneous qualification in indian context. Eng. Prod. Des. Edu. Postgrad. Des. Edu. **1**, 711–716 (2012)
19. Young, L.: Right-brained decision support systems. ACM SIGMIS Database **14**, 28–36 (1983)

Part II
Eco-Design, Sustainable Manufacturing, Design for Sustainability

Designing Technology, Services and Systems for Social Impact in the Developing World: Strong Sustainability Required

Gavin Melles, Blair Kuys, Ajay Kapoor, James Rajanayagam, Joseph Thomas and Aswhin Mahalingam

Abstract In the post-Bruntland era of sustainable development, design has taken a decided turn towards social impact although shows limited understanding of sustainable development. Socially responsible design (SRD) in developing countries builds on discourses of human-centred design, sustainability and social impact. Despite its successes, design approaches, however, remain firmly embedded in the mainstream sustainable development (MSD) discourse of market environmentalism, populism and ecological modernization. This has led to short term successes and failures of technology innovation, which a broader and deeper understanding could have helped avoid. Beyond Life Cycle Analysis (LCA) and eco-design for sustainability are development demands that question economic, social and environmental common sense, demanding systemic approaches to innovation and policy and governance change for sustainable livelihoods in BoP contexts. Focused on material and object constraints, industrial and product design is intellectually weak in this respect and divorced from actively participating in larger informed debates although new models, such as industrial ecology suggest new possibilities. From end-of-pipe pollution control to deep green ecology, the ladder of sustainable development ideologies position technology innovation relative to human and ecological well-being differently. Taking such an agenda seriously will demand radical change to design curricula to better integrate social impact economics and innovation. This paper proposes that sustainable development and social entrepreneurship and not sustainability and philanthropy should be the framework of choice for design's theoretical and practical contribution to development in BoP. Contributing to this discussion are outcomes and experiences from a joint IITM (Madras) and Swinburne University (Australia) collaboration.

G. Melles (✉) · B. Kuys
School of Design, Swinburne University, Melbourne, Australia
e-mail: gmelles@swin.edu.au

A. Kapoor
School of Engineering and Industrial Sciences,
Swinburne University, Melbourne, Australia

J. Rajanayagam · J. Thomas · A. Mahalingam
Centre for Innovation and Social Entrepreneurship, IIT Madras, Chennai, India

A. Chakrabarti (ed.), *ICoRD'15 – Research into Design Across Boundaries Volume 2*,
Smart Innovation, Systems and Technologies 35, DOI 10.1007/978-81-322-2229-3_8

Keywords Socially responsible design · Sustainable development · Education

1 Sustainable Development: Design Avoidance and Confusion

Design's recent turn to social impact in developing countries has emerged at a time the concept of sustainable development was increasingly being defined in terms of the triple bottom line (TBL) of economic, environmental and social constraints. Although acknowledging this discourse, design schools until recently have shown limited understanding of sustainable development (SD) and TBL for defining constraints on technology-driven (and service) innovation, preferring the vagueness of sustainability and the concrete but limited application of life cycle analysis (LCA) and similar tools as technical synonyms [1]. This limited perspective on sustainable development is due also to writers such as Walker [2] describing sustainable development as a 'myth', preferring it seems to limit 'hands on' design to material constraints and convention. Walker, we believe, is wrong! His claim seems to mean that SD is intellectually, politically and economically complex and challenging [3]. Thus, it will require more than the typical 'hands on' studio creativity of designers, a challenge that Fry [4] is happy to accept and develop into a program for design. We subscribe in principle to Fry's position focusing our discussion on technology innovation.

The goals, targets and measures of the Millennium Development Goals (MDGs) in 2000 regarding targets for 2015 placed sustainable development on the global agenda. In the current discussion about renegotiating the MDGs for the period 2015 and beyond, there is broad recognition that the needs based goals and targets for the Bottom Billion [5] are no longer fully adequate to the aspirations of Bottom of the Pyramid (BoP) households. Beyond survival needs these households wish to develop sustainable livelihoods by engaging in markets [6]. In this context of the triple-f, post-financial crisis and post-Washington consensus [7] of hyper-collective action [8], new institutional economic arrangements, including hybrids such as social entrepreneurship [9–11], are offering hope for development and innovation in the 'rising' South, including India.

2 Socially Responsible Design (SRD): Design's Response?

Despite much rhetoric to the contrary, sustainability education still has an indifferent record in Australia [12]. Industrial design and related fields have preferred the tangible demands of sustainability, e.g. materials choice [13] rather than (sustainable) development as the theme of choice. One discussion of these broader ethical, political, social and economic issues and the limits of current design approaches has been that of Tony Fry [4]; but translating this 'philosophy' into practice has had little impact it seems.

Although terminology varies, design has developed a broad set of human-centred design perspectives and tools for urban and rural contexts which, following Cooper [14], I will call socially responsible design (SRD). These developments are a positive response to the call by Margolin and Margolin [15] for a model of social design. Social innovation driven by human-centred SRD approaches has become a flagship for agencies like IDEO [16]. Co-design and participatory approaches have for some time been part of the new landscape of design practice, and are seen to be core to achieving social innovation [17]. Morelli [18] among others, sees the new social turn in design as viewing 'clients as co-producers'; an oblique reference perhaps to private-public engagements. Cooper [10] suggests, however, that despite all the activity, the ability of design to account for say economic and social tradeoffs and context is still limited. In general, we believe the broader ideological, economic and social concerns of sustainable development are poorly understood in design schools.

More recently more mainstream development concepts such as Sen's capabilities [19] have been proposed as a better guide to technology engagement [20]. However, such a philosophical position, while interesting, does not capture how technology plays different roles in the promotion of sustainable livelihoods for the predominantly rural poor in agricultural and transforming economies [21]. The rhetoric and reality of industrial design (education) remains conventional. In this paper we offer another framework—the ladder of sustainable development—to promote a better understanding of the continuum of positions along which strong and weak sustainability are positioned, and in which technology innovation has different roles. There is room for design to challenge convention and at least at the post-graduate level encourage designers to see innovation more broadly.

3 Weak and Strong Sustainability

Since the Bruntland Commission [22] formulation of Sustainable Development (SD) as a balanced focus on human and conservation needs with a focus on the present and future, a mainstream discourse of sustainable development (MSD) has been established. Adams [23] describes this as a mix of ecological modernization, market environmentalism and populist discourses of man and nature, including 'romantic' notions of nature and humanities relationship. MSD presumes that economic growth can be maintained through unrestrained technology innovation or varying degrees of voluntary and enforced regulation. Separating this 'weak sustainability' model from strong sustainable design is a belief about the substitutability of nature by human capital, i.e. that human needs and economic growth demand we sacrifice nature to a greater or lesser extent to those needs, often through technology innovation. In contradistinction, strong sustainability argues that a critical natural capital exists that cannot be sacrificed or at one extreme—deep ecology—that nature has an intrinsic value and no substitution at all is possible.

It is this dividing line between strong and weak sustainable development that Baker [24] has developed into a 'ladder of sustainable development' with implications for technology, civil-state relations, economic, policy and governance, and normative principles for a continuum stretching from end-of-pipe technocratic to deep ecology approaches. Thus, strong sustainability favours changes in patterns and levels of consumption, has a focus on non-material aspects of growth, aims to maintain critical natural capital and biodiversity, heighten local self-sufficiency in the context of globalized markets, and favours democratic participation and bottom up community structures; a good example is Tim Jackson's [25] thesis about Prosperity without Growth. On this basis, LCA, recycling and other mainstream tools and processes in design education when coupled with understanding of the regulatory environment are at best weak sustainable development proposals. They neither understand nor question existing economic models and propositions about growth, and view technology as a tool for regulated human development. Approaches in design also rarely rise above the isolated project level—ignoring broader TBL goals and policies, and therefore achieving at best short-term piecemeal change [26].

Market environmentalism takes the market as the essential mechanism for regulating human-nature interactions, aims to reduce the role of the state, promote the deregulation of markets, and extend market relations into the domain of society–environment interactions. It is an approach that believes in unrestrained economic growth, rejecting alternatives, such as zero growth or other attempts at a more equitable distribution of resources and opportunities. Businesses are encouraged to take voluntary measures to protect the environment and also to enter into partnerships with environmental groups, as highlighted in the Rio Agenda 21. Ecological modernization, meanwhile, is reformist and regulatory while supporting capitalism and conventional growth although seeing technological improvements as essential in the environmental performance of companies and individuals. Proponents of ecological modernization acknowledge that environmental degradation may accompany the operation of free markets, but they argue that the solution lies in the self-corrective potential of capitalist modernization. It is an approach which is popular with policy makers, as it integrates a role for the state, appeals to the green vote, and is a strong voice in the Bruntland document. In the following Table 1 have selected from Bakers table some of the relevant categories to this discussion.

4 Weak Sustainability in Design Education

Design education seems firmly embedded in weak and pollution control sustainability commitments to regulation, end of pipe solutions and other aspects of market environmentalism and ecological modernization. This is mixed with populist discourses about the value of nature albeit contradicted by oxymorons such as sustainable

Table 1 The ladder of sustainable development

SD model	Principles	Governance	Technology	Policy	State-Society
Ideal model	Needs (not wants), biophysical limits, no substitution of nature; bioregionalism; self-sufficiency	Decentralisation of political, legal, social and economic institutions	Labour-intensive, appropriate, green technology; valuing work	Principles priority to environment; socialization of SD norms	Bottom-up community structures; equitable participation
Strong SD	Principles in international law, consumption changes; non-material growth; 3rd world focus on equity	Partnership and shared responsibility at multi-levels of governance	EcoMod of production; mixed labour and capital intensive technology	Sector level integration of environmental principles; green planning and design; SD indicators; green accounting	Democratic participation; open dialogue; alternative futures envisaged
Weak SD	More rhetorical than real principles; decoupling, reuse, recycling, LCA and management; natural capital substitution	Some institutional reform and innovation; global regulation	End-of-pipe technical solutions; mixed labour and capital intensive technology	Pollution addressed at source; some policy coordination across sectors	Top-down initiatives; limited dialogue; elite participation
Pollution Control		Command-and-control state led regulation of pollution	Capital intensive technology automation	End-of-pipe approach to pollution management	State and economic interests dialogue

Source Baker [24, pp. 208–209]

packaging or consumption, see [27]. Existing claims about socially responsible or social design show this mix of mainstream populism, market environmentalism and ecological modernization (MSD), i.e. provide no detail nor show awareness of alternative economic scenarios or anthropocentric emphases e.g. [15, 18]. Populism rather than principled strong SD arises because the technology-driven commercial mindset of design is intellectually unable to cope with the economic, ideological and scientific arguments about strong sustainability and the ideal model is an explicit rejection of design's raison d'être. We argue here not to replace weak mainstream sustainability with stronger versions but to make clear what the full array of positions offers. Design may wish to contribute to sustainable development by adopting mainstream approaches but should not limit itself to these.

Technology, service and system innovation from a strong sustainability perspective will question existing economic and growth models, advocate policy and institutional change to create enabling conditions for sustainable livelihoods, adopt a global perspective on macroeconomic interdependencies, e.g. trade, of developed and developing countries. The fundamental weakness of design education has been the emphasis on sustainability and not sustainable development—the latter is a concept inviting true engagement with TBL. One of the models having some success in India is social entrepreneurship, stepping in where market failures for the poor have failed [28].

5 Implications for Design Curriculum

What would taking sustainable development, reverse innovation, BoP and other factors imply for programs of socially responsible design at the postgraduate level, particularly for engineers and designers? Stanford Graduate School has offered since 2008 a two quarter class entitled Entrepreneurial Design for Extreme Affordability, which addresses some of the economic issues mentioned here. IIT Madras, meanwhile, like other schools in the region, has developed a minor program in social entrepreneurship and innovation, which encourages technology-oriented students to consider all three 'legs' of TBL in proposing technology and service innovations. The implications from both programs are clear, multidisciplinary input into sustainable innovations, which may then be further leveraged in research-industry incubation cells, requires fieldwork and engagement and consideration of the economic, ecological, and social constraints facing SD.

A recurring feature of the literature and recent conversations with Faculty at IIT Madras involved in innovation for social impact was the fact that too many innovations of this sort were not sustainable in the relevant rural environments as they lacked understanding of the socio-cultural and economic constraints of household decision making [29]. One example discussed was the failure of affordable composting toilets whose uptake was poor due to cultural preference for open defecation and rejection of the idea of toilets being installed inside houses. Other more public examples include the failure of the one laptop per child (OLPC) campaign [30].

Students in the North need recognize that the South is innovating without expert assistance from the North and under constraints that produce 'reverse innovations' now selling back to OECD and industrial nations [31, 32]. Education programs integrating the relevant social, economic and environmental concerns need to be developed or modified along the lines of models already in existence in economies in transition (EIT), such as India; it is not apparent that the developed 'West' has figured this change out and begun to remodel its relationships with economies of transition on cooperation; there is widespread recognition that the future is being written in the South not the North [33].

6 Conclusion

Waage [34] also suggests the time is ripe for product design to develop a coherent plan to address the plethora of issues associated with sustainability. In a similar vein Spangenberg et al. [35] suggest that Design for Sustainability (DfS) is far more than eco-design and conventional approaches, noting also how little inroad the movement has made in design education and practice. DFs principles in fact sit at the borderline of weak and strong sustainability, largely repeating ecological modernisms mantra with some broader aspirational statements. Most recently a philosophy of capability-centred design [20] has been added to the landscape of design approaches [36] although there is no evidence of its application. This will not be a simple task as scholars at TU Delft, reporting on an attempt to integrate sustainable development across the engineering curriculum can testify [37]. Ehrenfield has argued for a deeper engagement with sustainability including but going beyond LC, which questions our assumptions about weak sustainability, consumption, etc. He notes 'industrial ecology suggests that societies built around principles derived from ecosystem properties and dynamics might be sustainable in the same sense that ecosystems are' [38].

This paper calls for frameworks that question taken for granted assumptions and the mainstream sustainability discourse that drives current approaches in design. This will mean teaching about ecological economics, systems thinking, and alternative concepts of wealth and value. Thus, building on the example of IITM, Stanford, and TU Delft, we suggest such a multidisciplinary curriculum will involve exposing students to: sustainable development principles, social entrepreneurship (and business modeling), frugal innovation, practical fieldwork strategies, and integrate start-up and incubation hubs with support from university and industry. Can design (education) transcend its modernist industrial roots? Innovation that is restricted to promoting the modernist modified neo-liberal agenda of growth through increased consumption and production is an agenda the rich North should re-examine as it is overtaken by the rising South. T-shaped graduates will study sustainable development, development economics, social entrepreneurship, and related topics to prepare them for the future (or present).

References

1. Melles, G., De Vere, I., Misic, V.: Socially responsible design: thinking beyond the triple bottom line to socially responsive and sustainable product design. CoDesign **7**(3–4), 37–41 (2011)
2. Walker, S.: Sustainable by Design: Explorations in Theory and Design. Earthscan Publications Ltd., London (2006)
3. Elliott, J.: What is Sustainable Development, 4th edn, pp. 8–56. Routledge; Taylor & Francis, London (2013)
4. Fry, T.: Design Futuring: Sustainability, Ethics and New Practice, p. 278. Bloomsbury Academic, London (2009)
5. Sumner, A.: Global poverty and the new bottom billion: what if three-quarters of the world's poor live in middle-income countries? IDS Work. Pap. **2010**(349), 01–43 (2010)
6. Melamed, C., Scott, L.: After 2015: Progress and Challenges for Development. Overseas Development Institute, London (2011)
7. Birdsall, N., Fukuyama, F.: The post-washington consensus-development after the crisis. Foreign Aff. **90**(2), 45–53 (2011)

8. Severino, J., Ray, O.: The end of ODA (II): the birth of hypercollective action. Working paper 218, Washington DC (2010)
9. Khanapuri, V.B., Khandelwal, M.R.: Scope for fair trade and social entrepreneurship in India. Bus. Strateg. Ser. **12**(4), 209–215 (2011)
10. Martin, B.R.L., Osberg, S.: Social entrepreneurship: the case for definition. Stanford Soc. Innov. Rev. **5**, 28–39 (2007)
11. Sonne, L.: Innovative initiatives supporting inclusive innovation in India: social business incubation and micro venture capital. Technol. Forecast. Soc. Change **79**(4), 638–647 (2012)
12. Ramirez, M.: Sustainability in the education of industrial designers: the case for Australia. Int. J. Sustain. High. Educ. **7**(2), 189–202 (2006)
13. Diegel, O., Singamneni, S., Reay, S., Withell, A.: Tools for sustainable product design: additive manufacturing. Sustain. Dev. **3**(3), 68–75 (2010)
14. Cooper, R.: Ethics and altruism: what constitutes socially responsible design? Des. Manag. Rev. **16**(3), 10–18 (2010)
15. Margolin, V., Margolin, S.: A 'social model' of design: issues of practice and research. Des. Issues **18**(4), 24–30 (2002)
16. Brown, B.T., Wyatt, J.: Design thinking for social innovation. Stanford Soc. Innov. Rev. Winter, 30–35 (2010)
17. Sanders, E.B.N., Stappers, P.J.: Co-creation and the new landscapes of design. CoDesign **4**(1), 5–18 (2008)
18. Morelli, N.: Social innovation and new industrial contexts: can designers 'industrialize' socially responsible solutions? Des. Issues **23**(4), 3–21 (2007)
19. Anand, S., Sen, A.: Human development and economic sustainability. World Dev. **28**(12), 2029–2049 (2000)
20. Oosterlaken, I.: Design for development: a capability approach. Des. Issues **25**(4), 91–102 (2009)
21. Ashley, C., Carney, D., Britain, G.: Sustainable Livelihoods: Lessons from Early Experience. DFID, London (1999)
22. WCED: Our Common Future. Oxford University Press, Oxford (1987)
23. Adams, W.M.: The dilemma of sustainability. In: Green Development: Environment and Sustainability in a Developing World, 3rd (edn), (Paperback), pp. 1–25. Routledge, London (2009)
24. Baker, S.: The concept of sustainable development. In: Sustainable Development, pp. 17–48. Routledge Taylor & Francis Group, London (2006)
25. Jackson, T.: Prosperity Without Growth. Earthscan, London (2009)
26. Dobson, A.: Environment sustainabilities: an analysis and a typology. Env. Polit. **5**(3), 401–428 (1996)
27. Lele, S.: Sustainable development: a critical review. World Dev. **19**(6), 607–621 (1991)
28. Asia Development Bank: India social enterprise: landscape report (2012)
29. Melles, G., Thomas, J., Kuys, B., Ranscombe, C.: Social entrepreneurship with design in Southern India: lessons for Australia. In: Design for Sustainable Well-being and Empowerment (2014)
30. Kraemer, K.L., Dedrick, J., Sharma, P.: One laptop per child: vision versus reality. Commun. ACM **52**(6), 66–73 (2009)
31. Zeschky, M., Widenmayer, B., Gassmann, O.: Frugal Innovation in Emerging Markets. Res. Manag. **54**(4), 38–45 (2011)
32. Australia-India Taskforce: Science technology innovation. VIC, Australia and India, Melbourne (2013)
33. UNDP: Human development report 2013. The rise of the south, New York (2013)
34. Waage, S.: Re-considering product design: a practical 'road-map' for integration of sustainability issues. J. Clean. Prod. **15**(7), 638–649 (2007)

35. Spangenberg, J.H., Fuad-Luke, A., Blincoe, K.: Design for sustainability (DfS): the interface of sustainable production and consumption. J. Clean. Prod. **18**(15), 1485–1493 (2010)
36. Sanders, L.: An evolving map of design practice and design research. Interactions **15**(6), 13–17 (2008)
37. Peet, D.-J., Mulder, K.F., Bijma, A.: Integrating SD into engineering courses at the Delft University of Technology: the individual interaction method. Int. J. Sustain. High. Educ. **5**(3), 278–288 (2004)
38. Ehrenfeld, J.R.: Searching for sustainability: no quick fix. Reflections Sol J. **5**(8), 1–13 (2004)

Understanding Consumers' Perceptions of Sustainable Products in India

Prabir Sarkar, Srinivas Kota and Bijendra Kumar

Abstract Understanding consumer perception of sustainable products is critical for product designers and sustainability enablers; for instance, policy developers can develop successful policies only through detailed understanding of consumers' attitude. Additionally, future sales of products can be predicted by understanding consumer's willingness of purchase and sustainable related policies. In this study, through an all encompassing, non-anonymous online survey, we tried to estimate the perceptual experience of sustainable products in the mind of consumers and predict its implications on their future purchasing patterns. The results show that most of the consumers are aware and interested in sustainable products, but majority of them are not practicing sustainable living. Majority of consumers are willing to spend extra for sustainable products, but we need to see whether they do it in practice or not. This work is a start in the direction of understanding consumer's perception towards sustainable products and develop support to help manufacturers; however, the results indicate that a detailed, and broader study is needed to gain insight into the diversified population perception and behavior towards sustainable products and develop design support.

Keywords Sustainable product · Consumer perception · Design for sustainability · Sustainable development

1 Introduction

Published in 1987, 'Our Common Future', also known as the Brundtland Report [1], from the United Nations World Commission on Environment and Development (WCED), describes sustainable development as, 'the development that meets the

P. Sarkar (✉) · B. Kumar
School of Mechanical, Materials and Energy Engineering, IIT Ropar, Ropar, Punjab, India
e-mail: prabir@iitrpr.ac.in

S. Kota
Department of Mechanical Engineering, BITS Pilani, Rajasthan, India

A. Chakrabarti (ed.), *ICoRD'15 – Research into Design Across Boundaries Volume 2*, Smart Innovation, Systems and Technologies 35, DOI 10.1007/978-81-322-2229-3_9

need of present without compromising the ability of future generation to meet their own.' India holds the second biggest population with one of the heaviest population density on the earth with diversified economic, social and cultural background. Thus, it is important to understand the perception of product users (consumers/buyers/customers) in India, so that appropriate strategies, methods, and regulations could be developed in India. Only few legislations are dealing with sustainability in product design and manufacturing are implemented in India and most often the emphasis is on the protection of the environment, in terms of pollution control, emission control, and public safety [2].

> Legislation alone does not guarantee sustainable development. The situation is made more difficult in India as required infrastructure for enforcement and implementation are lacking. Rules can work as a driver towards conservation: for instance, those that are inbuilt in the permission system force industries to take environmental issues into consideration. Also, India is a signatory to many of the international treaties bearing directly on environmental protection and sustainable development and takes an active role in prosecuting polluters [2].

Sustainable product development is critical for developing countries such as India. India is an ideal place for marketing variety of goods and services due to ever-increasing population, growing number of middle class with high purchasing power, high GDP growth, and a large variation in society. However, depleting resources, lack of proper policies, end-of-life infrastructure, and inflation causing an increase in commodity prices, creating landfills and pollution; hence, the goods and services offered often become unsustainable. To avoid this situation, manufacturers are increasingly putting a lot of emphasis on developing products that are both economic and environmental friendly. Developing such products often requires substantial resource allocation for the research and development of new technologies, new policies, corresponding infrastructure and controlling mechanisms. This calls for an in-depth understanding of various stakeholders' perspectives on sustainability and sustainable products to enable them in allocating appropriate resources on various sustainability enabling technologies.

The success of a product in the market depends on various factors including the availability of the product, willingness of buyers to buy the product, and prevailing policies that affect product, thus its purchase. Makower and Pike [3] wrote in the 'Strategies for the Green Economy,' that consumers really care about 'Green,' however, there is a chasm between 'green concern' and 'green consumerism.' To close that gap, companies need to communicate more effectively, and make sure their messages and marketing are pitch-perfect [3].

This work will help in better understanding of the views of various stakeholders about sustainability and sustainable products in India. This will pave way for devising various strategies with the current and future policies to develop more sustainable products/services. Moreover, companies can gain competitive advantage over their competitors by targeting their efforts in a focused manner by knowing the regulations and the wishes of the user in advance, which we have tried to capture in this work through an online survey.

2 Literature Review

Consumers have a major role to play in addressing the climate change and energy challenges as large number of current and future products depend on the use of some form of energy; and, the user behaviour is going to determine the amount of energy consumption and associated environmental, societal, and economic impacts. Mah et al. [4] conducted a survey in Hong Kong to know consumer perception about smart grid technologies, where consumer welcomed smart grid technologies and had a preference for energy saving, energy efficiency, and renewable energy. They mentioned that more attention was needed to improve the relation between consumer, the government, and utilities [4].

In any company, some of the stakeholders may wish to see things done differently—a desire to change practices to be more "sustainable". Short et al. [5], conducted a questionnaire based investigation to understand the eco/sustainable design in manufacturing companies of Sweden and United Kingdom, and discussed management of risk and discussed on the risks associated with taking 'Design for Sustainability' initiative as a design method or as a company strategy.

Consumer perception about the carbon footprint of food product were surveyed by Hartikainen et al. [6]. Their study shows that the term 'product carbon footprint' was familiar to many, however, due to misunderstanding of its meaning, only 7 % linked it to greenhouse gas emission associated with product and only 5 % linked it to climate change. Therefore, it was clear that consumer needs more education to better understand carbon footprint. If this was the situation in developed country like Finland then what could it be in a developing country like India.

A review was written by Schleenbecker and Hamm [7] on consumer perception about organic product characteristics. Market needs to be informed about interested consumers in organic products to serve them better according to their actual need. van Doorn and Verhoef [8] explore the reasons behind consumer's willingness to pay for organic food and investigate whether it differs between virtue and vice food. They found that because of growing environmental problems, food safety issues, and increasing obesity rates, many consumers' desire healthier, less processed natural foods that are less harmful to humans and environment.

Tseng and Hung [9] developed a green product instrument based on the basic attributes of product quality, attributes for measuring environmental performance and eco-certification of green products. They found that many green products in the marketplace do not fulfil consumers' expectations and creates a gap between expectations and perceptions. Meise et al. [10] provided suggestions to retailer on what kind of sustainability related information should be provided to consumers.

Co-operative product development processes can enhance the emergence and diffusion of sustainable product innovations by combining different forms and bodies of knowledge. Hoffmann [11] mentioned that changes in production and consumption patterns are crucial elements of sustainable development. User integration in product development as well as communication between product manufacturers and users could lead to sustainable product development and consumption. Ulaga and Chacour

[12] identified customer is becoming very important to supplier. They constructed a multi-item model of customer-perceived value and developed a marketing strategy.

Petiot and Yannou [13] measured the consumer perception for a better comprehension, specification and assessment of product semantics. They provided a tool for designers to understand and specify the semantic part of the need; it rates and ranks the new product prototypes according to their closeness to the specified "ideal product". Baker and Ozaki [14] investigated the marketing influence on the consumer purchase decision of greener product. They identified that consumers were not exposed enough to green product and they suggested the greater use of marketing and brands to promote and sell products that are environmentally friendly and function effectively. This shows the need for identifying the consumer's current state and then devise strategies to make them aware of the salient features of the sustainable products.

Pujari [15] investigated the influences of market performance on companies through a survey of environmental new product development (ENPD) projects in North America.

Hou et al. [16] identified 27 sustainable practices in remediation of limited understanding of actual sustainable behaviour being adopted and the determinants of such sustainable behaviour. A survey was used to rank and compare them in the US (n = 112) and the UK (n = 54). It was found that US and UK practitioners adopted many similar sustainable practices. Comparing the two countries, they found that the US adopted innovative in situ remediation more effectively; while the UK adopted reuse, recycling, and minimizing material usage more effectively. Customer competitive pressure was found to be the most extensively significant external force.

From the above literature review we can infer that different researchers conducted surveys to know the user perception about various products and services, but in the context of sustainable product development only few i.e. Hoffmann, Petiot and Yannou, and Baker and Ozaki were involved. There were differences in user perception in different countries and that lead us in conducting a survey in India. We found that a gap that still exists in understanding user perception about sustainable products and how they would use it during its life and thereafter. We tried to fill that gap through understanding the consumers' perception of sustainable products.

3 Aim and Methodology

The analysis of the responses aims to provide answers to the following questions. First, what are the consumers' perceptions of sustainable products/services, environment, pollution, sustainability, eco-products/eco-services, energy efficiency, eco-labels, and society? Second, what are the current usage patterns for different products/services and would the users inclined to pay extra for sustainable products? We hope to answer these questions with appropriate justification based on the responses and the size of the responded population.

The research methodology to understand the perception of sustainability in the minds of the users (who uses the product), manufacturers (who creates the product), media persons (who advertises the product) and policy developers (who develops the policies for markets/products) consists of the followings:

1. An extensive review of literature to find existing works on the understanding of users' perceptions of sustainability (Sect. 2)
2. Design an online survey that was used to collect responses from customers of varied strata of Indian society (Sect. 3)
3. Analysis and discussion of the survey responses (Sect. 4).

Even though we have sent out this survey to more than 700 people belonging to different strata of Indian population (educated), we received about 86 responses (n = 86) which include, users(68), policy makers(0), environmentalists(2), industrialists(8), social workers(6), and media persons(2). The respondents of the online survey include people with different educational backgrounds: 32 % graduates, 45 % Masters, 15 % doctorate and 8 % others. 62 % of them are of engineering background, whereas 15 % are from science, 8 % of management, 5 % arts, 2 % commerce and the rest 7 % with other background. Most of them are of the age group of 18–30 years (80 %) and in the remaining, 12 % are of 31–40 years, 5 % are of 41–50 years, 2 % are of 51–60 years, and 1 % are of 61+ of age group. 88 % respondents are male and 12 % respondents are female, and among them 59 % are from cities, 22 % of towns, 16 % of Metro cities and the rest are from villages (3 %) as shown in Table 1.

Table 1 Respondents' details

Gender			Living place			Education		
Male	76	88 %	Metro City	14	16 %	Post Graduate	38	45 %
Female	10	12 %	City	50	59 %	Graduate	27	32 %
			Town	19	22 %	Doctorate	13	15 %
			Village	2	2 %	Other	7	8 %
Age group			Area of expertise			Organization		
18–30	68	80 %	Science	13	15 %	Central government	29	35 %
31–40	10	12 %	Arts	4	5 %	State government	3	4 %
41–50	4	5 %	Engineering	53	62 %	Public sector unit	13	15 %
51–60	2	2 %	Management	7	8 %	Private—Multinational company	15	18 %
61+	1	1 %	Commerce	2	2 %	Private—Micro, Small and Medium Industry	9	11 %
			Others	6	7 %	Self employed	15	18 %

4 Details of Survey Questions, Results and Interpretation

The entire survey can be accessed online [17] and here we summarise the findings. Additionally, we discuss in detail the inferences that can be derived from the responses of this survey.

4.1 Consumers' Awareness of Sustainability

The first question is about the awareness of sustainability—respondents were asked whether they have heard the term sustainability. As shown in Fig. 1, 72 respondents (86 %) agree that they are aware, while only six respondents (7 %) replied it as 'no,' and another six respondents (7 %) were not sure about it. Next, regarding the question on consumers' interest on becoming more sustainable, as shown in Fig. 2, most of the consumers (75 %) are interested to be more sustainable. However, this is not surprising for us, since as per Table 1, respondents were educated and thus are exposed to sustainability, and so we would expect them to have a strong concern for sustainability.

Next, consumer perception of achievability of sustainable development is captured in Fig. 3, which indicates that most consumers firmly believe that sustainable development is something that is achievable. More than 75 % of them agreed that sustainable development is definitely achievable. Additionally, respondents do

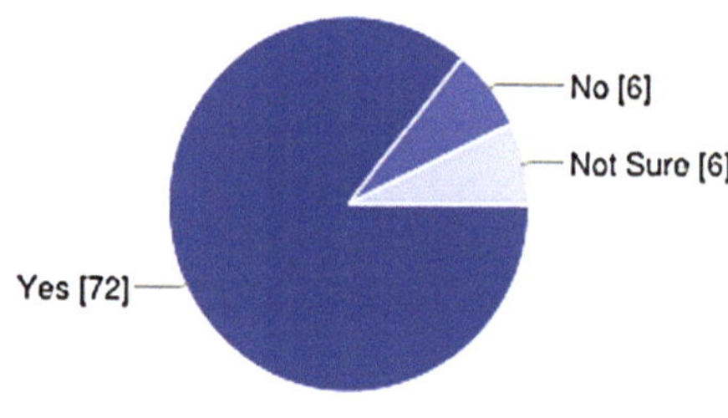

Fig. 1 Consumers' awareness of sustainability

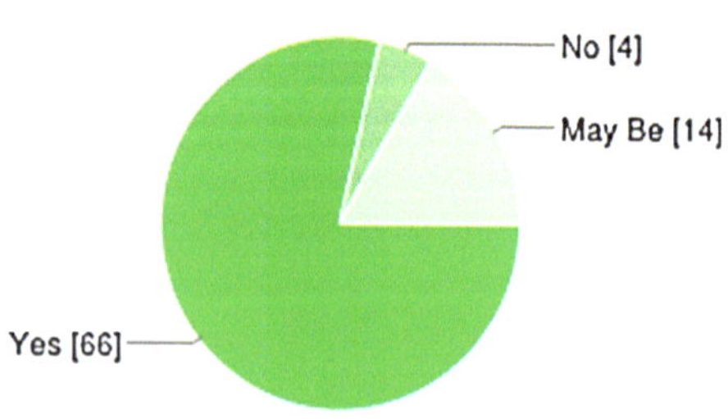

Fig. 2 Consumers' interest to be sustainable

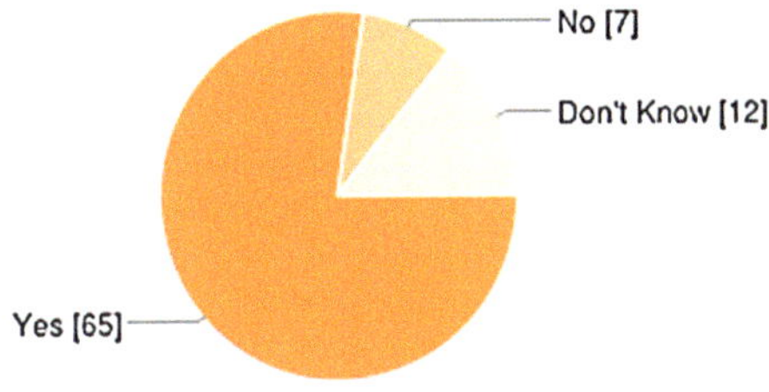

Fig. 3 Consumers' perception about sustainable development achievement

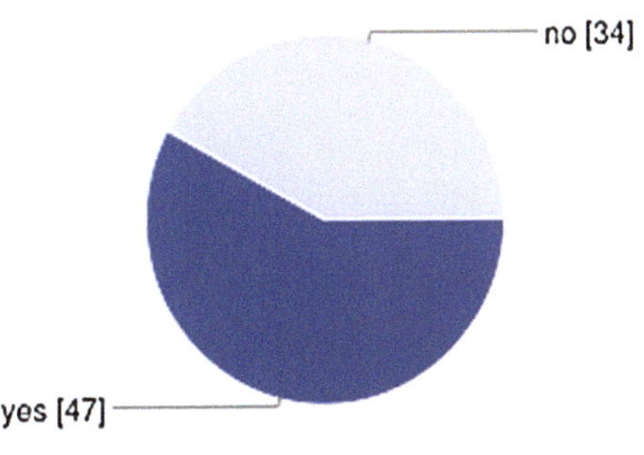

Fig. 4 Consumers' perception about separate shop for sustainable product

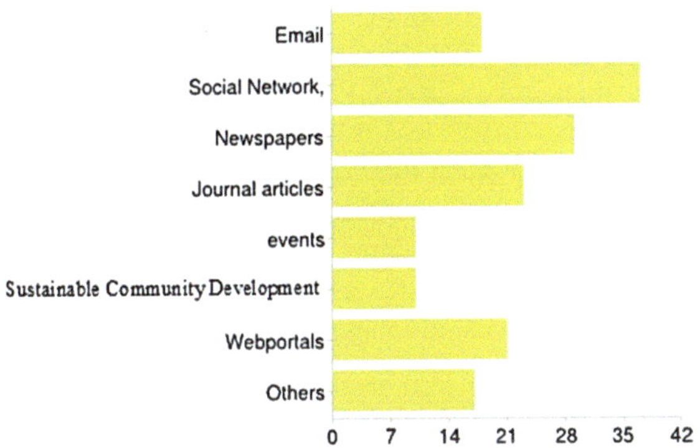

Fig. 5 Different sources of information on sustainability

indicate they wish for dedicated shops for purchasing sustainable products, as shown in Fig. 4.

Figure 4 indicates that 58 % respondent supports the idea of having dedicated shops for sustainable products.

Later, we asked consumers about the ways that they use to access information on sustainability. Most of them indicated that the social network is a major source of information for them, while other different means of receiving information on sustainability are also critical as shown in Fig. 5.

From Fig. 5, we can see that majority of the respondent were accessing the information about sustainability via social network (22 %), newspapers (18 %),

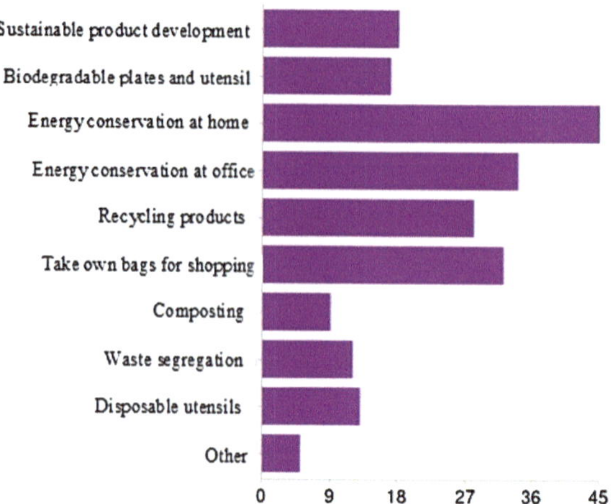

Fig. 6 Consumers' activities to achieve sustainable development

journal article (14 %), and Web portals (13 %). From this, we conclude that social network and newspaper play a major role in sustainable development.

When respondents were asked about their activity at home or in organization towards achieving sustainable development, the majority of them indicated that they take active initiatives for energy saving at home (21 %), and office (16 %), whereas, only 15 % people agreed to take own bags for shopping and 13 % people said they are involved in recycling of products. Since we received very less responses on the ideas of composting, disposable utensils, waste segregation and sustainable product development as shown in Fig. 6, it could indicate that people were not aware about these practices and thus we might need to encourage people to practice these in future. A more detailed survey needs to be conducted to see the problems faced by the consumers in adopting sustainable way of living. This shows that people say some things and practice different things.

4.2 Consumers' Perception About Sustainable Product Buying

When respondents were asked questions whether they are willing to purchase sustainable products with an extra payment; the responses were scattered. Majority of respondents 61 % indicated that they could pay extra for sustainable products. Whereas 27 % respondents indicated that sustainable initiatives should not increase the cost of products and 12 % respondents ask for discounts on purchase and use of sustainable products as shown in Fig. 7.

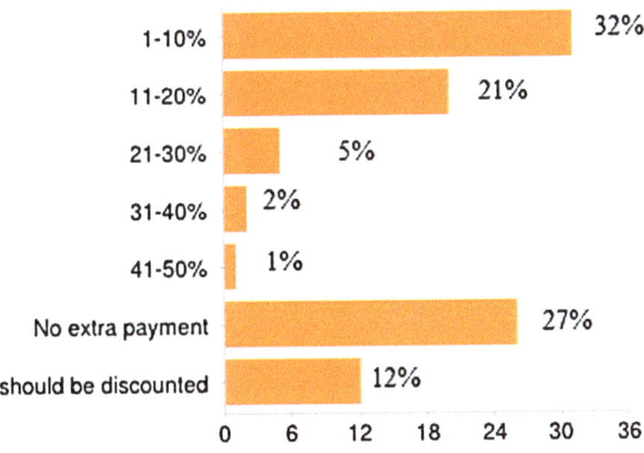

Fig. 7 Consumers' perception on extra payment for sustainable product

4.3 Consumer Perception About Three Pillars of Sustainability

Environment, Economy and Society are the three pillars of sustainability, when we asked if consumers' responsibilities helps to make a difference towards strengthening these pillars as shown in Table 2, a majority of respondents agreed about their importance, as shown in Fig. 8. When people were asked about their responsibilities to help make a difference on economic (unemployment, inflation, local economy etc.), Environmental (waste, resource consumption, water use etc.) and social (safety, education, health, etc.) issues, the majority of the respondent agreed to make a difference on Economy (72 %), Environmental (83 %) and the Society (82 %) as shown in Table 2. In addition, we note from Fig. 8 that for the consumer's perception on the importance of three pillars of sustainability, the majority of people (51 %) think that all the three pillars are important.

Table 2 Consumer's responsibility to make difference on three pillars of sustainability

	Economy		Environment		Society	
Strongly disagree	8	10 %	8	10 %	5	6 %
Disagree	12	14 %	2	3 %	9	11 %
Slightly disagree	3	4 %	3	4 %	1	1 %
Slightly agree	10	12 %	5	6 %	2	3 %
Agree	36	**43** %	37	**47** %	41	**51** %
Strongly agree	14	**17** %	24	**30** %	22	**28** %

4.4 Consumer Perception on Government Role in Sustainable Development

The majority of the people (38 %) believes that government is insensitive to sustainable development and are not taking obligatory initiatives towards achieving sustainable development; whereas, 32 % people believe that government need to be more sensitive and take more initiatives towards achieving sustainable development and the remaining 30 % respondents are not aware of such initiatives, as shown in Fig. 9. Figure 9 shows the consumers' perceptions of how government is sensitive to sustainable product development. Among the respondents, 72 % people said that the government should provide financial support to companies that are producing sustainable products as shown in Fig. 10. From Fig. 11 we can infer that 47 % people say India does not have sufficient sustainability related standards.

We believe that since the majority of the respondents are with an engineering background; their responses are more inclined towards making product more sustainable. 30 % of them agreed to use recyclable materials for making products more sustainable, whereas, 18 % said that designing energy efficient products is a good idea, and only 14 % were in favor of designing long lasting products, as shown in Fig. 12. These results show that only recycling, energy efficiency and long lasting products are perceived as sustainable products around 38 % of people did not perceive this also. We need to make people aware of sustainability as a holistic approach rather than piecemeal approach.

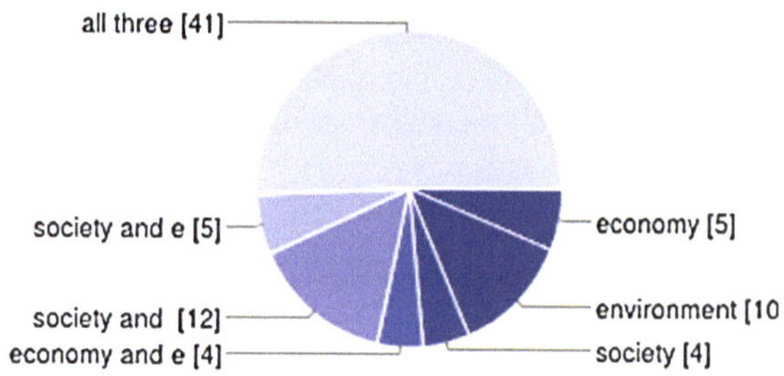

Fig. 8 Consumer perception of importance of three pillars of sustainability

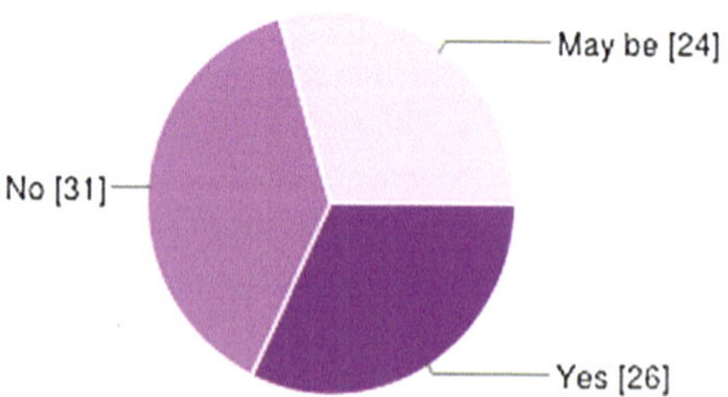

Fig. 9 Consumer perception on government sensitivity to sustainable development

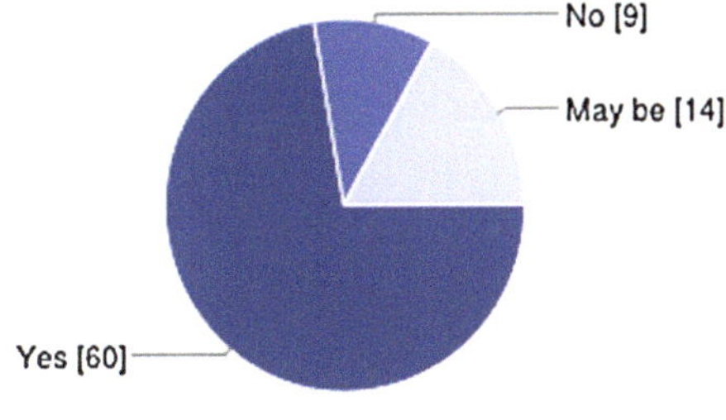

Fig. 10 Consumers' perception of financial support from government

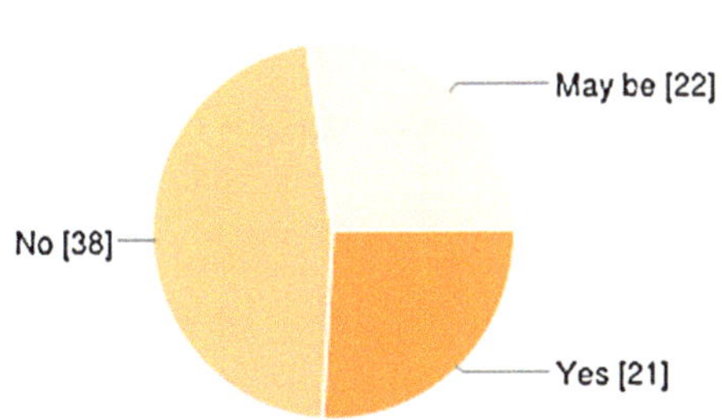

Fig. 11 Consumers' perception of having a sufficient sustainability standard in India

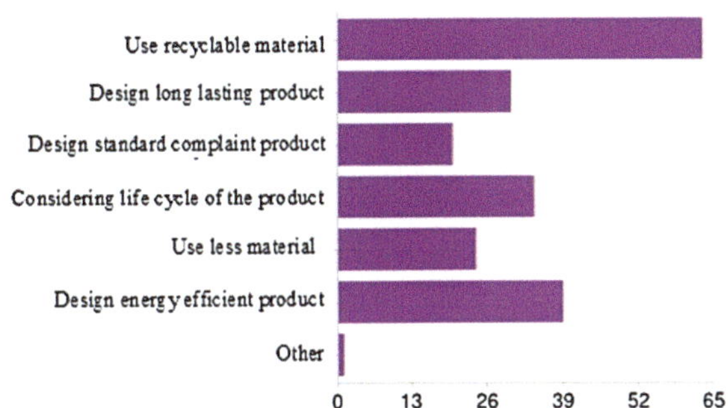

Fig. 12 Consumer perception on making product more sustainable

We propose that people from the media need to pay more attention to publicize the sustainability related issues, as we saw most of the people accessing the information about sustainability from either social networking or newspapers. Newspaper is a powerful medium to make most of the population aware of sustainability as 63 % respondents indicated that sustainability should be the focus of today's newspapers. As shown in Fig. 14, the respondents also indicate the willingness to take the initiative to make people around aware about sustainability. We infer from Fig. 13 that, 48 % respondents agree that current products available in India are not sustainable, 37 % are not sure about this and only 16 % said that the current products are sustainable. As people are not exactly know what sustainability means these results can be misleading.

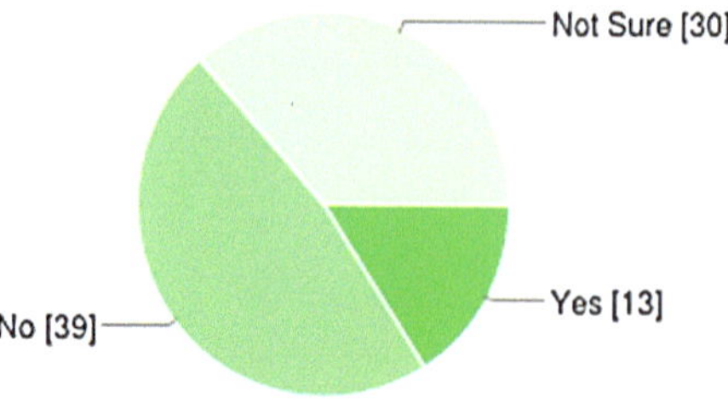

Fig. 13 Consumer perception of current product available in India

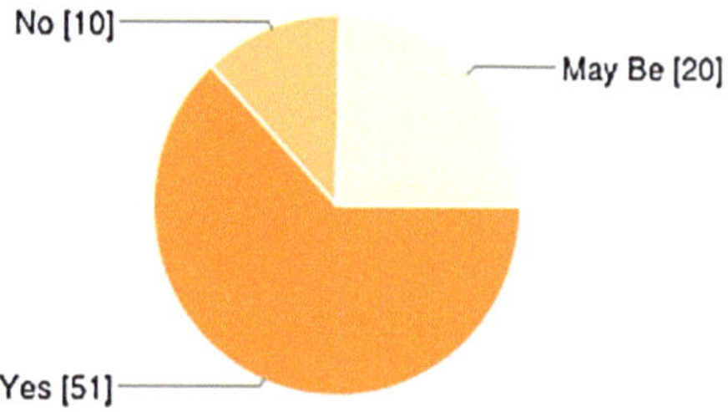

Fig. 14 Consumer perception about media focus on sustainability issues

5 Conclusion

This paper deals with the consumers and stakeholders perception of sustainability and sustainable products, and is based on the survey responses of 86 Indian individuals consisting of consumers, media persons, policy makers and industrialists from different places in India. This current survey is a preliminary one whose results will be used to design and conduct a more comprehensive survey to take diversified consumers opinion for devising strategies on sustainable products. The majority of the respondents indicated strong concern about sustainability and expressed interest to become more sustainable but the detailed analysis shows that there is a lacking in terms of proper knowledge and the daily practices. Consumers do think that India does not have sufficient sustainability related standards and government should take initiatives for sustainable product development by applying stronger policies, strict environmental laws, and educate people about the environment and sustainability. Respondents also thought that the government should provide financial support to industries that are producing sustainable products. People employed by social network and media need to pay more attention to advertising sustainability related products in those media, as we saw that most of the people accessing the information about sustainability from either social networking sites or newspapers. The majority of respondents indicated that sustainability should be the focus of today's newspapers. Many respondents agree to take personal initiatives to make other people aware about sustainability, however most of them are not practicing sustainable living. We believe that more policies should be developed related to recyclable materials and energy conservations because, in general, people were more aware about recyclable material and energy conservation. This study shows that being aware doesn't mean people are going to practice, there are lot of other factors which influence their purchasing behaviour and usage behaviour. A detailed study needs to be done to cover all segments of the society and to uncover the most significant factors which influence sustainable product purchase and use.

Acknowledgments Authors are thankful to all the respondents from different places of India. Additionally, we acknowledge the support provided by Google® for conducting the online survey.

References

1. Our common future, Chapter 2: Towards Sustainable Development—A/42/427 Annex, Chapter 2—UN Documents: Gathering a Body of Global Agreements [Online]. Available: http://www.un-documents.net/ocf-02.htm. Accessed 20 May 2014
2. Johnson, A., Gibson, A.: Drivers of Sustainability in Design. In: Sustainability in Engineering Design, Elsevier, pp. 345–371 (2014)
3. Makower, J., Pike, C.: Strategies for the Green Economy: Opportunities and Challenges in the New World of Business. McGraw-Hill, New York (2009)
4. Mah, D.N., van der Vleuten, J.M., Hills, P., Tao, J.: Consumer perceptions of smart grid development: results of a Hong Kong survey and policy implications. Energy Policy **49**, 204–216 (2012)
5. Short, T., Lee-Mortimer, A., Luttropp, C., Johansson, G.: Manufacturing, sustainability, ecodesign and risk: lessons learned from a study of Swedish and English companies. J. Clean. Prod. **37**, 342–352 (2012)
6. Hartikainen, H., Roininen, T., Katajajuuri, J.-M., Pulkkinen, H.: Finnish consumer perceptions of carbon footprints and carbon labelling of food products. J. Clean. Prod. **73**, 285–293 (2013)
7. Schleenbecker, R., Hamm, U.: Consumers' perception of organic product characteristics. A review. Appetite **71**, 420–429 (2013)
8. van Doorn, J., Verhoef, P.C.: Willingness to pay for organic products: differences between virtue and vice foods. Int. J. Res. Mark. **28**(3), 167–180 (2011)
9. Tseng, S.-C., Hung, S.-W.: A framework identifying the gaps between customers' expectations and their perceptions in green products. J. Clean. Prod. **59**, 174–184 (2013)
10. Meise, J.N., Rudolph, T., Kenning, P., Phillips, D.M.: Feed them facts: value perceptions and consumer use of sustainability-related product information. J. Retail. Consum. Serv. **21**(4), 510–519 (2014)
11. Hoffmann, E.: Consumer integration in sustainable product development. Bus. Strategy Environ. **16**(5), 322–338 (2007)
12. Ulaga, W., Chacour, S.: Measuring customer-perceived value in business markets: a prerequisite for marketing strategy development and implementation. Ind. Mark. Manag. **30** (6), 525–540 (2001)
13. Petiot, J.-F., Yannou, B.: Measuring consumer perceptions for a better comprehension, specification and assessment of product semantics. Int. J. Ind. Ergon. **33**(6), 507–525 (2004)
14. Pickett-Baker, J., Ozaki, R.: Pro-environmental products: marketing influence on consumer purchase decision. J. Consum. Mark. **25**(5), 281–293 (2008)
15. Pujari, D.: Eco-innovation and new product development: understanding the influences on market performance. Technovation **26**(1), 76–85 (2006)

16. Hou, D., Al-Tabbaa, A., Guthrie, P.: The adoption of sustainable remediation behaviour in the US and UK: a cross country comparison and determinant analysis. Sci. Total Environ. **490**, 905–913 (2014)
17. https://docs.google.com/forms/d/13Tude_cAWWg5aa4M5IGyFZ824dkjconDC39UiR0JnKY/viewform?sid&c=0&w=1&token&usp=mail_form_link

The Use of Sugarcane Bagasse-Based Green Materials for Sustainable Packaging Design

L. Pereira, R. Mafalda, J.M. Marconcini and G.L. Mantovani

Abstract This paper simulates the environmental impact of a design alternative choice based on materials selection. Applying an interdisciplinary research approach, we have used lab results of a nanocomposite material obtained from the mixture with sugarcane bagasse in order to simulate the carbon footprint and energy consumed in a Solidworks Sustainability design experiment. This practice can offer useful insights both for material and design engineers demonstrating quantitative and qualitatively the implications of eco design alternatives in early life cycle, including the life cycle of a new material.

Keywords Nanocomposites · Sugarcane nanofibers · Design alternatives · Product life cycle · Packaging

1 Introduction

The aim of this paper is to analyze the linkages between properties of green materials and its impact on sustainable design. Particularly, we are interested in learning more on the use of sugarcane bagasse cellulose fibers in the production of composite materials at the nanoscale and its applications in packaging design [1, 2].

In order to evaluate this new material, we have used an interdisciplinary approach applying the results obtained by researches on materials and design. We tested the materials specification inputs on software SolidWorks Sustainability to estimate green attributes such as carbon footprint and total energy consumed, knit

L. Pereira (✉) · R. Mafalda · G.L. Mantovani
Center for Engineering, Modeling and Applied Social Sciences,
Federal University of ABC, Santo André, Brazil
e-mail: luciana.pereira@ufabc.edu.br

J.M. Marconcini
National Nanotechnology Laboratory for Agriculture,
Embrapa Agricultural Instrumentation, São Carlos, Brazil

A. Chakrabarti (ed.), *ICoRD'15 – Research into Design Across Boundaries Volume 2*,
Smart Innovation, Systems and Technologies 35, DOI 10.1007/978-81-322-2229-3_10

together with life-cycle thinking. Translated into the practice of design, they help evaluate the relative greenness of different material as an instrument for design alternatives, ensure a high product performance with the lowest possible environmental impact.

The remainder of this paper is structured as follows: In Sect. 2, we briefly discuss related work about design for sustainability and eco-materials for packaging design. In Sect. 3, we outline the proposed methodology for achieve the research results. Section 4 presents the results, while the last section is dedicated to conclude remarks.

2 Design for Sustainability

In the interest of meeting market demands, new product development should not be focused only on technical and economic performances [3]. Consequently, strategies and development methods to promote products as sustainable as possible have been incorporated in product design. In this context, to have a broader view of the product's environmental impacts, an innovative approach which extends to the end of the product's useful life and retirement have been developed, in opposition to the conventional approach, where the product's sale is considered as the final analytical step [4]. Life cycle engineering (LCE) emerged in response to the need to develop life cycles causing the lowest possible environmental impacts, while still offering economic viability.

2.1 Eco-materials for Packaging Design

Well-designed packaging uses only as much of the right kind of material as necessary to deliver what is required. As we can see in Fig. 1, as packaging is reduced, the range of scenarios under which product losses occur rises, until eventually a point is reached where the increase in product loss exceeds the savings from the use of less packaging material. Any reduction in packaging beyond that point is a false benefit, since it increases the total amount of waste in the system.

Consequently, there has been an increasing demand for identifying biodegradable packaging materials and finding innovative methods to make degradable solutions. Biodegradation is the process by which carbon-containing chemical compounds are decomposed in the presence of enzymes secreted by living organisms. There is a list of biodegradable materials used in different packaging applications [6].

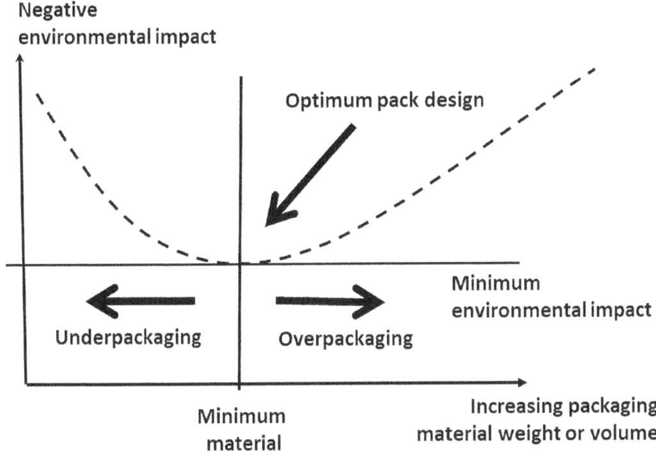

Fig. 1 Optimal packaging. *Source* The Consumer Goods Forum Sustainability Pillar [5]

2.2 Sugar Cane Bagasse Properties Advantages

In our review, we also were looking for themes or lessons that should be taken from the literature on the properties of sugarcane bagasse [7]. Some of these themes deserve special attention for its packaging applications, as summed in the next paragraph:

- Fibrous residue of sugarcane after crushing and extraction of its juice, known as 'bagasse', is one of the largest agriculture residues in the world Versatility of sugarcane residue usages; through its conversion inclusive but not limited to paper, feed stock and biofuel.
- Analysis of SCB indicates that its main constituents are cellulose, hemicellulose, lignin, ash, and wax. This composition of SCB makes it an ideal ingredient to be applied and utilized as reinforcement fiber in composite materials for the purposes of creating new materials which possess distinct physical and chemical properties. These in turn are desired for anticipated performances based on preset objectives.
- In comparison with other natural fibers, SCB is favored in the manufacturing of high quality green products given its low production cost. This is mainly attributed to the abundant availability of raw materials from the sugar processing plants and its low pre-treatment costs [8].
- Within the natural fibers group, and after appropriate modifications and manufacturing procedures are applied, SCB displays improved mechanical properties such as tensile strength, flexural strength, flexural modulus, hardness, and impact [9, 10].
- It can be mixed with gelatin, starch and agar to produce tableware packaging material.

3 Methodological Approach

In this work a proposal for an innovative analysis of sugarcane based materials for packaging design is presented. The sustainability analysis has been carried out using sustainability module of SolidWorks software package.

3.1 How a Cellulose Nanofiber Is Obtained?

Following the experiments performed by [1, 11, 12], the Polylactide Acid (PLA) used was supplied by NatureWorks with density 1,240 kg/m^3 (ASTM 1505); gel permeation chromatography (GPC), spectroscopic characterization (FT-IR), thermogravimetric analysis (TGA) and differential scanning calorimetry (DSC) were used to characterization.

3.1.1 Nanofiber Production Using Chemical Treatment

In order to obtain the cellulose nanofibers (CnF), the sugarcane bagasse was bleached with a hydrogen peroxide solution as a way of separating cellulose from lignin, the natural glue that holds cellulose fibers together and stiffens plant stems. The addition of a sulfuric acid 60 % (v/v) at 45 °C to an aqueous solution under for half an hour mechanical agitation allowed the production of CnF. Then the suspension was centrifuged and dialyzed with cold water until a pH (6) was reached.

3.1.2 Characterization of the Nanofibers

The atomic force micrograph (AFM), scanning electron microscopy (SEM) and transmission electron microscopy (TEM) were used to investigate the morphology and size of the dispersed structures.

As usual during development of new materials, structural analysis is performed in order to predict its behavior. These measures are then used in engineering design. In the case of this paper, the CnF specifications used to calculate the environmental impact on SolidWorks was the one with PLA 2.5 % nanofibers, as pictured in Table 1. The reason is based on the set of properties considered most satisfactory for packaging design.

3.1.3 Environmental Life Cycle Assessment Software

A life cycle assessment (LCA) is the comprehensive analysis of the direct and indirect environmental impacts which are attributed to a product, material, service or process, from raw material extraction to waste management. LCA has been used

Table 1 Material specification used for structural analysis and environmental impact calculation

Sample	Density	Elasticity	Rigidity	Tensile strength	Yield strength
PLA/2.5 % CnF	1,240 kg/m^3	10.3×10^8 N/m^2	5.6×10^6 N/m^2	21.84×10^6 N/m^2	17.94×10^6 N/m^2

Source Adapted from [4]

by design teams to explore the environmental footprint of different material, enabling better decision-making in product development, influencing eco-design and validating environmental benefit claims.

SolidWorks Sustainability is a diagnostic tool that will identify strengths and weaknesses in design process and predict the likelihood of sustainability thorough data collection in regards to resource requirements as well as air, water and ground emissions from manufacturing facilities [13, 14].

4 Environmental Impacts of Sugarcane Nanofibers Packaging Design

As part of the design process, construction of structures depends on the combination of material selection, structural analysis, and optimization. Therefore, computational simulation tools are important intermediary between design and realization [15]. Computer-Aided Design enables analysis of stress, strain patterns, and, more recently, the impact of design alternatives on the environment. Thus, in the early stages of the design is possible to get reliable estimates on the performance of products during their life cycles. As a result, in order to solve problems, designers must visualize the scenarios to identify the positive and negative effects on the system.

4.1 Modeling Plates with Sugarcane Nanofibers Material for Packaging

The chosen packaging concept is based on optimization of geometric characteristics. Rectangular prisms appear to maximize volume, or how much a box can hold inside it, while minimizing surface area, which determines the amount of material necessary to build it. Taking this into account, two plates were modeled to simulate multiple details of cellulose nanofibers material, which may be used to build several package types. The plates were 0.50 m long, 0.25 m wide and 0.03 m thick. The geometric structures represent hypothetical patterns that can be used to assist in an exploratory search for a suitable behavior for a particular resistance pattern you would expect from a product and at the same time to simulate the material reduction that occur in corrugated plates. As a result, plate 1 has a surface area equal to 0.221 m^2 and volume of 0.314×10^{-3} m^3 On the other hand, plate 2 has a surface area equal to 0.194 m^2 and volume of 0.269×10^{-3} m^3.

4.2 Sugarcane Nanofibers Plates' Performance

Analyzes from sugarcane nanofibers plates experiments have considered that the plates are glued to the four corners, forming a rectangular prism box. The reason is because the base and the lateral faces of a box is supposed to be the most requested. We also consider that the bottom of the box undergoes the action of uniformly distributed loads, describing a situation that best represents a real use. In addition, we applied loads estimated in the order of 390 N/m^2 to the faces of the plates, while the values for the physical properties of material were estimated from the density and elastic modulus obtained experimentally, as described in Table 1.

Once assigned the material properties of the plates, the aim of the experiment was to determine whether the tension at the same point exceeds a certain critical value called the yield stress. This state is characterized when the equivalent von Mises stress reaches the critical value. The von Mises equivalent stress is typically used to analyze the tension across the plate when it is subjected to a load condition. In this case, the critical values were obtained from the results of tensile testing. In Fig. 2, we show the distribution of stress in the plates 1a and 2b. In plate 1, the maximum stress reached 1.1×10^6 N/m^2 while in plate 2 the maximum stress reached 1.18×10^6 N/m^2. These values are 10 % lower than the maximum tension

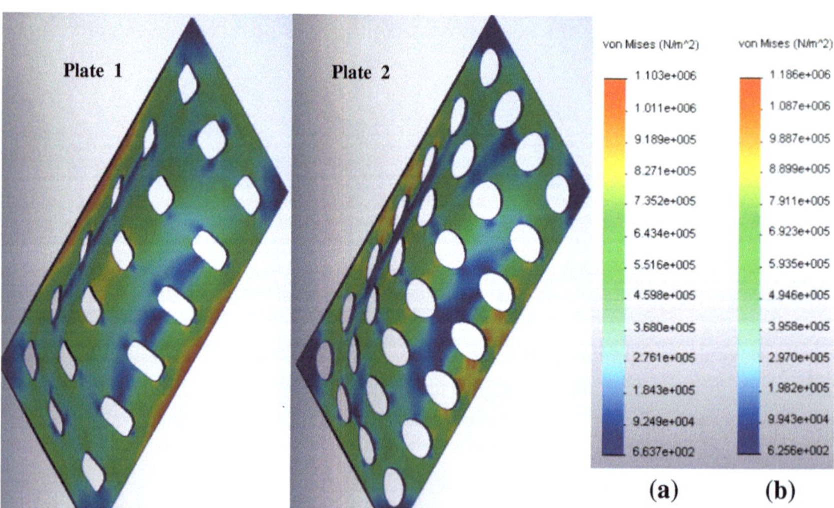

Fig. 2 Tension distributions in modeled plates. *Source* Authors

supported by the material. This suggests that the amount of materials used in the plates can be diminished in order to improve its mechanical properties. Moreover, the plates may resist to greater applied forces.

The differences in values shown in plate 1 and plate 2 are due to differences in geometric pattern, which may lead to different results in mass and surface area, implying a distinct load carrying capacity. The geometric pattern of the plate 2 results in greater strength compared with the plate 1.

4.3 Sugarcane Nanofibers Corrugated Box Design Sustainability Performance

In this paper, the hypothetical design of a corrugated box was considered in order to study sustainability outcomes. The analysis looks at a product's life cycle, which encompasses material production, manufacturing, product use, end-of-life disposal, and all of the transportation that occurs between these stages.

It was also considered that the area of production and consumption is South America. A region of production's choice determines the energy sources and types of technologies that will be used as well as the stages of production present in the product life cycle. Together with the consumption region, these values are used to estimate the environmental impact associated with transport the product from the production site until it reaches the final consumer. In this study, the distance was estimated in 1,600 km considering the whole life cycle and a half-life of about 1 month. We ran the simulations applying these fixed parameters and estimating values for manufacturing energy consumption, material recycling, incineration and waste destined for landfill.

Regarding energy consumption data to manufacture the material, they are estimated since there are not actual data for the production of PLA in the literature. This occurs due to the fact it is still an experimental material. However, given its production depends on mechanical mixing and drying processes, these values were estimated between 1.5 and 2.0 kWh to produce 0.456 kg of material. The environmental impact factors applied in this analysis are: carbon footprint, air acidification, water eutrophication, and total energy consumption.

4.3.1 Experiment 1

The first experiment considered a consumption of 1.5 kWh for the production of 0.456 kg of product. It was also found that 80 % of the material can be recycled and 0 % of it will be incinerated while 20 % will go to a landfill.

4.3.2 Experiment 2

The second experiment considered a consumption of 2.0 kWh for the production of 0.456 kg of product. It was also found that 70 % of waste can be recycled and that 22 % of them will be incinerated and that 8 % will go to a landfill.

4.3.3 Comparing the Scenarios for CnF Use

In general, the sustainable performance of CnF observed in the two experiments show that, as expected, levels of sustainability depend on variables such as energy consumption, rate of product recycling waste after use, the amount of product final waste which is incinerated and the amount of waste destined for a landfill.

In the first experiment, as observed in Fig. 3, items that are intensive in energy consumption are both the production of material and the boxes. Likewise these are

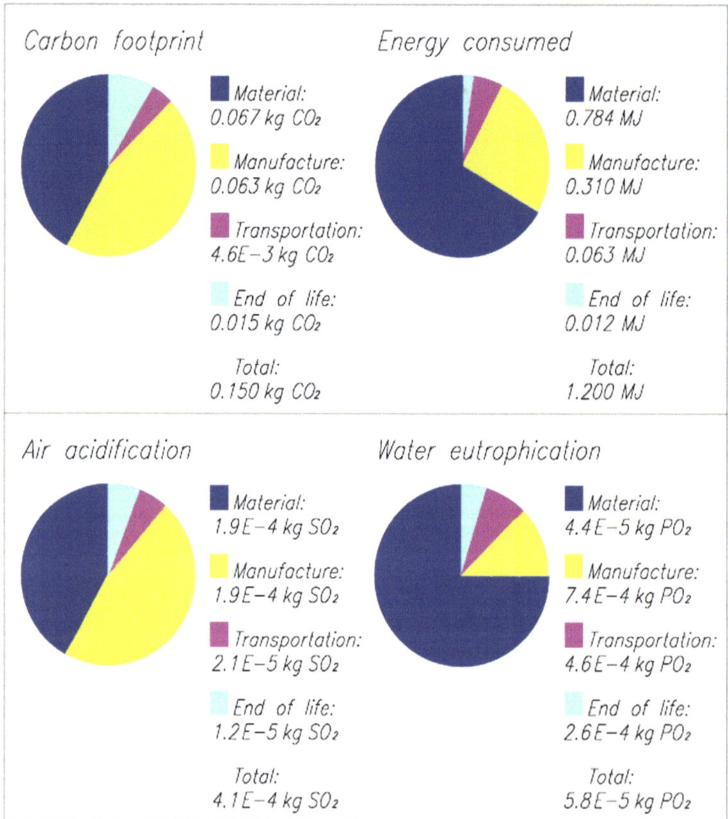

Fig. 3 Environmental impact of experiment 1. *Source* Authors

the items that emit more CO_2 in the atmosphere. For the same reason, these two items are also the most relevant to air acidification and water eutrophication.

In the second experiment (Fig. 4), the amount of energy used is 33.3 % larger than in the previous one. Therefore it has achieved the lower sustainability index because the amount of energy used has a direct relationship with the amount of CO_2 emitted into the atmosphere. This level of energy consumption, 2.0 kWh resulted in the lowest condition of sustainability: 70 % recycling, 22 % incineration and 8 % allocation to a garbage dump. The total energy consumption during the product life cycle can reach 1.3 MJ, the highest value obtained in the two experiments.

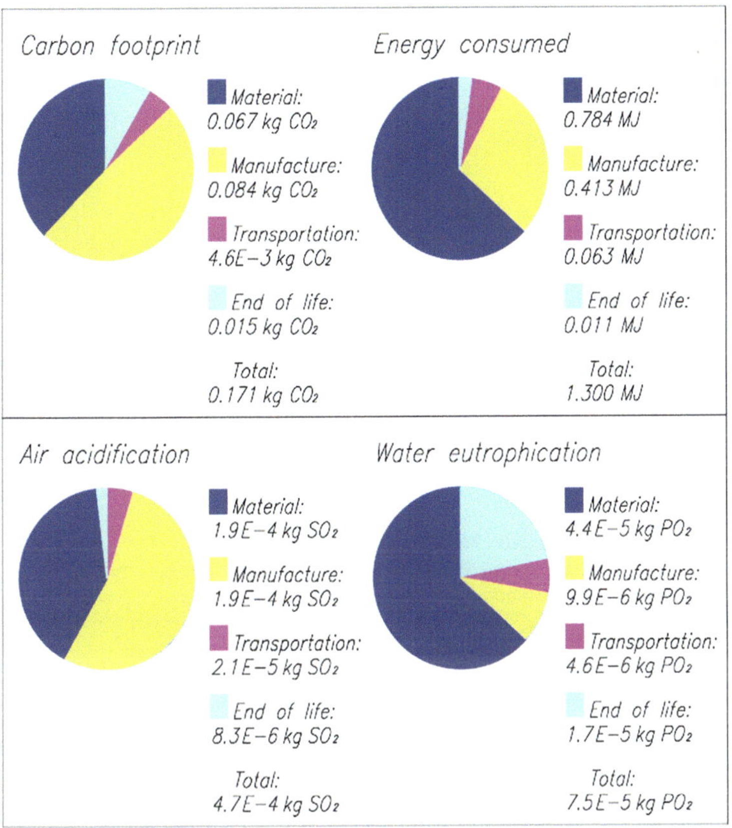

Fig. 4 Environmental impact of experiment 2. *Source* Authors

5 Preliminary Conclusions

Sustainability is a comparative process, which mean that in a situation where the emission of greenhouse gases and energy consumption are at a certain level, we must incorporate practices that reduce these values as lower as possible albeit slowly. The CAD tools available have great power to foster sustainable practices as now designers have the option to set the percentage of recycled material as well as take into account the complete life cycle.

In this case, we were able to evaluate the environmental impact of a packaging design that simulates the use of a nanocomposite based on sugarcane bagasse. These analyzes include the sustainable design practices should account for levels of power consumption compatible with product life cycle. As a result, encouraging the reduction of energy consumption is the most important practice of sustainability and it should be practiced by the society as a whole. The second most important practice is to apply techniques for recycling and reuse of waste as a way to prolong the life of the raw materials taken from nature. In the long run, it may decrease the need for exploitation of raw materials, preserving the nature and offering more time for recovering.

As part of future research, we intend to do more experiments with other samples of PLA as well as other natural fibers so we could have a more complete view of its impact on environment, resulting in better design decision.

Acknowledgements This research was partially supported by grants #2014/20971-3; #2010/12119-4 São Paulo Research Foundation (FAPESP).

References

1. da Silva, R.G., Oliveira, F.B., Marconcini, J.M., Mattoso, L.H.C., Mantovani, G.L.: Obtenção e estudo das propriedades mecânicas de nanocompósitos poliméricos biodegradáveis de poli (ácido lático) com nanofibras de celulose. Rede Nanotecnologia Aplicada ao Agronegócio 303–305 (2012)
2. Taekema, J., Karana, E.: Creating awareness on natural fibre composites in design. In: International Design Conference (DESIGN12), pp. 1141–1150. Dubrovnik, Croatia (2012)
3. Alting, L.: Life cycle engineering and design. CIRP Ann.Manufact. Technol. **44**(2), 569–580 (1995)
4. Ribeiro, I., Peças, P., Silva, A., Henriques, E.: Life cycle engineering methodology applied to material selection, a fender case study. J. Clean. Prod. **16**, 1887–1899 (2008)
5. The Consumer Goods Forum Sustainability Pillar: A global language for packaging and sustainability. A framework and a measurement system for our industry. The Consumer Goods Forum Report. Paris. http://globalpackaging.mycgforum.com/allfiles/GPP_FinalReport_170610.pdf (2009)
6. Mahalik, N.P., Nambiar, A.N.: Trends in food packaging and manufacturing systems and technology. Trends Food Sci. Technol. **2121**(3), 117–128 (2010)
7. Loh, Y.R., Sujan, D., Rahman, M., Das, C.A.: Sugarcane bagasse—the future composite material: a literature review. Resour. Conserv. Recycl. **25**, 14–22 (2013)
8. Maya, J.J., Sabu, T.: Biofibres and biocomposites. Carbohydr. Polym. **71**(3), 343–364 (2008)

9. Cao, Y., Shibata, S., Fukumoto, I.: Mechanical properties of biodegradable composites reinforced with bagasse fibre before and after alkali treatments. Compos. A Appl. Sci. Manuf. **37**(3), 423–429 (2006)
10. Luz, S.M., Del Tio, J., Rocha, G.J.M., Gonçalves, A.R.: Cellulose and cellulignin from sugarcane bagasse reinforced polypropylene composites: effect of acetylation on mechanical and thermal properties. Compos. A Appl. Sci. Manuf. **39**(9), 1362–1369 (2008)
11. Corrêa, A., Teixeira, E.D.M., Carmona, V., Teodoro, K., Ribeiro, C., Mattoso, L., Marconcini, J.M.: Obtaining nanocomposites of polyamide 6 and cellulose whiskers via extrusion and injection molding. Cellulose **21**, 3–11 (2014)
12. Teixeira, E.C.M., Bondancia, T.J., Teodoro, K.B.R., Corrêa, A.C., Marconcini, J.M., Mattoso, L.H.C.: Sugarcane bagasse whiskers: extraction and characterizations. Ind. Crops Prod. **33**, 63–66 (2011)
13. Solidworks sustainability: http://www.solidworks.com/sustainability/sustainable-design-guide/appendix-c-the-hannover-principles.htm
14. Klöpffer, W., Grahl, B.: Life Cycle Assessment (LCA)—A Guide to Best Practice, 440 p. Wiley VCH, Weinheim (2014)
15. Ermolaeva, N., Kirill, S., Kaveline, G., Spoormaker, S.J.: Materials selection combined with optimal structural design: concept and some results. Mater. Des. **23**, 459–470 (2002)

Does the Ecomark Label Promote Environmentally Improved Products in India and What Experiences Can Be Drawn from the Nordic Ecolabel?

Jakob Thomsen and Tim C. McAloone

Abstract Ecolabels are used to give consumers information about the environmental impact of products and thereby give an informed choice and an incentive to the consumer to choose the ecolabelled product. The environmental effectiveness of ecolabels ultimately relies on consumers' willingness to pay extra for a product with a lower environmental impact than alternative products. Various ecolabels exist throughout the world, but the difference between the successful and the non-effective ecolabels is large. This paper compares a best practice example of an ecolabel with a less successful ecolabel—each belonging to a different region of the world. Firstly we analyse, the ability of the Indian "Ecomark" label to promote environmentally improved products in India. The Ecomark scheme was launched in 1991, however the Ecomark label currently remains a nonstarter, with little awareness and no real consumer demand. India's Ecomark scheme has been analysed in several studies, all of which indicate that the scheme has had some flaws from the outset regarding its basic structure and execution. With a point of departure in these existing studies and further empirical insights from ecolabelling experts in Europe, this paper seeks to identify the differences between the Ecomark label and the second ecolabel under analysis—the much more successful "Nordic Ecolabel" from Scandinavia. The structural differences, selection process for criteria and the demand and awareness of the two ecolabels are analysed, in order to create suggestions of how to learn from the schemes and ensure knowledge transfer, regarding best practice.

Keywords Ecolabel · Ecomark · Nordic ecolabel · Ecodesign

J. Thomsen (✉) · T.C. McAloone
Department of Mechanical Engineering/Section of Engineering Design and Product Development, Technical University of Denmark (DTU), Lyngby, Denmark
e-mail: s112978@student.dtu.dk

A. Chakrabarti (ed.), *ICoRD'15 – Research into Design Across Boundaries Volume 2*,
Smart Innovation, Systems and Technologies 35, DOI 10.1007/978-81-322-2229-3_11

1 Introduction

The growing concern for the environment that has been experienced over the past decades has seen a rise in the number of consumers, with a demand for products that have been proven to be less environmentally harmful. Ecolabels provide consumers with an opportunity to distinguish between the most environmentally improved products and the rest of the market. Around the world, ecolabel schemes are being used to provide consumers with environmental information about products. Ecolabels are a market-based tool, designed to give the consumer the possibility to choose less environmentally harmful products, on the one hand, and to give the manufacturer a way of differentiating its products, on the other. Environmental goals are reached, because consumers are willing to choose ecolabelled products over non-labelled products. Consumers, who are willing to pay extra for an environmentally improved product, are often referred to as green consumers.

The first ecolabel to be released was the Blue Angel, established in Germany in 1977 [1]. Since then—and especially since the early 1990s, numerous national ecolabels have emerged throughout both the developed and developing countries, but the effectiveness of these varies considerably from country to country. Some ecolabels are considered successful, with both a demand and a supply of green products encompassing them, where in other cases the ecolabels are not being used.

International Standards Organisation (ISO) has defined three categories of ecolabels [2]:

- Type 1—Environmental labels (ISO 14024)
- Type 2—Self-declared claims
- Type 3—Environmental declarations for the comparison of products (ISO 14025)

The purpose of this paper is to identify the differences between the Nordic Ecolabel and the Ecomark label and make suggestions that could be adopted by the Ecomark scheme to improve the efficiency of the label. The two ecolabels are both Type 1 ecolabels (voluntary label, but with externally reviewed criteria). A Type 1 ecolabel follows a cradle-to-grave approach from raw material extraction, manufacturing and use, through to disposal. This means that the entire product life cycle is evaluated, by use of a Life Cycle Assessment (LCA) [3] and if the product meets the requirements set by the label, it will be a candidate for an ecolabel. The material created to fulfil these requirements is subject to external review, in order to be eligible for an actual ecolabel. The two ecolabels under study in this paper have similar goals [4, 5], but their influences on the market, within which they operate, are quite different. The Ecomark scheme has faced difficulties since it was launched in 1991 and still remains a nonstarter [1].

2 Methodology

This article bases its discussion and argumentation on the basis of three main sources: existing studies of the Ecomark label; articles and reports on ecolabelling in general; and further empirical insights from ecolabelling experts in Europe. The articles and reports have been used to gain knowledge about which elements influence the success of an ecolabel, in particular the Nordic Ecolabel and the Indian Ecomark label.

3 The Ecomark Scheme

In 1991 the Government of India introduced the ecolabel scheme Ecomark. The scheme was introduced to make it possible for consumers to identify environmentally improved products. The Ministry of Environment and Forests (MoEF) constituted two committees to manage the scheme. Firstly an *Inter-Ministerial Steering Committee* was established and given the responsibility of selecting product categories, creating awareness of the scheme through promotion, and formulating future strategies for the scheme. Secondly, a *Technical Committee*, constituted in the Central Pollution Control Board (CPCB), was established to specify the criteria for products, set up sub-committees for specific product categories with external experts, and evaluate the environmental impact of the specific products [1]. The Bureau of Indian Standards (BIS) evaluates and certifies submitted proposals of candidate products to receive the Ecomark label, and decides whether a product with the Ecomark label can keep the label or not.

To obtain the Ecomark label the company must pay a fee (500 Rupees equal to 6.16 € at the time of writing), fill out predescribed application forms and open its production facilities to inspections and testing of the product [6]. Expenses from testing and inspection are bourn by the applying company. The licence to hold the Ecomark label is awarded for a duration of one year, with the possibility of a two year extension, based on the performance of the product in preceding years.

4 The Nordic Ecolabel

The Nordic Ecolabel was established in 1989 by the Nordic Council of Ministers as a way of increasing demand and supply for green products. The member states of the Nordic Ecolabel are: Iceland, Norway, Sweden, Finland and Denmark.

The label is administered by a third-party control agency, *The Nordic Ecolabelling Board*, which sets the same criteria for products across all the Nordic countries. The member states each have a local organisation, which takes care of the daily tasks related to the label. In Denmark the Nordic Ecolabel is operated and administered by a non-profit organisation called *Ecolabelling Denmark*. This organisation is a part of the *Danish Standard Foundation*, which has been appointed to handle the operation of the ecolabel by the Danish Ministry of the Environment.

The Nordic Ecolabel has standards for over 60 product groups and over 5,000 products have been granted licence to use the label [7]. To obtain the ecolabel there is an application fee (2,000 € at the time of writing) and once the licence has been granted, there is an additional annual fee, based on the product or service.

5 Problem Analysis

Both labels are Type 1 ecolabels with the similar goals, but the two labels have experienced very different rates of success. The Ecomark scheme has so far not experienced success [1, 8, 9], whereas the Nordic Ecolabel is considered to be a large success, with market coverage in some product categories of up to 50 % [7]. Different analyses on the Ecomark label have suggested several reasons for the lack of success of the label. These reasons are examined in the following sections and compared to how the same problems are being tackled by the Nordic Ecolabel.

5.1 Structural Differences

The selection of criteria and operation of the two ecolabels are quite different. In the Indian example the organisation behind the Ecomark label is split into a three-tiered system. The Steering Committee controls both promotion of the label and selection of product categories. This committee consists mostly of state officials and a maximum of five non-officials, selected by the Central Government to represent the industry, consumer groups or other NGO's. Of these five non-officials at least two should represent consumer groups [8].

The Technical Committee consists of directors from various technical institutions and representatives from consumer groups and industry [8]. The Technical Committee identifies the criteria upon which the products should be evaluated.

The Ecomark scheme has been criticised for having some structural problems. The majority of state officials in the Steering Committee means that these carry a great responsibility for the progress of the scheme. Due to the fact that progress relies on state officials, their willingness to promote the label has been an important factor. The Inter-Ministerial structure of the Steering Committee has resulted in the participation of the members being on a voluntary basis [8].

According to Mehta, the Ecomark label has been considered by the CPCP (which is where the Technical Committee is constituted) to be additional work, dictated from the MoEF, because the Ecomark was not originally under their purview [8]. Additionally the BIS has been criticised for its lack of commitment to implementing the Ecomark label. This stance is supported by the fact that the BIS, which is the certifying agency, has no mention of the scheme on its webpage [1]. The abovementioned structural problems indicate that the Ecomark system has not been given the sufficient attention by the main bodies in charge of its implementation and execution.

It seems that the tried and tested structure of the Nordic Ecolabel could be the example to follow, if the Ecomark system, were to consider restructuring with a view to being more efficient, representative and ultimately successful. In the literature reviewed about the two ecolabels' structures, a series of recommendations are discussed by many of the authors. In the following we present our collation of the considered recommendations towards adjusting the structure of the Ecomark system, which we have chosen to present in a normative manner.

An ecolabel board should be established, similar to the Nordic Ecolabelling Board, and a control agency should also be established, similar to Ecolabelling Denmark. The ecolabel board should consist of representatives from industry and consumer groups as well as environmental NGO's. The MoEF should appoint the representatives on recommendations from the different interest groups. The ecolabelling board should identify product categories and verify the specific requirements for the products. The control agency should be a separate part of the BIS and should make recommendations for product criteria, test and verify products, and award the label. The control agency should consist of people assigned outside the central government to be responsible for the ecolabel instead of giving officials extra assignments. The scheme would have an executive power in the BIS and a legislature power in the ecolabel board, with a combination of the Steering and Technical Committee. The point of making an ecolabel board is to make sure that the stakeholders are being involved in the development of the product criteria. By establishing a separate control agency, testing and verifying products would not be regarded as additional work and this would ensure commitment from the employees involved. The above changes would come at an additional cost to the Indian government, meaning that there would have to be political will to invest in the Ecomark scheme.

5.2 Demand and Awareness

For an ecolabel to be effective both industry and customers need to know what the ecolabel represents. Analysis suggests that there is a clear lack of awareness of the Ecomark scheme [1, 8, 10].

The lack of awareness from both industry and consumers means that companies do not consider the Ecomark label as a way of differentiating their products. Several paper mills were granted the Ecomark label, but chose not to use it [10]. The fact that companies that originally applied and paid for the ecolabel, but later decided not to use it is a clear indication that promotion is needed, to make the scheme both understandable for the end customer and more attractive for a manufacturer to use the label. In contrast, surveys from northern Europe show that 94 % of the population recognize the symbol of the Nordic Swan (the logo of the Nordic Ecolabel) and are aware that it is an environmental label [11]. This level of awareness creates a much larger incentive for companies to file an application for the Nordic Ecolabel, because the label ensures a clear differentiation of their products from the rest of the market.

The largest difference between the Nordic Ecolabel and the Ecomark label is, of course, the markets within which they operate, and the creation of ecolabels is often based upon the local demand for environmental improved products. A Gallup survey in the beginning of the 1990s shows that consumers in India were willing to pay a higher price for a environmentally improved product [12], but according to Dilip Biswas, the former chairman of the CPCB, this is not the case: "*Consumers' attitude of 'buy cheap' rather than 'buy green' is a major impediment in promotion of Indian Ecomark*" [13]. The quote is of course part of one person's explanation of why the Ecomark was unsuccessful, but his conclusion is supported by the paper *Identification of Enviro-Eco Indicators for Environmental Sensitivity*, which suggests that the reasons why people in India are more likely to buy low quality cheap products—and sometimes also hazardous products—might be to lack of money, knowledge, and/or education [1]. A survey in 1998 also showed that only some consumers in the higher income group were willing to pay a higher price for an environmentally improved product, provided the price were not higher than 10 % [8].

According to Grossman and Krueger there is a threshold limit value for when countries will experience an improvement in demand for environmentally improved products, which is estimated to be around US$8,000 per capita income [14]. In India the income per capita is approximately US$1,570 compared to approximately US$61,110 in Denmark [15]. The differences in income indicate that the two labels clearly have very different bases for creating demand for ecolabelled products. However a study in Denmark found that an increase in income for both high and low income groups would increase consumers' probability of buying ecolabelled products, instead of non-labelled [7]. With the expected income rises in India over coming years, this might increase consumers' willingness to choose ecolabelled products.

To create an incentive for companies to apply for the Ecomark, it is necessary to create a demand for environmentally improved products; this could be created by the Indian government adopting a green procurement strategy. The MoEF initiated in 2009 the development of a green public procurement, based on the Japanese model [16]. Government-related institutions constitute nearly 80 % of buyers of goods [1], but in order to create a demand for environmentally improved products, there must be a way for governmental institutions to differentiate between alternative products. It is exactly here, where the Ecomark label can play its part. With the government demanding Ecomark labelled products, an incentive for companies to supply labelled products would be created. By combining green public procurement with the Ecomark label the government would help to kickstart the Ecomark label. This positive effect could be transferred to the private consumer market; studies from Denmark show that a greater availability of ecolabelled goods makes it more likely for consumers to choose ecolabelled goods [7].

The analysis made on both demand and awareness suggests that a promotion campaign of the Ecomark scheme is necessary. Both the industry and consumers are currently unaware of the existence of the Ecomark label. Studies show that gender, age and education level have an influence on the consumer's willingness to pay for an ecolabelled product. For example, studies in different places in the world—including Denmark and India—show that female buyers are more likely to choose an ecolabelled product [7, 8]. The results from these studies should be used in order to try and create a demand for products with the Ecomark label.

5.3 Criteria for Products

The Ecomark label has been criticised for setting standards, that did not fit the Indian market [9]. In some cases the criteria have been set too high, which would make the production for an Indian company much more expensive. The Nordic Ecolabel avoids this problem by setting (and adjusting) the label criteria, so that the top third of the market are able to meet the environmental standards. This means that the product is differentiated from the majority of the market, but the standards are still possible to reach. As soon as more than a third of the market exceeds the environmental demands set by the Nordic Ecolabel, the criteria are revised and adjusted.

In the Nordic Ecolabel criteria for a given product group are decided and also given an expiry date. The expiry date is normally set to four years after approval. The Nordic Ecolabel then makes sure that the criteria are revised, one year before they expire and that the companies possessing the label are given notice about changes in the criteria. Newly approved criteria are evaluated six months later, to check whether the level of demand has been set adequately, too high or too low. This strategy makes the development of criteria very transparent to both industry and green consumers. The most suitable product groups are chosen from a so-called "RPS" (relevance, potential, steerability) strategy. For a product to be considered it needs to have: *relevance* with respect to specific environmental problems it relates to; *potential* to have a significant environmental impact, if a label is created for this category of product/service; and *steerability*, related to how the product, activity or problem might be affected by the requirements of the Nordic Ecolabel [17].

The fact that the Ecomark scheme has been criticised in several studies for setting requirements too high indicates that requirements should be revised. The RPS strategy used by the Nordic Ecolabel would seem to be a sensible strategy for the Ecomark to adopt, in order to revise the focus for its criteria for various product groups.

Especially the fact that a product must live up to both an environmental and a quality standard has been criticised [8]. Due to the fact that the product has to comply with BIS' quality standards means that many companies will be put off making environmental improvements towards the Ecomark label, simply because it becomes too expensive to meet both standards [1, 9]. The BIS has over 1,700 standards, but according to BIS themselves, only 1,300 standards are recognised by industry, in practice [8]. If it is required to live up to standards that are not recognised in industry, companies will be likely not to apply for the Ecomark label. The Nordic Ecolabel also has standards in both environmental impact and quality. The fact that the Ecomark label consists of environmental and quality standards means that even more requirements are necessary to fulfil for a company applying for the label. If it does not add value to the consumer's opinion of the product, one consideration could be to split the label into two parts—one for quality and one for environmental performance. On the other hand it could be used as a marketing tool if used correctly. If the Ecomark label is associated with high standards in both environment and quality, the green consumer might be willing to pay more for a product with the Ecomark label.

6 Discussion

The Ecomark scheme has existed for over 20 years, but still has not gained any real influence on consumers or manufacturers. The fact that the Ecomark scheme still remains a non-starter has raised questions on whether a Type 1 ecolabel really is what India needs to promote environmentally improved products, just now. Studies have suggested that another kind of label might fit the Indian market better [8, 9]. The cradle-to-grave approach might be difficult to understand for manufacturers and consumers in an emerging economy and the Type 1 ecolabel requires a lot of credibility in the ecolabel.

The Ecomark has to compete with ecolabels that are easier to implement and understand, such as self-claimed ecolabels (Type 2). Type 2 ecolabels give claims about one environmental factor, where the product is doing well, but do not consider the whole product life cycle. These labels rely on the consumer to be able to make their own judgements about which product is ultimately the most environmentally improved. In the absence of a dominant ecolabel, Indian industry has made many different environmental claims. An independent study shows, that multiple products used misleading environmental claims to make the product seem more environmentally improved than it really is [8]. Changing the Ecomark label to a Type 2 would mean that the scheme would use its resources to verify these claims. This would be quite expensive and would leave the green consumer with the responsibility of determining, which product is less harmful to the environment. If the goal is to create a market for environmentally improved products, this is not the most effective path to take.

It has been suggested to make a grading of the Ecomark label, in order to make it easier for people to distinguish between products. This approach is often used in performance-based ecolabels where, for example, one product has been certified with a label that indicates that it is more energy efficient than other products in the category. In 2006 India founded a performance-based ecolabel on energy efficiency called the “BEE Star Label” (Bureau of Energy Efficiency). The BEE Star Label has been quite successful, considering that it is a new ecolabel [9]. These kinds of labels have the advantage that every product can be labelled instead of just the most environmentally improved product. Making a grading within the Ecomark would make it possible to differentiate products within the label, but it would also leave it to the consumer to make a choice, how green they want to be. This will create competition between the individual products in the label and will make the decision of how the product is graded much more sensitive. With the Type 1 ecolabel as it is, the consumer knows that he or she is getting one of the most environmentally efficient products available and the competition is between the products with the ecolabel and the products without the label. From a consumer’s point of view it would make the decision even more difficult, if products within the Ecomark label were graded.

What do ecolabels mean for product development? There seems to be high congruence between countries with strict environmental demands and successful ecolabels, and the amount of consideration made in product development towards environmental improvements. Scandinavian and other northern European countries (e.g. Germany, France, Holland) fall into this category. In these countries the majority of existing methodologies and tools regarding ecodesign, life cycle assessment and eco-innovation originate. It seems, therefore, that ecolabelling and ecodesign activities are closely connected. For many years there have been large efforts to ensure tools and methods for ecodesign (often with improvement goals similar or identical to ecolabel criteria). More recent research efforts seem to be moving away from “yet another tool”, towards understanding how to ensure the right conditions for ecodesign in a company, studying typical environmental performance indicators and understanding how to help companies become more mature in their ecodesign processes. It seems clear that with ambitious yet realistic ecolabels in place at a country level, the role of ecodesign, from product to product, becomes more focused, thus guiding the tools, methodologies and processes of ecodesign itself.

7 Conclusion

The Ecomark scheme started in 1991 with the intentions to influence Indian consumers to buy environmentally improved products, but it in order to make this impact the Ecomark label needs a revision. The three-tiered system has proven to be inefficient and way too bureaucratic. The current system’s dependence on state officials, the lack of involvement of the BIS and lack of awareness of both industry and consumers makes it unattractive to apply for the Ecomark label. To make the Ecomark an efficient ecolabel, the responsible stakeholders must be committed to the progress of the ecolabel. This will require a political willingness to invest in the ecolabel and provide a kickstart to ensure its success.

The Ecomark's ability to promote environmentally improved products relies greatly on the organisation behind the label to be efficient and well-structured. To ensure active involvement in the Ecomark scheme we propose that an ecolabelling board and a separate control agency be constituted. The ecolabelling board will make sure all stakeholders are involved in the development of criteria. The ecolabelling board should be constituted of people representing environmental NGO's, industry interest groups and consumer interest groups. A control agency such as Ecolabelling Denmark will consist of a group of people with technical expertise and people with expertise in sales and marketing. The agency would have the responsibility of promoting the label and taking care of technical aspects, such as setting requirements and evaluating products. This would create a clear job description for both organisations and the ecolabel would not be part of another organisation, where it is not the main priority.

Even though these initiatives would improve the functionality of the Ecomark scheme, it would still face large challenges in promoting environmentally improved products. The price sensitivity of consumers in India will still be a big factor and will make it difficult create demand for ecolabelled products. Green public procurement could be a way of kickstarting the Ecomark label by creating a demand of products holding the Ecomark label, but to make any real difference, a private market demand must exist. This demand can only come from a rise in consumers' willingness to pay for ecolabelled products. The willingness of consumers relies partly on the credibility of the label and partly on their awareness, ensured by campaigns and descriptive marketing. The Ecomark scheme should therefore make an effort to enforce the credibility and awareness of the label. There is no doubt that India has great potential to contribute to greatly improved products, seen from an environmental perspective. Even though the Ecomark label is currently not successful, many important considerations and decisions have been made. Given a re-framing, revision and kickstart of the Ecomark scheme, we believe that the scheme can be developed to contribute greatly to sustainable development.

References

1. Raman, S.D.B., Waghe, U.P.: Identification of enviro-eco indicators for environmental sensitivity. Int. J. Eng. Res. IJER **2**(5) (2013)
2. International Institute for Sustainable Development.: The ISO 14020 series [Online]. Available http://www.iisd.org/business/markets/eco_label_iso14020.aspx. Accessed 25 Aug 2014
3. Baumann, H.: The Hitch Hiker's Guide to LCA: An Orientation in Life Cycle Assessment Methodology and Application. Studentlitteratur, Sweden (2004)
4. Central Pollution Control Board: Objectives of the scheme [Online]. Available http://www.cpcb.nic.in/objectives_scheme.php (2008). Accessed 25 Aug 2014
5. Nordic Ecolabel: Mission for nordic eco-label [Online]. Available http://www.nordic-ecolabel.org/about/the-mission/. Accessed 20 Aug 2014
6. Central Pollution Control Board: How to obtain the licence to use ecomark. [Online]. Available http://www.cpcb.nic.in/licence_ecomark.php (2008)

7. Brouhle, K., Khanna, M.: Determinants of participation versus consumption in the Nordic Swan eco-labeled market. Ecol. Econ. **73**, 142–151 (2012)
8. Mehta, P.S.: Why was India's Ecomark Scheme Unsuccessful? CUTS Centre for International Trade, Economics & Environment, Jaipur, India, ISBN 978-81-8257-084-9 (2007)
9. Raghupathy, L., Henzler, M.P., Chaturvedi, A. , Arora, R., Eisinger, F., Strasser, C.: Green electronic products in India—lessons from the BEE star label and the Ecomark scheme. In: Proceedings of Electronics Goes Green 2012+, ECG 2012—Joint International Conference and Exhibition (2012)
10. Srivastava, S.: Ecomark scheme finds few takers in industry. Business Standard, 2007 [Online]. Available http://www.business-standard.com/article/economy-policy/ecomark-scheme-finds-few-takers-in-industry-107081601086_1.html. Accessed 25 Aug 2014
11. Nordic Ecolabel.: Recognition of the Nordic Ecolabel [Online]. Available http://www.nordic-ecolabel.org/about/. Accessed 25 Aug 2014
12. Shams, R.: Eco-labelling and environmental policy efforts in developing countries. Intereconomics **303**, 143–149 (1995)
13. Biswas, D.: The response to the Ecomark scheme has been hesitant if not different [Online]. http://expressindia.indianexpress.com/fe/daily/19990808/fex08026.html. Accessed 25 Aug 2014
14. Grossman, G.M., Krueger, A.B.: Economic growth and the environment. Q. J. Econ. **110**(2), 353–377 (1995)
15. The World Bank: GNI per capita [Online]. http://data.worldbank.org/ (2014). Accessed 25 Aug 2014
16. Centre of Excellence for Sustainable Development: Green Public Procurement Guidelines in India [Online]. Available http://www.ecoindustrialparks.net/live/hrdpmp/hrdpmaster/hrdp-asem/content/e8451/e8981/e41200/e41953/e41954/eventReport41965/Presentation-Ms. ShaguftaKamran.pdf (2012). Accessed 25 Aug 2014
17. Nordic Ecolabel: Nordic ecolabelling steps [Online]. Available http://www.nordic-ecolabel.org/CmsGlobal/Downloads/EcolabellingStepstowardsSustainability.pdf (2001). Accessed 25 Aug 2014

Taki, the Community (Sustainable) Sensory Garden

Design as Generator (DAG) Part I

Phebe Valencia and Martin L. Katoppo

Abstract Design as Generator (DAG) is an idea turning design to be a generator for others especially within communities. In the world of DAG, design should be practical and applicable as it served its own goal: to enhance the quality of human life. If we talk about the quality of human life, it could not be separated from its environment. Human has amazing ability to shape its environment and fill it with their daily habits which will then develop into a culture that shows their identity. Thus we propose that environmental concern awareness is needed to be built within the community in relation with its sustainability issues. *Taki* (shortened for *Taman kita,* means 'our garden') community (sustainable) sensory garden will be one of the first community projects of DAG. As a garden, *Taki* will not just aiming for visual pleasantly garden but rather it will be aiming on building the community awareness and the place for inducing children creativity. It will be using the idea of designing sensory garden which is useful for all members of the community to stimulate and to provide experiences for heightened sight, smell, hearing, touch, and taste. *Taki* will be located inside the suburban kampong of Pondok Pucung, South Tangerang, Indonesia; surrounded with a mega modern residential real estate. The specific location of it will be on an empty disarray lot owned by local and being used by members of its community as garbage dumping and burning site. This paper will emphasize on: (1) how to design for, with or even by the community; (2) how to design and build a sensory garden for community; (3) how to make the design dwell (sustain) in the community—with its main activity of building awareness and inducing environmental sustainable way of life. The design process of *Taki* aim for larger purpose: community empowerment through design. For this purpose in the end of the paper, authors will try to propose a research design method deemed suitable.

P. Valencia (✉) · M.L. Katoppo
Pelita Harapan University, Tangerang, Indonesia
e-mail: Phebe.valencia@uph.edu

M.L. Katoppo
School of Architecture, ITB, Bandung, Indonesia

A. Chakrabarti (ed.), *ICoRD'15 – Research into Design Across Boundaries Volume 2*,
Smart Innovation, Systems and Technologies 35, DOI 10.1007/978-81-322-2229-3_12

Keywords Community (sustainable) garden · Design as generator (DAG) · Sensory garden · Design (for, with and by) community

1 Mission and Purpose

If we talk about the quality of human life, then it could not be separated from its environment. Current trends showed awareness of environmental sustainability is increasing. Designing a garden for community is one way to protect and creatively enhancing the quality of natural environment. *Taki* (shortened for *Taman kita,* means 'our garden') community (sustainable) sensory garden will be a design project for, with or by the community. It short range plan mission is "to engage the community and their needs through inducing sustainable way of life in Kampong Pondok Pucung, South Tangerang, Indonesia", while it long range plan mission is "to enhance Kampong Pondok Pucung community quality of life, thus to empower the community through design activities (research and action) inducing creativity and providing innovation pipe lines within them to ensure its sustainability". This is where the idea of Design as Generator (DAG) comes in. DAG is an idea turning design to be a generator for others especially within communities. In the world of DAG, design should be practical and applicable as it served its own goal: to enhance the quality of human life [1].

Taki will be DAG prototype and example of how design can be a generator that generates others. *Taki* as a garden will not just aiming for visual pleasantly garden but rather it will be aiming on building the community awareness and the place for inducing children creativity. It will be using the idea of designing sensory garden which is useful for all members of the community to stimulate and to provide experiences for heightened sight, smell, hearing, touch, and taste. Sensory garden also can serve many functions such as teaching, socializing, and healing. The garden will not only be educational and fun, but it will be benefitting the children with their interaction with nature in all aspects of their development: physical, mental, moral and emotional [2].

The project goals are:

- To provide communal space for gathering and maintaining the warm connectivity within the Kampong's community.
- To provide stimulating garden environments for the children to play creatively and imaginatively.
- To provide environmental friendly and sustainable application system that is easy and accessible for everyone, in which later should promote awareness within the community.

2 The Design Team

The planning and design team consisted of DAG members (Martin L Katoppo—architect and lecturer with his wife Ruth Euselfvita Oppusunggu-architect and a mother of two, they lives in the Kampong where *taki* garden has been proposed; Rakyan Tantular-architect and environmentalist; Phebe Valencia-interior designer and lecturer, and; Ade Maradona SW-graphic designer) [3].

The stakeholders are the community of Kampong Pondok Pucung, South Tangerang, Indonesia consist of the natives who usually owned the land and the temporal dwellers; the local informal leaders; the formal or local institutional or government representatives; the housewives and mothers; and especially the children (age 3–12 years old). Children in the age of 3–7 years old are beginning to explore, while children in the age of 5–12 years old are beginning to explore lots, woods, ditches and other interesting places around their home [4].

3 The Design Process

Taki begins as an idea, emerged as a continuing research on how design influence and changing people perception towards their improvement and later their empowerment. The first experimental project was a dwelling designed by DAG members within Kampong Pondok Pucung that uses many features that promote sustainable way of life. The dwelling intentionally designed to influence and generates inquiry from the Kampong's community. Because of the responses were pleasing, DAG members then thought that this intention of influencing community toward their improvement through design should serve broader context. Thus the idea of designing public facilities in a form of community (sustainable) sensory garden emerged, giving birth to *Taki*.

As stated above in *Taki's* mission, the garden at first should urge community engagement and later pushing the community to empower themselves through design activities and intervention. With this in mind, DAG team then planned the design process phase, which are: (a) Observation and Interaction phase—DAG team observed intently and built up relationship with the entire stakeholder within the community, especially children (age 3–12 years old) to absorbed and experienced their everyday lives. The main goal of this phase was to identify the problems (explained in Sects. 3.1 and 3.2); (b) Designing phase—DAG team then will design the garden based on findings in the first phase (explained in Sects. 3.3 and 3.4); (c) Action phase—DAG team will arrange action plans to engage *Taki's* design ideas to the community members, while in the same time gathering feedbacks from them (explained partially in Sect. 4) and; (d) Delivering and the making of *Taki* for, with and by the community (for the further research).

3.1 Site Inventory and Analysis

The garden location is in Kampong Pondok Pucung, South Tangerang, Indonesia. The kampong is mainly a remaining native kampong surrounded by concrete walls built by the developer of mega modern real estate adjacent to it. The kampong itself in contrast to the mega modern real estate, is far of being well developed as it is a self-growing housing area through informal and disarray setting. It has no proper roads, sewer system, utilization, sanitation, and garbage disposal system, thus resulted on community ignorance or at best tried conventionally managed it (i.e.: burning the garbage, made a traditional embankment for human waste disposal, etc.) (Fig. 1).

The kampong inhabitants profile comprise of the natives (few clan of families who owned most of the remaining lands, private and rental houses or rooms) and the temporal dwellers (most are from Java's region). From the observation DAG team found that both of the kampong inhabitants, adults and children, are dealing poorly towards garbage disposal system, healthy and appropriate environment. On the other hand, the observation also showed that relationship between the kampong inhabitants still shows the traditional and warm connectivity towards each other. They socialize in front of their terrace, on the road with passerby, mingled in front of small kiosks, gathered in mosque and prayer groups, engaged in every community celebration (i.e. wedding, circumcision party, watching temporal big screen movie, etc.), maintained night watch activities, and many other communal gathering activities. The young ones, the children always and love to play, doing outdoor activities wherever there are open spaces, let it be on the road, empty disarray lot or even in someone's yards (Fig. 2). Interestingly, both attitude and values mentioned above are absolute opposite if compared to the modern society living in the mega modern real estate neighboring the Kampong.

The specific location of *taki* garden will be on an empty disarray remaining lot owned by one of the native families in the kampong. The location is strategic as it is located in the junction between neighborhood association and the main kampong

Fig. 1 Kampong Pondok Pucung, a self-growing housing area through informal and disarray setting without proper infrastructure *Source* Katoppo and Oppusunggu 2011–2014

Fig. 2 Kampong Pondok Pucung social-community and children activities. *Source* Katoppo and Oppusunggu 2012–2014

street that leads to the mega modern real estate. The owner built five rental rooms for the temporal dwellers on the lot and leaving the remaining empty for now. In the future the remaining vacant lot will be built as another rental rooms. In the meantime, this remaining 13×14 m^2 vacant lot is just an empty disarray lot in a wreck condition. The lot was used mostly as garbage burning and dumping sites, place to dry the washed clothes, children playing, vehicle parking, temporary activities (i.e.: celebration tent for wedding, traditional musician performer stage, temporal big screen movie, etc.). Although seemed strategic (one of the kampong's center and intersection between many places) the place, instead of contributing positive value to the Kampong's environment, is actually contributing more to the already deteriorated quality of the whole Kampong's environment (Fig. 3).

The background was essential in the preliminary design discussion, including the lot status that is still being negotiated with the owner and the public official until now. DAG team proposed to only borrow the lot meanwhile it empty and not being used by the owner. The owner is still not certain lending the lot for public use, especially in the form of garden. His main concern that he might lost the legal ownership of the lot. Suppose that he gave his permission on it, the owner needed assurance that the lot will always be his and that he still can do anything he wants to the lot whenever he needed it especially to build another rental rooms.

Fig. 3 *Taki's* lot. *Source* Katoppo and Valencia 2013

3.2 Program Development and User Needs Identification

When designing a sustainable garden we should have and consider the following [5, 6]:

1. **Think holistically**: knowledge of the larger context ensures the connectivity of the garden, thus ensuring its sustainability—such as: observe and maintaining the people activities and habits. Sustainable garden should provide social spaces for people to interact and build social capital and educational component that promotes stewardship. Garden should be able becoming place that build the quality within its communities.
2. **Be Yourself**: create a unique experience that visitors could not imagine finding anywhere else.
3. **The detail features** of Sustainable Garden design: (a) **Resource Conservation**. The garden should make best of the site's potentials; (b) **Storm Water Management**. The garden should manage its water usage and recycled it, including the rain water; (c) **Waste Reduction**. The garden should provide garbage disposal system that put forward the 3R principle: reduce, reuse, and recycle; (d) **Social Capital**. The garden should foremost deals with accessibility for every types of its user, including disable user. It should also design social gathering place, incorporate interpretive signage of the site history and sustainable design principles, provide bicycle and pedestrian linkages, and the possibilities of exhibiting public art activities; (e) **Low Maintenance**. The garden in this sense should be easy for everyone and sustainable for its communities to be environmentally friendly maintained.
4. **The Challenges and Opportunities**: (a) **Funding**. As in every sustainable design projects, the initial costs higher than building just the usual garden, while the maintenance an operating costs in the long run will be much more effective and efficient, especially if the garden connectivity to its communities intertwined; (b) **Changing perception**. This comes up throughout the design process until the garden operates and with everyone that is involved in it. It is imperative that everyone is involved because this will create a sense of public ownership and stewardship, thus the sustainability of the garden will also be ensured; (c) **Collaborate**. Use professionals with a wide variety of backgrounds and perspectives because collaboration generates rich and dynamic experiences.

When designing garden for children, we have to consider: (1) their age group (in our case varied between 3 and 12 years old), and; (2) what are they going to play. Children are always active. They have broad imagination spheres, love to get along with each other, like to do experiments and while their attention absorbed, they also like playing in peace and quiet. Therefore playground for children should considered physical games, creative games, social games, sensorial games and resting and reflection space. The most recent trend of designing garden is the consideration of having chances stimulating all the 5 senses of its user, whether the adults or the young ones. The approach of having 5 senses stimulating garden is usually a combination between making use of natural resources and middle to high technology. The key concept of 5 senses garden is interaction between all the elements of the garden and its user. The 5 senses garden should encourage its user to actively participate and interact with acoustic, visual, tactile and olfactory phenomena while at the same time learning how their own actions elicit certain responses [7].

DAG team identified Kampong Pondok Pucung community essential needs to be projected unto this garden, which are: (1) **For the environment**, the lack of public facilities and proper infrastructure resulted in the ignorance attitude towards the living and natural environment from the inhabitants or the community itself. The presence of sensory garden that induce sustainable way of life awareness should be appropriate; **For the adults**, they need communal space for gathering—a place for community engagement; (2) **For the children**, they need space to creatively play. In this sense providing a garden serves special purposes. Garden means play for children, while children mean the future. Thus by designing and proposing garden, we will also be investing to new character and awareness to the future generation and by that to the future of their dwelling place. Playing activities, therefore, are invaluable 'dress rehearsals' for their future lives and garden is the best space to be the place for this specific and driven-purpose activity.

3.3 The Design

Taki, the community (sustainable) sensory garden will provide [8]:

1. **Communal gathering space** will be place at the best spot facing the main road as the gate and the symbol of the garden. It will be built as a simple movable stage and roof construction using bamboo or used wood and steel as the materials. It will be in a form of amphitheater. The floor will use scraped roofs or knitted bamboo woven. This space ensures any temporary celebration activities usage.
2. **Children playground** will be the heart of this garden. The playground designed with vast green grass formed mildly leveled and contoured stimulating adventurous sense to the children. It served as parts. The green grass will be planted on the top of movable and elevated simple used wooden planks or used steel construction.
3. **The 5 senses interactive and greenery walls** will be located and using the walls from the existing rental rooms. The designed walls will not just be interactive equipment for children through the 5 senses stimulating materials, but also interactive equipment for adults through its greenery walls on the upper side of the 5 senses interactive walls. The walls also served as place where the adults still can hang their washed clothes in the morning before the children play in the early evening.
4. **The water and garbage disposal management system**. The water management system will use the existing water tank modified to be also able catching the rainwater, and re-use it for the garden maintenance and for the water feature

that children can play with. The garbage disposal management system will use separate bins for types of garbage, *biopori*, recycling bins, *takakura* bins, small garbage sorting workshops and creative area where adults or children could play with recycled and non-hazardous garbage. Both systems will also act as interactive educational tools for the adults and the children, targeting on building up their awareness on the issues and as a trigger for their creativity especially responding to their own dwellings.

5. **Playing equipment** (i.e.: sliding and climbing) will be made from recycled materials.
6. **Stimulating Path** will use the existing layer of the lot combined with materials that can stimulate the tactile experience, such as: stone pebbles, scraped roof tile, used tiles, etc.; while also considering the safety for its user especially the children.
7. **Accessibility of the garden** will be from every corner of the lot, as it designed without fence or walls. The garden will also provide: vehicle and bicycle parking.

It should be noted that *Taki* is designed as modular parts using knock down system as a response to the lot's status of ownership concerned by the owner.

3.4 Design Implementation

See Figs. 4, 5, 6, 7, 8, 9, 10 and 11.

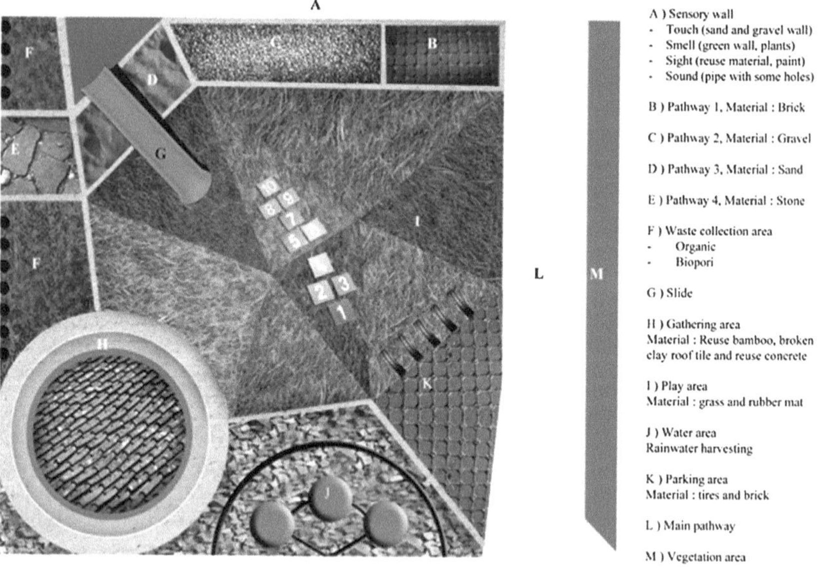

Fig. 4 *Taki's* plan. *Source* Katoppo and Valencia 2013

Fig. 5 *Taki's* park1 (*before*). *Source* Katoppo and Valencia 2013

Fig. 6 *Taki's* park1 (*after*). *Source* Katoppo and Valencia 2013

Fig. 7 *Taki's* park2 (*before*). *Source* Katoppo and Valencia 2013

4 Reflections

'The needs to turn *Taki* garden into collaborative and engaging participatory process'. *Taki* design is the results of the team intent and deep observation, as well as interaction with the community, especially with the children. However the design presented in this paper is still in its preliminary stage on its process to be shared and discussed with the community. We believe that to complete the cycle of

Fig. 8 *Taki's* park2 (*after*). *Source* Katoppo and Valencia 2013

Fig. 9 *Taki's* park3 (*before*). *Source* Katoppo and Valencia 2013

Fig. 10 *Taki's* park3 (*after*). *Source* Katoppo and Valencia 2013

sustainability of *Taki*, we must giving it back to its community that will share *Taki* from its impregnation, to its becoming and later growing infatuation, deviate, stray and depart towards the quality of life enhancement of this particular community itself and of course the quality of the environment. The recent development of *Taki* took form of an action-event that is arranged and carried out by the DAG team, called 'Color Your Kite' day held on the previous mid-April, 2014. This action-event targeted for the kampong's children, was aim at gathering them in playful activities, while in the same time became learning activities and campaign for raising awareness on how they should contribute on keeping their kampong clean and healthy. The action-event also became DAG team research activities for

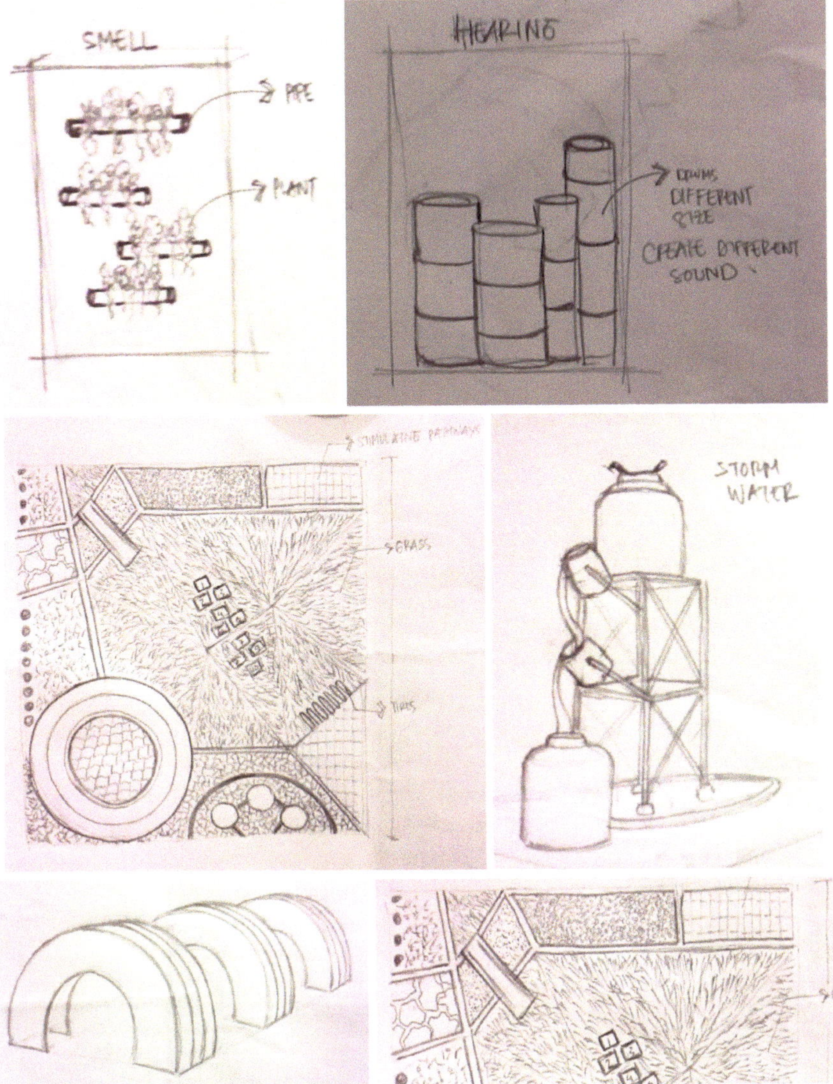

Fig. 11 Development sketches. *Source* Valencia 2013

gathering more data especially on the community everydayness, having a chance to interact with the children and the parents–mothers that accompany their children, in a more casual manner. 'Color Your Kite' day was actually planned as one of the preliminary action-events towards the needs to turn *Taki* garden into collaborative and engaging participatory process in its awake [9]. The action-events series are also served as preliminary research and introductory notion towards *Taki* greater goal, **in which showing how design activities** (both within the context of research

Fig. 12 'Color your kite' day. *Source* Valencia 2014

and action) **can induce creativity within the community and then later empowering them** (Fig. 12).

While undertaking the research and action activities throughout *Taki* design process, DAG team founded, developed and built a specific research design method, adopting a mixed methods model developed by Creswell and Clark (2007), namely Sequential Embedded Experimental Model [10]. The proposed mixed methods design will allow quantitative and qualitative methods to be used complementary in design research, taking the researcher along quantitative sides within the experimental design innovation done through prototyping and effects measurements during design intervention; and along qualitative sides as experience appreciations within participatory activities among every stakeholders during the whole process of design, done through Field Action Research. The proposed mixed methods will give new values to design, in virtue of its innovation and its response to social context. **Ultimately, through design, people will share knowledge production activities that emancipate and empower everyone** (Fig. 13) [11].

We realized that *Taki* community (sustainable) sensory garden is still inundated with shortcomings and far from being neither perfect nor comprehensive especially as a sustainable design intervention which goals are to promote awareness, induce creativity and push empowerment to happen within the community. However, initial steps must be taken and we believe it is worth to pursue, because we believe that investing on earth and in the future generation is worth the life itself.

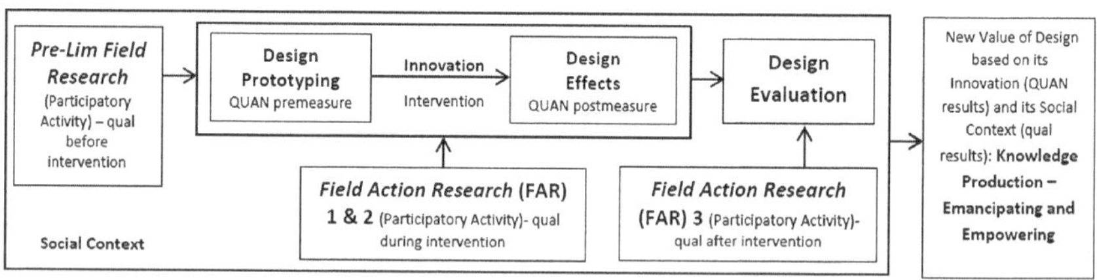

Fig. 13 Proposed mixed methods for architectural or design research *Source* Katoppo and Sudradjat 2014 translation based on Creswell and Clark 2007: 68–Fig. 4.2b

Acknowledgments The Authors would like to thank all the people of Kampong Pondok Pucung, Tangerang, Indonesia, especially the mothers and children; the DAG team and www.dagedubrag.org.

References

1. Katoppo, M., Oppusunggu, R.: Design as generator, Artepolis 4. In: International Conference, vol. 1, pp. 131–140. School of Architecture, ITB, Bandung (2012)
2. Tai, L., Haque, M.T., McLellan, G.K., Knight, E.J.: Designing Outdoor Environment for Children-Landscaping Schoolyards, Gardens and Playgrounds, p. 24. McGraw-Hill, New York (2006)
3. DAG team. www.dagedubrag.org/ about DAG
4. Tai, L., Haque, M. T., McLellan, G.K., Knight, E.J.: Designing Outdoor Environment for Children-Landscaping Schoolyards, Gardens and Playgrounds, p. 15. McGraw-Hill, New York (2006)
5. Byler, T., Goltsman, M.I.: Designing in the green: An approach To sustainable park design, p. 32. http://www.cprs.org/membersonly/Spring08_GreenDesign.htm (2008)
6. Tai, L., Haque, M. T., McLellan, G.K., Knight, E.J.: Designing Outdoor Environment for Children-Landscaping Schoolyards, Gardens and Playgrounds, p. 71. McGraw-Hill, New York (2006)
7. del Alamo, M.R.: Design for Fun: Playgrounds, pp. 266–298. Page One Publishing, Links International, Spain
8. Katoppo, M. and Valencia, P.: Taki Community park. In: International Conference on Creative Industry (ICCI), pp. 57–62, Surabaya (2013)
9. DAG team.: Discover phase—'color your kite' day", +Acumen and Human centered design for social innovation Project workshop report. http://www.slideshare.net/DesignAsGenerator/human-centered-design-for-social-innovation-3rd-week-170414 (2014)
10. Creswell, J.W and Clark, V.L.P.: Designing and Conducting Mixed Methods Research, p. 68. Sage Publication, Thousand Oaks (2007)
11. Katoppo, M., Sudradjat, I.: Combining participatory action research (PAR) and design thinking (DT) as an alternative research method in architecture. In: Artepolis 5 International Conference, School of Architecture, vol. 2. ITB, Bandung (2014)

Empowerment for Chhattisgarh Craft Clusters

Parth Shukla and Satyaki Roy

Abstract Chhattisgarh, a state which has almost 41 % of its land area under forest cover is one of the major contributors to the handicrafts sectors inhibiting art forms like Dhokra, Terracotta, Bamboo work, Wrought iron etc. The main contributors to these crafts sectors in the state are the tribes residing at remote locations in the state. There are many families or groups who have been practicing these art forms for generations, are now drifting to other means to earn their livelihood. A report from UN says that the no of artisans working in India has been reduced by 30 % in past 3 decades. Therefore in our work we try to address the issues pertaining to the state of the crafts and understand what technological interventions could be brought in, to make the crafts self-sustainable. In this paper we try to bring to forefront of our attention that the traditions of the handmade crafts have to be empowered for its sustainability and improvement of living conditions of the artisans.

Keywords Handicraft · Artisans · Dhokra · Wrought Iron · Terracotta · Current state · Dependency · Ignorance

1 Introduction

Craft is traditionally associated with creation of artefacts as a source of experience and emotion. Crafts are bound to be sensibilities of material and material understanding, making and haptic perception. Crafts have helped to shape our country and history. To lose them would be losing a part of our culture that can never be replaced. We should be conscious of the interlocking nature of the world that we live in and must realize that the species are dis-appearing, that environments are getting damaged. Now we should also realize the value of our own human cultural heritage. Crafts is like a golden thread that links our past, present to future and this a

P. Shukla (✉) · S. Roy
Indian Institute of Technology, Kanpur, India
e-mail: getparth@iitk.ac.in; parthpathfinder@gmail.com

A. Chakrabarti (ed.), *ICoRD'15 – Research into Design Across Boundaries Volume 2*, Smart Innovation, Systems and Technologies 35, DOI 10.1007/978-81-322-2229-3_13

timely reminder that the diversity and individuality of craft mirrors the qualities of the people themselves. Crafts plays an important role in educating people about the environment and local countryside, widening the learning perception of the world/ environment around them and grounding them well with it.

Handicrafts are a major stakeholder of home accessory market which includes handcrafted, semi-handcrafted and machined goods. It is the 2nd largest employer in rural sector after Agriculture. In most of the occasions the artisans or the craftsmen are unaware of the changes in the market due to rapid modernization, also the behavior of the consumers have become unpredictable now. With globalization penetrating to every corner of the society there is a need to realize the value of crafts, before it's too late and there wouldn't be anything left but legacy of our cultural heritage.

1.1 Literature Review

Jatin Bhatt in *Philosophy and Practice of Crafts and Design* says that handcrafted products form an important part of creative cultural industries, they also have the space to counter the techno-aesthetic dominance, for crafts independently present, to the patron of sustainable practice, a connect and concern with material and environment [1].

Dak has listed out reasons for failure of village crafts such as poor designs, low quality materials, and inefficient market approaches; consequently these approaches are then exploited by machine made urban craft units, who apparently lockout village crafts from global markets [2].

Chamikutty [3] comments that, today the sector carries the stigma of inferiority and backwardness and is viewed as just the items of '*decorative, peripheral and elitist*'.

Ms. Divecha explains that to realize the full potential of the crafts sector, the gaps, overlaps and challenges in the value chain must be overcome to create an ecosystem that enables crafts to thrive. She identifies that philanthropy has a "*key role to play in supporting the creation of such an ecosystem*" [4].

2 Objective

(i) To understand the state of the craft in Chhattisgarh
(ii) To understand the components of the value chain
(iii) Sustainability for the craft

3 Research Methodology

This research paper is based on exploratory or qualitative design research. The qualitative method was selected because of the lack of previous research on craft clusters of Chhattisgarh. In this research the author has tried to understand the real world situation as it happened naturally without influencing the environment or the subject of his study. The author also conducted several interviews and recorded the data based on his observations. People/artisans were interviewed on the basis of their craft practice.

The Dhokra art in Chhattisgarh has been practiced for over 300 years now. The major employment for this sector comes from tribes. They usually rely on their surroundings for raw materials and are rooted to the environment. The typical process of Dhokra involves following five steps (i) making the model out of clay to the near actual size of the artefact, (ii) a very thin layer of fine clay, (iii) actual design for the artefact by wax, (iv) layer of very fine clay, (v) another layer of clay mixed with fodder and hay. The artisans doesn't have access to modern facilities like fine sand or clay, binding materials or high efficiency furnaces. They tend to use whatever they can find nearby them like for making the casting surface smooth they used the fine clay which is found near the river banks or water bodies. For binding material they use a mixture of clay, hay and fodder. This mixture helps clay retain its shape when it is being baked in the furnace for melting out the wax. The Bell metal casting is also known as Lost Wax casting process, as the wax is melted out to create a cavity where the metal can be poured into take the desired form, hence the name Lost wax casting process.

The actual designs which are made in the artefacts are made from wax, this wax is actually a mixture of wax (by-product of honey) + camphor + tar in the quantities 50:25:25. The camphor and tar are added to give the wax strength and retaining shape at the same time to melt out when heat is applied. The wax is beaded in the form of threads of different sizes, fine tools are used to create the shape out of the layer of wax. Like the wax which they make by themselves, they melt the metal using open hearth furnace which requires large amount of wood as fuel and take time to complete the melting of metal as a large part of the heat is lost during the process. They melt 20–25 kg of metal at the same time as it uses time and energy both. The molds of the artefacts have been prepared prior to melting the metal, at a time 8–10 artefacts are created depending upon the size of the design. Once the metal has been poured in, it takes a day or two to solidify completely and to cool down, once its cooled down the outer covering of clay is been hammered softly to break apart, revealing the artefact.

The artefact is then required to undergo brazing and polishing to make the finish better and removal of excess material.

The Wrought Iron craft or Lohkala doesn't have so many process but its more labor intensive. In Lohkala, the metal sheet is being hammered constantly to match the design and the dimensions also constantly heating it in between so that the machinability is maintained. Once the desired shape has been achieved the intricate designs were made by a fine chisel.

In the Terracotta craft the main ingredient or the essential component is the clay, so the quality of the product depends greatly on how the quality of clay is and how much impurities has been filtered out. The artisans of Kumharpara follows the traditional process of making the artefacts i.e. the hollow part of the products are spun on the wheels to create the desired geometry and rest of the fine intricacies are done totally by hands. So the skill of the artisans also plays a major role in the consistency of the product. Once the product is made it's baked for certain period of time to give it strength. The artisan working in this cluster were mostly ladies and have been involved in this profession since their childhood, but not all of their next generation adopted the same profession, some of them moved to other professions as well.

The artisans of Kondagaon belt in CG always try to associate the artefacts that they produce to their mythological events or their god/goddess. A close interview revealed what are their thought process behind creation of the artefacts the way they are. Their major quantity of goods are sold in the National Craft fairs that they attend and other methods of selling goods include selling in the local market, giving it to traders and also the emporium which is opened by the govt. to promote the sale of the craft. The terracotta artisans in order to attract more consumers have started making a specialized wall tiles. These tiles are made up of mixture of clay and other binding agents. The typical value chain which has been observed in all of the artisans of Kondagaon belt is fairly long when it comes to reaching the end consumer. They give their finished products to the middlemen which in-turn deliver their products to the traders in markets of different cities, then it reaches the end consumers. There are SHG's and NGO in the belt but they are also not able to reach the outer market so easily, they issue they face is related to packaging, transporting etc.

The author also visited the Bolpur belt of West Bengal which is famous for different art forms existing there. Bolpur inhibits art forms like Dhokra (Dariapur), Shola pith (Surul), Terracotta (illambajar) for a comparative study of the craft practices and patterns of the artisans.

The artisans of Bolpur enjoys good connectivity from the major markets of WB, making them little bit aware of the needs of the market as there is huge inflow of tourists in Bolpur due to presence of Shantiniketan. For Shola Pith artisans the raw material required (i.e. Indian cork, glue etc.) is easily available and the quality of the materials is fairly constant. They use indigenous tools as it suits them well to work with tools that they've made according to their specific requirement. The literacy level of artisans are higher when compared to the artisans of CG. There is a cluster which comprises of only women who work for Dhokra art form. To sustain their livelihood they visit nearby market, weekly markets and also source it through local stores and emporiums themselves. The artisans of Dariapur who are famous for their intricate designs and very light weight castings, are now loosing on their skills i.e. no of artisans working for this craft (Dhokra) are decreasing now, many of their family members are turning towards different professions.

4 Result

The current state of the artisans and the value chain are the two major focus areas where the author has stressed during the research. The following are the key findings of the research:

(i) The literacy is very low
(ii) Quality of artefacts produced varies with time and location
(iii) Avg. income distribution for per Rs. 100 paid by consumer is (Fig. 1)
(iv) Percentage of artisans drifting away from their professions (Fig. 2)
(v) Percentage source of income for artisans (Fig. 3)

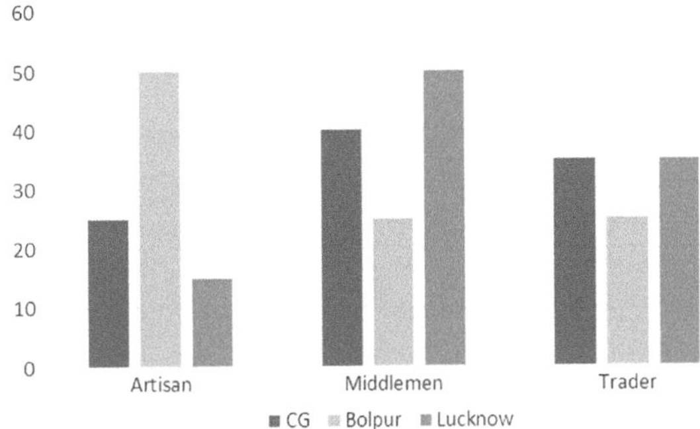

Fig. 1 Avg. income distribution

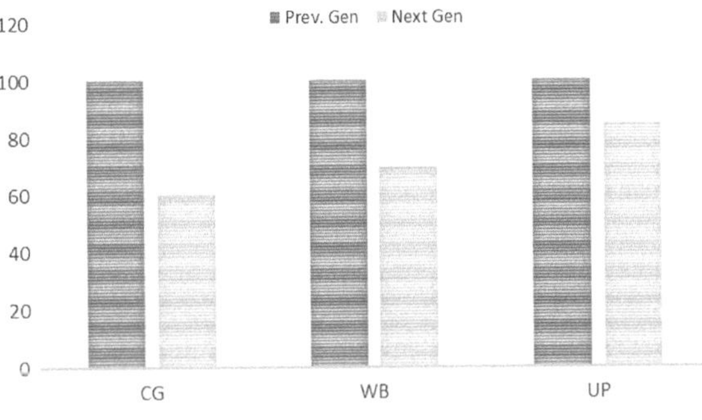

Fig. 2 Percentage retention of artisans

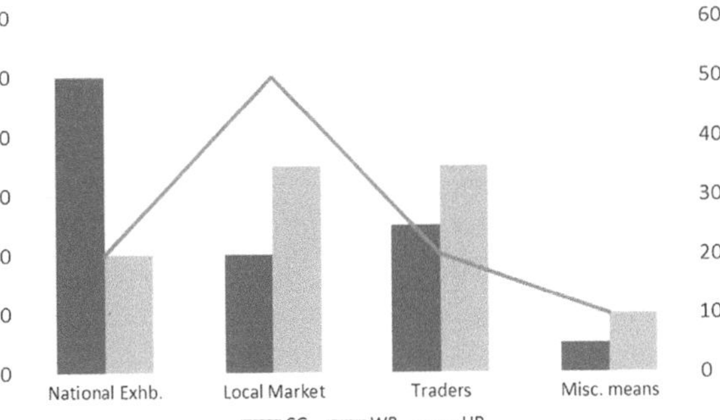

Fig. 3 Percentage source of income for artisans

5 Discussion

Since raw materials are the essential components required for the process of making a product are taken or procured locally, the quality of goods produced depends upon the quality of materials and subject to their availability. Same item or goods when produced in different regions separately varies greatly in terms of craftsmanship, material quality and fineness. When compared to the Dhokra items of both the place the quality of work in Bolpur is better than that of CG, because the artisans of Bolpur have access to modern tools etc. unavailability of modern tools leads to higher production time. The low literacy rate among the artisans hampers their growth in big way. Since, a very long channel exist which has to traversed by the product to finally reach the consumer market, only a fraction of the actual selling price of the product reaches the artisans, a major share is been taken by the middleman and the traders or entrepreneurs. The craftsman who are versed with technology are using it as a medium to diversify their business and make more money. Since their children see them earning meagre amount to sustain the livelihood of the family, they are turning to different fields of profession in search of work and money, hence drifting away from the cultural heritage of workmanship. The issue at hand is not to increase the productivity of the artisans (although that is also needed), but the focus is how to make them aware of the markets and changing needs of the consumers. If their production rate increases, they would have too much products to stock and very less market to consume them. For ex-if the artisans continue to build the artefacts as they were till now, there is very less possibility that the same would be widely accepted by the people, reason being their main stakeholders are the middleclass people who don't have the luxury of having very open or big spaces in their home now a days. The size of the homes are getting smaller and smaller, if an artefact would be able to serve more than just one purpose then logically it would make more sense in buying that, rather than just consuming space.

The other thing which has a very big potential in this sector is its merger with the e-commerce industry. If we could propagate the work of artisans by means of the e-commerce players of the industry, we would be giving them a big boost in terms of market exposure. The Indian e-commerce industry is one of the fastest growing sectors in the world. The early we tap into it the better would be long time results. Some of the people in the higher ranks within the govt. supported the thought of the merger. This is where the designers have a big potential to contribute to sustainability of the craft.

For artisans of CG the National Craft Exhibition remains the main source of income for them. They usually attend 3–4 exhibitions in a year and try to build the contact there so that they can get orders from different places.

Because of the close vicinity of major markets in Bolpur, the shola pith and Dhokra are more organized. Also the Artisans of Bolpur have higher literacy levels so they are aware of the trends and role of middleman is reduced thereby maximizing their profit share in the final cost of the products.

6 Conclusion

Since there has been a steep decline in the number of artisans in India due to lack of earning, which suggests that the market for the crafts needs to be expanded and the consumers and the artisans needed to be brought closer by minimizing the role of middleman. If middleman's share of the product pricing directly goes to the artisans, there would be a big leap in terms of monetary gain that they'd experience. The more ambitious option which could be taken would be to educate the artisans to such a level that they become conversant with the current needs and trends of the society and change their designs accordingly so as to give the user options to buy from, hence ending the saturation which is prominent as of now.

References

1. Bhatt, J.: Philosophy and practice of crafts and design. In: Seminar Magzine. http://www.india-seminar.com/2007/570/570_jatin_bhatt.html (2007)
2. Dak, T.M.: Rural Industrialisation: Challenges and Responses, pp. 23–24. North Book, Delhi (1989)
3. Chamikutty, P.: Problems facing the Indian artifacts sector. http://social.yourstory.com/2013/06/crafting-a-livelihood-a-snapshot-of-the-indian-artifacts-sector-2/
4. Shivakumar, G.: To the aid of crafts. http://www.thehindu.com/features/friday-review/art/to-the-aid-of-crafts/article5003961.ece

Sustainable Supply Chain in Product Development

Srinivas Kota and Kirthi Bandi

Abstract There exists an urgency for organizations all over the world to consider sustainable supply chain strategy in product development. Every firm needs to provide a safe working environment and reduce its ecological footprint as well as ensure returns to its investors. This calls for cross-functional integration, better decision making and performance improvement. Every stakeholder in the supply chain is responsible to ensure its' sustainability. Several incentives can be found in adopting a sustainable supply chain management strategy. Sustainable supply chain management pushes the system towards forethought, exploration and data analysis in business so as to measure up to the expectation of customers and flourish in the long run. The focus of this study has been on why a company requires sustainable supply chain management and how it would benefit by implementing a sustainable supply chain management strategy. The core aspect was to analyze the advantages of sustainability in a supply chain and how everybody would be motivated to work towards it. Main outcome from the research is that stakeholders in different domains are more or less similar but the challenges and motivations of stakeholders are different in different domains. It is required to consider the domain specific characteristics for inclusion of sustainable supply chain in product development. A conceptual strategy is proposed to include sustainable supply chain in product development.

Keywords Supply chain · Stakeholders · Product development · Sustainability

1 Introduction

A supply chain is a sequenced network of resources and activities that support the production or distribution of a good or service. It is characterized by different stages and is essentially directional. The vital activities of a supply chain are manufacturing,

S. Kota (✉) · K. Bandi
Department of Mechanical Engineering, BITS Pilani, Pilani, Rajasthan, India
e-mail: srinivas.kota@pilani.bits-pilani.ac.in

A. Chakrabarti (ed.), *ICoRD'15 – Research into Design Across Boundaries Volume 2*,
Smart Innovation, Systems and Technologies 35, DOI 10.1007/978-81-322-2229-3_14

demand forecasting, purchasing, customer relationship management and logistics. Supply chains may be interlinked between companies too. A supply chain accounts for more than 50 percent of the total expenses for manufacturing companies. Supply chains have become complex and insubstantial due to globalization. A sustainable supply chain is no longer a viable option but an indispensable necessity for the success of an organization [1].

Supply Chain Management (SCM) is the co-ordination of business functions within an organization and with its partners in order to provide goods and services in order to fulfill customer demand responsively, efficiently and sustainably. It encompasses three major aspects, namely, Strategy, Planning and Operations. Strategy signifies building a supply chain for the good or service to be provided. Planning involves leveraging the supply chain to aid medium term goals. Operations are surveillance and assistance to the supply chain plan. There exists an urgency for organizations all over the world to implement a sustainable supply chain strategy. Every firm needs to provide a safe working environment and reduce its ecological footprint as well as ensure return to its investors [2].

The focus of Supply Chain Management has been a blend of market responsiveness and cost strategy in line with the firm's competitive plan. The rising trend is to include sustainability viewpoints in SCM. Sustainability means to keep the process alive and going. Analysis of a supply chain model is required to review our decisions on sustainability. This involves expounding the problem, stating assumptions, collecting data, building the structure and obtaining results. Metrics and logic play an important role in measurement. A considerable amount of imposition is necessary to implement Sustainable Supply Chain Management (SSCM). This could be done through voluntary participation of the company, internal incentives or validation by extraneous agents [3].

A supply chain expects a blend of enthusiasm and commitment from all the links of the chain, that is, the stakeholders. The key to a successfully run a supply chain is to maximize the positive inputs of the each participant and reduce potential inimical impact [4]. The first step in stakeholder management is to identify the stakeholders and the influence they exert on the supply chain. Work effectiveness, trust-building and loyalty could be developed by motivating the stakeholders.

Consolidation of the existing work to define sustainable supply chain is being carried out. The stakeholders responsible are identified throughout the supply chain across different domains. Each stakeholder has certain unique traits and interests and the ones which are relevant to sustainability were found. For example, a supplier has to be responsible for information flow downwards as he is at the starting point whereas distributors count on mutual trust and commitment for information exchange. Encouraging as well as discouraging factors were distinguished to aid motivators and reduce barriers. The idea is to develop a strategy to analyze the advantages of sustainability in a supply chain and support each stakeholder to maximize the advantages.

2 Objectives and Methodology

2.1 Objective

- To understand the importance of sustainable supply chain management
- To obtain a comprehensive view of the incentives of all the stakeholders in a supply chain

2.2 Methodology

The study of Sustainable Supply Chain Management (SSCM) necessitates knowledge of the purpose. The objectives are fulfilled by the following methodology.

- Literature review on need for sustainable supply chain management
- Study of SSCM and its forms in different domains
- Study on impetus of participants in the supply chain
- Analysis of benefits of stakeholders in adopting a SSCM strategy

An assessment of the research conducted till date about the need of SSCM has been summarized in Sect. 3. The requirements of a supply chain vary according to the industry. The intricacy of the chain will change with the size of the business and the quantity of goods manufactured. A review of the scenario in some major domains has been given in Sect. 4 and Sect. 5. A detailed analysis of the motivating factors of the stakeholders has been elaborated in Sect. 6. Major challenges for sustainable supply chain in different domains is discussed in Sect. 7. Suggestions for following up a sustainable supply chain strategy have been proposed in Sect. 8.

3 Need for Sustainable Supply Chain Management

Supply chains should be consciously managed. Sustainable practices in supply chain management provide the ability for businesses to make better decisions to reduce costs, improve productivity, support growth and take effective long term decisions.

Supply chains incur expenses to transfer information, produce goods, store them, transport them and pay for services. The total cost of the supply chain tends to increase due to inflating capital costs and freight charges especially in global businesses. However, planning saves time, money and energy. In case of internationally spread supply chains, every stakeholder performs a wide range of functions such as order fulfillment, international procurement, acquisition of information technology, manufacturing, faster and reliable delivery of products and customer service. Supply Chain Management involves extensive research and data analysis to perform all the activities efficiently. Secondly, the whole supply network could improve its ability to meet expectations of users in terms of quality and supply chain practices.

Sustainable supply chain management deals with the dialogue companies create with suppliers in order to prevent violations of fundamental human rights and international environmental standards. The three basic dimensions to incorporate sustainability in the supply chain are society, environment and economy [5]. The fluctuation in the world economy, increasing levels of carbon footprint and unfair employee policies call for remodelling the existing supply chains all over the world.

Most companies recognize that compliance with emissions reporting requirements is only one aspect of more responsible environmental and social stewardship. Organizations are now working collaboratively across the supply chain so as to cut costs, manage risks better, generate new sources of revenue and boost their brand value.

4 Supply Chain in Different Domains

A thorough literature review of different case studies in various domains has been conducted and the salient points have been summarized in this section. The various domains are fast moving consumer goods, aerospace, information technology, pharmaceutical, automobile and small & medium enterprises (SMEs).

Fast moving consumer goods companies like Unilever targets growth in emerging markets and emerging movement of highly conscious consumers to achieve sustainability in its supply chain. Owing to its vast product range, this domain needs to have a clear idea of its' supply chain, and key elements in the supply chain have to be identified. This domain focuses on cost of the product, sustainable return on investments, consumers' willingness to pay and product differentiation to figure out areas for implanting sustainability [6].

In ***Aerospace domain*** companies like Boeing, lean supply chain management is the core policy. It focuses on process improvements to reduce the cost and lead time. Higher responsibility is transferred to the suppliers as the manufacturers concentrate on efficient design and assembly technologies. There is a trend of dependency on third party organizations for maintenance activities due to high expectations in the logistics part of the supply chain [7].

Information Technology projects have always given importance to the technology adopted at every point of the supply chain. The role of stakeholders has been undermined. Besides the strategic factor and engineering ergonomics factor, the effective involvement of stakeholders can aid in improving the supply chain. A complete remodelling is suggested for efficiency.

Pharmaceutical domain has key stakeholders as manufacturers, wholesale distributors, pharmacies and pharmacy benefit managers with pharmaceutical manufacturers being the most influential of the lot [8]. The pricing of the drug is highly scrutinized and holds the interest of every stakeholder.

Automobile domain gives emphasis on supply chain management and considers managing cost and efficiency as pivotal points. Owing to the wide spread in the location of customers, it relies on an established distribution network. Lead time is very little as the competition is very high and need most production processes as assembly line based. Third party logistics is the preferred option of most companies [9]. Currently, the industry is focusing on sustainable supply chains due to pollution and global warming.

In ***Small and Medium enterprises***, the focus is to co-ordinate together to achieve their targets and at the same time retain their identity. Collaboration is necessary due to the numerous stakeholders in the supply chain [10].

5 Stakeholders with Respect to Different Domains

Derik H.T. Walker defines stakeholders in the article, Stakeholders and the Supply Chain in the book [11] as 'Individuals or groups who have an interest in some aspect of rights or ownership in the project, and can contribute to, or be impacted by either the work or the outcomes of the project'. A comprehensive but not exhaustive list of the important stakeholders in each of the above discussed domains is given below:

Fast Moving Consumer Goods

- Manufacturers
- ***Distributors***
- Consumers

Aerospace

- ***Suppliers***
- Manufacturers
- Maintenance operators
- Distribution network
- Customers

IT Sector

- Shareholder
- Staff
- Government
- Competitor
- ***User***

Pharmaceutical Industry

- ***Manufacturers***
- Distributors
- Pharmacies
- Consumers

Automobile

- ***Vendors***
- Assembly operators or manufacturers
- Logistic providers
- Distribution network (Depots, Forwarding Agents, Dealers, Sub-dealers)
- Customers

Small and Medium Enterprises

- Co-ordination amongst all groups

From the above list we can see that there are different names given for similar stakeholders but ***suppliers***, ***manufacturers and distributors*** and ***consumers*** were present in all domains with emphasis on a particular or a group of stakeholders in different domains. The influencing stakeholder in each domain is given in bold.

6 Motivation of Key Stakeholders

As aforementioned, the stakeholders characteristics and interests have been explained below:

6.1 Suppliers

A supplier in a supply chain is an enterprise that contributes goods and services to the next link in the chain which could be a company or a group of individuals. A supplier, also known as a vendor, provides the inventory or raw material to a manufacturer.

6.1.1 Characteristics

- A vendor needs to meet the minimum requirements of customers. An efficient supplier would make sure that the material is available in surplus.
- A vendor should be flexible enough to accommodate the needs of a customer. A responsive supply strategy would ensure the sustenance of the vendor.
- A supply chain starts with the vendor in most cases. As a forerunner in the chain, the supplier needs to be responsible to corroborate information flow down the supply chain. Communication amongst stakeholders is a crucial factor in supply chain management.
- A supplier should strive to be innovative. The market connects with a creative supplier to increase the sale of goods.

6.1.2 Motivating Factors

- Diversified Customer Base [12]—A high level of competition amongst manufacturers will push vendors towards responsible sourcing. More number of buyers would mean higher profits.
- Concentrated Supplier Base—If suppliers are shared by manufacturers, they would be compelled to provide the goods on time periodically. A diffuse supply base will induce callousness in the vendor.
- Inflexible Supply Chain—The supply chain structure has a significant impact on the suppliers to source efficiently. A manufacturer willing to allow for shortage of material supply would be a deterrent to effective sustainable SCM.
- Manufacturer's role—The manufacturer being the next link in the supply chain can aid by proper payment of bills, providing certain contract security, issuing public statements of recognition through business practices like supplier evaluation and supplier performance management.

6.2 Manufacturers and Distributors

A manufacturer is an entity that produces goods through a process involving raw materials, components or assemblies. A distributor is an intermediary who makes the product available for use by a consumer. Manufacturers and Distributors, being the middle men in the supply chain work together to meet consumer needs.

6.2.1 Characteristics

- Manufacturers and distributors give high priority to the requirements of a user. Regular checks on product quality, agile response to demand and accelerated innovation are some of their strategies.
- They work on expanding into popular markets and new domains. A good reputation is necessary for their prosperity.
- These stakeholders control sourcing and procurement costs. They cut down expenditure and aim at minimizing risks.

6.2.2 Motivating Factors

- Cost Savings and Efficiency Optimization- Lower costs of components and complete utilization of resources is a motivator to stakeholders. Manufacturers are extremely conscious about incurring costs.
- Sustainable Supply Chain Integration—Strategy formulation should involve information sharing across the supply chain. Lack of knowledge on sustainability

and understanding of best practices mislead stakeholders. A holistic approach in planning is beneficial.
- Effective Marketing—The focus of distributors is to sell their goods successfully. They choose suppliers willing to help them to become renowned. Suppliers who educate the distributors regarding the product and keep them informed about industry trends are in a position to encourage them.
- Mutual trust and commitment—Business relations between stakeholders will continue on based on performance measures, however, certain amount of reliance is required from both sides.

6.3 Consumers

A consumer, generally the last link in the supply chain, obtains goods or services for direct use or ownership. He pays to consume the goods produced. He is the motivating factor for most other links of the supply chain.

6.3.1 Characteristics

- Low prices—The monetary value of a commodity is extremely important to the user. He expects to get the best deals from the lowest prices.
- Proper working standards—Consumers hope that their products were manufactured in a healthy and positive environment. They assume minimum working conditions to be available at the product workplace.
- Access to product information—Users demand to be informed about the product quality and potency to protect themselves from unfair trade practices.
- Extreme purchasing behavior—A consumer would go to any extent to satisfy his requirements. He might not be ready to spend 5 % more on a normal commodity but may be willing to spend 20 % more on a luxury commodity to please himself.

6.3.2 Motivating Factors

- Ethical factor—A product made by workers who have been provided with the basic amenities would be preferred by a consumer over a product made by exploitation of labour. For instance, nobody would like to buy crackers made by child workers during the festival season.
- Advertisement—Marketing by distributors influence the purchasing behavior of users. An appealing advertisement can make a lot of difference to the preferences of a consumer.

- Feedback—Consumers expect responsiveness from the higher end of the supply chain. Manufacturers sensitive to the suggestions and complaints of a consumer will do well in their business.
- Safety—Consumers are concerned about the safety and security that a product has to offer. Most people would prefer healthier food given a choice.
- Environmental impact—The present day consumer is interested to buy an environmentally friendly product as he understands the importance of protecting his surroundings.

7 Challenges in SSCM

Domain specific problems of the chosen sectors have been enumerated:

- Consumer goods industry
  - FMCG inventory with the company is more than the required level [14]
  - Improper handling of different varieties of goods
  - Poor performance of distributors and logistic providers
- Aviation industry
  - Timely delivery of the plane parts by the suppliers is not efficiently coordinated
  - Maintenance of the airplanes is a huge challenge in this monopolistic industry
- IT industry
  - Unrealistic assumptions about technology solutions
  - Information flow across the supply chain is not regulated
- Pharmaceutical industry
  - Regulatory framework established by the government is not favourable
  - Adulteration and counterfeiting of products by anti-social elements
  - Outdated manufacturing sites
- Automobile industry
  - Too long material lead times leading to obsolescence [15]
  - Lack of sustainability in terms of capacity and liquidity
- Small and medium scale enterprises (SMEs)
  - Short product lifecycle is a deterrent to opportune working of the supply chain [13]
  - Lack of understanding of full capabilities of suppliers and service partners

8 Adopting a Sustainable Supply Chain Management Strategy

A highly effective supply chain operation focuses on sustainability issues in trade and marketing. An organization needs to connect its competitive strategy with its' supply chain strategy using sustainability.

The factors affecting an organization's supply chain are demand, supply and the societal conditions. A Sustainable Supply Chain Strategy (SSCS) should be unambiguous to augment learning within the corporation. Supply chain managers should be able to influence the employees to work in this direction. Increase in research and development will help the company to identify and evaluate the shift in policies in the future. It is also important to gauge the economic angle of an operation. The challenge in adopting a SSCS is to establish pragmatic targets balancing social, economic and environmental ideas and the company's mission.

The following approach is proposed to achieve a sustainable supply chain:

- **Assess the existing SC**—The existing supply chain of the company has to be evaluated so as to understand prospective intra and inter organizational goal variances. The current practices, resources and capacity need to be judged.
- **Assess the business environment**—Special importance has to be given to the input resources, namely, fuel, energy, shareholders and stakeholders.
- **Measure the effect** of the supply chain on environment, economy and society.
- **Redesign or modify the supply chain strategy incorporating the changes required for sustainability**—A collaborative approach should be adopted

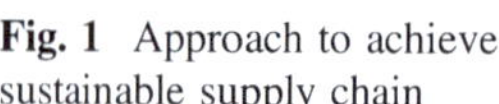
Fig. 1 Approach to achieve sustainable supply chain

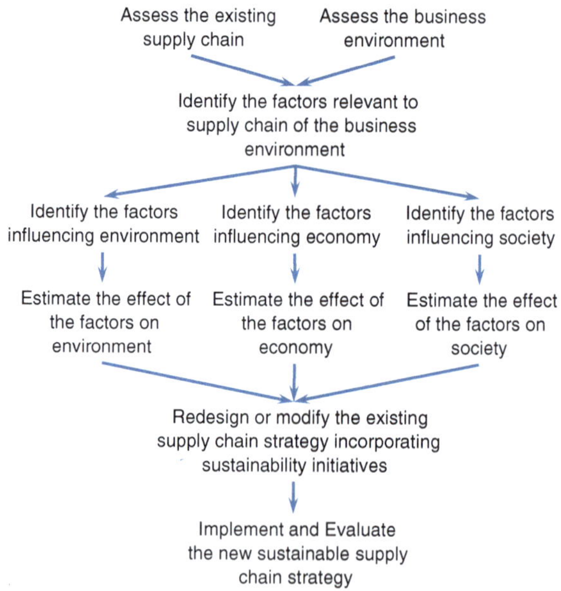

while remodelling. A shift to aggressive strategies will ensure long term benefits.

- **Implement and Evaluate**—Use of a sustainable supply chain measure to assess the productivity, development of technology, infrastructure and energy resources along the supply chain. Figure 1 shows the approach in detail.

9 Conclusion

This paper has sought to establish the exigency for organizations to adopt a sustainable supply chain strategy. The factors which encourage important stakeholders to participate in such a scheme have been enlisted. Complex patterns of customer demand and global competition make commodity prices unpredictable. Companies incur a lot of expenditure on logistics and transportation. Consumer expectations about product quality are also high as they have up to date information. Higher level issues such as downturns in the global economy, shortage of critical resources, political instability, regulatory requirements and natural disasters are major setbacks to the supply chain. Cost pressure, increasing volatility of customer and environmental concerns are only few of the challenges. Further scope of research would include examining the roadblocks of the stakeholders in carrying out their functions and providing solutions to domain specific problems.

References

1. Gupta, S., Palsule-Desai, O.D.: Sustainable supply chain management: Review and research opportunities. IIMB Manage. Rev. **23**, 234–245 (2011)
2. Beske, P., Land, A., Seuring, S.: Sustainable supply chain management practices and dynamic capabilities in the food industry: a critical analysis of the literature. Int. J. Prod. Econ. (2013) (accepted in Dec 2013)
3. Cholette, S.: Sustainable Supply Chain. San Francisco State University USAS, Basics Decision Sciences (2011)
4. Borade, A.B., Bansod, S.V.: Domain of supply chain management—a state of art. J. Technol. Manage. Innov. **2**(4) (2007)
5. Cetinkaya, B., et al.: Sustainable Supply Chain Management—Practical Ideas for Moving Towards Best Practice. Springer, Berlin (2011)
6. Dhawan, E., Goodman, E., Harris, S., Mitchell, C.: Unilever and its supply chain: Embracing radical transparency to implement sustainability, 15.915. S-Lab Final Report, May 12, 2010. Available at accessed on 10 Mar 2014.http://mitsloan.mit.edu/actionlearning/media/documents/s-lab-projects/Unilever-report.pdf
7. Cizmeci, D.: An examination of Boeing's supply chain management practices within the context of the global aerospace industry. Thesis, Massachusetts Institute of Technology, Engineering Systems Division (2005)
8. Supply Chain Design for the Pharmaceutical Industry. http://www.llamasoft.com/wp-content/uploads/2013/05/DS-LLamasoft-for-Pharma-US.pdf (2013). Accessed on 10 Mar 2014

9. Rao, R.V.: SCM in two wheeler industry—a study on hero Honda and Bajaj auto SC practices. Int. J. Res. Commer IT Manage. **2**(3) (2012)
10. Wolf, Julia: Sustainable supply chain management integration: a qualitative analysis of the German manufacturing industry. J. Bus. Ethics **102**(2), 221–235 (2011)
11. Derek, H., Walker, T., Rowlinson, S.: Procurement Systems: A Cross-Industry Project Management Perspective. Taylor & Francis, London (2007)
12. Guo, R., Lee, H., Swinney, R.: The impact of supply chain structure on responsible sourcing. In: Proceedings of M&SOM Sustainable Operations SIG Conference, INSEAD, Fontainebleau, France, 28 July 2013
13. Marchese, K., Paramasivam, S.: The Ripple Effect, How Manufacturing and Retail Executives View the Growing Challenge Of Supply Chain Risk. Deloitte Publication, New York (2013)
14. Raghuram, G.: Integrated Supply Chain Challenges for FMCG and Organized Retail. IIM Ahmedabad. http://www.scribd.com/doc/234745267/2006-G-Raghuram-s-Presentation-at-RAI-Retail-ITW-1-Ppt (2006). Accessed on 10 Mar 2014
15. Naude, M.J., Badenhorst-Weiss, J.A.: Supply chain management problems at South African automotive component manufacturers. S. Afr. Bus. Rev. **15**(1) (2011)

Use of MFF and Concurrent Engineering to Develop a Sustainable Product—Radical Redesign of Flushing System

Nikola Vukašinović, Žiga Zadnik and Jože Duhovnik

Abstract The paper shows the approach to radical product redesign of sanitary flushing equipment which is conservative market niche product. The approach demonstrates use of the Matrix of Functions and Functionalities (MFF) and elements of concurrent engineering to achieve a significant reduction of necessary resources during whole product's lifecycle: production, use as well as its disposal. This approach led to a considerable reduction of number of different parts and materials, as well as total number of parts and overall mass of the product. The result was a product proposal, which fulfills the same functional requirements as exiting products using different principal solutions. This change of exploited working principles opened new possibilities for a significant reduction of necessary amount of flushing water, thanks to the energy stored in the new flushing device. The proposed solution was experimentally confirmed also with working prototype.

Keywords Concurrent engineering · MFF matrix · Radical innovation · Sustainability · Design for manufacturing · Design for assembly

1 Introduction

1.1 The Project

In 2011 we were approached by an international company that produces sanitary equipment including traditional market segment of toilet flushing systems and devices. Due to the market demand, which calls for compatibility of different toilet elements across different models and even manufacturers, this market segment is strongly regulated and directed by different international and national standards

N. Vukašinović (✉) · Ž. Zadnik · J. Duhovnik
LECAD Lab, Faculty of Mechanical Engineering, University of Ljubljana, Ljubljana, Slovenia
e-mail: nikola.vukasinovic@lecad.fs.uni-lj.si

A. Chakrabarti (ed.), *ICoRD'15 – Research into Design Across Boundaries Volume 2*, Smart Innovation, Systems and Technologies 35, DOI 10.1007/978-81-322-2229-3_15

(e.g. EN 14055:2006 or EN 997:2004). However, the company recognized that it is difficult to remain competitive in global environment in such conservative market niche without having constantly innovative product portfolio. Due to our common successfully finished projects in the past, they came to us expressing a wish for development of a new approach which would solve open problems concerning their existing toilet flushing systems. Water consumption, flushing effectiveness and noise are just major of them. Since the company gave us only few basic limitations and demands regarding the product (e.g. necessary compatibility of the new system with existing infrastructure in majority of building in the region), we were able to draft the new product development process from the very beginning, opening the opportunity to radically reconsider traditional toilet flushing principles. However, the development time from the agreement to the prototype which would demonstrate working principles was quite short, limited to 6 months.

For that reason we decided to arrange product development phases in a way to implement elements of concurrent engineering, especially for developing, designing and detailing different components of the concept. This approach resulted in reduction of different components of the concept and in expanded product portfolio with different valve variants which could be used for various other purposes as well.

This project was also good opportunity to test newly developed method for a systematic and computer aided development of design concepts called Matrix of Functions and Functionalities. The MFF method was developed to solve problems of managing large number of concept solutions emerging from traditional morphological matrices [1] based on a mathematical model and predetermined conditions instead of relying only on the designer's intuition as in traditional methods.

Our hypothesis claims that the MFF method can be useful during the development process of radical design or radical redesign of products. The second hypothesis claims that it is crucial to identify and modify the main functions of the system in order to achieve radical (re)design of a system or a product.

2 Method

2.1 Concurrent Engineering Product Development

The basic premise for concurrent engineering (CE) revolves around two ideas. The first is the idea that all the elements of a product's preliminary design, from function to functional requirements, should be taken into careful consideration and, based on the parallelization of tasks, be carefully managed [2] The second idea is that the subsequent design activities should all be occurring at the same time, or concurrently. The overall goal is that the concurrent nature of these processes significantly increases the productivity and the product quality due to established two-way communication among concurrent activities [2–4] (Fig. 1).

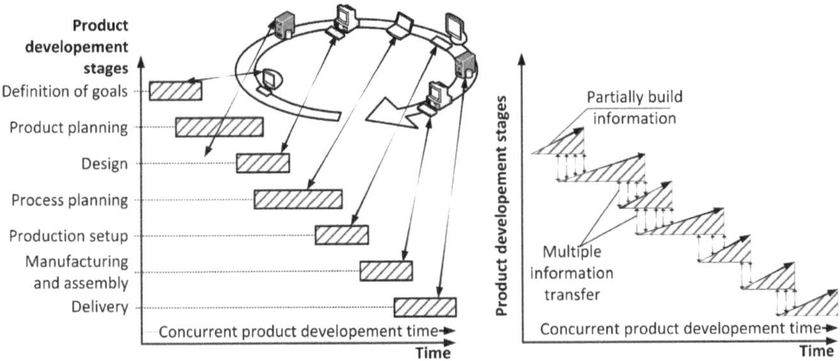

Fig. 1 Concurrent product-development stages during time [5]

Research and development activities within the product-development process have their own characteristic and distinctive features, dominated by unpredictability, creativity, mentality, and abstraction [3]. As a result of these features it is difficult to thoroughly describe, develop, and implement the design process, especially in the initial phases of application development [6]. In order to reduce the time and costs involved in solving, arranging, and improving the functional requirements, there is an increasing demand for new methods that would enable the concurrent and parallel solving of objectives [7].

2.2 Matrix of Functions and Functionalities

The basic morphological matrix [1] represents the basis for the development of the MFF model. With a small number of rows and columns, the morphological matrix can yield a large number of solutions, which often makes them poor and unsuitable.

For this reason, the MFF is based on the interaction between the functional requirement and the functionality, and is developed according to specific criteria, while we are striving for a solution that would be, in relation to the morphological matrix, built and determined on the basis of a mathematical model and predetermined conditions, and therefore, not only on the designer's intuition.

The MFF is a method that is a tabular representation of the bindings/links between the functional requirements and functionalities (Fig. 2) [8, 9]. It can be devised if the key elements are known, such as the initial functional requirements and functionalities. The functional requirements are derived from the market requirements and represent the most important attributes of the requested system, that is, functions, while the functionalities are represented by technical systems [10] or shape models which either in part or as a whole fulfill the required functions.

The functional requirements: The functional requirements or the desired market requirements (Fig. 2) represent the first possible information and abstract desires

FUNCTIONAL REQUIREMENTS	FUNCTIONALITIES/SOLUTIONS						
	TS_1 →	← TS_2 →	← TS_3 →	← … →	← TS_j →	← … →	← TS_m
Functional requirement - FR_1 ↓ [Suggested solution R_{FR_1}]	$X_{FR_1 TS_1}$	$X_{FR_1 TS_2}$	$X_{FR_1 TS_3}$	…	$X_{FR_1 TS_j}$	…	$X_{FR_1 TS_m}$
Functional requirement - FR_2 ↑↓ [Suggested solution R_{FR_2}]	$X_{FR_2 TS_1}$	$X_{FR_2 TS_2}$	…	…	…	…	$X_{FR_2 TS_m}$
Functional requirement - FR_3 ↑↓ [Suggested solution R_{FR_3}]	$X_{FR_3 TS_1}$	…	…	…	…	…	⋮
⋮ ↑↓	⋮	…	…	⋱	⋮	⋱	⋮
Functional requirement - FR_i ↑↓ [Suggested solution R_{FR_i}]	$X_{FR_i TS_1}$	…	…	…	$X_{FR_i TS_j}$	…	⋮
⋮ ↑↓	⋮	…	…	⋱	⋮	⋱	⋮
Functional requirement - FR_n ↑ [Suggested solution R_{FR_n}]	$X_{FR_n TS_1}$	$X_{FR_n TS_2}$	…	…	…	…	$X_{FR_n TS_m}$

Fig. 2 Mathematical MFF model

about a future product (e.g. the intake of air and the ease of suctioning). Functional requirements are similar to functions, except that they do not have known solution in the form of functionality; therefore, they represent only abstractions and desires. They can be derived from various sources, using various methods, e.g. QFD [11, 12]. The benefit of applying QFD into this process is the possibility to get well structured and weighted functional requirements, however, we should be careful not to prejudice some of the functionalities/solutions used in the QFD to early.

The functionalities: Functionalities are presented as possible realizations or technical systems. Each functionality is an individual unity, outwardly represented by a main title (e.g. Blower or Wire) while inwardly it contains precisely defined functions with the corresponding parameters. As such, the described functionalities represent a source of information for finding answers to functional requirements [13–15]. Functionalities could be found either intuitive, based on designers experience and knowledge, or systematically, based on knowledge databases, catalogues or using methods for searching and generating principal technical solutions, e.g. TRIZ [16, 17].

2.3 Implementation

In the initial step we had to analyze existing solutions and standards of concealed flushing units. We identified their main and supportive functions and functionalities. This analysis revealed the main functional requirement (level 1) to be flushing effect that is the function of two variables: water quantity and water energy:

$$FR_1 = f(V_{fl},\ E_{fl}) \tag{1}$$

The other functions (deriving from functional requirements) were identified as supportive functions (level 2 and higher) that are necessary to establish the working flushing system. They could be optimized and rearranged, but would provide only minor advantage over the original concept.

This deductively leads to the need for the radical redesign, which can be obtained only by significant modification on the level of the main functions. From Eq. (1) it is easy to identify the idea, that reduction of water usage needs to be compensated by increasing the other component, which is the energy stored in the water.

The further research focused on finding potential sources to increase the water energy. In the past the increase of water potential energy was mostly used for these purposes, i.e. putting the reservoir on a higher level above the toilet bowl. However, we identified another energy source, which is mostly neglected in standard flushing units; this energy source is the pressure within the water pipeline system. This huge amount of energy is being annihilated every moment when water is being spilled from the pipeline system into the atmospheric water container. Further research showed, that standardized pipeline pressure in Central Europe is between 4 and 10 bars. Therefore we decided to design a system that would work on 4 bars of pressure to fit with most of the pipelines.

The further step was the rearrangement and reduction of supportive functions before designing a working prototype. We identified three functional requirements that need to be fulfilled in order to satisfy the main functional requirement of the system: pressurized water container, filling unit and water release unit. Filling unit consists of several (level 3) functional requirements: pressure reduction to 4 bars, back-flow preventing, automatic opening and closing. We also identified (level 3) functional requirements for water release unit: 2-mode working (small flush–big flush); automatic closing after the flushing action, which can be associated to the pressure drop in the container; easy electronic or manual activation.

Since we had to develop both valve units anew, the research focused on similarities between both units trying to develop as many common parts as possible. This led us to use concurrent engineering methods, to drag parallel development of both units, sharing knowledge, research results and ideas during the development process of both units. Although the functions of two units had nothing in common, we managed to develop the concepts, which have two of their largest components in common.

3 Results

According to the before described procedure we attempted to develop a substitute design of a flushing unit which could be mounted in place of existing concealed flushing units models using standard company's manual activation panels while only minor mounting modifications would be necessary. The result of the final concept is depicted on Fig. 3.

Fig. 3 3D CAD model of flushing system concept that was radically (re)designed using MFF and concurrent engineering design approach

The final resulting concept consists of water pressure tank with 4.5 L total volume (3.0 L of water, 1.5 L for pressurized air) with three valve attachments on its bottom part. First valve from the left (Fig. 3) is intake valve that limits the maximum water container pressure and volume and prevents backflow. The other two are water release valves; central release valve for full flush while the right valve serves small flushing operation. We decided for two separate valves to reduce complexity of the concept since the water release valves differ only in preset spring stiffness that closes the valve when appropriate pressure drop occurs in the water container.

All valves are permanently submerged to provide reliable operation, to avoid noise, which would be generated if water would be spilling from height onto the bottom of the container in the filling phase of the operation; to reduce the volume of the concept and to simplify the production process.

The comparison of the new concept and the company's best-selling concealed flushing model (Table 1) showed several great improvements of the new concept compared to the standard concealed flushing systems. Since most of the concealed flushing systems are built similarly, we assume similar advantage also towards other models of various manufacturers.

Table 1 Comparison between LIV-Duo concealed toilet flushing system and flushing system concept that was radically (re)designed based on MFF and concurrent engineering approach

	LIV-Duo	Concept	Difference
Filling unit			
Mass (g)	200.83	44.68	−156.15
Number of parts	12	7	−5
Number of different materials	8	3	−5
Water release unit			
Mass (g)	248.02	51.84	−196.24
Number of parts	7	11	+4
Number of different materials	4	4	0
Water container			
Mass (g)	2,905.85	425.40	−2,480.45
Number of parts	8	4	−4
Number of different materials	4	2	−2
Total			
Mass (g)	3,354.70	599.03	−2,755.67
Number of parts	28	33	+5
Number of different parts	28	15	−13
Number of different materials	8	4	−4

The advantage in terms of reduced mass per unit as well as in term of number of different materials being used was obtained at all three main functions of the system: filling, water storage and water release.

Besides that, we reduced the total number of parts per two of three units: filling unit and water container, while release unit gained 4 more parts compared to the release unit of standard flushing system.

On the level of a whole system we saved 2,755.67 g or 82 % of mass compared to the existing flushing unit, the total number of used parts increased for 5 items, however, the total number of different used items was reduced by almost 50 % due to the use of common elements for different units and parallel use of two water release valves.

For environmental reasons, to improve recyclability of the product at the end of its lifecycle, it is recommended to reduce the number of different used material. The number of different used material in new concept was reduced for 50 % compared to the reference flushing system.

3.1 Prototype Testing

At the final stage we built functional prototype (Fig. 4) of the system, to demonstrate the working principles and functionality of the solution. The prototype was

Fig. 4 Working prototype of the concept in its testing environment

used for various tests to evaluate filling times, emptying time, and flushing performance according to valid European standards for flushing systems (bowls and flushing units).

The prototype testing results showed functionality of the radically developed concept. All components performed as predicted, providing reliable filling, emptying and pressurized water-storing functions. Besides that, both components proved to be reliable, not causing noticeable hydraulic shocks. Their modular and simple construction thus promises various additional possibilities of usage.

The flushing effect test was performed on standard off-the-shelf toilet bowl with syphon volume of 1.4 L and toilet entrance and rim space of 0.8 L. Testing was done based on two standard testing procedures: (1) flushing 12 dry toilet paper pieces that are separately thrown in the bowl; (2) flushing 50 standardized plastic balls (SIST EN: 997:2003 standard). We tested concepts with two different flushing water quantities: 3.0 and 4.6 L. Testing procedures also included reference measurement with classical flushing unit (3 and 6 L container). Another measurement was performed, using the same flushing unit, but with released flushing water accelerated with additional applied energy by injecting small amount (0.39 L) of pressurized water directly from pipelines, through Venturi nozzle at the exit of the water container.

However, the testing results were only partially successful. The result of toilet-paper flushing with 3 L of water were better on the new concept compared to standard flushing system, but significantly worse than using standard 6-L flushing system. We suspect that the use of standardized toilet bowl which is designed for flushing with a sudden wave of large quantity of water is the reason for that.

Namely, 0.8 L of water spills within the bowl rim, which means there is only 2.2 L of effective flushing water available. This amount of water must immediately replace all 1.4 L of contaminated water in syphon containing also paper pieces, while this is almost impossible due to bowl's geometry. Therefore only large quantity of water can do this job.

Further tests with plastic balls standard testing method indicated that using pressurized water with standard bowls could even reduce the flushing effect compared to reference measurement, which was significant at 3-L flushing quantity. However, by increasing amount of pressurized flushing water to 4.6 L, the flushing effect has almost reached standard 6-liter flushing at atmospheric pressure.

4 Conclusions

The results of this research showed that the use of MFF method could support the development process of radically redesigned products. In this case it was particularly helpful to shift paradigm of thinking that water energy instead of water quantity can be used as flushing unit. It led to the realization that traditionally designed flushing cisterns literally annihilate enormous amount of pressurized water energy from the pipelines.

The prototype analysis showed several advantages including solving numbers of unresolved problems occurring with traditional flushing units. However, the new approach opened additional problems, which have to be resolved before the concept will become commercially attractive.

The benefits of the results are: significant noise reduction due to the closed container system and simplified and submerged water inflow; smaller number of necessary different components; smaller total number of used components; reduction of total weight; reduction of different used materials; reduction of water consumption.

The main identified but unresolved problems are the following: exploitation of water kinetic energy for flushing demands short water transport paths to the flushing area in order to reduce energy losses. Therefore, the water cistern must be placed near or integrated into the toilet bowl, while upgrade of existing systems according to valid ISO and EN standards was demonstrated as ineffective.

The additional problem of toilet bowls designed according to ISO and EN standards is in being effective only if sufficient amount of water is being used for flushing. Therefore it is impossible to use them effectively with less than 4.5 L flushing water due to large quantity of contaminated water in their siphons (1–2.5 L) that needs to be replaced during each flushing.

Therefore, we recommend using only specially designed toilet bowls with water reservoir integrated within the bowl. Development of such bowl can be large design and engineering challenge, but also an opportunity for a complete and integrated product system with the highest added value on the market.

References

1. Zwicky, F: The morphological method of analysis and construction (Courant anniversary volume), pp. 461–470. Interscience Publishers, New York (1948)
2. Duhovnik, J., Zargi, U., Kušar, J., et al.: Project-driven concurrent product development. Concur. Eng. Res. Appl. **17**(3), 225–236 (2009)
3. Duhovnik, J., Kušar, J., Tomaževič, R., et al.: Development process with regard to customer requirements. Concur. Eng. Res. Appl. **14**(1), 67–82 (2006)
4. Prasad, B.: Concurrent Engineering Fundamentals: Integrated Product and Process Organization, vol. 1. Prentice-Hall PTR, Upper Saddle River (1996)
5. Duhovnik, J., Starbek, M., Prasad, B.: Development of new products in small companies. Concur. Eng. Res. Appl. **9**(3), 191–210 (2001)
6. Kurtoglu, T.A.: Computational approach to innovative conceptual design. Ph.D. Thesis, The University of Texas at Austin, Austin (2007)
7. Hales, C., Wallace, K.: Design research reflections—30 years on. In: Maier, A.M., Mougaard, K., Howard, T.J., McAloone, T.C. (eds.) Proceedings of the 18th International Conference on Engineering Design—ICED 11, Lingby/Copenhagen, Denmark, 15–19 August 2011, vol. 2, pp. 163–172. Technical University of Denmark (DTU), Design Society (2011)
8. Karakasic, M., Zadnik, Z., Kljajin, M., Duhovnik, J.: Functional structure generation within multi-structured matrix forms. Tech. Gaz. **17**(4), 465–473 (2010)
9. Zadnik, Z., Cok, V., Karakasic, M., Kljajin, M., Duhovnik, J.: Modularity solutions within a matrix of function and functionality (MFF). Tech. Gaz. **18**(4), 471–478 (2011)
10. Hubka, V., Eder, W.E.: Theory of Technical Systems. Springer, Heidelberg (1988)
11. Lai, K.C., Ming, L.W.: Quality function deployment: literature review. Eur. J. Oper. Res. **143** (3), 463–497 (2002)
12. Büyüközkan, G., Çifçi, G.: A new incomplete preference relations based approach to quality function deployment. Inf. Sci. **206**, 30–41 (2012)
13. Zadnik, Z., Starbek, M., Duhovnik, J.: Enhancing preliminary design within concurrent engineering using the matrix of functions and functionalities. Concur. Eng. Res. Appl. **20**(4), 275–285 (2012)
14. Zadnik, Z., Karakasic, M., Kljajin, M., Duhovnik, J.: Function and functionality in the conceptual design process. J. Mech. Eng. **55**(7–8), 455–471 (2009)
15. Karakasic, M., Zadnik, Z., Kljajin, M., Duhovnik, J.: Product function matrix and its request model. Strojarstvo **51**(4), 293–301 (2009)
16. Hua, Z., Yang, J., Coulibaly, S., Zhang, B.: Integration TRIZ with problem-solving tools: a literature review from 1995 to 2006. Int. J. Bus. Innov. Res. **1**(1–2), 111–128 (2006)
17. Altshuller, G.: 40 Principles: Extended Edition. In: Shulyak, L., Clarke, D. (trans.) Technical Innovation Center, Sr. Worcester, MA (2005). ISBN 0-9640740-5-2

Biogenic Domestic Waste—Exploring Select Dimensions of Socio Technical Innovation Using Design Probe

Amit Kundal, Jayanta Chatterjee and Shatarupa Thakurta Roy

Abstract The aim of this research paper is to explore select dimension of social innovation through design that communicates and signals for behavioral change towards behavioral transformation of citizens/consumers. The study focuses on waste utilization by developing a case for an innovative product-service business system specifically designed for campus communities/high-rise buildings/and urban residential colonies. The study probes an economic solution for large scale composting and design driven customer education on the matters of sustainability and increased environmental responsibility. Urban Citizens behavior towards organic waste is studied to design a self/citizen driven a product-service business system where composting form the core. As waste handling in many Indian cities suffers from various social taboos, so a design probe based research method is adopted to generate preliminary data and insight through direct and indirect observation. This research approach helped to generate empathic design ideas and to explore theories regarding consumption innovation. The paper outlines the emerging theory of societal design and understanding of second life for the biogenic domestic waste and how to convert it into a new product for a new market and how waste can be an incentive for generating social innovation. This project situates design as a core component in the social process of transitioning to a novel system of composting. Initial results support that design has a critical role to play in facilitating social learning in systemic innovation.

Keywords Design · Social innovation waste · User behavior · Waste management · Biogenic waste · Composting · Probe

A. Kundal (✉)
Design Programme, IIT Kanpur, Kanpur, India
e-mail: akundal@iitk.ac.in; amitkundal@gmail.com

J. Chatterjee
IME Department, Design Programme, IIT Kanpur, Kanpur, India

S.T. Roy
Department of HSS, Design Programme, IIT Kanpur, Kanpur, India

A. Chakrabarti (ed.), *ICoRD'15 – Research into Design Across Boundaries Volume 2*, Smart Innovation, Systems and Technologies 35, DOI 10.1007/978-81-322-2229-3_16

1 Introduction

The socio-cultural and technological problems that designers need to perceive engage with and respond to are growing in complexity and scope. India's growth story has created the escalating problem of domestic and industrial waste. The main driver for domestic waste is the rapid urbanization that is slated to change India from a largely rural to a majority urban country in the next decade.

"Wastes" [1] are materials that are discarded after use at the end of their intended life span. Waste management [2] is a collective activity involving segregation, collection, transportation, re-processing, recycling and disposal of various types of wastes. Sustainable waste management involves managing waste in an environmentally sound, socially satisfactory and a techno economically viable manner. Sustainability of waste management is the key to providing a continuous and effective service that satisfies the needs of all the stakeholders and end users. Sustainable waste management can be achieved through strategic planning, institutional capacity building, fiscal incentives, techno-economically viable technologies, public-private partnerships, community participation and such others. Strategic planning needs to be based on the local needs with the formulation of long-term goals.

2 Opportunity for Socio Techno Design Investigations

Food Waste is a complex problem that has economic, social and environmental aspects. From the economic perspective, food has a value that can be exchanged with any other goods in the global market. According to Worldwide Responsible Accredited Production

(WRAP) [7], UK citizens annually throw away food costing 12 billion €, of which 68 % can be classified as avoidable. According to the UN Food and Agriculture Organization (FAO) nearly a fourth of its 1.1 billion population hungry, India indeed is the world's hunger capital. For another example to show the importance of food waste in economic terms, 2.2 million TJ (equal to Switzerland's total annual energy consumption) was embedded in food wasted in the USA in 2007 [6]. To summarize, it can be stated that addressing the food waste problem of domestic kitchens can create multiple advantages from social, environmental and economic perspectives.

2.1 Overview of Design Research Method

The past decade has seen tremendous interest in design and "design research" from a multitude of perspectives in industry and academia. Indeed, the term "design research" has become part of the common vernacular in the field of design and is increasingly used to describe a myriad of possible approaches, perspectives, philosophies and methods. Research is generally defined as a systematic investigation that establishes novel facts, solves new or existing problems, proves new ideas, or develops new theories. It is primarily associated with the search for knowledge, especially in the sciences and technological fields. Design, in contrast, deals with the act of planning and communicating a course of action to others, usually through the creative exploration of an area of interest. Charles Eames defined design as "A plan for arranging elements in such a way as to best accomplish a particular purpose [4]." The term "design research" combines these two reasonably well-understood areas of practice, research and design, resulting in a seemingly meaningful merger roughly equivalent to the investigation of knowledge through purposeful design (Figs. 1 and 2) [3, 5].

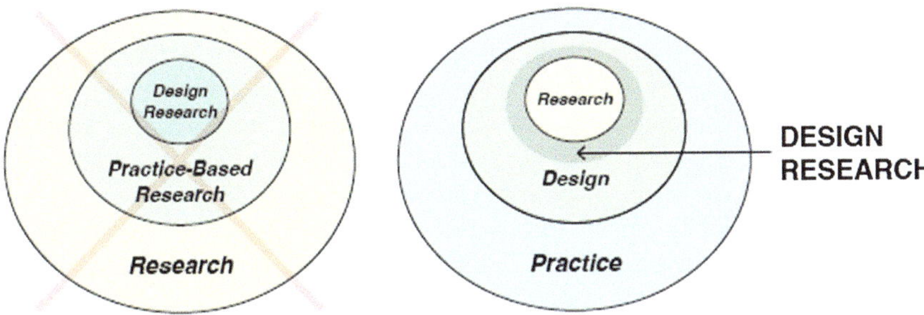

Fig. 1 Design research is not a "kind" of research. Rather, research is a practice, and it is part of design practice [6]

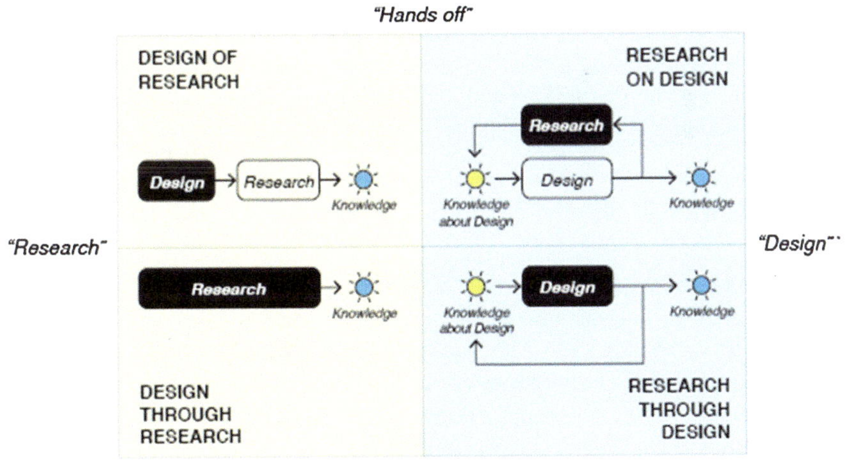

Fig. 2 Design through research—research through design [3]

2.2 Assumptions Developed as We Explored

Better-designed containers/kitchen bins will encourage segregated garbage collection.

Effective collection in the domestic unit will enhance the efficiency of the community system. The composting route has a higher probability of success as opposed to biogas etc. because user can directly participate in co creating the solution.

2.3 Investigating Through the Medium of Design Probe

Design probes are an approach of user-centered design for understanding human phenomena and exploring design opportunities. The literature and the experiences from several probe studies suggest three features describing the probes in this book (Mattelmäki 2005). The first emphasizes the user's active role in recording the material. The second and the third show similarities with the characteristics of applied ethnography [8]. Probes are based on user participation by means of self-documentation. The users or potential users collect and document the material, working as active participants in the user-centered design process. They explore new opportunities rather than solve problems that are known already. Probes are meant to support both the designers and the users in their interpretations and creativity. They are used to ask the users to experiment, express and explicate their experiences. The open probe assignments are both descriptive and explorative. The openness and room for interpretation also involve the expectation of a surprising or unexpected result.

The probe process intending dialogue emphasizes discussion at both tuning in probing and interpretation stage. Meeting, workshops, brainstorming sessions encourage the dialogue. In this research probes are used throughout the design process to facilitate discussions and dialogue with the users and potential users.

2.4 Research Study Design and Guidelines

Active participation of a user is more energetic in a physical space. The idea was to install a set of 3 composters in the vicinity of IIT Kanpur campus and involve some families to use the same. This active participation of user will help in generating key observation and point to further investigation.

The Design probe pedagogy and case study format based on the composting pilot installed at IIT Kanpur faculty apartments. The purpose of this is

1. To study the food waste disposal habit of apartment dwellers.
2. To study the food waste handling inside the kitchen and at the community site.

3. Develop composting product design and user interface alternatives for the Indian context.
4. To understand the relationship between Indian household and domestic food waste generated by them.
5. To study weather the co creation of devices and artifacts with waste handlers from the plate to the final dumping bin can lead to efficient and effective designs.

Rather than focusing barely on how to persuade users to recycle in a correct way, the goal is to reach deeper understanding of people's approach to food waste disposal.

Throughout the entire research, a user centered design (UCD) approach was applied. There are interconnected studies that were conducted to answer these research questions.

3 Research Study Analyses

In Study I, 26 participants were observed and conversations were initiated and recorded over a period of three months to explore their perceptions and attitudes towards food waste, revealing the most wasted food types and reasons for food wastage. The findings of Study I were clustered under four phases of food handling: acquisition, preparation, consumption and storage. With the help of study I, the food waste problem in domestic kitchens and user behavior became clear.

The attitudes that we have towards purchasing food can influence the way we shop and how much food we waste at home. Most respondents indicated that they feel guilty when they waste food (82 %). Although most respondents (79 %) claimed they thought about how much food they would use, just 57 % indicated that they are careful about buying foods that they know will be used. This suggests a disparity between thinking about how much will be used and following through with the corresponding purchasing behavior.

Most respondents (70 %) indicated they only buy the amount of fruit and vegetables they need, while 15 % indicated they buy the best value fruit and vegetables even if it is more than what they need.

Individuals can waste food intentionally and unintentionally. Study was devised to discern the behaviors of individuals falling into these two categories. However, some individuals may not be aware of what they waste, whilst others may not waste any food at all. Candidates who answered "no" to this question were not selected as participants for further Study, even though they might not have been aware of their waste behavior. Furthermore, for such candidates, it is hard to determine whether they really do not waste food or whether they are not aware of their food waste behaviors. We decided to instead to simply exclude from the study the candidates who answered "no" to "do you waste food?" We tried to divide participant's behaviors into intentional and unintentional food wasting. Afterwards, the participants were asked several questions to learn whether food wastage is really a problem according to their perception.

These questions were as follows:

- Do you remember what kind of food did you throw away last week?
- How can the food waste problem be solved? Do you think that it can be solve with regulations and education? Or, can the help of products or services solve it?
- Do you think that food waste is our individual problem? Can we solve it by changing our behaviors such as planning skills?

After these questions were asked, participants were requested to answer several questions about their acquisition, preparation, consumption and disposing behaviors, in order to gather explicit knowledge about the issue.

These results were helped to designate the Study II, a generative session with three users and two designers, which mainly aimed to understand latent and tacit information about the possible solution ways and reasons behind the food waste problem. The results of Study II were used for creating a set of criteria for design solution, which could minimize food waste, was drawn-up.

4 Best Practices on Food Waste Prevention

The Paradigm shift in the waste and resources management sectors, that will actually achieve significant and measurable prevention and ultimately contribute to sustainability, has as prerequisite the implementation of effective strategies. A range of waste usage prevention strategies, targeting households, with varying degrees of success, has been applied in several EU countries. The strategies applied fall into three categories with different levels of engagement of central or local authorities:

(i) Diffusion of information;
(ii) Promotional campaigns; and
(iii) Setting regulations.

A range of best practices for food waste usage prevention is presented and briefly analyzed below, to feature good examples and stimulate waste prevention.

4.1 "Love Food Hate Waste" Campaign, WRAP, UK

This has proved to be one of the most successful food waste related awareness campaigns. Over the first two years of its application (from its launch in 2007–2009) in the UK, it contributed to the prevention of approximately 137,000 tonnes of food waste. The campaign aims at raising awareness, personalizing the issue of food waste usage prevention, and developing "encourage and enable" actions (http://lovefoodhatewaste.com).

4.2 Réduisons nos Déchets, National Authority 2005, France

The awareness campaign "Réduisons nos Déchets" (Reducing our Waste) was developed by the ADEME, the French Environmental and Energy Management Agency, in order to provide information to households about waste generation usage and prevention. This campaign also participates and promotes the European Week for Waste Reduction (EWWR) initiative (http://www.reduisonsnosdechets.fr).

5 Next Practices for Food Waste

5.1 Levels of Design

As Don Norman explain in his book Emotional Design that Human behavior towards everyday objects and the environment is very complex. Each of the three levels of designs Visceral Design, Behavioral Design and Reflective Design defines the experience of the engagement.

5.1.1 Visceral Design

Visceral is what nature does. We humans evolved to coexist in the environment of other humans, plants, landscapes, weather and other natural phenomena. As a result we are exquisitely tuned to receive powerful emotional signals from the environment that get interpreted automatically at the visceral level. When we perceive something as 'pretty', that judgment comes directly from the visceral level.

5.1.2 Behavioral Design

Appearance doesn't really matter but performance does. What matters are four components of good design: Function, understandability, usability and physical feel. If a well-designer product misses the target when it comes to fulfilling its purpose, it deserves to fail. What if a peeler doesn't peel?

5.1.3 Reflective Design

We all worry about the image we present to others. You avoid things because 'it wouldn't be right' or you buy things so support a cause you prefer. Attractiveness is visceral, beauty is reflective. Reflective level operations often determine a person's overall impression of a product. You think back, reflect upon its total appeal and experience of use.

Table 1 The three parts to consider for a successful composting system in India

Inside	Outside	Awareness
Problem: waste separation	Problem: a composter that is suitable for India	Problem: people lack awareness to what composting is. This leads to problems with waste separation. Once educated, people embrace the idea and are excited about the concept
Discussion: a bio bin to promote awareness and encourage waste separation Design can provide a direction of the use of the bin through pictures and illustrations. Materials like charcoal powder can be used to reduce smell from the bin. Color-coding in many cases has been adapted successfully for encouraging segregation, Use of space in reference to indian kitchen layout is and important area to focus	Discussion: design to fit India's climate. Less insulation, cheaper materials and manufacturing, and wheels. A Caretaker to look after composting process and monitor waste separation	Discussion: a composting campaign that has a recall value with posters, stationery and shirts to constantly remind people about composting

6 A Step to Sustainable Composting Solution for Indian Consumer

Based on the results from the previous chapter a three-way solution for Indian market was envisaged. Indian household are not used to self motivated segregation for successful composting activity, hence it was important to construct entire system around the composting so as to ensure that composting becomes a part of everyday activity (Table 1).

7 Conclusions

Food waste is a stream with multiple social, economical and environmental implications generated throughout the life cycle stages of food. Among those stages, the role of the households in the generation of food waste is very important. Therefore, in order to demonstrate its potential, quantification of the household food waste benefits is required.

Composting is often implemented for the wrong reasons. It can make large profits, may be it will not solve all solid waste management problems. Incentives, such as the converting you waste into an economic inflow rather that out flow to are needed to be addressed to set up composting projects, which can be sustained on a long-term basis. Composting should be considered as part of a larger educational strategy for educating Indian consumer about the process and benefits, so it can be adopted within the culture of Indian community supported with appropriate technologies based on market opportunities, economic feasibility, and social acceptance. Cost effective and sustainable composting is possible within a household. Participation and cooperation from many stakeholders is required, including national governments, municipalities, local communities, waste generators, and the private sector. Government and other bodies should support and encourage community based, private sector, and municipal composting initiatives by:

- Providing technical assistance on composting techniques
- Developing guidelines for the implementation of low-cost facilities
- Evaluating loans and other financial support
- Allocating land for compost facilities on a long term lease basis
- Establishing and enforcing compost quality standards
- Regulating and monitoring the performance of compost operations
- Promoting the use of compost through public awareness campaigns

8 Limitations and Future Direction of Study

Same methodology can be applied in understanding bigger and complex systems like hostels; hotels and new probes can be developed to enhance the process and solution domain. Understanding the sociological and cultural attitude of people towards waste and reduction of source. After the compost is created from biogenic domestic waste an economically feasible system has to build for sustainability.

References

1. Ahmed, K., Jamwal, N.: (2000). In: Zurbrcugg, C., Drescher, S., Patel, A., Sharatchandra, H.C. (2004): Decentralized composting of urban waste—an overview of community private initiatives in Indian cities. Waste Manage. 24:655–662 (2004)
2. Akolkar, A.B.: Status of Solid Waste Management in India, Implementation Status of Municipal Solid Wastes, Management and Handling Rules 2000. Central Pollution Control Board, New Delhi (2005)
3. Daniel, F.: Why Research oriented Design Isn't Design oriented Research. In: Proceedings of Nordes: Nordic Design Research Conference, May 29–31, Copenhagen, Denmark (2005)
4. John, N., et al.: Eames Design: the Work of the Office of Charles and Ray Eames. New York: Abrams, H.N. (1989)

5. Jonas, W.: Research through DESIGN through research a problem statement and a conceptual sketch. In: Proceedings of DRS international conference, Lisbon (2006)
6. Kirshenbaum, S.R.: Waste not, want less. New Scientist (2010)
7. Quested, T., Johnson, H.: Household Food and Drink Waste in the UK: a report containing quantification of the amount and types of household food and drink waste in the UK. Report Prepared by WRAP (Waste and Resources Action Programme), Banbury (2009)
8. Sanders, E.B.: Virtuosos of the experience domain. Paper presented at IDSA education conference, London (2001)

Design for the BOP and the TOP: Requirements Handling Behaviour of Designers

Santosh Jagtap, Andreas Larsson, Anders Warell, Deepak Santhanakrishnan and Sachin Jagtap

Abstract The base (BOP) and the top (TOP) of the world income pyramid represent the people living in poverty and the people from developed countries, respectively. In the approach of business development combined with poverty alleviation, the design of products for the BOP plays an important role. There is an urgent need to develop an understanding of the process of designing products for the BOP. Requirements handling is an important ingredient of a design process. This research, using a protocol study, examined the differences between the requirements handling behaviour of designers when they design a product for the BOP and TOP markets. We found differences between their requirements handling behaviour in terms of their attention to different topics of requirements, and their handling of solution-specific and solution-neutral requirements.

Keywords Base of the pyramid · Product design · Requirements · Protocol analysis

1 Introduction

The world income pyramid can be divided into three segments—top, middle, and bottom. The top segments (i.e., 'Top of the Pyramid'—TOP), includes people from developed countries. The middle segment consists of the rising middle class from

S. Jagtap (✉) · A. Larsson · A. Warell
Department of Design Sciences, Lund University, Lund, Sweden
e-mail: snjagtap22@gmail.com; santosh.jagtap@design.lth.se

D. Santhanakrishnan
Indian Council for Research on International Economic Relations, New Delhi, India

S. Jagtap
Maharashtra State Board of Secondary and Higher Secondary Education, Pune, India

A. Chakrabarti (ed.), *ICoRD'15 – Research into Design Across Boundaries Volume 2*, Smart Innovation, Systems and Technologies 35, DOI 10.1007/978-81-322-2229-3_17

developing countries. The bottom segment, generally called the 'Base of the Pyramid' (BOP), consists of the poor people. About two-fifths of the world population can be categorized as poor [1].

1.1 Design for BOP Markets

In recent years, a poverty reduction approach that combines business development with poverty alleviation has received attention [2]. In this approach, the poor at the BOP are considered as producers and consumers of products. Design of products is an important ingredient of this market-based approach. Furthermore, some universities have begun to offer courses and/or design projects in the area of the design for the BOP.

Design research is important in understanding and improving design practice and education [3]. However, design researchers have given little attention to the field of the design for the BOP. Most of the design research has been carried out in the context of developed countries and relatively affluent markets [4, 5]. There has been little empirical examination of the design for the BOP. This limits our ability to develop tools and methods for improving current practice and education of design for the BOP.

In our previous research [5], using a protocol study, we explored the differences between the design processes for the BOP and TOP markets. Specifically, we investigated the strategies (i.e., problem driven, solution driven strategy) used by the designers. In this paper, based on the data from the protocol study, we examine how designers handle requirements in designing products for the BOP and TOP markets. We have discussed the implications of the findings for design practice.

1.2 Requirements Handling in Design

Chakrabarti et al. [6] found that the main ingredients of the design process are: requirements (i.e., problems), solutions, information, and strategy (i.e., plan of action to progress through the design process). In the design process, requirements and solutions co-evolve [7].

Based on the analysis of designers' activities, Nidamarthi [8] found that the designers use tentative solutions to enhance the understanding of the initial requirements. He also observed that these solution-generated requirements (i.e., solution-specific requirements) played an important role in the problem solving process. Restrepo and Christiaans [9], based on their empirical studies of designers, have characterised requirements depending on their specificity—'solution-specific' and 'solution-neutral'.

In his experiments with designers, Nidamarthi [8] found that requirements do not get fulfilled if they are ignored. In the experiments, the designers were asked to

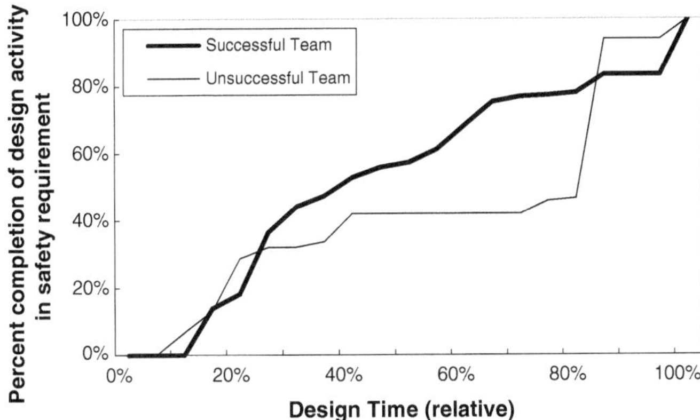

Fig. 1 Distribution of activity related to the requirement regarding 'safety' by two teams—adopted from [8]

consider some safety aspects (see Fig. 1). In this Fig. 1, any horizontal segment indicates inactivity by the design team in the related requirement for that duration. The design team, which provided enough attention to the requirements regarding safety aspects, was successful in satisfying safety-related requirements.

2 The Protocol Study

The details of the protocol study are presented in our previous research [5]. However, for easy reference, we briefly present the protocol study in this paper. In total, eight designers individually participated in the study. These designers were divided into two groups, namely BOP and TOP groups/sessions. In a laboratory setting, four designers (BOP designers) solved a design problem for the BOP, and four other designers (TOP designers) solved the same problem for the TOP. These eight designers were Masters students in 'Industrial Design' or 'Product Design'. Before this protocol study, we ensured that the BOP and TOP designers had prior experience of working on university-based design projects for the BOP and the TOP, respectively. We believe that excepting this difference in experience of working on university-based design projects, the designers in the BOP and TOP sessions are fairly similar. It is therefore likely that the differences in the design processes for the BOP and TOP markets are mainly due to the differences in these markets. There can be some differences in these design processes due to the difference in the BOP and TOP designers' degree of familiarity with the respective contexts. We have discussed these issues later in Sect. 4. Our experimental arrangement was pragmatic, and the findings gained through this research are useful in terms of their implications for design practice. These implications are discussed in Sect. 4.

The formulated design problem needs to be applicable for the BOP and TOP markets. We created the design problem as follows.

> A highly contagious and deadly disease called 'anthrax-d5' is spreading across (…). This disease is transmitted only through contaminated food and water. A person infected with this disease needs to be hospitalized in order to save his/her life. The spread of this disease is such that the existing healthcare infrastructure (i.e., available number of hospitals) is inadequate to hospitalize and treat the large number of infected people. There is an urgent need to erect a number of temporary shelters that can be used as hospitals. For (xxx), where the 'anthrax-d5' is spreading at an enormous rate, design such a temporary shelter that can be used to hospitalize 5 infected people (per shelter). Each shelter also needs to accommodate basic healthcare facilities and healthcare staff consisting of 1 nurse. The time to install this shelter must be less than 2 h. The shelter also needs to withstand different types of weather conditions.

In the above problem, in the case of the BOP sessions, (…) was replaced by 'a cluster of BOP communities in a developing country' and (xxx) by 'the cluster of BOP communities'. In the TOP sessions, (…) was replaced by 'a city in a developed country' and (xxx) by 'the city in the developed country'.

The audio recordings were transcribed. The transcripts were divided into segments, with each segment corresponding to a single thought, expression, or idea. The coding scheme consisted of four major categories, borrowed from the coding scheme successfully implemented and developed by Chakrabarti et al. [6]. The four major categories are: 'requirement', 'solution', 'information', and 'strategy' (see Table 1).

As shown in Table 1, for the segments that were classified under the 'requirement' category, we coded the type of requirement (i.e., solution-specific or

Table 1 Coding scheme

Category	Description (example)
Requirement	Designer deals with a requirement ("That needs to include…", "I am assuming this should be…")
Req. type	
Solution-specific (*SR*)	A requirement that is specific to any of the designer' solutions (The designer, in relation to a specific solution, dealt with the following requirement, "The outside of it should be of leak-proof material to protect from rain.")
Solution-neutral (*NR*)	A requirement that is not specific to any of the designer's solutions ("The solution needs to be as cost-efficient as possible.")
Req. topic	Categorization of a requirement based on its topic (e.g., materials, geometry, aesthetics, ergonomics, etc.)
Solution	Designer deals with a solution ("Let's put cloth on inside…", "So, this is efficient to…")
Information	Designer deals with information ("Developed countries have…", "This is actually not accurate information of…")
Strategy	A plan of action for proceeding through the design process ("I will start by just taking some notes about what this task is.")

solution-neutral), and the topics of requirement (e.g., materials, geometry, etc.). Some topics of requirements were borrowed from Pahl and Beitz [10] and Dwarakanath and Blessing [11], and some topics evolved during the coding process.

We measured the reliability of the coding process by calculating the percentage agreement between two coders. Due to resource limitations, two out of the eight protocols (i.e., two transcripts) were coded by the researcher and one coder. The average inter-coder reliability was above 85 %.

3 Results

3.1 Specificity of Requirements

Figure 2 shows the average percentage of segments according to the requirements-specificity in the BOP and TOP sessions. In the BOP and TOP sessions, there are differences between the occurrence percentages of these two types of requirements. The average percentage of segments associated with the solution-specific requirements (SRs) is higher in the TOP sessions as compared to that in the BOP sessions (48.0 and 32.2 %). In contrary, the average percentage of segments associated with the solution-neutral requirements (NRs) is higher in the BOP sessions as compared to that in the TOP sessions (67.8 and 53.1 %).

Figure 2 also shows that, in the BOP sessions, there is a substantial difference between the average percentage of segments classified into SRs and NRs. The designers in the BOP sessions dealt more with the NRs than with the SRs (67.8 and 32.2 %). While the designers in the TOP sessions dealt more with the NRs than

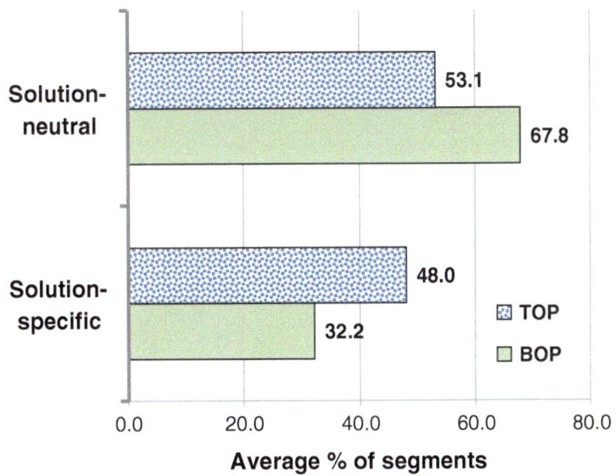

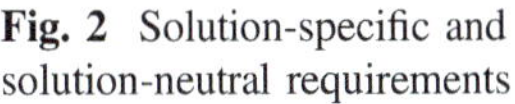
Fig. 2 Solution-specific and solution-neutral requirements

with the SRs (53.1 and 48.0 %), the difference between the average percentage of segments under these two types of requirements is small in the TOP sessions (53.1 and 48.0 %) as compared to that in the BOP sessions (67.8 and 32.2 %).

3.2 Topics of Requirements

The BOP and TOP designers mainly dealt with requirements related to geometry and installation (see Fig. 3). As compared to the TOP designers, the BOP designers dealt more with requirements from the topics-materials (12.5 and 6.7 %), users (13.3 and 6.5 %), energy/power (6.3 and 4.3 %), and costs (3.6 and 1.7 %). As compared to the TOP designers, the BOP designers paid little attention to the requirements related to aesthetics (0.4 and 5.6 %), ergonomics (4 and 10.5 %), information provision (0.4 and 2.6 %), supply chain/logistics (2 and 7.3 %), healthcare (8.6 and 13.8 %), and hygiene (5.4 and 12.4 %). The designers from both BOP and TOP sessions have not considered maintenance requirements. While the TOP designers have considered the requirements about forces and production, the BOP designers have not considered requirements from these topics.

The above differences between the BOP and TOP designers can be due to the differences between the TOP and BOP markets (e.g., the poor physical infrastructure in the BOP markets, low income of the BOP people, etc.). Furthermore, the differences in the occurrence percentages of requirements from different topics indicate the degree of importance the designers have placed on these topics. For example, the findings suggest that, as compared to the TOP designers, the BOP designers have placed more importance on the requirements regarding materials, energy/power, costs, etc. and less importance on the requirements regarding aesthetics, ergonomics, information provision, hygiene, etc.

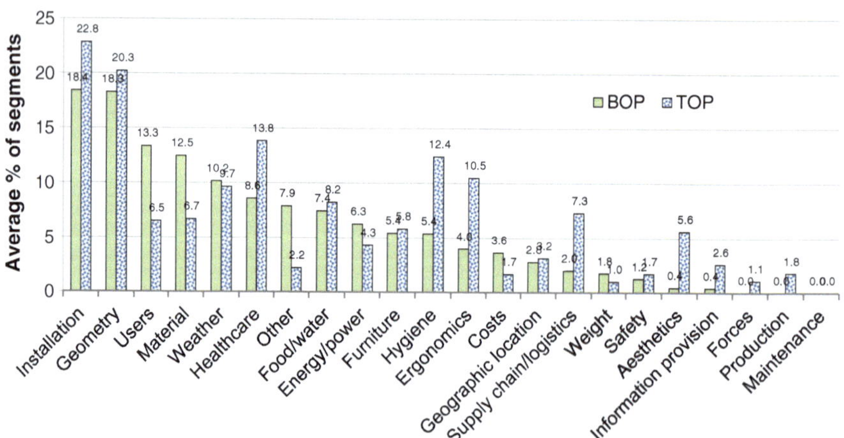

Fig. 3 Topics of requirements

		Q1	Q2	Q3	Q4
SR	BOP	1.9	10.5	11.9	8.0
	TOP	7.2	9.7	21.1	10.0
NR	BOP	39.2	19.9	6.6	2.0
	TOP	35.2	16.4	0.7	0.9

Fig. 4 Distribution of SRs and NRs (average % of segments)

3.3 Distribution of Solution-Specific and Solution-Neutral Requirements

For each designer, the timeline of the design process was divided into four equal quarters, namely Q1, Q2, Q3, and Q4. We counted the number of segments corresponding to solution-specific requirements (SRs) and solution-neutral requirements (NRs) in each of these quarters. Figure 4 shows the average percentage of segments for SRs and NRs in each of the quarters in the case of the BOP and TOP sessions. The coloured bars in this figure are drawn by using the conditional formatting facility of the Microsoft Excel.

From Fig. 4, the following observations can be made. Throughout the process, the occurrence percentage of SRs is higher in the TOP sessions than in the BOP sessions, except for Q2. In Q2, the occurrence percentage of SRs is slightly higher in the BOP sessions than in the TOP sessions (10.5 and 9.7 %). In Q1, the occurrence percentage of SRs is considerably higher in the TOP session than in the BOP session (7.2 and 1.9 %). This indicates that the TOP designers engaged in activities associated with solutions from the beginning of the process (i.e., in Q1).

Figure 4 also shows that, in the TOP and BOP sessions, from Q1 to Q4, there is a gradual decrease in the occurrence percentage of segments associated with NRs. However, there are some differences between these two sessions regarding the occurrence percentages of these NRs along the timeline. The designers in the TOP sessions considered the NRs mainly in the early phases of the process (i.e., in quarters Q1 and Q2). On the contrary, the designers in the BOP sessions dealt with these requirements throughout the process as can be seen from the average percentage of segments in Q3 (6.6 %) and Q4 (2 %).

4 Discussion, Conclusions, and Limitations

In this research, we used a protocol study to compare requirements handling behaviour of BOP and TOP designers. In comparison to the TOP designers, the BOP designers predominantly handled NRs than SRs. The BOP designers handled NRs throughout the process with an emphasis in the early phases. NRs are not specific to a solution, and the higher handling of NRs suggests that a designer is engaged more in the clarification of the design objectives that the final design solution needs to meet. These findings indicate that the BOP designers engaged more in the clarification of the objectives than the TOP designers. A reason can be greater unfamiliarity with the design task in the BOP sessions. The BOP designers had experience of working on university-based design projects for the BOP. However, they come from the middle to upper middle class strata of the society, and therefore it is likely that they did not experience the context of the poverty/BOP. Consequently, they had less direct knowledge of the BOP. The TOP designers' familiarity with the TOP was relatively higher as they come from middle to upper middle class strata of the society.

There are differences between the BOP and TOP sessions in terms of different types of requirements considered by the BOP and TOP designers. The BOP designers have placed more importance on the requirements related to materials, energy/power, costs, etc. and less importance on the requirements regarding aesthetics, ergonomics, information provision, hygiene, etc. These differences can be attributed to the obvious differences between the BOP and TOP markets (e.g., the physical infrastructure, the income of the BOP people is meagre, etc.).

The TOP designers have paid attention to the requirements regarding aesthetics, ergonomics, information provisions, hygiene, and supply chain/logistics. The BOP designers have not paid enough attention to these requirements despite the importance of these requirements in the BOP. For example, the BOP people, in general, are semiliterate or illiterate, and therefore the requirements regarding information provision are important in the BOP. Also, the BOP people can have preferences regarding aesthetic qualities of products. Furthermore, the requirements regarding supply chain/logistics are important in the BOP.

We propose the following reasons for the BOP designers' less attention to the requirements regarding aesthetics, ergonomics, and information provision. The BOP designers placed more importance on some requirements (e.g., materials, energy/power, costs, etc.); and they thus perceived other requirements as less important. Another reason can be that there was higher degree of unfamiliarity with the design task in the BOP sessions. Furthermore, the designers might tend to think that the BOP people mainly have basic survival needs, and they might tend to give little attention to their other needs (e.g., their aesthetic preference). However, the BOP people can have other needs besides the basic survival needs. Van Kempen's [12] experiments in Bolivia revealed that the poor people can consume status products before satisfying their physiological needs.

The BOP designers paid less attention to the requirements regarding aesthetics, ergonomics, etc. The requirements, which are given less attention, do not get fulfilled [8]. There are examples of 'real life' BOP design projects where the requirements regarding aesthetics and ergonomics were not taken into account, and that caused in the unacceptance of the products by the BOP people. These 'real life' projects are from different sectors such as healthcare and access to clean drinking water [13, 14]. Consider for example a product, namely 'LifeStraw', which is specifically designed for providing clean drinking water to the BOP people. LifeStraw is a water filter in the form of a tube, and can be used by one person (see

Fig. 5 Lifestraw (*Source*—https://www.wikipedia.org/; "LifeStraw use" by Edyta Materka from London, United Kingdom)

Fig. 5). Water passes through a filter when a person sucks it up. According to Starr [14], requirements regarding ergonomics and symbolic meaning of a product were not considered in the design of the LifeStraw, and therefore the BOP people may not like using product. He states, "People don't really like squatting over dirty water and sucking it up in a straw… it's a lot of work to suck it up through a filter." Another example of a product that is not widely used by the BOP people is 'PlayPump' [14]. The PlayPump uses the energy of children at play to operate a water pump. However, in the design of this product, the requirements regarding ergonomics were not taken into account. It can be difficult to operate the PlayPump [15].

Lockwood [13] has explained why condoms are not widely used in the DR Congo. HIV is a serious problem in the DR Congo, and aid agencies have distributed low-priced condoms in the country. However, a few people are using them. According to Lockwood [13], a reason behind this is that the requirements regarding the symbolic meaning of a product were not considered and implemented in the design of condom-packaging. The packaging of condoms, distributed by the aid agencies, shows pictures such as a wife and a husband, and a 'red ribbon' that reminds people about HIV. This type of packaging-design does not motivate people to use condoms.

These examples of 'real life' BOP design projects and the findings of our research suggest that there appears to be a tendency not to pay enough attention to the requirements regarding aesthetics and ergonomics in the design of products for the BOP despite the importance of these requirements in the acceptability of products by the BOP people. A potential implication of these findings is that designers need to overcome the above tendency, and that they ought to consider such requirements in the design of products for the BOP.

There are some limitations to this research. The results are based on the design task that is not a genuine 'real life' design task. The designers worked individually in contrast to genuine design projects that are, in general, carried out by a team. The process of thinking aloud while solving a design problem may affect the design process. While the sample size in our study is small, the experiment provided sufficient data to observe overall trends and observations. We believe that it is important to validate the results of this research in studies of real design projects using ethnographic methodologies. It would also be interesting to study the differences between the design processes for the BOP and TOP when designers are given the same set of product specifications. For example, designers can be given the same product specifications and then can be asked to design and develop solutions to meet those specification for the BOP and TOP markets.

Acknowledgments This work was partly financed by VINNOVA within the Product Innovation Engineering program (PIEp). We are grateful to the participating designers. This paper is an enriched version of some part of the paper—Jagtap et al. [5].

References

1. Karnani, A.: Fighting Poverty Together: Rethinking Strategies for Business Governments and Civil Society to Reduce Poverty. Palgrave Macmillan, Basingstoke (2011)
2. Prahalad, C.K.: The Fortune at the Bottom of the Pyramid: Eradicating Poverty through Profits. Wharton School Publishing, Upper Saddle River (2004)
3. Blessing, L.T.M., Chakrabarti, A.: DRG, A Design Research Methodology. Springer-Verlag London Limited, London (2009)
4. Jagtap, S., Larsson A.: Design of product service systems at the base of the pyramid. In: International Conference on Research into Design (ICoRD '13), Chennai, India (2013)
5. Jagtap, S., et al.: How design process for the base of the Pyramid differs from that for the top of the Pyramid. Des. Stud. **35**, 527–558 (2014)
6. Chakrabarti, A., Morgenstern, S., Knaab, H.: Identification and application of requirements and their impact on the design process: a protocol study. Res. Eng. Design **15**(1), 22–39 (2004)
7. Suwa, M., Gero, J., Purcell, T.: Unexpected discoveries and S-invention of design requirements: important vehicles for a design process. Des. Stud. **21**(6), 539–567 (2000)
8. Nidamarthi, S.: Understanding and Supporting Requirement Satisfaction in the Design Process. In Engineering Department. University of Cambridge, Cambridge (1999)
9. Restrepo, J., Christiaans H.: Design requirements: conditioners or conditioned? In: International Conference on Engineering Design, Stockolm (2003)
10. Pahl, G., Beitz, W.: Engineering Design, 2nd edn. Springer, London (1996)
11. Dwarakanath, S., Blessing, L.: Ingredients of the design process: a comparison between group and individual work, in analysing design activity. In: Cross, N., Christiaans, H., Dorst, K. (eds), Wiley, Chichester (1996)
12. Van Kempen, L.: Status Consumption and Poverty in Developing Countries. VDM Publishing, Saarbrücken (2009)
13. Lockwood, A.: Selling condoms in the Congo, TED talk. (2011)
14. Starr, K.: Design for (Real) Social Impact, IIT Design Research Conference (2010)
15. PlayPump: Available from http://en.wikipedia.org/wiki/Roundabout_PlayPump (2014)

A Sustainable Design Method Acting as an Innovation Tool

Jeremy Faludi

Abstract Product companies generally see sustainability as a burden limiting their design process, similar to cost or safety limits. A method for sustainable design was created, attempting to turn sustainability from a burden into an innovation tool with inherent business value. The method combines creative whole-systems thinking with quantitative sustainability metrics. It facilitates innovation by the creation of visual whole-system maps that encourage more thorough and more radical brainstorming. It facilitates sustainability by using quantitative measurements, such as life-cycle assessment or point-based certification systems, to set priorities and choose final designs. The method has been anecdotally tested in classes at four universities, and many of the companies partnering with these classes have said the students provided both sustainability and feature /functionality benefits. This paper also compares the method to Lindahl's nine recommendations for being useful to engineering designers. Thus there is at least anecdotal evidence that the design method may turn sustainability from a burden into an innovation tool. Future studies should compare the method against industry-leading innovation and green design methods.

Keywords Sustainable design · Innovation · Design method · Design process · Systems thinking · Life-cycle assessment · Design fixation

1 Introduction

Sustainability is one of the critical problems of our time, affecting all people on Earth in some way, as well as affecting economies and ecologies. Green architecture is an increasingly established industry, with three billion square feet of LEED-certified buildings alone [1], but green product design is decades behind, with few sustainable alternatives to most consumer products. Behrisch et al. bemoaned the lack of studies

J. Faludi (✉)
Department of Mechanical Engineering, University of California, Berkeley, USA
e-mail: faludi@berkeley.edu

A. Chakrabarti (ed.), *ICoRD'15 – Research into Design Across Boundaries Volume 2*,
Smart Innovation, Systems and Technologies 35, DOI 10.1007/978-81-322-2229-3_18

on how often green design techniques are even used in industry [2] or advertised as capabilities [3]. Sustainable design has been studied extensively, but most studies have found that companies perceive it to be a burden on designers and engineers, due to the extra time and money it requires [4–6]. Many eco-designers have been touting sustainability as an innovation tool for years—a conference named "Sustainable Innovation" is in its 19th year [7]. However, academic studies showing its advantages for innovation are scarce to nonexistent. Collado-Ruiz's quantitative study shows the opposite—that eco-design methods limit design creativity [8].

This does not have to be the final word, however. Different design methods can be assumed to drive different outcomes, or else they would not be used. Some methods are actually not design but just sustainability assessment, such as life-cycle assessment ("LCA"), or assessment and strategizing, such as Waage's "road map" [9]. Other methods focus almost entirely on idea generation, such as Biomimicry [10]. Most methods combine ideation, strategizing, and assessment together, e.g., The Natural Step [11], the UNEP/TU Delft Design for Sustainability manual [12], Factor Ten [13], and various certification systems like Cradle to Cradle [14] or EPEAT [15].

Some green design strategies have unquestionably caused innovation in the market (in the sense of radical design change): for example, turning a product into a service. Product-service-systems have been great financial successes for Xerox, Electrolux, Castrol, and many others [16], showing that economically valuable innovation can come from sustainable design. Product-service-systems are not appropriate for every product category, but perhaps a design method prioritizing both sustainability and innovation could suggest valuable solutions for any product category.

2 Background

Why is it important for a sustainable design method to also be an innovation tool? Most companies only use sustainable design when it is required by government regulation, reduces legal liability, or saves money, rather than prioritizing sustainability in itself [17]. Of these, only "saving money" sees sustainability as an investment rather than a burden, but all three motivations have direct economic value for companies [18]. Innovation is a direct economic value, which is also seen as an investment by companies [19, 20]. Therefore, sustainable design as innovation should be pursued for three reasons: First, because innovation is valued by companies, and can thus promote sustainable design in cases where it does not obviously save money, reduce legal liability, or solve regulatory compliance. Second, because sustainable design usually requires companies to think very differently about their products—a natural driver for innovation. Third, because innovative companies are market leaders, and leaders shifting towards sustainability could pull whole industries along with them.

Lindahl [21] studied engineering designers to find their requirements for sustainable design tools, and listed nine recommendations: #1. Easy-to-understand benefits. #2. Easy to understand process. #3. Adjustable to different contexts. #4.

Low setup time. #5. No need for simultaneous cooperation. #6. Low need for data. #7. Visualization of results. #8. IT-based (use dedicated software). #9. Give direction, not a result.

The Whole-Systems and Life-Cycle ("WSLC") method was created for the Autodesk Sustainability Workshop [22], a free online set of training resources for students, professionals, or professors. While quantitative assessments of learning outcomes have not been studied, anecdotal evidence suggests it may be an effective learning platform [23–26]. In the Sustainability Workshop, the WSLC design method is a keystone. It frames later videos and readings on specialized sustainable design strategies such as material choices, design for disassembly, reducing material use, and energy efficiency.

3 The Whole Systems + Life-Cycle Design Method

The WSLC method combines creative whole-systems thinking with quantitative sustainability metrics in a four-step process: First, designers /engineers /business managers start with the existing product or service, and visually map its whole system. This includes all major physical sub-systems, life-cycle stages, inputs and outputs, customer use, and the connections between these nodes. Second, this system is quantitatively analyzed, using LCA or point-based certification systems (e.g., Cradle to Cradle or EPEAT) to find the worst environmental impact(s). The worst impact becomes the top-priority goal for sustainable redesign. The design team then ranks the top-priority sustainability goal(s) along with the project's top business goal (s) such as cost and functionality, and writes quantitative metrics of success for all goals. Third, the team brainstorms on their top-priority sustainability goal, using the visual system map to facilitate the brainstorm. This helps designers be more thorough by ensuring there are alternatives to every component or step in the system map, and helps designers generate more radical ideas by encouraging them to skip steps/eliminate components in the system. Finally, quantitative environmental impact estimates (LCA or certification scores) are used to judge the brainstorm results. These estimates are very imperfect, but are still better than untrained guesswork—engineers, designers, and managers are not environmental scientists, so quantitative estimates will be better than their intuition of where eco-impacts lie. The WSLC design method's two divergent, creative stages balance its two convergent, analytical stages to turn sustainability from a limitation into a jumping-off point for exploration. A video describing the method can be seen at http://sustainabilityworkshop.autodesk.com/products/whole-systems-and-lifecycle-thinking . Some illustrations from educator materials are included here as examples.

For Step 1, Fig. 1 shows a possible whole system map for a refrigerator. In the center is the product with all its major parts: doors, insulation, cooling coils, compressor, etc. The life-cycle of the refrigerator is shown from bottom left to bottom right, going from raw materials to landfill. Since refrigerators exist to prevent food from spoiling, the food's life-cycle is shown from top left to top right.

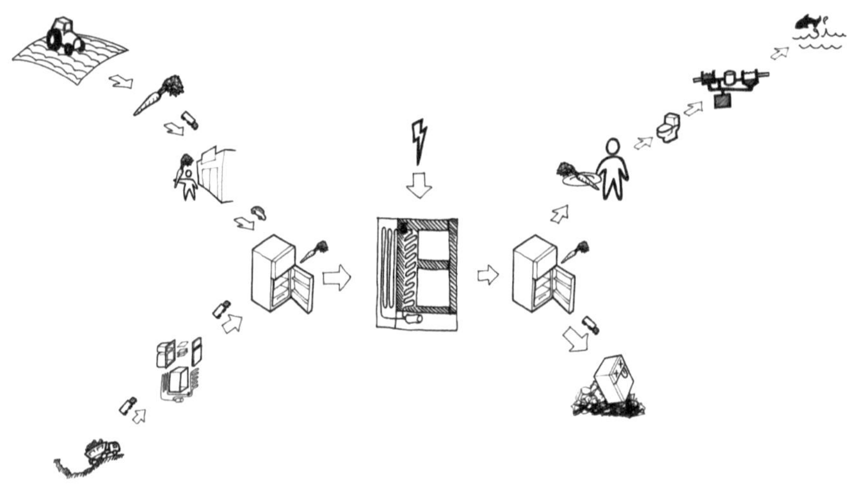

Fig. 1 WSLC Step 1: an example whole-system map for a refrigerator

Next to the cutaway drawing of the product in the center, the adjacent nodes in the map show how it is used: the user opens and closes the refrigerator door to insert and remove food. The refrigerator's other significant input is electricity, shown at center top. There is not necessarily a "right" system map—it cannot capture the immense complexity of the entire world, it is merely a model to aid design teams in focusing on all the aspects of a design they can influence. Therefore the design team should tune the system map's completeness and level of detail to their purposes. More complete is generally better.

For Step 2, Fig. 2 shows the quantitative sustainability assessment using LCA. Step 2 starts with Step 1's whole system map and either calculates an LCA of the whole system, or of a subset of it chosen by the design team to only include things they feel they can influence. Step 1 also helps clarify the functional unit for Step 2's analysis. More expansive LCAs give a more complete picture, but require more

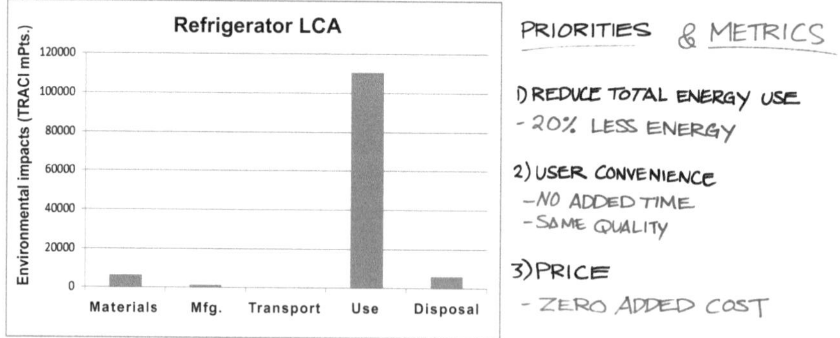

Fig. 2 WSLC Step 2: an example life-cycle assessment for a refrigerator, with the resulting sustainability priority and metric along with business priorities and metrics

time and expertise to perform. LCAs may be done by an engineer on the team, or by an outside department/consultant, using software such as SimaPro, GaBi, Sustainable Minds, or even paper-based systems if budget is limited. It may be a pre-existing assessment. The LCA should use a methodology that measures a broad variety of sustainability impacts and combines them into a single score; e.g., EcoIndicator [27] or ReCiPe [28]. This makes it easy to identify top sustainability priorities at a glance. Systems other than LCA can be used instead: for instance, scorecards such as EPEAT and Cradle to Cradle include both environmental and social impacts, and provide simple scores. Once the top sustainability priority is identified by one of these means, it is listed in the design specification, ranked alongside business priorities. These business priorities and metrics will generally be pre-existing, set by team managers or executives. Sustainability rarely ranks first, but it should be given a place in the list. Concrete metrics to measure success are decided upon here, for later accountability to the vision set forth in the design spec. Metrics for success depend on the team—an ambitious team might aim for 80 % reduction of energy use, while a modest team might choose 20 %. Metrics are not strictly needed but are helpful–they help teams decide when they are "done", versus when they need to keep generating more ideas.

For Step 3, Fig. 3 shows brainstorming off of the whole system map created in Step 1. (Note that Step 2's LCA may have narrowed the boundaries chosen by the

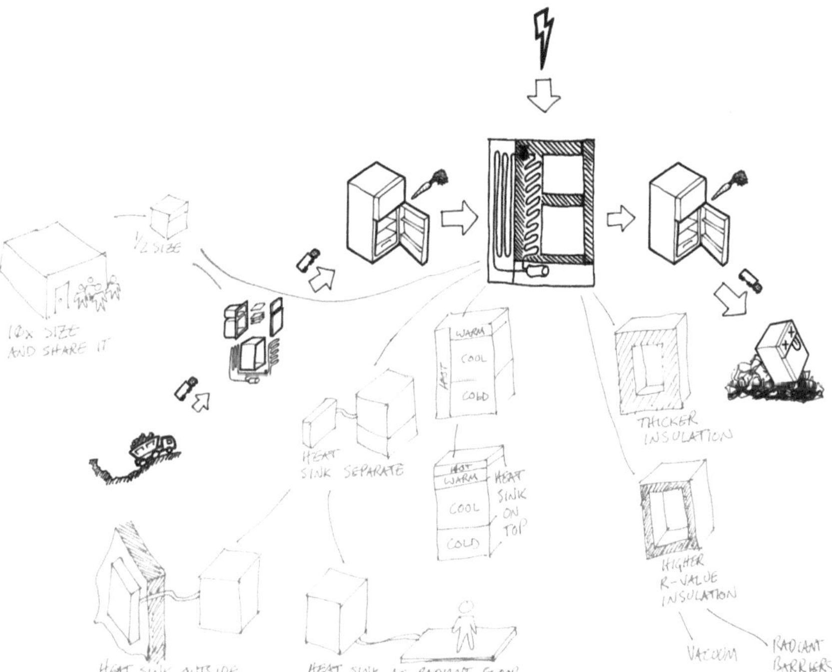

Fig. 3 WSLC Step 3: brainstorming off a whole-system map shows which parts of the system have new ideas and which do not, encouraging thoroughness in ideation

design team as being within their scope of influence—the map here does not include all the nodes of the original whole system map in Step 1.) Step 2's top sustainability priority is the "problem statement" that Step 3's brainstorm generates solutions for. Designers can avoid the classic brainstorming trap of fixating on certain solution types [29, 30], simply by brainstorming off of every node on the map. They immediately and constantly see what components of the product or steps in the system they have new ideas for, and which do not have new ideas. By requiring new ideas for every single node in the system map, brainstormers can break fixation to consider more varied solutions and ensure they are not missing opportunities for innovation in forgotten parts of the product's system. Visual messiness can be alleviated by having several copies of the system map (see Fig. 4 for a second page with more ideas). This is especially easy when teams are mapping and brainstorming with software. Most of the text in Figs. 3 and 4 is not readable here, but it is not necessary to understand the results of the brainstorm—the point is that brainstormers can see where they do and do not have new ideas in the system.

Figure 4 shows the second value of brainstorming off the system map: designers can see when they have eliminated a component or skipped a step in the system. The more steps they skip with an idea, the more radically innovative that idea is.

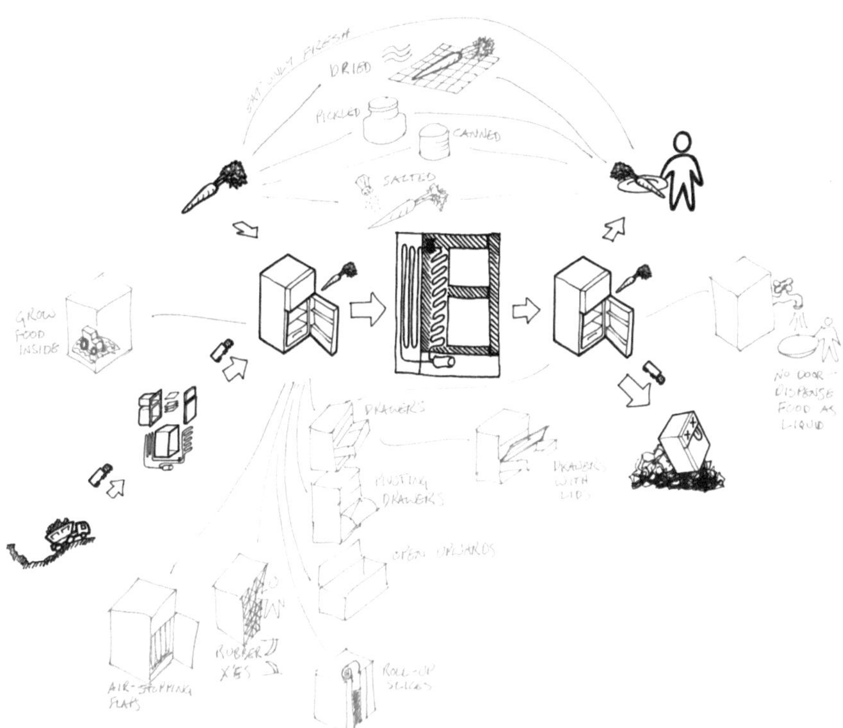

Fig. 4 WSLC Step 3 continued: brainstorming off a whole system map also shows when an idea eliminates components or skips steps in the system, encouraging more radical ideation

Teams can intentionally brainstorm to skip more steps for more radicalism. In this example, several ideas (salting food, pickling food, etc.) actually eliminate he refrigerator entirely. (Note the electricity part of the system was removed here to make the illustration compact.) These ideas may or may not be improvements, but that will not be judged here, it will be judged in Step 4.

For Step 4, Fig. 5 shows the design team's favorite few ideas from Step 3 being evaluated with LCAs to find a "winner" or "winners." The team chooses the number of top ideas by balancing convenience (fewer ideas) against thoroughness (more ideas). A team engineer or external consultant performs these assessments by estimating the reduction or increase in product energy use, materials, change in materials, etc. for the different design ideas. The estimated bill of materials, transport, and usage have their impacts calculated the same LCA software as in Step 2, using the same scoring methodology and functional units. The assessments should provide a single score for each design idea and the original product, so they can easily compared as in this graph. Figure 5 shows a student team using both single-score and CO_2e measurements; it is simpler to only use single-score, but two measurements can help show uncertainty. Again, as in Step 2, sustainability measurement systems other than LCA can be used (e.g., Cradle to Cradle, EPEAT, etc.) as long as they have clear scores to show at a glance which ideas are the biggest winners.

Design ideas from Step 3 that do not significantly improve the score from the original product (such as "A", "C", and "E" above) should be thrown out in Step 4; likewise with design ideas that do not meet the business metrics from Step 2. This leaves only top-performing ideas for both the environment and the business. A faster but less rigorous alternative to this analysis is to simply measure ideas against Step 2's metrics. Remember that quantitative analyses of early-stage design ideas are merely guesses, and there will be large uncertainties. Differences of a few percent, or even perhaps 20–30 %, may be illusory. Ideally teams should only choose final design ideas that are very clearly large improvements. If none of the available ideas show large improvements, Step 3 may need to be repeated for more ideas.

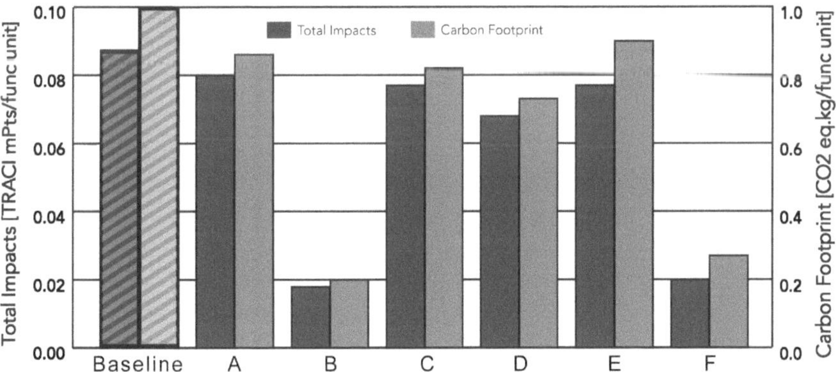

Fig. 5 WSLC Step 4: choosing between estimated LCAs of the top six design ideas ("A"–"F") by comparing them to the original product ("Baseline")

The WSLC design process can be done once or can be done iteratively, to address multiple issues or to drill down into specific parts of the product's system. A system map can be made once and used for many brainstorms, or many maps can be made at different scales. The method is intended to be flexible so teams can shape it to their needs.

4 Anecdotal Trials

The WSLC method has been used in classes at Minneapolis College of Art and Design, California College of the Arts, Emily Carr University of Art and Design, UC Berkeley, and elsewhere. In addition, universities such as University of Calgary in Canada, India Institute of Technology Kharagpur, University of Hongik in Korea, and many others have used the online video summary of the method in engineering and design classes. Hundreds of thousands of people from countries all around the world have viewed the video. The method has been applied to consumer electronics, clothing, outdoor gear, furniture, and kitchen appliances. In theory it can be used for any product category. Partner companies for anecdotal trials have included Motorola, Steelcase, Hamilton Beach, Cascade Designs, Oboz, Panasonic, Pacific Outdoor Gear, Anthro, Rayne Longboards, and others. Several of these companies have commented on its usefulness. Panasonic executives said they were "surprised and delighted how the students' ideas were not just improvements for sustainability, but were innovations for the functionality and aesthetics of the product as well." A Hamilton Beach executive said "Your process was as valuable as the actual concepts it produced."

5 Discussion

There are many green design methods; the most popular ones listed in an informal survey of practitioners were biomimicry, Cradle to Cradle, LCA, TU Delft's/UNEP's method [12], and "systems thinking." Comparing these, we find: TU Delft's method has tools to help teams prioritize sustainability, ideate, and evaluate results, much like WSLC. It has more tools related to business processes, but is more complicated (a nine-step process) and has less rigor in sustainability evaluation (uses team opinion rather than quantitative modeling). Its ideation tool is standard brainstorming, it does not have unique tools as biomimicry, WSLC, or systems thinking have. Biomimicry is an excellent ideation tool, and can also provide inspiration for measurement, but it does not have well-developed tools to measure sustainability like LCA or Cradle to Cradle, nor to ensure that they are prioritized in the design process, as WSLC or TU Delft's do. Cradle to Cradle the book is just a design philosophy; Cradle to Cradle the certification standard is a very well-developed tool to measure the sustainability of a product or service, and provides specific prescriptive suggestions, but it does not have ideation tools as biomimicry, WSLC, or systems thinking do. LCA is a measurement tool, useful for setting priorities and deciding between design options, but it does not provide ideation tools. "Systems thinking" is often poorly defined, but two concrete methods are Donella Meadows's "12 leverage points" [31] and Rocky Mountain Institute's Factor Ten Engineering Principles [13]. These are useful for driving radical ideation and can help set goals, but do not provide concrete metrics like LCA or Cradle to Cradle. The Whole Systems + Life-Cycle method was an attempt to join these for the best of both worlds. It also adds visual mapping of the system, to aid designers and engineers in boundary-setting and ideation.

While the WSLC method does appear promising, it has limitations. It requires the company to have at least one team member, or a consultant, who is conversant with a sustainability measurement system (LCA is preferred, but point-based certification systems such as EPEAT or Cradle to Cradle may also be used). It requires some training to learn how to create and use the system maps most effectively, and as with any design method, it does require time and effort to deploy.

The WSLC method always satisfies Lindahl's recommendations for engineer-friendly sustainable design tools in points #1, 2, 3, and 7, and can be used in ways that fit #4, 5, 8, and 9 to a greater or lesser extent; it does not fit #6. Specifically:

1. Its benefits are easy to understand: it should simultaneously improve product sustainability and innovation by focusing teams on their highest sustainability priorities, then driving more thorough and radical brainstorming on those priorities.
2. Its process is easy to understand, a simple four steps describable in five minutes.
3. It is adjustable to different contexts by letting the team choose the depth and breadth of the system map, choose the complexity of the priorities list, and choose whether to use LCA or other scoring systems.
4. Its system-mapping has no setup time other than finding a whiteboard or large paper to draw on. LCA does take significant time to perform, but it can be done beforehand, or certification checklists take little time to estimate and are part of the design thinking process, priming team members for issues to consider.
5. It is designed for simultaneous cooperation, but this is not a hard requirement. System maps can be collaboratively drawn asynchronously, brainstorms can be asynchronous, and LCA is best done individually and asynchronously.
6. It does require data for the LCA, and for the product's bill of materials in the system map. The method could be used without data, but its effectiveness will likely be reduced.
7. It is a fundamentally visual method, both in the system mapping and the graphing of LCA results (or other scores).
8. Its LCA portion is IT-based, using specific LCA software; however, its actual design ideation portions generally use whiteboard or paper. System mapping and brainstorming can (and often have been) performed using online collaboration software such as Google Docs, but these tools are not specific to it.

9. It is intended to give a result, not just a direction, but this result can be quite general. Scoring of final design ideas often shows that one category of ideas (e.g., product-as-service) performs much better than other categories (e.g., substituting a material with a greener material). Engineers and designers can use this as a direction rather than a fully-specified result.

6 Conclusion

The Whole-Systems and Life-Cycle method has been anecdotally shown to drive both sustainability and innovation. It clarifies sustainability priorities by using LCA or other measurements to find the product system's highest eco-impacts, and encourages accountability by choosing "winning" ideas based on their estimated improvement of these measurements. It encourages innovation by making the brainstorming process both more thorough and more radical through the use of system maps. While its effectiveness has not been deeply studied, evidence from student teams on industry projects suggests that it drives environmental improvements as well as creativity in product features and business. Companies have found the results valuable, and the method was found to meet between four and eight of Lindahl's nine recommendations for engineer-friendly design tools. This innovation may provide another important economic incentive for companies to practice sustainability, in addition to cost savings, lowered liability risk, and regulatory compliance.

Future studies should perform field trials of this method versus others, not only to compare which perform best, but also to find the valuable elements of each method. There may be other values that companies perceive in the methods, along with or even above innovation. Engineering and design educators are encouraged to try the method, as are industry engineering and design teams, to see if their results match those described here, as well as to suggest improvements. Together we can turn sustainability from a burden into an innovation tool that provides direct value to companies whether or not they prioritize sustainability itself.

Acknowledgments Thanks to Adam Menter and Dawn Danby of Autodesk for their support of the method.

References

1. US Green Building Council, Three Billion Square Feet of Green Building Space LEED-Certified [Press Release]. Accessed 14 April 2014 from http://www.usgbc.org/articles/three-billion-square-feet-green-building-space-leed%AE-certified
2. Behrisch, J., Ramirez, M., Giurco, D.: The use of ecodesign strategies and tools: State of the art in industrial design praxis. In: Proceedings of the ERSCP-EMSU Conference, Delft. 25–29 Oct 2010

3. Behrisch, J., Ramirez, M., Giurco, D.: Representation of ecodesign practice: international comparison of industrial design consultancies. Sustainability **3**(10), 1778–1791 (2011)
4. Bansal, P.: The corporate challenges of sustainable development. Academy. Manage. Executive. **16**(2), 122–131 (2002)
5. Hauschild, M., et al.: From life cycle assessment to sustainable production: status and perspectives. CIRP Ann. Manufact. Technol. **54**(2), 1–21 (2005)
6. Millet, D., Bistagnino., et al.: Does the potential of the use of LCA match the design team needs?, J. Clean. Prod. 15(4): 335–346 (2007)
7. Centre for Sustainable Design: Sustainable Innovation 2014 conference website. http://cfsd.org.uk/events/sustainable-innovation-2014/. Accessed on 14 May 2014
8. Collado-Ruiz, D., Ostad-Ahmad-Ghorabi, H.: Influence of environmental information on creativity. Des. Stud. **31**(5), 479–498 (2010)
9. Waage, S.A.: Re-considering product design: a practical 'road-map' for integration of sustainability issues. J. Clean. Prod. **15**(7), 638–649 (2007)
10. Benyus, J.: Biomimicry: Innovation Inspired by Nature, Harper Perennial (1997)
11. Cook, D.: The Natural Step: Towards a Sustainable Society, Green Books (2004)
12. Crul, M., Diehl, J. C.: Design for Sustainability: A Practical Approach For Developing Economies, UNEP/Earthprint (2006)
13. Lovins, A., et al.: Factor Ten Engineering Design Principles, Rocky Mountain Institute (2010)
14. MBDC, LLC.: Overview of the Cradle to Cradle Certified Product Standard. Accessed 14 May 2014 from http://www.c2ccertified.org/images/uploads/C2CCertified_V3_Overview_121113.pdf (2012)
15. IEEE Standards Board, IEEE 1680: Standard for Environmental Assessment of Personal Computer Products, Including Laptop Personal Computers, Desktop Personal Computers, and Personal Computer Monitors, IEEE (2006)
16. Baines, T.S. et al.: State-of-the-art in product-service systems. In: Proceedings of the Institution of Mechanical Engineers, Part B: Journal of Engineering Manufacture, vol. 221, no. 10, pp. 1543–1552 (2007)
17. Hart, S.L., Ahuja, G.: Does it pay to be green? An empirical examination of the relationship between emission reduction and firm performance. Bus. Strategy. Environ. **5**(1), 30–37 (1996)
18. Konar, S., Cohen, M.A.: Does the market value environmental performance? Rev. Econ. Stat. **83**(2), 281–289 (2001)
19. Kim, W.C., Mauborgne, R.: Strategy, value innovation, and the knowledge economy. Sloan. Manag. Rev. **40**, 41–54 (1999)
20. Roos, J. (Ed.): Intellectual capital, Macmillan (1997)
21. Lindahl, M.: Engineering designers' experience of design for environment methods and tools–Requirement definitions from an interview study. J. Clean. Prod. **14**(5), 487–496 (2006)
22. Menter, A.: Autodesk sustainability workshop: advancing the practice of sustainable engineering through education. In: 2011 IEEE International Symposium on Sustainable Systems and Technology (ISSST) pp. 1–6 (2011)
23. Eggermont, M., Rosehart, W. B.: Integration of desired learning outcomes in a first year design course. In: Proceedings of the Canadian Engineering Education Association (2012)
24. Faludi, J.: Sustainable design education through modular online content. Soc. Info. Technol. Teacher Edu. Int. Conf. **2012**(1), 3599–3601 (2012)
25. Schlecht, L. L.A.: Impact of prototyping resource environments on idea generation in product design, PhD Thesis, Massachusetts Institute of Technology (2013)
26. Unknown,: Reviews and Resources, J. Edu. Sustain. Develop. 6: 161 (2012)
27. Goedkoop, M., Spriensma, R.: The Eco-Indicator99: A Damage Oriented Method For Life Cycle Impact Assessment, Pré Consultants (2001)
28. Goedkoop, M., Heijungs, R., et al.: Recipe 2008. A Life Cycle Impact Assessment Method Which Comprises Harmonised Category Indicators at the Midpoint and the Endpoint Level, Pré Consultants (2009)

29. Cardoso, C., & Badke-Schaub, P.: Give design a break? the role of incubation periods during idea generation, DS 58-2. In: Proceedings of ICED 09, the 17th International Conference on Engineering Design, Design Theory and Research Methodology, vol. 2, pp. 24–27 (2009)
30. Cardoso, C., Badke-Schaub, P.: The influence of different pictorial representations during idea generation. J. Creative. Behav **45**(2), 130–146 (2011)
31. Meadows, D.: Leverage points. Places to Intervene in a System. The Sustainability Institute, Hartland, Vermont, USA (1999)

Green Is the New Colour for Menstruation. *Uger* Fabric Pads Show the Way Through a Sustainable Perspective

Lakshmi Murthy

Abstract Early woman managed menstruation in many ways. These included isolating herself when bleeding, using leaves or cloth and other techniques. These used to be environmentally friendly systems, causing no harm. As societies evolved, newer ways of managing emerged, giving rise to menstrual products that are no longer sustainable. Waste now includes menstrual debris, which is an environmental concern. We examined menstrual products used by women in South Rajasthan through a sustainable perspective. We developed a new product, the *Uger* fabric sanitary napkin to address some of the gaps we found. For assessing sustainability, we developed a diagrammatic representation for sustainable menstrual management, the PASS diagram. We assessed all these products through PASS using qualitative data from studies. We found *Uger* to be a sustainable option. The PASS diagram has limitations, but when improved in future, it can potentially be a local powerful tool for communities to assess menstruation management sustainability.

Keywords Menstruation · Sustainable · Reusable · Diagram

1 Introduction

Humans, as others in the animal kingdom, hunted, gathered, reproduced and passed on. The next generation carried on in much the same way. When communities began to settle, it changed the way life was managed, it was no longer the simple hunting—gathering sequence, it was replaced by complexities. In the 21st century, it is no longer easy to be simple. "Complexity is failed simplicity" [1]. In this study we examine the issue of menstruation, a body expulsion specific to women. We look at design systems or products that exist for this biological phenomenon. Earlier

L. Murthy (✉)
Industrial Design Centre, Indian Institute of Technology, Bombay, India
e-mail: ellemurthy@gmail.com

A. Chakrabarti (ed.), *ICoRD'15 – Research into Design Across Boundaries Volume 2*,
Smart Innovation, Systems and Technologies 35, DOI 10.1007/978-81-322-2229-3_19

periods were managed in simple ways which aligned with nature causing no imbalance to the environment. The systems were sustainable. Period management today has been commoditized, dominated by disposable products. An impact caused by this are huge volumes of visible menstrual waste. This can no longer be ignored, clearly, menstruation management has become unsustainable. Can we re-examine menstruation management products and systems and will such an assessment assist in moving from an unsustainable to sustainable scenario? The study tries to find some clues looking at menstruation management from sustainable perspectives.

2 Background

2.1 Menstruation and Management

The human body expels saliva, sweat, gas and many other products. Menstruation or periods is one such physiological phenomenon. "A period is the part of the menstrual cycle when a woman bleeds from her vagina for a few days. In most women, this happens every 28 days or so" [16]. Early woman managed menstruation using leaves, moss, mud or sand [5]. When clothing evolved, periods were managed differently. When garments became old or torn, the stronger parts of the fabric were recovered from it, converting this into absorbent material. The early nineties saw inventions, the sanitary napkin, the tampon and the menstrual cup [5]. The disposable sanitary napkin revolutionized the way menstruation was managed. Today there are many varieties of branded napkins to choose from and more and more women show preference for disposable [10].

2.2 Menstrual Management in the Study Area

Three types of menstrual products are currently used by girls and women in the study area—Southern Rajasthan, India. The first is cloth recovered from old garments,[1] the second the *Time Piece*[2]—TP and the third, the Disposable Sanitary Napkin[3]—DSN. Infrastructure and social circumstances govern and influence both selection of a menstrual product and the way products are managed. Most homes do not have latrines, jungles and fields are used for defecation and urination. Changing

[1] Cloth is recycled from old garments petticoats, turbans, veils, towels, bed sheets, saris, loin cloth.

[2] *Time Piece*, is a rectangular piece of fleece or poly acrylic fibre. The non woven material, is available in dark shades, the fabric was tested using standards for identification of fiber—ASTM D276 and identification of dyes as per AATCC 161-2012.

[3] DSN—Disposable Sanitary Napkins—branded or locally produced.

of the menstrual product is done in the *nyora* or barn where livestock are tied or at home at a time when men folk are out of the house. Alternatively, men are asked to wait outside as they change. All this has an impact on health as women are not able to change as often as they need to. Additionally there are many superstitions; one strong notion is that men turn blind if they see menstrual blood. Dark coloured fabrics are thus preferred over light colours so that residual blood stains remain hidden; this colour selection means the product is never completely washed out. Subsequently this is hung in a dark corner, away from the sight of men, with no sunlight reaching it. These practices compromise health. We also found that reusable products are losing value, as users see reusable as an old fashioned method. There is aspiration and migration from homemade products to those available in the market, is a common trend. A need was thus felt for an alternate product that would be a healthy option, providing an environmental friendly and sustainable menstrual management item that would give the user dignity.

2.3 Menstruation Products and Impact on Environment

Used cotton menstrual rags, when buried by users, degrade easily, within six months. Rags when burned, give off a small amount of ash like residue going back into earth. Disposable sanitary products available in the early part of this century, were made from materials that were bio degradable. These were sustainable systems. Recently all this changed, impact on the environment was realized when used disposable products started becoming visible. Environmentalists saw that debris washed up on coast lines were ear buds, diapers, tampons, applicators and sanitary napkins, made up of non degradable raw materials [17]. The modern non woven sanitary napkin was an innovation that had clearly ignored long term impact on environment. The highly absorbent disposable napkin available today, is composed of non degradable material such as polymers, polyacrylate gel and other plastic derivatives [19]. This along with other waste has choked landfills [2]. In India, menstrual waste has started to become a problem, [15, 18].

Designers have ignored sustainability issues, designing instead for planned obsolescence [11]. There is clearly a need to relook at menstruation management from a sustainable perspective, from the point view of the product, its management and its long term impact on environment.

3 Objective of the Study

The study was designed:

- to measure or assess sustainability in existing menstrual management products
- to develop a new menstrual product and assess this for sustainability

4 Method

4.1 Developing a Diagram for Assessing Sustainable Menstruation Management Called "PASS"

A sustainability diagram was developed, the "PASS" diagram to assess menstruation management sustainability. Additionally this would assist a user to make informed choices during menstrual product selection. This is discussed in Sect. 5.

4.2 Designing a Menstrual Product—Uger Pads

A new menstrual product was developed and introduced to users in the study area. This is discussed in Sect. 6

4.3 Assessing Sustainability of Menstrual Products Using the "PASS" Diagram

We had earlier studied all menstruation products that are in use and the practices that govern the use.[4] These products were then assessed and graded using "PASS". The new menstrual product was also assed using PASS. We discuss this in Sect. 7

5 PASS—A Diagram for Assessing Sustainable Menstruation Management

"Sustainable" is defined in dictionaries as "capable of being continued with minimal long-term effect on the environment" [20]. Sustainable communities mean—well being of the community from the perspectives of surroundings, wealth, health and social-cultural contexts [7]. Researchers have represented sustainability through visuals, for example the Venn diagram model [12, 13], which has three aspects to sustainability—social, economic and environmental. The four pillar model [21] includes culture. Culture takes into account community values and community expressions [6, 8] (Fig. 1).

[4] Studies were done through 5 home visits, visiting spaces used by women for toilet needs, conducting 3 group discussions with 68 young women and girls. All girls and women are from schedule caste families, age group 16–50 years. Additionally we interviewed 3 doctors.

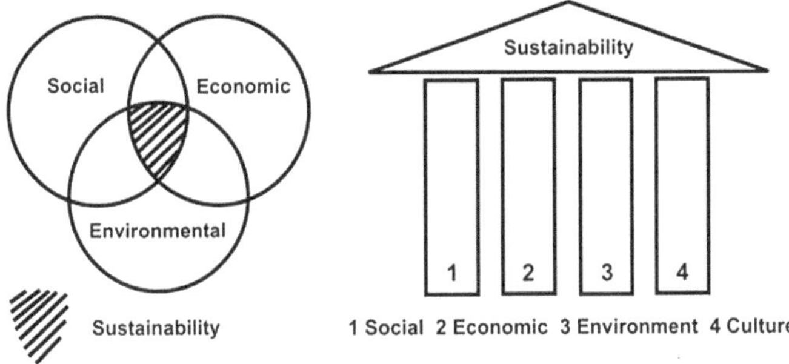

Fig. 1 Sustainability, Venn diagram model, 4 Pillar model. *Source* Author generated

We saw menstruation management as sustainable if it included 4 aspects—environmental, economic, social and health. This aligned with the 4 pillar model as opposed to the Venn model. We defined "Sustainable Menstruation Management" or SMM as—*a method of managing menstruation, by the practice and use of products and systems that maintain balance between environmental, economic, social and health aspects, causing no harm to the user or to the larger community*. The four pillars represented environment, economic, social and health, together holding up a roof, the roof representing menstruation (Fig. 2).

Ideally when all four pillars attain equal height a fully sustainable menstrual management situation can be achieved. For different levels of sustainability each pillar was demarcated into a scale as described in Fig. 3. A product could be measured using this scale, guided by a simple qualitative analysis based on the studies conducted.

To make the model relevant to the study area each pillar was translated into Hindi language and renamed. Environment Pillar—*Paryavaran*—named P—to

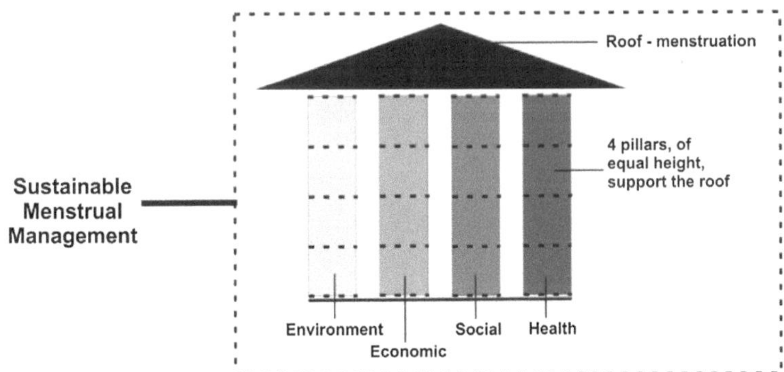

Fig. 2 Diagram of SMM model. *Source* Author Generated

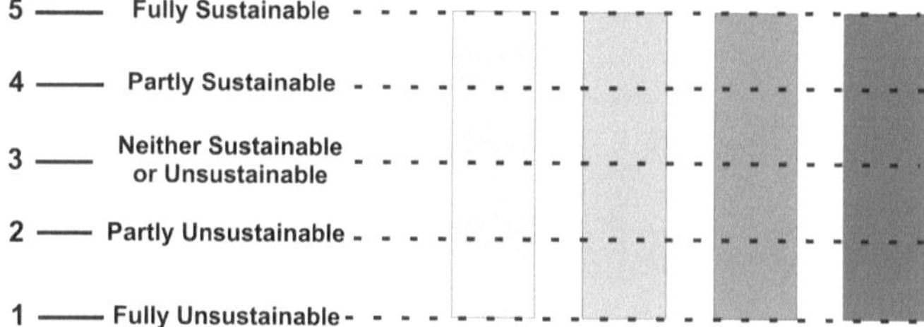

Fig. 3 Scale of 1 to 5 with description is assigned to each pillar

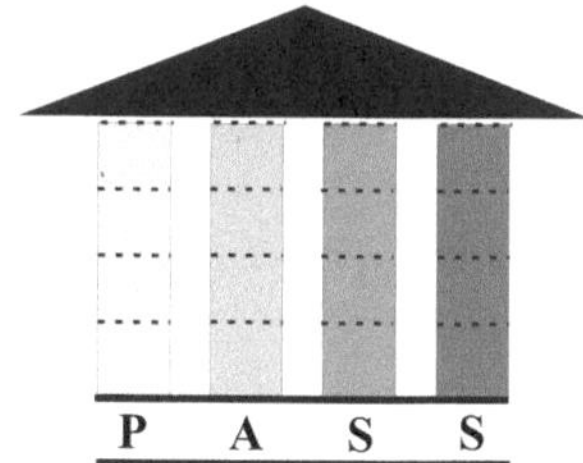

Fig. 4 PASS Diagram—with Score 20–100 % SMM—Source Author generated

consider LCA,[5] WF[6] and MD.[7] Economic Pillar—*Arthic*—named A—to consider production cost, cost to user. Social Pillar—*Samaj*—named S to consider infrastructure, social acceptance, attitude toward a practice or product having direct bearing on individual user and community. Health Pillar—*Swasth*—named S to consider negative or positive health impact caused by the use of the product both on the user and on the community.

"PASS"—also worked as a pun—"pass" and "fail" are common terms used in the area to sanction or reject anything, hence the acronym was appropriate. When all four pillars reach full height, Pillar P = 5, Pillar A = 5, Pillar S = 5, Pillar S = 5, the score is added. 5 + 5 + 5 + 5 = 20, an ideal score, denoting the product is 100 % sustainable (Figs. 4 and 5).

[5] LCA Life Cycle Assessment assess a product at all stages production, consumption, till disposal.

[6] WF Water Footprint assesses total amount of fresh water, direct, indirect water use by producer and consumer.

[7] MD Menstrual Debris means residue, remains of menstrual product after use and disposal.

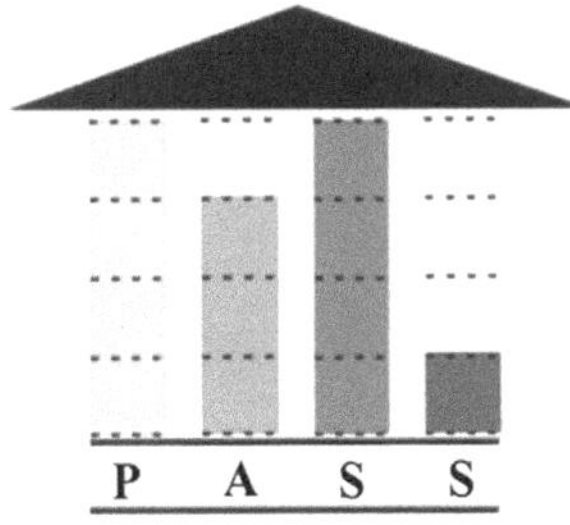

Fig. 5 PASS Diagram—with Score 16–80 % SMM—Source Author generated

6 Designing a Menstrual Product—*Uger* Pads

Our understanding from Sect. 2 guided the brief for the design of a menstrual product. Reusable cloth pads appeared to be a solution. We consulted with *Eco-femme* at *Auroville*, *Tamil Nadu*, the pioneers of cloth pad making in India [4]. We took inspiration from their product, both the design and the philosophy of reuse—reusing places lesser burdens on environment. Thus cotton fabric[8] was selected, cool to skin, a non irritating raw material, causing no menstrual debris as the material biodegrades

- The design was made to manage different volumes of discharges,
- Colour of fabric that would be next to the skin was kept white to help user identify differently coloured abnormal discharges
- It was styled, to be fixed to underwear, to have a comfortable contour and size

The pads designed were named *Uger*[9] Pads, in two styles, for light flow and heavy flow. A Self Help Group of the NGO produced these cloth pads (Figs. 6 and 7).

7 Using the PASS Diagram for Assessing Menstrual Products

7.1 PASS for Cloth

Pillar P—Score 5—LCA and WF—Cloth is harvested from old used cotton garments, and takes on a new life, no cost for production and processing. MD—there is no menstrual debris, the fabric, bio-degrades or turns to ash if burned.

[8] Interviews with gynecologists—cotton cloth will not potentially cause a health problem. Problems occur due to poor personal hygiene and improper care of menstrual cloth.

[9] *Uger* means “New Beginning” in *Mewadi*, the language of Sothern Rajasthan.

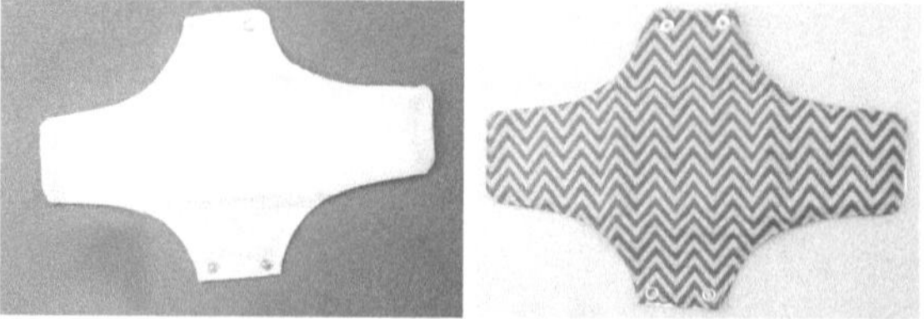

Fig. 6 Light pad designed for light flow, colour. *Source* Author generated

Fig. 7 Pads for heavy flow have were designed with towel inserts

Pillar A—Score 5—There is no manufacturing cost, no purchase cost. Maintenance cost, a cake of soap and 6–8 buckets of water per cycle are insignificant, as this has been considered a reproductive right [9] to maintain menstrual hygiene.

Pillar S—Score 3—Society's attitudes to privacy compromises sustainability. Very few private spaces for women are made. Society sanctions negative practices such as the use of dark menstrual products, endorsing poor hygiene and maintenance. Users are seeing home made products as old fashioned.

Pillar S—Score 4—Use of a poorly maintained dark cloth can lead to reproductive infections. Colour of discharge is never identified in time, treatment seeking is delayed. As there is no menstrual debris from this product, community health is also maintained.

7.2 *PASS for TP*

Pillar P—Score 3—LCA—Fabric is a non woven polymer based product, it is not biodegradable. WF—Water used for production was not calculated but assumed to be much lower than cotton production[10] [3]. MD—At the end of the product's life there is menstrual debris, but of a lower volume as compared to DSNs—user will

[10] It takes 2,720 litres of water to produce one cotton T-shirt.

dispose 8–10 TPs per year or 400 TPs in lifetime. This disposed material will pelletize[11] and enter waterways if just discarded. If it is burned, pungent fumes are toxic.

Pillar A—Score 4—Cost to individual user is low, hence affordable. Maintenance cost is less than half cake of detergent soap, 4 buckets of water per menstrual cycle.

Pillar S—Score 3—TP has social sanction due to its dark colour, superstition is managed, it is perceived to be very convenient as it dries very quickly, uses less water and detergent as compared to cloth or *Uger* pads.

Pillar S—Score 1—Nature of material in TP is reported to increase vaginal temperature, leading to discomfort and health problems such as itching, burning, fungal infection. The porous product traps detergent causing contact dermatitis in some users.

7.3 PASS for DSN

Pillar P—Score 2—LCA—We found most raw materials in the product to be non biodegradable—made up of polymers, non woven material and polyacrylate gel. Cellulose while biodegradable remains trapped in between polymer layers in the napkin, causing it to be preserved, remaining intact in a landfill. WP—Water used in production was not calculated but cellulose production and bleaching requires water. Overall water consumption was understood to much lower than cotton production. MD—Very high amount of menstrual debris, a user will throw away approximately 100–120 napkins at the end of each year, 3,000–5,000 napkins in a life time. There will be no biodegradation, only increase of waste in the landfill.

Pillar A—Score 2—Purchase cost to individual user is high, but no maintenance cost.

Pillar S—Score 3—Increasing accepted for convenience. Due to high cost, the napkins are thus worn for very long periods. This is a machine made product [23] and not labour dependent.[12] It generates no local employment and hence fails in an SLCA Frame work Social Life Cycle Assessment framework [22].

Pillar S—Score 3—Individual users present with itching, ulcers due to allergy to materials in the product. Often users wear the same napkin for 12 h to save money, this causes severe itching and other health problems.

---

[11] Materials in *Time Piece* photodegrade, they break down and enter waterways causing pollution.

[12] It takes a machine only 40 s to make one sanitary napkin.

7.4 *PASS for Uger*

Pillar P—Score 4—For LCA and WF—Cotton cloth is water intensive, with huge cultivation and processing cost. Maintenance cost is similar to cloth. MD—significantly very less menstrual debris as compared to DSNs. User will throw away 6 *Uger* pads after 12–14 menstrual cycles and 250 Uger pads in a life time. *Uger* pads biodegrade.

Pillar A—Score 4—The cost of each pad is Rs 80 very high as compared to the cost of a TP or DSN. But life time cost to user is less over all.

Pillar S—Score 3—light colored pads are not easily accepted. However many users saw *Uger* as an upward social move from Cloth to TP to Uger.

Table 1 Table showing product sustainability through PASS diagram

Product	Image	PASS diagram	SMM score
Cloth	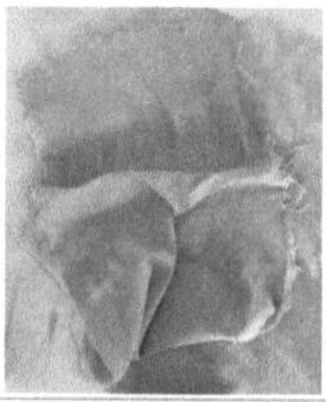	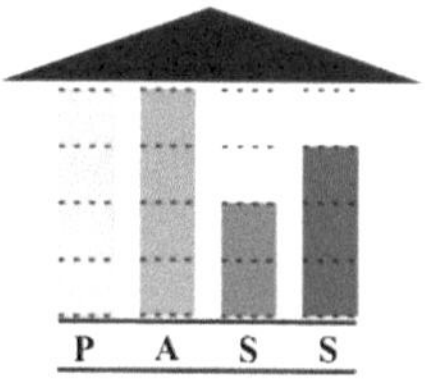 P A S S	5 + 5 + 3 + 4 = 17 85 %
Time piece	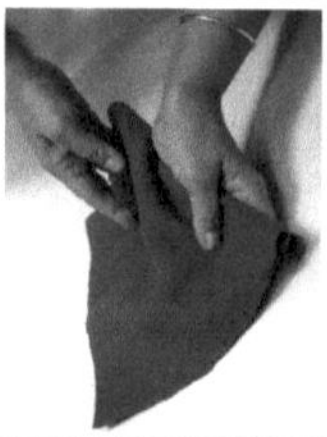	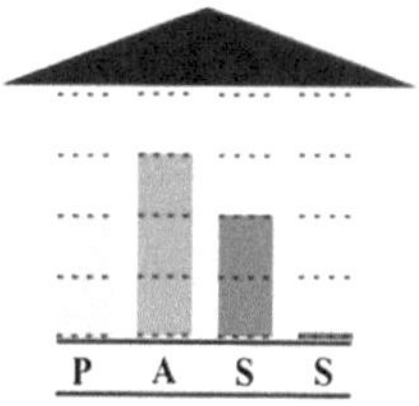 P A S S	3 + 4 + 3 + 1 = 11 55 %
Disposable sanitary napkin	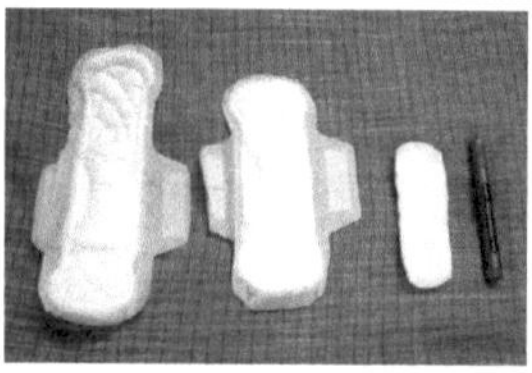	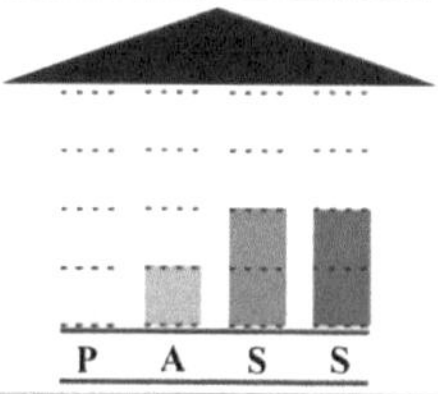 P A S S	2 + 2 + 3 + 3 = 10 50 %
Uger pads	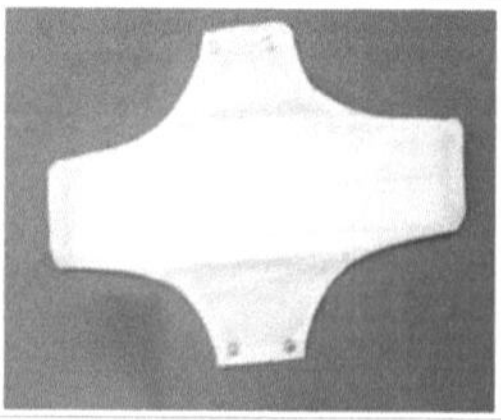	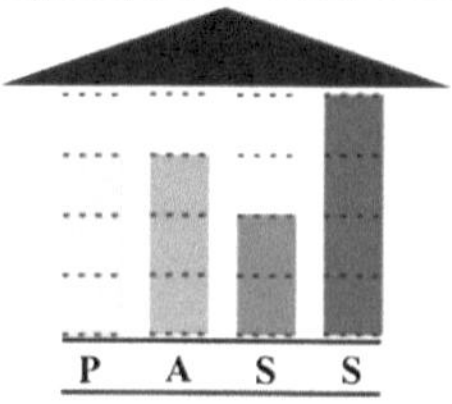 P A S S	4 + 4 + 3 + 5 = 16 64 %

Pillar S—Score 5—No health problems were reported from using *Uger*. Different trials have been going on in the field since 18 months. The pad design has white coloured fabric next to skin, allowing user to easily identify abnormal discharges and seek medical advise early. No debris from this product, community health is also maintained.

7.5 Summary of PASS Scores

See Table 1.

8 Conclusion

8.1 Limitations of the PASS Diagram

There are many limitations to the PASS diagram. The first drawback is that the grading has been done qualitatively by the researcher based on experiences and studies from the field areas. If this same diagram is used by another individual from a different background, it may be assessed differently, hence SMM scores will not necessarily be consistent, accuracy stands challenged. The second drawback is—the PASS diagram has not been tried by users in the field area. This test will determine user friendliness of the diagram and will understand whether the diagram can actually be used to make choices as this was one of the intentions.

8.2 Strengths of the PASS Diagram

Giving choices to the user from a sustainable perspective is not a new concept, Retailers such as Target in the USA [14] have begun to use indexing and scoring systems designed for users to make choices.

Scoring menstrual products from a sustainability lens is relatively a newer area. The PASS diagram in its current state provides that first step to assessing sustainability of a product, with scores that can be simply computed. In Table 1 from the score calculations, we can see that no menstrual management product can truly be declared as a fully sustainable product.

All products will have drawbacks when seen from different aspects. For example from an environmental perspective “P”, when the four products are assessed, cloth from old garments appears to be the most sustainable. It puts the least environmental negative load when compared to DSN which has the highest load. TP on the other hand puts some environmental load but is still significantly less when compared to DSN. Because *Uger* is made from BT cotton fabric,[13] cultivation is highly water intensive putting the product lower on the environment score. Similar analysis can be done through different pillars of the PASS diagram.

8.3 Strengths of Uger Pads Based on PASS Diagram

As economics in a family improve the move toward store bought menstrual products is a reality. Cost to environment and user health will be an important consideration. *Uger* pads is an option that is taking the middle path, it is trendy and has taken into account health and environment factors. Additionally *Uger* pads provides livelihood, pads are handmade generating employment within the community. Some factors will go against *Uger*, cost, resistance to using white and inconvenience of maintaining. Creating awareness and advocacy around sustainability can catalyse behaviour change and influence product selection. Users can be made aware that long term investments are required in a menstrual management product, sustainable from all four aspects, environmental, economic, social and health. Design thus has a huge role to play in keeping environment safe and in the current scenario sustainability has to be kept central to creating and manufacturing.

Acknowledgments The researcher gratefully acknowledges the support from (1) Aniruddha Joshi, Industrial Design Centre, Indian Institute of Technology, Bombay; (2) Deepti Panchratna and H.B.N. Murthy, for going over drafts of the paper; (3) NGO, Jatan Sansthan, Rajsamand District, Rajasthan—for sponsoring ongoing research; (4) People who willingly gave the researcher their valuable time; (5) Gynecologists, V. Ramakrishnan and V. Pendse; (6) NGO, Ecofemme, Auroville, Tamil Nadu—for sharing their work with us; (7) Design Department, Bansathali University, Tonk, Rajasthan, for laboratory testing of fabrics.

References

1. Bono, D.E.: Simplicity. Penguin Books, p. 42, (1998)
2. Chapman, J.: Emotionally durable design, Earthscan, p. 17, (2007)
3. Cotton and Water, (n.d).: http://ejfoundation.org/cotton/cotton-and-water. (April 23, 2014)
4. Ecofemme Washable Cloth Pads, (n.d).: http://ecofemme.org (May 14, 2014)
5. Finley, H. (n.d).: Museum of menstruation (mum.org) (Oct 6, 2013)
6. Four pillars of Sustainability, (n.d).: http://sustainablekingston.ca/community-plan/four-pillars-of-sustainability (May 2, 2014)
7. Green Communities.: http://www.epa.gov/greenkit/sustain.htm (2012). Accesses 4 Jan 2014
8. Hawkes, J.: The fourth pillar of sustainability cultural essential role in public planning. Common Ground Publishing Pvt Ltd (2001)
9. Human Rights, (n.d).: http://www.unfpa.org/rights/rh.htm (March 4, 2014)

[13] *Uger* products are currently made from BT cotton.

10. Juyal. R, Kandpal, S.D., Semwal, J., Negi, K.S.: Practices of Menstrual hygiene among adolescent girls in a district of Uttrakhand. Indian J. Community Health. **24**(2), (2012). http://www.iapsmupuk.org/journal/index.php/IJCH/article/view/173. Feb 8, 2013
11. Landes, L.: Resist planned obsolescence or accept the financial consequences. http://www.forbes.com/sites/moneybuilder/2012/11/06/resist-planned-obsolescence-or-accept-the-financial-consequences/ (2012). Accessed 26 May 2014
12. Lozano, R.: Envisioning sustainability three-dimensionally. J. Cleaner Prod. **16**(2008), 1838–1846 (2008)
13. Mann, S.: Visualising Sustainability. http://computingforsustainability.com/2009/03/15/visualising-sustainability/ (2009). Accessed 23 May 2014
14. Moore, J.: Target using scoring system to look at product sustainability. http://www.startribune.com/business/231211911.html (2013). Accessed 29 August 2014
15. Murthy, L.: It's time to take the bull by the horns—menstrual product debris can be reduced by using Uger fabric washable pads. To be published—Conference Design for Sustainable Wellbeing and Environment (June 2014)
16. Periods, (n.d).: http://www.nhs.uk/conditions/periods/pages/introduction.aspx (March 6, 2014)
17. Potts, T., Hastings, E.: Marine litter, issues, impact and actions. http://computingforsustainability.com/2009/03/15/visualising-sustainability/ (2011). Accessed 24 May 2014
18. Sabnis, V.: Used sanitary napkins, condoms choking city, Mid day, Jan 15, 2013. http://archive.mid-day.com/news/2013/jan/150113-pune-used-sanitary-napkins-condoms-choking-city.htm (2013). Accessed 8 Aug 2013
19. Sanitary Napkin Properties, (n.d).: http://textilelearner.blogspot.in/2012/05/sanitary-napkins-properties-of-sanitary.html (May 13, 2014)
20. Sustainable, (n.d).: (http://www.thefreedictionary.com/sustainable (Feb 4, 2014)
21. Sustainable Development. http://rosia-montana-cultural-foundation.com/sustainable-development (2013). Accessed 1 May 2014
22. UNEP Guidelines.: Guidelines for social life cycle assessment of products (2009)
23. YouTube.: How sanitary pads are made. https://www.youtube.com/watch?v=IfLOpvWGj80 (Jan 6, 2013). Accessed 22 April 2014

Sustainable Machining Approach by Integrating the Environmental Assessment Within the CAD/CAM/CNC Chain

Hery Andriankaja, Julien Le Duigou and Benoît Eynard

Abstract This paper reports on a sustainable machining approach based on dynamic environmental assessment integration through the CAD/CAM/CNC chain. To achieve this objective, the framework for supporting the environmental assessment features key environmental indicators and simplified LCA database. Key environmental indicators are composed of 'environmental metrics' and 'environmental impacts'. To implement the environmental assessment, cutting data are first extracted from STEP-NC file, a standard extending STEP to machining process data. Then, through the environmental metrics evaluation, the metrics quantities (Input/output flows of the machining process) are translated into environmental impacts by the help of the simplified LCA database. The obtained environmental profile is used by the CAD/CAM entity as a decision-aid, providing a better understanding of the environmental impacts of their machining technologies.

Keywords Sustainable machining · STEP-NC · Environmental metrics · Life cycle assessment · Environmental impacts

1 Introduction

Regulations [1, 2] and high energy prices have been identified as potential sources of motivation on energy consumption and emissions reduction within manufacturing firms. Industries are confronted with the challenge of designing both sustainable products and manufacturing processes. However, the sustainable manufacturing actions have been systematically conducted up to now with a focus on reducing both the production lead time and the costs, all which in turn may lead to a reduction in

H. Andriankaja (✉) · J. Le Duigou · B. Eynard
Department of Mechanical Systems Engineering, Roberval Laboratory - UMR CNRS 7337, Université de Technologie de Compiègne, CS 60319, 60203 Compiègne Cedex, France
e-mail: hery-tsihoarana.andriankaja@utc.fr

A. Chakrabarti (ed.), *ICoRD'15 – Research into Design Across Boundaries Volume 2*, Smart Innovation, Systems and Technologies 35, DOI 10.1007/978-81-322-2229-3_20

energy consumption. In addition, environmental assessments are carried out beyond the product design (CAD) and manufacturing processes design (CAM). Thus, they are not yet integrated into the design of manufacturing processes, and this post-design environmental consideration even lead to solutions which are increased in production costs [3]. Integrating environmental aspects earlier in the design of the production system is a promising approach to set up a new model of sustainable manufacturing.

In this paper, we report to an on-going research work which stands for integrating environmental assessment within the CAD/CAM/CNC chain based on the STEP-NC standard. The STEP-NC (STEP-compliant Numerical Control) file is a new ISO standard (ISO 14649) to create machining programs that can be shared between many entities, thus allows integrating CAD requirement data with CAM process data. In Sect. 2, a literature review is provided and the Sect. 3 stands for a new proposition to integrate the environmental assessment within the CAD/CAM/CNC chain. The paper concludes with a discussion about the findings and future works to develop and validate a prototype tool for supporting the sustainable machining approach.

2 Research Background

Over the past four decades, the environmental burdens due to industrial activities became a global problem increasingly important and are currently a great societal challenge [4, 5]. The concept of sustainable manufacturing was appeared and was declined in various Anglo-Saxon labels (for instance environmentally conscious manufacturing, green manufacturing …). This section is devoted to the summary of relevant methods for sustainable manufacturing, in order to identify their concept and the nature of environmental indicators they use.

The simulation of a production system is initially based on the Discrete Event Simulation approach [6–8]. However, DES-based software tools available in commerce such as Plant Simulation, Witness, AnyLogic Enterprise Dynamics…only offer a very limited functionality to support an environmental analysis. Consequently, many researches have emerged in the development of specific approaches from these software tools.

Thiede et al. [9] have classified these approaches into three paradigms (Fig. 1) according to the implementation of the environmental assessment.

In Paradigm A, the DES module is used to simulate production systems and typical variables such as the state of the machine or the manufacturing duration. The outputs would be treated in an external evaluation tool where data are converted into environmental aspects (for instance energy consumption). Paradigm B regroups approaches structured with a dynamic coupling of the DES with an assessment tool. This allows a more detailed analysis of environmental characteristics in considering existing interdependencies between different sub-systems or

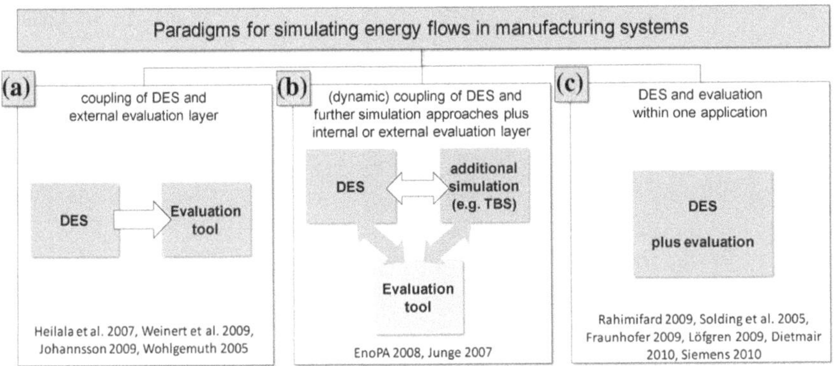

Fig. 1 The three paradigms for coupling environmental assessment with DES [9]

processes. In Paradigm C, the environmental concerns are implemented within a single DES application.

By analogy to the simulation of the production systems, the simulation of machining parameters at the CAD/CAM level seems an interesting place to integrate environmental impacts assessment of the machining operations. To address this issue, the proposed approaches aimed at optimizing environmental factors such as the duration of an operation, the quantity of removed chips, the use of cutting fluids, and energy consumption [10–12].

For measuring the energy impact of using a NC machine, Kong et al. [13] developed a model based on the cutting path planning via the NC code to estimate the energy consumption and greenhouse gases emissions resulting from the machining process behavior. NC code controls the motion of NC machine and the machining parameters, including the list of cutting tools, the specification of the machine and the geographic information on the power grid mix (electric energy). The entries are then processed to estimate the time required to run the NC code and GHG emissions resulting from the operation on the machine tool during the execution of NC code. The outputs of this model characterize the performance, versus the emission of greenhouse gases from the considered operation (Fig. 2).

Recently, Zhao et al. [14] used the information models of design and manufacturing processes, including the STEP-NC standard. According to Zhao et al., STEP-NC is the most appropriate structure and must be enriched with data on the

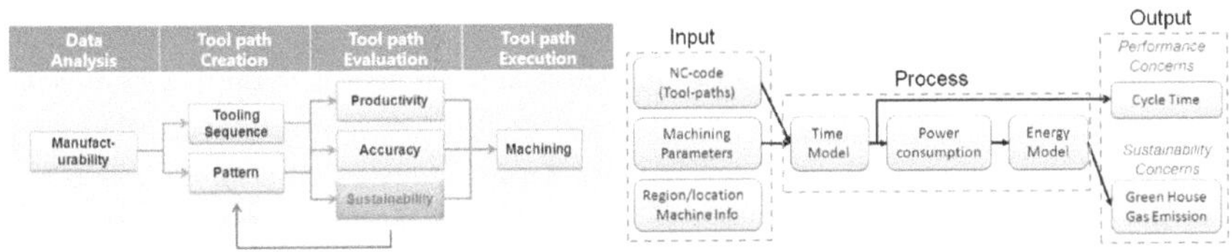

Fig. 2 Tool path planning simulation incorporating the sustainability concern [13]

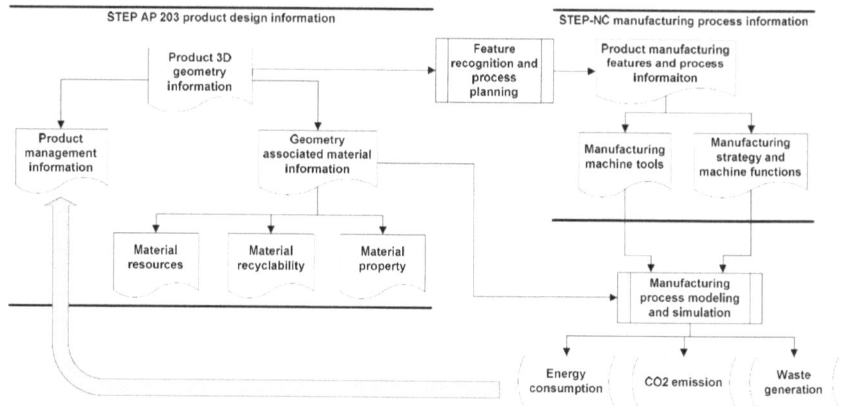

Fig. 3 STEP and STEP-NC information models for integrating environmental aspects [14]

environmental aspects of products and processes, as shown in Fig. 3. With the proposed model, Zhao et al. aim at integrating information related to LCA into the parameters of the design and the product development process planning. According to the interactivity of the CAD/CAM/CNC chain thanks to STEP-NC, the environmental knowledge and data accumulated during the manufacturing of the products or other life cycle stages (for instance end of life) can be returned to the design stage in terms of experience feedback to improve the environmental performance of new products.

In terms of integration, the methods presented above can be grouped into two main categories: (1) approach by coupling environmental tools with the software tools for simulation of the production system, and (2) approach by integrating environmental evaluation through the CNC chain. The first approach encompasses the entire production system and seems easy to achieve regarding the market availability of the DES-based software tools. However, the question arises on the appropriation of the environmental tools embedded in DES by the CAD/CAM engineers. Moreover, one wonders if the management of the entire production system simulation does not devote extensive material, human and financial resources due to the extent of the scope to be taken into account. Indeed, the production system in a holistic vision [9] includes not only flows contributing to environmental impacts and costs of the production, but also flows interacting with the infrastructure which provide the power supply, computers and peripherals and all other conditions necessary for the production. On the other hand, the evaluation of environmental impacts through the NC simulation seems more accessible by the CAD/CAM engineers. However, this approach is focused solely on the machining system. Then, the risk of pollution transfer must be monitored. However, conducting environmental assessment still remains a challenge for industrials although universal tools exist. The main problems are: (1) to collect data from products/processes and to appropriate environmental tools; (2) to ensure the representativeness of the model in spite of the intangibility of the environmental assessment through certain machining parameters (for instance, the variation of global warming potential impact when the depth of cut varies). These two barriers hinder to set up a pertinent approach for reflecting the progression/regression induced by a machining system in terms of environmental impacts. The question we will answer in this article is then: how to develop and integrate the environmental assessment within the CAD/CAM/CNC chain?

3 Research Proposal

3.1 CAD/CAM/CNC Chain Integrating Environmental Assessment

STEP and STEP-NC have been developed by the ISO committee as a basis for the product design, the machining standard, and the integration of different machining and control parameters. STEP-NC is a new standard for exchanging data between CAD/CAM system and NC machine. It is conceived to replace the traditional G-codes programming (ISO 6983) to allow intelligent control of NC machines and increased industrial productivity. STEP-NC file provides the CAD/CAM/CNC chain a detailed and structured interface for integrating machining data. Machining data are mainly composed of various information such as the forms (features) to machine, the type of tools to use, the type of operations to be performed (working steps), the process plan (work plan). By providing this level of information to the machining system, STEP-NC allows an unified environment for information exchanges between the product design and the machining processes planning or control. It thus promotes the establishment of a bidirectional information flow sent throughout the CAD/CAM/CNC chain.

Machining characteristics are the essential inputs for performing the environmental assessment within the CAD/CAM/CNC chain. The new proposal consists on enabling a dynamic environmental evaluation of the machining system. The input data are progressively filled by the users (mainly CAM engineers) at the rhythm of the data availability in STEP-NC. The assessment tool (eco-assessment) provides the environmental profile of the evaluated machining technique. The acquired knowledge from the environmental assessment results must be capitalized in MPM and will be translated into guidelines for further enhancing actions to reduce the environmental impacts of next machining techniques to implement. The overall scheme of the new proposal is shown in the Fig. 4.

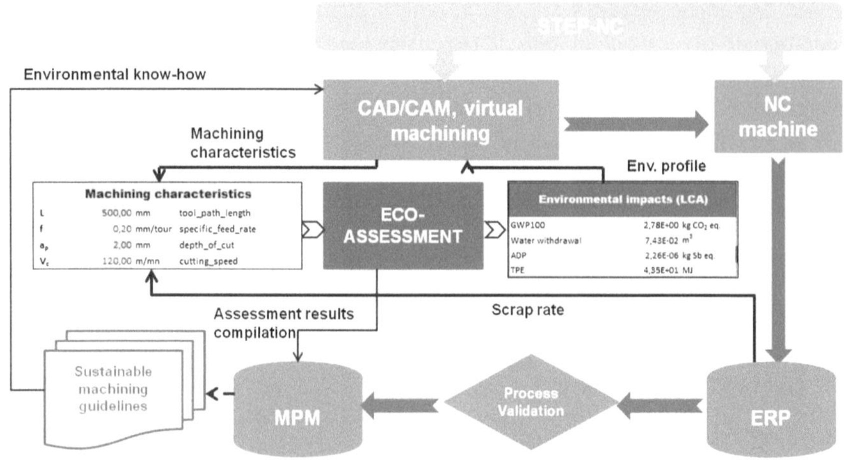

Fig. 4 CAD/CAM/CNC-chain integrating environmental assessment

3.2 Key Environmental Indicators

Defining key indicators to assess the environmental profile of a production system is a delicate exercise. Various indicators are available to assess the environmental performance of products. Seiffert [15] clearly distinguished 'environmental aspects' with 'environmental impacts' and illustrated the correlation between the two. For example, energy consumption (an environmental aspect) contributes to the depletion of natural resources (an environmental impact). On the other hand, Tam [16] distinguishes environmental indicators according to their industrial usefulness. The first category concerns the indicators used in environmental reports, such as Global Warming Potential (GWP), total primary energy, eutrophication, etc. The second category includes the indicators used in the context of environmental strategies such as energy use reduction, etc. The third category encompasses functional indicators that cannot be measured per unit of product, but rather by the function that this product meets such as the amount of electrical energy consumed per X hours for Y machining operation, etc. The first category of indicators addresses more global aspects and cannot report on specific requirements such as the legislation targets. These indicators are obtained from the realization of life cycle assessment (LCA). In the following sections, these will be identified as 'environmental impacts'. The second and third categories seem closer to the current practice, or in any case, can report on specific aspects of interest to the manufacturing industries. In the following sections, these will be called 'environmental metrics'. The selected indicators for supporting the new proposal are summarized in the Table 1.

Table 1 Key environmental indicators

Indicators	Type	Unit	Method for calculation	Source of data
Cutting energy	Env. metrics	E_c (kW, MJ)	$E_c = P_c * T_c = (K_c * A_D * V_c) * T_c$	CAM (expected cycle duration);
			P_c: cutting power; T_c: cutting duration; A_D: chip section; K_c: specific cutting force; V_c: cutting speed	STEP-NC (Touch_Probing); STEP-NC (Machining_operation);
Cutting fluid		Qualitative	Dry cutting	STEP-NC (technology); *Remark: Environmental impacts (from LCA modeling) of cutting fluid consumption depend on the nature of fluid*
			Oil based	
			Water based	
			Cryogenic	
Workpiece material		Qualitative	N.A	STEP-NC (workpiece)
Cutting tool material		Qualitative	N.A	STEP-NC (tool)
Amount of chips		Q_c (g)	Rough_workpiece_mass-machined_workpiece_mass;	STEP-NC (workpiece)
			$Q_c = \rho * Q * T_c$;	
			Q (cm3/mn): removed material flows;ρ : density of the material; Tc: cutting duration	
Tool depreciation rate		T_u (%)	$T_u = (T_c/T_o) * 100$	STEP-NC (tool)
			Allocation ratio from the cutting duration T_c over the tool lifetime T_o	
Scrap rate		T_r (%)	$T_r = (N_r/N) * 100$	ERP
			Allocation ratio from the number of failed pieces N_r over the number N of machined pieces in a lot serial production	
Global warming potential	Env. impacts (LCA)	GWP ($kgCO_2$ eq.)	$\sum_i Q_i \times EF_i^E$	Ecoinvent v2.2 database (EFs extraction); Impact 2002 + (evaluation method)
Water withdrawal		m^3		
Abiotic depletion potential		ADP (kgSb eq.)	Q_i: quantity of process ***i***; EF: Env. factors of the impact ***E*** of an unit of ***i***	
Primary total energy		PTE (MJ)		

Sustainability assessment of a machining system is based on a comprehensive evaluation of the inputs and outputs of the process. Machining process interacts with the environment through energy consumption, material and other consumables flows. LCA has emerged as the most objective tool for evaluating the environmental impacts of a product or a process. Environmental assessment with LCA consists on translating input/output flows of a process into impacts indicators. To integrate LCA within the eco-assessment tool, a simplification of its comprehensive feature is needed.[1] (EFs) are extracted from an LCA software package by modeling and evaluating a reference unit of a process $\boldsymbol{i}$. Then, following an allocation rule, an impact $\boldsymbol{E}$ for the process $\boldsymbol{i}$ is linearly correlated to the corresponding environmental impact factor EF_i^E (see Table 1).

3.3 Implementation

The cutting data are first collected from STEP-NC to complete the environmental metrics. Additional data from the ERP system are needed for evaluating the scraps rate metric. The second step deals with the environmental impacts evaluation through the completed environmental metrics, thanks to the simplified LCA database. The environmental profile of the machining system is then provided to the CAD/CAM entity that generates ideas for improvement if the environmental impacts are found critical. The undertaken improvement actions consequently induce changes in the CAM outputs for the STEP-NC. Meanwhile, the metrics and impacts evaluation results are capitalized in the MPM system to consolidate an environmental knowledge-base. Further, guidelines for sustainable machining can be established based on the collected environmental knowledge for enhancing continuous improvement actions (Fig. 5).

Moreover, this new proposal foresees two levels of actions for achieving dynamic environmental assessment. The first level of actions consists on trial-and-error actions for evaluation/improvement of the environmental profile. It corresponds to the first implementation (introduction stage) of the environmental assessment within the CAD/CAM/CNC chain. The second level of actions consists on a decision-making to select a more environment-friendly machining technique. This kind of action can only be achieved when a sufficient background from anterior machining simulations are available in the MPM system. This level corresponds then to the routine implementation (maturity stage) of the environmental assessment.

[1] Simplification of the comprehensive feature of LCA for its integration in design and development of product/process is a relevant research area in ecodesign.

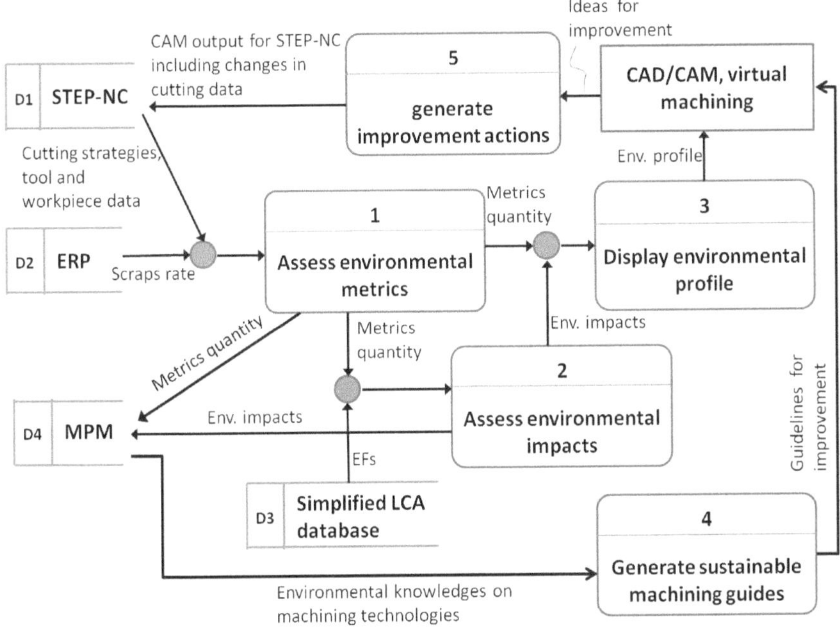

Fig. 5 Data flow diagram showing the implementation of the environmental assessment

4 Conclusion

Machining is one of the important material removal processes in manufacturing. Therefore, a study of the environmental impact of a machining system contributes greatly to sustainable manufacturing. The research described in this paper stands for a new approach which will provide a dynamic environmental assessment through the virtual environment of machining operations. The framework for supporting environmental assessment within the CAD/CAM/CNC chain has been developed. This framework mainly features key environmental indicators and their evaluation methods supported by the eco-assessment tool. To implement the environmental assessment, cutting data are first extracted from STEP-NC file. Then, through the environmental metrics evaluation, the metrics quantities (Input/output flows of the machining process) are translated into environmental impacts by the help of the simplified LCA database. The obtained environmental profile can be analyzed to aid decision-makers in CAD/CAM entity by providing a better understanding of the environmental impact of their machining technology. A prototype of the eco-assessment tool is under development. Then, it must be validated by the industrial partners prior its integration within the software system ANGEL.[2] STEP-NC file is

[2] ANGEL (**A**telier **N**umérique co**G**nitif int**E**ropérable et agi**L**e).

a new standard which is still in development. Then, the future of this study depends on the maturity of STEP-NC and its effective implementation in manufacturing firms. Nonetheless, it is hoped that the research reported in this paper can help the manufacturing firms have a better understanding of the environmental issues affecting the machining system.

Acknowledgments This work is done in the French FUI project ANGEL. We also thank all consortium partners for their contribution during the development of ideas and concepts proposed in this paper.

References

1. Commission of the European Communities: Communication from the commission to the council and the European Parliament. Integrated Product Policy, Building on Environmental Life Cycle Thinking, Brussels (2003)
2. Directive 2008/1/EC of the European Parliament and of the Council concerning "integrated pollution prevention and control" (2008)
3. Kaebernick, H., Kara, S., Sun, M.: Sustainable product development and manufacturing by considering environmental requirements. Robot. Comput. Integr. Manuf. **19**, 461–468 (2003)
4. Erkman, S.: Industrial ecology: an historical view. Cleaner Prod. **5**, 1–10 (1997)
5. Jovane, F., et al.: The incoming global technological and industrial revolution towards competitive sustainable manufacturing. CIRP Ann. Manufact. Technol. **57**, 641–659 (2008)
6. Pegden, C.D., Shannon, R.E., Sadowski, R.P.: Introduction to simulation using SIMAN. McGraw-Hill, New York (1995)
7. Banks, J.: Discrete-Event System Simulation. Pearson, Upper Saddle River (2010)
8. Law, A.M.: Simulation Modeling and Analysis. McGraw-Hill, Boston (2007)
9. Thiede, S., Seow, Y., Andersson, J., Johansson, B.: Environmental aspects in manufacturing system modeling and simulation, State of the art and research perspectives. CIRP J. Manufact. Sci. Technol. **6**, 78–87 (2013)
10. Thiriez, A., Gutowski, T.: An environmental analysis of injection molding. In: Proceeding of the IEEE International Symposium on Electronics and the Environment, pp. 195–200. (2006)
11. Neto, B., Kroeze, C., Hordijk, L., Costa, C.: Modelling the environmental impacts of an aluminium pressure die casting plant and options for control. Environ. Model Softw. **23**, 147–168 (2008)
12. Dietmair, A., Verl A.: Energy consumption modeling and optimization for production machines. In: IEEE International Conference on Sustainable Energy Technology, pp. 574–579. (2008)
13. Kong, D., Choi, S., Yasui, Y., Pavanaskar, S., Dornfeld, D., Wright, P.: Software-based tool path evaluation for environmental sustainability. J. Manuf. Syst. **30**, 241–247 (2011)
14. Zhao, Y.F., Perry N., Andriankaja H.: A manufacturing informatics framework for manufacturing sustainability assessment. In: Re-engineering Manufacturing for Sustainability, pp. 475–480. Springer, New York (2013)
15. Seiffert, M.E.B.: Environmental impact evaluation using a cooperative model for implementing EMS (ISO 14001) in small and medium-sized enterprises. J. Cleaner Prod. **16**, 1447–1461 (2008)
16. Tam, E.K.L.: Challenges in using environmental indicators for measuring sustainability practices. J. Environ. Eng. Science. **1**, 417–425 (2002). (Published on the NRC Research Press Web site at http://jees.nrc.ca/)

Idea-to-Market: Product Innovation Sustainability in Challenging Health-Care Context

Dipanka Boruah and Amarendra Kumar Das

Abstract The challenge of sustainable development is now recognized worldwide. There are large numbers of products locally innovated to suit particular context. However these are never available in the market in absence of their commercial manufacture that can have been mutually beneficial to innovators and users and also to Micro, Small and Medium Enterprises (MSMEs). Design aspect concerns emergency medical transportation services to marginalized people of the society in rural areas. A case in hand is rural health service in rural areas of India; National Rural Health Mission (NRHM) has launched boat clinic to make health services available to the rural population. But many places in North Eastern India are totally inaccessible by road. In this respect, design of amphibian featured new tricycle ambulance was initiated and the acceptance by the beneficiaries were studied and the research works has given very positive results and this work has established the vital contribution of design for bringing this innovation to market.

Keywords Contextual innovation · Sustainable development process · Design and technology transfer

1 Introduction

Design for Sustainability (D4S) is referred to as sustainable product design, which is a globally recognized method for companies to improve profit margins; product quality, market opportunities, environmental performance, and social benefits (UNEP). Canada Governments have invested heavily in the clinical and economic promise of health innovation and express increasing concern with the efficacy and

D. Boruah (✉) · A.K. Das
Department of Design, Indian Institute of Technology Guwahati, Guwahati, Assam, India
e-mail: b.dipanka@iitg.ernet.in

A.K. Das
e-mail: dasak@iitg.ernet.in

A. Chakrabarti (ed.), *ICoRD'15 – Research into Design Across Boundaries Volume 2*, Smart Innovation, Systems and Technologies 35, DOI 10.1007/978-81-322-2229-3_21

efficiency of health innovation systems [1]. In this case in hand is rural health service in rural areas of India; Government has launched National Rural Health Mission (NRHM) to make health services available to the rural population. But many places in North Eastern India are totally inaccessible by road. As a result, the health services of the people living in these areas are badly affected. Although government has introduced boat clinic, people cannot reach these in emergency, since patients cannot be carried by road to these. In contrast to high-tech innovation and development, it is considered that [2] (or appropriate) technologies may be considered effective. Appropriate technology is small scale, energy efficient, low-cost, radical, environmentally sound, labor intensive, and controlled by the local rural community [3]. In this context once an idea or concept is visualized by the designer, product developers' focus on lean product development strategies to materialize the innovation and produce it in a faster and cheaper rate for the consumer market. But producing cheaper and faster products alone, does not guarantee the success of the products in the commercial market. Understanding factors that affect a product's acceptability in the market and developing new product design strategies accordingly may form an important step in meeting this challenge. Victor Papanek [4] pointed out the designers' responsibilities with respect to major social and environmental needs. Due to globalization, a product which is developed in one part of the world is introduced in another but their acceptability with the local consumer community remains a debatable issue. Many a time, such globalized products end up as failures in local markets due lacking of issues addressed by designers, which the local consumers look for. To address this issue attention should be paid to develop successful marketing strategies for rural innovations that are hardly introduced in the market due to faulty design issues. In this aspect, designers should collaborate with rural innovators to provide them with strategies that would make their product a commercial success.

1.1 Sustainable Product Innovation Process

Design in the simplest form is defined as initiation of change in man- made things [5]. Product design deals with conversion of ideas into reality (tangible products) and, as in other of human activity, aims at fulfilling human needs.

Design development of rural ambulance is a new approach to vehicle design that fulfills the requirements of sustainable development for rural that. In contrast to the standard automobile design methodology of western manufacturers who place emphasis on comfort and luxury (from a developed perspective), this shifts the emphasis towards safety on emergency situation, functionality and affordability—the key criteria for the demographic, that is aimed to serve through this process. Advantages of rural ambulance are:

- Generates less pollution during its manufacture and disposal compared to fully motorized mode of transportation.
- Provides for income generation for vast unemployed youths of the country.
- It can easily move on un-surfaced road due to its low weight.

- It uses better transportation between remote area and hospital instead of boat and thela (hand gurney).
- This ambulance is manufactured using fiber reinforcement plastic (FRP) to decrease weight.

It is easy to repair, due to easy maintain availability of material and manpower (experts' supervision); it makes this product appropriate to the context and sustainable.

A new approach to rural ambulance interventions is emerging. It requires a more holistic understanding of the mobility and access needs of the rural communities than has traditionally been the case in rural roads. It is a demand-led (or Human-centered), approach with an emphasis on the needs expressed by affected rural people of riverine areas. In this context, rural medical transport is more broadly seen as an input into successful rural livelihood strategies.

1.2 Methodology Followed

In tune with the above objectives, design development of a tricycle rickshaw for passengers in the Indian context [6] addressed how a design to be a successful in Indian context when designing for the Indian population, a design will be right context (Indian Context) and meet the need of Indian population in their context when this culture, tradition and way of leaving is addressed. This helped in better visualization of the design (Fig. 1). The major constraint considered at this stage is to initially tape drawing. A functional model was carried out to experiment various factors such as form, height, width and proportion for an able bodied person. This model would be foot pedal one operated so that before putting the concept model to actual use by patient, rider, attendant and spaces for medical first-aid equipment. This can be tested by able bodied person for various factors like appropriateness of the design, stability of Ambulance during operation of road and water in riverine area. The other constraints considered for prototyping was the selection of part from, existing tandem bicycle-tricycle rickshaw and stretcher of rural health care centre. The focus is on small enterprise, in many cases tiny enterprise for its manufacturing system management.

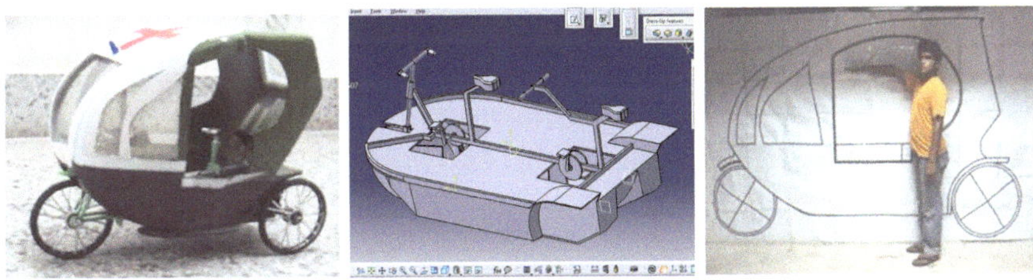

Fig. 1 Scaled model of the ambulance (*left*), CAD modeling of Catamaran type hull (*middle*) and 1:1 scaled tape drawing (*right*)

1.3 What is Meant by Sustainable Innovation for Health?

Technology for health is broader than health technology. For example, technologies for health could increase agricultural output in low-income countries such as the foot-operated treadle pump which improves health by reducing hunger and malnutrition. Likewise, technologies that improve road safety such as motorcycle crash helmets contribute to public health, but are not classified as health technologies. Such design and technologies improve health facility, but they are not usually the main concern of a health system in rural areas.

1.4 Global Health Needs and Contribution of Design Technology

Assessment of the contribution of technology to meeting global health needs is not easy. Evidence shows the benefits of health technology in high-income countries, 33 but worldwide and in low-income and middle income countries the benefits are not well studied. The 1999 World Health Report34 (taking a very broad view of technology; every advance not related to income or education) suggested that worldwide "half the gains in health between 1952 and 1992 result from access to better technology".

2 Analysis and Prediction

Considering the prevalence of tricycle rickshaw across India, a small development in this field would serve a large population that users this modes of transportation. With this view, for experimentation in this proposed research work, human powered amphibian tricycle vehicle is design and development considered.

The Ambulance (it's made from locally available materials), is a human gurney designed to carry sick or injured people across rural terrain. Made from locally available resources; its unique design that provides a lifeline for many people with no direct access to hospitals.

2.1 Detail Design Process Planning for Every Part, Sub Assembly and the Final Assembly

The concept evolved in the primary design, is the basis for the detailed design phase [7]. It is aimed at furnishing the engineering description of a tested and producible design. Prior to the detail design phase, the design is greater flexibility. This provides for accommodating changes in concept without greater financial burden. Concepts being exploratory in nature, flexibility are essential in seeking revel an adequate range of possible solutions.

First step in planning the production process, detailed process planning required for every part, sub assembly and final assembly was carried out. A sample process sheet is given below:

Each process sheet contains a sequential list of operations those are performed to produce the particular part. Specification for the tools and machines required are provided in the process sheet is shown in Fig. 6, PS-1. This step was particularly important and design features that lead to difficulties in production.

2.2 Design for Assembly (DFA) Guidelines [8] and Design for Manufacturing

Design for assembly entails making attachment direction and methods simpler, for example, making attachment easy by using snap fits instead of machine screws. It involves application of attachment time and complexity models and basic rules and tables based on simplified time studies. This takes into consideration the DFA guideline for wheel and propels shaft drive and nonetheless cost effective. To address such trade- offs actual concept costing methods must be explored. Then numerical comparisons can be made between designs with more complexes versus designs with more complex piece parts. Initial prototype of the rural ambulance has used existing boat and is fitted with tandem tricycle structure for the seat and propulsion. In the interim version of amphibian vehicle, the entire body structure was combined into one unit. The eliminated substantial number of components and fasteners, but for the ease of manufacture this was broken down to following individual parts.

- Catamaran Type hull in FRP
- Integrated with propulsion system
- Integrated with transmission system (ITS)

2.2.1 Catamaran Type Hull in FRP

The above aim led to the concept that was tested initially for its feasibility with the aspect of buoyancy principle and aerodynamic principle. A small boat with Catamaran type hull supported with 3 wheels added like a tricycle rickshaw is conceived and designed to enable it to travel over land and water. It can travel over the road and through the water with comfort and grace. The driving mechanism from foot pedaling is similar to a common form of tandem tricycle for able bodied user for self-travel and the stretcher placed between rider and attendant. In addition to use maximum effort when required the rear person can use both his hands and legs to

Fig. 2 Wooden pattern using plywood (*left*) and making of FRP component from wooden mould to FRP die (*middle*) and temporary bicycle chassis arrangement

propel the vehicle, facilitate by incorporating an exerciser as shown in Fig. 5 with synchronous lower limb and upper limb movement.

Based on actual testing of the products for various attributes, the design was finally modified for mass manufacturing using FRP moulding etc. The idea was for local assembly and manufacture through SMEs production facilities, a FRP component manufacture in various stages. Thus decision making was fairly easy task. The production planning phase for amphibian ambulance involves multiple steps.

These are however similar to mass production industry such as production industry and in particular to amphibian vehicle industry.

The preliminary design of the amphibian vehicle was intended to establish an overall concept for the project and served as guideline for detailed design.

In this attempt, it is assumed that if a new looks CAD modeling, Tape drawing (Fig. 1 right) based vehicle having sufficient safety with full overhead enclosure suitable for users, there is sufficient demand for it to be manufactured commercially while NRHM boat clinic are also available. Pattern making using plywood; the first step in making an FRP component to be made in FRP for face side, made in materials that can be easily curved. Based on expertise available, it can be made in plaster of paris (POP) or wood suitable for curving.

The second step (Fig. 2 left) for making FRP component is fabrication of the moulds itself is fabricated in FRP. The process of mould making starts after pattern is ready. The pattern is to be coated with releasing agent i.e. Polyvinyl Alcohol (PVA).

The final step (Fig. 2 middle) is making of the components using the moulds. First, a coat mixed with requires colour of pigment is applied of the mould. Final component, a surface mat with smooth surface and low GSM is first laid and then based on the size and type of component, 400 GSM mat is used. Once the resin set, which normally take 3–4 h with 2 % of catalyst, the component can be taken out and process repeated for components (Fig. 3).

2.2.2 Integrated Propulsion System

To propel the vehicle a chain drive mechanism with is used. Front and rear rider applying forces through individual set of pedal are transmitted to a jackshaft held on

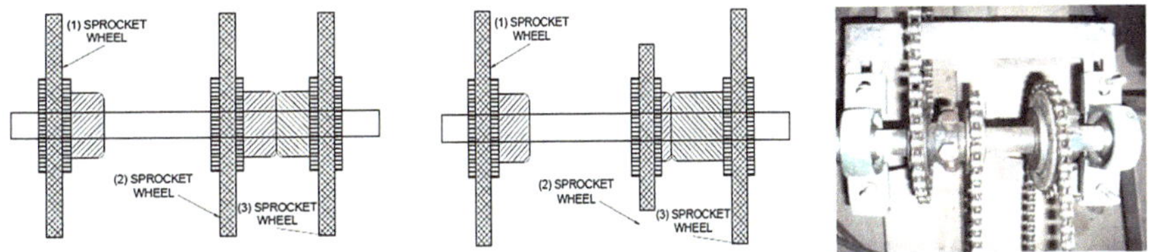

Fig. 3 CAD drawing of sprockets before being machined (*left*), after machined (*middle*) and after machined prototype (*right*)

a gear hub with the help of bearing. One of the jackshaft are mounted three sprocket which are connected to front and rear crank and the rear axle which driving the vehicle.

2.2.3 ITS-Gear Hub

The gear hub was made using Iron Oxide metal sheets of dimension 50 mm × 5 mm. A V-shaped form of design was proposed to produce cost consumption of metals and resource. The metal sheets were cut according to the dimensions and are welded to the mainframe. The gear hub has to be strong enough to hold the sprocket gear along with the jackshaft and the bearing house.

But after assembling the chain into the sprockets and the crank had some alignment problems causing the chain to break when the pedal is rotated at speed higher than a specified speed.

The distance between the two sprockets is reduced by machining the bolted section of one of the sprocket hub and fixing tightened with the same set of bolts on the second sprockets. This modification is chosen because the free play of both the sprocket is on the same side and therefore the motion of sprocket will not restrict the motion of the other.

2.2.4 ITS-Idler Gear (Fig. 4)

Idler sprockets can be used to reduce the slack in the chain. In amphibian vehicle design the idler was used to compensate for the imposer placement of the rear axle gear housing causing the chain from the jackshaft to the sprocket of the rear axle to clash with the chassis. The idler is mounted on the mainframe with the help of metal clamp. The clamp was L- shaped and was welded the mainframe. The idler was bolted to clamp. The chains were finally mounted on the sprockets on the jackshaft and on the rear axle and checked for alignment of the sprockets and crank wheel. The idler was also checked for its motion to be the sprockets. The elliptical tandem bike mechanism was mounted on the T- section welded to the mainframe. T-Section houses the solid rod which accommodated the sleeves on which the

Fig. 4 View of Rear wheel sprocket (*left*) and Idler gear (*right*)

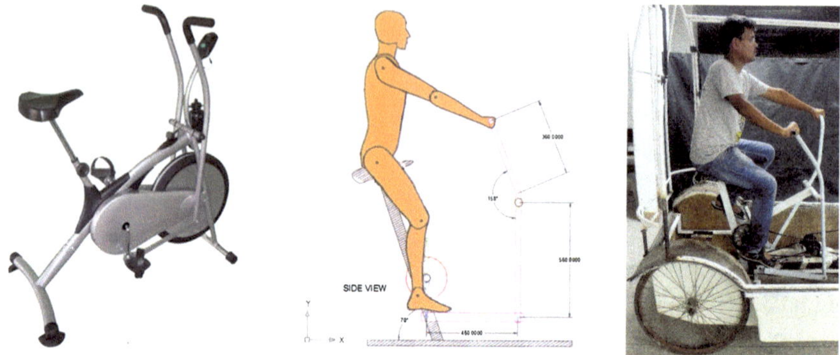

Fig. 5 Upper limb and lower limb of cross trainer exercise system (*left*), CAD drawing of cross trainer embedded with the ambulance (*middle*), prototype (*right*)

handlebars perform it's to and fro motion. The connecting rod was connected to the handlebar at one end and to the pedal spine at the other (Fig. 5).

The degrees of freedom included in this upper limb model are those considered most significant in accounting for motion of the upper limbs performing arm cranking tasks. Arm powered mobility aids and recreational vehicles for the disabled community have been the focus of a number of ergonomic, physiological and biomechanical [9] Chassis (Fig. 6).

Chassis and mainframe is an integrated piece made out of following components.

- 3 pieces of mild steel 50 × 25 mm rectangular sections of 2 mm thick.
- Body bottom socket- out sourced from market.
- 1 piece of mild steel 40 × 5 mm flat bar rectangular section of 2 mm thickness.
- 2 numbers reinforcement elements of mild steel 40 × 5 mm flat bar of 150 mm in length for bearing block attachment bracket.
- 4 numbers U-shaped brackets of 3 mm thickness mild steel plate for fixing rear end of side frame
- 1 piece 19 mm solid rod for fixing the front end of the side frames.

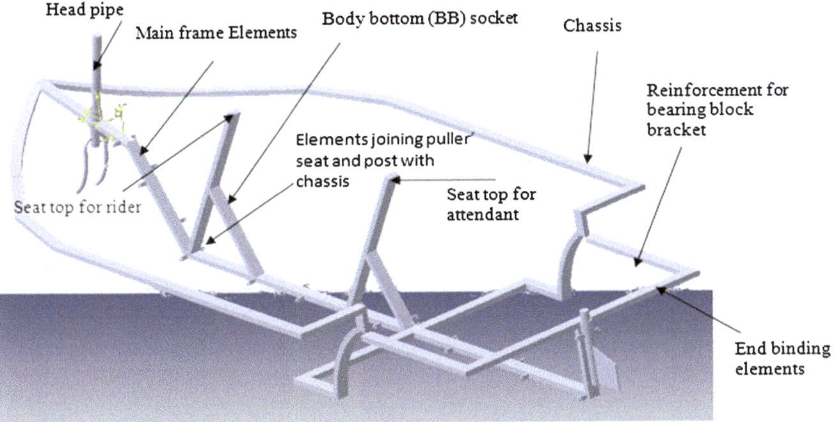

Fig. 6 PS-1 CAD modeling of integrated mainframe and chassis of rural ambulance

2.3 Design and Human Factor Consideration

The overall structure protects the ambulance puller as well as passenger from the elements of the nature (sunshine, rain etc.) through the weight and centre of gravity of the rural ambulance remain similar weight of human powered three wheel ones.

Incorporate human dimension for ergonomic seating arrangement [10]. The seat and back rest is maintained at around 100° (Figs. 7, 8, 9, 10 and 11).

Fig. 7 CAD modeling of integrated mainframe (*left*) and cross trainer mechanized transmission system (*right*) of the ambulance

Fig. 8 Head frame with chassis rod (*left*), Integrated propulsion joint (*middle*), front head frame of the FRP hull (*right*)

Fig. 9 View of catamaran type hull (*left*) and hull water testing (*right*)

Fig. 10 Side view (*left*) and front view and door opening system (*right*)

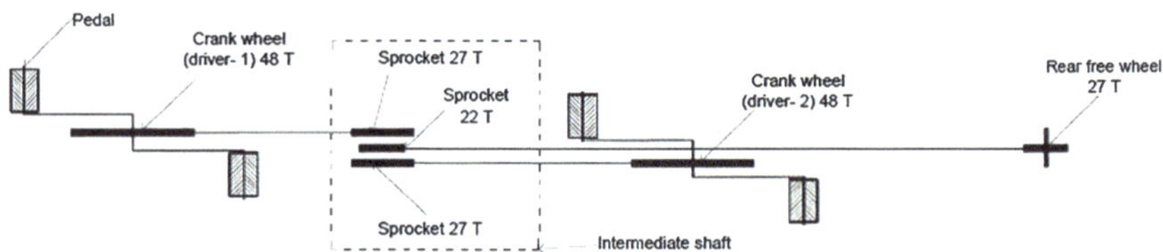

Fig. 11 Schematic reduction of layout of gear and pinion for gear reduction

2.4 Design of Tools and Fixtures

Whenever a product is designed for commercialization, manufacturing ease needs to be taken into consideration. To ease the manufacturing of multiple units, standard fixtures are to be designed. Second step in planning the production process is Design of tools and fixtures for the Production process. Second step in planning the production process is Design of tools and fixtures for the Production process.

2.5 Buoyancy Calculation and Land Testing

Total Volume of single hull: (from FRP catamaran type hull)

=Part1 + Part2 + Part3 + Part4

=(45,937,500 + 408,240,000 + 42,548,000 + 7,753,350)

=504,478,850 mm^3

=504.478 L = 0.504 m^3

Error Percentage = (Actual value − Theoretical value)/Actual value

=|(0.481–0.504)/0.481| = 4.78 %.

The sizes of hull dimensions (as per design) is 2,420 mm × 560 mm × 450 mm

Buoyant force (F.B) = V × v × SP of water

where V = volume of the hull

v = % of the object submerged in water.

Specific weight of water (SP) = 62.4

For two hulls

For 100 % = 0.481 m^3 × c2 = 0.962 m^3

For 80 % = (0.962/100) × 80 = 0.7696

Buoyant force, FB = 0.7696 × 0.80 × 62.4 = 38,418.432°N (without person)

For 3 persons (rider + attendant + patient) weight = 75 kg/each = 225 kg.

Number of revolution of crank wheel given by = S1 rpm

Number of revolution of sprocket wheel fixed with intermediate axel given by = S1

Number of teeth of crank wheel given by T1 = 48 T

Number of teeth of sprocket wheel at rear axle given by T2 = 27 T

Then, S1/S2 = T2/T1

Since, S1 is 1, S2 = T1/T2 = 48/27 = 1.78 i.e. for every rotation of the pedal by rider, the pedal by the rider, the rear wheel will rotate by 1.78 turns. Here gear ratio is T1/T2 = 48/27 = 1.78

Considering the diameter (d) of the rear wheel as 71 cm (28″), in one pedaling, the tri wheeler will travel a distance of S2 × π × d = 1.78 × 3.14 × 71 = 397 cm or 3.97 m.

To travel 1 km, numbers of pedaling required will be = 1,000/3.97 = 252 times.

For two stage gear reduction through introduction of intermediate shaft

Number of revolution of crank wheel given by = S1 rpm

Number of revolution of sprocket wheel fixed with intermediate axel given by = S1 rpm

Number of teeth of crank wheel given by T1 = 48 T

Number of teeth of driven sprocket wheel at intermediate shaft given by T2 = 22 T

Number of teeth of driver sprocket wheel at intermediate shaft given by T3 = 27 T

Number of revolution of sprocket wheel fixed to rear axle given by S3

Number of teeth of sprocket wheel fixed to rear axle given by T4 = 27 T

Then, S1/S2 = T2/T1

Since, S1 is 1, S2 = T1/T2 = 48/22 = 2.18
Again, S2/S3 = T4/T3 or S3 = S2 × T3/T4 = T1/T2 × T3/T4 (replacing S2 with T1/T2)
S3 = 48/22 × 27/22 = 2.18 × 1.22 = 2.65
For every rotation of the pedal by rider, the pedal by the rider, the rear wheel will rotate by 2.65 turns. Here gear ratio is T1 /T2 × T3/T4 = 48/22 × 27/22 = 2.65
Considering the diameter (d) of the rear wheel as 71 cm (28″), in one pedaling, the tri wheeler will travel a distance of S3 × π × d = 2.65 × 3.14 × 71 = 593 cm or 5.93 m.
To travel 1 km, number of pedaling required will be = 1,000/5.93 = 168 times

3 Participatory Approaches in Development and Transfer of Technology to Small Enterprises

3.1 Planning and Production Process

In planning and production, the process was planning for production personnel. Initial training was done in the Department of Design, Indian Institute of Technology Guwahati as part of design and technology transfer. Being a technology transfer initiative, it will important to impart proper process to the SMEs for future work development their capacity to deliver a successful product will all possible assistant including support in successful implementation and manufacturing process [10].

In this case Small and Medium Enterprises (SMEs) from Guwahati in North East Region in India The design methodology adopted for this purpose was found to be simple to understand by the employee of the SMEs while transferring design and technology and thus helps in subsequent design process by the target manufacturer in a conventional way.

3.2 Design and Technology Transfer

The Institute of Transportation and Development Policy (ITDP) is one of the few organizations all over the world associated with the tricycle rickshaw development specifically in the developing world (www.itdp.com, July, 2006). In view of it the universal access to public health care services Govt. of Assam of North East India under National Rural Health Mission (NRHM, 2011–2012) joined hands with an NGO named Center for North East Studies and Policy Research (C-NES) under Public Private Partnership (PPP) as it is presently providing preventive and primitive services in the riverine areas.

In the next step, in design and technology transfer process and implementation will through participation of the designer with SMEs throughout this type of rural design innovation and development and its implementation and results will visible in terms of product sustainability and indigenous.

4 Conclusion

The research work transpires a need to develop indigenous design with Indian need and context. The small enterprise can benefit from the participatory approach by building its capacity to design new products for rural remote areas people.

It is observed that, success of the research work is based on the active participation from the small enterprises including NGO. The model of contextual design and development, appropriate technology transfer and manufacturing approach in participation from small enterprises is found to be effective and can be successfully extended to this product.

References

1. Miller, F.A., Sanders, C.B., Lehoux, P.: Imagining value, imagining users: academic technology transfer for health, innovation. Soc. Sci. Med. **68**, 1481–1488 (2009)
2. Schumacher, E.F.: Small is Beautiful: Economics as if People Mattered. Harper & Row, New York (1973)
3. Boruah, D., Das, A.K.: Design education: empirical investigations of design theory in practice in specific context. In: Proceeding of the E&PDE International Conference on Engineering and Product Design Education, vol. DS75, 20(4), pp. 781–786. Design Society, Antwerp (2012)
4. Papanek, V.J.: Design for the Real World: Human Ecology and Social Change, 2nd edn, completely rev. Thames and Hudson, London (1985)
5. Boothroyd, G., Peter, D., Winson, K.: Product Design for Manufacture Assembly. Marcel Dekker Inc., New York (2002)
6. Nadkarni, S.: Design crossroads in India: the challenge admist confusion. Conference Proceeding of 'Nest Wave at Nagoya', Japan (1985)
7. Christopher, J.J.: Design Methods. Wiley, Van Nostrand Reinhold, New York (1992) (print)
8. Martin, D.A., et al.: Human Upper Limb Dynamics, pp. 151–163. Elsevier Science Publishers B.V., North-Holland (1989)
9. Chakrabarti, D.: Indian Anthropometric Dimensions for Ergonomic Design Practice. National Institute of Design, Ahmedabad (1997) (print)
10. Chitale, A.K., Gupta, R.C.: Product Design and Manufacturing. Prentice-Hall of India, New Delhi (1999)
11. Das, A.K.: Integrated process design using rapid prototyping technology and rapid tooling in concurrent engineering approach, advance in material manufacturing science and technology. Trans Tech Publication Ltd., Switzerland, Material Science Forum, pp. 471–472 (2004)

An Interface Between Life Cycle Assessment and Design

Praveen Uchil and Amaresh Chakrabarti

Abstract Results from a full Life Cycle Assessment (LCA) of a product contain complex information about its constituent components and materials; subsequent manufacturing processes, emissions, and potential environmental impacts. Understanding such LCA information in detail can help designers make robust decisions related to reducing environmental impacts of the product through appropriate choice of components, materials and processes during Eco-Design. A key resource for LCA information in a Full LCA tool (also called Detailed LCA, and henceforth referred to as LCA tool) is its databases. However, LCA tools have often been criticised for not being useful to designers. Among the various underlying causes, the issues with visual representation of LCA information and usability of LCA tools have gained minimal research attention. In order to address these issues, we adopt an information visualization approach. Information visualization is an interdisciplinary research area focused on computer supported development of visual representations of complex information in order to render it interpretable by humans. In this paper, we discuss the potential of using information visualization techniques as an interface for LCA tools for educating designers about the likely environmental impact of their design decisions. We also discuss potential benefits of using a novel interface developed by us using interactive, multi-view based visualization techniques in understanding context sensitivity of LCA information. The proposed interface is part of an ongoing research effort for developing user friendly interactive visual representations for LCA. Ultimately, the proposed interface is intended to enhance designers' capabilities for developing environmentally benign product life cycles.

Keywords Life cycle assessment · Information visualization · Ecodesign · Usability

P. Uchil (✉) · A. Chakrabarti
Centre for Product Design and Manufacturing, Indian Institute of Science, Bangalore, India
e-mail: praveen@cpdm.iisc.ernet.in

A. Chakrabarti (ed.), *ICoRD'15 – Research into Design Across Boundaries Volume 2*,
Smart Innovation, Systems and Technologies 35, DOI 10.1007/978-81-322-2229-3_22

1 Introduction

Life Cycle Assessment (LCA) results of a product contains comprehensive information about the complex interactions between materials and processes and the environment during the entire life cycle of the product. However, use of LCA tools for design has been critiqued for factors such as its outcomes being too complex to interpret, too time consuming to implement, the tools being not easy to use and not in alignment with designers' requirements, etc. [1–3]. Although there have been several efforts to improve adoption of LCA by the design community through various approaches such as integration with CAD systems [4], development of single scores [5] and improvement of LCA interfaces [6], many issues still persist, either due to their complexity or due to the lack of adequate, systematic effort [7]. In this paper, we discuss the feasibility of addressing some of these outstanding issues through the application of techniques adapted from the domains of decision making [8], information visualization [9] and user-centered design [10].

2 Objective and Methodology

The objective of this paper is to highlight information visualization issues in LCA tools and to discuss the potential of using recent techniques from information visualization for addressing some of these issues. Information visualization issues can exist in LCA either due to issues with the content, or with the form of the information representation used.

2.1 Research Questions and Methodology

The following research questions are asked:

1. What is the role of information visualization in LCA-driven decision-making? This is addressed by reviewing literature on decision-centered design e.g. [8], and by application of information visualization e.g. [9] and user centered design e.g. [10].
2. What are the design decisions taken during ecodesign? What among them can be supported by LCA?
   This is addressed by reviewing literature on models of designing e.g. [11], and by application of LCA in decision making e.g. [12].
3. What are the visualization issues faced by current LCA tools?
   This is addressed by reviewing literature on requirements of support tools for ecodesign e.g. [1] and analyzing user interfaces and visualization aspects of LCA tools from the perspective information visualization principles [9].

4. Which information visualization techniques can potentially address the issues identified?
   This is addressed by designing alternative representations using existing techniques of information visualization that have been found suitable in addressing similar issues in other domains.

3 Research Outcomes

3.1 Role of Information Visualization in LCA

A decision is a commitment to use resources; therefore, it deserves serious emphasis [8]. Design as a problem solving activity involves generating and refining information punctuated by decision-making [8]. The activities involved in ecodesign are similar to those in design [4]. We adopt the decision thinking framework for design developed by Ullman [8] as it succinctly represents the relationships between decisions and information (Fig. 1).

A decision is a conclusive piece of information obtained as an outcome of interpretation facilitated by an internal representation. Representation is visual encoding of data in terms of various shapes and relationships. Internal representation is a tacit entity that drives decision making, consisting of prior knowledge and knowledge obtained from external (LCA) representations. Visualization methods can accelerate internal representation by presenting information in an appropriate format or structure or by helping users find, relate and consolidate information, helping them to form an appropriate internal representation [13]. Interaction is a dynamic means through which a decision maker alters the LCA representations to obtain insight into the LCA information. LCA information refers

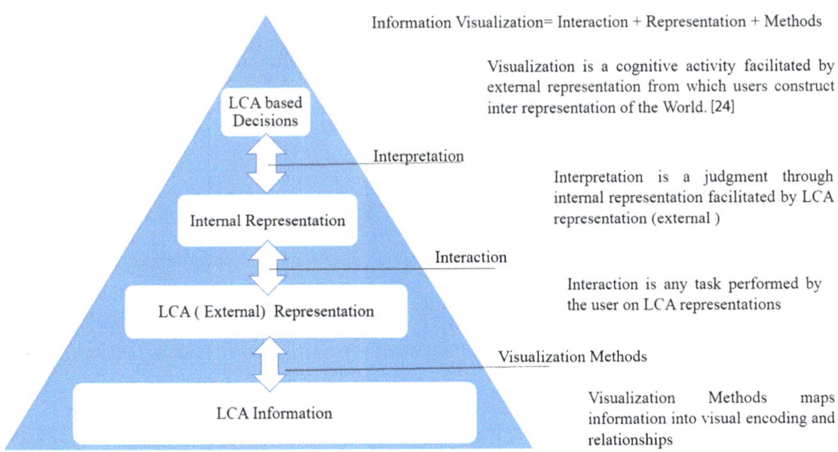

Fig. 1 The value of information visualization

not only to the computed outcomes of life cycle impact assessment models, but to all information required to make sense and assess credibility of the outcomes [7]. The outcome of the research discussed above is an extension to the information visualization framework proposed in our earlier paper [7].

3.2 A Decision Centred View of Ecodesign for Structuring Information Space

Table 1 shows a generalized decision framework according to the integrated model of designing proposed by Srinivasan and Chakrabarti [11], which can be used to represent the decisions driven by LCA. For example, decisions on evaluating product alternatives can be represented using "evaluate" decision, and decisions on identifying environmental target can be represented under "Generate" decision, in Table 1. The list of LCA applications in decision-making is adopted from Gloria et al. [13]. The decision typology compiled in Table 1 can be useful for developing decision-specific interfaces, since the typology provides a basis for grouping the methods and identifying the tasks necessary to represent information as per the information requirement of a decision.

3.3 Review of Information Visualization Approaches for LCA

Through a review of literature [12, 14–17], we identify that although many authors argue for using an information visualization approach in LCA, none of them benchmark the performance of the proposed approaches against current visualization in LCA tools, nor do they explore the most suitable visualization techniques for a given LCA decision. Addressing each of these gaps is a formidable challenge, because there are no formalized methods or criteria to identify the visualization issues, nor is there a concrete basis for prescribing the most suitable visualization for a given task. Moreover, lack of any concrete theory for visualization [18] and lack of objective evaluation methods pose further challenges for assessing the value of novel, alternative, information visualizations against existing ones.

Table 1 A decision typology for LCA

General decision category	GEMS frame work	LCA based application example
Analysis	Evaluate	Product/process/materials alternatives
Synthesis	Select	Energy materials emission audits
		Eco labelling (marketing) information
		ISO standards, Legislatory requirements
	Generate/modify	Environmental targets
		Product/process alternatives

3.4 Issues in a Current LCA Visualization

Put simply, a visualization issue is an example of bad design. A visualization issue in a decision support system may arise due to non-adherence to the principles of information visualization. In this section, we highlight various information visualization issues using a hypothetical example of LCA of a coffee machine design adapted from an example project used in a commercial LCA software tool. The most frequent application of LCA, as identified by a survey of LCA practice pertains to the decision of evaluating design alternatives [19]. We assume LCA information of an existing coffee machine is available for the designer. The objective of the designer involved in redesign of the coffee machine is to evaluate whether Plastic (Poly-propene) makes a better alternative for the housing of a coffee machine.

Figure 2 presents a 2D stacked bar chart representation as an outcome of an analysis for supporting the decision task comparative evaluation of product alternative. In order to understand the issues in this representation, it is necessary to consider the perspective of task sequence involved in generating the above representation in current LCA tools, and how accurately the representation reflects the truthfulness of the impact captured by the LCA results. On these lines of thought, we discuss visualization issues against the following requirements of information visualization.

(a) Accuracy and Insight: Results of LCA are context-specific [18]. Therefore, an accurate LCA representation should provide insight into the context-specific nature of the outcomes of the LCA. Currently, context-specific dimensions, such as functional unit, system boundary etc., are not explicitly represented in Fig. 2 representation. The above results, which considered only manufacturing phase in the life cycle analysis, may lead the designer to conclude Plastic as a

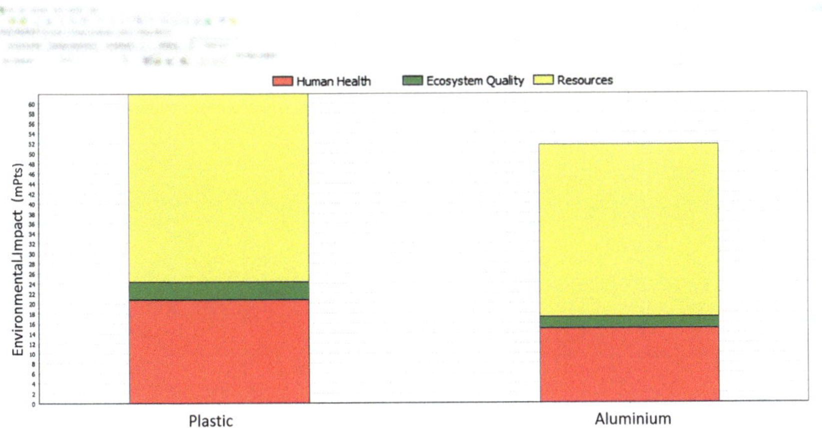

Fig. 2 A sample of LCA representation in a commercial LCA tool

worse alternative to Aluminium—the current constituent in the coffee machine. However, when the same analysis is repeated by changing the system boundary (e.g. including the material extraction processes in this case) reveals opposite results as shown in the top left chart in the Fig. 2. Relying on the representations without understanding the contextual parameters (in this case the exclusion of material extraction phase) could lead to incorrect decisions.

(b) Ease of Use: A commonly used heuristic for ease of use is the number of operations required to accomplish a task [20]. Generating the above representation requires the user to perform the following tasks in the current LCA software: Open the project > Product stages > Coffee machine > Assembly > Analyze > Calculate > Single Score > Bar Chart. The large number of operations as listed here indicates a poor ease of use.

(c) Performance: Performance in decision making can be assessed as the inverse of the amount of time taken for arriving at a decision. Performance is dependent on ease of use and adequacy of information. As the user has to search for the methodological dimensions within the current tool, this is likely to be time consuming.

3.5 *An Alternative Visualization*

In order to address some of the above mentioned issues, we propose an alternative visualization (Fig. 3) that uses a multi-view representation technique to display higher dimensional data. A multi-view representation technique uses two or more distinct representations to display a piece of information [21]. A multi-view

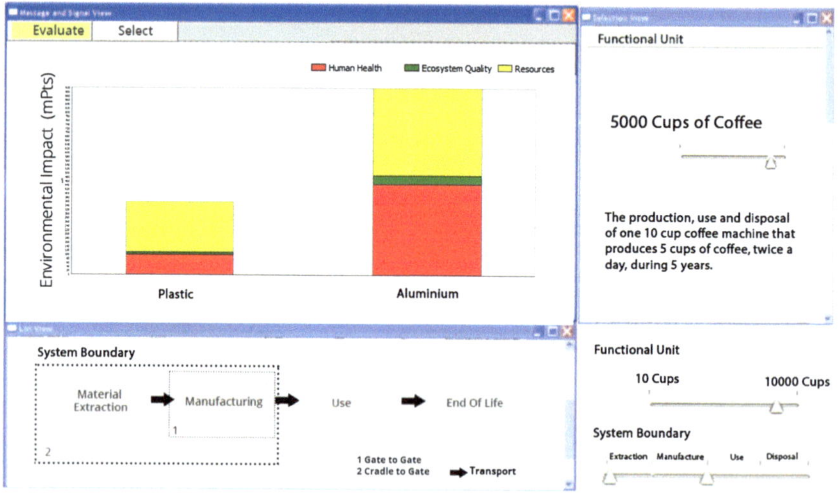

Fig. 3 An example of a multi-view visualization technique

representational technique is used in scenarios where interpreting information involves navigating complex data sets and identifying the relationships among various data fields required for decision making [22]. To facilitate interaction among such complex data sets within a multi-view approach, we propose integration of dynamic sliders on the interface.

The proposed representation is likely to be more effective than the current representation in the following aspects:

(a) Accuracy and Insight: The multi-view representation should enable linking of contextual data to LCA results through provision of dynamic sliders and distinct views for methodological dimensions. Thus, users can gain insight into the context-specificity of the information, by visualizing the effect of changing methodological parameters (using dynamic sliders) on LCA results. For instance, in the coffee machine, multiple representations of the same model can facilitate the user to perform 'what-if' analysis by quickly and interactively changing the system boundary from 'gate to gate' to 'cradle to gate', as indicated in the bottom view in Fig. 3. Users can interpret the results of Plastic being better than Aluminium for the Housing of the Coffee machine is contextual, and that often there is no single absolute answer to evaluation of design alternatives. Thus by making context-specificity explicit, the above representation more accurately represents the LCA model, and facilitates the user to gain insight into the dynamic nature of the LCA.

(b) Performance: North and Shneiderman [23] observes that multiple views offer improved user performance and identification of unforeseen relationships. The proposed interface reduces search for information and task sequence for a user to elicit the methodological dimensions by representing required dimensions (For instance: Functional unit and system boundary) explicitly using multiple representations.

(c) Ease of Use: The task sequence for generating the above representation in proposed prototype is the following: Open LCA results > Evaluate (Highlighted in Yellow Marker in Fig. 3). Any further interaction, if required can be made using dynamic sliders. Thus a smaller number of operations is required, indicating improved ease of use.

3.6 Discussion

The proposed representation shows a tentative, alternative visualization framework for evaluating product alternatives using LCA. Further studies are required to identify the most suitable representation for each given task. The proposed representation needs to be experimentally validated for the benefits it claims over existing tools. In our earlier paper [7], we had discussed the limitations of 2D bar charts in terms of representational attributes such as dimensionality, uncertainty and interactivity. This paper discusses the limitations of the visualizations from a relatively more tangible (i.e. empirically assessable) set of attributes such as ease of use, performance, perceived insight and accuracy, thus providing a more convincing approach for demonstrating the value of information visualization in LCA.

4 Conclusions

This paper discusses as to how interpretation of LCA information is hindered by the way LCA information is visualized. We highlighted some of the limitations of current LCA representations, and proposed application of higher dimensional techniques, such as multi-view and dynamic sliders, for improving this situation. The proposed visualization is intended to improve the adoption of LCA information in design by reducing the time taken to arrive at decisions, as well as by improving the credibility over the decisions through providing insight into the LCA information. Future work includes empirically evaluating the effectiveness of the proposed visualization against current LCA representations.

References

1. Kota, S., Chakrabarti, A.: Understanding the needs of designers for developing environmentally friendly products. In: International Conference on Research into Design (ICoRD '09) (2009)
2. Millet, D., Bistagnino, L., Lanzavecchia, C., Camous, R., Poldma, T.: Does the potential of the use of LCA match the design team needs? J. Clean. Prod. **15**(4), 335–346 (2007)
3. Bhander, G.S., Hauschild, M., McAloone, T.: Implementing life cycle assessment in product development. Environ. Prog. **22**(4), 255–267 (2003)
4. Kota, S.: An interactive support for developing environment friendly product life cycles. PhD thesis (2009)
5. Goedkoop, M., et al.: The Eco-indicator 98 explained. Int. J. Life Cycle Assess. **3**(6), 352–360 (1998)
6. Rio, M., Reyes, T., Roucoules, L.: A framework for ecodesign: an interface. Int. J. Eng. 121–126 (2011)
7. Uchil, P., Chakrabarti, A.: Communicating life cycle assessment results to design decision makers: need for an information visualization approach. In: DS 75-5 Proceedings of the 19th International Conference on Engineering Design (ICED13) (2013)
8. Ullman, D.G.: Toward the ideal mechanical engineering design support system. Res. Eng. Des. **13**(2), 55–64 (2002)
9. Card, S.K., Mackinlay, J.D., Shneiderman, B. (eds.): Readings in Information Visualization: Using Vision to Think. Morgan Kaufmann, Los Altos (1999)
10. Norman, D.A.: The Design of Everyday Things. Basic Books, London (2002)
11. Srinivasan,V., Chakrabarti, A.: An integrated model of designing. J. Comput. Inf. Sci. Eng. **10**(3), 031013 (2010)
12. Otto, H.E., Mueller, K.G., Kimura, F.: Efficient information visualization in LCA. Int. J. Life Cycle Assess. **8**(4), 183–189 (2003)
13. Börner, K., Chen, C., Boyack, K.W.: Visualizing knowledge domains. Annu. Rev. Inf. Sci. Technol. **37**(1), 179–255 (2003)

14. Otto, H.E., Mueller, K.G., Kimura, F.: Efficient information visualization in LCA: approach and examples. Int. J. Life Cycle Assess. **8**(5), 259–265 (2003)
15. Otto, H.E., Mueller, K.G., Kimura, F.: Efficient information visualization in LCA: application and practice. Int. J. Life Cycle Assess. **9**(1), 2–12 (2004)
16. Otto, H.E., Mueller, K.G., Kimura, F.: Interactive visualization for the comparative analysis of life cycles in complex product design. In: ds 31 Proceedings of ICED 03, the 14th International Conference on Engineering Design, Stockholm (2003)
17. Ragnerstam, E.: Enhanced interactivity in charts: visualization of life cycle assessment results. Skolan för datavetenskap och kommunikation, Kungliga Tekniska högskolan, Stockholm (2010)
18. Norris, G.A., Yost, P.: A transparent, interactive software environment for communicating life-cycle assessment results: an application to residential windows. J. Ind. Ecol. **5**(4), 15–28 (2001)
19. Gloria, T., Saad, T., Breville, M., O'Connell, M.: Life-cycle assessment: a survey of current implementation. Environ. Qual. Manage. **4**(3), 33–50 (1995)
20. Nielsen, J.: Usability Engineering. Elsevier, Amsterdam (1994)
21. Wang Baldonado, M.Q., Woodruff, A., Kuchinsky, A.: Guidelines for using multiple views in information visualization. In: Proceedings of the Working Conference on Advanced Visual Interfaces, pp. 110–119. ACM (2000)
22. Ainsworth, S., Van Labeke, N.: Using a multi-representational design framework to develop and evaluate a dynamic simulation environment. In: International Workshop on Dynamic Visualizations and Learning, Tubingen, Germany (2002)
23. North, C., Shneiderman, B.: A Taxonomy of Multiple Window Coordinations. University of Maryland. Technical Report #CS-TR-3854 (1997)
24. Spence, R., Press, A.: Information visualization (2000)

Part III
Design for X (Safety, Manufacture and Assembly, Cost, Reliability etc.)

Three Dimensional Form Giving of *Kundan* Jewellery—A Parametric, Cluster Based Approach to Jewellery Design and Prototyping

Parag K. Vyas

Abstract Traditionally jewellery in India is not only ornaments but acts as a last line of defense of a family in a moment of need. It is condensed wealth, available to women of a family that can be converted to liquid cash if the family faces extremely trying times. *Kundan* is a style that is embellished with chips of diamond or a variety of colored gemstones such as emeralds and rubies. Such engaging and intricate works are intense both in terms of time and money. A jeweller's work starts with a hand drawn sketch. This is for initial approval of design. Presently it is done in an old fashioned way that is iterative and tentative in nature as there is no methodical approach. Often there are reworking leading to client's dissatisfaction and it is difficult to detect and correct things absence of a systematic procedure. As articles of *Kundan* are commissioned as one piece off, each project is in a sense prototyping work. This study takes a cluster based approach to design that is methodical in nature. It is expected to benefit both the client and jeweller by providing a tool in the early stages of design. It also facilitates concurrence of views by providing a digital prototype, a three dimensional modeling of gold framework, eliminating reworking at the most critical, final stages of manufacturing. The individual fitting of complex components can also be seen and corrected, pre-empting counterproductive rework in advanced stages of manufacturing.

Keywords Prototyping · Cluster · Design tool · Jewellery · Manufacturing

---

P.K. Vyas (✉)
Grau Bär Designs, Indore, India
e-mail: paragvyas01@gmail.com

A. Chakrabarti (ed.), *ICoRD'15 – Research into Design Across Boundaries Volume 2*, Smart Innovation, Systems and Technologies 35, DOI 10.1007/978-81-322-2229-3_23

1 Introduction

Jewellery in India is synonymous with bullion metals, preferably gold or silver. The reasons for these metals being preferred for jewellery are both historical and social customs as often the bulk users are women.[1] The affinity of Indian populace to gold can be attributed to deep roots dating back to times when India was a collection of several kingdoms coexisting side by side. In those days, often the family from which a girl was given was from a different kingdom or state in today's context. Once the girl left her father's house, contact back home was scarce and visitations few. Therefore, it was a custom to give a share of the girl's right in paternal property in the form of jewellery, at the time of the wedding, an important part of her trousseau. Sons of family in turn married women from other republics and their women brought with them their share in their paternal property, in a similar form.

Thus, women carried their wealth with them in a compact form of 'jewellery' that was portable and held cash value across diverse states and ethnicities. Even to date the cash value of gold holds fairly consistent across countries and continents. Gold and in turn jewellery made from gold, also has a universal appeal in tribal, rural and/or urban settings.

Kundan is a style that is embellished with chips of diamond or a variety of colored gemstones, such as emeralds and rubies. Such engaging and intricate works are intense both in terms of time and money. The motifs are distinctively Indian and identifiable at a glance amongst other contemporary styles and trends.

The process of *Kundan* setting is intensive labor and difficult to mechanize. In such situations research inputs can make most meaningful contributions at various manufacturing stages and to overall domain of jewellery.

2 Background Study

2.1 Formal Aspects

The form of a typical *Kundan* necklace is a collection of clusters made of smallest semantic units. These clusters are arranged around a curve, best fitting the circumference of the neck, for frontal observation and appreciation (Vyas and Bapat 2010; [1: 168]). There is a slider over strings for finer adjustments and comfort over nape of neck as shown in Fig. 1. This is a typical configuration that allows women to wear it neatly fitting around the neck and present article for a visual appreciation by observers.

[1] Largely the quantity of gold held by a family was in individual possession of women, where they enjoyed absolute control over this material possession. This gold was also protected by a strict code of conduct in society where it was considered beneath a family to use it as collateral even in most trying situations.

Fig. 1 A *Kundan* article made of form clusters worn for visual appreciation form front

2.2 Design Aspects

The exquisiteness of this design is that the removal of a few clusters will not change the overall appearance of the design in a perceptible manner.[2] These traditional aspects make an important part of design parameters and technical details that must be incorporated in modern processes.

The mechanical construction of a *Kundan* form cluster begins with a tracing of a single line artwork on a paper. This artwork, that stands in for a crude drawing, is sent to the jeweller to be used as reference. The jeweller is expected to design and incorporate details, this requires interpretation of the sketch which may and does differ from person to person.

The maker subsequently takes a strip of gold of suitable dimensions and bends it in a form of a motif using a pair or tweezers, sometimes taking help of specially formed mandrels. A collection of these bent forms are arranged over the reference sketch and checked for visual similarity. Adjustments are made by tweaking and teasing strip with fine tipped tweezers until a satisfactory conformance is achieved. This composition is then put together over refractory material and spot soldered in key locations as shown in Fig. 2. This latticed structure is thereafter put on a flat plate of adequate thickness and soldered on the base. Excessive material is removed by using a piercing saw.

The gemstone is set from the visible front using a foil of very high purity, almost 24 Carat gold foil. For the purpose of setting, a very thin foil (rolled in a mill) is used. At such fineness and low thickness gold becomes very pliable. Parts and pieces of this foil are pressed in cavities left around the gemstone, partly set with setting paste as shown in Fig. 3. By pressing layer after layer of gold, space around gemstone is slowly filled with solid gold. A pointed tool that gives good pressure and negotiates tight corners does this work creating flush stone close setting [2: 614]. Gold surface

---

[2] This typical feature, made it possible for the owner to discreetly sell (or place as collateral) a portion of this condensed wealth without showing signs of distress in peer group or society which is very watchful of such occurrences in a family.

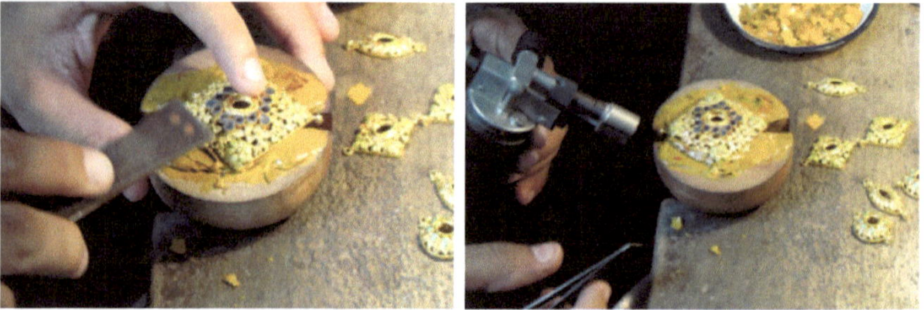

Fig. 2 An article being made showing two different stages of the overall process

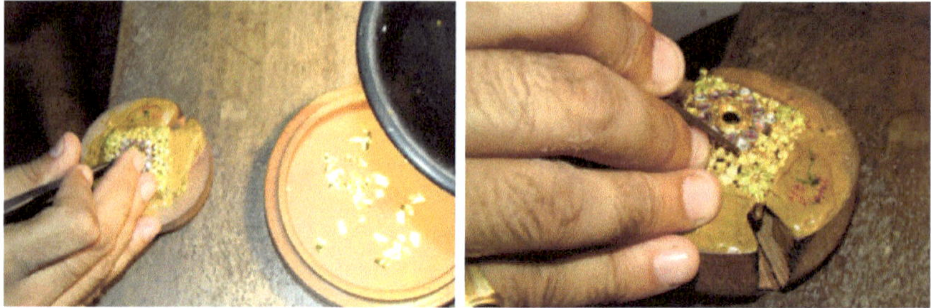

Fig. 3 *Kundan* setting process using very thin gold foils

holding gemstone in place is evened out using an engraver that leaves behind a lustrous surface. This style of setting was different from the western style of prong or claw setting brilliant cut diamonds and formally cut gemstones [3: 169, 4: 174–188].

Backside of *Kundan* jewellery is embellished with hard enamel typically red green and red blue colors. This makes back side of jewellery equally pleasurable to look at. Backs often outshine the front [5: 21], a feature that makes *Kundan* jewellery wearable from both sides

3 Literature

Gold ornaments, as rings and such smooth mono-form articles for casting can be visualized using diverse CAD software's. Exhaustive modeling techniques are available as technical details in form of setting, insitu-casting of gemstones, linkages, jewellery fasteners and findings (Fig. 4). Literature is silent on digital modeling techniques for *Kundan*, hence this subject gains importance.

First interaction of an object with the observer is based on overall form (Vyas 2007; [6]). It is therefore of importance that the overall form of jewellery is presented to a present or prospect client for visual appreciation. Presently, for

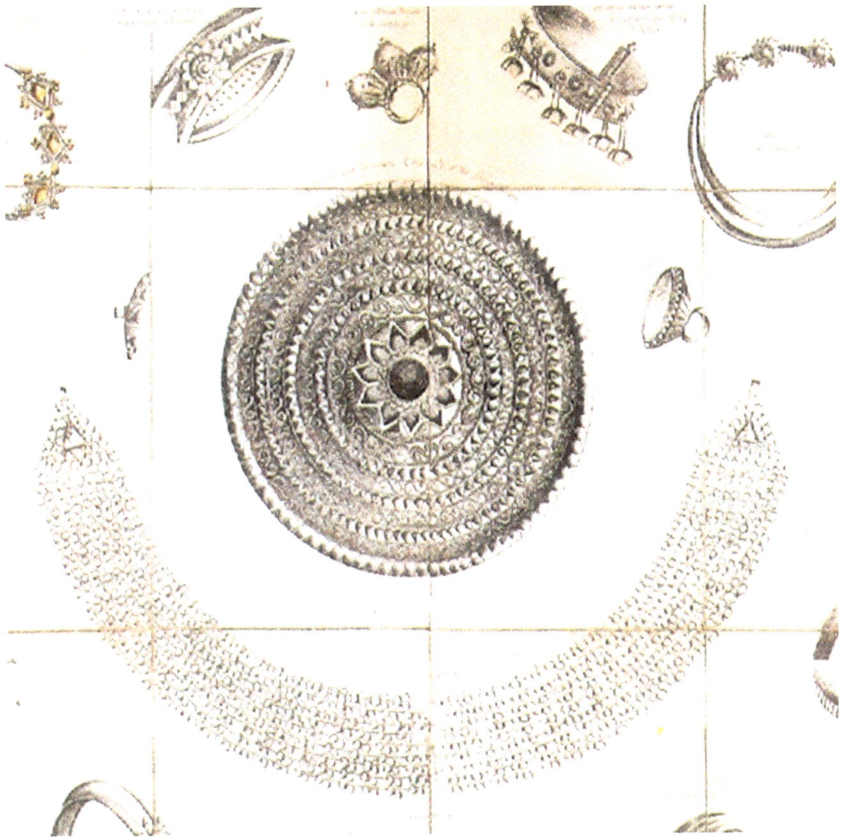

Fig. 4 Sketch of conventional *Kundan* articles showing a neck piece prominently

articles of *Kundan* jewellery it is done in a traditional way by presenting a sketch or rendering as shown in Fig. 4 that has changed very little over years [7]. This is to seek a broad agreement on shape, size and overall appearance of the form with reference to the client's neck [1]. This work is iterative in nature and often takes several trials and errors to have satisfactory results.

There exist chasms that need to be bridged by quick modeling techniques for communication with the client. This is an important aspect of jeweller client communication. Expression by clients need to be understood and translated correctly by jeweller for construction of a masterpiece. What happens in reality often is a compromise between 'what you see and what you get' to accept. Literature is silent on this issue.

Few relevant articles are reviewed for their contributions and limitations in Table 1.

Table 1 Reviewed articles

Sr. No.	Title of paper/Article	References	Contribution	Keywords	Limitations
1	Rapid prototyping and tooling technology in jewellery CAD	Wannarumon and Bohez [11]	Investigates computer-aided design and rapid prototyping technologies in jewellery design and manufacturing	Computer-aided design (CAD), rapid prototyping (RP), rapid tooling, stereo lithography apparatus (SLA), jewellery design, jewellery making	The gap as appears is too wide to be bridged by practising jewellers by adapting. On the other hand when academicians approach the issue they cannot fathom the formal details
2	A parametric feature-based CAD system for reproducing traditional jewellery	Stamati and Fudos [12]	Computer aided design (CAD) systems, reproducing traditional jewellery	Feature-based design, traditional jewellery, voxels, CAD, scaling algor	Though the paper throws good light on lithography and other techniques in general, particularly product specific applications need to be explored by jewellers
3	A parametric voxel oriented cad paradigm to produce forming components for stretch formed jewellery	Gulati and Tandon [13]	A unique kind of jewellery, called stretch formed jewellery is discussed	Jewellery, CAD, parametric, voxel, rapid prototyping	
4	A parametric voxel based unified modeller for creating carved jewellery	Gulati et al. [14]	This paper presents a parametric voxel based jewellery modeller for designing and creating carved jewellery	Parametric, voxel, jewellery, CAD, rapid prototyping	The papers throw light on modelling techniques but limited by academic approach. Firstly carving applicable to wax and materials suchlike for replication. Secondly, fractal based application being theoretical in nature, may pose practical difficulties in adaptation for a technique suited for forward integration with manufacturing
5	An aesthetics driven approach to jewellery design	Wannarumon [15]	Proposes a computer-based design tool to automate art form generation used in jewellery design	Computer design support, evolutionary art, fractal, jewellery design	

To further strengthen the reviews, the India International jewellery show from year 2007–2011 were routinely visited to gain insights. To observe and learn by inspection about *Kundan* and its various aspects of form and construction details. Visits to such venues have an added merit of getting in touch with domain experts and practitioners. It was further realized that an informal approach with moderate interest in subject begets good information. Formal approach and recording devices, including cameras, trigger activation of a defensive mode in people and often interviews are abruptly terminated.

4 Methodology

With a thorough background study and a clear understanding of the subject matter, a systematic step by step process was adopted. To devise a quick modeling technique for presentation to clients, a visualization aid, which can subsequently be used for prototyping as well, motifs were methodically documented and converted to vector artworks [8]. These vector artworks, in the form of smallest semantic units were the foundation for subsequent modeling.

For the purpose of understanding Modeling follows three stages,

- Selection of Smallest Semantic Units—Starting with selection of a number of smallest semantic units that gel with each other as shown in Fig. 5. Some motifs are used as central units while the others are better suited to support the focal member or (Varadrajan 2008; [9: 768]) on outer perimeter.
- Form clusters—The chosen semantic units are arranged in a manner to have a meaningful form cluster [10: 118]. This form cluster is visually checked for an appeal and aesthetically pleasing appearance, if needed adjustments if any are made to the composition of the form cluster. Merely following geometry is not adequate, a poetic expression, a rhythmic movement that leads the eye is the essence of creation of form clusters.
- Complete articles—For the creation of an article such as a choker, the form cluster is placed centrally to a principle circle, with the apex of the form cluster just touching on a point of contact. This cluster is then checked for size with

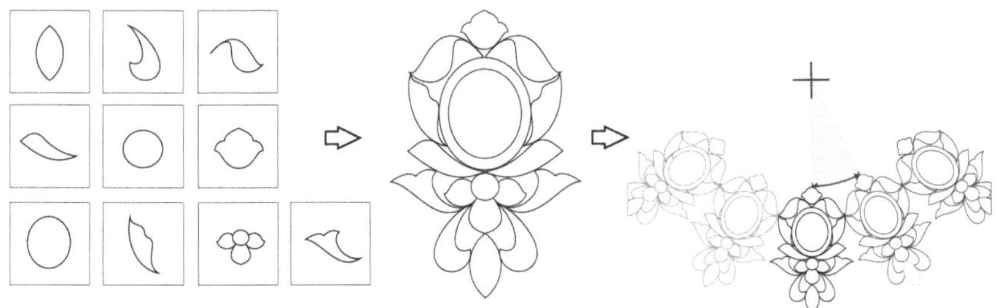

Fig. 5 Semantic units that gel with each other, a typical example

respect to the circle. There are two parameters to be controlled at this stage: firstly the size, secondly the cluster pitch. This pitch is the linear distance between two adjoining clusters.

5 Investigational Setup

A subject was chosen from a particular client profile, from amongst those who volunteered for an experiment to test the procedure on how it works. The designs were presented for selection in the form of a portfolio and the client finally chose one as seen in Fig. 6 showing the development process.

This design was first rendered in color to seek concurrence of views on choice of color stones and in what manner combinations will appear on setting.

An estimation was possible due to a much precise, to the scale drawing now being available. This accuracy was reflected in estimation and precise estimations could be given, within a short period of time.

In comparison to conventional methods, it takes a few days to week for such estimations owing to several reasons. Few major reasons are: time taken in scaling and re-drafting or re-sketching. In this case scaling and adjustments were done in the presence of the client. Pictures with the client in session are not produced following a request for anonymity. However, the article is photographed with gracious permission and is reproduced.

The three dimensional modeling also helped to determine suitable soldering locations. This was especially helpful at the shop floor where intricate piercing details of base plate could be visualized before actual soldering. The progressive development pictorial depiction is self illuminating with Fig. 7. These renderings were very useful for shop floor work and as a result no reworking was required.

The actual piece as it came after prototyping is produced with clients permission. It can be observed through the developmental process that every step has a check and correct visual feedback. The article took just one third time for manufacturing as a result of methodical approach as there were no iterations or amendments required.

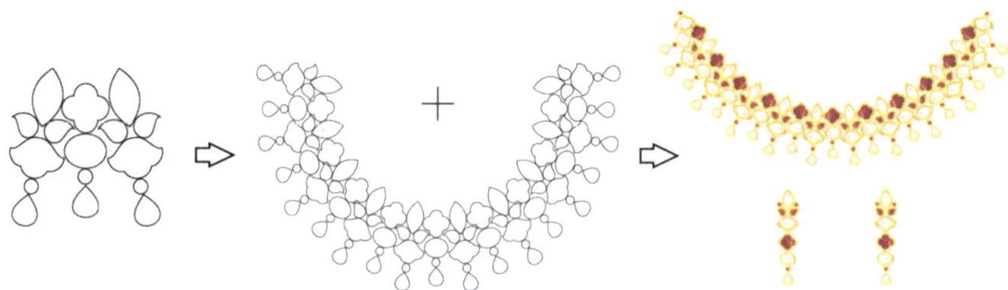

Fig. 6 Creation of necklace selected by client for initial approval

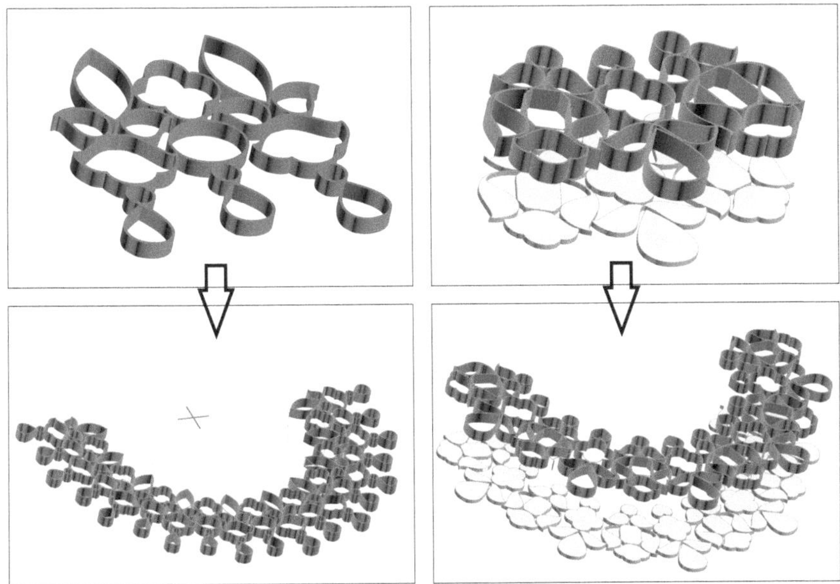

Fig. 7 Three dimensional view of cluster showing bent stripes

6 Implications

6.1 The Managerial Implications Are as Follows

1. Considerable choices become available to the client in the initial stages of design due to use of technology as clusters can be made in front of the client by picking and placing.
2. Changes, variations and detailing become prompt and instantaneous. Various transformations such as scale, skew, rotation and translate can be applied to individual units as well as the whole cluster.
3. The desired results can be seen in three dimensions from various angles on screen before implementation avoiding reworks and revisions.
4. It is beneficial for the jeweller to have a much clearer mental picture before work commences. Tricky corners, nooks and crevices can be planned and avoided with better visual appearance in final piece.
5. The article can be made in a systematic manner, with little or no reworking. In practice jewellers keep sufficient monetary margins to counter such cases. As with this technologically sound approach there is little need for those safety margins in quotations. The savings in terms of time and money is beneficial for overall client jeweller relationship.

6. The savings can be equitably shared by all stakeholders, a portion can be extended to client himself as a discount.
7. Better time schedules for delivery can be met. Often a delay is expected and accepted in practice as of now.

6.2 The Academic Implications Are as Follows

1. The process can be used to sensitize a new client, making them aware of detailed aspects of *Kundan* using virtual models.
2. Apprentice Jewellers can be trained on formal aspects of *Kundan*, reducing learning time by conventional methods.
3. An interface with SLA and FDM technologies can be devised with ease. This will be beneficial for limited edition batch productions which currently are not plausible.
4. Mechanical detailing can be greatly improved as location of hinge points and fixing details can be easily adapted and incorporated.
5. Estimation protocols can be written making spot estimations possible, both for amount of gold and number and size of gemstones.

7 Limitations

The tool is tested only few times so far owing to very expensive nature of basic material. Though it is working satisfactorily, wider tests need to be done. Also, incorporating stones in modelling process remains a challenge because chips are not as homogeneous as formally cut gemstones. However, an indicative place holder can be modelled.

8 Conclusions

The use of technology is most effective on two fronts: Firstly, in a better three dimensional modelling that helps at various stages of design and manufacturing process. This is a clear improvement over conventional sketches. Secondly, time and efforts are saved as iterations and reworking is successfully avoided.

The study can be later integrated with diverse design platforms in practice. This will give practicing designers a good collection of ready to use forms in their designs.

References

1. Chakrabarti, D.: Indian Anthropometric Dimensions, for Ergonomic Design Practice. National Institute of Design, Ahmadabad (1997)
2. Untracht, O.: Jewelry Concepts and Technology. Doubleday & Company Inc., Garden City, New York (1985)
3. Pagel Theisen, V.: Diamond Grading ABC Handbook for Diamond Grading. Rubin & Son Bvba, Antwerpen, Belgium (1993)
4. Snowman, A.K.: The Master Jewelers. Thames & Hudson Ltd., London (1990)
5. Sharma, R.D., Varadarajan, M.: Handcrafted Indian Enamel Jewellery. Lustre Press, Roli Books, India (2008)
6. Yarbus, A.: Eye Movement and Vision. Plenum Press, New York (1993)
7. Balakrishnan, U.R., Kumar, M.S.: Jewels of the Nizams. India Book House Pvt. Ltd., India (2006)
8. Vyas, P., Bapat, V.P.: Identification and Classification of Semantic Units Used in Formation of Patterns in Kundan Jewellery, A Methodical Approach, pp. 59–72. Design Thoughts, IDC, IIT Bombay (2010)
9. Vyas, P., Bapat, V.P.: Investigation of Form Clusters Made of Smallest Semantic Units and Patterns they Create as Building Blocks of Kundan Jewellery, pp. 766–774. Research into Design, Research Publishing, India (2011)
10. Adorno, T.W.: Aesthetic Theory, 6th edn. University of Minnesota Press, USA, p. 118 (2008)
11. Wannarumon, S., Bohez, E.L.J.: Rapid prototyping and tooling technology in jewelry CAD. Comput. Aided Des. Appl. **1**(1–4), 569–576 (2004)
12. Stamati, V., Fudos, I.: A parametric feature-based CAD system for reproducing traditional jewelry, vol. 37(4). Department of Computer Science, University of Ioannina, pp. 431–449 (2005)
13. Gulati, V., Tandon, P.: A parametric voxel oriented CAD paradigm to produce forming components for stretch formed jewelry. Comput. Aided Des. Appl. **4**(1–4), 137–145 (2007)
14. Gulati, V., Singh, H., Tandon, P.: A parametric voxel based unified modeler for creating carved jewelry. Comput. Aided Des. Appl. **5**(6), 811–821 (2008)
15. Wannarumon, S.: An aesthetics driven approach to jewelry design. Comput. Aided Des. Appl. **7**(4), 489–503 (2010)
16. Vyas, P.: Understanding Features of Form in Fine Jewellery, by Overt Eye Moment Studies-Catch People Watching Jewels, pp. 59–72. Design Thoughts, IDC, IIT Bombay (2007)

Design and Implementation of a Line Balance Visualization and Editing Tool

Rahul Sharan Renu and Gregory Mocko

Abstract The objective of this research is to develop a knowledge representation for managing and modifying assembly line balance information through visualization tools. Additionally, a software tool, the Line Balancing Visualization and Editing Tool (LVET), is developed that enables assembly line balancing experts to view and edit line balance knowledge. The tool is integrated with external line balancing algorithms and assembly line constraints. This knowledge representation and associated tool provide a graphical representation of workstation layout, processes assigned to workstations, and tool and time constraint information. The tool enables complex line text based results to be visualized graphically. The tool is validated by checking for consistent and correct representation of information from an algorithm generated line balance.

Keywords Knowledge representation · Assembly line balancing · Visualization · GUI · Constraint visualization

1 Introduction

An assembly line consists of a number of workstations that can be arranged in different ways. Some of these arrangements include U-shaped assembly lines, parallel assembly lines and two-sided assembly lines [1]. A base product is placed on top of a moving transportation mechanism, such as a conveyor, at the first station. At every workstation, activities are performed on this base product. These activities include assembling parts on the base product, checking the parts that have been assembled and placing parts on the base product for subsequent assembly. This research was conducted in perspective of an automotive assembly line and in such a setting, the activities that are to be performed at workstations are dictated by assembly process sheets (see Fig. 1) [2].

R.S. Renu · G. Mocko (✉)
Department of Mechanical Engineering, Clemson University, Clemson, SC, USA
e-mail: gmocko@clemson.edu

A. Chakrabarti (ed.), *ICoRD'15 – Research into Design Across Boundaries Volume 2*,
Smart Innovation, Systems and Technologies 35, DOI 10.1007/978-81-322-2229-3_24

Fig. 1 Example process sheet

Process Sheet Number: 1234567	**Title:** Example Process Sheet	
Variants: A1, A2, A3, C3		
Work Instructions: 10 Get bracket; 20 Align bracket to holes on body-in-white; 30 Insert screws		
Time Studies:		
10 Get and place bracket	ABC23	15
20 Insert screws	DEF25	10
Part Number: 7890123		

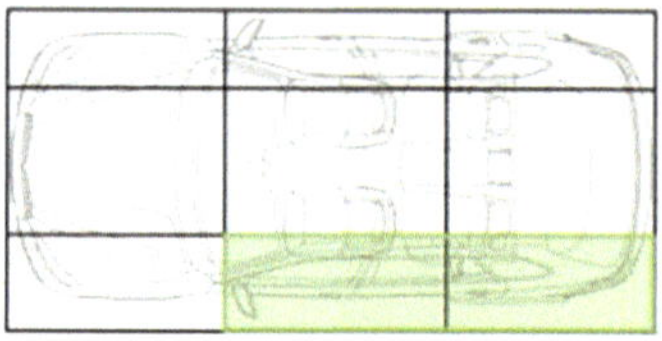

Assembly line balancing involves assigning process sheets (see Fig. 1) to workstations based on a variety of objective functions. These objective functions may include minimize time, maximize worker utilization, or minimize number of workstations. In addition, line balancing is subject to a number of constraints.

The most significant assignment constraint comes in the form of precedence. Precedence relations dictate the order in which assembly tasks must be performed. Another assignment constraint is the (non-portable) tools that are required for assembly operations. Tasks that require these tools must be assigned to the station at which the tools are present. The optimization considerations coupled with the assignment constraints make the generation of a feasible assembly sequence a challenging task.

Market trends, strong competition and shorter product lifecycles have forced companies into manufacturing products that are highly customized [1, 3–6]. Another requirement for companies is to reduce lead times and maintain the Just-In-Time (JIT) manufacturing principles [7]. This necessitates a reduction of batch sizes. It also implies a need to move away from mass production strategies and move towards mass customization strategies. Boysen et al. [1] identifies the problems associated with balancing mixed-model assembly lines within the JIT environment:

- Different variants imply varying cycle times and the possibility of reduced worker utilization.
- Tasks have to be assigned to workstations such that there is no intrusion of line side associate workspaces.

These complexities make line balancing a challenging and time consuming task and warrant the need to have computer supported line balancing. Attempts have been made to completely automate the line balancing process [1, 3, 8, 9]. However the process of line balancing has proven to be hard to completely automate. Capturing and codifying expert knowledge has proven to be a challenge [10]. Researchers have found it difficult to include all necessary constraints in line balancing algorithms to eliminate the need for human input. It has been recognized that the assembly line balancing problem has “inherent problem complexity and diverging problem settings” [10]. There is a need for a decision support tool that will assist in performing semi-automated line balancing.

There is a need to leverage and apply information from the line balancing algorithms as well as information from assembly line balancing experts. It has been observed that manually editing outputs of line balancing algorithms is unintuitive and error-prone (see Fig. 2). This further warrants the need to have a decision support tool for semi-automated assembly line balancing. There have been efforts to visualize line balance information [11–13]. However, these efforts contribute towards the visualization of the statistics associated with the line balance algorithm output. Some efforts have also contributed towards creating a virtual environment; often called *virtual factories*. There remains a need to create a representation of tasks and tools assigned to workstations to allow line balance experts to make necessary editing capability that balances visualization and the immersive virtual reality representation.

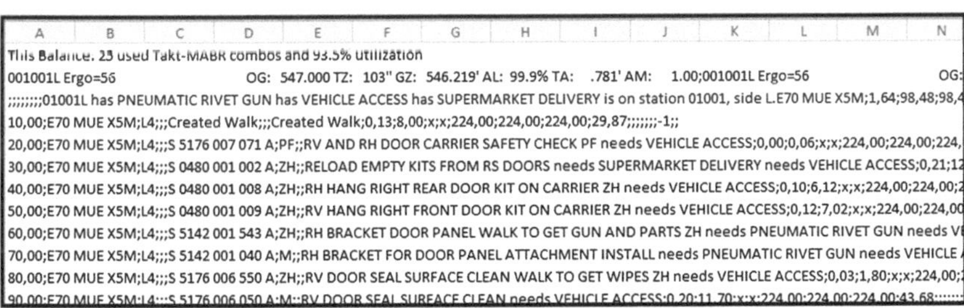

A B C D E F G H I J K L M N
This Balance. 23 used Takt-MABR combos and 93.5% utilization
001001L Ergo=56 OG: 547.000 TZ: 103" GZ: 546.219' AL: 99.9% TA: .781' AM: 1.00;001001L Ergo=56 OG:
;;;;;;;;01001L has PNEUMATIC RIVET GUN has VEHICLE ACCESS has SUPERMARKET DELIVERY is on station 01001, side L.E70 MUE X5M;1,64;98,48;98,4
10,00;E70 MUE X5M;L4;;;Created Walk;;;Created Walk;0,13;8,00;x;x;224,00;224,00;224,00;29,87;;;;;;;-1;;
20,00;E70 MUE X5M;L4;;;S 5176 007 071 A;PF;;RV AND RH DOOR CARRIER SAFETY CHECK PF needs VEHICLE ACCESS;0,00;0,06;x;x;224,00;224,00;224,
30,00;E70 MUE X5M;L4;;;S 0480 001 002 A;ZH;;RELOAD EMPTY KITS FROM RS DOORS needs SUPERMARKET DELIVERY needs VEHICLE ACCESS;0,21;12
40,00;E70 MUE X5M;L4;;;S 0480 001 008 A;ZH;;RH HANG RIGHT REAR DOOR KIT ON CARRIER ZH needs VEHICLE ACCESS;0,10;6,12;x;x;224,00;224,00;2
50,00;E70 MUE X5M;L4;;;S 0480 001 009 A;ZH;;RV HANG RIGHT FRONT DOOR KIT ON CARRIER ZH needs VEHICLE ACCESS;0,12;7,02;x;x;224,00;224,00
60,00;E70 MUE X5M;L4;;;S 5142 001 543 A;ZH;;RH BRACKET DOOR PANEL WALK TO GET GUN AND PARTS ZH needs PNEUMATIC RIVET GUN needs V
70,00;E70 MUE X5M;L4;;;S 5142 001 040 A;M;;RH BRACKET FOR DOOR PANEL ATTACHMENT INSTALL needs PNEUMATIC RIVET GUN needs VEHICLE
80,00;E70 MUE X5M;L4;;;S 5176 006 550 A;ZH;;RV DOOR SEAL SURFACE CLEAN WALK TO GET WIPES ZH needs VEHICLE ACCESS;0,03;1,80;x;x;224,00;2
90,00;E70 MUE X5M;L4;;;S 5176 006 050 A;M;;RV DOOR SEAL SURFACE CLEAN needs VEHICLE ACCESS;0,20;11,70;x;x;224,00;224,00;224,00;43,68;;;;;;

Fig. 2 Snippet of output from an automated line balancing algorithm

2 Frame of Reference

The literature studied in this research is divided into three categories. First, the use of line balancing algorithms has been reviewed. The next section presents the use of some of these algorithms in line balancing tools. These line balancing tools also provide statistical visualization of line balance data. In the last section, research based on virtual, immersive environments that allow the evaluation of the feasibility of product designs and line balances, has been reviewed. There have also been some efforts towards finding a middle ground between the two categories.

2.1 Line Balancing Algorithms

Assembly line balancing algorithms have been researched for over 50 years [14–16]. Due to the complex nature of the problem, metaheuristic approaches have dominated [5, 10, 17–26]. These metaheuristic approaches have included both, simple local search algorithms [1, 10, 22, 24–26] and also algorithms with search strategies that have the ability to learn [17–21].

The research works cited above all make simplifying assumptions to reduce the complexity of the assembly line balancing problem. The research presented by Hyun et al. [17], Falknauer and Delchambre [20], Vilarinho and Simaria [23], McMullen and Frazier [26] and Lapierre et al. [24] consider only a limited set of objectives and constraints while balancing an assembly line. Goncalves and colleagues [18] assume that the line balancing is being performed for a line that has not been set up yet. By making this assumption, they disregard permanent tooling constraints (some tools are built into the ground and can't be moved). This constraint is vital while rebalancing assembly lines that have already been set up. Research presented by Sabuncouglu et al. and Vilarinho et al. assume that the assembly line produces only one variant of a certain product. This again is a simplifying assumption does not hold true in a mixed-model assembly environment.

Some researchers have implemented their line balancing algorithms into commercial software packages. These software not only perform assembly line balancing, but also provide statistical reports of line balance data. The next section reviews the literature relevant to such research efforts.

2.2 Assembly Line Balancing Tool

Torenli [11] has developed a line balancing tool—Probalance® which takes the required processes, available resources, the stations to which certain resources have been allocated and available time as input and provides the user with a balance that

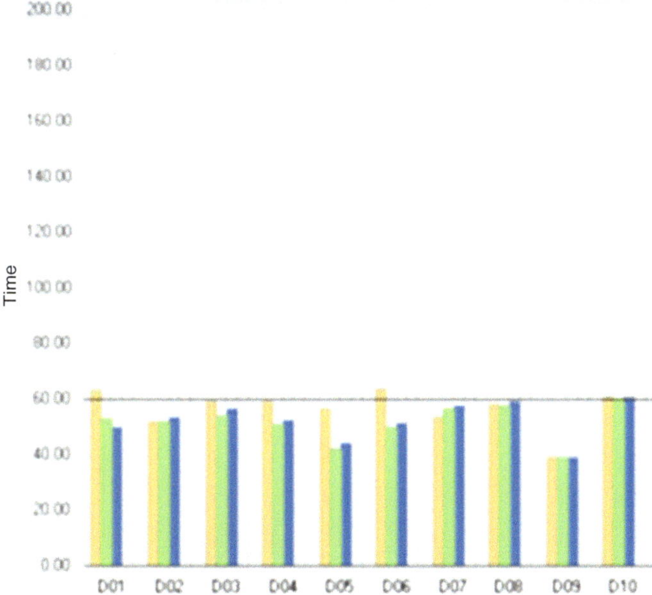

Fig. 3 Line balance data visualization (*Source* [11])

is editable. This balance is visualized by way of traditional bar graphs. The visualization lacks, in terms of presenting work station resources to the user. While editing assignments of work content to work stations, i.e. while editing line balances, it is essential to have knowledge of the resources present at the stations to which process are intended to be moved. The bar graphs presented by Torenli (see Fig. 3) do not present the user with work station resource information. Rentschler and colleagues [12] have presented a method to balance an assembly line by taking input data in the form of a spreadsheet. The method proposes to display the output information as a combined bar and line graph.

2.3 Use of Virtual Reality for Assembly Line Planning

Wenbin and colleagues [13] present a Production Engineering-Oriented Virtual Factory. While their approach models and visualizes the details of an entire assembly line, the goal for the tools that they have presented is to foster concurrent manufacturing systems design. The tools presented do not focus on visualization of an algorithm generated assembly line balance.

Zhong and Shirinzadeh [27] have also presented a 3D virtual factory modelling methodology. They have recognized that discrete event simulation approaches to production make several simplifying assumptions that limit their output from being implemented without further modifications (Fig. 4).

Fig. 4 Virtual reality to model and visualize assembly lines (*Source* [27])

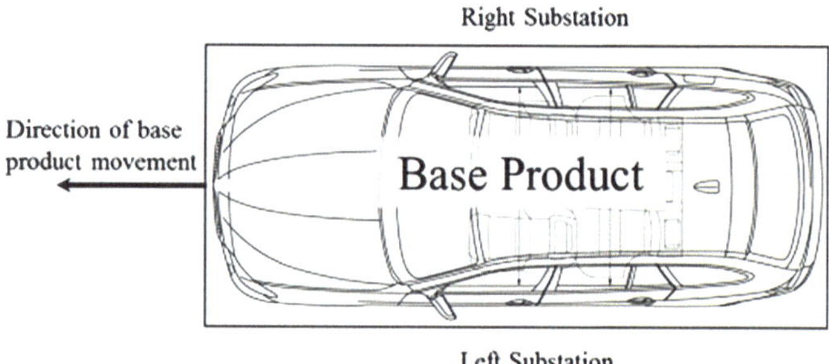

Fig. 5 Division of workstation into *left* and *right substation*

Also, they recognize that the outputs from discrete event simulations are not easily interpretable by non-experts leading to the need for a visualization system. While the above listed shortcomings of discrete event simulation are shared with line balancing algorithms, Zhong and colleague's virtual factory does not visualize line balance data. Hong et al. [28] have also developed a 3D virtual factory to assist in visualizing and optimizing the outputs obtained discrete event simulations.

Fan and colleagues [29] have developed an assembly line balancing algorithm and a visualization method. However, they do not delve into the details of the latter. Also, their tool does not provide information regarding tools present at stations. This approach visualizes the data better than statistical approaches. It is also not as immersive as 3D virtual environments.

The performance gaps identified in this section have led to establishment of requirements for an assembly line balance visualization tool. These requirements are presented in the next section.

While several research efforts have presented tools to perform line balancing and present line balance information as bar charts and/or virtual reality environments, there is a need for tools and methods which will allow assembly line planners to visualize and edit line balances, workstation layouts and constraints associated with each station. The visualization of information as bar and line graphs is not intuitive and does not allow for the station-related information to be visualized. On the other end of the spectrum, virtual reality is extremely immersive, computationally intensive and leads to high cognitive load. A middle ground between these two approaches must be developed and linked with an algorithm-based assembly line balancing software.

3 Design of Visualization Tool

Automated assembly line balancing algorithms do not capture all the knowledge necessary to obtain the most useful line balance. However, these algorithms can be used to obtain an initial balance that will need review and revision from an expert assembly line planner. This warrants the need to visualize the assignments made by the line balancing algorithm and allow experts to make necessary changes. It is important to note the tool is a not a line balancing tool and does not implement line balancing algorithms, but rather imports line balancing data, provides functionality to visualize the data, and enables the planner to make minor changes and additions to the line balancing data.

Modifying line balancing data, often representing using text, is a challenging task for the planner. A snippet of the output from an automated line balancing tool implemented in Microsoft Excel is shown in Fig. 2. This figure shows the unfavorable user interface for editing the line balancing output. The design method proposed by Pahl et al. [30] is used. Requirements presented in the following section will dictate the development of a tool that eliminates this issue.

3.1 Requirements

The requirements stated below were formulated to guide the design and development of a tool that is capable of addressing the research gap identified in Section 2. The tool must:

1. Have capabilities to pre-process outputs of automated assembly line balancing algorithms.
2. Visually represent the tools and available time present at each station.
3. Visually represent the process sheets assigned to each station.
4. Not disallow any assignment that the user would like to make.

5. Inform the user of any infeasible assignments that have been made either by the algorithm or by the user.
6. Should be a network-based tool.

The last requirement stems from the need to encourage the use of collaborative process planning. Requirement four ensures that the tool does not defeat its own purpose. The objective of visualizing and allowing edits to algorithm generated line balances is to capture expert knowledge. This objective will remain unfulfilled if the tool imposes hard constraints on the manual re-assignment of process sheets to stations.

3.2 Knowledge Representation

Representation of the knowledge that the tool will have to manage is the first step in the development of a tool that will fulfill the requirements stated in the previous section. shows the knowledge representation in the form of an Entity-Relationship (ER) diagram [31]. The following points enumerate provide a textual description of the knowledge representation.

While manually editing line balances it is important that the planner be aware of:

- tools present at the workstation to which he/she intends to move a process sheet, and
- available time at the destination workstation (Fig. 6).

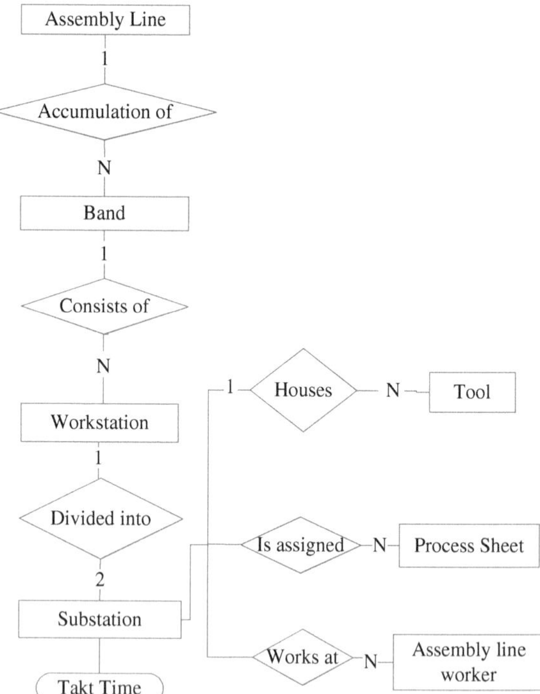

Fig. 6 Knowledge representation for assembly line data

Fig. 7 System design

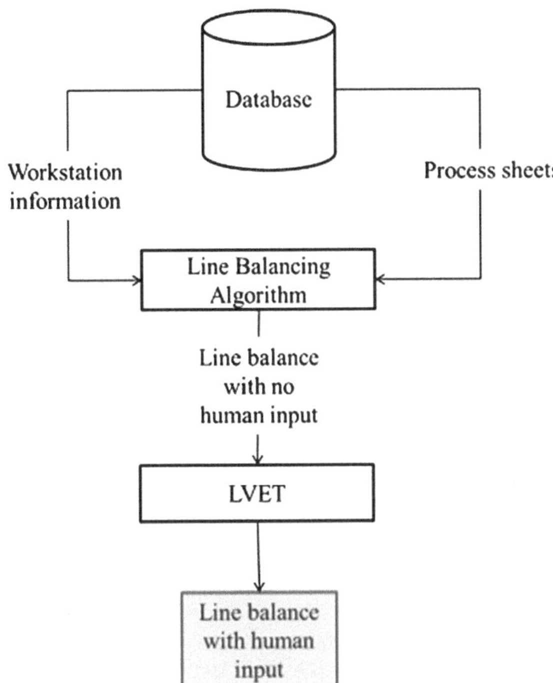

3.3 *System Design*

The functioning of the visualization tool is found in Fig. 7. The line balancing algorithm requires the process sheets that have to be assigned to workstations, and workstation information (such as presence of tools, type of carrier that the base product moves on, available platforms for workers to stand on) to perform line balancing. The output from the line balancing algorithm is one which has no human input. The visualization tool takes this output from the line balancing algorithm, extracts the required information and generates a visual representation of the information that users are allowed to edit.

4 LVET Implementation

It is conceptualized that when the user selects a station all the process sheets assigned to that station will be displayed. Also, the time available at that station will be displayed to the user along with the list of constraints that are considered. The user will be notified of any broken constraints.

4.1 Pseudo-guide for Implementing Visualization Tool

Although information about the product variants are being captured, it is decided that the visualization tool will only consider the line balances for one single variant alone. The information model was developed to capture, compute and display information regarding only two constraints, time and tooling. The pseudo code developed for LVET is the following four steps.

1. Extract information from automated line balancing algorithm's output
2. Compute time taken by process sheets at a station
3. Match process sheet tool requirement to station tool availability
4. Highlight infeasible assignments

4.2 Visualization Tool User Interface

The automated line balancing algorithm's output contains all the information that is needed for LVET. This includes availability of tools at stations, station time and tool requirement of process sheets. A tool was created within Microsoft Excel 2010 by programming in VBA. This tool parses through every line of the automated line balancing algorithm's output and extracts all the necessary information. The tool creates Comma Separated Value (CSV) files. The database is populated with the information from these files by accessing a webpage that has been created for this purpose alone. Once the information has been uploaded to the database the user can access LVET's main page where the pseudo code is implemented using HTML and PHP.

The bottom most frames show all the stations within the band under consideration. Those stations that have violated constraints are differently colored to alert the user. Above this frame, the process sheets that are assigned to this station are listed. The user can select any number of these process sheets and move them to the desired station by making an appropriate selection in the "Move to station" section. The top most frames displays the tools present at either zone of the selected station.

5 Validation of the Visualization Tool

The Line balance Visualization and Editing Tool (LVET) works on the standard output of an automated line balancing algorithm. In order to test and validate the consistency and accuracy of this tool, three outputs from the automated line balancing algorithm were input to the LVET as test cases. The input data to the automated line balancing algorithm was data from one band in the assembly line of a large scale automobile OEM. Table 1 shows the requirements of the tool.

Table 1 Knowledge representation elements

Entity	Description
Assembly line	An assembly line is an accumulation of bands
Band	A band is a collection of workstations
Workstation	The physical space on an assembly line where assembly operations occur. For line balancing purposes, a workstation is divided into substations (see Fig. 5)
Substation	These are virtually demarcated spaces within a workstation. Tools are housed at substations. Assembly line workers perform assembly operations from substations. Process sheets are assigned to substations

Table 2 List of requirements of the visualization tool

Description	Satisfied?
Have capabilities to pre-process outputs of automated assembly line balancing algorithms	Yes
Visually represent, in a human readable form, the tools and available time present at each station accurately	Yes
Visually represent, in a human readable form, the process sheets assigned to each station accurately	Yes
Not disallow any assignment that the user would like to make	Yes
Inform the user of any infeasible assignments that have been made either by the algorithm or by the user	Yes
Should be a network-based tool	Yes

Validation of the tool checks if these requirements are met. It is important to note that although the output from the automated line balancing algorithm contained data of several variants of cars, the LVET only visualizes the data pertaining to one single variant. The data that is presented by the LVET is compared manually to the data contained within the output of the automated line balancing algorithm (Table 2).

The line balancing algorithm assigned 359 process sheets to twenty four substations. The LVET output accurately displays each of these 359 process sheets in their respective substations. A subset of five process sheets is shown textually in Fig. 8 from the line balancing algorithms and graphically in Fig. 9 through LVET.

The three requirements were found to be satisfied for all the three automated line balance algorithm outputs that were visualized. The tool causes color changes of the station boxes based on the constraint violations accurately and consistently.

- The LVET accurately visualizes the line balance information. The tool allows users to move process sheets from one station to another and notifies the user of time and/or tooling constraints being violated.
- Testing and validation of the tool also helped recognize opportunities for improvement of these tools.

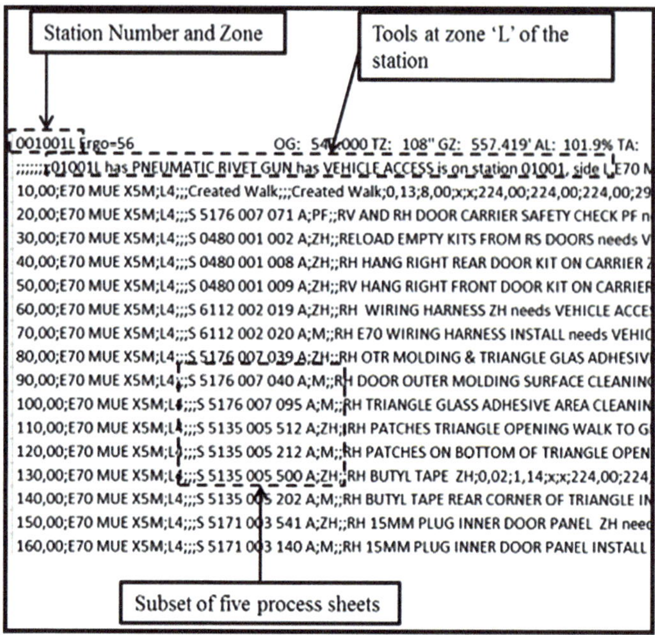

Fig. 8 Snapshot of automated line balancing tool output [32]

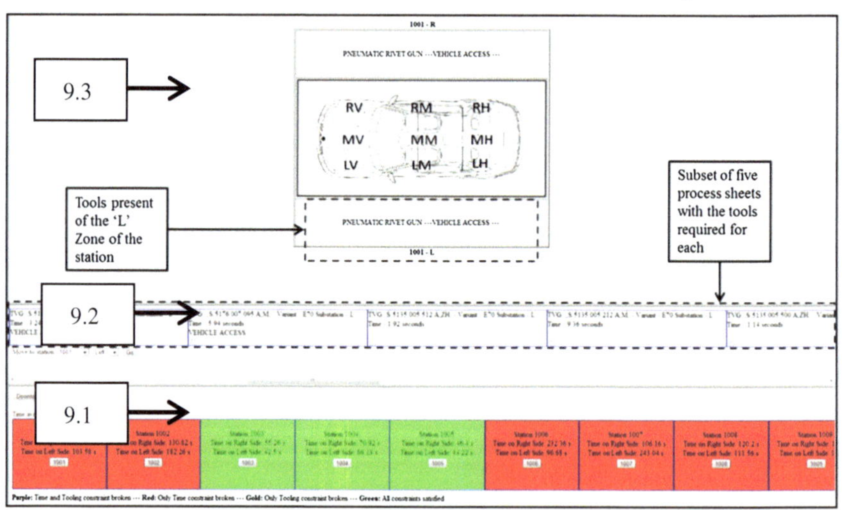

Fig. 9 Snapshot of the web interface of LVET [32]

The user interface of the visualization tool is shown in Fig. 9. This interface has three sections. Section labelled 9.1 lists workstations that have been assigned process sheets by the line balancing algorithm. In this section, the color of the box representing a workstation will change depending upon satisfaction of tooling and time constraints. When a particular workstation is selected, the list of process sheets that have been assigned to this workstation is displayed in Section 9.2. If a process sheet violates a tooling constraint, then its color is changed. Users are allowed to select and move process sheets from one station to another in this section.

When a workstation is selected in Section 9.1, a graphical representation showing the tools are each substation is generated in Section 9.3. This will allow users to verify the tools present at each station and allow for better informed decisions while moving process sheets from one workstation to another.

6 Conclusions and Future Work

The knowledge representation and tool developed in this research provide the following contributions:

1. **A knowledge representation to store line balance information**: This framework (Fig. 2) presents a hierarchical representation of the knowledge associated with line balancing. This framework has been used, in this research, to develop a relational database. Tests conducted show that this knowledge representation captures all line balance information for a single variant of the product. It is therefore concluded that while the knowledge representation presented is a good starting point, further research is required to ensure that the knowledge representation is made more comprehensive. A research opportunity exists to utilize this framework to develop an ontological representation in order to allow for collaborative research efforts in ensuring that the knowledge representation is correct and complete.
2. **A tool to interpret algorithm-generated line balance information**: Apart from the lack of an established knowledge representation for assembly line balance information, a need was identified to establish a method to interpret information from algorithm-generated line balances. The method (shown in Section 3.1) fills this gap and the pseudo code presented should be used to implement other similar tools. The pseudo code should also be extended to ensure that any other forms of constraint violations are also detected and presented to the user.
3. **Visualization scheme for line balance information**: Commercially available software tools visualize statistics associated to line balances. The tool developed in this research provides a more intuitive graphical rendition of line balance information. The stations are laid out as they would be on an assembly line at the bottom of the screen. On the top half of the screen, the tools present at the station are displayed on either the left or right side depending on their respective substations.

The research presented in this paper has the following limitations which future research must address:

1. The visualization tool can computationally parse the output of one specific line balance algorithm. In order for this tool to be extensible to other algorithms, a general parser must be developed.
2. The tool must be further developed in order to be able to visualize line balance information of all variants of the product.
3. Testing must be performed to determine the cognitive load on users of this tool.

References

1. Boysen, N., Fliedner, M., Scholl, A.: A classification of assembly line balancing problems. Eur. J. Oper. Res. **183**(2), 674–693 (2007)
2. Rychtyckyj, N., Klampfl, E., Rossi, G.: Application of intelligent methods to automotive assembly planning. IEEE Int. Conf. Syst. Man Cybern. 2479–2483 (2007)
3. Merengo, C., Nava, F., Pozzetti, A.: International Journal of Balancing and sequencing manual mixed-model assembly lines. Int. J. Prod. Res. **37**(12), 2835–2860 (1999)
4. Bukchin, J., Dar-El, E.M., Rubinovitz, J.: Mixed model assembly line design in a make-to-order environment. Comput. Ind. Eng. **41**(4), 405–421 (2002)
5. Vilarinho, P.M., Simaria, A.S.: A two-stage heuristic method for balancing mixed-model assembly lines with parallel workstations. Int. J. Prod. Res. **40**(6), 1405–1420 (2002)
6. Gerritsen, B.H.M.: IT innovations and their impact on industrial design and manufacturing. In: Proceedings of TMCE 2008 Symposium, Izmir, Turkey, pp. 63–78 (2008)
7. Funk, J.L.: Just-in-time manufacturing and logistical complexity: a contingency model. Int. J. Oper. Prod. Manage. **15**(5), 60–71 (1995)
8. Scholl, A., Becker, C.: State-of-the-art exact and heuristic solution procedures for simple assembly line balancing. Eur. J. Oper. Res. **168**(3), 666–693 (2006)
9. Arnold, P., Rudolph, S.: Bridging the gap between product design and product manufacturing by means of graph-based design languages. In: Horvath, I., Rusak, Z., Albers, A., Behrendt, M. (eds.) Proceedings of TMCE, Karlsruhe, Germany, pp. 985–998 (2012)
10. Altemeier, S., Helmdach, M., Koberstein, A., Dangelmaier, W.: Reconfiguration of assembly lines under the influence of high product variety in the automotive industry—a decision support system. Int. J. Prod. **48**(21), 37–41 (2010)
11. Torenli, A.: Assembly Line Design and Optimization. Master of Science Thesis, Chalmers University of Technology, Gothenburg (2009)
12. Rentschler, D., Stevens, D.: Assembly Line Balancer. U.S. Pat. **5**(177), 688 (1993)
13. Wenbin, Z., Juanqi, Y., Dengzhe, M., Ye, J., Xiumin, F.: Production engineering-oriented virtual factory: a planning cell-based approach to manufacturing systems design. Int. J. Adv. Manufact. Technol. **28**(9–10), 957–965 (2006)
14. Bowman, A.E.H., Jun, N.M., Bowman, E.H.: Assembly-line balancing by linear programming. Oper. Res. **8**(3), 385–389 (1960)
15. Held, M., Karp, R.M.: A dynamic programming approach to sequencing problems. J. Soc. Ind. Appl. Math. **10**(1), 196–210 (1962)
16. Gutjahr, A.L., Nemhauser, G.L.: An algorithm for line balancing problem. Manage. Sci. **11**(2), 308–315 (1964)
17. Hyun, C.J., Kim, Y., Kim, Y.K.: A genetic algorithm for multiple objective sequencing problems in mixed model assembly lines. Comput. Oper. Res. **25**(7–8), 675–690 (1998)

18. Goncalves, J.F., De Almeida, J.R.: A Hybrid genetic algorithm for assembly line balancing. J. Heuristics **8**, 629–642 (2002)
19. Sabuncuoglu, I., Erel, E., Tanyer, M.: Assembly line balancing using genetic algorithms. J. Intell. Manuf. **11,** 295–310 (2000)
20. Falknauer, E., Delchambre, A.: A genetic algorithm for bin packing and line balancing. In: International Conference on Robotics and Automation, pp. 1186–1192 (1992)
21. Rubinovitz, J., Levitin, G.: Genetic algorithm for assembly line balancing. Int. J. Prod. Econ. **41**(1–3), 343–354 (1995)
22. Chiang, W.: The application of a tabu search metaheuristic to the assembly line balancing problem. Ann. Oper. Res. **77**, 209–227 (1998)
23. Vilarinho, P.M., Simaria, A.S.: ANTBAL: an ant colony optimization algorithm for balancing mixed-model assembly lines with parallel workstations. Int. J. Prod. Res. **44**(2), 291–303 (2007)
24. Lapierre, S.D., Ruiz, A., Soriano, P.: Balancing assembly lines with tabu search. Eur. J. Oper. Res. **168**(3), 826–837 (2006)
25. Erel, E., Sabuncuoglu, I., Aksu, B.A.: Balancing of U-type assembly systems using simulated annealing. Int. J. Prod. Res. **39**(13), 3003–3015 (2001)
26. Mcmullen, P.R., Frazier, G.V.: Using simulated annealing to solve a multiobjective assembly line balancing problem with parallel workstations. Int. J. Prod. Res. **36**(10), 2717–2741 (1998)
27. Zhong, Y., Shirinzadeh, B.: Virtual factory for manufacturing process visualization. Complex. Int. **12**, 1–22 (2005)
28. Hong, Z., Soon, T.H., Sivakumar, A.I.: Digital models for manufacturing process visualization. Proc. Int. Conf. Integr. Logistics **2**, 113–122 (2001)
29. Fan, W., Gao, Z., Xu, W., Xiao, T.: Balancing and simulating of assembly line with overlapped and stopped operation. Simul. Model. Pract. Theory **18**(8), 1069–1079 (2010)
30. Pahl, G., Beitz, W., Feldhuesen, J., Grote, K.H.: Engineering Design: a Systematic Approach. Springer, London (2007)
31. Pin-shan, P.: The entity-relationship unified view of data model. ACM Trans. Database Syst. **1** (1), 9–36 (1976)
32. Renu, R.S.: Decision Support Systems for Assembly Line Planning: Modular Subsystems for a Large-Scale Production Management System. Clemson University, Clemson (2013)

Identifying and Utilizing Technological Synergies—A Methodological Framework

Michael Roth, Matthias Gürtler and Udo Lindemann

Abstract As customers' expectations and requirements steadily increase, the utilization of synergies provides a way of achieving cost savings and improving product design. However, a systematical approach to employ synergies that addresses all levels of abstraction is missing. This paper identifies the various interpretations of the term "synergies" and defines it in the context of engineering through a detailed literature review. The paper develops a seven-phase methodological framework to systematically support the identification and utilization of technological synergies. Unlike existing methodologies it extends the scope to all levels of abstraction. The paper describes the framework and the main challenges in each of the seven phases. Two industrial applications underline the applicability and advantages of the framework.

Keywords Synergy · Technology transfer · Methodological framework · Innovation

1 Introduction

Customers' expectations and requirements are steadily increasing [1]. Thus, companies are under pressure to evermore fulfil each customer's individual demands. This leads to an increasing number of variants while the number of sales for each decreases [2]. In achieving this degree of innovation and individualization while maintaining competitiveness, the identification and utilization of synergies plays an important role. Synergies are one of the three ways of increasing the value of an enterprise [3]. Moreover innovative products require the synergetic integration of different disciplines [4].

M. Roth (✉) · M. Gürtler · U. Lindemann
Institute of Product Development, Technische Universität München, Munich, Germany
e-mail: michael.roth@pe.mw.tum.de

A. Chakrabarti (ed.), *ICoRD'15 – Research into Design Across Boundaries Volume 2*, Smart Innovation, Systems and Technologies 35, DOI 10.1007/978-81-322-2229-3_25

Utilizing synergies leads to numerous advantages, e.g. the reduction of development efforts and the spread of innovations to all branches of the company. However, while the identification and utilization of synergies on the operational and executive levels is common, on the technological level only the synergetic use of components and modules has been widely adapted (e.g. as platforms and standards). When this happens, mostly direct component transfer is mostly followed. But as requirements and restrictions usually differ, the achievable success is limited. Outside of component transfer, only a few applications of synergies in the design of technical products are published and an applicable systematic approach is missing. In general, the applications focus on the "pull" of technological solutions in order to solve internal problems. Since technology "push" is harder to realize [5], it remains out of scope. Thus, the research question we focus on in this paper is how technological synergies can be systematically identified and utilized in order to improve the product design process and products themselves.

In this paper we first provide an overview of synergies and the interpretations of this term as well as of exemplary utilization approaches. To resolve the differences in terminology, this paper then provides a clear and consistent definition and categorizes synergies. This is followed by the analysis of the synergies' properties within product development and we present existing approaches in detail. A methodological framework for the identification and utilization of technological synergies is derived and applied within two industrial case studies. It is the first generic and systematic methodology for identifying and utilizing technological synergies at all abstraction levels and it provides the basis for further research. The paper concludes with the discussion of the findings and applicability of the developed methodology as well as recommendations for further work.

2 State of the Art and Definitions

2.1 Synergies in the Context of Engineering Design

The term "synergy" is used across multiple disciplines and has taken on a wide variety of meanings and interpretations [6]. We screened journals and conferences from engineering and innovation management to clarify the meaning in engineering design.

The origin of "synergy" is the Greek word "synergia", which can be translated as collaboration [7]. Ansoff first introduced the term "synergy" in the context of economics in 1965. He describes synergies as the effect, of the combined output of different entities exceeding the sum of their single outputs [8]. This effect is often called the "superadditive principle". Since then, the term synergy has been widely used in economics, usually keeping in line with the superadditive principle [3]. According to these definitions, synergies are for example classified into material and immaterial synergies or distinguished by their place of appearance (internal or external) [3].

In engineering design no concept to structure or to classify synergies has been published so far. Based on our literature review and discussions with practitioners, we identify communalities and differences of the definitions. Their clustering results in the following three groups of interpretations in the context of engineering.

2.1.1 Class A: Organizational Synergies

We categorize synergies resulting in effects on organizational or administrational levels in the class of organizational synergies. This interpretation resembles the definition of synergies in economics. Starting from [8] extensive research has been done and various approaches, strategies and methods supporting the use and fostering of synergies on administration levels have been published. For example Hagedoorn [9] analyse and review modes of inter-firm cooperation and the associated technology transfer.

2.1.2 Class B: Interdisciplinary Synergies

According to our definition, interdisciplinary synergies describe effects achieved by the combination of knowledge and concepts from different disciplines within one product. The most common occurrence of these synergies are mechatronic products. In his definition of mechatronics Harashima integrates the aspect of synergies as the synergistic integration of mechanical engineering with electronics [10]. Tähemaa et al. [11] take up this interpretation and define synergy in mechatronic or interdisciplinary products as "[…] the effect of suitable integration when the whole is more than the sum of its parts". Moreover, they state, that synergies do not occur automatically but are a result of systematic design. Examples of the systematic utilization of synergies are published in [7, 11]. Both works increase the integration level of the disciplines during the product design and thus maximize achievable synergies.

2.1.3 Class C: Technological Synergies

In the third class we gather the remaining synergies on technological level. We define these as the effects resulting out of the collaborational use of technological knowledge, functions, principles, concepts or components. Depending on the abstraction level, the depth of research and the usage of these synergies strongly vary. While the utilization of technological synergies within components is established (e.g. the use of catalogues, modules and databases), the use of synergies on more abstract levels is not common [12]. Only few specialized approaches exist, for example the design of a feeder by using synergies on functional level [12]. Yet, fundamental approaches for the identification and utilization of technological synergies on all abstraction levels are not available.

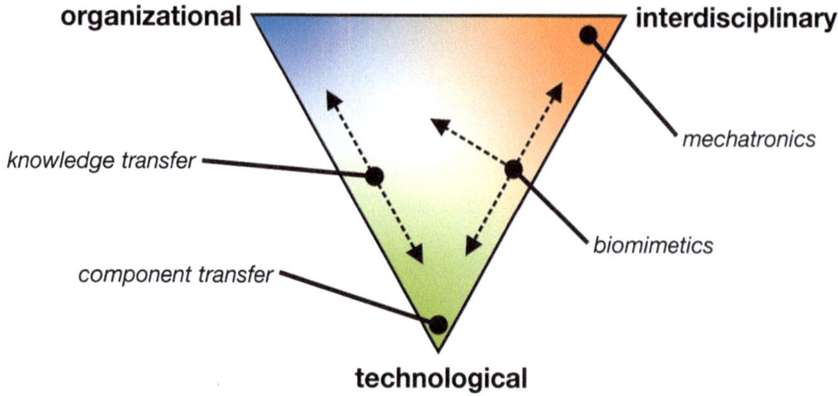

Fig. 1 Classification of synergies with exemplary fields and their categorization

2.1.4 Differentiation and Transition of Synergy Classes

Within the three classes of synergies no clear boundaries can be drawn. Figure 1 depicts the classes and illustrates the fluid transitions between them. Mostly the usage of synergies cannot be reduced to one class and involves at least one other synergetic aspect. The approaches of biomimetics provide an example of this [13]: on one hand, the simple transfer of principles from biology to engineering design can be assigned to the class of technological synergies. On the other hand, synergies on organizational levels or interdisciplinary synergies can also be addressed by using more sophisticated approaches.

To align with our goal of improving the usage of synergies in engineering and based on the learnings of failed attempts, we in this paper narrow our focus on technological synergies.

2.2 Technological Synergies and Their Properties in Engineering Design

Technological synergies in engineering design can be described by specific features. Basically, synergies are applicable for both, the design of new products and the improvement of existing technical systems [12]. However, as technological synergies are defined as the effects resulting out of collaboration, they can manifest positively and negatively [11]. In mechatronic products with interfaces between the components, negative synergies can for example, manifest as incompatibilities of components [11]. Hence, the goal of identifying and utilizing synergies is to achieve an overall positive synergetic effect. Another important property of technological synergies is the place of appearance of their positive effect. In case it is achieved or aspired at only one of the collaboration partners, we call it unidirectional synergy. If both partners of the collaboration achieve positive effects, we consider this a bidirectional synergy.

In order to use technological synergies, technology transfer is one basic concept. We understand technology transfer according to Taschler's definition as "[…] the managed, interpersonal, and systematic process of passing control of a technology from one party to its adoption by another party […]" [5]. Thus, the utilization of unidirectional synergies in its core resembles a sole technology transfer.

2.3 Identifying and Utilizing Technological Synergies in Engineering Design

The common approaches of using synergies within product design are mainly located on the level of components, modules, code, etc. Established concepts are common parts and type series [12]. But also construction catalogues, modularization or platform strategies address the technological synergies on component level. Manifold approaches, methods and tools are published in this field [1].

On higher abstraction levels only few approaches have been developed. Thus, better support is needed in order to utilize synergies in engineering design. To classify and structure the levels of abstraction we decompose products into the three levels of functions, principles/behaviour and realization.

To conduct knowledge transfer on functional level in engineering design for example, Formanek and Hosnedl use a 5-step model based on the Theory of Technical Systems [14]. According to general procedural models for design engineering, they define the steps "arise of a need", "specification of the problem", "searching for solutions (incl. other branches)", "evaluation of properties and decision" and "elaboration of documentation" [12]. While the process of technology transfer is well supported by their approach, synergies on all levels of abstraction as well as the properties of technological synergies and associated challenges are not in their scope.

Potential challenges within the utilization of technological synergies on all of these abstraction levels for example, are the comparison and harmonization of restrictions and boundary conditions as well as the matching of requirements. Moreover social aspects like the "not invented here" syndrome have to be considered [15].

Thus, the detailed research question we address is how the systematic identification and utilization of technological synergies on all abstraction levels in engineering design can be supported, whilst the properties and characteristics of these synergies are considered.

3 Methodological Framework to Identify and Utilize Synergies

The identification and utilization of technological synergies in engineering design can be reduced to the basic problems of product development. Therefore, a wide range of established procedural models is available, for example the Generic Product Development Process [16]. In comparison, the "Munich Procedure Model" (MPM) supports both the planning and reflection of development processes, by providing general guidance throughout the problem solving [17].

To support the identification and utilization of technological synergies [12] show the advantages of a procedural model driven approach. As the basic problem does not differ, we adapt these models. As being closer to the generic aspect of problem solving, we focus on the MPM.

Based on the MPM, we developed the seven-step framework depicted in Fig. 2. An important aspect during the procedure are iterations. In our framework we define two types of iterations: The decision-based iterations are performed according to the acting in step seven. If the assessment and decision does not lead to satisfactory results, going back to one of the phases will help to achieve suitable results. Contrary the iterations performed within the other steps are called integrated iterations. They are integrated in the identification procedure and should be performed during the whole process.

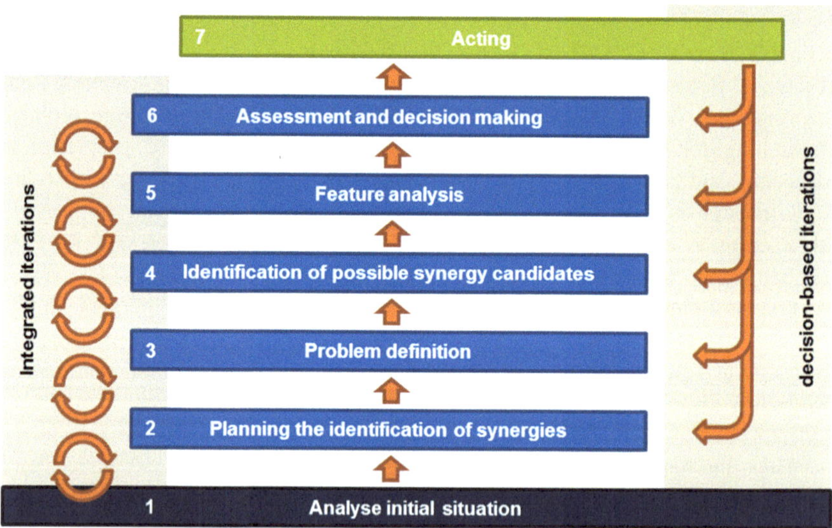

Fig. 2 Methodological framework to identify and utilize technological synergies

3.1 Analyze Initial Situation

Prerequisite of the identification and utilization of synergies is the understanding of the contemplated system. Based on the system overview, developers can deduce problems and potentials of improvement within the current product design. If the process is push-motivated, innovative solutions and concepts have to be identified instead. Suitable methods in this phase are e.g. functional or structural modelling and expert interviews.

3.2 Planning the Identification of Synergies

In the second phase the exploration and identification has to be planned. Important aspects are the exploration areas, the abstraction levels, the direction of desired synergies, specific requirements and other restrictions.

The basic decision of the planning phase defines the area of the exploration. It has to be considered, if the searching activities are limited to internal partners only or if external partners and companies are included. Likewise the included disciplines, branches and partners need to be defined. As the exploration should be limited to a manageable scope, the experience and intuition of the developers are essential input factors.

Another important aspect during this phase is to define the properties of the aspired synergies. According to the given definition in Sect. 2.3, we propose the structuring in levels of abstraction. The identification of potential synergies can be restricted to one level only, but a search on all levels is possible as well. The direction of the aspired synergies is the second property to be determined in the planning phase. Possible objectives are unidirectional synergies with both, pull- and push-oriented motivation or bidirectional synergies.

In addition to the properties, specific restrictions and requirements of the explored problem fields and areas strongly influence the identification process and thus, influence the quality of results. They are an important element of the planning phase as, based on them, the assessment and integrated iterations can be efficiently performed.

3.3 Problem Definition

In this phase the goal and resulting objectives are specified and the requirements are documented. According to the definition of possible negative synergies it is important to include all effects and results of the aspired synergies. Therefore, the identification of stakeholders in the context of potential synergies is essential and specific requirements and restrictions have to be included. Same, requirements which directly address the synergetic effect are an important aspect of the problem definition. Suitable methods in this phase for example are stakeholder analysis and functional modelling.

3.4 Identification of Possible Synergy Candidates

In order to ensure the manageability of the synergy identification we propose a brief screening of the solution space which is defined by the branches and areas included into the exploration. Additionally a preselection support the structuring of the solution space. Therefore components, principles, solutions or combinations with expected synergy potential have to be defined. They are in the following denoted as synergy candidates. This step, thus includes a rough feature analysis of the possible synergy candidates. However, a detailed analysis is not necessary. Depending on the amount of identified possible synergies, relevant candidates are selected.

3.5 Feature Analysis

With the selected synergy candidates a detailed feature analysis has to be conducted. Depending on the chosen partners as well as the directions of the identification and aspired utilization, different methods apply. Outside of workshops or the evaluation of documents, experiments and benchmarking are applicable options. In case of aspired bidirectional synergies in this phase a strong collaboration is essential.

3.6 Assessment and Decision Making

Main prerequisite for the generation of positive synergies is the fulfilment of requirements. Therefore, a mapping and evaluation of features and requirements has to be conducted. Additionally in the context of synergies, the estimated duration of the synergy is an important criterion. For instance, different product lifecycles in two branches can limit the period of achievable synergetic effects to the next product change.

Using these considerations the synergy candidates and their overall effect can be assessed. Following the assessment, ranking, and the final selection of candidates can be conducted and candidates with bad ranking are dropped. Suitable methods vary from point assessment to scenario techniques.

3.7 Acting

Depending on the decision in the previous phase, the action can veer towards two different directions. If promising synergy candidates are identified, the implementation of these can be initiated. In this case multiple candidates are to be realized, possible interferences between these candidates have to be analysed in order to ensure the maximum positive overall synergetic effect. This applies for both, the utilization of unidirectional and bidirectional synergies. The subsequent implementation and application complies with usual and well known development projects.

If the assessment leads to a non-satisfactory decision, a decision-based iteration is the proposed option. Depending on results and restrictions a conscious iteration of the whole procedure starting over from one of the seven phases applies.

3.8 Integrated Iterations

Complementing the previously introduced decision-based iterations, the integrated iterations are an essential part of the presented procedure. The permeability between each of the phases allows improving the efficiency and quality of the identification. At any point during the identification of synergies a need of refinement or adjustment e.g. through changes in boundary conditions can occur. Thus, previous phases have to be repeated, extended or refined. The approach of integrated iterations improves the primarily linear proceeding by adding the required flexibility and generating the consciousness of the relevance of these adjustments.

4 Application of the Methodological Framework

The developed methodological framework was evaluated by two industrial applications. Both are located in a company from the mobility sector. The main driver for the search of synergies is saving costs. The company incorporates two branches with similar functionalities. The first branch A is characterized by mass production and a large number of sales. The other branch B applies an engineer-to-order (e2o) strategy with small batches. Thus, the company expects potential synergies by the transfer of technology from branch A to B. The identification process in both cases is "pull" motivated. Past attempts of direct component transfer failed, particularly due to differing operating conditions and differing lifecycles of the components.

4.1 Synergies on Component Level

One central component in both branches is a special valve. The main function is the adjustment of the input pressure and its transduction to the output pressure, dependent on the control pressure. With this function and boundary conditions, the initial situation is clarified and the problem definition is conducted. In the planning phase the identification of synergies is limited to the identification of unidirectional synergies on component level. Quickly possible candidates of technological synergies are identified and chosen. However, the first analysis of boundary conditions and restrictions reduces the options to two candidates. By applying functional modelling, we detect incompatibilities within the components. Thus, in detailed experiments and calculations the features of the components are determined. The comparison of features and requirements allows prioritising one of the two remaining candidates. As not all requirements are satisfactorily fulfilled, adjustments have to be made. Thus, the decision is to design and develop a concept for these product adjustments, which is performed in the acting phase.

This example demonstrates, why previous attempts of component transfer failed. The boundary conditions and requirements in the branches vary, so that adjustments have to be done. Thus, reduced durability or increased life cycle costs need to be accepted.

4.2 Synergies on Higher Abstraction Levels

The previous case once more pointed out the motivation for the search for synergies between these branches on higher abstraction levels. Thus in another application we applied the methodology on synergies on higher abstraction levels. During problem definition and the planning phase, we use functional modelling, interviews and scenario techniques in order to understand the system of interest as well as possible developments in the near future. In a screening with experts five synergy candidates are identified and analysed. Their features are documented in a specially developed feature form. Based on this, we conduct the assessment in a workshop using a two-aspect assessment. The fulfilment of requirements is rated in a score evaluation, while the cost-benefit-ratio is estimated within a portfolio. In this process one major challenge manifested: As the degree of fulfilment of requirements is roughly determined, the exact assessment on higher abstraction levels is not possible. Thus, in many aspects a score range dependent on possible implementations is recorded. These uncertainties influence the cost-benefit-ratio and complicate its assessment. Nevertheless, promising candidates are identified. To reduce uncertainty, our decision is to more closely examine the candidates in a second feature analysis and to develop first transfer concepts. The acting phase here implies both, the early stage of technology transfer and a decision-based iteration to the feature analysis.

Moreover we observe barriers within the company, such as the "not invented here" syndrome. The designers struggle to accept the other approaches or principles and we often face the statement "that won't work" without funded reasoning.

5 Discussion and Conclusion

The classification of synergies supports the coordination of activities connected to synergies. Yet the three categories have to be validated in further research.

Failed attempts show, that there is a need for a systematic identification and utilization of technological synergies in product design. In scientific literature this need is only addressed rudimentarily. Thus, we present a methodological framework to support the identification of candidates for technology transfer and synergy utilization.

The framework has to be replenished with fitted methods for each phase in order to improve efficiency and the quality of the outputs. Especially bidirectional stimulation and social issues should be considered. Synergy identification usually is pull-motivated, thus incentives for a technology-push and for mutual bidirectional exchange have to be researched. Based on this methods and tools to improve this aspect can be developed. Moreover a method and process to implement and to operationalize the identification of synergies within product design is required. This can sustainably improve the efficiency of development and the grade of innovation of products and systems.

Processes to improve the coordination in application are also needed. Therefore, the awareness of advantages of bidirectional synergies has to be strengthened. Ongoing research of Open Innovation and Change Management can provide the required support.

Social aspects represent a major challenge during the utilization of synergies. As real-life examples, the industrial case studies show that social barriers such as the "not invented here syndrome" are a risk. In Open Innovation these phenomena subject of ongoing research [18]. Thus, strategies, tools and methods to prevent or diminish these effects within the product development process need to be researched.

References

1. Lindemann, U., Reichwald, R., Zäh, M.F.: Individualisierte Produkte-Komplexität beherrschen in Entwicklung und Produktion. Springer, Berlin (2006)
2. Piller, F., Stotko, C.M.: Mass customization und Kundenintegration - Neue Wege zum innovativen Produkt. Symposion Publishing, Düsseldorf (2003)
3. Zorn, C.: Erfolg von verwandten Produktdiversifikationen in gesättigten Märkten - Eine Untersuchung der Automobilindustrie in Deutschland. Logos, Berlin (2012)

4. Felgen, L., Deubzer, F., Lindemann, U.: Complexity management during the analysis of mechatronic systems. In: 15th International Conference on Engineering Design (ICED'05), pp. 2562–2575 (2005)
5. Taschler, D.R., Chappelow, C.C.: Intra-company technology transfer in a multinational industrial firm. J. Technol. Transfer **22**, 29–34 (1997)
6. Foss, N.J., Iversen, M.: Promoting synergies in multiproduct firms. Institut Industriøkonomi og Virksomhedsstrategi, Handelshøjskolen i København, Copenhagen (1997)
7. Kaljas, F., Reedik, V.: On using the DSM technology approach to synergy-based design of interdisciplinary systems. In: 15th International Conference on Engineering Design (ICED'05), pp. 498–499 (2005)
8. Ansoff, H.I.: Corporate Strategy—An Analytic Approach to Business Policy for Growth and Expansion. McGraw-Hill, New York (1974)
9. Hagedoorn, J.: Organizational modes of inter-firm co-operation and technology transfer. Technovation **10**(1), 17–30 (1990)
10. Harashima, F., Tomizuka, M., Fukuda, T.: Mechatronics—what is it, why, and how? IEEE/ASME Trans. Mechatron. **1**(1), 1–4 (1996)
11. Tähemaa, T., Kaljas, F., Reedik, V.: Assurance of Synergy and Competitive Reliability at Mechatronic Systems Design. Mechatronic Systems 2002, Elsevier, Oxford, pp. 797–802 (2003)
12. Formanek, J., Hosnedl, S.: Transfer of knowledge among different branches with use of the theory of technical systems. In: Proceedings of Design 2004, Design Society, Glasgow, pp. 1021–1026 (2004)
13. Bar-Cohen, Y.: Biomimetics: Biologically Inspired Technologies CRC. Taylor & Francis, Boca Raton (2006)
14. Hubka, V., Eder, W.E.: Design Science—Introduction to Needs, Scope and Organization of Engineering Design Knowledge. Springer, Berlin (1996).
15. Smilor, R.W., Gibson, D.V.: Technology transfer in multi-organizational environments: the case of R&D consortia. IEEE Trans. Eng. Manage. **38**(1), 3–13 (1991)
16. Ulrich, K.T., Eppinger, S.D.: Product Design and Development. McGraw-Hill/Irwin, Boston (2004)
17. Lindemann, U.: Methodische Entwicklung technischer Produkte - Methoden flexibel und situationsgerecht anwenden. Springer, Berlin (2009)
18. Enkel, E., Kausch, C., Gassmann, O.: Managing the risk of customer integration. Eur. Manag. J. **23**(2), 203–213 (2005)

Examples of Poor Design from an Engineering Expert Witness and Consultant

Swaminathan Balachandran

Abstract As an engineering expert witness and consultant, the author has participated successfully in about twenty six cases that involved personal injuries or death. Seven of the above cases are briefly outlined to illustrate how inadequate design and/or analysis of components, parts, subsystems and products resulted in death of users or severe injuries to users. One example is a liquid sprayer that was designed incorrectly, without regard to applicable standards. In addition, neither design failure mode and effects analysis (FMEA) nor process FMEA was conducted. This product resulted in the death of one user. This paper will discuss this case and summarize what designers can and must learn from it. The rest of the six cases will be presented briefly to illustrate what designers can learn from each case. The focus will be on design principles, instructions to users, applications of human factors in design, and designing for safety of users.

Keywords Poor design · Design principles · Design priorities · Design liability issues

1 Introduction

The most common definition of the engineering design process is that it involves a series of steps that lead to the development of a new product or system as outlined in Fig. 1 [1].

This paper deals with design of specific products to serve the needs and requirements of targeted consumers or the general public. The design scope of such products depends on the goals of design and may focus sequentially on the

---

S. Balachandran (✉)
Department of Mechanical and Industrial Engineering, UW-Platteville, Platteville, WI, USA
e-mail: balachas@uwplatt.edu

A. Chakrabarti (ed.), *ICoRD'15 – Research into Design Across Boundaries Volume 2*, Smart Innovation, Systems and Technologies 35, DOI 10.1007/978-81-322-2229-3_26

following design activities: system-level concept design, subsystem-level preliminary and detailed design, assembly-level preliminary and detailed design, and component-level preliminary and detailed design. At each of these levels specific design criteria, principles, and design elements may be identified, explored, and applied. The evolution of design will be influenced by the experiences of the design team including composition of the team, their history of successful designs, knowledge of available materials, expertise with manufacturing processes, understanding of customer behavior, market characteristics, product and process lifecycle details, etc.

The successful design of products requires a multidisciplinary team that is capable of addressing many ergonomic, manufacturing, lifecycle, environmental, social, financial, marketing, maintenance, sustainability, universal design, and other market constraints. The design of a specific product is an iterative process [5] in which the product evolves from a preliminary design to final design in a spiral shaped sequence of the steps that were identified in Fig. 1 so that constraints are satisfied and the final design satisfies the ten principles of excellent design [2]. This paper reminds designers to be aware of legal aspects of engineering design.

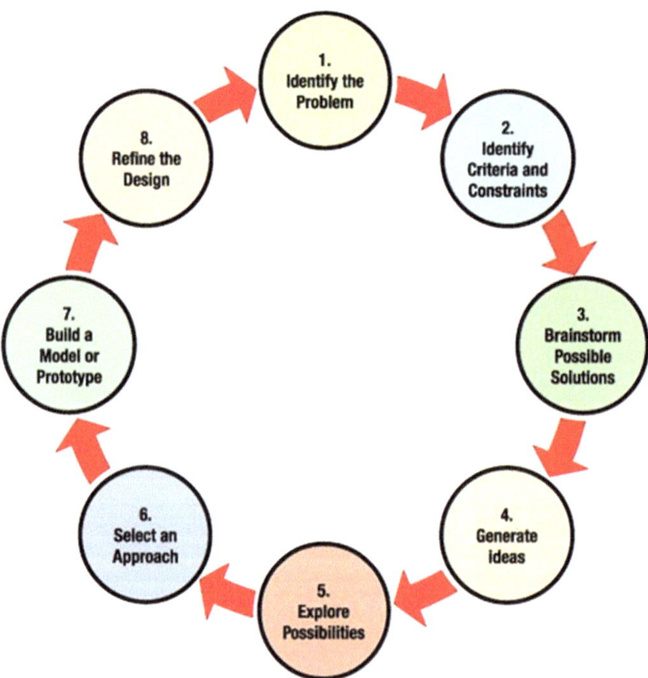

Fig. 1 Steps in engineering design process (*Source* [1])

2 Basic Information About Personal Injury and Loss Cases

The civil law in the United States evolved to resolve disputes regarding product liability, negligence, intentional acts, and breach of contract in tort cases. In personal injury or loss cases, the person who suffered an injury is called the plaintiff or petitioner. Plaintiff's attorney may take up a personal injury case for an hourly fee, one flat fee, or a contingency fee. The plaintiff's attorney may first send a demand letter to the defendant(s) or respondent(s) who is the person(s) or entity being sued. As soon as a defendant gets the demand letter that sets forth the plaintiff's version of the facts surrounding an alleged personal injury or loss and the monetary amount of damages being sought, the defendant may hire an attorney and respond to the demand letter on a timely basis in consultation with the attorney. If the response to the demand letter does not satisfy the plaintiffs, they may instruct their attorney to file a petition (pleading or complaint) against the defendant. The lawsuit begins when the plaintiff's attorney files the pleading, complaint, or petition with the appropriate court.

A petition is a document that officially requests a court's assistance in resolving a dispute. The complaint or petition lays out the theory of the case, what facts support that theory and what remedy is requested. The person filing the lawsuit has to pay a filing fee to the court. The petition will identify specifically the plaintiff and the defendant. In addition, it will describe the matter it wishes for the court to decide. Included in the complaint against the defendant will be the plaintiff's suggestion for resolution of the issue. When the petition has been filed with the administrative clerk of the court, the lawsuit officially begins. Once the complaint is filed with the court, the court (and/or the plaintiff, through counsel) will notify the defendant of the plaintiff's charge, and will include a copy of the complaint in the notification. Upon receipt of the complaint the defendant needs to respond in writing within the time specified in the notice from the court. The following section provides the anatomy of a civil lawsuit and discusses the roles of an expert witness.

3 Anatomy of a Case

Most attorneys will have an interview with the client before deciding to take up a case. The interview will assist them to evaluate client's legal rights against those who may be responsible for the personal injury loss. The next phase will be the investigation to identify witnesses, collect data, estimate loss, and preserve evidence. The attorney will then try to negotiate and settle the case with insurance representatives and/or lawyers without the necessity of filing suit because lawsuits may have unpredictable outcomes. In order to win a favorable judgment, the plaintiff's attorneys must convince a jury that the wrongdoer was careless, the product was unnecessarily or unexpectedly hazardous, or the service was substandard. The attorneys representing the defendant will try to defeat the case, or diminish its value substantially, by criticizing the plaintiff as having been careless, inattentive, or negligent in putting oneself in the predicament that led to injury. The right to a jury in a civil action is guaranteed by the 7th amendment to the U.S. Constitution. The parties may agree to not seat a jury and to let the judge act as a fact finder. This called a "bench trial". A vast majority of cases do not go to trial. Settlements can be reached at any stage of a case, even after the start of a trial or even after a verdict. It is common to settle a case while a matter is on appeal.

Frequently, settlement negotiations do not succeed unless a law suit is filed and all parties are convinced that the plaintiff is serious in pursuing the claim to a just conclusion, by jury trial if necessary. To file a lawsuit, plaintiff's attorneys draft a formal legal paper called a complaint, petition, or pleading and file it with the appropriate court. The defendants respond in a legal pleading called an answer. They usually deny each allegation in the complaint and set forth their defenses. Defendants, in return, often make a number of motions to request more information to be placed in their pleading.

While the suit is pending, the discovery phase of a civil lawsuit begins. In this phase, both parties seek to learn the facts necessary to best support their position. This can include answering questions via interrogatories or depositions, requests for records or other evidence, and sometimes visiting the scene of the incident that caused the complaint. The plaintiff in the lawsuit has the burden of proving the allegations set forth in the petition. This is the responsibility of proving to the trier of facts (judge or jury) that a particular view of the facts is true. A deposition is a formal statement, under oath, given in response to questioning by the lawyers in the case. Defense attorney will depose the plaintiff, witnesses, and also the expert witness. The plaintiff's attorney will depose the defendant, witnesses, and also the expert witness. Depositions are very important means of getting evidence.

Most injury cases do not go to trial and it is impossible to predict when a trial will take place. A request for trial cannot be made until the end of the pleading stage. Defendants will usually try to do everything to delay the trial. About 30 days before the trial, a pretrial conference takes place. At this conference before the judge, each side names all of its witnesses and lists all of its exhibits. The trial will be tried before six jurors selected at random. Each side is entitled to challenge potential jurors for selection, with six selected at the end.

The trial phase of a lawsuit may consist of the following sequence of events: Presentation by plaintiff, Presentation by defendant, Plaintiff's rebuttal, Summation by both parties, Judge's instructions about the applicable law and procedures to the jury, Jury deliberation, Verdict, Judgment or Award, and Appeal of verdict and/or award. In the trial each side tries to persuade the judge or jury that his or her version of the facts and points of law should prevail. Either side has the right to appeal a verdict or award. In the personal injury area, it is common for a losing defendant to appeal the size of the award if it is considered to be excessive.

In a case with the title Smith v. Jones, the first name is usually that of the plaintiff, or the person who initiated the lawsuit and the second name is that of the defendant, or the person against whom the lawsuit has been brought. There may be cases with multiple plaintiffs or defendants, but only one name will be used in the official title of the case. The name that is picked may be the person most primary to the case or in alphabetical order and it will be in uppercase letters. Title of a criminal case will involve only one proper name, but the title of a civil case will contain two names. Following title will be the citation to the case, which will be the volume number, the name of the book, and the page number. If there is a string citation it will all appear there. Sometimes instead of the citation, there will be the docket number that denotes when the court heard the case.

4 Expert Witness and Consultant

Because of the complexity of scientific, engineering, and technical issues in some civil cases, the responsibility for a personal injury or loss is, sometimes, unclear and expert witnesses may be needed to explain and interpret the evidence to trier of facts (judge and/or jury). They may be hired as expert witness by the attorney for the plaintiff or the defendant. The role of an expert witness is the same whether he/she works for the defendant of the plaintiff: assist concerned parties to understand the facts of the case. Unlike other witnesses who may only testify about facts based on first-hand experience or knowledge, the Rules of Evidence allow experts to give opinions based on professional theories and assumptions if that testimony will assist the trier of fact to understand the evidence.

Experts in several areas such as forensics, medicine, allied health, aviation, and other fields are needed to help collect, analyze, and interpret data or evidence. The State Bar of Wisconsin defines responsibility of the expert witness as having a commitment to investigate, evaluate, analyze the facts, and advice attorneys on behalf of their clients. Expert witnesses are usually paid well for their time and expertise. The compensation may depend of the field of expertise, number of years of experience, ability to communicate well, education, training, etc. A testifying or consulting expert witness may be involved in the following phases of a lawsuit: Investigative Phase, Discovery Phase, and the Trial. Most of the work performed by an expert witness occurs before the trial stage of litigation as 90 % of all litigations are settled before the trial.

A professional may be considered an expert witness in some specific fields by their education, experience, skill, and training. An expert witness is distinguished from an average person by the expertise in a particular field that is sufficient for others to rely upon his/her specialized opinion. Attorneys for the plaintiff and defendant may have their own expert witness and an expert opinion may be rebutted in all phases of a case. Having an advanced degree, being registered as a professional engineer, having publications in journals and presentations at conferences, teaching courses on topics related to the case, and possessing skills to succeed as an expert will be advantageous. Excellent communication skills and ability to explain technical information in a way everyone can understand will be helpful. Candidates searching for work as an expert witness may create a website and post credentials available for review, list publications, and list areas of expertise. Another course of action will be to join a service that helps lawyers find expert witnesses. Finally, local law firms may be contacted to check whether an expert witness may be needed in any current case. The following section on billing rates may be used to develop a summary of compensation for expert witness work and can be left with the law firms along with contact information and available times for such work.

The discovery and trial phases of a case require more expert witness work such as developing a list of documents to be requested from the defendant, draft or interpret interrogatories, advise attorneys of the need for additional tests or information, educate attorneys and clients understand theories, data, and technical language, list of questions to be submitted to the opposing side to gather information, visit to the site of injury, research, data collection, analysis, tests, etc. It will be necessary to keep a detailed log of all activities along with the actual time for each category of work. An expert witness should neither testify on anything outside of one's area of expertise, nor accept any case where there is a conflict of interest. It will be important for an expert witness to answer all questions honestly and not guess on an answer as the stakes are high in legal cases.

5 Defects in Engineered Products

Engineers practicing in the United States have seen increasing societal expectations for safe products and safe services. Engineers have responded by continually improving safety characteristics through design and manufacturing capabilities. Engineers are expected identify the physical and functional requirements of the end product which are necessary to satisfy the requirements of customer's intended use of product, foreseeable misuse of the product, and environment in which product will be used [3].

In product liability and personal injury cases, strict liability in torts for products may be due to manufacturing defect(s) or design defect(s). The liability can be assigned regardless of whether the defendant has been negligent or has been careful. Defects can be created by manufacture, assembly, design, lack of adequate warning labels, inadequate instructions, etc. A product that had defects when it left the defendant's hands can be deemed to be unreasonably dangerous due to its defective condition. If the defect(s) caused physical and economic harm to the plaintiff, then the defendants will be held liable. In most cases, strict liability in torts may relieve the plaintiff of responsibility for unforeseeable misuse, abuse, alterations, etc. Engineers are expected to apply accepted standards of care for the product, its design, its manufacture, its assembly, instructions for use, and associated warnings [4].

Designers can prevent defective design of a product through simultaneous and concurrent engineering concepts, conducting the design failure modes and effects analysis (DFMEA), designed experiments to optimize design, testing, etc. the claim is that the product functioned as it was designed but the design was negligent. Similarly, manufacturing process design must be preceded by process failure modes and effects analysis (PFMEA). Once a hazard, which is a condition that may cause injury or death, is identified in a product, the design engineer must follow the following accepted design priority recognized by all design engineers in reducing the possibility of the dangerous condition of the product causing injury or death:

- The defect(s) and resulting dangerous condition must be designed out of the product if such can be done without destroying the utility of the product. If this cannot be done then:
- The defect(s) and resulting dangerous condition of the product must be guarded against. If this dangerous condition of the product cannot be guarded against then:
- A warning label and or instructions must be designed in accordance with acceptable standards used effectively to educate users about the dangerous condition of the product and consequences of improper use.

6 Examples of Poor Design of Products and Processes

The author has worked as expert witness and consultant in about thirty cases during the past two decades. About four of these cases were not undertaken by the attorney on the advice of the author and the rest of the cases were either settled out of court or won after cases went to trail in a court of law. The following is a brief summary of poor designs from the most interesting cases.

1. Mr. Bertram Wright v. CorrectRide Mower Inc.
   - Inadequate ergonomic clearances for user
   - Incorrect ergonomic design parameters
   - Confusing operation and/or maintenance instructions
   - Inadequate warning labels
2. Mr. Stephen Gould v. Hideout Bar and Restaurant
   - Improper design of facilities and not satisfying applicable fire codes
   - Lack of specific instructions for periodic inspection of controls and sensors
   - In adequate automatic safety devices, alarms, valves, etc.
   - Not recommending material safety data sheets for treatment of exposure to chemicals
   - Not using a fire protection system that is suitable for expected fire type
3. Ms. Holly Jane Morchoen v. Automatic, Inc.
   - The design of the machine was defective and unreasonably dangerous, as the manual specifies in the daily clean-up procedures that the conveyor be set at slow speed for cleaning.
   - There was no machine guard to prevent the operation of the machine when operator's hand is near pinch points.
   - The design of the machine was defective and unreasonably dangerous, as the emergency controls to stop the conveyor and the machine were not designed taking into account human factors engineering principles.

- There were no danger, warning, or caution signs at potential hazard locations to prevent unsafe tasks and educate user about consequences of such acts.
- There were no detailed instructions for operation, maintenance, repair, and cleaning.

4. Ms. Clairol R. Jameson v. Sneko Gas Inc.
   - Designer did not specify frequency of maintenance activities to avoid malfunctioning
   - Designer did not incorporate signals from critical components/subsystems to inform operator about operating condition and whether they are working in the normal range
   - Designer did not provide easy shutoff control
   - There were no material safety data sheets for fluids normally used in the system
   - There were no adequate warnings

5. Design of Perfect Sprayer Inc. v. Sebastian Cailiff
   In this case a worker was using a pressurized two liter canister to spray concrete forms with oil so that concrete can be poured. When the oil was not spraying well due to cold weather on that day, he decided to warm up the oil with heat from a torch. Unfortunately, the canister exploded, the fragments caused traumatic injuries to the worker, and he died on the spot. Engineering expert witness and consultants were hired by plaintiff's attorneys to collect and analyze engineering data about the design of the product and facts about the injuries sustained by the plaintiff. The expert witness and consultant analyzed the product design using human factors engineering, safety engineering, and mechanical engineering design principles. In the following list, DFMEA refers to design failure modes and effects analysis and PFMEA refers to process failure modes and effects analysis. The following list is based upon the documents that were supplied by the company that designed and manufactured the spray canister, responses from the defendants to interrogatories, depositions, and other court documents. The product design drawings and specifications are considered to be proprietary information and cannot be published. The following is a summary of conclusions from the expert witness and consultant and these were the most important factors that lead to the verdict in favor of the plaintiff. This case took about 3 years and required about 50 hours of engineering analysis by the expert witness and consultant.
   - Design was done by one person who was not an engineer
   - There were no formal design process and/or design documents
   - Design was not analyzed in detail. There were no test records
   - No DFMEA and PFMEA; There was no control plan
   - Designer did not follow standards for burst pressure, safety valve, gages, etc.
   - The warning label for poorly designed (Inadequate markings—color, font, size and safe operating pressure, etc.
   - Label suggested wrong or incorrect uses. Designer did not anticipate all possible uses.

6. Design of Machine to Trim Drop Ceiling Grid Supports: Gem Tools v. David Hammond
    - No engineer on design team, no formal design process and design documents
    - Improvised design while fabricating and assembling the Machine
    - Firm that subcontracted work did not conduct design reviews
    - No analysis and no test records
    - No DFMEA and PFMEA; No control plan; Designer did not follow standards
    - Poor warning label
    - Sought grinding wheels of large diameter after major design was completed
7. Harvey P. Leonard v. Parsec Brick Co., WI
    - Controls on the machine did not have clear markings to open or close moving elements and operator's head could not be easily released from machine members
    - Machine requires operator had to work in a confined space, but applicable standards were not followed

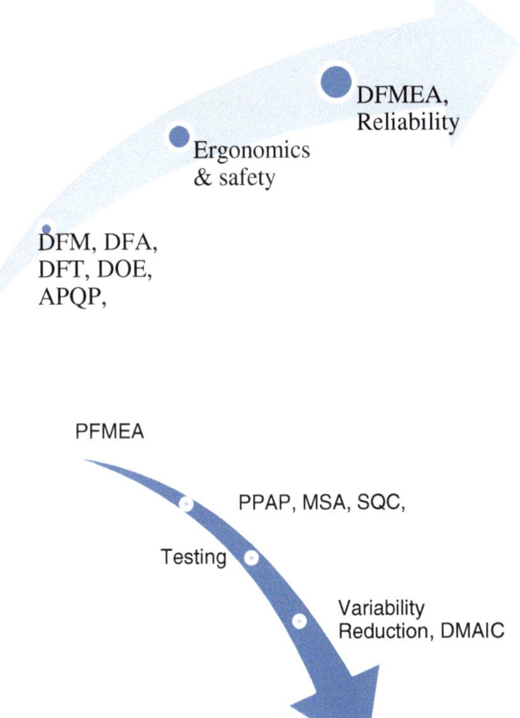

Fig. 2 Steps in design (*top arrow*) and manufacturing (*bottom arrow*) to reduce defects. *DFM* Design for manufacturability, *DFA* design for assembly, *DFT* design for test, *DOE* design of experiments, *APQP* advanced product quality planning, *DFMEA* design failure modes and effects analysis, *PFMEA* process failure modes and effects analysis, *PPAP* production part approval process, *MSA* measurement system analysis, *SQC* statistical quality control, *DMAIC* define, measure, analyze, improve, control

7 Conclusions

An attorney's perspective of product design and product liability are very useful to designers. The above examples illustrate that product failures occurred because the design team did not follow applicable standards or use commonly available analysis methods in the design phase. In some examples that were presented above, the designer did not have a formal engineering education and may not have been aware of the design process and commonly available design and analysis tools. In addition, in some of the examples, properly designed warning labels were not used when some design constraints could not be guarded against or designed against. The most important conclusion from the above examples and a decade of work of the author in lawsuits is that designers must diligently follow and document all steps in the design or manufacturing process [1, 5] to attain a six-sigma quality level as shown in Fig. 2 [6]. The successful design of products requires a multidisciplinary team so that performance, aesthetics, ergonomics, and safety aspects of design are not only addressed, but are also well documented [6]. The design team must be familiar with risk analysis and must take sufficient care in reducing the risk of product failure and resultant loss or injury to users.

References

1. Engineering Design Process, NASA, 7 Feb 2008, http://www.nasa.gov/audience/foreducators/plantgrowth/reference/Eng_Design_5-12.html#.U2WYP6L0-1U. Accessed 03 May 2014
2. Karissa, R., Rams, D.: 10 Principles of "Good Design", 09 Jan 2012. ArchDaily, http://www.archdaily.com/?p=198583. Accessed 03 May 2014
3. Nichols, SP.: A Design Engineer's View of Liability in Engineering Practice: Negligence and Other Potential Liabilities. Online Ethics Center for Engineering 10/4/2006 National Academy of Engineering, www.onlineethics.org/Topics/ProfPractice/PPEssays/designnichols.aspx. Accessed 03 May 2014
4. Woodson, W.E.: Human Factors Engineering for Forensic and Safety Specialists. Lawyers and Judges Publishing Company, Inc., Tucson, (1998). ISBN 0-913875-40-6, Chapter II, p. 19
5. Hayes, C.C., Akhavi, F.: Creating effective decision aids for complex tasks. J. Usability Stud. **3**(4), 152–172 (2008)
6. http://www.freescale.com/webapp/sps/site/overview.jsp?code=QUALITY_0_DEFECTS

Design for Method Study—Work Measurement: Do We Need It?

Sangarappillai Sivaloganathan and Rana Yanis

Abstract DFX is a methodology for the definition and implementation of range of insights and demands which allow the development of an optimal and mature product in the aspects of reliability, manufacturing, assembly, service, transportation, usability, ergonomics, environment and more. It has been rightly argued that introduction of the DFX principles at the early stages would enable the designer to take into account the different 'X contexts' that would impact in his/her design. In this context Design for Method Study—Work Measurement is a suitable DFX method for design choice in high volume low value products. Use of PMTS to determine the assembling time and use of the time for design choice is proposed in this paper. The method is illustrated with two designs of the three pin plug.

Keywords Method study · PMTS · Assembling time estimation

1 Introduction

Design for Manufacturing is a philosophy and mind set aimed at designing parts and products that can be produced more easily and economically. It is a practice that emphasizes manufacturing issues throughout the product development process with a view to influence the design. Substantial amount of work, adding significant value, has been done in design for manufacture and assembly over the past few decades.

If the basic recommendations of "Reduce" and "combine" the number of parts (as outlined in Design for Assembly) are strictly followed, one is bound to arrive at embodiment with least motion and time in assembly. However it appears that they have not been strictly adhered to and products coming to the market suggest that the design of products, especially the high volume low value products, can be sub-

S. Sivaloganathan · R. Yanis (✉)
Department of Mechanical Engineering, UAE University, Al Ain, UAE
e-mail: 201180946@uaeu.ac.ae

A. Chakrabarti (ed.), *ICoRD'15 – Research into Design Across Boundaries Volume 2*, Smart Innovation, Systems and Technologies 35, DOI 10.1007/978-81-322-2229-3_27

stantially improved. This necessitates the question 'Is there a need for a DFX to address this issue as a specialist design method or as a tool'.

The purpose of motion study is to find the greatest economy of physical effort with due regard for safety and human aspect. It analyses the basic hand, arm and body movements of workers as they perform work. Time study on the other hand is the recording of the time needed to do a certain amount of work in a certain way. It is tied in directly with the specific method of doing the work and is good only for that method. May be because of this, manufacturing time has not been considered as an explicit reason for the choice of the design of products like the three pin plugs discussed in the case study. But the use of predetermined motion and time systems, PMTS, can help to estimate the time needed to manufacture or assemble the product made with the proposed design and this time can be used as a basis for concept selection and product evolution.

This paper proposes that, the use of 'Estimated Assembly Time' as a criterion for concept and embodiment selection, can improve the design significantly and reap the benefits of high labor efficiency and demonstrates it with an example. It uses a five stepped methodology to estimate the assembling time of a task that is repetitive and of relatively short duration, and performed many times. It analyses the design of two different types of electric plugs to illustrate the effectiveness of the method.

2 Literature Survey

Literature survey for the project has been carried out under two headings (i) design process and (ii) motion and time study. Under the design process heading, how design process has evolved into a 'Design and Debug' practice to accommodate Design for X and how DFA has been developed as part of the design process, is explored. This leads to the discussion, which identifies the place for motion and time study as part of DFA or DFX methodology. In the motion and time study part the development of method study and work measurement are looked briefly and motion study in its present form is explored.

2.1 Design Process

In systematic design the design process is described as a sequence of activities or as a sequence of stages in the passage of converting an abstract set of requirements into the definition of a physically realizable product. Design methods are the tools and techniques that can be used at different stages of the design process. Past few decades saw a flurry of new design methods and the number of the design methods used in a design project depends on the size and complexity of the product being developed.

Cross [1] observes that the originators of the 'design methods movement' realized that there had been a change from the craftwork of preindustrial design to the mechanization of industrial design, and modern industrial design had become too complex for intuitive methods. During the early development, engineering design was seen as the integration of the sciences and technologies to achieve the functionality required to meet the societal need. Demand for short time-to-market, reduction of cost and customer satisfaction during the life of the product drove the design process towards concurrent engineering.

Concurrent Engineering [2] is the systematic approach to the integrated simultaneous design of both products and their related processes including production. Advent and the ever-increasing use of computers in design further enhanced the design process. The design process became a kind of 'Design and Debug' process where an original concept, developed and detailed to meet the company's aspirations and customer needs, is being evaluated and optimized from several perspectives. These perspectives include design for manufacturability, design for procurement, design for supply, design for portability, design for reusability, design for interoperability, design for performance, design for regulatory compliance, design for reliability, design for serviceability and design for sustainability, and the list is ever increasing. These all are captured under the title 'Design for X' or DFX. It is a methodology for the definition and implementation of range of insights and demands which allow the development of an optimal and mature product in the aspects of reliability, manufacturing, assembly, service, transportation, usability, ergonomics, environment and more [3].

Several DFX methods, each investigating one or few particular aspects have been and are continuing to be developed and among these only a handful are chosen for application in any design process. The definition and choice of suitable DFX methodologies supporting rapid evaluation of design alternatives and the resulting design changes is an important problem [4]. Raffaeli et al. [5] argue that introduction of the DFX principles at the early stages would enable the designer to take into account the different 'X contexts' that would impact in his design. The argument forwarded by this paper is that assembly cost estimation using PMTS should be used as a DFX method for design selection.

It may be argued that the method outlined here needs the details such as parts list and workstation design which are well downstream in the product development process. However knowing that they ought to be considered as a DFX context for concept and embodiment selection will facilitate the designer to adhere to concurrent engineering in its true sense.

Design for Manufacture, DFM, and Design for Assembly, DFA, were started by Boothroyd and Dewhurst in the 70s and has led to enormous benefits such as simplification of products, reduction in assembly and manufacturing costs, improvement of quality and reduction of time to market. Kuo et al. [6] summarize the DFA method developed by Boothroyd and Dewhurst [7] in the following way:

1. Through the use of basic criteria, the existence of each separate part is questioned and the designer is required to provide the reasons why the part cannot be eliminated or combined with others.
2. The actual assembly time is estimated using a database of real-time standards developed specifically for the purpose.
3. A DFA index (design efficiency) is obtained by comparing the actual assembly time.
4. Assembly difficulties, which may lead to manufacturing and quality problems, are identified.

Discussion: The 'Design and Debug' approach and the DFX sets of design methods harmoniously fit together as a design methodology that can lend itself for the design of several kinds of products by permitting appropriate choices of DFX tools. The DFA methodology also benefits the design of better products with a broad estimate of the assembly times. However the assembly times for small products like the three-pin plug considered in the case study section of this paper depend heavily on the specific workstation design and the method of assembly.

This group of products can greatly benefit if a method that considers the workstation design, assembly method and the assembly time for the specific method, is employed during the design process to guide the design towards a more suitable one for the method.

2.2 Motion and Time Study

Motion study may be defined as the study of the body motions used in performing an operation, to improve the operation by eliminating unnecessary motions, simplifying necessary motions, and then establishing the most favorable motion sequence for maximum efficiency. Once the favorable motion sequence is defined the time taken for the process can be determined by time study.

Time study is often referred to as work measurement and it involves the techniques of establishing an allowed time standard to perform a given task, with due allowance for fatigue and for personal and unavoidable delays. Basically, motion study is the foundation for time study. The time study determines the time to do the job according to a certain method and is valid only so long as the method is continued.

The origin of motion and time study was due to Frank and Lillian Gilbreth. It is claimed that '*Frank developed an adjustable scaffold to keep his workers in level with the wall they built so as to eliminate stooping. He arranged mortar and bricks to eliminate 'reaching'. He simplified the labor process so that a bricklayer could repetitiously grab a brick and trowel full of mortar simultaneously, swivel, and simultaneously deposit mortar in the furthest tier of bricks and the brick in the next closest. Thus he claimed to reduce the bricklayers' motions from as many as 18 to as few as 4 and a 1/2*' [8]. Gilbreth identified seventeen basic elements of motions named by him as 'Therbligs', which are the same to all pieces of manual work and the study of them is called micro-motion study. The body motions should be recorded as they are performed for a micro-motion study to yield accurate results. Pigage and Tucker [9] advocate the collection of the following details during the study:

1. The general information about the job
2. The workplace description
3. The conditions and environment surrounding the workplace
4. The methods used by the operator (Can be given by an operation chart)

In order to consider whether the information is complete or not the following two questions should be answered:

1. Can the job be reproduced from the methods description?
2. Does the description include everything that the worker does?

2.2.1 Predetermined Motion Time Systems, PMTS

A predetermined motion time system (PMTS) may be defined as a procedure, that analyzes any manual activity in terms of basic or fundamental motions, (of limbs) required for performing it. Each of these motions is assigned a previously established standard time value and then the timings for the individual motions are synthesized to obtain the total time needed for performing the activity. PMTS had its origin with the Therbligs but there are more than fifty systems according to Groover [11]. Each of them has their own grouping of motions as the basis for the analysis. The PMTS application steps are as follows:

- Synthesize the method that would be used by a worker to perform the task. The method should be described in terms of the basic motions consisting the task, based on a predefined workplace layouts and set of tools.
- Retrieve the normal time value for each motion element, based on the work variables and conditions under which the element is performed. Sum the normal times for all motion elements to determine the normal time for the task.
- Evaluate the method to make improvements by eliminating motions, reducing distances, introducing special tools, introducing simultaneous right and left hand motions and so on.

2.2.2 The PMTS System MOST

The basic MOST has three activities (i) General Move with four sub activities (ii) controlled move additional three sub activities and (iii) tool use with eight additional sub activities. With these sub activities a sequence model is built for each task according to the model. The model consists of a letter and a suffix. The suffixes

Table 1 Activity models for basic MOST actions

Activity	Sequence model	Sub-activity
General move	*ABGABPA*	A—action distance
		B—body motion
		G—gain control
		P—placement
Controlled move	*ABGMXIA*	M—move controlled
		X—process time
		I—alignment
Tool use	*ABGABP* ***Operation*** *ABPA*	F—fasten
		L—location
		C—cut
		S—surface treat
		M—measure
		R—record
		T—think
		F—fasten

sometimes called indexes when multiplied by 10 gives the time taken for the motion in TMUs. One TMU is 0.036 s. The entire details are given in Table 1. Guide tables are available in text books [11] to obtain values for the indices.

3 Methodology

The methodology proposed requires the development of the assembly process before firming up the embodiment design. This falls in line with the concurrent engineering philosophy. It has five stages (i) establishing the process map (ii) designing the workstation (iii) designing of the tooling and equipment (iv) breaking down the task within the workstation and tooling conditions established into motion elements as defined by MOST and (v) estimate the time to complete the task.

Understanding the Design: At this stage the process designer should have a complete parts list and understanding of their geometries with respect to sizes so that he can identify the right sizes and shapes of bins. The products considered are relatively small but produced in large quantities and thus a small change in assembly time can lead to large savings. It therefore justifies the detail designing of parts and assembling workstations. Also this will enable him/her to decide the supply of the components to the assembling workstation.

Establishing the Process Map: Process Map describes the assembling process as a sequence of basic motions. It is a simple flow chart indicating the actions that are carried out in sequence. Thus in this context a process is defined as the sequence of activities that add value to one more inputs to produce outputs. Process maps provide a detailed picture of the process by showing the steps in the process. Often they are drawn as a two hand maps showing the activities performed by the right and left hands.

Designing the Workstation: This is a hypothetical arrangement of the workstation where the worker will sit and perform the task. It shows the positions of all components and tools. It is an arrangement according to the normal and maximum working area dimensions. There are several guidelines on the arrangement of workstation [11].

Design of Tooling and Equipment: This again is a hypothetical arrangement of the various hand and semi-automated tools that could be used if this arrangement and the design are chosen for final implementation.

List down the Motion Elements according to MOST Classifications: These are essentially the activity descriptions and the motion sequence models. The indexes representing the time should be allocated according to the guidelines.

Establish the Time: The total time required for the task can be obtained by summing up the indexes and multiply the sum by 10 to get the time in TMUs. The TMUs have to be multiplied by 0.036 s to get the time in seconds. In short this requires all the minor details of the assembling process. This estimated assembly time could be used as the basis for choosing the final embodiment.

4 Case Study

In this case study the designs of two different plugs used for domestic appliances were considered. They both were bought in the market for similar prices.

4.1 Description of the Designs and Process Maps

Plug A has 11 different types of parts and each plug assembly contains 15 parts. The tasks involved in assembling it can be enumerated in the following way:

1. Reach the plastic plug base, grasp it and position in vice holder. Rotate vice handle to secure plug base in holder. "10 Seconds process time".
2. Reach earth and neutral pins, grasp using both hands and insert them in plug.
3. Reach live pin and live outlet, grasp using both hands and insert them in plug.
4. Reach neutral screw, grasp, insert in neutral pin and tighten it by right hand.
5. Reach live screw, grasp, insert in live pin and tighten it by right hand.
6. Reach earth screw, grasp, insert in earth pin and tighten it by right hand.
7. Reach fuse, grasp it and insert in plug.
8. Reach the base in Vice with left hand, grasp it and rotate the vice handle with the right hand "10 s process time".
9. Reach and grasp cable grip and place in plug.

10. Reach and grasp cable grip screw 1, place it in plug and give three turns by fingers. Reach and grasp screw driver then tighten the cable grip screw 1.
11. Reach and grasp cable grip screw 2, place it in plug and give three turns by fingers. Reach and grasp screw driver then tighten the cable grip screw 2.
12. Reach and grasp top plastic cover and place on the top of plastic base.
13. Reach and grasp closing screw, insert in plug, grasp screwdriver and tighten the closing screw.
14. Reach and grasp paper label and then place on the plastic base of the plug (Figs. 1 and 2).

4.2 Work Station Design for Plug A

The workstation design shown is for the sequence of operations by both hands as shown in Fig. 3. The plastic base and cover being big were given bigger bins. They are placed within the normal working range of the assembler. The hand tool is given a designated area. The workstation design shown in Fig. 5 is made to scale so that the assembly times can be calculated for design A using a PMTS. Table 3 shows the estimation procedure (Fig. 4, Table 2).

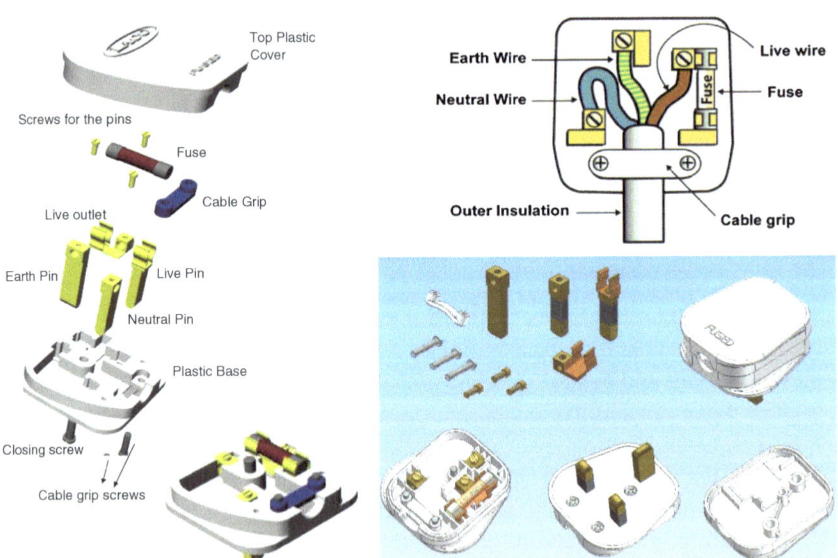

Fig. 1 Constituent parts plugs

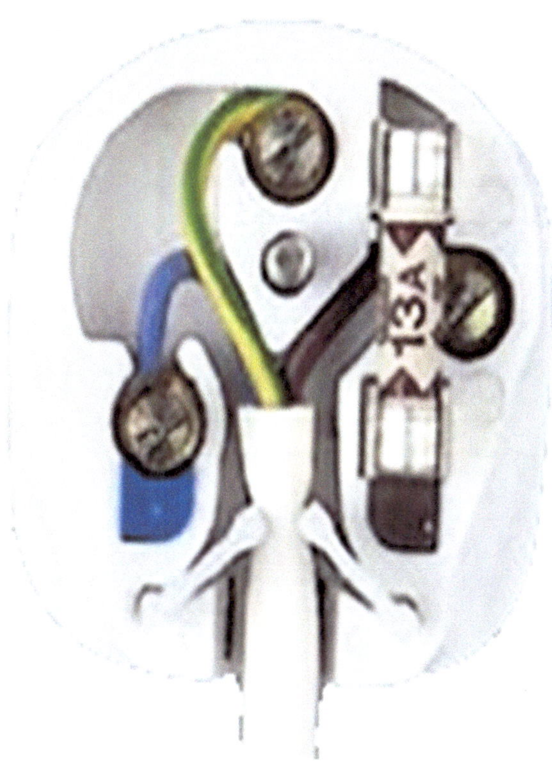

Fig. 2 Constituent parts of plug 2

Motion Elements according to MOST Classifications and their Time for Type A Plug,

$$\begin{aligned}\text{Total time} &= 340 + 40 + 40 + 60 + 320 + 320 + 320 + 330 \\ &\quad + 40 + 160 + 160 + 60 + 160 + 40 = 2{,}390 \\ &= 2{,}390 \times 0.036 = 86.04\ \text{s}\end{aligned}$$

4.3 Analysis of Plug B

Plug B has 11 different types of parts and each piece contains 14 parts. Activities for both hands in the assembling process are shown in Fig. 4. The tasks involved in assembling it can be enumerated in the following way:

1. Reach the plastic base, grasp it and position in vice holder then rotate handle to secure plug in holder. "10 s process time".
2. Reach earth and neutral pins, grasp both using both hands and insert them in plug.
3. Reach live pin and live outlet, grasp both using both hands and insert them in plug.

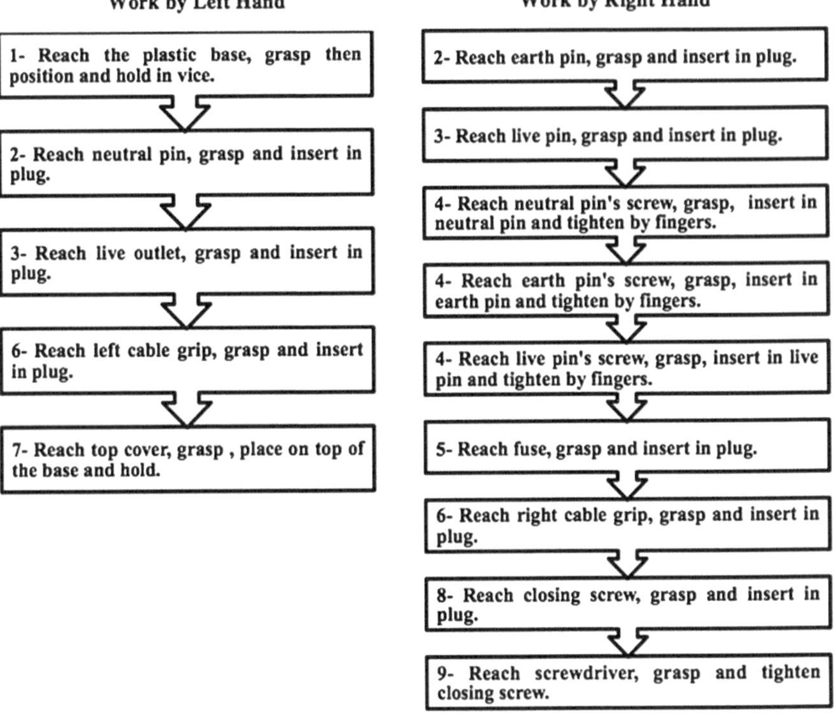

Fig. 3 Activities by the left and right hands

4. Reach neutral screw, grasp it and tighten by fingers.
5. Reach live outlet screw, grasp it and tighten by fingers.
6. Reach earth screw, grasp it and tighten by fingers.
7. Reach fuse, grasp it and insert in plug.
8. Reach the plastic base with left hand, grasp it then rotate the vice handler to loosen the base. "10 s process time".
9. Reach cable grips with both hands, grasp them and insert in plug.
10. Reach and grasp top plastic cover then place on the top of the plastic base.
11. Reach and grasp closing screw, insert in plug, grasp screwdriver and tighten the closing screw.
12. Reach and grasp paper label then place on the plastic base of the plug.
13. Motion Elements according to MOST Classifications and their Time for Type B Plug

$$\begin{aligned}\text{Total Time} &= 340 + 40 + 40 + 320 + 320 + 320 + 60 \\ &\quad + 330 + 40 + 60 + 160 + 40 = 2{,}070 \\ &= 2{,}070 \times 0.036 = 74.52 \text{ s}\end{aligned}$$

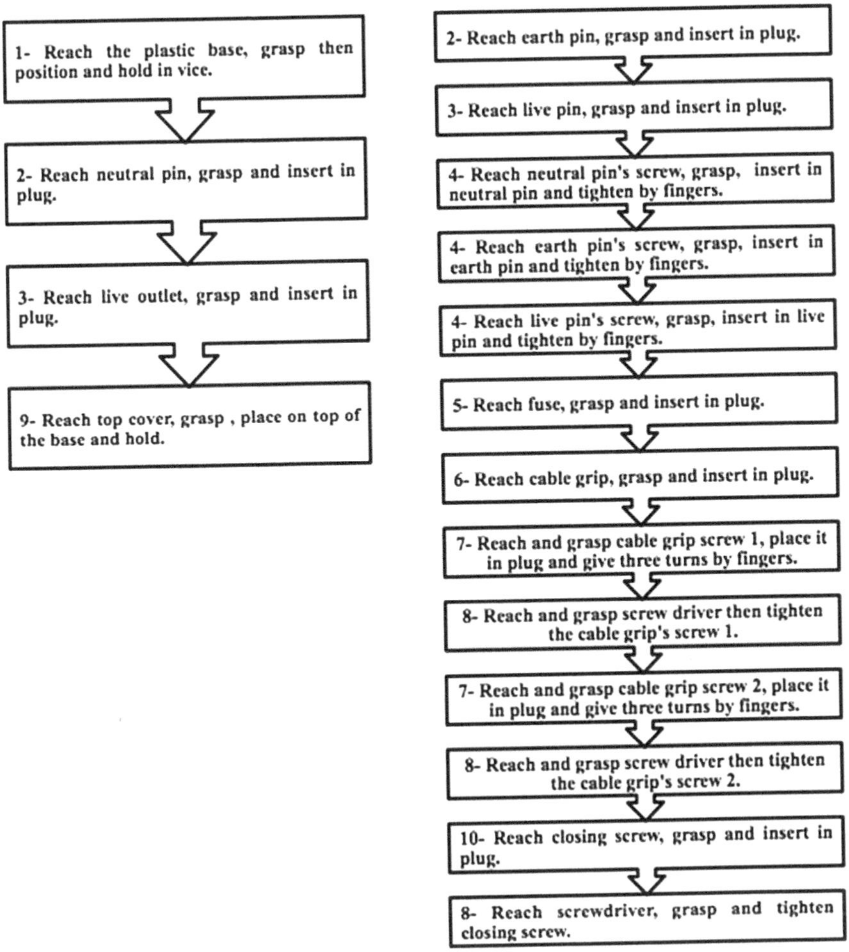

1- Reach the plastic base, grasp then position and hold in vice.
2- Reach neutral pin, grasp and insert in plug.
3- Reach live outlet, grasp and insert in plug.
9- Reach top cover, grasp , place on top of the base and hold.
2- Reach earth pin, grasp and insert in plug.
3- Reach live pin, grasp and insert in plug.
4- Reach neutral pin's screw, grasp, insert in neutral pin and tighten by fingers.
4- Reach earth pin's screw, grasp, insert in earth pin and tighten by fingers.
4- Reach live pin's screw, grasp, insert in live pin and tighten by fingers.
5- Reach fuse, grasp and insert in plug.
6- Reach cable grip, grasp and insert in plug.
7- Reach and grasp cable grip screw 1, place it in plug and give three turns by fingers.
8- Reach and grasp screw driver then tighten the cable grip's screw 1.
7- Reach and grasp cable grip screw 2, place it in plug and give three turns by fingers.
8- Reach and grasp screw driver then tighten the cable grip's screw 2.
10- Reach closing screw, grasp and insert in plug.
8- Reach screwdriver, grasp and tighten closing screw.

Fig. 4 Activities by the left and right hands

7L
1L
6L
3L
2L
8R
6R
4R
9R
2R
3R
5R
PLUG A
9L
1L
3L
2L
10R
4R
6R
7R
8R
2R
3R
5R
PLUG B

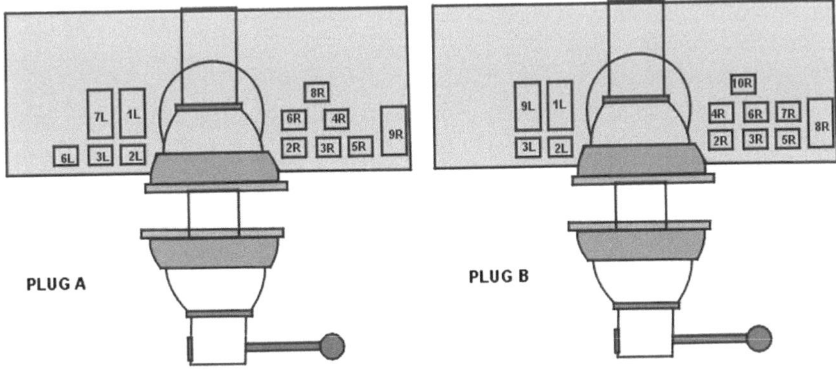

Fig. 5 Workstation design for plug A & B

Table 2 Task sequencing and time calculations for plug A

S. No.	Activities and sequence model	TMUs
1	Reach the plastic base, grasp it and position in vice, rotate handle to secure plug in holder. "10 s process time" $A_1B_0G_1M_3X_{28}I_1A_0$	340
2	Reach earth and neutral pins, grasp using both hands and insert them in plug $A_1B_0G_1A_1B_0P_1A_0$	40
3	Reach live pin and live outlet, grasp using both hands and insert them in plug $A_1B_0G_1A_1B_0P_1A_0$	40
4	Reach earth screw, grasp it and tighten it by hand $A_1B_0G_1M_1X_{28}I_1A_0$	320
5	Reach neutral screw, grasp it and tighten it by hand $A_1B_0G_1M_1X_{28}I_1A_0$	320
6	Reach live screw, grasp it and tighten it by hand $A_1B_0G_1M_1X_{28}I_1A_0$	320
7	Reach fuse, grasp it and insert in plug $A_1B_0G_1A_1B_0P_3A_0$	60
8	Reach the base in vice with right hand, grasp it and rotate the vice handle with the left hand 10 s process time $A_1B_0G_1M_3X_{28}I_0A_0$	330
9	Reach and grasp cable grip and place in plug $A_1B_0G_1A_1B_0P_1A_0$	40
10	Reach and grasp cable grip screw 1, place it in plug and give three turns by hand; reach and grasp screw driver then tighten the cable grip screw 1 $A_1B_0G_1A_1B_0P_3F_{10}A_0B_0P_0A_0$	160
11	Reach and grasp cable grip screw 2, place it in plug and give three turns by hand; reach and grasp screw driver then tighten the cable grip screw 2 $A_1B_0G_1A_1B_0P_3F_{10}A_0B_0P_0A_0$	160
12	Reach and grasp top plastic cover and cover the plug $A_1B_0G_1A_1B_0P_3A_0$	60
13	Reach and grasp closing screw, insert in plug, grasp screw driver and tighten the closing screw *A1 B0 G A1 B0 P3 A0*	160
14	Reach and grasp paper label and then place on the plastic base of the plug $A_1B_0G_1A_1B_0P_1A_0$	40

5 Discussion and Conclusions

In systematic design process the designer considers the DFX contexts and navigates the design process with due consideration of downstream issues as advocated by Concurrent Engineering. Low value high volume products coming to the market suggest that their designs can be substantially improved and subsequently their labor cost significantly. A method has been devised to investigate a design for estimating assembly times. It has been applied on the designs of two commercially available three pin plugs.

The design of Plug A demands more assembling operations and time because of the complex features. Comparatively the design of plug B has features which can be assembled easily. This becomes visible with the assembling time. The saving in time may appear small but becomes very significant when large quantities are involved.

Table 3 Task sequencing and time calculations for plug B

S. No.	Activities and sequence model	TMUs
1	Reach the plastic base, grasp it and position in vice holder then rotate handle to secure plug in holder. "10 s process time". A1 B0 G1 M3 X28 I1 A0	340
2	Reach earth and neutral pins, grasp both using both hands and insert them in plug. A1B0G1A1B0P1A0	40
3	Reach live pin and live outlet, grasp both using both hands and insert them in plug. A1B0G1A1B0P1A0	40
4	Reach neutral screw, grasp it and tighten by hand. A1B0G1M1X280 I1 A0	320
5	Reach Live outlet screw, grasp it and tighten by hand. A1B0G1 M1X280 I1 A1	320
6	Reach earth screw, grasp it and tighten by hand. A1B0G1 M1 X280 I1 A2	320
7	Reach fuse, grasp it and insert in plug. A1B0 G1A1B0P3 A0	60
8	Reach the plastic base with left hand, grasp it then rotate the vice handler to loosen the base. "10 s process time". A1 B0 G1 M3 X28 I0 A0	330
9	Reach cable grips with both hands, grasp them and insert in plug. A1B0G1 A1B0P1 A0	40
10	Reach and grasp top plastic cover then place on the top of the plastic base. A1B0 G1 A1B0P3 A0	60
11	Reach and grasp closing screw, insert in plug, grasp screwdriver and tighten the closing screw. A1B0G1 A1B0P3 A0	160
12	Reach and grasp paper label then place on the plastic base of the plug. A1B0G1A1B0P1 A0	40

The example suggests that Design for Method Study—Work measurement is an appropriate design method to make the design simpler to assemble where appropriate. In general there are several small products, which are produced in large quantities, are ideal candidates for the DFX tool Design for Method Study—Work Measurement.

Since submitting the draft of this paper the method has been applied on some other designs and similar results have been obtained.

5.1 Conclusion

A method showing how PMTS based Method Study—Work Measurement can be effectively used as an appropriate DFX tool has been established. Its effect in small products made in large quantities has been demonstrated by considering the assembly process for two plugs having different design features. Further studies with similar products have to be conducted to confirm the general applicability of the method.

References

1. Cross, N.: Designerly ways of knowing: design discipline versus design science. Des. Issues **17**(3), 49–55 (2001)
2. Gunasekaran, A.: Concurrent engineering: a competitive strategy for process industries. J. Oper. Res. Soc. **49**(7), 758–765 (1998)
3. http://www.tefen.com visited on 29th May 2014
4. Linderman, U.: A vision to overcome 'chaotic' design for X processes in early phases. In: Proceedings of the 16th International Conference on Engineering Design, vol. 1, pp. 231–239. Paris (2007)
5. Raffaeli, R., Mengoni, M., Germani, M.: A software system for 'design for X' impact evaluations in redesign process. J. Mech. Eng. **56**(1), 77–717 (2010)
6. Kuo, T.C., Huang, S.H., Zhang, H.C.: Design for manufacture and design for X: concepts. Appl. Perspect. Comput. Ind. Eng. **41**, 241–260 (2001)
7. Boothroyd, G., Dewhurst, P.: Product Design for Assembly. Boothroyd Dewhurst Inc., Walefield, RI (1987)
8. Price, B.: Frank and Lillian Gilbreth and the Manufacture and Marketing of Motion Study, 1908–1924. Bus. Econ. Hist. **18**, 88–98 (1989) (Second Series)
9. Pigage, L.C., Tucker, J.L.: Motion and Time Study. Bulletin of University of Illinois, Institute of Labor and Industrial Relations **51**(73) (1954)
10. Dagdeviren, M., Eraslan, E., Celebi, F.V.: An alternative work measurement method and its application to a manufacturing industry. J. Loss Prev. Process Ind. **24**, 563–567 (2011)
11. Groover, M.P.: Work Systems. Pearson Education International, New Jersey (2007)

Identification of Distinct Events in an Assembly by Automatically Tracking Body Postures

B. Santhi, Amaresh Chakrabarti, B. Gurumoorthy and Dibakar Sen

Abstract Identification of distinct events in an assembly process plays a vital role in assigning difficulty of assembly to specific events within the process. Literature addresses identification of events in assembly by analyzing videos, which is a manual process. We propose an approach to identify these events automatically using a method based on tracking of body postures and time. We use this method for an assembly exercise designed to introduce progressively more reach difficulty, and was performed in laboratory settings to track all limb data in time. The tracked data was used as an input to identify distinct events in an assembly. We use a technique we call slope method to smoothen the input data. Using this method helps in identifying distinct events involved in an assembly task. The result is validated by identification of events by observation of videos and correlation of these events with those identified automatically using the proposed method.

Keywords Assemblability · Assessment · Difficulty

1 Introduction

Assembly costs are significant, and their reduction is a major goal in design for manufacturing. One of the recent trends in the early design phase to reduce assembly process costs is to improve assemblability. Identification of distinct events in an assembly plays a vital role in assigning difficulty of assembly to specific events of the assembly process. 'Assemblability' refers to the ease of assembling a product from its parts. Assemblability evaluation is applied by product designers for estimating the degree of difficulty of assembly. It helps in reduction of process time and production costs. Existing methods for analyzing assembly rely mainly on

B. Santhi (✉) · A. Chakrabarti · B. Gurumoorthy · D. Sen
Virtual Reality Lab, Centre for Product Design and Manufacturing,
Indian Institute of Science, Bangalore, India
e-mail: santhi@cpdm.iisc.ernet.in

A. Chakrabarti (ed.), *ICoRD'15 – Research into Design Across Boundaries Volume 2*,
Smart Innovation, Systems and Technologies 35, DOI 10.1007/978-81-322-2229-3_28

manual video analysis, which finds its application in various areas [1–7]. This technique usually consumes a large amount of time, and its accuracy can be questioned [4]. Our goal is to resolve these limitations by developing an automated process for identifying events using automated, real-time capture and analyses of postural and time data. Currently electromagnetic trackers have been used for this purpose. For this purpose, we conducted an experiment of known difficulty which was progressively increased as the experiment progressed. This has been carried out in a laboratory set up. The objective is to propose and apply a method to automatically track human postural and time data associated with the operators involved in an assembly process, and to identify distinct events in the assembly process. The following section details the experimental procedure, results and their analysis.

2 Background

2.1 Assemblability

Assemblability as defined by Lee and Yi [8] is the probability of successfully assembling product parts. The definition, however, does not take into account cases where parts can be successfully assembled, but with varying degrees of difficulty, rather than only the worst case scenario where products are not possible to be assembled. Fujimoto and Ahmed [9] defines in terms of part ease of gripping, positioning and inserting parts in an assembly process. A definition which covers all cases of varying difficulty is 'Assembly-Operation difficulty', which uses a fuzzy number between 0 and 1 to represent the degree of difficulty [10]. Yet another definition, one that deals with complexity, time and quantity is 'Assembly requirement'; it covers assembly of individual parts/sub-assembly of parts and substances into final assembly of often high complexity and of a given quantity in a given period of time [11]. Notwithstanding all these definitions, however, a definition that directly deals with ergonomic aspects of the difficulty associated with assembly seems to be missing. We define 'Assemblability' as the ease of assembling a product from its parts. Traditionally, assemblability is assessed by using design for assembly (DFA) methodologies [11–14]. Literature review indicates that manual assemblies are assessed primarily by two methods: Ergonomic postural analysis [15–21] and Predetermined time systems [22, 23].

2.2 Electromagnetic Trackers

An electromagnetic tracker (EM) allows several body parts to be tracked simultaneously and will function correctly even if objects come between the source and the detector. In this type of tracker, the source produces three mutually perpendicular electromagnetic fields. The detector on the user's body then measures field attenuation (the strength and direction of the electromagnetic field) and sends this information back to a computer. The computer triangulates the distance and orientation of the three perpendicular axes in the detector relative to the three electromagnetic fields produced by the source. The detector must be within a restricted range from the source or it will not be able to send back accurate information, so the user has a limited working volume [24, 25].

2.3 Video Analysis

Research community uses video recording for various studies. Some of the related studies of work are discussed in this section. Shinde and Jadhav [1] used video recording for study of assembly workstations in welding shops to improve these workstations by applying ergonomics principles. Montakarn [2] used video based analysis for postural analysis in electronic assembly. Aggarwal and Cai [3] in their review on human motion analysis used video streaming for recognition of human activity and tracking of human motion. Mares and Senderska [4] explained the pros and cons of using video recording for assembly analysis. Bart et al. [5] used markers for measuring trunk movement in 2D video analysis. Sean et al. [6] developed a streaming tool called SAPPHIRE for analysing video for computer aided diagnosis and computer aided surgeries. Rodrigo et al. [7] implemented a computer vision technique to automatically detect ancillary gestures produced by clarinetists during musical performances.

3 Scheme of Experiment

This section provides details on tracking of human posture and time and the experimental setup used for the purpose of identifying distinct events.

3.1 Tracking of Human Postures Using Electromagnetic Trackers

To conduct this study, Polhems—Liberty [24] EM tracking system with 8 sensors was used, Fig. 1. Touted by its manufacturers as being one of the most accurate tracking systems in its range and it does not requiring a direct line of sight. Each tracker records its position in xyz format and orientation in yaw, pitch and roll angle format.

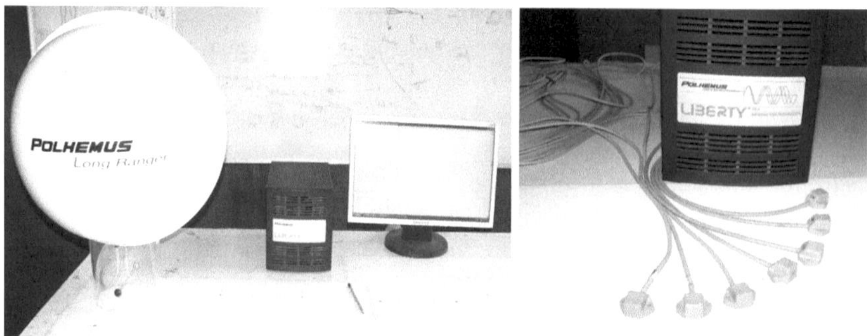

Fig. 1 Polhems—liberty tracking system

The joint angle for each limb was found by using data from two trackers. Each tracker orientation was represented by its rotation matrix from the global reference frame (GRF), for the determination of various limbs joint angles, as given by the Eq. (1),

$$\left[{}^{A}R_{B}\right] = \left[{}^{0}R_{A}\right]^{-1}\left[{}^{0}R_{B}\right] \tag{1}$$

where $[{}^{A}R_{B}]$—rotation matrix, 'A'—reference frame, 'B'—rotated frame.

Further computation ensured that even if the two trackers had different orientations with respect to GRF, they assigned the tracked values parallel to the original orientation.

3.1.1 Calibrating Home Position

In order to use the tracking device for a particular individual, the system was calibrated before the start of the experiment by connecting all eight sensors to the subject and recording the ideal or home position. The ideal or home position for the subject used in the experiment is shown in Fig. 2, and is described as follows.

The subject should stand straight, keeping his head looking straight to face the front side of the wall, holding his shoulders back, with his arm held in attention position and his torso at inline position with the head. At this position, the subject was instructed to stand for a few seconds such that his position is recorded and assigned to the system for further tracking. After this process, the subject was given a signal to "start", by the researcher, to carrying out the assembly exercise. After that, the subject performs the exercise by uttering the word "start" at the beginning and "stop" at the end.

Fig. 2 Home position for calibration of device

3.1.2 Locations of Trackers on the Body

The trackers were fixed on the subject in the locations as explained below and as in the schematic representation in Fig. 3:

- Tracker-1: at the base of the subject's neck, on the last cervical vertebra.
- Tracker-2: between the bicep, triceps and deltoid muscles of left hand.
- Tracker-3: on the ulna near the elbow of left hand.
- Tracker-4: in the lumber region of spine.
- Tracker-5: between the bicep, triceps and deltoid muscles of right hand.

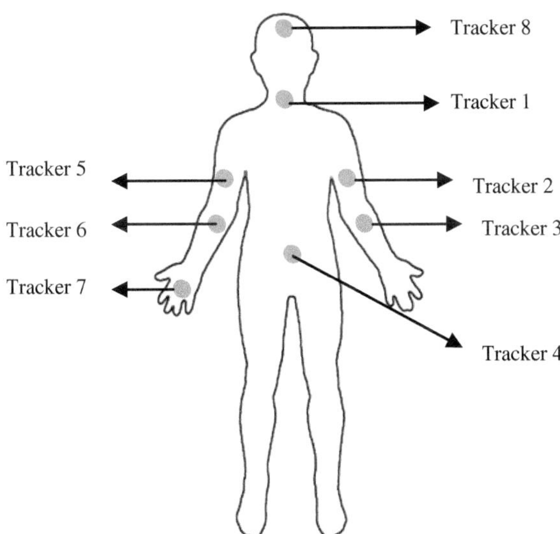

Fig. 3 Locations of tracker on the body

- Tracker-6: was attached on the ulna near the elbow of right hand.
- Tracker-7: on the dorsal surface of the right hand.
- Tracker-8: on the cranium.

Each tracker was marked with the number indicated as above for further reference and to avoid any mistake in connection with the system and subject. As a subject perform the experiment, a pair of trackers was used to capture postural angles for each limb/part of the body. The angles of a limb were captured as follows. The head angles were tracked by placing Tracker-8 anywhere on the cranium and taking roll, pitch and yaw with respect to Tracker-1. Torso was tracked by placing Tracker-4 on the belt, and taking the angles of Tracker-1 with respect to it. Right upper arm was tracked by placing Tracker-5 anywhere between the bicep, triceps and deltoid muscles and taking reference from Tracker-1. Right lower arm was tracked by placing Tracker-6 attached on the ulna near the elbow and using its reading against that of Tracker-5. Right wrist was tracked by fixing Tracker-7 on the dorsal surface of the hand and taking reference from Tracker-6. Similarly the other side of the hand i.e. left hand was tracked.

3.2 Experimental Method

An assembly process designed to vary reach difficulty was set in the laboratory to conduct the experiment, such that the level of difficulty progressively increased as the assembly process progressed. The assumption was that if the type (i.e. reach) and relative degree of difficulty (i.e. progressively more) were known beforehand, it would be easier to judge as to which limb or part of the body were responsible for the difficulty.

For this experiment, the tracking system using trackers 1–8 was mounted on the subject as explained in Sect. 3.1.2. The experimental setup is shown in Fig. 4. Though the devices are capable of giving accurate and reliable data but skin

Fig. 4 Experiment on reach difficulty

movement cannot be avoided during the experiment. So, we ensured that the sensors are attached in a way that minimizes slipping of sensors due to skin movement during motion. For this purpose we made additional arrangements to fix the sensors on the body. We attached the sensors using adjustable velcro strips; during motion, the sensors were observed for any slippage, which if were corrected by adjustment of the velcro strips. Then the experiment was carried out as described below.

The subject was asked to move freely in the wired condition. If any discomfort in the movement was noted, corrective action was taken by the researcher by fixing the trackers conveniently on the subject. In this experiment, the subject was asked to attain the various ‘reach’ points as shown in Fig. 4, and was instructed to fix a bolt and nut at each of these locations.

The subject was asked to maintain ideal or home position as detailed in Sect. 3.1.1 for few seconds, while data was gathered for calibration. Then the signal from the researcher to start was given, and the subject started the experiment by uttering the word “start”. During the experiment data for each limb was recorded automatically with the help of electromagnetic trackers and was stored in a file for further analysis. For each data gathered from the trackers, programs were written to calculate the joint angles of each joint and output them in a single line, starting with the time followed by the joint angles of left upper arm, right upper arm, left lower arm, right lower arm, left wrist, right wrist, head and torso. The next data gathered was stored in a new line, thus providing a continuous stream of data for all limbs, head and torso angles. This data was also automatically written to a file for later, more detailed analysis without the necessity for any manual manipulation. A sample output of the streamed data from an experiment is shown in Table 1.

Also, a video recording of the exercise was carried out to cross-check the data gathered as well as for validation of results by comparison. At the end of the

Table 1 Sample output data of automatically streamed values of body postures

Time shown in device	Right upper arm			Left upper arm			Right lower arm		
	Θ1	Θ2	Θ3	Θ1	Θ2	Θ3	Θ1	Θ2	Θ3
1,150,360.00	0.81	0.61	0.94	4.33	5.18	4.14	3.77	−3.00	−0.44
1,150,480.00	−1.01	2.50	−1.77	3.38	5.84	6.17	10.98	−9.74	−0.14
1,150,610.00	−2.65	4.69	−5.51	−0.75	6.22	6.44	20.39	−18.44	−0.74
1,150,720.00	−4.43	6.23	−10.63	−6.82	6.26	6.13	32.50	−27.83	−3.61
1,150,830.00	−4.04	7.87	−12.26	−11.95	7.02	6.20	44.90	−35.10	−9.33
1,150,940.00	−1.57	7.61	−11.12	−15.75	8.89	6.54	53.60	−38.44	−13.98
1,151,050.00	1.50	6.64	−8.48	−19.11	10.14	6.44	58.46	−39.72	−16.87
1,151,160.00	4.73	6.15	−5.00	−21.68	9.69	5.81	60.43	−39.71	−17.77
1,151,270.00	6.77	5.02	−2.68	−24.43	8.99	3.59	62.83	−40.25	−18.92
1,151,380.00	7.90	4.00	−1.73	−27.31	8.32	0.85	65.05	−40.78	−20.18
1,151,480.00	8.48	3.48	−0.25	−28.63	8.17	−0.73	64.68	−40.82	−20.39
1,151,590.00	8.83	1.25	2.34	−29.72	9.13	−2.83	63.35	−40.52	−19.53
1,151,700.00	9.68	−1.23	7.17	−30.71	10.70	−5.30	63.23	−40.73	−19.63

experiment, feedback from the subject was obtained in terms of the various tasks as well as events that he felt he performed, and also in terms of any discomfort he felt during these.

4 Analysis of Tracked Data

During this stage, comparisons were made between tracked and plotted data. This experiment sought to validate the hypothesis that 'an event is characterized by gross movements at its beginning and end', where an event is defined as 'a complete set of tasks in an assembly which is not repeated'. For example, an event was following a sequence of tasks like reach, pick, position, align, pick, align and screw. These set of sequence of operations was called event. In this work, gross and fine motions are taken respectively as motions where the angular change is relatively large or small over a period of time in an assembly process. Figure 5 shows the tracked values of torso for the experiment in Sect. 3.2. This figure plots time along the x-axis and angle of torso (as recorded using electromagnetic trackers) along the y-axis. The blue, red and green colour lines in the graph respectively indicate yaw, pitch and roll angle variations of the torso during the experiment. In a similar manner, angles related to all the other body segments (right upper arm, left upper arm, right lower arm, left lower arm, head, right wrist) were obtained and plotted for analysis.

Subsequently, tracked data was analysed in using two methods—the Method of area under the curve (MoAC) (see Fig. 6), and the Method of slope (MoS) (see Fig. 7). In MoAC, the blue points in the figure are the indicators of area under the curve for twenty tracked values of the torso data that we gathered using the trackers. Similarly, for other body segments, graphs were plotted. In MoS, the same data was first refined for consequent increase/decrease in value for each limb while ignoring three intermediate values; then the slope of the curve was determined. The analysed data was plotted along with square root of the three angles of torso.

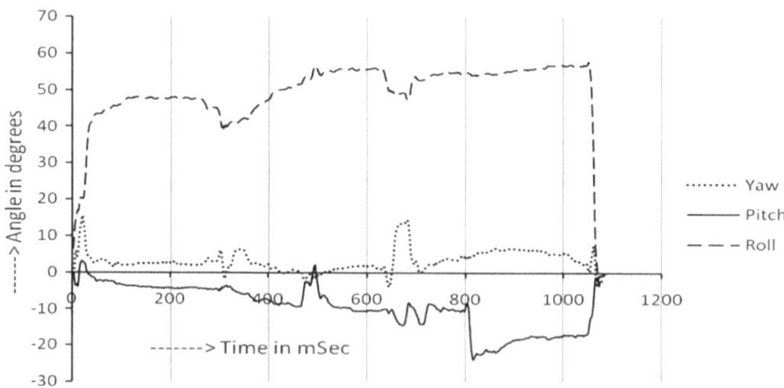

Fig. 5 Tracked angle data for torso

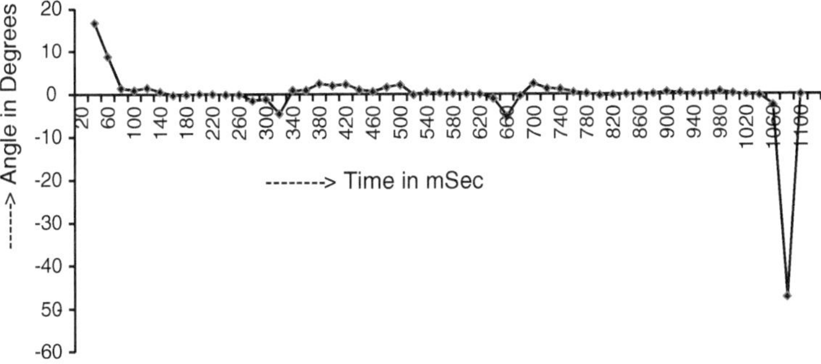

Fig. 6 Tracked data for torso by MoAC

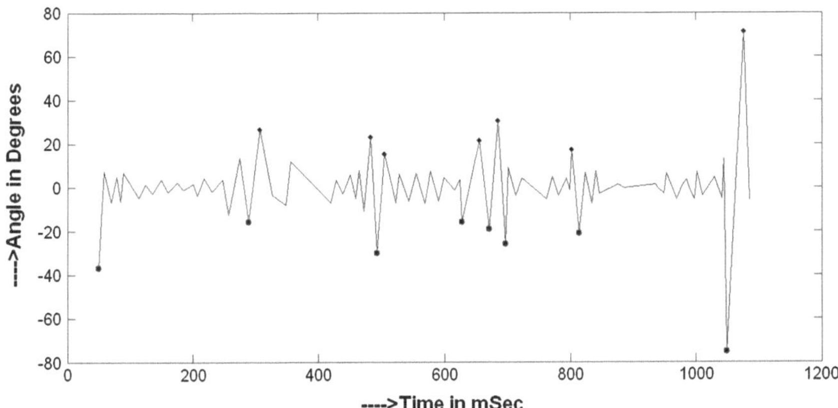

Fig. 7 Tracked data for torso by MoS

5 Results and Discussion

From the study, the following can be observed,

- The graphical outputs support the hypothesis that angular variation is more where gross motions occur and is less when the fine motions occur. Referring to Fig. 5, if the angular changes in these values are more over a period of time in a region, such regions were denoted as having gross motion. For example, for the green curve in Fig. 5, the Y-value drastically changes from 0°–44° during 1–40 ms; this is an example of gross motion. On the other hand, if angular change is less (<5°/ms) it is seen as a fine motion region; for example, for the same curve in Fig. 5, the angular value changes by 5° between 40 and 284 ms and this region would be classified as having fine motion.

- By comparing with experiments, analyzed data in MoAC (Fig. 6) helps in identifying the distinct tasks in an assembly operation. The experiment for which this method was tested had three events, e.g. reaching one point, say Reach 1, was considered to be an event. The output from this method indicates similar distinctions beyond a certain minimal threshold point. In Fig. 6, a change in configuration is indicated by a change in value for more than a value of ±2. There were 4 points in the curve at x-values (30, 320, 670, 1,080) above the value of ±2; if we ignore the initial point, which signifies bringing the body postures from the home position, there are 3 points in the curve, which match with the time span of the three events (i.e. reach1, reach2, reach3) in the assembly exercise. So, the hypothesis is validated.
- The data analysed using MoS (Fig. 7) helps to distinguish between the tasks within an event in an assembly exercise. This method captured several points within the three events. When checked with the video recording, each was found to be an important part of an event. The curve in Fig. 7 indicates the slope value of tracked data with round and diamond marks. The round (20, 280, 480, 620, 640, 680, 800, 1,020 point values in the x-axis) and diamond (300, 470, 490, 630, 650, 790, 1,040 point values in the x-axis) marks were an indication of the slope value of more than +15 or less than −15 degrees. These are the indicators of the various tasks within an event that were important for further analysis.

The above study can be further improved in several ways:

1. The data captured for the other limbs can also be analysed using the same methods i.e. MoAC (Fig. 6) and MoS (Fig. 7).
2. The data can be further analysed for determination of difficulty involved in an assembly task.

Also, this experimental methodology can be extended to address other sources of difficulty, like Vision and dexterity, also occurring during assembly.

6 Conclusions

A novel attempt, towards identification of distinct events by automatically tracking body postures, is reported in this paper. The results indicate a possible means for automatically segregating an assembly process into various tasks and events by two methods: MoAC and MoS. Also, it is noticed that the tracked data can be distinguished into 'fine' and 'gross' motions by the concept of no (or little) change per unit time as fine, and distinct, large change per unit time as gross. This work, we hope, will form the basis for automatically assessing difficulty of assembly in real time.

Acknowledgments Mr. Kalyan Ramana for participating in assembly exercise, and Mr. N. Madhusudhanan, for helping us to set the device for conducting the experiment, is acknowledged.

References

1. Shinde, G.V., Jadhav, V.S.: Ergonomic analysis of an assembly workstation to identify time consuming and fatigue causing factors using application of motion study. Int. J. Eng. Technol. **4**(4), 220–227 (2012)
2. Chaikumarn, M.: Variability in activities and postures during electronic assembly work assessed by video-based analysis system: a preliminary study. Master thesis, Human Work Science (2001)
3. Aggarwal, J., Cai, Q.: Human motion analysis: a review. Comput. Vis. Image Underst. **73**(3), 428–440 (1999)
4. Mares, A., Senderska, K.: Experience with the application in the manual assembly
5. Dingenen, B., Malfait, B., Vanrenterghem, J., Verschueren, S.M.P., Staes, F.F.: The reliability and validity of the measurement of lateral trunk motion in two-dimensional video analysis during unipodal functional screening tests in elite female athletes. J. Phys. Ther. Sport **15**, 117–123 (2014)
6. Stanek, S., Tavanapong, W., Wong, J., Oh, J.H., Nawarathna, R.D., Muthukudage, J., de Groen, P.C.: SAPPHIRE: a toolkit for building efficient stream programs for medical video analysis. J. comput. Methods Programs Biomed. **112**, 407–421 (2014)
7. Seger, R.A., Wanderley, M.M., Koerich, A.L.: Automatic detection of musicians' ancillary gestures based on video analysis. J. Expert Syst. Appl. **41**, 2098–2106 (2014)
8. Lee, S., Yi, C.: An Analytic Approach to Assemblability Analysis. In: International Conference on Robotics and Automation IEEE, pp. 1484–1489 (1998)
9. Fujimoto, H., Ahmed, A.: Entropic Evaluation of Assemblability in Concurrent Approach to Assembly Planning. In: 4th International Symposium on Assembly and Task Planning, pp. 306–311. IEEE (2001)
10. Zha, X.F., Lim, S.Y.E., Fok, S.C.: Integrated intelligent design and assembly survey planning: a survey. Int. J. Adv. Manfact. Technol. **14**, 664–685 (1998)
11. Boothroyd, G., Dewhurst, P., Knight, K.A.: Product Design for Manufacturing and Assembly. Marcel Decker Inc., New York (1994)
12. Rempersad, H.K.: The House of DFA, pp. 312–318. IEEE (1995)
13. Sturges, R.H.: A quantification of manual dexterity: the design for assembly calculator. J. Robot Comput. Integr. Manufact. **6**(3), 237–252 (1989)
14. Whitney, D.E.: Mechanical Assemblies. Oxford University Press, Oxford (2004)
15. McAtamney, L., Corlett, E.N.: RULA: a survey method for the investigation of work related upper limb disorders pdf assembly planning. J. Appl. Ergon. **24**(2), 91–99 (1993)
16. Feyen, R., Liu, Y., Chasn, D., Jimmerson, G., Joseph, B.: Computer-aided ergonomics a case study of incorporating ergonomics analysis into workplace design. J. Appl. Ergon. **31**, 291–300 (2000)
17. Kadefors, R., Forsman, M.: Ergonomic evaluation of complex work a participative approach employing video computer interaction exemplified in a study of order picking. Int. J. Ind. Ergon. **25**, 435–445 (2000)
18. Hignett, S., McAtamney, L.: Rapid entire body assessment. J. Appl. Ergon. **31**, 201–205 (2000)
19. Kee, D., Waldemar, K.: LUBA an assessment technique for postural loading on the upper body based on joint motion discomfort and maximum holding time. J Appl. Ergon. **32**, 357–366 (2001)
20. Massaccesi, M., Pagnotta, A., Soccetti, A., Masalib, M., Masiero, C., Greco, F.: Investigation of work related disorders in truck drivers using RULA method. J. Appl. Ergon. **34**, 303–307 (1985)
21. Santos, J., Sarriegi, J.M., Serrano, N., Torres, J.M.: Using ergonomic software in non-repetitive manufacturing processes. Int. J. Ind. Ergon. **37**, 267–275 (2007)
22. Laurig, W., Kühn, F.M., Schoo, K.C.: An approach to assessing motor workload in assembly tasks by the use of predetermined motion time systems. J. Appl. Ergon. **16**(2), 119–125 (1985)

23. Laring, J., Forsman, M., Kadeforsa, R., Ortengren, R.: MTM-based ergonomic workload analysis. J. Ind. Ergon. **30**, 135–148 (2002)
24. http://www.polhemus.com/?page=Motion_LIBERTY
25. Stolkin, R. (ed.): Scene Reconstruction, Pose Estimation and Tracking, pp. 530. I-Tech, Vienna (2007). ISBN 978-3-902613-06-6

Validation of Methodology and Tool for Design for Adaptability in Accomplishment of Project Objectives

V. Srinivasan, Phillip Schrieverhoff, Cristina Carro Saavedra, Matthias Gürtler and Udo Lindemann

Abstract A methodology and a tool for Design for Adaptability (DfA), developed in a project comprising academic and industrial partners, are validated to assess if they accomplish the objectives of the project. The DfA methodology and tool, developed by the academic partners, are applied in diverse use-cases by the industrial partners. A questionnaire is developed to collate the opinions of the industrial personnel who use the methodology and tool. An analysis of these opinions shows that: (a) the methodology and tool have high genericness, usability, tailorability, scalability and cost-effectiveness, and (b) designs and products developed using the methodology and tool have reduced lifecycle cost, shortened cycle time, extended lifespan, increased value to stakeholders and reduced consumption of natural resources and energy in their manufacturing. Based on the opinions of the industrial personnel, the objectives of the project are accomplished. Merits and demerits of the validation approach are also discussed.

Keywords Adaptability · Methodology · Tool · Validation

1 Introduction

Fricke and Schulz [1] identify adaptability, flexibility, robustness and agility as the four aspects of changeability. Adaptability is defined as the ability of a system to be modified by an agent located outside the system to make the system conform to changes in its environment [2]. Since adaptability is a measure of extrinsic ability, it is associated with expending a combination of effort, cost and time, in effecting a change [3]. In other words, if a change in a system requires lesser effort, lower cost and shorter time, then the system has higher adaptability than another system, which requires greater effort, higher cost and longer time to be changed. Hashemian [4]

V. Srinivasan (✉) · P. Schrieverhoff · C.C. Saavedra · M. Gürtler · U. Lindemann
Institute of Product Development, Technische Universität München, Munich, Germany
e-mail: srinivasan.venkataraman@pe.mw.tum.de

A. Chakrabarti (ed.), *ICoRD'15 – Research into Design Across Boundaries Volume 2*, Smart Innovation, Systems and Technologies 35, DOI 10.1007/978-81-322-2229-3_29

identifies the following types of adaptability: (a) based on what is adapted—general and specific adaptability, (b) based on who is adapting—producer adaptability or design adaptability and user adaptability or product adaptability, and (c) depending on the reversibility of the effected change—sequential and parallel adaptabilities. Design for adaptability (DfA) is a relatively new design paradigm to create designs and products which can be adapted to change at any stage of their life [5]. Adaptable designs and products are claimed to have several advantages, namely: (a) savings in product development time and costs due to better chances of reuse, (b) reduction in amount of waste produced due to higher chances of reconfiguration and re-use instead of disposal or recycling, (c) reduction in design and manufacturing efforts to achieve the required functions, (d) higher value to stakeholders over the entire product lifecycle, (e) longer lifespans due to the provisions to upgrade and reconfigure [6, 7]. Several efforts [2–4, 6, 7] have been undertaken to understand and support design for adaptability. In this paper, we specifically focus on validation of a methodology and tool for DfA, which are developed in a project, to check if the objectives of the project are fulfilled.

2 Earlier Work and Goal

Architecting Manufacturing Industries and Systems for Adaptability (AMISA) is a recently concluded project funded by the European Commission under the 7th Framework, with two academic partners [Technische Universität München (TUM), Germany and Tel Aviv University (TAU), Israel] and six industrial partners [Israel Aerospace Industries (IAI), Israel; MAG IAS, Germany; MAN, Germany; Optoelectronica (OPT), Romania; TetraPak Packaging Solutions (TPPS), Italy; and Tecnologias de Telecommunicaciones and Informacacion (TTI), Spain] covering domains of aerospace, machine tools, automotive, optical, packaging and telecommunication industries. The objectives of the project are as follows: (1) to develop a generic, tailorable, quantitative and usable methodology for architecting systems for optimal adaptability to possible future changes in stakeholder needs and technology development; such systems must exhibit better cost-efficiency, longer lifespan, reduced cycle time, and provide more value to stakeholders; (2) validate and prove the methodology by means of real-life pilot projects, and provide concrete evidence that the methodology is generic, tailorable, scalable, usable and cost effective; (3) show by the end of the project that reconfiguring systems designed for adaptability, yields savings of 20 % in cost, cycle time, or combination of cost and cycle time; (4) show by the end of the project that lifespan of systems designed for adaptability increases by 25 %; (5) show that the systems yielding more service for a longer duration will exhibit the following qualitative benefits: (a) during the manufacturing process, the overall usage of natural resources and energy consumption as well as the overall pollution and byproduct waste will be reduced, and (b) adaptable systems will be more amenable to sustained evolving regulatory framework (i.e. environmental, health, safety, etc.). In this project, to support DfA, a methodology and a tool are developed by the academic partners, TAU and TUM. More information of the project is available in the AMISA website [8].

2.1 DfA Methodology

The DfA methodology is described in detail in [3, 9]. The purpose of the DfA methodology is to help identify if an adaptation can be made efficiently and effectively, how and where in the system can the adaptation be performed, and when and at what cost should the system be adapted. The DfA methodology comprises the following broad stages: problem formulation, criteria-based modeling, engineering change scenario assessment, cost evaluation, and option value estimation. Formulation of problem helps to clarify ways with which adaptability can provide value to stakeholders, and assist in selection of appropriate and subsequent steps in the methodology. To estimate the evolution in future needs and technology for identifying relevant business cases and key parameters, TRIZ technology forecasting can be used together with expert interviews and scenario methods. In criteria-based modeling, relevant criteria for the redesign of the architectures of systems are identified. These criteria comprise constraints from the systems and are the basis for modeling the systems. Systems are modeled using Design Structure Matrix and Multiple Domain Matrix and the relevant elements of systems are identified. This model and the structural metrics are used for analyzing the architecture of the system and to identify potential improvements. In the engineering change scenario assessment, based on the model of the system and structural metrics, alternative architectures with potentially higher adaptability are developed and assessed using the criteria and key parameters defined in the earlier step. In the cost estimation, all the costs (development, production, assembly, upgrade, etc.) associated with the alternatives are estimated. Systems with higher adaptability have additional costs i.e., option cost (OC) and upgrade cost (UC). OC is accounted in advance when designing and offering an adaptable system. UC occurs only when an upgrade is performed. In the option value estimation, the expected customer value of the developed architectures is determined. The evolution of the key parameters is estimated using TRIZ technology forecasting method. Two approaches are used for estimating option values: (a) Black-Scholes method for components and congregated systems and (b) Monte-Carlo simulation for system architectures.

2.2 DfA Tool

The DfA tool is explained in detail in [9, 10]. The tool is an interactive and partial implementation of the DfA methodology on computer. The tool assists in the following stages of the DfA methodology: criteria-based modeling, engineering change scenario assessment, cost evaluation, and option value estimation using Black-Scholes approach.

2.3 Goal

The goal of this research is to validate the DfA methodology and tool to check whether or not they assist in the accomplishment of objectives of the AMISA project.

3 Research Approach

To validate the DfA methodology and tool, they are used by the industrial partners to develop adaptable designs and products in their respective industry for around 20 months. The following are the details of the industrial use-cases: (a) Robotic vehicle localization system at IAI; (b) Fiber placement machinery at MAG; (c) Differential lock and powertrain at MAN; (d) Hologram production line at OPT; (e) Cap applicator at TPPS; and (f) Solid state power amplifiers at TTI. Before using the methodology and tool, the designers and engineers involved in the development of designs and products are provided a documentation of the methodology and tool. In addition, a number of workshops are held with the designers and engineers, separately for each industry, to guide them through the various stages of the methodology and tool in their application to the respective industrial cases. The application of the DfA methodology and tool is monitored in the group workshops, which are organized every 3 months and involves participation of all the project partners. Since the various variables in the objectives are difficult to be assessed during the duration of the project (36 months), the opinions of the designers and engineers, based on their experience of developing designs and products using: (a) the DfA methodology and tool, and (b) prior to the development of the DfA methodology and tool, are used as a means to assess the fulfillment of the objectives. These opinions are collected using a questionnaire and the responses to the questionnaire are analyzed in the context of the extent of fulfillment of the project objectives.

4 Design of Questionnaire

To collect the opinions of designers and engineers from the industrial partners based on their use of the DfA methodology and tool, an online-based questionnaire is developed, see [11]. The questionnaire contains twenty-six questions divided into three sections. Most of the questions, barring a few, are multiple-objective type with only one option to choose from among the many. The objective of the three sections is: (a) to understand the background of the respondents and their organizations, (b) the status before the AMISA project, and (c) the effect of use of the DfA methodology and the tool. The questionnaire is pilot tested: (a) with the group at TUM and, (b) with a smaller group of industrial partners. In these pilot studies, the response to the questionnaire is collected, analyzed and shared with the respondents. Based on their feedback, a few questions are modified to improve the understanding and elicit relevant responses.

5 Findings

In this section, important findings from the response to the questionnaire are reported. Twenty-one designers and engineers, who used the DfA methodology and tool, from the six industrial partners responded to the questionnaire (see Fig. 1). In the following part of this section, their responses are analyzed in the context of accomplishment of the project objectives. Since all the industrial partners use a combination of the DfA methodology and tool to develop designs and products, a combined analysis of the methodology and tool is performed.

Figure 2 shows the respondents' rating of the genericness, usability, tailorability, scalability and cost-effectiveness of the DfA methodology and tool in a qualitative

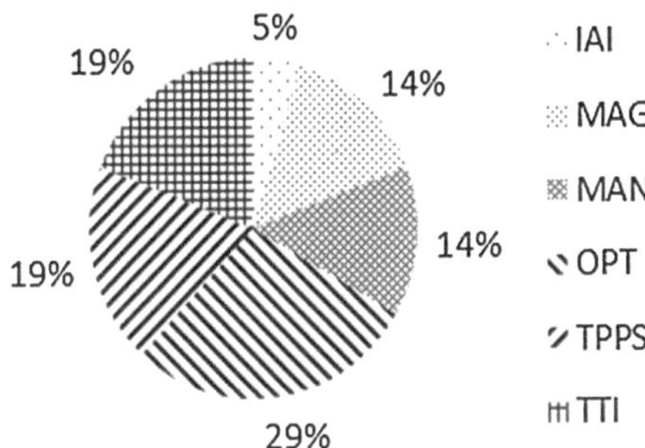

Fig. 1 Percentage distribution of respondents to questionnaire

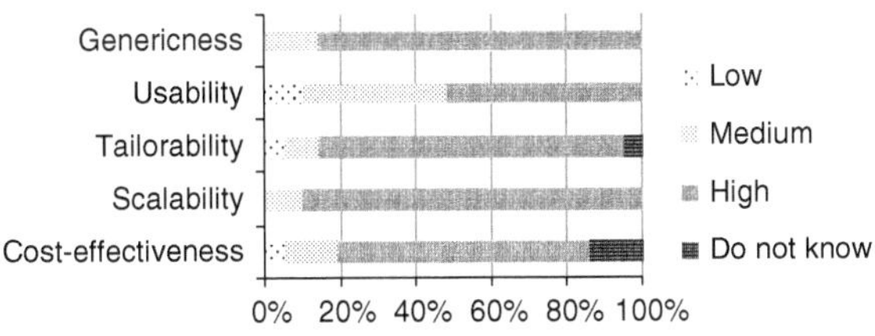

Fig. 2 Genericness, usability, tailorability, scalability and cost-effectiveness of DfA methodology and tool

scale as low, medium, high or do not know. Note that genericness, usability, tailorability, scalability and cost-effectiveness have been defined in the questionnaire after clarifying them in earlier workshops. The horizontal bar-graphs show the trends of the parameters. For instance, 5 % of respondents rate the cost-effectiveness of the combination of the DfA methodology and tool as low, around 15 % rate it as medium, close to 65 % rate it as high and the remaining did not know. Overall, a majority of the respondents reported that the combination of the DfA methodology and tool has high genericness, usability, tailorability, scalability and cost-effectiveness.

In the findings henceforth, a distinction is made in the effects on: designs and products. Designs refer to all kinds of outcomes of designing such as instructions, sketches, clay models, prototypes, etc. except product which can be taken directly to the market. This is done to distinguish design adaptability from product adaptability as done by Hashemian [4], and the ensuing benefits. Figure 3 shows the respondents' rating of the effect of designs (top horizontal bar) and products (bottom horizontal bar), developed using the DfA methodology and tool, on their lifecycle cost, in terms of decrease, remaining constant, increase or do not know. From the graphs, it is observed that a majority of the respondents feel that the lifecycle cost decreases. Amongst those who report decrease in lifecycle cost, the average decrease in lifecycle cost in designs and products is found to be approximately 19 % (see Fig. 4) and 16 % (see Fig. 5), respectively.

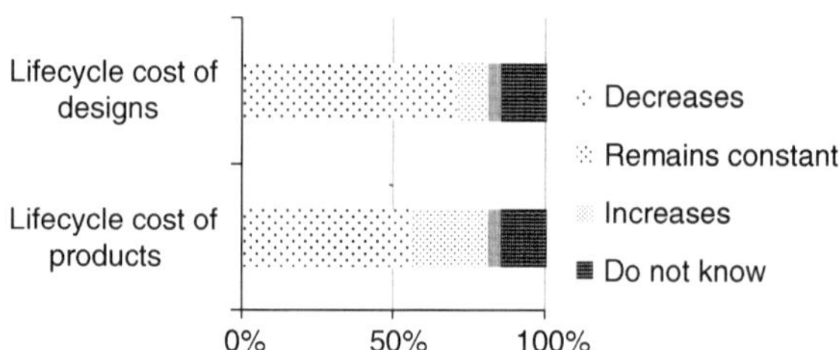

Fig. 3 Effect of designs and products developed using methodology and tool on lifecycle cost

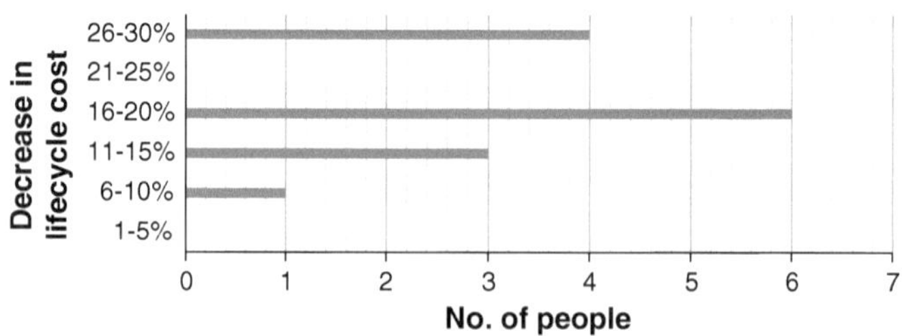

Fig. 4 Decrease in lifecycle cost of designs

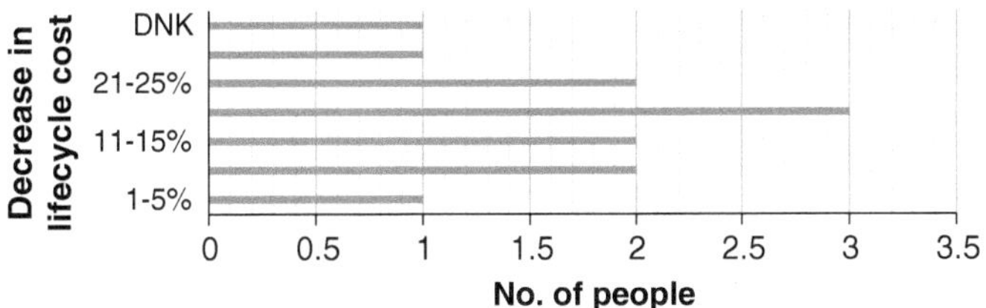

Fig. 5 Decrease in lifecycle cost of products

The rating of the influence of the designs and products developed using the DfA methodology and tool on their cycle time, in terms of decrease, remaining constant, increase or do not know, is shown in Fig. 6. A majority of the respondents (about 60 % for designs and 50 % for products) opine that the cycle time decreases for both designs and products. For this population, the average reduction is found to be 20 % (Fig. 7) and 17 % (Fig. 8) for designs and products, respectively.

Figure 9 shows the respondents' rating of the impact of designs and products, developed using the DfA methodology and tool, on their lifespan in terms of increase, remain constant, decrease or do not know. A majority of the respondents (71 and 60 % for designs and products, respectively) report that the lifespan increases. Among these groups, the average increase is found to be 20 % (Fig. 10) and 17 % (Fig. 11) for designs and products, respectively.

The rating of the effect of designs and products developed using the DfA methodology and tool on the value to stakeholders, in terms of decrease, remain

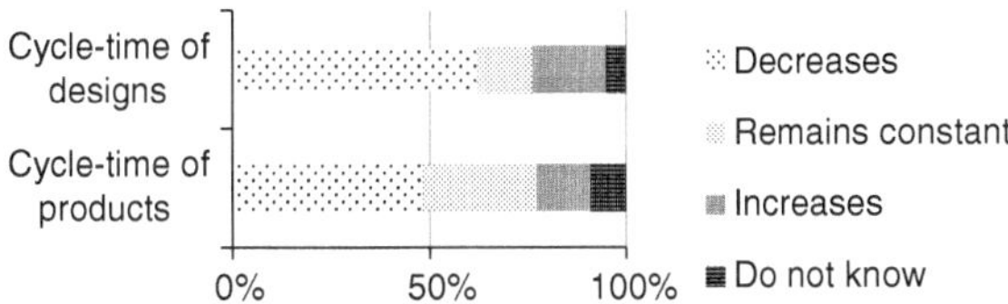

Fig. 6 Cycle time of designs and products

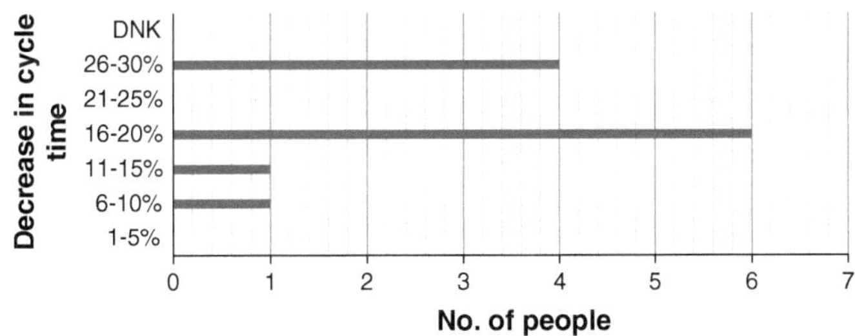

Fig. 7 Decrease in cycle time of designs

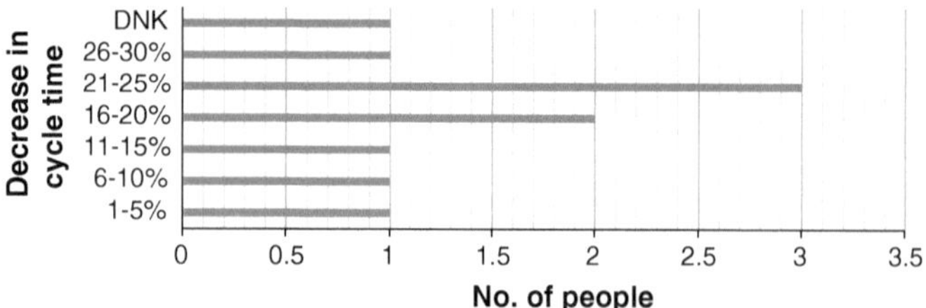

Fig. 8 Decrease in cycle time of products

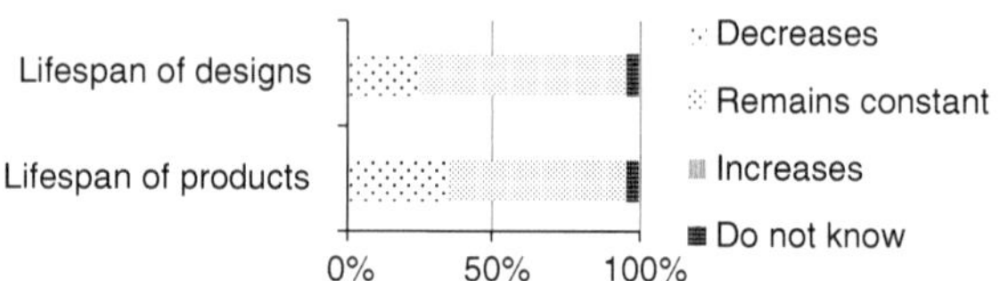

Fig. 9 Lifespan of designs and products

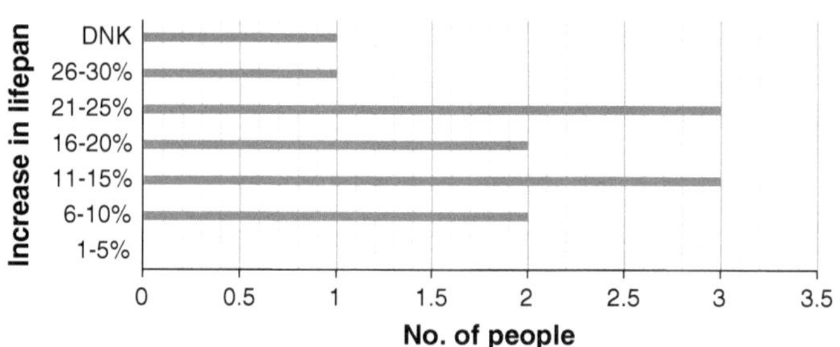

Fig. 10 Increase in lifespan of designs

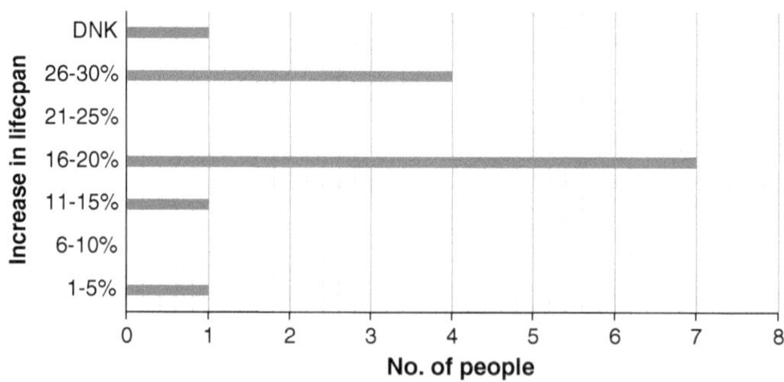

Fig. 11 Increase in lifespan of products

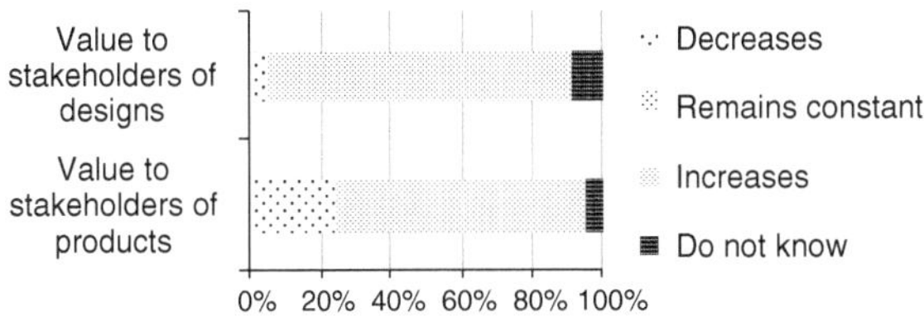

Fig. 12 Value to stakeholders of designs and products

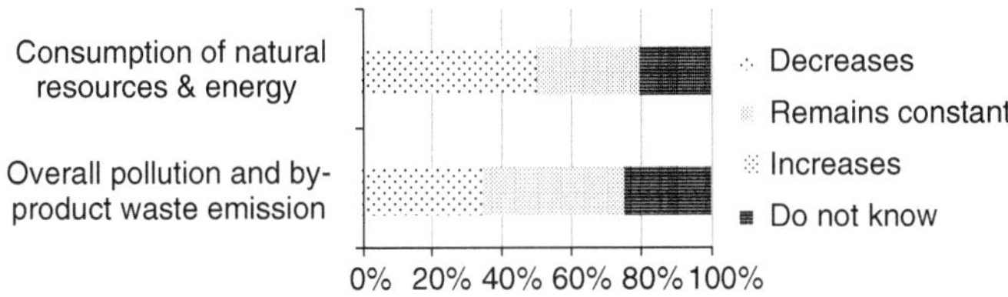

Fig. 13 Consumption of natural resources and energy, and overall pollution and by-product waste emission

constant, increase or do not know, is shown in Fig. 12. The graph shows that a majority of the respondents (close to 70 % for designs and 85 % for products) feel that the value to stakeholders of designs and products developed using the DfA methodology and tool increases.

In Fig. 13, the rating of the influence of designs and products developed using the DfA methodology and tool on the consumption of natural resources and energy (top horizontal bar) and overall pollution and by-product waste emission (bottom horizontal bar), during the manufacturing process, is shown. Majority of respondents (50 %) feel that the consumption of natural resources and energy decreases. About 40 % respondents feel that designs and products developed using the DfA methodology and tool have no influence on pollution and by-product waste emission and only 35 % feel that this value decreases, during the manufacturing process.

To verify the responses to the questionnaire, all the trends shown in this section are also shared with all the respondents, their causes and potential effects are discussed in a meeting.

6 Discussion

The following is a summary of the important findings based on the questionnaire study:

(a) The methodology and tool developed for DfA are found to be generic, usable, scalable, tailorable and cost-effective.

(b) The designs and products developed using the DfA methodology and tool are observed to have, on average:

i. a cost saving of 19 and 16 %,
ii. a cycle time savings of 20 and 17 %,
iii. an increase in lifespan by 20 and 17 %,
iv. an increase in value to stakeholders,
v. a reduced consumption of natural resources and energy, and no influence on overall pollution and byproduct waste emission, both during manufacturing.

Thus, most of the objectives of the project are found to be accomplished based on the responses to the questionnaire.

As seen from the results, the DfA methodology and tool help develop better designs than products. This could be due to the following reasons: (a) the methodology and tool address designs better than products, (b) addressing adaptability in designs is better than in products, and (c) benefits in designs are perceived easier than in products due to the shorter observation time.

Although the objectives of the AMISA project are accomplished based on the perceptions of the designers and engineers who used the DfA methodology and tool, the assessment of this accomplishment has the following constraints. The accomplishment of most of the objectives, with the exception of Objective 2, needs a longer duration, which is beyond the funding period of 36 months of the project. This is especially relevant for variables such as lifecycle cost, cycle time, lifespan, savings in usage of natural resources and energy, etc. of designs and products, which cannot be measured easily within the funding period of the project. Another issue is that developing measures for these variables or measuring these in terms of other variables that can be assessed in shorter duration is beyond the scope of the project. Therefore, it is concluded that the objectives of the AMISA project are accomplished in short-term but need to be monitored over longer duration.

In this validation, a combination of case-study and questionnaire is used. As explained earlier, the DfA methodology and tool are used by the industrial partners within their settings to develop designs and products, constituting the case-study. The opinions of the designers and engineers, who used the DfA methodology and tool, are collected using a questionnaire developed for this purpose. This combination of case-study and questionnaire is proven useful considering the resources (including time) available in the project for the validation. In this study, only a limited number of people with several years of industrial experience used the DfA methodology and tool, in uncontrolled settings in industry to solve six real-time tasks by developing designs and products that are intended to be used. Apart from the project constraints, no other time constraints are imposed on the designers and engineers, and overall, the methodology and tool are used for around 20 months. This study is interventional, involving the combined use of the DfA methodology and tool, and performed on a non-comparative basis. However, the designers and engineers are requested for comparative opinions in terms of changes in designs and products developed: (a) before introducing the methodology and tool and (b) by using the methodology and tool. Questionnaire is used as a data-collection method because it also enables the collection of data from the past (i.e., prior to the development of the DfA methodology and tool) and the background of the participants. It also facilitates data collection to be performed within the resources, including the allotted time of the project. The merits of the methodology and tool are assessed in terms of the perceived performances of designs and products developed using the methodology and tool, and compared with the performances of designs and products (performing the same functions) developed before introducing the methodology and tool. The involvement of researchers in the application of the methodology and tool is minimal and is restricted to assisting the designers and engineers when they run into impediments. In other validation studies, comparative or non-comparative interventional studies with several groups of experienced and/or novice designers in controlled settings (such as laboratory), solving various less real design tasks in limited time duration are used. The choice of variables in the validation of a support is dictated by the requirements and constraints of the project.

On a positive side, the combination of case-study and questionnaire enables validation to be performed within the resources and time allotted in the project. It also enables collecting data from the past against which the benefits of development using the DfA methodology and tool can be compared to. On the other side, first, the fulfillment of the project objectives is assessed using the opinions of designers and engineers of the perceived performances of designs and products. Second, this is a short-term, therefore, preliminary validation checking the accomplishment of the objectives, which need longer time to be accomplished. Third, this validation is a qualitative assessment of quantitative objectives. Fourth, it is possible that the designers and engineers are biased by the objectives of the project, which are known to them earlier, and therefore, responded in accordance to their accomplishment.

7 Conclusions

The validation of the DfA methodology and tool based on: (a) their application in diverse industrial cases and (b) the perceptions of the industrial personnel, who used the methodology and tool, collected using an online questionnaire, shows that the objectives of the AMISA project are accomplished in short-term. The objectives need to be monitored over a longer term to prove the efficacy of the methodology and tool.

References

1. Fricke, E., Schulz, A.: Design for changeability (DfC): principles to enable changes in systems throughout their entire lifecycle. Syst. Eng. **8**(4), 342–359 (2005)
2. Engel, A., Browning, T.: Designing systems for adaptability by means of architecture options. Syst. Eng. **11**(2), 125–146 (2008)

3. Kissel, M., Schrieverhoff, P., Lindemann, U.: Design for adaptability—identifying potential for improvement on an architecture basis. In: NordDesign 2012, Aalborg, Denmark (2012) (Digital proceedings)
4. Hashemian, M.: Design for adaptability. Ph.D. Thesis, Department of Mechanical Engineering, University of Saskatchewan, Canada (2005)
5. Gu, P., Hashemian, M., Nee, A.Y.C.: Adaptable design. CIRP Ann. Manuf. Technol. **53**(2), 539–557 (2004)
6. Gu, P., Xue, D., Nee, A.Y.C.: Adaptable design: concepts, methods, and applications. Inst. Mech. Eng. Part B: J. Eng. Manuf. **223**(11), 1367–1387 (2009)
7. Li, Y., Xue, D., Gu, P.: Design for product adaptability. Concurrent Eng. **16**(3), 221–232 (2008)
8. AMISA website_1. http://www.amisa.eu/. Accessed in May 2014
9. AMISA website_2. http://amisa.eu/index.php/downloads/130-amisa-methodology-guideline. Accessed in May 2014
10. AMISA website_3. http://amisa.eu/index.php/downloads/129-dfa-tool-manual. Accessed in May, 2014
11. Questionnaire in SurveyMonkey. https://www.surveymonkey.com/s/AMISA_Questionnaire. Accessed in May 2014

Part IV
Enabling Technologies and Tools (Computer Aided Conceptual Design, Virtual Reality, Haptics, etc)

Effectiveness of Tangible and Tablet Devices as Learning Mediums for Primary School Children in India

Anmol Srivastava and Pradeep Yammiyavar

Abstract A study was conducted in two primary schools of Guwahati, India to understand how tangible devices and tablets impact the learning process. Data was collated from field notes, interviews, audio logs and recorded videos. Children between ages 6 and 7 years were introduced to tablets and tangible devices and asked to describe their learning experience after the designed usage. The data was analyzed to understand how these children imbibed the learning experience. Framework of Multiple Intelligence (MI) and Kolb's Learning Cycle was used in categorizing the resulting learning observed in the children. This paper presents a qualitative analysis of the study to explore if new paradigm based on combination of MI theory and Kolb's Cycle can emerge to determine the effectiveness of screen based devices and tangible learning aids.

Keywords Tablets · Tangible devices · Experiential learning · Multiple intelligence

1 Introduction

With increasing penetration of screen-based devices like mobiles and tablets resulting in shrinking of conventional classroom space, several questions arise regarding their utility, usability and effectiveness as learning aids. The most common cited reason being the accompanying distraction that students face while learning on such devices. Students tend to deviate, knowingly or unknowingly, from their learning regime and move towards entertainment content within these devices (e.g. preloaded games or videos), which any way seem more interesting than the topic of learning. Since these devices are beginning to be considered as a disruption, they raise a vital question of how soon should children be inducted into

A. Srivastava (✉) · P. Yammiyavar
Department of Design, Indian Institute of Technology Guwahati, Guwahati, India
e-mail: anmol.srivastava@iitg.ernet.in

A. Chakrabarti (ed.), *ICoRD'15 – Research into Design Across Boundaries Volume 2*, Smart Innovation, Systems and Technologies 35, DOI 10.1007/978-81-322-2229-3_30

digital devices usage as part of their educational activities. As there is a vast difference between their use in schools by children and adults in college, it also leads us to the question of how soon and how much should we bring the use of Information and Communication Technology into education system and how gradually should it be brought in?

Technology, on the other hand, is facilitating embedded intelligence into children's physical world of toys and learning aids at a rapid rate. In this paper, such learning aid are referred as tangible devices and tablets indicates digital communication and computing devices including large screened mobile phones.

Various studies [1, 2] suggest that tangible devices help in enhancing learning effectiveness. Research studies in the domain of tangible interactions indicate pedagogical links between device use among children and their level of learning. Since learning takes place by acquiring knowledge through various sources and experiences, it is important to understand how technology assists in shaping education. The purpose of this paper is to (a) understand how use of educational tangible devices and screen-based communication mediums, such as tablets, leads towards experiential learning; (b) find how these devices are effectively imbibed by children while learning; (c) find the usability and effectiveness value of tablets and tangible devices in primary school education; (d) to seek answer to the question—what age is appropriate to introduce such devices?

The underlying posit for taking up these studies is the fair assumption that there is a limit to learning when predominantly screen based digital devices are involved. Children often tend to lose focus and attention when such devices become more of a challenge rather than mediums of instruction. Bigger the challenge more is the probability of children preferring to escape into games and music preloaded in such devices.

2 Literature Survey

Kolb [3] contributed an important theory of learning based on experience. His model divided the learning cycle into two experiences. Concrete Experience (CE)—feeling and Abstract Conceptualization (AC)—thinking, forms the grasping experience, Reflective Observation (RO)—watching and Active Experimentation (AE)—doing, forms transforming experience.

Beard and Wilson [4] presented a model linking experiential part of learning to intelligences. It consisted of three main components, (1) External Environment, such as outdoors, VLEs, lecture halls, etc.; (2) Sensors, like eyes, nose, ears, nerves, mouth, intuition, etc.; and (3) Internal Environment consisting of emotions, intelligences and various learning.

A study [5] done on tangible and augmented reality user interface systems provides a theoretical framework based on Kolb's learning model for cognitive development. It also explains the relation of physicality and embodied cognition on learning.

According to Howard Gardner's MI theory [6], intelligence consists of eight distinct types to which a person can relate their internal strengths and capabilities. These intelligences reflect as bodily-motions and use of other senses in children while they learn. Thus learning can be analyzed by observing these as markers.

Tablets are being adopted in educational settings at a pace quicker than research studies being published to validate and prove their pedagogical effectiveness. However, several works e.g. [5, 7] suggest that use of such digital devices promotes an involved and engaging experience for students. Various types of learning styles based on Kolb's learning cycle and modality theory have been studied [8] which suggest that if tablets are to be introduced, they should promote active-learning environment in class, which it lacks. A study [9] also mentioned about the distraction caused by tablets in classrooms.

Studies above indicate that consideration of educational epistemology is important before introducing tablets in primary schools. However tablets have a high potential for visual representation, which can be utilized for developing reflective observation and abstract conceptualization in students. Thus, integration of tangible devices and tablets can provide a concrete learning experience along with active experimentation. This paper presents a model by merging Kolb's learning cycle and MI theory, which can be used to qualitatively measure the effectiveness of Tangible devices and Screen-based learning mediums and can be helpful in designing them.

3 Methodology

Two prototypes were conceptualized and fabricated as shown in Fig. 1 for this study. First prototype was a musical application (app) installed on a Tablet. The app interface consisted of eight colored circles, which played musical notes when touched (Fig. 1a). The second prototype had eight colored capacitive sensors placed equally on the circumference of a bicycle wheel. When the sensors were touched or pressed, musical notes were generated through computer connected to it. The studies comprised of mixed activities that were assigned to different groups of students. All the activities conducted were video recorded and analyzed using the technique mentioned in literature [10] of interaction analysis.

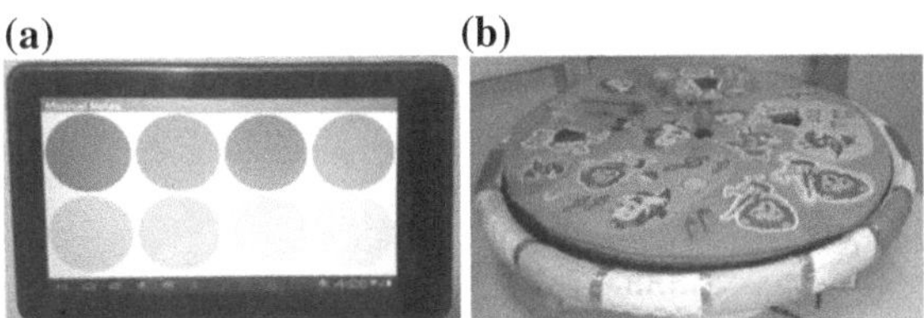

Fig. 1 **a** Musical notes app on tablet. **b** Round piano prototype

Twelve teachers and 25 children from 2 primary schools of Guwahati, India participated in the study.

The first two activities observed learning in children. Two groups of children of age 6-years from different schools were formed. These consisted of (1) five children having no prior exposure music lessons, (2) nine children who had taken music lessons. Musical notes were arranged sequentially on the tablet and randomly on the round piano. These groups were first taught musical notes on tablet and asked to reproduce them on the Round piano. This was done to make sure that learning had indeed happened.

In third activity, musical patterns in both prototypes were arranged sequentially. The devices were let loose in a class full of children. Physical interaction of five children of age 7-years that interacted with the prototypes for longer duration was considered for analysis. Children's behavior while they interacted with these prototypes were observed.

In fourth activity, time taken by children to learn on tangible device and tablet was measured. Six children, having prior exposure to music lessons of age 7-years from same class was chosen. Three of these children were taught musical notes on the tablet. Other three were taught on round piano.

4 Findings and Analysis

Analysis of collated/observed data was done based on the qualitative study of the field notes, audio logs, interviews and recorded videos. The following section highlights these findings.

4.1 Analysis of Interaction

Tables 1 and 2 give descriptions of children's interaction with Round piano and tablet along with the types of intelligences they used, as observed by authors. Dashed white circles have been used to highlight important actions in the pictures.

Table 1 The images provide a description of an atypical interaction analysis of tangible device let loose amongst children

Picture/clip	Description
	The boys exploring the device by touching stickers on top of it. One of them, tries to peep inside the device to find out how it works. Intelligence observed: musical, kinaesthetic, naturalist, interpersonal, spatial
	A boy experimenting with the device using his handkerchief by pressing colored sticker. Others watch him try it. Intelligence observed: naturalist, spatial

Table 2 An atypical interaction analysis of children interacting with musical app installed on tablet

Picture/clip	Description
	Children tapping the colored buttons of app interface to explore musical notes being played. Intelligence observed: musical, naturalist, kinaesthetic (limited to finger tapping), spatial
	A boy blocks the hand of another to interact with tablet. See dashed white round circle in the picture. He told others "we'll play it one-by-one". Intelligence observed: interpersonal

The round piano prototype encouraged collaborative activity amongst children. They were inquisitive to find how the device worked and enjoyed pressing the rubber tire and knocking it at various places to create sound. They experimented with the devices by touching the sensors with the handkerchief to check if it would still play musical notes. It was observed that children guided their class fellows by giving them direction to press stickers. All other videos were analyzed in similar way.

The children were excited to use the tablet and attempted to tap the screen and play with it. They tried to block each other's hands sometimes to interact with the device. In one video, it was observed that a girl held the hand of another boy and pushed it aside saying, "Let me try to play". Figure 2 (Left) highlights one such case.

Few children decided to take turns for interacting with the tablet. This strategy did not work for long as one or the other would break this flow and then everyone started tapping randomly. An interesting observation was that one of the boys held the finger of another and helped him to play the required pattern, see Fig. 2 (Right). This suggests that children can sometimes perform collaborative activity while interacting with the tablets.

The children were approximately engaged for 5–6 min in each cycle to interact with the round piano and 7–9 min to interact with the tablet.

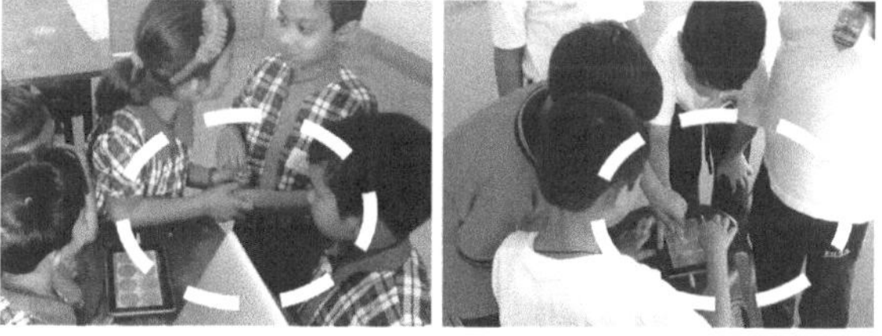

Fig. 2 A girl blocking the hand of the boy to have control over tablet (*left*). A boy assisting another boy to play the pattern of musical notes as we had asked (*right*)

Table 3 Response of the children when asked to describe their experience learning experience on these prototypes

Participants	Responses
B1	"I liked round piano as it was fun to play with…"
B2	"I liked tablet because we can install games on it and sound was clear. But I liked running around and playing with round piano."

4.2 Experience

This section highlights children's responses after they interacted with prototypes. The children were asked to describe which device they liked and why? Mixed responses were received from them. Two such responses have been shown in Table 3. The participants have been labeled as B1, B2… Bn, where (n = 12).

Most of the children reported that they liked the "sound" that was played from these prototypes. Many of them liked both the devices. Those who liked the round piano emphasized that they liked to "press" the tire and the sound it made. Those who liked the tablet told that they liked its sound, while one them said that he liked "touching" the screen as he could "see" what was happening.

4.3 Effectiveness of the Devices

Figure 3 shows the time taken by the participants to learn the musical notes. A1, A2 and A3 are the participants who learned from the Tablet. B1, B2 and B3 are the participants who learned from Round piano. Task 1 was to reproduce a simple musical pattern and Task 2 was to reproduce a complex musical pattern like a nursery rhyme.

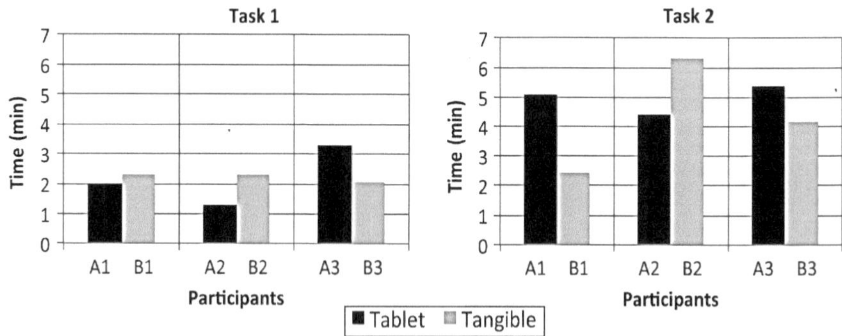

Fig. 3 Comparison of learning effectiveness of round piano prototype and tablet

4.4 Usability

Face-to-face semi-structured interviews of 12 teachers from primary schools were conducted. It enquired about the use tablets and tangibles devices in primary classrooms. The interview aimed at finding the motivation of teachers to use tablets, self-motivation of the students, lessons that can be imparted using tablets and tangible devices and their willingness to continue the usage of devices. The responses of the teachers have been labeled as T1, T2…Tn, where n = 12.

4.4.1 Tablets

None of the teachers had used tablets in the classrooms, but said that they utilized the smart-class, which provided them with all the audio-video aids for teaching. They stated that introduction of tablets would be "interesting" for students since they are familiar with mobiles and play games on them. Table 4 shows a few verbatim excerpts from the interview of the some teachers.

The teachers showed interest towards tablets and were aware of its capabilities as a teaching aid. They opined that students would be motivated to use them. However, they felt tablets would be a "distraction" for primary class children. They suggested introducing these devices from "class III or class IV" onwards when children start maturing. They posited that as teacher they would operate these devices rather than allowing primary school children to operate them.

4.4.2 Tangible Devices

On enquiring about teaching materials, teacher stated that they use learning materials, like workbooks, color pencils, chart papers etc. They said that toys are commonly used at pre-school level but not in primary. This could mean teachers associate and restrict toys as learning aids to pre-primary school level. Toys are seldom used for primary and higher levels. All 12 teachers said that such devices are good for "initial stages" and are "interesting to use for teaching". Some replied that the students were "inquisitive" to use this device (round piano) and showed "willingness to learn" from it. Four teachers said "we would like to have screen and audio-visual aids and round piano tangible device as compared to making use of

Table 4 Response of two teachers regarding the introduction of tablets in primary classes

Participants	Responses
T1	"…Students will be motivated to learn from them as they already play games on it. It will be better if these are introduced in class III or class IV, maybe higher, … but not for class I and II as they will only play games on them…"
T2	"…It will be better if tablets are introduced in class IV and class V as they become mature. In class II and I, they are too young…"

Table 5 Response of two teachers regarding the introduction of tangibles devices in primary classes

Participants	Responses
T3	"…children can learn by themselves…good this is that while one performs, the other watches…if this device is integrated with tablet, it will be very good…"
T4	"It will be good for initial stages of children and will be useful and interesting for teaching…"

black boards alone". However, most of them suggested that the round piano device should be more "engaging" for children. Table 5 shows the response of two teachers T3 and T4.

From the interviews, it was inferred that teachers were willing to use such tangible devices in classroom teaching. It was observed that teachers supported use of such devices in primary and pre-schools. It was pointed out that these devices should provide higher engaging experience for children or else they are likely to lose interest in them after one or two rounds of interaction. They also suggested that integrating tangible devices with screen-based devices could prove very helpful in teaching and learning.

5 Inferences

The video analysis reveals that the children utilize their senses and bodily-motion more when they interact with the tangible device. They try to explore these devices and collaborate with one another while playing with them. On the other hand, interaction with tablets is limited to individual finger taps on screen unless inter-device communication such as Bluetooth is made available. Since children tried to block each other's hand while interacting with tablet in groups, it can be inferred that tablets and mobiles, by their nature of interaction, encourage solitary absorption of self as opposed to collaborating with many.

It was also observed that the children took more time to interact with the Tablet as compared to the tangible device. Thus tangible device has to match the level of attention of the child to sustain their interest. This can be described in form of a relation (1),

$$\text{Length of Interaction} \propto \textit{Ability to interact} \tag{1}$$

Children also reported that they liked pressing the tangible device and running around it. Therefore, it can be inferred that sensory-kinaesthetic experiential value of physical contraptions with embedded intelligence, is much more as compared to the screen-based tablets.

The results indicate that tablets can be introduced in schools from class III onwards, i.e. from the age group of 8 years and above. Children in primary classes,

5–7 years of age, often tend to get distracted by the pre-loaded contents in the tablets such as games and videos. However, if tablets and tangible devices are integrated together, they can extend attention and experiential involvement increasing their usefulness as learning aids.

6 Suggested Framework

According to Piaget, children between the ages of 2–6 years begin to develop reflective observation and by 7–11 years abstract conceptualization starts developing [3]. Therefore, tablets can help in providing reflective observations in children through various visual representations, which in turn can lead them towards abstract conceptualization. On the other hand tangible devices can provide physicality, which leads to embodiment effect and learning [5]. Thus tangible devices help in providing concrete experience much more than the tablets. A combination of these leads to active experimentation. Intelligence is multifarious and dynamic in nature and is continuously interacting with our conscious and our surroundings. Since educational toys and games, when designed properly, embody various intelligences [6]; it is possible to involve various intelligences in children while they interact with the above devices. According to Piaget, intelligence is a continuous process of accommodation and assimilation tending towards cognitive growth; therefore it always leads to knowledge creation thus elevating the learning process. Figure 4 shows our conceptualized model of intelligences overlaid on Kolb's learning cycle stages. Intelligence is a vector normal to the learning stages, which have been represented on a Cartesian plane. Whenever a person moves from one stage to another, the intelligence vector acts perpendicular to it thereby accommodating and assimilating new knowledge in an individual. The type of intelligence that has to be nurtured at each stage of this cycle has to be embodied in the tangible devices, tablet apps and the activity to be performed using them. Thus it is possible

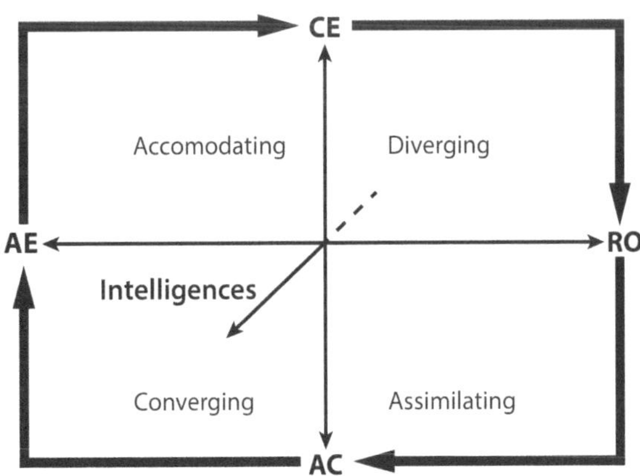

Fig. 4 Model proposed in this paper based on overlapping Kolb and multiple intelligence frameworks

to distribute the type of intelligences equally among the different quadrants of the learning stage to foster learning. Using the terminology from mathematics, the proposed model can be viewed analogous to vector cross-product or vector multiplication. Here, if different learning styles are considered as vector quantities, their cross-product will yield a perpendicular vector called intelligence. However, this is only an analogy and does not follow rules of vector algebra strictly.

Equation 2 shows an analogy proposed of the conceptualized model with the vector cross product:

$$\mathrm{LS_m} \times \mathrm{LS_n} = \mathrm{I_k} \tag{2}$$

where, *LS* are learning stages, *I* is intelligence, *m* and *n* are the type of learning stages and *k* is type of intelligence. Therefore, if body-kinaesthetic (*BK*) and musical (*M*) intelligences needs nurturing in a divergent learner ($m = CE$, $n = RO$), then substituting these parameters in Eq. 2 yields an analogy as represented in Eq. 3.

$$\mathrm{LS_{CE}} \times \mathrm{LS_{RO}} = \mathrm{I_{BK,M}} \tag{3}$$

Designers and educators can use this proposed model as a reference to conceptualize and design learning aids as well as activities for students so as to match their individual learning styles. This model requires further development.

7 Conclusion

The above study indicates that tablet-based application provide a way for the learners to conceptualize and visualize information. Tangible objects assist them to utilize all their bodily senses and motion—as recommended in the Theory of MI. Introducing screen—based mediums alone in primary classes can tend to distract the children from their normal learning regime, which may have future consequences for the children.

The study recommends introduction of tablets after the age of seven years in formal education. However, tablets can be used in collaboration with tangible objects embedded with intelligence, especially to aid visual representation component in learning content. A model based on combination of Learning and Intelligence theories has potential as a design heuristic.

Acknowledgments We are thankful to all the children and teachers, who became a part of our study. We would also like to state that due consent was taken to record videos and photographs.

We are thankful to Dr. Shobha Shashidhara of Ankura foundation, Bengaluru in giving her expert inputs on teaching practices of primary school children and sharing experimental devices conceived and used by her in her institution.

References

1. O'Malley, C., Fraser, D.: Literature review in learning with tangible technologies. Report for NESTA Futurelab, NESTA Futurelab (2004)
2. Antle, A.N.: Designing tangibles for children: what designers need to know? In: Extended Abstracts on Human Factors in Computing Systems (CHI EA07), pp. 2243–2248. ACM, New York (2007)
3. Kolb, D.A.: Experience as the Source for Learning and Development, pp. 20–38. Englewood Cliff, Prentice-Hall, New Jersey (1984)
4. Beard, C., Wilson, J.P.: Experiential Learning: A Best Practice Handbook for Educators and Trainers. Kogan Page, London (2007)
5. Chen, R., Wang, X.: Conceptualizing tangible augmented reality systems for design learning. In: Proceedings of Design Computing and Cognition (DCC08). Springer, Berlin (2008)
6. Becker, K.: How Are games educational? Learning theories embodied in games. In: Digital Games Research Association (DGRA05), p. 4 (2005)
7. Tse, E., Marentette, L., Ishtiaque Ahmed, S., Thayer, A., Huber, J., Mühlhäuser, M., Kim, S.J., Brown, Q.: Educational interfaces, software, and technology. In: Extended Abstracts on Human Factors in Computing Systems (CHI EA12), pp. 2691–2694. ACM, New York (2012)
8. Theys, M.D., Lawless, K., George, S.: Tablet computers and the traditional lecture. In: Frontiers in Education (FIE05) (2005)
9. Elbert, J., Code, J., Irvine, V.: iPad on practicum: perspective of a student teacher. Arbutus Rev. **4**(1), 1–18 (2013)
10. Jordan, B., Henderson, A.: Interaction analysis: foundations practice. J. Learn. Sci. **4**(1), 39–103 (1995)

Analysing the Innovation Growth of Robotic Pets Through Patent Data Mining

Teo Kiah Hwee, Mohan Rajesh Elara, Ricardo Sosa and Ning Tan

Abstract This study aims to illustrate the importance of patent data mining as a tool to analyse the innovation growth of robotic pets across time, countries and patent types. Such an approach differs very much from the traditional academic strategies which often technology centric focusing on validating incremental technical contributions with respect to a pet robotic system. To conduct this study, patents of robotic pets were retrieved and analysed. A period of 5 years, from 2009 to 2014, was defined as the time of reference to which the patents were extracted from the relevant patents database for analysis. The patent office of which the patents are registered and patent types were also used as indicators to further segregate the patents found in the database. The search gave a total return of 82 patents that were documented. Analysis done found that distribution trends indicates an overall increase in technology advancement made by companies, academic institutions and individuals from various countries to robotic pets. The results summarized in this paper is set to greatly benefit the policy makers in government agencies, and entrepreneurs to effectively manage the growth of this emerging market of robotic pet products.

Keywords Patent data mining · Pet robot · Robotic pet · Technology advancement · Assisted therapy · Growth

T.K. Hwee
Singapore Institute of Management University, Clementi, Singapore

M.R. Elara · R. Sosa · N. Tan (✉)
Singapore University of Technology and Design, Dover, Singapore
e-mail: ning_tan@sutd.edu.sg

A. Chakrabarti (ed.), *ICoRD'15 – Research into Design Across Boundaries Volume 2*, Smart Innovation, Systems and Technologies 35, DOI 10.1007/978-81-322-2229-3_31

1 Introduction

Animal Assisted Therapy (AAT) is a form of therapy that is deployed to help patient improve their social, emotional and/or cognitive functions. This is especially so in an aging population. However there are various difficulties when deploying such a therapy to a patient. As living animals are used in this therapy, certain issues such as safety of patient, maintenance for the animal and allergies to the animal's fur hinders the success of such a therapy. Therefore, in order to remove the difficulties and still be able to achieve the same success, the Artificial Animal Assisted Therapy (AAAT) was introduced. Thus this is also where robotic pets comes into play as they are designed and built to behave like the animal that it is supposed to represent.

Over the years many robotic pets have been developed [1, 2] and placed into the market [3, 4]. The different varieties and features allow users to choose the type that suits their purpose. However, in AAAT, one of the main purpose of the use of robotic pets is to provide companionship. Through companionship, it hopes to achieve the goal of the therapy. In order to come up with more variations of robotic pets and to have improved features on existing ones, innovation is one main ingredient that can't be missed. There are many firms in the world that deals with the creation of robotics. All of these firms compete to show that their technology in robotics are more superior to the rest.

If innovation is an important element of the success and future of robotic pets, it is imperative that studies must be done to reflect the importance of innovation to senior management of robotics firms that being innovative is one of the main aspect to ensure success in robotic pets. Such an effort towards mapping innovation growth through patent analysis has yielded positive results in other industries [5]. In order to show that innovation is an important fact, this paper will provide a study by gathering patent data related to robotic pets to depict how patent data mining can be used to explain the importance of innovation in the robotic pets industry and to show what does the future holds for the industry.

2 Study Methodology

To conduct the study in this paper, patent data were collected from the Google Patent database. Patents were gathered from different countries between the period of 2009 and 2013. Multiple keywords were also used, such as, pet robot, robotic pet, companion pet and companion pet robot, to gather relevant data on robotic pets. Manual analysis will then be done to make sure that the patent results are relevant to the subject matter and are within the scope of study. After doing individual searches on each keyword, the results show that a total of 82 patents that were filed and published over a period of 5 years from 2009 to 2013 and were relevant to robotic pets.

3 Results and Discussions

3.1 General Overview of Robotic Pet and Tools Derived from Global Patent Information

The results of 82 patents from the patent data search were compiled into a summary dataset and is illustrated in Table 1.

From the dataset, we can see that a total of 222 patents were filed and/or published during the period of 2009–2013. 28 patents were filed in 2010, which make it the year with the most patents being filed. The general overview of the whole dataset shows that there is a good spread of patent activities that span across the 5 years. This also indicated that the technology growth of robotic pet is on a steady pace.

3.2 Annual Patenting Activities

Using the results obtained from the search, an analysis was done based on the patents filing dates and publish dates. The two analysis were plotted in a graphical illustration to show the distribution of patents being filed and patents published over the 5 years. This analysis will provide the study with a better understanding of patent activities that occurred during the 5 year period.

From Fig. 1, we can observed that there was a steady number of patents being filed in the year 2009, with 27 patents and year 2010, with 28 patents. From this we

Table 1 Summary of dataset found on robotic pets

Time frame	2009–2013
Number of patents filed and published	82
Peak year	2010 (28 patents)
Top country	Japan
Patent offices	6
Years	5

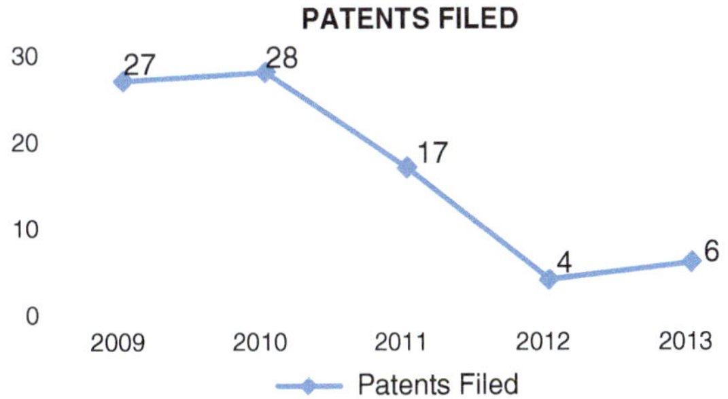

Fig. 1 Distribution of patent filed over time

can infer that the innovation of robotic pet technology has been explored even before 2009 and 2010. The filing dates of a patent are only an indication of when the patent application was submitted. It doesn't indicate that the technology was invented and explored then. Therefore, we know that inventive activities of robotic pets came about way earlier than the 5 year period of study. Despite the fact the study only covers a period of 5 years, it is sufficient for us to say that from the year 2009 onwards, the technology of robotic pet have showed great progress. Although according to Fig. 1 there is a rather steep decline in patent applications from 2011 onwards till 2012 and only slightly pick up on 2013, it doesn't mean that there was a decline in overall patent activities and technology growth decline. If we look at Fig. 2, it is observed that there was a peak in patents being published in the year 2011 and followed by a steady spread in 2012 and 2013 as well. Therefore, we can infer that there is no sign of a decline of patent activities from 2010 onwards. In fact, we can say that the two charts correspond to each other as from the manual analysis done on the search results, we observed that most of the patents filed during the peak period of 2009 and 2010, were then published in 2011 onwards.

In both Figs. 1 and 2, it can be observed that there was a pick-up of patent activities in 2013 for both patents filing and patents publishing. From this, it can be inferred that such growth in patent activities could only mean that companies are looking into bringing robotic pet technologies to a greater height, exploring new means to of improving existing robotic pet technologies. To support this claim, a search was done on the current market trend on robotics in general. From the search results, it was observed that researchers believed that there will be an influx of robotics product over the next 10 years and robotic pets such as the SONY AIBO dog will be more visible as well.

In this modern era where new technologies are surfacing at a very fast pace, many technology companies will want to have their creation and invention protected so as to avoid technology infringement from other companies. Therefore, companies put up patents for their technologies to acquire the protection that patents provide. All these are part of the Intellectual Property Rights (IPR). An increase in patent activities also means that companies are more aware of the IPR and are more afraid of their technologies being copied. With their technology

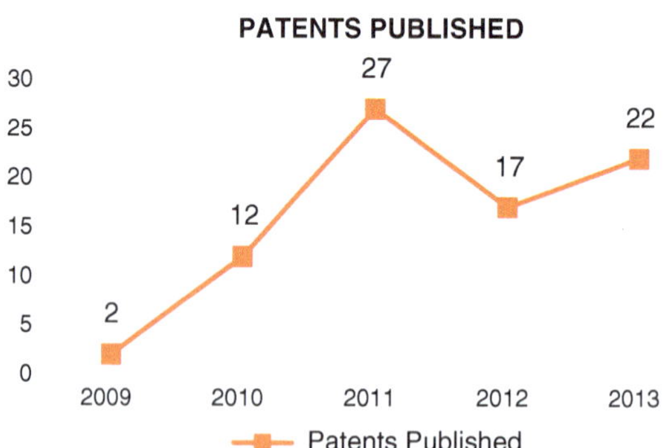

Fig. 2 Distribution of patents published over time

protected, they have every right to the technology and will be ready to file a law suit with any companies that are deemed to have infringed on the patented technology. A very good example of IPR at play would be on two of the largest smart phone companies, Apple Inc. and Samsung, where they have sued each other on multiple occasions on patent infringements. However, it is not an easy task to determine if a patented technology has been infringed as two different technologies could be very similar to a staggering point but at the same time are distinct in their own ways [6].

The increase in patent activities could mean that there is a higher demand in robotic pets by the market. Existing robotic pets would also need to improve to provide better features to accommodate to the demands of consumers. Many consumers are seeking robotic pets for features far more advance than before. Robotic pets that are currently in the market have features that far surpass those in the past. Robotic pet can now be used on different fields like therapeutic, companionship or military applications. Robotic pets like PARO the seal developed by the National Institute of Advance Industrial Science and Technology (AIST) is an example of a therapeutic robotic pet that administer animal therapy to patients in different environment [7].

3.3 Geographical Distribution of Patents

The understanding of the geographical spread of the patents activities will provide a better observation on the trends of companies acquiring patent protections on their technology to give a better insight of where their products would likely be marketed. Therefore, a graphical illustration was drawn to plot the geographical spread of patent activities in 6 different patent offices over the 5 years. Through the graph plotted, it will be able to show which patent office has the most number of patents filed in each of the 5 years.

Figure 3 illustrates the patent activities of different patent offices over the period of 5 years based on the publication date. The 4 trends were based on 6 choices available on Google Patent, of which, two patent offices, Germany and Canada were not shown on the graph because no patents were filed under the 2 patent offices. Also, International is not a patent office by itself. Patents results found under the International tab are formed by patents filed under patent office such as Japan or South Korea patent offices that has a much smaller patent database that is not an available choice in Google Patent.

From Fig. 3 we can observed that patents filed under the Europe patent office has remained at a steady pace over the 5 years. In the case of United States and China, it can be observed that patent activities started increasing from 2010 onwards, reaching a peak in 2011, then followed by a gradual decrease in 2012 onwards. However, for the case of International, where patents are filed and published in patent offices from Japan, Taiwan and Korea, with Japan having the most patents published. It can be observed that the trend was on a steady pace throughout the first 4 years and then followed by a steep increase in 2013.

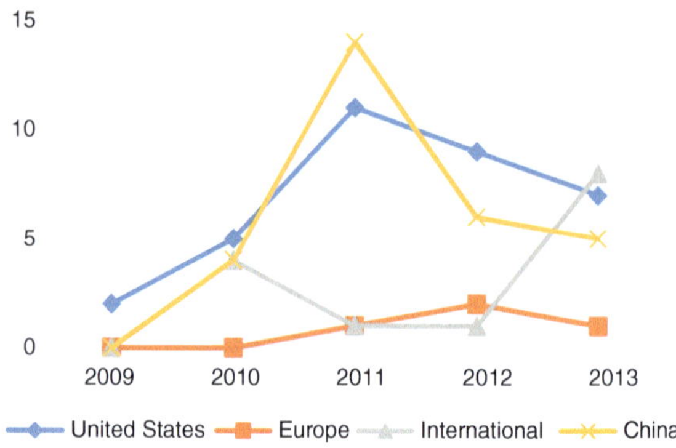

Fig. 3 Geographical distribution of patent activities

One reason for the different trends to show such a behavior is because, most companies, even those that are not of US origins used to file their patent applications with the United States patent office making them US or patents. This shows that the robotic pets market were dominant in the United States. However, the decline in US patents and increase in Japanese or Korean patents indicate that companies of these countries are slowly moving the market back into their own countries as well, allowing growth not just in United States, but within their own countries as well. From this, it can also be observed that quite a number technology companies have origins from Asian countries like Japan and Korea. For the case of China, the observation made was that there was a rather steep decline on patent activities in 2012 after its peak at 2011 with the highest number of patent activities recorded. Chinese technology companies have always been trying to develop their own inventions, trying to compete with foreign markets. This would explain the peak in patent activities in 2011. The decline in patent activities could mean that China is slowly moving its R&D focus out of robotic pets and more on other form of technology like smart phones. Thus, lesser patents activities were recorded in recent years.

Each patent office has their own patent system and each system provide different types of patents. Typically, a patent office will provide three types of patents, design patents, plant patents and utility model patents. Design and plant patents are available in every patent system in the world, whereas utility model patents are available in 77 countries with extensive usage in some European countries, Japan, China and South Korea. It is therefore, important to consider the available patent system in these patent offices when analysing their patent activities.

Utility model patents are usually more sought after because of the protection it provides for the utility and functionality of the invention. Most companies will want to prevent their competitors from coming up with a product that differs in their own design, but has the same functions. This type of patent is especially suited for small and medium enterprises (SME). To many SMEs, acquiring patent protection for their technology is a very stringent and costly process. However, they are not able to do away with patent protection for their technology. This is where utility model patents becomes a suitable choice for SMEs. The process to acquire a utility model patent is much less stringent as compare to other forms of patents. The cost of filing for such patents are also significantly lower. A utility model patent can be obtained 6–8 months upon application. This is well suited for SMEs who makes “minor” improvements and adaptations to existing products, who may not have met the inventive step requirement of a patent [8].

3.4 Comparative Analysis of Patent Output

3.4.1 Patents Filed in United States

Out of the 36 patents filed in the United States patent office from 2009 to 2013, only 19 % of the patents filed were by domestic assignees or applicants. The majority of patents filed in the United States patent offices were of foreign origins. Among the foreign applicants, Netherlands, Israel and China filed 3 % of the total patents, Hong Kong with 8 %, both South Korea and Taiwan filed 14 % and Japan who file the biggest percentage of patents at 36 %. Further segregation of the patents based on the type of applicants also indicated that 81 % of total patents filed are by commercial manufacturers, and 19 % are filed by individual inventors. Figure 4 illustrates the distribution in the United States patent office.

3.4.2 Patents Filed in Europe

All of the 4 patents filed in the Europe patent office over the 5 years are all of foreign origins. Of which 3 (75 %) patents are filed by commercial manufacturers, only 1 (25 %) patent is filed by an individual inventor.

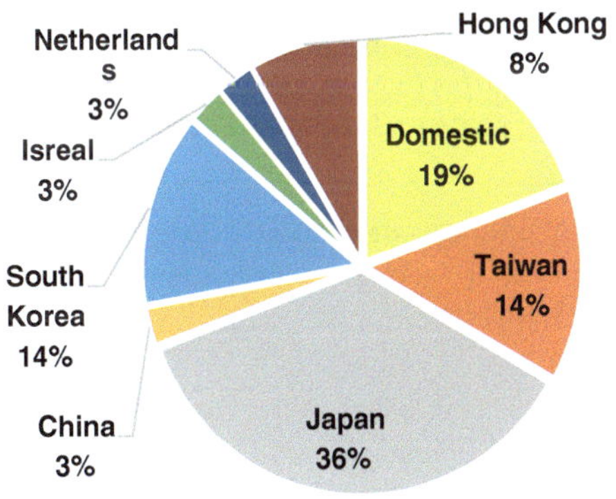

Fig. 4 Distribution of patent activities in United States patent office

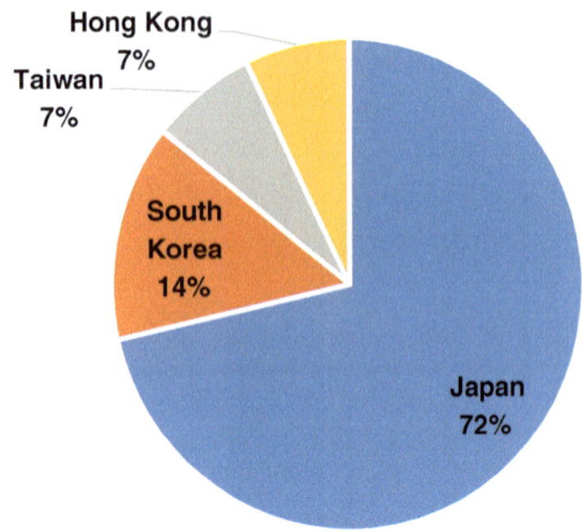

Fig. 5 Distribution of patents filed internationally

3.4.3 Patents Filed Internationally

Of the 14 patents that are filed internationally, 72 % of patents are filed in the Japanese patent office. 14 % of the patents are filed in the South Korean patent office. Both Taiwan and Hong Kong patent offices have 7 % each of the total patents filed internationally. 79 % of the total patents filed internationally are filed by commercial manufacturers, 21 % are by individual inventors. Figure 5 will illustrate the distribution of patents that are filed internationally.

3.4.4 Patents Filed in China

The total number of patents filed in China patent office 29. Of which, 20 (69 %) patents were filed by commercial manufacturers which included foreign applicants mainly from Japan. The remaining 9 (31 %) patents were filed by domestic inventors.

3.5 Technology Analysis

By understanding what kind of technology relevant to robotic pets are being patented, it will allow the study to better understand the direction of technology advancement for robotic pets. To understand this, technology analysis is done using the type of patents being filed in the different patent offices. 76 out of the 82 patents found were listed as patent of inventions, which according to United States Patent and Trademark Office (USPTO) is also referred to as the utility patent [9]. The remaining 6 patents are listed as design patents. Figure 6 illustrate the distribution of the patents based on their patent types.

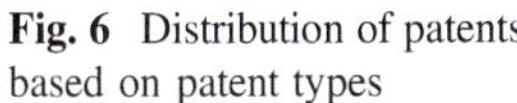

Fig. 6 Distribution of patents based on patent types

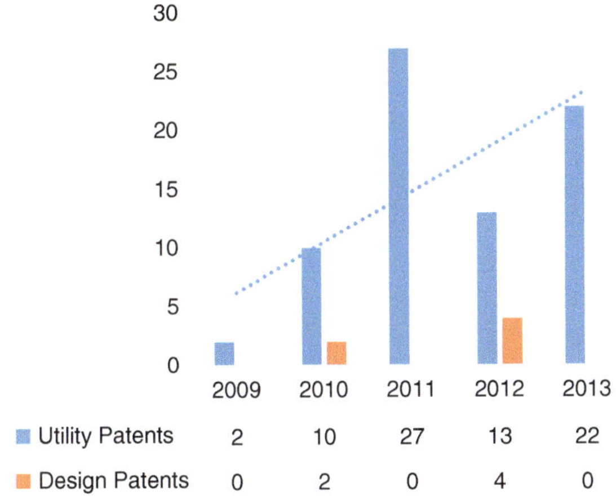

From this, it can be observed that over the given time frame of 5 years from 2009 to 2013, there are more utility patents filed as compared to design patents. The difference between the two is quite significant. The most number of utility patents filed were made in 2011. Through this observation, it can be said that technology advancement have been increasing steadily every year. Although there was a drop in 2012, it quickly pick up in 2013. This shows that technologies companies are constantly exploring into new technology advancements of robotic pets.

4 Conclusion

In conclusion, we can observe that the overall patent activities related to robotic pets have generally increased over the given time frame of 5 years from 2009 to 2013. There are a few factors behind the increase in patent activities over the years. One of the main factors that causes the increase of patent activities could be that there is an increase in demand for robotic pets by consumers. With the advent of computing, sensing and miniaturized electronics hardware, there is an increasing trend to accommodate more features into robotic pets and extend to new innovative application scenarios. From the patent analysis, we could validate that trend as the application domain seems to have shifted from purely entertainment to elderly companionship and therapeutic scenarios. Also, it can be witnessed that a vast majority of pet robot related innovations arise from Japan and/or Japanese companies. Patents activities made by Japan in the United States patent office alone took up 36 % of the total patent activities made throughout the 5 years, let alone the number of patents filed by them in other patent offices.

This paper presents our results from patent data mining experiences to analyze innovation growth of robotic pets over time, countries, and patent types. The work presented here is novel and differs entirely from the traditional technology centric validation studies in core robotics domain. With large number of national level initiatives across several developed countries towards developing robotics market aimed at handling aging societies, the results presented in this paper is expected to be highly useful for policy makers and entrepreneurs in managing this new market segment of robotic pet products. Future work would include deeper analysis of core technology orientation of considered countries, their commonalities and difference over years. Another possibility of future work would include study of correlation between the preferred morphological form and cultural orientation of the nations considered.

References

1. Sheba, J.K., Mohan, R.E., Garcia, E.A.M.: Easiness of acceptance metric for effective human robot interactions in therapeutic pet robots. In: 7th IEEE Conference on Industrial Electronics and Applications (ICIEA), pp. 150–155. IEEE (2012)
2. Tan, Y.K., Wong, A., Wong, A., Dung, T.A., Tay, A., Kumar, D.L., Dat, T.H., Ng, W.Z., Yan, R., Tay, B.: Evaluation of the pet robot CuDDler using godspeed questionnaire. In: Kobayashi, J.B.H., Abdulrazak, L.W.B., Mokhtari, M. (eds.) Inclusive Society: Health and Wellbeing in the Community, and Care at Home, pp. 102–109. Springer, Berlin (2013)
3. Saint-Aimé, S., Le-Pevedic, B., Duhaut, D., Shibata, T.: EmotiRob: companion robot project. In: The 16th IEEE International Symposium on Robot and Human interactive Communication. RO-MAN 2007, pp. 919–924. IEEE (2007)
4. Friedman, B., Kahn Jr, P.H., Hagman, J.: Hardware companions?: what online AIBO discussion forums reveal about the human-robotic relationship. In: Proceedings of the SIGCHI Conference on Human Factors in Computing Systems, pp. 273–280. ACM (2003)
5. Dulakakhoria, S., Jana, T.: Mapping innovation growth in the sports industry through patent data mining. J. Intellect. Property Rights **18**, 410–418 (2013)
6. Kuester, J.R., Horstemeyer, S.A., Santos, D.J.: New frontier in patents: patent claims to propagated signals. A. J. Marshall J. Comput. Info. Law **17**, 75 (1998)
7. Wada, K., Shibata, T.: Living with seal robots—its sociopsychological and physiological influences on the elderly at a care house. IEEE Trans. Rob. **23**(5), 972–980 (2007)
8. Beneito, Pilar: The innovative performance of in-house and contracted R&D in terms of patents and utility models. Res. Policy **35**(4), 502–517 (2006)
9. U.S. Patent and Trademark Office: Types of patents. Retrieved from http://www.uspto.gov/web/offices/ac/ido/oeip/taf/patdesc.htm (2013)

Studies in Application of Augmented Reality in E-Learning Courses

Mannu Amrit, Himanshu Bansal and Pradeep Yammiyavar

Abstract Previous Studies have indicated that specific concepts in chemistry education require visuospatial skills by students. This paper presents an attempt to extend Augmented Reality technology to school textbooks—particularly Chemistry. A qualitative study conducted with five high school chemistry teachers in India with the aim to identify existing pedagogical patterns related to Solid State Chemistry taught in high schools in India and attempt to design a device based application that will aid students' learning. The results were found to be concurring with existing literature in chemistry learning. On the basis of this study, design suggestions to be considered while designing AR based solutions in the field of solid state chemistry are proposed. Incorporating these suggestions, we then conceptualized and developed an AR application for mobile and tablet devices based on a popular platform. This application uses standard XII NCERT textbook images as markers/reference to augment dynamic 3-dimensional content. User testing of the application indicated its acceptance.

Keywords Augmented reality · E-Learning · Solid-state chemistry models

1 Introduction

One of the challenges of modern chemistry education is that while learning one has to address multiple levels of representation which requires interdependent, networked thinking of the students [1, 2] and establishing relationships between these multiple levels can be challenging for novice learners [2]. In our study, we aimed at addressing learning difficulties of Solid State chemistry and similar concept followed by proposing an Augmented Reality application design.

Solid State chemistry which is taught as the first topic in the 12th year of school (standard XII in high school chemistry in India) involves several concepts with

M. Amrit (✉) · H. Bansal · P. Yammiyavar
Department of Design, Indian Institute of Technology, Guwahati, India
e-mail: mannuamrit@gmail.com

A. Chakrabarti (ed.), *ICoRD'15 – Research into Design Across Boundaries Volume 2*,
Smart Innovation, Systems and Technologies 35, DOI 10.1007/978-81-322-2229-3_32

3-dimensional visualization of atoms and molecules. Other than spatial visualization skill, students also require to constantly switch between micro and symbolic levels to understand the concepts. Educational research of Potter et al. [3] concluded that spatial ability influences academic performance in engineering. Every student in the classroom doesn't necessarily have a good spatial visualizing ability. Previous studies have shown that students with lower visuospatial abilities are unable to perform well in solving spatial and non-spatial chemistry problems [4, 5].

Augmented Reality (AR) is defined as any system which combines real and virtual, is interactive in real time and is registered in three dimensions [6]. Literature [7] suggests that such immersive Augmented Reality tools increase motivation, contributes to better learning, and enhances the educational experience for students. AR offers an alternative way to see the abstract chemistry world and allows students to interact with the system and discover knowledge by themselves through "sensorimotor feedback" [8]. Overall, AR as an educational medium provides a great alternative environment for students to learn. There have been research projects like Studierstube and Construct 3d etc. [9, 10] towards application of AR in education. Chen [11] investigated student perceptions while using AR based tool and physical models to learn about amino acids. In contrast with these, we followed a user centric design process in the development of this e learning tool, by taking inputs from teachers and students before and after building the prototype respectively. We aimed at development of AR based e-learning tool for Solid State chemistry whose features, interactions, topics were chosen based on user expectations and needs, making our tool more useful and usable.

2 Application Design Methodology

2.1 Topic Chosen

We chose Solid States, which is the first chapter in Chemistry book of class 12th according to NCERT [12] course curriculum. This chapter deals with 3d arrangement of atoms of crystalline metallic, non-metallic elements and ionic and covalent compounds which need the students to understand the concepts sub-micro and symbolic level at the same time. More importantly, it requires students to visualize the atomic arrangement in 3d space which deals with Visio-spatial thinking capability of the students.

2.2 User Need Analysis

The primary user of our application is the student of 12th class studying Solid state chemistry. The secondary user was identified as the Teacher who teaches the chapter. With the aim to identify the problem points and additional needs of

Fig. 1 Teachers under interview in their respective school environment

teachers and students in terms of their learning/teaching and understanding the underlying concepts, a user survey was conducted (Fig. 1). Semi-structured in-depth interviews with five higher secondary class chemistry teachers currently involved in teaching 12th class students, were conducted. Out of these five teachers, 3 were interviewed in person and remaining two were interviewed through email dialogue as they were located in a distant place.

The interview questionnaire had six questions which are reported below under each question heading. In the semi structured interview having terms such as "what is the difference …" or "Do you feel there is difference ….." was adopted.

2.3 *Questions Categorized and Insights from Interviews*

(a) Difference between Solid States and other chapters:
Responses to this question were found to be quite consistent for all five interviewed teachers. They describe Solid States chapter as more demanding in terms of 3 dimensional visualization and imagination for students. One teacher said, "As solid state involves 3d concepts, it requires more visualization and imagination skills of the students". According to another teacher: "It gives help to understand 3-D structures of metals and Ionic Compounds. Visualization in 3-D is required." These feedback gives support to our assumption that there is need of 3d visualization aiding for students in Solid States and nurture our motivation to design a Augmented Reality based tool for the same.

(b) Division of chapters into different modules and sub-topics:
Based on the dialogues and discussions that were recorded and analyzed the following insight emerges.
As some teachers are more focused towards teaching school syllabus whereas other are focused towards teaching entrance exam syllabus, there are slight differences across teachers in the content and the modules in which the content is divided. Even though, there is similarity in terms in terms of teaching core

concepts of the chapter: different layer wise 3 dimensional arrangement of atoms, unit cells of Face Centered Cubic (FCC) and Hexagonal Closed Packing (HCP) and tetragonal and octahedral voids. We also asked from some of the teacher's most important points of emphasis during teaching the chapter. These insights helped us to choose spatial arrangement of atoms in unit cells and voids formed inside them as content for AR based pedagogical tool to start with.

(c) Relatively difficult topics to teach and learn:
Teachers find it difficult to make student visualize and understand the spatial arrangement of particles in 3d space in his/her imagination. One teacher informed, "For students it is difficult to understand 3d crystalline structure and where and how different voids are present inside the structures." From different structures couple of teachers found Hexagonal cubic packing relatively difficult to visualize and so to teach.
Solid States chapter contains other concepts as well e.g. Voids, Cation-Anion Ratio, Coordination number. There are numerical problems in these concepts. These concepts are associated with and extension of basic concepts of 3d structure arrangement and unit cells. According to one teacher, "Once 3d arrangement of atoms is clearly understood by the student, everything else falls into place and becomes easier." This resulting insight motivated us to start conceptualizing an AR application with spatial arrangement of atoms in unit cells and voids as instructional content to be resolved.

(d) Text book alone is insufficient:
Most of the teachers admire prescribed text books (NCERT) because of the content and instruction design. Though It helps students understand the crystalline structure with the help of colorful 2d figures. Students do not find it sufficient in terms of depth of content and its effectiveness in providing a clear 3d visualization of structures and lattices inside the void. One teacher stated, "Text books are good and there are some diagrams and explanations for 3d concepts but these are not clearly not sufficient." Teachers often search and refer to multiple books that have rich illustrations. The student however does not have access to such resources when studying on their own.

(e) Use of additional tools:
Although few teachers used to take help from physical ball and stick models, they expressed the shortcomings of such tools when probed further. 3d physical models could be difficult to make, store or carry and demonstrate. According to one teacher, "It is time consuming to make slides or use 3d models. There is non-availability of 3d models in market." Also, these models are just static 3d representation of one state of lattices. The models are not dynamically changing which is required for a student to understand the concepts. Animations are often relied upon but a teacher shared his views, "Unfortunately the videos and models are not very useful and user friendly so they also do not provide much help for teachers." Insight regarding use of animations that emerged is as follows: Rewinding and replaying the

animations is not of much help if the student fails to grasp the underlying concept the first time itself. Instead, we hypothesized that if we can have the visualization of the 3-D structure illustrate how a structure is formed step wise, it will help. It should be handy and simple to use." It was interesting to note that most teachers mentioned the analogy of classroom space to teach arrangement of atom in cubic unit cell and sharing among different unit cells.

3 Design Suggestions

In [13], authors have suggested five principles for designing chemistry visualization tools that help students understand concepts and develop representational skills through supporting their visuospatial thinking. On the basis of these principles and insights form teachers' interviews, we propose following suggestions to be considered while designing AR application for chemistry learning:

(i) Step wise instruction: One of the teacher told that visualization of 3d molecular structure formation should be step wise to make it clearer. In his fifth principle [13] also suggested to reduce cognitive load by making information explicit and integrated. Hence, rather than one 3d model for one concept multiple models arranged in appropriate flow through which student can step wise navigate, are advisable. We implemented this navigation using two on-screen GUI buttons, next and previous.

(ii) Content delivery style: Most of the previous works in AR based e-learning tools are solely relies on visual 3d content [9–11] whereas according to [14], animation and narration should also be presented in a coordinated way to ameliorate the learning of low spatial ability students. We incorporated background audio instruction and animated 3d models in our tool.

(iii) Touch based interaction: As the marker based detection technique requires image always to be present in front of the camera, it is not possible to see bottommost surface of the 3d model just by tilt interaction. Also, part of image can come out of the camera view during device tilting, causing disappearance of the 3d model. With the increasing popularity of touch based devices, users have become well versed with touch interactions and it should be used consistently in AR applications as well [15]. We added touch interaction to rotate 3d models along with tilt.

(iv) Direct interaction with content: Wu and Shah [13] proposed that there should be visible referential link among representations provided by the visualization tool. One of the biggest advantages of AR based e-learning tool is that there is direct connection between content in the tool and content in the textbook. Following this, we included virtual button feature in the tool which allows students to switch to a particular subpart of the content by moving the finger over image related to that subtopic in textbook page.

4 Development and Design

4.1 Development Platforms

Initially, D'Fusion studio [16], a GUI based cross platform SDK for building AR applications was explored. Scenario intelligence programming is done using Lua script. 3D rendered objects can be directly imported from Autodesk 3ds Max and Maya using exporters provided in its developer package. We were successful in augmenting 3d molecular structure over a black and white pattern marker. We also tried adding interactivity to it by changing the rendered supplement when two markers are brought nearby. However, during the course of our exploration with D'Fusion studio, we found certain technical issues such as flickering of 3D models and adding interactive elements etc. which compelled us to look for alternatives such as another tool named as Vuforia.

Vuforia [17] by Qualcomm is an Augmented Reality Software Development Kit (SDK) for mobile devices that enables the creation of Augmented Reality applications. Tracking in Vuforia is quite more stable in comparison with D'fusion. AR application can be developed by importing Vuforia SDK into Unity 3d [18] which is a cross-platform game engine with a built-in IDE and used to develop video games for web plugins, desktop platforms, consoles and mobile devices. Apart from providing Image tracking capabilities, Vuforia also gives developers the flexibility to add interactions through buttons, gestures, animation, sound etc. in the mobile application by programming in C sharp and Java script in Unity 3d. For generating 3D models, we used SketchUp [19] which can be easily imported in Unity 3d.

4.2 Application Features and GUI of Developed Solution

Virtual buttons in AR interfaces are developer-defined rectangular regions on image targets that trigger an event when touched or occluded in the camera view. Such buttons provide an intuitive means of interaction since the users are directly using the content (on paper/surface) to navigate/as a button rather than on screen buttons.

The graphical user interface of Augmented Reality Apps is primarily simple because a major chunk of screen space is dedicated to the camera for easy viewing. Any additional content that needs to be shown to the user is subsequently placed on layers above the camera layer. In this application, we have used two GUI buttons to allow users to navigate/toggle between different views of the same 3D model (Fig. 2). The models are placed in a chronological order—i.e. the next view of the model is obtained from the previous view.

To assist learning and provide instruction, audio feedback was added into the application to guide users through the flow of the application as well as help in instruction. A pause button to turn off these instructions has also been provided on the GUI in our design. We also incorporated animation in 3D models in cases where

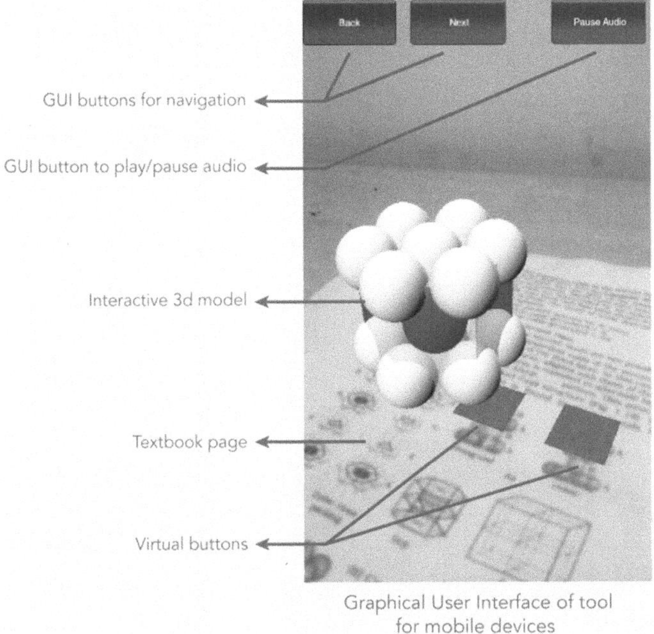

Fig. 2 Labeled GUI of AR application for mobile devices

we needed to portray positioning of one layer of atoms over another. We showed through animation how one layer fits over another and forms a complete lattice. This was based on insights gained from Teacher's interviews.

Another key feature of our application was on screen interactions. Users could rotate the models as per their convenience by swiping on the screen, in the direction of rotation. The swipe rotation also included inertia so that the rotation looked more natural i.e. upon swiping in a particular direction, the model rotated for a particular angle and then came to a smooth stop based on the speed of the swipe. A single tap on the rotating model also brought it to a halt.

We divided our teaching content into two modules, based on the content finalized through feedback from our qualitative research. Both these modules were on separate pages on the NCERT textbook, and thus, had separate image trackers. Content related to these modules was displayed on the tablet/mobile device upon successful tracking of the respective text book page (image tracker) by the camera of the tablet/mobile device.

The developed modules are:

(a) Understanding 3D Closed Packing Structure:

 (i) Hexagonal Close Packing
 (ii) Cubic Close Packing

Fig. 3 (**a**, **b**, **c**) Different GUI screens of AR application for tablet devices. **a** First 3d model: face centered cubic; **b** Octahedral void (OV) selected and navigated to edge centered OV; **c** Switched to tetrahedral void and zoomed by moving device closer to marker

(b) Understanding Voids:

  (i) Tetragonal voids
  (ii) Octahedral voids

Both these modules have similar information architecture. The difference is only in terms of content—i.e. 3D models and related audio feedback. The working of the application can be understood through the following steps with reference to Fig. 3.

(a) User is reading the NCERT textbook and comes across the concept of 3 dimensional closed packing.
(b) User turns on the application on his mobile/tablet.
(c) The home screen of the application is essentially live feed from the camera of the device. The user points the device to that particular page of the NCERT book.
(d) A 3D model is augmented on the device with audio feedback. Virtual buttons to toggle between hexagonal close packing and cubic close packing are also augmented on the device. This 3D model, consists of two layers of atoms in which placement of second layer, is shown through animation. The first layer is white in color while the second is in green. Different colors are used to differentiate between orientation of layers and easy understanding.
(e) The user points/touches the desired concept to be explored on the NCERT book.
(f) The virtual button triggers the animation and placement of third layer which is augmented on the textbook.
(g) User can also toggle between different 3D models of the selected concept (hexagonal close packing or cubic close packing in case of module 1 and tetragonal void or octahedral void in case of module 2) using on screen GUI buttons (back and next).

5 Testing

The prototype developed was subjected to a limited pilot testing involving high school students (Fig. 4). Qualitative testing by heuristics and response gathering was done. Five students of class 12 were involved in this qualitative testing. Some key insights from the testing are as follows:

Fig. 4 School students using the developed prototype on a mobile device

- All 5 Students were happy to see such a learning content. They wanted similar 3D explanations to all the examples in the chemistry textbook.
- Wow factor and non-familiarity with technology was found to be the major driving force behind initial feedback which was very positive.
- One student wanted content to be broken down to even smaller steps (atom joining atom instead of layer joining layer).
- When students were provided with prototype not having touch interactions, they pointed that rotation and movement of 3d models should be accessible through touch gestures as well.

6 Discussions and Conclusion

The initial intent of exploring the potential of new emerging technologies such as Augmented Reality in learning applications was achieved. From the Teachers' as well as students' responses in the limited Qualitative survey and study it can be inferred that A R holds tremendous potential to aid understanding and learning especially of concepts involving spatial characters and features. In our study, we aimed at addressing learning difficulties of Solid State chemistry and similar concept followed by proposing an Augmented Reality application design.

We submit that in the context of non-availability of online internet access widely in the hinterland of a country like India, AR can become a very economical tool to bring three dimensional learning experience to the student on a device such as a mobile or tablet independent of the net connectivity and based on simply pointing out the device towards a text book.

Acknowledgments We would like to thank all the teachers and students of Kendriaya Vidalaya, IIT Guwahati Campus for their help in agreeing to the survey and discussing the prototype solution. We would like to extend our sincere thanks to our teachers at Kendriya Vidyalaya, Oriental Tutorials and Concept Education, Guwahati; FIITJEE and Vidyamandir Classes, New Delhi for their valuable insights and support throughout the project. We would like to thank the participants for their informed consent to use their pictures. We acknowledge respective IPR of platforms and products which we used for academic research purposes on terms of fair use: (a) Unity 3d, (b) Total Immersion D' Fusion, (c) Qualcomm Vuforia, (d) Trimble Sketch Up.

References

1. Johnstone, A.H.: Macro- and micro-chemistry. Sch. Sci. Rev. **64**, 377–379 (1982)
2. Johnstone, A.H.: Why is science difficult to learn? Things are seldom what they seem. J. Comput. Assist. Learn. **7**(2), 75–83 (1991)
3. Potter, C., Van der Merwe, E.: Perception, imagery, visualization and engineering graphics. Eur. J. Eng. Educ. **28**(1), 117–133 (2003)
4. Bodner, G.M., McMillen, T.L.B.: Cognitive restructuring as an early stage in problem solving. J. Res. Sci. Teach. **23**(8), 727–737 (1986)
5. Carter, C.S., LaRussa, M.A., Bodner, G.M.: A study of two measures of spatial ability as predictors of success in different levels of general chemistry. J. Res. Sci. Teach. **24**(7), 645–657 (1987)
6. Azuma, R.T.: A survey of augmented reality. Presence **6**(4), 355–385 (1997)
7. Pantelidis, V.S.: Reasons to use virtual reality in education. VR Schools **1**(1), 9 (1995)
8. Shelton, B.E., Hedley, N.R.: Exploring a cognitive basis for learning spatial relationships with augmented reality. Tech. Inst. Cogn. Learn. **1**, 323–357 (2004)
9. Szalavári, Z., et al.: Studierstube: an environment for collaboration in augmented reality. Virtual Reality **3**(1), 37–48 (1998)
10. Kaufmann, H., Schmalstieg, D., Wagner, M.: Construct3D: a virtual reality application for mathematics and geometry education. Educ. Inf. Technol. **5**(4), 263–276 (2000)
11. Chen, Y.: A study of comparing the use of augmented reality and physical models in chemistry education. In: Proceedings of the 2006 ACM International Conference on Virtual Reality Continuum and Its Application, pp. 369–372. Hong Kong, China, 14–17 June 2006
12. National Council of Educational Research and Training (NCERT): http://www.ncert.nic.in/index.html
13. Wu, H.K., Shah, P.: Exploring visuospatial thinking in chemistry learning. Sci. Educ. **88**(3), 465–492 (2004)
14. Mayer, R.E.: Multimedia Learning. Cambridge University Press, Cambridge (2001)
15. Nielsen, J.: Ten Usability Heuristics (2005)
16. Total Immersion D' Fusion Studio: http://www.t-immersion.com/products/dfusion-suite/dfusion-studio
17. Qualcomm Vuforia SDK for Android: https://developer.vuforia.com/resources/sdk/android
18. Unity 3d, free version: https://unity3d.com/unity/download
19. Sketch Up by Trimble: http://www.sketchup.com/download

Mobilising Design for Development: An Analysis of a Human-Centered Design Process Used for a South African mHealth Student Project

Fatima Cassim and Nina Honiball

Abstract This paper explores the use of a human-centered design process by university students to develop effective communication solutions using mobile technology as part of their communication design training. To this end, the paper presents Diamobi, a mHealth student project that proposes an educational mobile system to help spread basic healthcare information related to diabetes within a South African context. The analysis is theoretically informed by the design research map developed by Elizabeth Sanders and considers contemporary design approaches and mindsets relevant to the field of social innovation. Through the analysis, the paper sets out to illustrate how the students employed contemporary design approaches such as participatory and user-centered design and related methods such as personas, mapping and user-journeys throughout the project. Furthermore, the paper discusses how an understanding and implementation of these different design approaches can help communication design students to be more mindful of their role as socially responsible designers when developing mobile communication solutions specifically within a developing context.

Keywords Design education · Human-centered design · Mhealth · Design for development

1 Introduction

The theme of the conference, Design Across Boundaries, alludes to the fact that contemporary societies are faced with a number of boundaries. Very often these boundaries are dictated by societies' and individuals' social and economic standings. In other instances, boundaries may also exist owing to geographic location (matched with a lack of adequate infrastructure such as transport). The effects of

F. Cassim (✉) · N. Honiball
Visual Arts Department/Information Design, University of Pretoria, Pretoria, South Africa
e-mail: cassim.fatima@gmail.com

A. Chakrabarti (ed.), *ICoRD'15 – Research into Design Across Boundaries Volume 2*, Smart Innovation, Systems and Technologies 35, DOI 10.1007/978-81-322-2229-3_33

these different types of boundaries can be quite severe and have implications on the access people have within these boundaries. Here, access refers to a wide range of rights and services that all relate to some extent to the well-being, development and flourishing of humans. Owing to these boundaries, increasing attention is currently being paid to the use and value of mobile devices to address developmental issues such as healthcare and education; for example, the World Health Organisation (WHO) [1], explored "mHealth: new horizons for health through mobile technologies" in a report of the same name in 2011. The small letter m is often used as a prefix to denote areas of practice and research that focus on mobile devices; for example, the use of mobile phones to address and achieve health outcomes, is referred to as mHealth. However, there is no standardised definition to date for the term mHealth.

The immediate communication potential of mobile phones is what differentiates it from their technological counterparts. According to Bentley and Barrett [2], mobile devices serve "as the default connection to people in our lives" and "this interaction is quite different from how people interact with their computers." As such, the new technology opens up innovative spaces for communication, has potential for civic engagement and action and requires complex and different interactions. Within the context of healthcare, the WHO [1] also recognises this stance by affirming that "with increased accessibility comes the possibility of greater personalization and citizen-focused public health and medical care." What this implies for designers is that they need to provide more participatory design solutions which put the power in citizens' hands to address and manage their well-being.

Within an education context, technological advancements and the changing nature of design products also impact on the way that design students need to be trained, especially when considering design for social change such as mHealth. As design educators, we believe that communication design students should not only embrace new technologies as a medium for social innovation but like interaction designers for example, they need to understand and adopt human-centered design methods while doing so in order to arrive at appropriate and meaningful design concepts and/or solutions. In light of this stance, this paper considers Diamobi, a design for development project by Information Design students at the University of Pretoria, as a case study. Diamobi, a mHealth design solution, proposes a mobile system to give South African citizens basic healthcare information related to diabetes. The case study is theoretically aligned by positioning it on Elizabeth Sanders' [3, 4] design research map which illustrates a variety of design tools and processes in order to better understand the expanding scope of contemporary design practice and research. Sanders has been an influential voice in design discourse owing to the many tools, techniques and methods she has introduced to inform and advance human-centered design practices. Her intention for the creation of the map was to understand the main characteristics of the respective design processes and to visualise their relationship/s with each other. For similar reasons, this paper refers to the map to first define human-centered design processes before it discusses their use and considers their value specifically for communication design students when they learn to design mobile communication solutions for developmental purposes.

2 Mapping Contemporary Design Processes in Theory and Practice

Contemporary design discourse and design practice have broadened significantly both from within the respective disciplines and outside them as well. No longer is design relegated to a secondary position where designers merely offer value on an aesthetic level. Nowadays designers are called upon for their intuitive and abductive ways of thinking in order to address many of the complex, social problems that societies are faced with. The call for socially responsible design has a long history but gained significant impetus during the 1970s with the publication of Victor Papanek's book, *Design for the real world* [5], which served as a seminal text for designers and encouraged them to consider the social and ethical dimension of their work. Since then, there has been an on-going shift towards socially and environmentally responsible design and this focus in design is currently widespread and traverses all design disciplines. What this implies is that today designers seldom work in isolation and often find themselves as part of multidisciplinary teams who design products and services that focus specifically on the needs of the end-user. Similarly, end-users have also become key players in the design process and this practice has facilitated and advanced social and human-centered design.

According to William Drenttel (in [6]), former president of the Winterhouse Institute, "social design defines a new kind of designer." Furthermore, Drenttel [6] believed that social design "needs to be expansively conceived beyond trained designers to include end users and social participants." There are two considerations which stem from his viewpoint. Firstly, social design is not a new discipline that has emerged. Instead, it is a way of thinking or a mindset that is altruistic in nature. Secondly, and not surprisingly, contemporary design is increasingly becoming a collaborative activity which not only considers users but also focuses on co-creation with end-users and other stakeholders. Depending on the nature of the interaction with end-users, there are varying design processes which can be used, such as user-centered design and participatory design. As stated in the introduction, these processes have been visually mapped by Sanders. It must be noted here that although Sanders is not the exclusive voice on the topic, the visual accessibility and holistic nature of her maps provide a point of departure for elaborating on and explaining these relatively new design practices for purposes of the paper.

The map (Fig. 1) comprises two axes: the y-axis represents approach and the x-axis represents mind-set [3]. The x-axis is a kind of continuum which moves from an expert mindset on the left to a participatory mindset on the right. According to Sanders, moving from the left to right represents "a shift in attitude from designing for users to designing with users" [3]. It is this shift in attitude which differentiates user-centered design from participatory design.

In user-centered design, positioned in the top left quadrant of the map (Fig. 1), a designer adopts the role of a "user advocate" by learning more about the user and translating this information in a way that is accessible and usable for other designers to improve their work and design contextually appropriate products [3]. Bentley and Barrett [2] share this viewpoint by defining user-centered design as a process that "emphasizes actual, real-world human experience as the primary and necessary inspiration for application design. Generally, it follows a recursive pattern from observation to design to implementation." As such, the end user is not a member of the design team and the team's focus is primarily on the end-product of the iterative design process. Participatory design on the other hand, positioned in the top right-hand corner of the map (Fig. 1), involves end-users to a greater extent and is a process of co-creation where the end-user is a key member of the design team [3]. Bjögvinsson et al. [7] indicate that there are two types of values that guide the practice of participatory design: the first value is that of democracy which favours "legitimate user participation" and the second value is the importance of the tacit knowledge of participants so that they can practically apply their intuitive thinking skills in the creation of meaningful and desirable products. These values have implications for the generative phase of the design process where initial product ideas begin to take shape. Although user-centered design and participatory design adopt different mindsets it should be noted that the two processes share common ground because they are both human-centered, largely pragmatic in nature and are applied in real world contexts. For purposes of this paper only these two design processes are discussed. However, this selection does not imply that these processes are superior to any others that appear on the map but it is clear that they are the two dominant processes across the four quadrants.

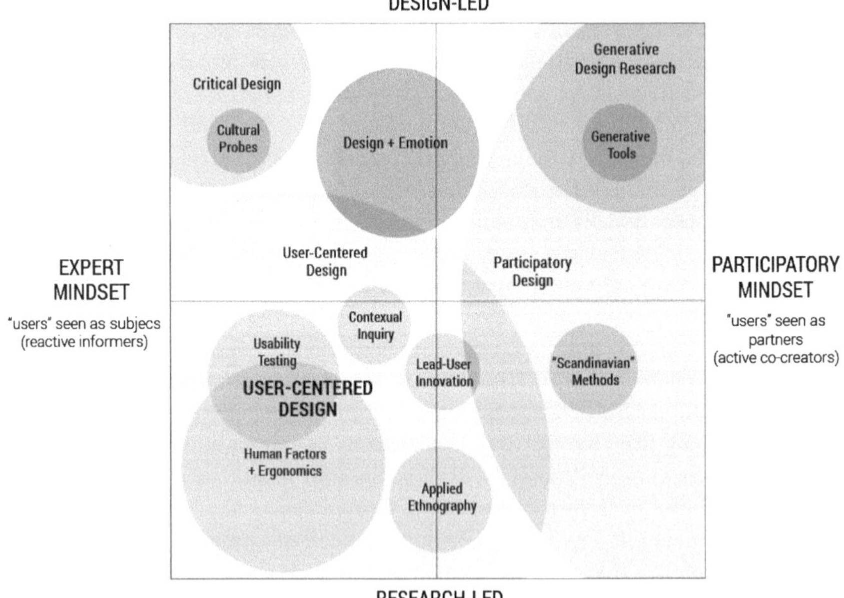

Fig. 1 "Map of design research" [4]

3 Case Study: Diamobi

Diamobi is a student project proposal for a mobile solution that offers mHealth information to patients suffering from diabetes. The project was completed with third year BA Information Design students at the University of Pretoria, South Africa, in August 2011. The project was implemented in an attempt to introduce communication design students to the field of interaction design by providing them with a real world opportunity to be placed at the front-end of a social innovation project. At the same time, the curricula-related project provided students with an opportunity to grasp the potential and value of designing for hand-held devices, specifically mobile phones. To include an industry partner, the project was delivered in conjunction with staff members from the Praekelt Foundation (a non-governmental organisation based in South Africa). The objective of the Foundation is to incubate, develop and implement various design solutions related to mobile technology in an attempt to improve the health and well-being of people living within limited means.

For pedagogic purposes, the project offered the Information Design students an opportunity to work in groups and engage with an iterative design process. The project consisted of three phases: phase one encompassed information gathering and visual research; phase two involved idea generation; and finally, the third phase looked at interface design and asked of students to carry out basic user-testing to introduce them to the practice of prototyping. Drawing on a human-centered approach to design, at different points during the project, students adopted both participatory and user-centered design mindsets as they are conceptualised by Sanders.

As a point of departure, the design brief invited students to identify a specific social issue related to healthcare in South Africa. Working in groups of 3–4 students, students were required to research the complexities surrounding family healthcare in a previously disadvantaged community in Pretoria and find a way to empower the community and/or health system in question through the use of low-cost mobile phones. Diamobi, conceptualised and designed by a group of four students is one successful example of the project. The authors acknowledge the limitation of only referencing one case study but owing to the scope of the paper, Diamobi was chosen to illustrate how contemporary human-centered design processes were put into practice by communication design students.

During the first phase of the project, the group of four students visited various state clinics in and around the city of Pretoria. The students were required to immerse themselves in the real context of their research. Not knowing much about the topic, students were reminded not to adopt an expert mindset but rather to approach the different people they met at the clinics as potential participants and co-investigators. Accordingly, students conversed with doctors, nurses and patients at the clinic, spending time at the clinic with the staff to observe how the clinic operated. As part of the project brief, students were expected to reflect regularly on their insights and experiences in order to identify a design challenge that would be appropriate to resolve with a mobile handset. Whilst doing their field research work, the students were also required to do secondary research and compile a short project report that looked at existing online literature related to challenges faced by state clinics in South Africa and identify any relevant case studies related to mobile technology and primary health care that could inform their thinking. In this way, students were provided with an opportunity to approach the project more holistically by combining secondary research with primary research through immediate engagement with a number of participants in a real world context. The following comment by one of the four students, speaks to this point: "One of the most important aspects we experienced on a daily basis was the confidence to simply ask for advice or information rather than endless 'googling' sessions."

After a week of rigorous field research and information gathering the students reached a turning point when they spoke to medical students working at the Chris Hani Baragwanath State Hospital in Johannesburg. One medical student specifically pointed out that patients often need to return to the hospital and queue for long hours to get basic health care information related to a particular chronic disease that they are suffering from. The students saw this as a research gap and after extensive brainstorming made an informed decision to use diabetes as their focal point and used this insight as the starting point to address the project brief. Were it not for their participatory mindset and the resulting engagement with the medical student, the students acknowledged that they would not have arrived at this significant point in their design process. Equipped with this new-found motivation to engage with relevant key players on the topic of diabetes, students also identified and contacted the Roche Diabetes Care Centre in Johannesburg. They engaged in e-mail correspondence with relevant members of staff at the Centre to help them further understand the various complexities related to diabetes and the students even considered The Centre as a possible client for the mobile solution they were developing.

Ultimately, the student's fieldwork and other research resulted in the conceptualisation of Diamobi, a design proposal for an mHealth application that would enable a variety of people to have easy access to basic healthcare information regarding diabetes. As a mobile solution, Diamobi needed to be a free mobile solution that could be implemented as an USSD system for basic feature phones and mobile application for smartphones, offering the end user the ability to select their local language and access the information in a format that is clear and accessible to them. "USSD stands for Unstructured Supplementary Service Data and is a Global System for Mobile (GSM) communication technology that is used to send text between a mobile phone and an application program in the network. Applications may include prepaid roaming or mobile chatting" [8]. The reason for this dual implementation was to allow the design solution to be available for people across the different income brackets common in South Africa.

During the second, ideation phase of the project, the project brief shifted to involve more user-centered design processes as those in the upper left quadrant of Sanders' research map. The four students were required to clarify their project aim and conceptualise their project as a mobile solution that could offer Diabetes patients access to a basic selection of information ranging from medical facts about diabetes, where to get treatment and support, information regarding diet and exercise and access to emergency numbers for help. During this phase, students were encouraged to use three design processes commonly used in the field of user experience and human-centered design, namely personas, mapping and user-journey design. Students created personas of potential users, based on people they had interviewed and met during their field-research phase to better understand the needs, motivations and technological constraints of the identified user groups. Technological constraints in this instance hints at user limitations, for example language barriers, as well as technological and financial limitations owing to a lack of access to a stable internet connection, limited airtime for making and receiving calls and some users having access to basic feature phones without internet capabilities. For the mapping process, students were asked to create a diagram of the existing situation that their service would address and indicate how their idea would work in conjunction with the existing process in question. This helped the students to finalise their ideation process. Once the mapping was completed, the students were required to work with a process or technique called user journey design where they visualised the entry-points and touch-points different users would need to have with their service (Fig. 2). Although new to the students, the use of these three design processes proved to be an invaluable learning experience as they required students to carefully plan and conceptualise their idea in order to arrive at an end-product that would be relevant, accessible as well as financially feasible.

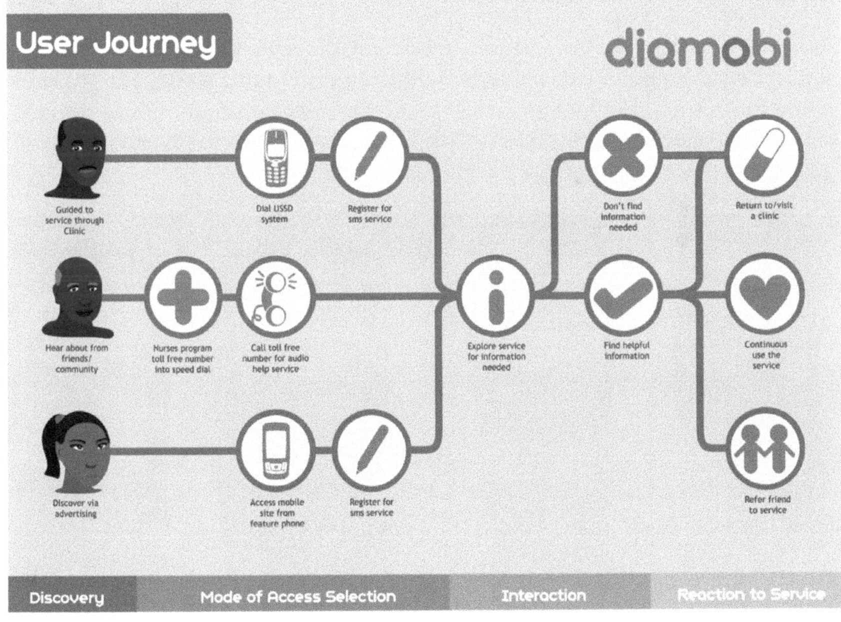

Fig. 2 Example of user journey design [8]

During the third phase of the design process, students were introduced to the process of wireframing to conceptualise their mobile solution for a range of mobile phones. Through observation and discussions students were given an opportunity to test the usability of these wireframes, using a method called paper-prototyping to experience first-hand how people would interact with their system. This process enabled the students to identify any challenges or stumbling blocks that their wireframes presented. It should be noted however that the project brief did not require students to develop real prototypes because as communication design students, they did not have the technical know-how and in addition, the three week timeframe of the project did not allow for such skills development and application. Parallel to the paper-prototyping process, students were guided through a traditional communication design phase to develop a proposed look and feel for the branding and user interface design of their product (Fig. 3). The design phase included coming up with a name for their service, designing a logo and doing a more polished design mock-up to visualise selected screens of their mobile design solution. After the completion of the project, all the student groups presented their ideas to a selection of staff members from the Praekelt Foundation, the objective being that the Foundation would select a winning idea. The Diamobi group was selected by the Foundation as being the most innovative and feasible design solution and the four students were offered a month long internship to work alongside strategists and engineers at the Praekelt Foundation's headquarters in Johannesburg to gain industry experience in this field of design.

From an education perspective the Diamobi project offers many insights, especially when considering the contextual complexities within which the project evolved and the opportunity that a design brief of this nature offered the learners, local health organisations and also future investors looking to identify relevant ideas to incubate for mobile platforms within the developing world. The project was a first for this group of design students and therefore significant in that it asked of students to stretch the application of their design skills beyond the mere making of

Fig. 3 User interface designs created for low- and mid-end phones [8]

the visual and placed a significant emphasis on participatory idea generation and putting the theory of human-centered design into practice in the "real world". In this sense, the project then became a vehicle for knowledge around a developing cause that invited the students to move from being passive observers to becoming active collaborators; the project even encouraged the students to engage more directly in the thinking, unthinking and rethinking of the social challenge at hand.

4 Discussion and Conclusion

In terms of its positioning on Sanders' research map, the Diamobi case study indicates the students' use of different human-centered design processes and methods to develop mobile solutions for change. One of the key observations, when considering the application of the theory to the students' practice, is that the quadrants on Sanders' map are not rigid as each of the design approaches and mindsets can be complementary rather than independent of each other. Sanders [4] states that "[i]t is difficult for many people to move from the left to the right side of the map (or vice versa), as this shift entails a significant cultural change." However, the case study suggests the fact that within an educational context a project brief of this nature enabled fluidity in practice where one mindset/approach informed and/or evolved into the other. Between the first two phases of the design process, the project moved from a participatory, research-led approach to a more user-centered, design-led approach. For the students, the shift was not so much a cultural change as a change in mindset about the different roles a designer plays throughout the process, from co-creator and facilitator to communication design expert (for purposes of the interface design). Although the extent of engagement with the end-users was limited within the timeframe of the project, it made students aware that human-centered processes do not disregard the people to whom solutions are aimed at. As such, the project instilled a greater sense of responsibility in students.

Students' reflections about their Diamobi project indicate that the projects gave the four students a good opportunity to see themselves not just as communication designers, but also as researchers and strategic thinkers and to understand the integral role that a human-centered approach plays in creating solutions that have social value. During the first part of the project, the strong focus on information gathering and research, together with the rigorous process of documentation (both visual and verbal) proved to be extremely valuable as it enabled students to identify a real need for which to design. It also taught the students the value of designing with people as opposed to designing for people. During the second and third phase of the project, the strong focus on user-centered design processes necessitated that students engage with new ideation and design tools such as personas, user journey design and wireframing to broaden their scope of practice. This also contributed to their learning curve because they were exposed to methods that are more commonly used within the context of interaction design and strategic thinking. This in turn opened up a new horizon for a number of students who found a strong inclination towards social design and/or interaction design as a potential career path. For those students who had a preference for research, the project sparked increased interest in contemporary design discourse on human-centered design, socially responsible design and design education.

In addition to the project having an impact on the students, there were teaching implications as well in that the communication design lecturers learnt from the teaching experience and were able to develop similar approaches in future developmental projects where students need to embrace and work with mobile technologies as a digital design medium. Technology continues to evolve and as such a teaching opportunity of this nature allowed the educators to keep the BA Information Design curriculum relevant to changing practices in the industry.

Another insight from the project is that the way it brought different people together effectively reflects the multidisciplinary nature of human-centered design processes. More specifically, the Diamobi project took the communication design students out of their comfort zone and asked of them to engage with people outside of their discipline. Although multi- and inter-disciplinary collaboration is inherent in contemporary design practice, implementing and managing such collaboration is often challenging within a South African design education context. Therefore, part of the success of the project discussed lies in bringing together key players related to the chosen topic (lecturers, students, health professionals, design professionals and members from the general public). This is supported by one of the student's reaction to the project; the student arrived at the clear realisation that asking for advice and collaborating is much more valuable than "googling" and working in isolation. The multidisciplinary focus of the project also resulted in a mutually beneficial output whereby the design students could offer an insightful design proposal to both Roche and the Praekelt Foundation, and in exchange the quality of the project's proposal was greatly informed by the input and support given to the students by both organisations respectively. In this way academia and industry overcame barriers by entering into a synergistic partnership and this engagement makes visible the innovation firm IDEO's belief that "[a]ll of us are smarter than any of us" [9]. In essence, this is the underpinning, collaborative ethos of human-centered design which students were meant to grasp through their experience.

In conclusion, this paper serves as a springboard to encourage and reflect on the incorporation of human-centered design approaches within a communication design curriculum to address developmental issues such as healthcare within a South African context. To this end, the research map by Sanders provides a useful starting point for students to visualise and understand various design mindsets and approaches to match them with their educational and ultimately, professional practice. Sanders openly calls for people to use her map and therefore new projects may in turn inform the map with new approaches and design tools with the intention to stay abreast of emerging trends and practices in design research. Ultimately, the intention is to nurture designers who pursue new technologies in responsible and accountable ways in order to affect change in the real world.

Acknowledgments The authors are grateful to Praekelt Foundation for lending their professional expertise and supporting the student project. The authors also express their gratitude to the subject coordinator and the students who gave us permission to use their project as a case study.

References

1. World Health Organisation: mHealth: New horizons for health through mobile technologies. [Online]. Available: http://whqlibdoc.who.int/publications/9789241564250_eng.pdf (2011). Accessed 5 May 2013
2. Bentley, F., Barrett, E.: Building Mobile Experiences. MIT Press, Cambridge (2013)
3. Sanders, E.: From user-centered to participatory design approaches. In: Frascara, J. (ed.) Design and the Social Sciences: Making Connections. Taylor & Francis, London (2002)
4. Sanders, E.: An evolving map of design practice. Interactions **15**(6), 13–17 (2008)
5. Papanek, V.: Design for the Real World. Collins, London (1974)
6. Shea, A.: Designing for Social Change: Strategies for Community-Based Graphic Design. Princeton Architectural Press, Princeton (2012)
7. Bjögvinsson, E., Ehn, P., Hillgren, P.A.: Design things and design thinking: contemporary participatory design challenges. Des. Issues **28**(3), 101–116 (2012)
8. Rouse, M.: USSD (Unstructured Supplementary Service Data), [Online]. Available: http://searchnetworking.techtarget.com/definition/USSD. Accessed 26 Nov 2013
9. Brown, T.: Change by Design. Harper Collins, New York (2009)

A Methodology for the Analysis of the Influence of Odours on the Users' Evaluation of Industrial Products

Marina Carulli, Monica Bordegoni, Umberto Cugini and Ding Weibin

Abstract The sense of smell has a great importance in our daily living: today pleasant odors are used to elicit positive emotions in users. In the marketing area, a lot of works have been done concerning the use of odors for communicating information for products as household cleaners and foods. In the area of Virtual Reality (VR) several researches have focused in presenting odors in virtual environments. The introduction of odors simulation in virtual environment could represent an easy and flexible tool for evaluating industrial products characteristics. This research work aims at evaluating in which way odors can influence the users' evaluation of products and if studies on the influence of odors on the users' evaluation of products in a VR environment and in a real environment can be comparable. For this purpose, an experimental framework has been defined, a wearable olfactory display has been developed and experimental testing sessions have been performed.

Keywords Olfactory display · Product evaluation · User experience · Multisensory interaction

1 Introduction

The sense of smell has a great importance in our daily life, also if today humans use more the senses of vision and hearing for interacting with the environment. Pleasant odors are used, usually in the marketing research area, to characterize environments, to improve or orient the person' mood, to elicit positive emotions in people

M. Carulli (✉) · M. Bordegoni · U. Cugini
Mechanical Engineering Department, Politecnico Di Milano, Milan, Italy
e-mail: marina.carulli@polimi.it

D. Weibin
School of Design, Politecnico Di Milano, Milan, Italy

A. Chakrabarti (ed.), *ICoRD'15 – Research into Design Across Boundaries Volume 2*, Smart Innovation, Systems and Technologies 35, DOI 10.1007/978-81-322-2229-3_34

and to communicate information about products as perfumes, household cleaners and food.

While at the present moment the *scent design* field is devoted to the design of personal or ambient fragrances, also industrial products not traditionally associated with odors, as household appliances, furniture, cars, and so on can particularly benefit from its introduction. In fact, the use of pleasant odors for communicating product characteristics to users could represent a competitive advantage. Specifically, odors can evoke positive feelings and differentiate these products from the others with very similar functionalities and appearance.

At the same time, in the area of Virtual Reality (VR), several studies have focused on presenting odors in virtual environments, by developing devices named *olfactory display*. These devices are controlled by a computer and generate scented air. Unfortunately, the olfactory displays developed so far have not found commercial success: they are often cumbersome or very limited in the quantity and number of scents that can be stored and generated. Nevertheless, the introduction of odors simulation in multisensory VR environment could represent an easy and flexible tool for increasing the realism of the experience in the evaluation of industrial products characteristics.

In the research presented in this paper the hypothesis is that the introduction of odors simulation in VR multisensory environments could allow for verifying and possibly quantifying the influence of odors on the users' assessment of products. Consequently, the research presented in this paper aims at evaluating in what way odors can influence the users' evaluation of products, and if the influence of odors on the users' assessment of products in a VR multisensory environment and in a real environment can be comparable. For this purpose, an experimental framework has been defined, a prototype of a wearable olfactory display has been developed and experimental testing sessions have been performed. Consequently, the data collected in the experimental testing sessions have been analyzed from a statistical point of view for identifying any possible differences and correlations in users' evaluation with the used scent typology. Also, these data have been compared to results of testing sessions carried out in a real environment, in order to understand if the influence of odors on the users' assessment of products in a VR multisensory environment and in a real environment can be comparable.

2 State of the Art

2.1 Odors in Marketing and Design

The sense of smell has a great importance in our daily life, because it is devoted to acquire and interpret chemical signals in the environment and supports some basic biological and behavioral functions, as dangers recognition, identification of food, and social communication [1]. However, today humans use more the senses of vision and hearing, while the sense of smell is only used to characterize environments, to improve the person' mood and to communicate and elicit positive emotions in people.

In fact, the regions of the brain involved in the olfaction process are the cerebral cortex, the thalamus and the frontal cortex and the limbic system. The cerebral cortex, the thalamus and the frontal cortex are involved in the cognitive processes of identification, recognition and measurement of olfactory stimulus, integrating and comparing it with other types of information (past experiences, stimuli from other senses, etc.).

The limbic system, which governs the unconscious and emotional components of perception, is responsible for the human olfactory memory. It is, therefore, possible that a perceived smell evokes memories, unconscious responses at the emotional level even before the olfactory stimulus is consciously perceived and recognized.

For these reasons, odors can improve a person's mood, as well as enhance learning activities, increase attention level, and are deeply evocative [2]. Concerning the link between smell and memory, many studies focused on the ability of scent signals to evoke autobiographical memories, or memories of people, products, environments and also scents met or experienced long time ago [3, 4]. Also, Herz [5] demonstrated that memories triggered by scents are considered as more emotional than those evoked by the other types of stimuli. Also, researches [6, 7] demonstrated that odors can impact on physiological parameters such as heart rate or skin conductance, and consequently can induce activation or relaxation states in people.

Odors can also influence moods: while pleasant odors can induce positive moods, unpleasant odors can induce negative ones [8]. This property has been applied in the marketing research area, in which a lot of works have been done on the use of odors for eliciting positive moods in people, and for communicating information about products as perfumes, household cleaners and food. Studies found out congruent odors in the environment can be used for improving the time spent by customers in store [9]. Moreover, the use of pleasant scents in the environment can enhance evaluations of products [10] and of stores [11] in comparison with evaluations in no scent conditions.

In the mass market, often the perceived quality of the products is associated with their odors [12]. Then, companies add odors to several commercial products, as household cleaners, to communicate to users the feeling of clean, and foods to supplement their organoleptic characteristics. In these cases, the addition of odors (such as artificial flavors) allows companies to standardize the taste, filling any gaps caused by the food characteristics, and the possible changes in food preparation.

On the opposite side, few research works and design practices have been carried out in the field of industrial products not traditionally associated with odors, but which can particularly benefit from their introduction. For example, some industrial products are linked, in the collective perception, to the odors of what they contain (as in the case of food packaging), or to the good or bad performances of the functions they are performing (as in the case of the food in a refrigerator, the odor of washed cloth in the case of a washing machine, the odor of a dishwasher or a vacuum cleaner) or the perceived quality of the product (the car interiors, the quality of leather garments, and so on). Nevertheless, usually for these products their smell is not specifically designed, but relies on the characteristics of the product (as, for instance, the materials that constitute it), or on what the product will generate naturally, and could be both pleasant and unpleasant. However, in the second case, an unpleasant smell could affect the product judgments by users.

On the contrary, the design and the use of pleasant odors for communicating product characteristics to users could represent a competitive advantage for these products. In fact, odors can evoke positive feelings and differentiate them from the others of the same category, thus impacting both in the conscious and unconscious levels of perception.

2.2 *Odors in Virtual Environments*

In the area of Virtual Reality and simulation, several studies have focused on presenting odors in virtual environments, by developing both ubiquitous and personal olfactory displays. An olfactory display is a device controlled by a computer that generates scented air and provides a human user with odors. Specifically, because humans sense odors through the air, the role of an olfactory display is to make scented air from odor materials in a stocked form (liquid, soaked in porous materials, encapsulated or gelled) with the desired components and concentration and to deliver the scented air to the human olfactory organ. There are various technologies used to construct an olfactory display, categorized by scent generation methods and scent delivery methods [1].

These technologies have been used in several research projects that focused on the development of olfactory display for specific purpose. For instance, Yanagida et al. [13] proposed an unencumbering olfactory display that conveys a clump of scented air from a remote place to the user's nose by using an air cannon. Wook et al. [14] developed an olfactory display based on a chemical container of temperature responsive hydrogel and controlled release of aroma by using a Peltier module to control the temperature. Yamada et al. [15] developed and evaluated two prototypes of wearable olfactory displays (the first prototype deals with odor in the gaseous state, while the second one uses odor in the liquid state) to present the spatiality of odor in an outdoor environment. Also, Yoshimura and Sakashita [16] developed an olfactory device for emitting a real perfume into the user's immediate area, when a mail with a key program is read via Internet. Narumi et al. [17] developed a "Pseudo-gustatory" display for creating a gustatory sensation. This display presents flavors by means of a cross-modal effect elicited by visual and olfactory Augmented Reality. Matsukura et al. [18] proposed an olfactory display that presents a virtual odor at an arbitrary position on a two-dimensional screen.

Finally, some companies have developed and tried to sell virtual olfactory displays for personal computer use [19]. Examples are AromaJet (www.aromajet.com), DigiScents-iSmell, ScentAir (www.scentair.com), and TriSenx. They use a number of odors stored in cartridge and, upon receiving a signal describing an odor, they release a mixture of these odors by using, for example, pumps. Unfortunately, due to the fact that personal olfactory displays developed so far try to simulate a great number of odors, these are often cumbersome or very limited (in the number and quantity of scents that can be stored and generated, in the distance that the scented air can cover, and so on), and they have not found commercial success.

Nevertheless, the introduction of odors simulation in VR multisensory environment could represent an easy and flexible means for evaluating characteristics of industrial products. In fact, the use of Virtual Prototyping in the product development process is more and more becoming a diffused practice for evaluating design solutions in advance and modify them, reducing the need to develop real prototypes [20]. Specifically, the use of multisensory approach, in which a combination of more sensory channels is used, allows the evaluation of multiple characteristics of the design solutions. For example, Bordegoni et al. [21] demonstrated the effectiveness of using a multisensory approach based on the combination of vision, sound and touch in a design review activity.

In the research presented in this paper we make the hypothesis that the introduction of odors simulation in VR multisensory environments could allow for evaluating the influence of odors on the users' assessment of products.

3 Evaluating the Odors Influence on the Users' Assessment of Products in a Multisensory Virtual Environment

The research presented in this paper aims at evaluating:

(a) in what way odors can influence the users' assessment of products, and
(b) if the influence of odors on the users' assessment of products in a VR multisensory environment and in a real environment can be comparable.

In order to perform this study, a VR experimental framework has been set up. First we intended to study if odors influence the users' assessment of products. The experimental framework consists of the visualization through a Head Mounted Display (HMD) of the virtual models of various types of a consumer product. Selected odors rendered by an olfactory display are added to virtual scene. Subjects, not aware of the presence of odors in the virtual environment, are asked to evaluate the virtual products, which are visually presented to them. This method, in which the only variable parameter is the odor, has been chosen to reduce the possible variables to take into account and that can impact on the experimental analysis. Moreover, the authors decided to use a 2 (pleasant versus unpleasant) X 2 (congruent versus incongruent) between-subject method. Also, the authors included a no-scent control condition for comparing users' evaluations in odor and odorless conditions. Consequently, five distinct groups of subjects were asked to perform the same testing session. After the tests, the collected data are analyzed for identifying any possible differences in users' assessments related to the use of odors and to the used odor typology. Besides, it is analyzed if the obtained results can be comparable with those of Bosmans [10]. According to this study, odors congruent with the product category can have a strong influence on consumers' product evaluations.

4 Wearable Olfactory Display

A wearable olfactory display to integrate in the multisensory virtual environment has been developed (Fig. 1). For what concerns the generation of odors, it has been decided to use the ultrasonic atomization method (in which fine particles from a liquid are generated by using ultrasonic energy), while the delivery of odors has been developed by using a direct injection method. The device consists of air cannons for the generation and the delivery of the odor. In each air cannon a small cylinder of porous cotton is placed and used as scented water cartridge. Commercial water-soluble flavors are used, diluted with water. The cylinder of porous cotton is drenched with the scented water and, under the effect of ultrasonic energy, a scented mist is generated and released in the air.

For obtaining a wearable olfactory display, air cannons have been miniaturized, and two of them have been placed in the external part of commercial earphones (one air cannon in each side of the earphones). The authors have chosen this solution to reach the user's nose with the scented air and, at the same time, to develop a non-cumbersome and comfortably wearable olfactory device.

The air cannons are controlled by means of a software application specifically developed. The architecture of the wearable olfactory display is shown in Fig. 2. The software application, on a tablet, sends the user's input via wireless to an Arduino board, which processes the input and sends to the air cannons the command of generating and emitting the odor.

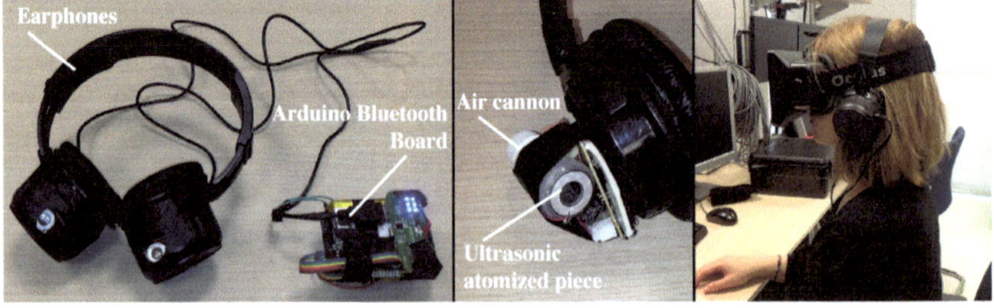

Fig. 1 The prototype of the wearable olfactory display and a user while performing the testing session wearing the Oculus HMD and the wearable olfactory display

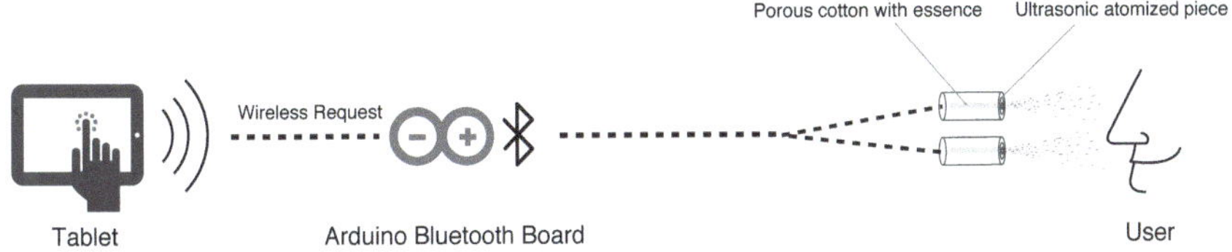

Fig. 2 The architecture of the wearable olfactory display

5 Experimental Testing Sessions

According to the defined experimental framework, several experimental testing sessions have been carried out. In particular, five different groups of subjects, each made up of 15 subjects, were asked to evaluate the digital models of three washing machines, while a specific odor (one for each group) was added to the virtual environment. In one case (the no-scent control condition) no odors were added to the virtual environment.

The VR multisensory environment consists of an Oculus Head Mounted Display (http://www.oculusvr.com/), through which subjects can visualize the digital scene, and the developed wearable olfactory display.

The digital scene is made up of the digital models of three commercial washing machines, placed in a virtual room, in which a picture of a household appliances store has been imported as background (Fig. 3). The shape of the washing machines and of their components are different, while their colors have been made uniform.

At the beginning of each testing session, the olfactory display was filled in with the water added with commercial water-soluble flavors. According to the experimental framework, a specific odor has been used for each group of subjects. The four selected odors are lavender (pleasant and congruent), orange (pleasant and incongruent), smoke (unpleasant and congruent) and anchovy (unpleasant and incongruent). In one case (the no-scent control condition) the olfactory display was filled in with plain water.

Fig. 3 The virtual environment with the 3D models and the background showed to the users

In total, 75 subjects were asked to perform the experimental testing session. The subjects were students of the Design and of the Mechanical Engineering Schools of the Politecnico di Milano. Their average age is 24.5 and 45 % of them are females. Each subject was received alone and subjects were randomly assigned to one of the five experimental conditions. An informed consent was read and signed by each subject; they were voluntary and did not receive monetary compensation. The subjects were not aware of the presence of odors in the multisensory virtual environment. In the informed consent form, odors were presented as "multi-sensory" items. Each testing session was made of a pre-test, a testing, and a post-test session, and lasted about 30 min in total.

In the pre-test session, which lasted about 5 min, subjects were asked to fill in a short questionnaire for obtaining background information about the subjects, including their age, sex and experience with HMD. Also, subjects were asked if they are smoker or not, if they suffer from any allergy, and if they had a cold.

In the testing session, subjects were asked to wear the HMD and the wearable olfactory display (Fig. 1), and then to look at the virtual scene displayed through the HMD. The testing session lasted about 5 min.

During the post-test session, which lasted about 20 min, qualitative data about the subjects' evaluation of the washing machine have been collected. In this case, subjects had to fill in three questionnaires dedicated to the assessment of the virtual washing machines. Each questionnaire was dedicated to a specific washing machine, and subjects were required to express assessments on the washing machine and on the shape of its main components, its perceived value, its perceived level of usability and washing performances, and so on. In these questionnaires, a six-point scale format was used.

6 Results Analysis

The collected data have been analyzed to verify the influence of the odors on the users' evaluation of products. For this purpose, for each question of the questionnaire, each washing machine and each odor (or no odor, for the no-scent control condition), the medium value of the collected users' evaluations have been calculated.

The analysis of the collected data demonstrates that the use of congruent and pleasant odor positively influences the users' evaluations of the product shape. The influence of odors on the evaluation of the shape of the washing machines has been evaluated by using two different questions, distant in the questionnaire, related to the aesthetic pleasantness of the washing machines (question number 1 "*How much do you like the shape of the washing machine?*" and question number 10 "*Does the shape of the washing machine fulfill your aesthetic sense?*"). In both cases the medium value of the users' evaluations in the test with lavender is the highest one. These results are confirmed by the question number 11, in which the subjects were asked to express their intention about placing the washing machines in a visible place of their house. In the 67 % of the cases, the medium value of users' evaluations after the test with lavender is the highest one.

For what concerns the influence of odors on the evaluation of the shape of the window and the interface of the washing machines, the medium value of the users' evaluations in the test with lavender is the highest in the 50 % of the cases, while in the 33 % of the cases the highest medium value has been collected in the test with orange.

For what concerns the users' evaluations of the perceived level of usability of the washing machines, in the 45.5 % of the cases the medium value of users' evaluations in the test with lavender is the highest one, while in the 22 % of the cases the highest medium value of users' evaluations has been collected in the test with smoke.

Regarding the users' evaluations of the perceived washing performances of the washing machines, it seems to not be related to the use of odors. In fact, the highest medium value of users' evaluations occurs similarly in the test with lavender (33 % of the cases), orange (33 %), smoke (22 %) and anchovy (22 %), while the odorless test in the 66 % of the cases collected the second highest medium value of users' evaluations.

Also, the collected data demonstrated that the users' evaluations on the perceived value of the washing machines is influenced by the use of congruent and pleasant odors: the medium value of users' evaluations after the test with lavender is the highest one.

Finally, for what concerns the influence of scents on the users' purchase intentions, the medium value of the users' evaluations in the test with the congruent and pleasant odor (lavender) is the highest in the 67 % of the cases.

These results match with those of Bosmans [10], also if the two studies present some differences: firstly, the testing sessions in the Bosmans' study have been carried out concerning ambient odors, while in this work odors are directly associated with the product. More, in the Bosmans' study only pleasant odors (and a no-scent control condition) have been used, while in this work the authors used both pleasant and unpleasant odors. Despite these differences, both in the Bosmans' study carried out in a real environment and in the present study, it is demonstrated that odors congruent with the product category can have a strong influence on users' product evaluation.

7 Conclusion

The research presented in this work deals with the benefits that the introduction of the *scent design* could bring to industrial products not traditionally associated with odors, as household appliances, furniture, and so on. In particular, while up to date the odor of these products is not designed, the design of pleasant odors for communicating product characteristics to users could represent a competitive advantage, evoking positive feelings and differentiate these products from the others of the same category.

Also, the authors make the hypothesis that the influence of the product odors could be evaluated in VR multisensory environments, in which odors simulation is introduced by using an olfactory display. In fact, the use of VR multisensory environments in design review activities has already proven its effectiveness, and the introduction of odors simulation could make the design review activities more engaging and complete.

Consequently, the research presented in this work aimed at evaluating in what way odors can influence the users' evaluation of products, and if the influence of odors on the users' assessment of products in a VR multisensory environment and in a real environment can be comparable.

The analysis of the data collected in the experimental testing sessions demonstrates that the use of congruent and pleasant odor positively influences:

- the users' evaluation of the shape of the product and product components;
- the users' evaluation of the perceived level of usability of the product;
- the users' evaluation on the perceived value of the product;
- the users' purchase intention.

Moreover, the presented results match with those obtained in a real environment [10], and demonstrate that studies on the influence of odors on the users' evaluation of products in a VR environment and in a real environment are comparable.

References

1. Nakamoto, T.: Human Olfactory Displays and Interfaces: Odor Sensing and Presentation. Tokyo Institute of Technology, Japan (2013). (Information Science Reference)
2. Porcherot, C., Delplanque, S., Raviot-Derrien, S., Le Calvé, B., Chrea, C., Gaudreau, N.: How do you feel when you smell this? Optimization of a verbal measurement of odor-elicited emotions. Food Qual. Prefer. **21**, 938–947 (2010)
3. Chu, S., Downes, J.J.: Proust nose best: Odors are better cues of autobiographical memory. Mem. Cogn. **30**(4), 511–518 (2002)
4. Willander, J., Larsson, M.: Smell your way back to childhood: autobiographical odor memory. Psychon. Bull. Rev. **13**(2), 240–244 (2006)
5. Herz, R.S.: A naturalistic analysis of autobiographical memories triggered by olfactory visual and auditory stimuli. Chem. Senses **29**(3), 217–224 (2004)
6. Alaoui-Ismaili, O., Vernet-Maury, E., Dittmar, A., Delhomme, G., Chanel, J.: Odor hedonics: Connection with emotional response estimated by autonomic parameters. Chem. Senses **223**, 237–248 (1997)
7. Bensafi, M., Rouby, C., Farget, V., Bertrand, B., Holley, A.: Influence of affective and cognitive judgments on autonomic parameters during inhalation of pleasant and unpleasant odors in humans. Neurosci. Lett. **319**, 162–166 (2002)
8. Rétiveau, A.N., Chambers, I.V.E., Milliken, G.A.: Common and specific effects of fine fragrances on the mood of women. J. Sens. Stud. **19**, 373–394 (2004)
9. Mitchell, D.J., Kahn, B.E., Knasko, S.C.: There's something in the air: effects of congruent or incongruent ambient odor on consumer decision making. J. Consum. Res. **22**(2), 229–238 (1995)
10. Bosmans, A.: Scents and sensibility: when do (in)congruent ambient scents influence product evaluations? J. Mark. **70**(3), 32–43 (2006)

11. Spangenberg, E.R., Crowley, A.E.: Improving the store environment: do olfactory cues affect evaluations and behaviors? J. Mark. **60**(2), 67–80 (1996)
12. Gatti, E., Caruso, G., Bordegoni, M., Spence, C.: Can the feel of the haptic interaction modify a user's emotional state?. World Haptics Conference (WHC) (2013)
13. Yanagida, Y., Noma, H., Tetsutani, N., Tomono, A.: An Unencumbering, Localized Olfactory Display. CHI (2003)
14. Kim, D.W., Nishimoto, K., Kunifuji, S., Cho, Y.H., Kawakami, Y., Ando, H.: Development of aroma-card based soundless olfactory display, electronics, circuits, and systems. ICECS (2009)
15. Yamada, T., Yokoyama, S., Tanikawa, T., Hirota, K., Hirose, M.: Wearable olfactory display: using odor in outdoor environment. IEEE Virtual Reality (2006)
16. Yoshimura, T., Sakashita, Y.: Development of a perfume emission system via internet. J. Comput. Chem. Jpn. **5**(4), 227–230 (2006)
17. Narumi, T., Nishizaka, S., Kajinami, T., Tanikawa, T., Hirose, M.: Meta cookie+: an illusion-based gustatory display. In: Shumaker, R. (ed.) Virtual and Mixed Reality, pp. 260–269. Springer, Berlin (2011)
18. Matsukura, H., Yoneda, T., Ishida, H.: Smelling screen: technique to present a virtual odor source at an arbitrary position on a screen. IEEE Virtual Reality, IEEE (2012)
19. Davide, F., Holmberg, M., Lundström, I.: Virtual Olfactory Interfaces: Electronic Noses and Olfactory Displays, Communications Through Virtual Technology: Identity Community and Technology in the Internet Age. IOS Press, Amsterdam (2001)
20. Wang, G.: Definition and review of virtual prototyping. J. Comput. Inf. Sci. Eng. **2**(3), 232–236 (2002)
21. Bordegoni, M., Ferrise, F., Covarrubias, M., Antolini, M.: Haptic and sound interface for shape rendering. Presence Teleoperators Virtual Environ. **19**(4), 341–363 (2010)

Skweezee-Mote: A Case-Study of a Gesture-Based Tangible Product Design for a Television Remote Control

Mehul Agrawal, Vero Vanden Abeele, Karen Vanderloock and Luc Geurts

Abstract Skweezees are soft objects filled with conductive padding, which are capable of detecting different squeeze gestures using electrodes dispersed all over the object. This paper presents a case-study on the design and development of a tangible product based on the Skweezee system, namely a cushion remote. Squeeze-based gestures for soft user interfaces have rarely been explored. Therefore, we have worked on establishing squeeze-based gestures for a soft cushion interface for controlling a television, by means of a user-centered approach. The user study has brought out appropriate gestures for controlling a cushion remote. A prototype was designed using these gestures. The end result is a cushion remote that uses the Skweezee system and a gesture set for a Skweezee-based cushion remote control.

Keywords Soft-User interface · Tangible user interface · Gestural study · User-centered design

1 Introduction

Tangible user interfaces are becoming part of our life with the emergence of wearable computing and the internet of things. Tangible User Interfaces give us the power to alter the coupling between physical and digital data [1] i.e. input and output can be computationally coupled. When a computer acts as a mediator between input and output, a wide array of input-output coupling are possible. Having the freedom to choose the input and coupling gives designers flexibility to generate a wide variety of interaction styles or gestures for product interaction. While this might be an

M. Agrawal (✉)
Department of Design, Indian Institute of Technology, Guwahati, Assam, India
e-mail: agrawal.mehul92@gmail.com

V.V. Abeele · K. Vanderloock · L. Geurts
E-Media Lab, KU Leuven, A. Vesaliusstraat 13, 3000 Louvain, Belgium

A. Chakrabarti (ed.), *ICoRD'15 – Research into Design Across Boundaries Volume 2*,
Smart Innovation, Systems and Technologies 35, DOI 10.1007/978-81-322-2229-3_35

opportunity for designers, the effort required from users while interacting with everyday products can be enormous. This brings forth an inherent need to design gestures in such a way that it reduces cognitive effort of the user.

Kirk et al. [2] highlight that the kinesthetic memory of moving a tangible object can help in reducing the risk of mode errors by increasing the awareness of performed actions. Gestures add a benefit of reduced cognition [3] to 3D manipulation of a tangible object [2]. However, Norman [4] points out that, using gestures can have side effects such as misrecognition and poor designs of the gesture-to-function mapping. To counter poor designs of the gesture-to-function mapping, it is paramount to adhere to a user-centered design process. In this paper we provide a case study where we designed gestures for a cushion remote based on the users' previous interaction with the product (in this case a cushion).

In this study, gesture detection for cushion-remote was achieved using the Skweezee system. Skweezees [5, 6] are soft objects that are programmed to recognize specific shape deformations. They are filled with conductive padding and can be deformed or squeezed by applying pressure. They have electrodes dispersed all over their surface and they detect the gesture or shape change by measuring the changing resistance between every possible pair of electrodes. The authors of the Skweezee system were curious about the different application areas of the Skweezee system, and what type of squeeze-based tangible interface could be built with the technology. To address this question, we followed a product design approach. In this paper we discuss this approach and the result, namely the cushion remote (Fig. 1).

We will first discuss related work with respect to squeeze-based gestures and highlight why we opted for a cushion remote. Next we will present our user-centered approach for studying the diverse gestures for interacting with the cushion remote. A Wizard of Oz technique [7] was employed to conduct this empirical study during which users were asked to control their television by means of the cushion remote. Next, a video analysis of gestures was conducted to identify the most appropriate gestures for the "Skweezee System". The gestures of all the users were screened and it was investigated which gestures were enacted by multiple users. These repetitive gestural interactions formed the basic repository for a gesture set for a cushion remote. They were coded into the form and the interaction of our "Skweezee System". Finally, a Skweezee Cushion Remote was prototyped using these set of gestures.

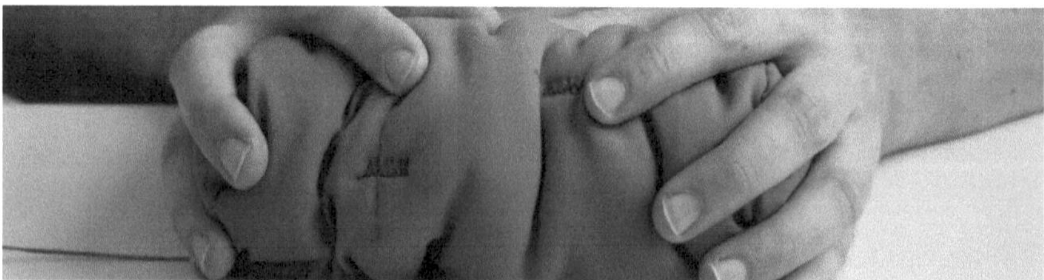

Fig. 1 Squeezing the Skweezee object [5]

2 Related Work and Probable Application Areas

The product design phase started with analyzing products similar to the Skweezee system. These products were then grouped according to similarity. These groups formed probable application areas for the Skweezee system. Thereafter, mind mapping was employed to identify the possible application area. We here discuss the applications that we believe are most relevant to the Skweezee system i.e. rehabilitation products, interactive and programmable soft toys and remote controls for house appliances.

2.1 Rehabilitation Product for Children

When suffering from chronic pain, it is hard to interact with hard objects. Nevertheless, this might be necessary for physical therapy. Children like plush toys, particularly if they seem alive by means of sound or movement [8]. Thus using a soft user interface for children suffering from chronic pain can be easier to interact with, along with being more likeable. The soft object can be used to attract children in interacting (squeezing, pulling, pinching, picking, etc.) with objects thereby promoting physical rehabilitation. Researchers from the Pain Medicine Care Complex [9] have employed games as a way to collect data from children suffering from chronic pain. Christian Schönauer et al. [10] used serious games with multimodal input for facilitating rehabilitation for chronic patients. Coupling the soft input from the Skweezee system with output in form of movement of a soft object can increase interaction with an object supporting rehabilitation among children. The natural benefit of the soft object designed with the Skweezee system can be that it is easy to interact with, and merging it with games can provide a motivational factor for children.

2.2 Interactive and Programmable Soft Toy

Pinoky [11] used a wireless-ring like device to animate a plush toy by moving its limbs. Storytelling Agent Generation Environment is an authoring tool by Marina Umaschi [12] that allows children to design their own wise storytellers to interact with. Using the gesture learning capability of the Skweezee system, an interactive toy companion can be designed such that it can be programmed by children. A normal toy can be augmented to an interactive toy by using the Skweezee system. For example, they can tell the teddy bear to produce a particular sound when they move the hand in particular way, or produce a certain sound when they hug them or squeeze their paws or ears. Another alternative version could be in a form of puppets. So a certain gesture can be associated in a form of certain sound or activity.

2.3 Cushion Remote

Researchers have employed various techniques for redesigning conventional remotes like using gesture-based interaction [13–15], smart phones [15], video game controllers [16] and even a cushion as a remote [17–19]. Vatavu [13] has used a set of referents containing frequently used commands for controlling the TV set: Starting-Up and Shutting-Down the TV, Next and Previous Channel, Increasing and decreasing the volume, Mute, Opening and Closing Menu, Accepting and Rejecting Option and Help. Bernhaupt et al. [20] said that although many buttons are included in today's remote controls, they just lead to puzzling effect. Vatavu [13] posits that this has led to new and shorter versions of the remote like the Pal Simple Remote or the Flipper [21], featuring 6 frequently-used buttons only (power, channel ±, volume ±, mute).

Although there have attempts [17–19] in re-designing present day remote controls into a cushion remote, none of them have employed a user-centered methodology to come up with gestures for the soft remote. Rather, they assigned different locations of the cushion as pressable locations without working out appropriate gestures for them.

Based on the preliminary findings, the "cushion remote" was found to be the most promising application area wherein the "Skweezee System" can be embedded, at least for this study. Remote controls have been there for a long time and have not changed much. The present design of remote is easily lost among cushions of sofa. Adding to it the hard feel of the plastic casing in present designs, this does not fit the comfortable environment (sofa, cushion and soft lights) of living room.

Therefore, it was decided to merge a TV remote with a soft sofa cushion and redesign its interaction based on user-designed gestures. We put forward the concept of the "cushion-remote" in which the cushion of your sofa would act as a remote control for the television. In this manner, the new cushion remote would have the added benefit of not getting lost in some corner of your living room as classical remote controls, but always be there conveniently located on your sofa.

3 User Study

As aforementioned, a user-centered approach was followed which involved users designing the interaction and look of the cushion. The aim of the study was to figure out the optimal gestures for the users, to investigate what kind of cushion, users would relate to and to understand how users' interaction with the cushion remote would change when the user is not looking on the cushion. 'Not-looking' was an extra condition put forward on the basis of the assumption that it would be an additional benefit if users could control the television, without having to look at the remote, but instead be able to focus on the television screen. 20 users participated in the activity, of which 10 users performed the activity with blindfold and 10 users performed the activity without blindfold.

A "Wizard of OZ" technique was employed for this user study. These 20 users were given a cushion and asked to perform various gestures related to some selected functions of the TV remote. Upon their gestures, the television magically responded as if the cushion remote truly worked. In reality, the researcher controlled the television.

All participants (both with and without blindfold) were asked to perform the gestures. Next, they were asked to participate in the design activity during which they designed the look of the cushion remote that can facilitate the gestures performed by them. Video recording was done for all the users while performing the tasks to facilitate later analysis.

3.1 Task 1

A cushion was kept on sofa which was placed facing the TV. The user had to use it to carry out a gesture corresponding to the selected functions of a classical remote, however this time using the cushion. A plain white square (50 × 50 cm) cushion was given to them to accomplish the task. The functions that users were asked to accomplish were Start TV, Stop TV, Volume Up, Volume Down, Next Channel, Previous Channel, Menu Open and Menu Exit (Fig. 2).

The first user group consisted of users without blindfold. This was done to investigate what gestures users would perceive as natural while interacting with a cushion. In this the users were allowed to see while performing the gestures. The second user group consisted of users who were blindfolded while performing the tasks. The blindfolded condition was introduced to understand what gestures were perceived as natural without looking at the cushion. Users were not allowed to see what gestures they were performing and so they were blindfolded (Fig. 3).

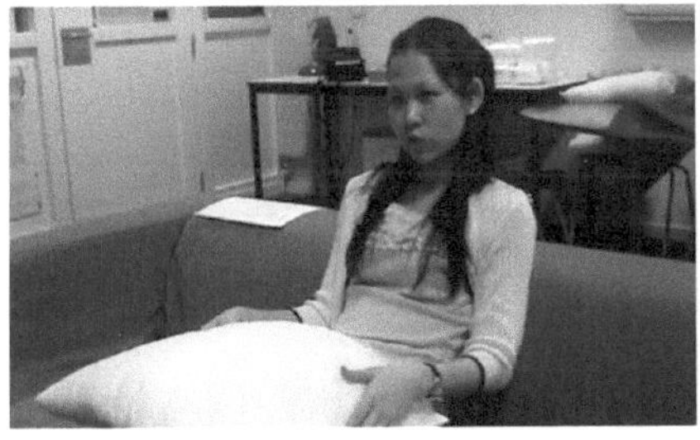

Fig. 2 Users performing tasks of controlling a television by interacting with a cushion, without blindfold (*left*) and with blindfold (*right*)

Fig. 3 Users crafting visual look of the cushion remote after performing 'Task 1'

3.2 Task 2

After task 1, users were asked to modify the plain white cushion using markers, colored papers and other basic arts, crafts and stitching material. Although they were given as much time as they needed to modify the cushion, most of these activities lasted for 15–20 min. This task was given to the users to learn about most common/dominant visual metaphors in the context of a TV remote (for given functions).

4 Analysis

Visually, many of the cushions borrowed looks from the existing TV remote control interface. Many cushions consisted of a button in the center with four arrows pointing outside from the center in each direction (top, bottom, left and right), or some variant of this design. Even during performing gestures most of the gestures performed by users were somehow associated with the position of them on the remote. However, 6 out of 20 users used gestures that took advantage of a cushion like twisting, pulling, pressing, squeezing, etc. (Fig. 4).

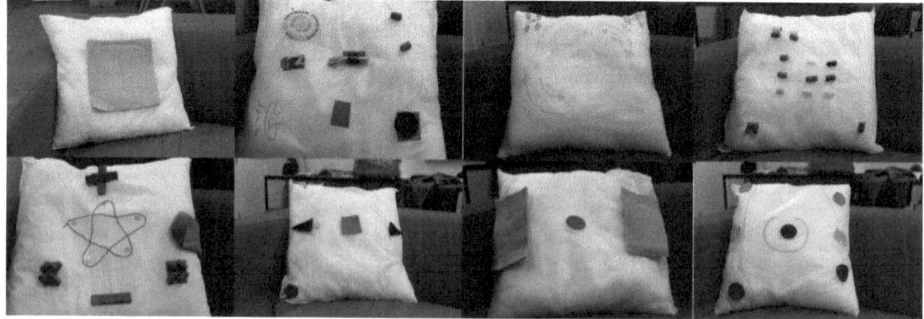

Fig. 4 Few outcomes of creation activity

Thus, it seems that the metaphor of a remote is printed strongly into the users' brain. Furthermore, analysis was conducted to identify the gestures deemed most appropriate by our users. Video of each user was broken down to keep only the 8 gestures of each user. This was followed by open coding; the gestures from the video were described on the basis of what we observed. Then these open codes were coded again (axial coding) to find the relationship between the task and the location of gesture, direction of gesture, and type of gesture. Within the scope of this paper we cannot discuss all three characteristics, but we will present the location of gesture in relation to the task being carried out.

Figure 5 shows that the center of the cushion was most preferred for opening the menu (9 out of 16) and an equal number of users preferred the center as an option for closing the menu. So the center of the cushion can act as toggle for opening and closing menu. 6 out of 14 users preferred using the Middle Left for going to previous channel, whereas the opposite end of it (i.e. Middle Right) was mostly preferred for going to the next channel. The bottom-center and top-center can act as target location for decreasing and increasing volume, despite having low agreeability on that as a location for performing gesture. The volume down button is the bottom-most button as compared to other functions (next channel, previous channel and volume up) in a conventional remote. This trend can be continued to the cushion remote by fixing the location of volume down gesture to bottom of cushion.

Either the television system can be in "On state" or "Off state" thus a common gesture like squeezing all of cushion can be used to toggle between these states.

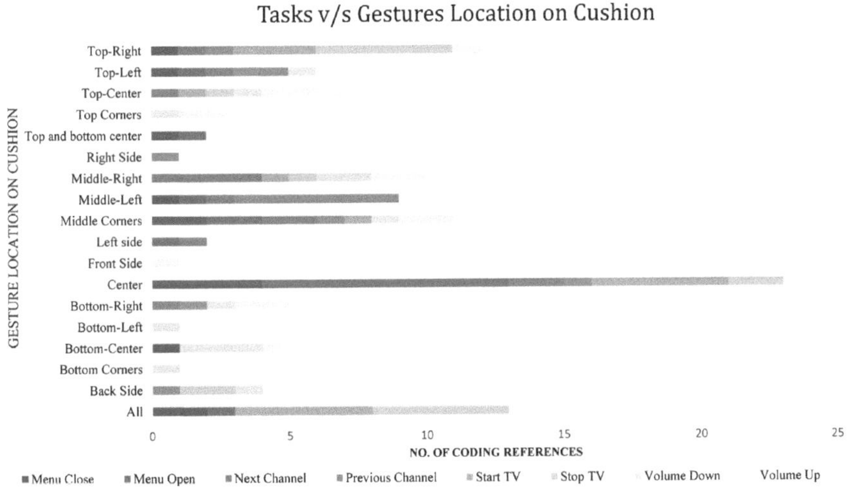

Fig. 5 A graph showing the number of users that used a certain gesture location for controlling with the cushion remote

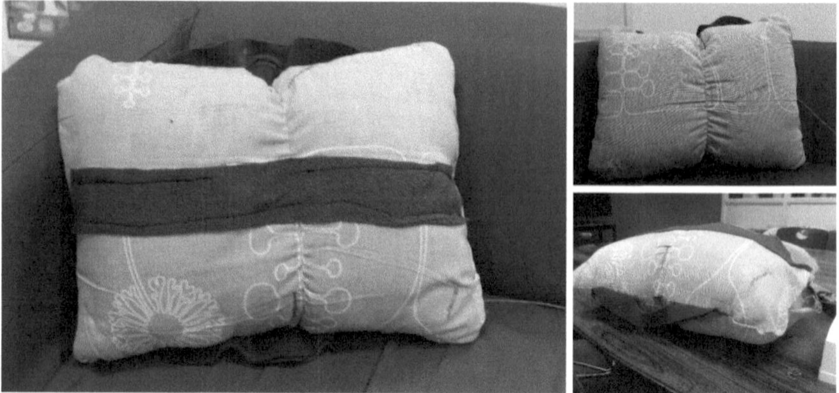

Fig. 6 Prototype of cushion remote

5 Prototyping

5.1 Important Considerations for Designing Gestures

In addition to the points from the user study, the gestures were filtered based on the following points:

- The location of the gesture on cushion should be in line with the user study.
- The gestures should be designed such that unintended gesture detection is minimal, for example the channel should not change when a user puts the cushion behind his or her back.
- Gestures should be such that they can be performed easily without looking at the cushion. Additionally they should be remember-able.

Based on these considerations, eliminations were done, in order to arrive at suitable cushion design and gestures set (Fig. 6).

5.2 Gestures Set and Prototype

Conductive cotton, conductive thread and conductive tape were used to make the prototype. For *starting and stopping the TV*, the gesture chosen was to squeeze the entire cushion. For *next channel*, we chose a squeeze middle-right or top-right. Top-right was included to have a clear detection, as pressing middle-right or top-right gave similar reading in the Skweezee system. The same applies for middle-left and top-left as a gesture location for *previous channel*. For *increasing the volume*, users need to pull the middle-top of the cushion away from the center, and for decreasing the volume, users need to push the middle-top of the cushion towards the center (Fig. 7).

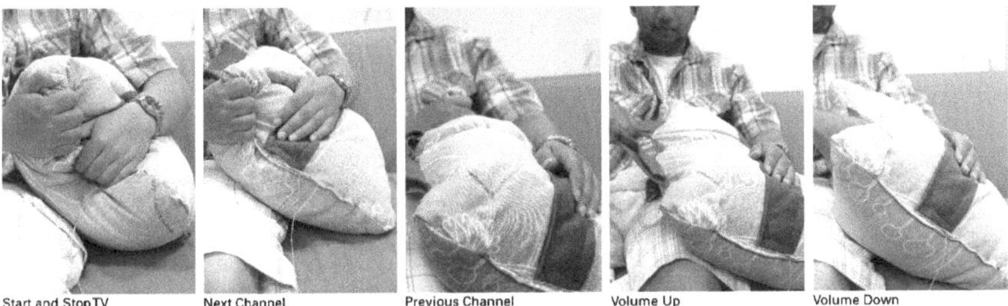

Fig. 7 Gesture sets for cushion remote (In order from *left* to *right*: starting and stopping TV, next channel, previous channel, volume up and volume down)

6 Limitations and Future Work

Our current work is limited to squeeze-based gestures for a cushion remote, a number of possibilities lie in other types of gestures, for example spatial movements of the cushion like waving the cushion, rotating it etc. A comparison can be done to understand the performance of users when interacting with a cushion remote whose gestures are designed using user study to cushion with user designed gestures. Furthermore, a research can be instigated to see variance of preference of gestures based on age, sex, technological literacy, etc.

7 Conclusion

We have employed a user-centered methodology to establish a gesture set for 6 functions of the television remote control, namely Starting and Stopping the TV, navigating to Next or Previous Channel and Increasing or Decreasing of Volume. These gestures employ squeeze-based interaction for cushion remote control. A prototype was designed using the same gesture set.

The methodology used here can be employed while designing interaction for a soft user interface. One important factor during our process was deciding on the target location as a primary aspect to work out the gestures. Other aspects like the movement of the gesture can become important when designing for other soft user interface because of the added possibilities of pulling, pushing, twisting, etc. as a possible interaction when interacting with soft interfaces compared to hard interfaces.

Acknowledgments We would like to thank KU Leuven for funding the internship. We are grateful to all the participants who helped in user-study. We would like to thank Dr. Debayan Dhar for reviewing the paper.

References

1. Edge, D.: Tangible user interfaces for peripheral interaction. Technical Report, (UCAM-CL-TR-733), Computer Laboratory, University of Cambridge (2008)
2. Kirk, D.S., Sellen, A., Taylor, S., Villar, N., Izadi, S.: Putting the physical into the digital: issues in designing hybrid interactive surfaces. In: Proceedings of HCI 2009 (2009)
3. Alibali, M.W., Kita, S., Young, A.: Gesture and the process of speech production: we think, therefore we gesture. Lang. Cogn. Proc. **15**, 593–613 (2000)
4. Norman, D.A.: Natural user interfaces are not natural. Interactions **17**(3), 6–10 (2010)
5. Geurts, L., Vanderloock, K., Vanden Abeele, V.: Skweezees: soft objects that sense their shape shifting. In: Proceedings of the 7th International Conference on Tangible, Embedded, and Embodied Interaction (2013)
6. Vanderloock, K., Vanden Abeele, V., Suykens, J.A., Geurts, L.: The skweezee system: enabling the design and the programming of squeeze interactions. In: Proceedings of the 26th Annual ACM Symposium on User Interface Software and Technology, pp. 521–530. ACM (2013)
7. Dahlbäck, N., Jönsson, A., Ahrenberg, L.: Wizard of Oz studies: why and how. In: Proceedings of the 1st International Conference on Intelligent User Interfaces, pp. 193–200. ACM (1993)
8. Schönauer, C., Pintaric, T., Kaufmann, H.: Full body interaction for serious games in motor rehabilitation. In: Proceedings of the 2nd Augmented Human International Conference, p. 4. ACM (2011)
9. Strommen, E., Alexander, K.: Emotional interfaces for interactive aardvarks: designing affect into social interfaces for children. In: Proceedings of the SIGCHI Conference on Human Factors in Computing Systems, pp 528–535. ACM (1999)
10. Meeting an unmet need: The burden of pediatric pain. Children's National Sheikh Zayed Institute for Pediatric Surgical Innovation. Accessed http://www.childrensnational.org/files/PDF/Sheikh_Zayed_Institute/Pain-Med-Care-Fact_sheet-ENG.PDF, 28 May 2014
11. Sugiura, Y., Lee, C., Ogata, M., Withana, A., Makino, Y., Sakamoto, D., Igarashi, T.: PINOKY: a ring that animates your plush toys. In: Proceedings of the SIGCHI Conference on Human Factors in Computing Systems, pp. 725–734. ACM (2012)
12. Umaschi, M.: Soft toys with computer hearts: building personal storytelling environments. In: CHI'97 Extended Abstracts on Human Factors in Computing Systems, pp. 20–21. ACM (1997)
13. Vatavu, R.D.: User-defined gestures for free-hand TV control. In: Proceedings of the 10th European Conference on Interactive TV and Video, pp. 45–48. ACM (2012)
14. Freeman, W.T., Weissman, C.D.: Television control by hand gestures. In: Proccedings of IEEE International Workshop on Automatic Face and Gesture Recognition, Zurich (1995)
15. Kühnel, C., Westermann, T., Hemmert, F., Kratz, S., Müller, A., Möller, S.: I'm home: defining and evaluating a gesture set for smart-home control. Int. J. Hum Comput. Stud. **69** (11), 693–704 (2011)
16. Vatavu, R.D., Point and Click Mediated Interactions for Large Home Entertainment Displays. Multimedia Tools Appl. **59**, 113–128 (2011). doi:10.1007/s11042-010-0698-5
17. Sugiura, Y., Kakehi, G., Withana, A., Lee, C., Sakamoto, D., Sugimoto, M., Igarashi, T.: Detecting shape deformation of soft objects using directional photo reflectivity measurement. In: Proceedings of the 24th Annual ACM Symposium on User Interface Software and Technology, pp. 509–516. ACM (2011)
18. Chu, L.W.J.: U.S. Patent No. D684,562. U.S. Patent and Trademark Office, Washington, DC (2013)
19. Swallow, S.S., Thompson, A.P.: Sensory fabric for ubiquitous interfaces. Int. J. Human Comput. Interac. **13**(2), 147–159 (2001)

20. Bernhaupt, R., Obrist, M., Weiss, A., Beck, E., Tscheligi, M.: Trends in the living room and beyond: results from ethnographic studies using creative and playful probing. Comput. Entertainment **6**(1), 5 (2008)
21. Flipper Remote. Accessed http://flipperremote.com/, 28 May 2014

Extraction of Gestures for Presentation: A Human Centered Approach

Surbhit Varma, Keyur Sorathia and Abdul Sameer Ashraf

Abstract Many discipline of multimedia and communication go towards ubiquitous computing, hands-free and touch-less interaction with computers. Gesturing is one possible method. In this paper, we present a study on gestures that are commonly used for manipulating presentations. Human centered design methodology was followed to identify appropriate gestures suitable to the context of classroom presentations. User study was conducted to identify commonly used functions and their associated gestures appropriate for classroom presentation. Identified gestures were benchmarked through memory test, stress test and semantic interpretation test. They were further prototyped and tested with users to measure intuitiveness, fun, distraction and stress. Results indicate that gestures identified through human centered design approach were natural, easy to remember and with minimal physical load. Participants were investigated using Wizard of Oz technique for their perceived experience of manipulating presentation through gestures. Results showed that participants did not get distracted or confused while presenting.

Keywords Gesture user interface · Gesture vocabulary · Human centered design · Classroom presentations

1 Introduction

Digital presentations have increasingly become a part of lectures and conferences. In order for the presenter to be comfortable, he needs to use certain devices like keyboard, mouse, remotes, laser light pen, Data gloves [6] etc. for manipulating the presentation remotely. These pose problems of wear and tear of the device, its compatibility and the fact that additional hardware is required. It also interrupts free interaction between the presenter and audience.

S. Varma (✉) · K. Sorathia · A.S. Ashraf
Design Department, IIT Guwahati, Guwahati, India
e-mail: varmasurbhit@gmail.com

A. Chakrabarti (ed.), *ICoRD'15 – Research into Design Across Boundaries Volume 2*, Smart Innovation, Systems and Technologies 35, DOI 10.1007/978-81-322-2229-3_36

Area of communication is moving towards ubiquitous computing, hands-free and no-touch interaction with computers where gestures have been a possible method of interaction, hence making such devices obsolete. In context of making the experience of giving presentations more convenient, many systems [3, 8, 16] using microphones, heterogeneous cameras and webs cameras have been prototyped. Also, gestures have been found to be combined with speech to facilitate manipulation in [6], "The Media Room" by Negroponte's Architecture Machine Group. But in a country like India, which has over 22 national languages and even larger number of dialects, speech based interaction systems are very challenging to implement. Additionally, user is always speaking during presentations and using speech would result in miscommunication with listeners and presentation content. In most of such systems gestures are defined based on easy gesture recognition but taking technology as backbone. This technical based approach can potentially result in gestures that are hard or impossible to perform and culturally unacceptable. It may also result in illogical mapping of functionalities and their associated gestures. Nielson in his research [13] gave two approaches of choosing the gesture- a technical based approach and a human behavior based approach.

Human behavior based approach uses human feedback to identify appropriate set of gestures. It considers ergonomic factors, learnability, memorability and intuitiveness. Our study is based on the human behavior approach. In phase 1 of the research, we sought to identify a candidate set of gestures that could be useful and usable across a range of contexts during a presentation. We decided upon the set of tasks for which we needed to find gestures. In phase 2, we asked participants to spontaneously generate gestures to perform given interaction tasks. We recorded the gestures performed by each participant and categorized most common gesture for all tasks. Upon analysis, a set of gestures was finalized. In the final phase of our study we performed a comparative assessment of intuitiveness of gestures, the ability of users to remember and perform gestures and the physical load felt during their performance. Finally, a prototype was made and tested with the participants.

2 Related Work

Gesture in its simplest definition is a motion of body that contains information [6]. Classification of different types of gestures according to various taxonomies provides a good general overview of gesture issues. A classification presented in [9] focuses on differentiation between Postures, Quasi Postures and Gestures. Driven by the increasing availability of low-cost sensing hardware, gesture-based input is quickly becoming a viable form of interaction for a variety of applications. Electronic presentations (e.g., PowerPoint, Keynote) have long been seen as a natural fit for this form of interaction. Similar work has been done on finding the input effect of gestures on presentation dynamics [1]. It is learned that gestures should not be algorithm-driven, rather they should be human based.

2.1 Gestures and Presentations

Various systems have been developed to enhance the experience of giving presentations. "Charade" [4] uses data-glove connected to Macintosh and issuing one-handed gestural commands. "Slideshow" in [7] uses a remote stick to capture hand movements. A real time system using thermal cameras and web camera was employed in [18]. Here, actions were performed on screen and pre-defined gestures were recognized. Another, solution using software and a web-cam was prototyped in [3]. Tasks were performed upon differences in finger detection.

Multimodal systems, which use speech along with gestures, have also been developed. In "The Media Room" [12] users were seated on a chair, wearing magnetic positioning sensing devices, in center of a room with a projection in front. User could say, "Create a blue square there (pointing at a screen location)", and a square was formed at the pointing location. Cognitive psychology gives reasons for integrating speech with gestures but with a little cognitive interference too [6]. Intelligent classroom [8] is a facility that enables interaction through multiple sensing modes and plan recognition. This classroom uses cameras and microphone to determine what speaker is trying to do and takes appropriate actions.

2.2 Approach for Finding a Gesture Vocabulary

Set of gestures in an interface is called gesture vocabulary. Neilson [13] proposes human based gesture approach, which investigates the people who are going to use the interface. The human based principles make the gestures such that they are easy to perform and remember, intuitive, metaphorically and iconic logical towards functionality and not physically stressing when used often. In order to achieve these principals it is necessary to consider usability theory and ergonomics. There are some important issues in choosing the set of gestures for the interface from a user-centered view [14]. Another approach, Technical based approach, is to find a gesture vocabulary by making it easy for recognition algorithm to recognize gestures by forcing functionalities on gestures, without being logical towards their semantic interpretation [13].

Wizard-of-Oz experiments have proven valuable in the development of gestures [5]. But it is important to keep in mind that gestures should be possible to recognize unambiguously by the system too. In [15] several experiments had been conducted to extract gestures and after considering trade-offs between system performance and human performance to design recommendations for derived 3D gestures.

Approach given by Neilson focuses on choosing a logical and ergonomic gesture vocabulary. It provides a procedure for a human based experiment for testing the found vocabulary. Procedure consists of tests, which verify the vocabulary for following attributes like Semantic interpretation, generalization, intuitiveness, memory, learning rate and stress of the gesture vocabulary. In addition to the tests performed the prototype was also being tested for the experience of such a system being developed. As the above literature analyzes, various systems are being developed for enhancing experience of presenting. The gestures implemented in this system were according to the ease of recognition by the system. Literature compels the need of defining a vocabulary, which should consider the heuristics so that the participant is content upon its interaction with the system. But it is reasonable to keep in mind that the gestures should be recognized by system hence considering trade-off between system performance with gestures and human performance is also important.

3 Methodology

Human centered design methodology proposed by [14] was used to define appropriate gestures. This approach investigates the people who will be using the interface. It is divided into two major stages. Stage I identifies appropriate gestures and stage II benchmarks these gestures based on semantic interpretation, memory and stress. Finalized gestures were evaluated using wizard of Oz technique in stage III.

3.1 Stage I: Identification of Appropriate Gestures

Stage I is further sub divided into 3 steps. Step 1 defines the correct set of functions used in existing system to manipulate presentations. These functions were presented to participants to provide user suggested gestures in step 2. User suggested gestures were further extracted in step 3 to benchmark them in stage II. Following subsections provide details of each step.

3.1.1 Step 1: Finding the Functions

The first task was to find out the major functions involved in system interactions for a classroom presentation. 14 students—10 male and 4 female, from Indian Institute of Technology Guwahati were observed while presenting their work during their classes. The participants were free to move around in the space available, while presenting. It was documented through video recordings and notes. Each presentation had 2 people involved; one giving the presentation and the other helped to control the system, as this would ensure the need of communication between the participant and operator.

All participants were familiar with classroom presentations and had presented at least twice prior to the study. Following list of functions was obtained at an initial stage:

Full screen, zoom in, zoom out, next slide, previous slide, pointing at an image/entity, scroll up, scroll down, play video, pause video, stop video, forward video, rewind video, increase volume, decrease volume. We eliminated few functions based on their frequency of use. For example, scroll up and scroll down were not used often as the software itself adjusts to the screen size. Similarly, we found that only 2 out of the 14 participants used "forward/rewind video" during their presentations. Finally, we selected 10 commonly used functions based on their frequency of use and requirement by the system, which were:

1. Full screen—to enlarge the presentation to full screen mode
2. Zoom in—to increase the presentation window size
3. Zoom out—to decrease the presentation window size
4. Next slide—to move to next slide
5. Previous slide—to move to previous slide
6. Play video—to play a video showcased in the slide
7. Pause video—to pause an ongoing video
8. Stop video—to stop an ongoing video
9. Increase volume—to increase volume of an ongoing audio/video
10. Decrease volume—to decrease volume of an ongoing audio/video

3.1.2 Step 2: Identification of Logical Gestures

The goal is to find gestures that represent the functions found in step 1 of study. 27 students (15 male and 12 female), from Indian Institute of Technology Guwahati were invited for study. These participants were different from those in step 1 with different course backgrounds. Each participant was taken to a presentation room and asked to stand before an empty screen without a mouse, keyboard or any tool. Each function was read out one-by-one and participants were asked to perform gestures appropriate to that function, intuitively.

3.1.3 Step 3: Extractions of Appropriate Gestures

Each video was analyzed to extract gestures performed in step 2. Table 1 describes the gestures observed for each function. Gestures were extracted based on their frequency for each function. Static posture and dynamics of gestures were also observed to identify if the information lies in static posture alone or in the dynamics too.

Gestures for some functions were easy to identify from frequency of a specific gesture being performed. For example, 19 out of 27 users performed *moving hand downward* gesture to explain 'Decrease volume'. Similarly, 'previous slide', 'play video' and 'increase volume' also had specific gestures performed by most of the participants.

Some functions did not have a stand out gesture. For example, for the full screen function, *'making a square' and 'moving hands apart'*, participants preferred both.

Table 1 Showcasing functions and their corresponding most frequent performed gestures in second row

Full screen	Play	Pause	Stop	Increase volume	Decrease volume	Next slide	Previous slide	Zoom in	Zoom out
Making a square	Pointing	Halt (one hand)	Halt (both hands	Move hands up	Move hands down	Move hand left to right	Move hand right to left	Move hands outward	Move hands in
D/14	S/9	S/16	S/20	D/19	D/19	D/11	D/19	D/12	D/14

In third row, the frequency of gestures (by 27 participants) and *S* static, *D* dynamic, represent the nature of gesture performed

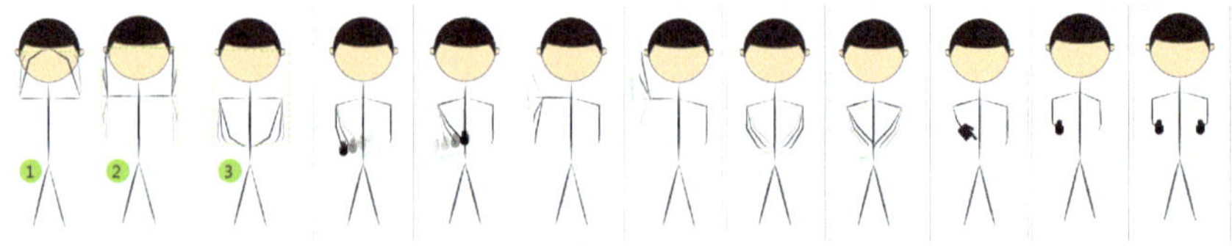

Fig. 1 Extracted final gestures. From *left*—full screen (making a *square*), previous slide (moving hand from *left to right*), next slide (moving hand from *right to left*), decrease volume (moving hand *downwards*, increase volume (moving hand *upwards*), zoom-in (moving hands *outwards*), zoom-out (moving hands *inwards*), play (pointing), pause (halt with 1 hand), stop (halt with 2 hands)]

Making a square was chosen as final gesture as *moving hands apart* was closely related to *move hand outward* of 'zoom-in' function. Implementing *moving hands apart* and *move hands outward* could have created confusion in users' mind as it was tough to identify the difference between them. *Move hands outward* was preferred due to its expressive nature for 'zoom-in'. *Pinch out and move hand forward* be omitted due to its secretive nature, as every user may perform pinch out in a different way. This could create difficulties in implementing it. Figure 1 showcases the finalized gestures for benchmarking.

3.2 Stage II: Gesture Benchmarking

Results obtained in step 3 of Stage I were further tested to benchmark based on various parameters. As proposed by [13], we conducted the following tests to benchmark the above parameters.

1. Guess the function—to identify semantic representation, generalization and intuitiveness of extracted gestures
2. Memory—to identify learning rate, memory and intuitiveness of extracted gesture vocabulary
3. Stress—to identify stress of extracted gesture vocabulary

Gesture benchmarking was conducted with 23 students—11 male and 8 female, different from those in stage I. Every participant participated in all 3 tests.

3.2.1 Step 1: Guess the Function

Participants were given a list of functions identified in step 1 of stage I. A deck of 10 cards with a function written on each card was used. Next, gestures extracted, in step 3 of stage I, were presented in form of a video on a 51″ television screen. Participants were asked to guess/pick the function on a card associated with demonstrated gesture video, one by one. Participants were not allowed to pick a pre-selected card for gestures that came later and were requested to select only from the remaining deck. This allowed us to identify total numbers of errors done by each participant. Table 2 showcases the sequence of videos presented to each participant.

Scores were calculated and analyzed at the end of the study. Score = number of errors per total number of functions. Mean, variance and standard deviation are shown in Table 3. Mean value was found to be an average of 3 errors for 10 functions. Overall, the functions were easy to guess and the gesture set is apt in its semantic interpretation, generalization and intuitiveness.

3.2.2 Step 2: Memory Test

Videos of the gesture vocabulary were showcased to each participant before the study. A slideshow was shown to participants, with functions written on each slide, in logical sequence was showcased with 4 s for each function, and participants were asked to perform the gestures associated with each function. Whenever participant made a mistake, gesture vocabulary was shown again and the slideshow was restarted. This process was continued till the users did all the gestures correctly. Scores were calculated, where Score = Number of restarts. Scores were calculated to present the mean, variance, standard deviation and range of the data. The mean variance and standard deviation are shown in Table 4.

Table 2 Showcasing sequence of gesture videos presented to participants

1st	2nd	3rd	4th	5th	6th	7th	8th	9th	10th
Making a square	Moving hand left to right	Moving hand right to left	Moving hands outwards	Moving hands inwards	Pointing	Halt with one hand	Halt with two hands	Moving hand up	Moving hand down

Table 3 Mean, variance and standard deviation of guess the function benchmark test

Mean score	Variance	Standard deviation
0.347	0.05	0.232

Table 4 Mean, variance and standard deviation of memory test

Mean score	Variance	Standard deviation
0.782	1.38	1.177

Result demonstrates mean value of 0.782 and range from 0 to 2, which means participants made minimum 0 and maximum 2 errors for 10 functions. This indicates lesser number of restarts. Hence the gestures were intuitive.

3.2.3 Step 3: Stress Test

All the participants were provided with a sequence of functions. Sequence was made assuming the number of times each function would be performed in an average 20 slide presentation with 1 embedded video. Following is the frequency of each function included in the test are showcase in Table 3 (Table 5).

Table 6 showcases mean value obtained from *stress test.* Results demonstrate that participants were relatively stressed while performing gestures for "zoom-in" and "zoom-out" functions. This may be due to two reasons; (a) it required use of both hands and (b) frequency was higher as compare to other functions that required using both hands, e.g. "stop" function. It was also observed that static gestures (e.g. play and pause) using one hand very relatively less stressful than dynamic gestures and static gestures with two hands. Similarly, dynamics of gestures played an important role in stress test. Participants performed longer gestures for "increase volume" and "decrease volume" in order to achieve maximum output. They were found more stressed despite using it just for once.

All gestures were found between *No problem* to *mildly tiring.* It indicates that gestures were not physically stressing and were easy to perform. Each gesture was benchmarked using above 3 tests. Results obtained in above tests were in favor of proposed gesture vocabulary. Hence, all extracted gestures were easy to remember,

Table 5 Sequence and frequency of functions used for stress test

Full screen	Next slide	Previous slide	Zoom in	Zoom out	Play	Pause	Stop	Increase volume	Decrease volume
2	11	4	4	3	4	3	1	1	1

After the test, a 5-point Likert scale was provided to each participant. Feedback was gathered based on physical stress they experienced while performing each gesture. Following ratings were provided for a 5-point Likert scale

1 No problem, *2* mildly tiring, *3* tiring, *4* annoying, *5* impossible

Table 6 Mean, variance and standard deviation for each function for stress test

Full screen	Next slide	Previous slide	Zoom in	zoom out	Play	Pause	Stop	Increase volume	Decrease volume
1.67	1.4	1.33	2	1.87	1.13	1.06	1.53	1.53	1.53

easy to recognize, intuitive and required less physical effort. All gestures extracted in step 3 of stage I was finalized. These gestures were further prototyped and evaluated with users.

4 Discussion

User study was conducted to extract gesture vocabulary. We conducted 3 tests to benchmark extracted gestures based on semantic interpretation, memory and stress. Initial results indicate that identified gestures have passed the tests, however it also initiates discussion on following topics.

4.1 Appropriate Sequence of Presented Gestures in Step 1

Sequence of presented gestures play an important role in finding accurate results. In this study, we presented *halt with one hand* prior to *halt with two hands* gesture, which resulted in higher number of errors for total functions. Participants chose "stop" function for *halt with one hand* instead of "pause". However, once both the gestures and associated functions were demonstrated to participants, they identified correct functionality for each demonstrated gesture. One user stated, "If I had been shown *halt with two hands* gesture first, I would have chosen stop for it. This would automatically lead me to choose "pause" for *halt with one hand* gesture".

4.2 Gestures for Opposite Functions

Our study identified opposite gestures for opposite functions. For example, "zoom-in" and "zoom-out" functions were represented using *moving hands apart* and *closing in both hands* respectively. Similarly, "previous slide" and "next slide" were presented using *moving hand left to right* and *moving hand right to left* respectively. It is found that such gestures may create confusion to identify the correct gesture to perform.

4.3 Prototype and User Testing

Final gestures were investigated using Wizard of Oz technique, with 10 participants—5 male and 5 female-different from those in stage I and II. Each participant was introduced to final gesture vocabulary and their associated functions. Moderators provided a presentation that contained 10 slides explaining details

Table 7 Mean values for each question from user testing

Questions	Q1	Q2	Q3	Q4	Q5
Mean	2.8	4	1.9	4.1	2.7

of mensuration using text, images and a video. Each participant presented all contents through defined gesture vocabulary. On the completion of task, participants were requested to fill a 5-point Likert scale (0—strongly disagree to 5—strongly agree). It aimed to understand ease of use, stress, distraction and confusion during presentation using gesture enabled presentation method. Following questions were asked to participants:

Q1. Performing gestures distracted you from presentation contents
Q2. Gesture enabled presentation method was fun to use
Q3. Gesture enabled presentation method was tiring to use
Q4. Gestures were easy to remember
Q5. The use of gestures created confusion during presentation

Table 7 showcases the results obtained, which were in favor of the proposed gesture vocabulary.

5 Conclusion

In this paper we presented a study upon gestures which were commonly used in a presentation. We have followed human centered design process to identify appropriate gestures, which were benchmarked as 1st stage evaluation and further tested with users. Gestures identified through human centered design were not only easy to remember but also fun to use. Body gestures are integral and natural when presenting, however users neither got distracted nor confused during their presentations. Frequency of use demonstrated increased stress for dynamic gestures using both hands. We have used wizard of Oz technique and evaluated with users. In future, we plan to develop this system, conduct a pilot study to evaluate based on users' and spectators' acceptance of gestures during high mental stress.

Acknowledgments We thank all the volunteers, and all publications support and staff, who wrote and provided helpful comments on previous versions of this document.

References

1. Adam, F., Michael, T., Richard, M.: Gesturing in the wild: understanding the effects and implications of gesture-based interaction for dynamic presentations. BCS'10, 230–240 (2010)
2. Baldwin, T., Chai, J.Y., Kirchhoff, K.: Communicative gestures in coreference identification in multiparty meetings, Proceedings of the 2009 international conference on Multimodal interfaces, pp. 211–218 (2009)

3. Barrage, A., Jamil, H., Ramadan, H.: Improving the presentation experience with an intelligent gesture based interface (2012)
4. Baudel, T., Beaudouin-Lafon, M.: Charade: remote control of objects using free-hand gestures, Comm. ACM, **36**(7):28–35 (1993)
5. Beringer, N.: Evoking gestures in SmartKom—design of the graphical user interface. Int'l Gesture Workshop, LNCS, vol. 2298, Heidelberg: Springer-Verlag, pp.228–240 (2002)
6. Buxton, B.: Chapter 14: Gesture Based Interaction, Haptic Input (2011)
7. Chen, Y., Liu, M., Liu, J. Shen, Z., Pan, W.: Slideshow: gesture-aware ppt presentation, IEEE International Conference on Multimedia and Expo (ICME), pp. 1–4
8. Franklin, D. Flachsbart, J. Hammond, K.: Improving human computer interaction in a classroom environment using computer vision. In: Proceedings of the Conference on Intelligent User Interfaces (IUI-2000), Northwestern University (2000)
9. Isenberg, T., Hancock, M.: Gestures vs. postures: gestural touch interaction in 3D environments of CHI'12, ACM (2012)
10. Jocob, O., Andrew, D., Meredith, R.: User-defined gestures for surface computing, CHI (2009)
11. Kurtenbach, G., Hulteen, E.: Gestures in human-computer communication. The Art and Science of Interface Design (1990)
12. Kayne, B., Danny, W., Christof, L., Robert, S.: A quantitative quality model for gesture based user interfaces. In: OZCHI'11, 28 Nov–2 Dec 2011
13. Negroponte, N.: The Media Room—Report for ONR and DARPA. MIT, Architecture Machine Group (1978)
14. Nielsen, M., Storring, M., Moeslund, T., Granum, E.: A procedure for developing intuitive and ergonomic gesture interfaces for man-machine interaction, Technical Report CVMT http://www.cvmt.dk/~mnielsen (2003)
15. Nielsen, M., Storring, M., Moeslund, T., Granum, E.: A procedure for developing intuitive and ergonomic gesture interfaces for HCI. Int'l Gesture Workshop, LNCS, vol. 2915, Heidelberg: Springer-Verlag, pp. 409–420 (2003)
16. Wright, M., Lin, C., O'Neill, E., Cosker, D., Johnson, P.: 3D Gesture Recognition: An Evaluation of User and System Performance. University of Bath (2011)
17. Xiang, C., Eyal, O., David, V.: Evaluation of alternative presentation control techniques. CHI (2005)
18. Zeng, B., Wang, G., Lin, X.: A hand gesture based interactive presentation system utilizing heterogeneous cameras, Tsinghua Science and Technology (2012)

Implementation of an Algorithm to Classify Discourse Segments from Documents for Knowledge Acquisition

N. Madhusudanan, Amaresh Chakrabarti and B. Gurumoorthy

Abstract The overall objective of this paper is to acquire diagnostic knowledge about aircraft assembly in an automated manner, in order to minimize issues from occurring in new, similar situations. This research uses documents, prepared by experts, as a source of knowledge. The first step of the process of knowledge acquisition is segmentation of relevant sections of documents. From many methods that currently exist for such segmentation and classification, one method, namely 'discourse analysis' is chosen for analyzing documents (with future knowledge considerations in mind). Using discourse analysis, entities from sentences are extracted to identify what is being discussed in a chunk of text. These entities are then compared to a domain knowledge base, such as an ontology, to see how (semantically) close the discussion is to the domain of interest. A method for such segmentation had been previously proposed, and is summarised here. This paper describes the efforts for partial implementation of this method. Computer-based tools are used for this implementation, such as Natural Language Toolkit, Boxer, and Ontologies. The Natural Language Toolkit is used for performing text processing, such as tokenization; Boxer is used for Discourse Analysis; Ontologies are used as a knowledge base for domain related terminologies. The method calculates a semantic score for each sentence against the terms taken from related domain ontologies. If the sentence has terms matching those in the ontology, that sentence is classified as being related to the domain of aircraft assembly. The implementation is then applied on test documents to evaluate its performance.

Keywords Text classification · Discourse analysis · Aircraft assembly

---

N. Madhusudanan (✉) · A. Chakrabarti · B. Gurumoorthy
Centre for Product Design and Manufacturing, Indian Institute of Science, Bangalore, Karnataka, India
e-mail: madhu@cpdm.iisc.ernet.in

A. Chakrabarti (ed.), *ICoRD'15 – Research into Design Across Boundaries Volume 2*, Smart Innovation, Systems and Technologies 35, DOI 10.1007/978-81-322-2229-3_37

1 Introduction

Documents are a useful source of knowledge because they capture the collective experience of domain experts. However, it is a challenge to extract knowledge represented in written language and represent it in a machine-understandable form.

The overall goal of the research presented in this paper is to automatically acquire diagnostic knowledge from documents. The intended domain of application of this knowledge is assembly of aircraft structures. This, it is argued, will help reuse knowledge for supporting design of better assembly processes, leading to less expensive aircraft structures with greater ease of assembly and productivity. It is not necessary that the input documents be only from the same domain. They may belong to peripheral domains such as manufacturing, hydraulics, ergonomics, and similar other domains. Some of the knowledge required for diagnosis may be directly available from documents. There is however, more knowledge that would complement the understanding of diagnostic knowledge in documents. For example, such complementary domain knowledge may indicate the meanings or domain-based synonyms for terms used in the diagnostic knowledge documents. Although some existing commercial tools [1] provide for effective manual search for knowledge, we aim at developing an automatic and virtual advisor. Our approach to understanding the above knowledge is to treat the documents as a one-way discourse between the authors and the reader [2]. Discourse analysis methods are planned to be used subsequently for knowledge acquisition. For the purposes of this research, only the textual part of a document is considered, without processing other parts of the document, such as figures and tables.

1.1 Role of Segmentation and Classification

In the process of understanding a large number of documents, the first step in our research is to identify documents that are relevant to the domain. This is applicable not only for entire documents but also for portions of a document. The process of segmenting out such relevant portions of a document is the problem being addressed in this paper. In particular, this paper describes the implementation and validation of a proposed method for document segmentation.

The need for segmentation and its use is explained in Sect. 2. A further elaboration of this is given in Sect. 2.1, followed by an explanation of the previously proposed method [2] for document segmentation. Details of the implementation are discussed in Sect. 4, with a discussion on the tools used, followed by the details of the validation of the implementation. This is followed a brief discussion on the conclusions of the paper and future work.

2 Text Segmentation and Classification

As mentioned in the previous section, in this work, the source of diagnostic knowledge for assembly is taken as documents. However there may be a vast majority of documents that are used in an organization. These may range from simple process documents such as standard operating procedures, to emails, to more complex documents such as process instruction sheets that are derived from a long chain of activities such as assembly planning. In order to avoid processing a large corpus of unwanted information, it is necessary to filter out the irrelevant sections of documents that may not include any knowledge pertaining to the domain of interest. Text segmentation and classification are intended to serve this purpose.

2.1 Coherence and Relevance

Segmenting text might imply different things, as seen in literature. It can be interpreted as one of many possible natural language processing applications that are trending today. It can be partly seen as a classification task [3, 4] and sentiment analysis, since the goal is to classify the text as either related to the domain of interest (aircraft assembly) or not related. It is also related to topic identification and topic segmentation [e.g. 5, 6].

There are two important parts to segmentation in our work [2]. Since the focus of this paper is only on the implementation of the proposed method, these will be only briefly discussed. The first part is *coherence*, which means that subsequent sentences of a discourse are related to one another through the topic of discussion. The second part is *relevance*, which addresses as to whether the topic of discussion belongs to the domain of interest, namely that of assembly of aircraft structures.

The combination of coherence and relevance of a segment of text in a document indicates that the segment is of interest in knowledge acquisition. Hence, further processing would be performed on such segments for acquiring required diagnostic knowledge.

3 Proposed Method for Discourse Analysis Based Segmentation and Classification

The algorithm for the proposed method was reported in [see 2] and repeated here as follows:

Algorithm 1

1. Given text from a document, tokenize it into sentences;
2. Resolve anaphora and pronouns on a per-sentence basis to obtain the Discourse Entity (DE) list for every sentence;

3. Segment the sentences which are both contiguous (i.e. within a certain distance) and share parts of their DE list;
4. Once the segments are recognized and marked, compare entities in the DE list to determine to how many of them relate to the assembly domain. The basis for comparison here are the terms (and their semantic neighbors) from one or more assembly ontologies;
5. For any segment, classify it as related to assembly beyond a certain semantic similarity e.g. Jiang-Conrath similarity for WordNet [7].

4 Implementation

In this section, implementation of the above method (with some modifications to the method) is discussed. The whole method could not be implemented at the time of writing; however, some suggestions as to how this can be done are given in the Future Work section (Fig. 1).

4.1 Tools Used

For implementing the proposed method, a combination of tools has been used. These are listed as below, with a short description and the purpose that each tool serves:

- Boxer and C&C Tools: This is a set of utilities [8] to interpret input text as discourses, provide a representation in the form of Discourse Representation Structures (DRS) [9] and resolve anaphora. However the anaphora resolution capabilities of Boxer are not practical, since it does not recognize anaphora very well [8]. Hence for the purpose of the work reported in this paper, anaphora resolution (i.e. *coherence*, as mentioned in Sect. 2.1) is not handled. The reason

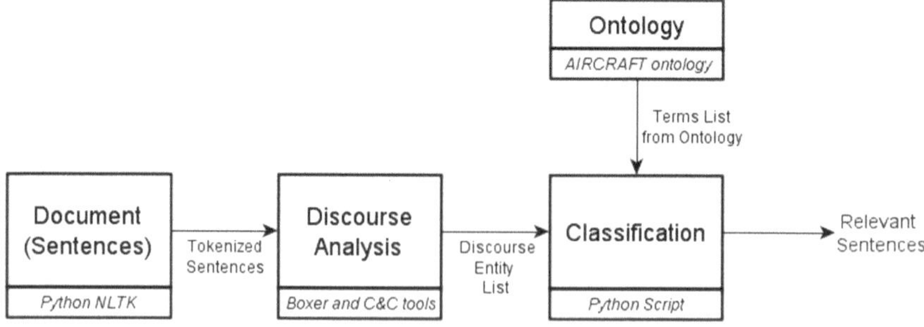

Fig. 1 Sequence of components of the implementation

for choice of this particular set of tools is its open availability and ability to perform discourse analysis, which we argue will aid in further knowledge acquisition activities.

- Python Natural Language ToolKit: It is a large Python language library [10] divided into modules according to functionality. It addresses specific tasks in Natural Language processing and understanding e.g. assign Part of Speech tags, find semantic relations among words, and tokenize input text. Amongst currently available toolkits, Python NLTK is open-source and the most well-known, which made it a natural choice for our purposes.
- Ontology related tools: In this case, the ontology editor, Protégé [11], was used only to manually read an existing ontology file (in OWL format), and to explore various levels of the ontology, if required.
- LaTeX: For printing out segmented text portions, with different colours to indicate the related and non-related segments for the classification (See Fig. 2). This was chosen as an aid for the researcher to clearly contrast related and unrelated sentences, avoiding the need to manually scan the entire text for results.

Figure 1 shows an overview of the proposed implementation. The implementation involved a Python script to first tokenize a given text into sentences. Then each of these sentences is processed individually, for the purpose of classification (since *coherence* has not been tackled in this paper). They are interpreted as individual DRSs, and for each the entities (which are marked as nouns by Boxer, e.g. indicated by n(), n_(), etc.) are extracted and listed as the relevant discourse entities. A list of aircraft related terms is already obtained from an AIRCRAFT ontology [12] and is available for reference. As of now, this list is derived only from the class hierarchy of the ontology. If there is a need for more terms for reference, these can be obtained from the object properties and the class descriptions.

where reliability and safety count : Semantic score = 0 [16]
A typical application for solid rivets can be found within the structural parts of aircraft : Semantic score = 1 [17]
Hundreds of thousands of solid rivets are used to assemble the frame of a modern aircraft : Semantic score = 1 [18]
Such rivets come with rounded (universal) or 100 countersunk heads : Semantic score = 0 [19]
Typical materials for aircraft rivets are aluminium alloys (2017, 2024, 2117, 7050, 5056, 55000, V-65), titanium, and nickel-based alloys (e.g., Monel) : Semantic score = 1 [20]
Some aluminum alloy rivets are too hard to buck and must be softened by annealing prior to being bucked : Semantic score = 0 [21]
"Ice box" aluminum alloy rivets harden with age, and must likewise

Fig. 2 Example of sentences classified using the implementation

4.2 Limitations in Anaphora Resolution

Since the anaphora resolution capabilities of Boxer—C&C tools are limited in that Boxer has high precision at the cost of recall [8], anaphora resolution has not been carried out in the implementation. Hence, referring back to the proposed method of segmenting discourses, only the relevance part of segmentation has been carried out using the suite of tools mentioned above. As a point for future work, it is proposed that the other part of implementation has to be separately handled using other tools, such as tools for coreference resolution (e.g. BART—(Beautiful Anaphora Resolution Toolkit) reference [13]). Such tools are potential sources for understanding and judging the coherence part. After this activity, both the coherence and relevance factors can be put together for arriving at the final segmentation of the given text.

4.3 Validation

Once the implementation of the proposed method for segmentation was completed, the next step was to validate the method on a set of test data. Since it was not clear whether a gold standard exists for a task comprising discourse segmentation and classification, documents which were available in the public domain were chosen as the test data set. These documents were manually segmented by experts and the researchers. In parallel, they were fed as input to the implemented method, so that a comparison between the two could be made.

Three documents were collected and shared amongst a test group consisting of Master's and Doctoral students and members of the project staff at the university. These documents consisted of a mix of sentences that did and did not relate to the aircraft domain. However, the respondents to these documents were mostly for one of these documents, namely a Wikipedia article for riveting [14]. The document had 177 sentences, from various domains including aircraft domain. However, the document did not contain any terms from the ontology other than the term 'aircraft'.

Figure 3 shows the responses of the subjects as to which sentences they thought were related to the aircraft domain, versus the classification made by the implemented method. The figure is a plot of the position of each sentence in the test document along the abscissa, and the subject's identity along the ordinate axis. At the top is the classification performed using the implementation of the DRS based method. Each vertical line in the plot indicates that, that particular sentence has been classified by the subject as being relevant to the domain of assembly. Note that there are clear clusters of sentences (with some exceptions) for all subjects, as well as for the DRS based method.

Among the other two documents, one was an article about the use of jigs and fixtures, while the other was the webpage content of a company specializing in machining. However, there were only two respondents for each of these documents. Their classification matched that of the DRS based classification by only 50 and

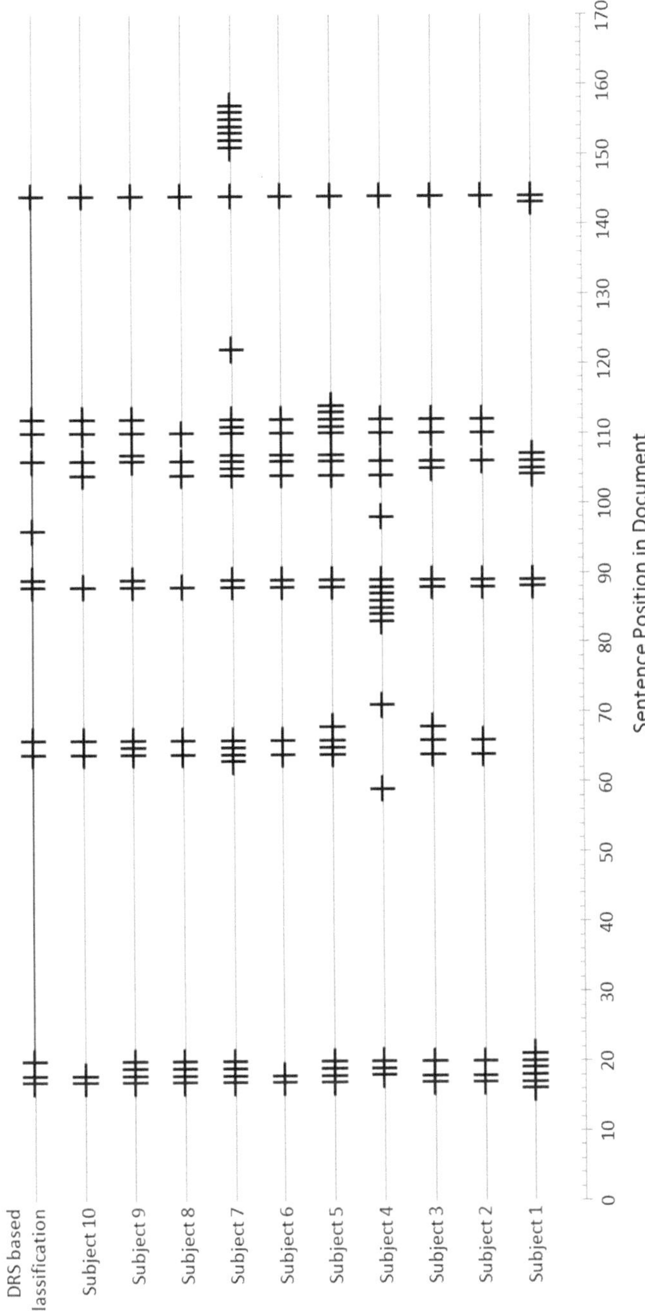

Fig. 3 User classification versus DRS based classification for one of the test documents

33 % respectively. However, for the latter case (33 % matching) one of the subjects had quoted reasons other than semantic relation, for classification (such as relating 'just-in-time' being an indicator of aircraft industry)—which was out of our purview.

5 Conclusion and Discussions

Domain knowledge or background knowledge is a major factor in variability in understanding relevance of a document to a specific domain, since the extent to which each subject understands domain related terms is variable. Although different subjects treated the text differently (e.g. some of them read some sentences together with other sentences) there were a set of sentences that were classified by almost all of the subjects. Also, domain knowledge seems to have helped some subjects with their inferencing, such as 'aerodynamic drag' for classification.

Another important observation is that coherence seems to be a factor for some subjects in judging that a set of sentences are related to the domain of assembly. For example, if two sentences separated by a sentence in between have been classified as related to aircraft domain by most subjects, some subjects have marked the in-between sentence also to be related. This indicates that they implicitly inferred the sentence to be related to these two sentences even though the sentence did not contain the term they were looking for.

Overall, the algorithm performed as expected on most data, with the exception of a few deviations either way (classified as true a sentence that subjects did not classify as true and vice versa), as mentioned above. On average, for the positive matches, it matched 82, 50 and 33 % respectively for the three documents used.

6 Future Work

Although the implementation of the algorithm has worked largely for the current set of test data, it is still one part of a larger piece of work. The challenge of identifying coherence and combining it with the current implementation for ascertaining relevance is required to successfully complete the segmentation and classification tasks. One possible means to identify coherence is to link sentences which have shared anaphora between them. These sentences can then be chunked together to represent coherent sentences which have intersecting sets of discourse entities.

This work also requires test data of a wider variety and heterogeneity to carry out stronger validation. The former refers to more terms *within* the aircraft domain, while the latter is important to cover the grey areas of classification between the aircraft domain and other, related domains. For example, 'tires' are common across multiple domains and are not exclusive to the aircraft domain.

Another remark about the current implementation of the classification (*relevance*) algorithm is the following. Currently the only measure for relevance is that there is at least one term from the ontology. However, this can be trivial, and when there are overlapping ontologies as mentioned in the previous paragraph, this may not always work. Hence, a more concrete measure has to be developed for explaining relevance. A possible method is to use unsupervised clustering methods where overlaps are present between aircraft-related and unrelated segments. Such methods might help in arriving at an average classification point to clearly demarcate the two categories.

Acknowledgments The authors convey their acknowledgements to all the participants in the classification study.

References

1. Website of IHS Goldfire: http://www.ihs.com/products/design/software-methods/goldfire/index.aspx. Accessed 18 Sept 2014
2. Madhusudanan, N., Gurumoorthy, B., Charkrabarti, A.: Segregating discourse segments from engineering documents for knowledge acquisition. In: PLM 2014—The IFIP WG 5.1 11th International Conference on Product Lifecycle Management, Yokohoma, Japan, 7–9 July 2014
3. Wijewickrema, C.M., Gamage, R.: An ontology based fully automatic document classification system using an existing semi-automatic system. IFLA WLIC 2013—Singapore—Future Libraries: Infinite Possibilities (2013)
4. Nyberg, K.: Document Classification Using Machine Learning and Ontologies. M.Sc. thesis, Aalto University, School of Science, Degree Programme of Information Networks, Jan 2011
5. Benno, S.: Topic identification: framework and application. In: Proceedings of I-KNOW '04, Graz, Austria, June 30–July 2 2004
6. Reynar, J.C.: Statistical models for topic segmentation. In: Proceedings of the 37th Annual Meeting of the Association for Computational Linguistics on Computational Linguistics. Association for Computational Linguistics (1999)
7. WordNet Documentation: http://wordnet.princeton.edu/wordnet/documentation/. Accessed on 18 September 2014
8. Bos, J.: Wide-coverage semantic analysis with boxer. In: Proceedings of the 2008 Conference on Semantics in Text Processing. Association for Computational Linguistics (2008)
9. Blackburn, P., Bos, J.: Representation and inference for natural language. A first course in computational semantics, working with discourse representation structures, vol. II. University of Saarland. (Unpublished manuscript)[AGBtM] (1999)
10. http://www.nltk.org/howto/wordnet.html. Accessed 26 June 2014
11. http://protege.stanford.edu/. Accessed 26 June 2014
12. Ast, M., Glas, M., Roehm, T., Luftfahrt eV.B.: Creating an Ontology for Aircraft Design. Deutsche Gesellschaft für Luft-und Raumfahrt-Lilienthal-Oberth eV, Germany (2014)
13. BART—co-reference and anaphora toolkit: http://www.bart-coref.org/. Accessed 26 June 2014
14. Riveting—Wikipedia, a free encyclopedia: www.en.wikipedia.org/wiki/Rivet. Accessed 04 June 2014

Part V
Design Management, Knowledge Management and Product Life Cycle Management

From Concept to Specification Maintaining Early Design Intent

Ricardo Sosa, Jun Bum Lee, Diana Albarran and Kevin Otto

Abstract Early design intent can be compromised as decisions transition in the design process from early concepts to prototyping and to the final specification. This paper presents a method to support designers in the decision making process in the design of innovative products. Bearing selection in kinetic design is used to demonstrate how implementation decisions made in the late stages can heavily impact the early design intent. An inductive research approach is adopted to analyze a collection of existing kinetic products with rotational joints across scales and domains. The resulting macroscopic decision trees demonstrate a way towards a new type of support systems that can be applied at the earliest stage of functional prototyping. An industry case is used to validate this two-stage knowledge schema representation. This method is viewed as a foundational step towards the development of future knowledge-based systems that help teams across disciplines develop creative solutions and maintain early design intent.

Keywords Knowledge representation · Design process · Kinetic design · Kansei

1 Introduction

In the new product development (NPD) process, design decisions are made -sequentially or concurrently- by professionals across various fields of expertise. Design intent refers to the reasoning behind decisions throughout the NPD process. In particular, early design intent is a key element of the 'fuzzy front end' [1, 2]. In industries such as consumer appliances, product designers often make early decisions on the look and feel, where these intentions may remain implicit or only

---

R. Sosa (✉) · J.B. Lee · K. Otto
Singapore University of Technology and Design, Singapore 131816, Singapore
e-mail: ricardo_sosa@sutd.edu.sg

D. Albarran
Out of the Box Thinking, Singapore 130017, Singapore

A. Chakrabarti (ed.), *ICoRD'15 – Research into Design Across Boundaries Volume 2*,
Smart Innovation, Systems and Technologies 35, DOI 10.1007/978-81-322-2229-3_38

partially specified in the design documentation. As a result, the design intent is often compromised in the chain of decisions, leading to innovativeness and implementation shortcomings [3]. This information loss is particularly noticeable in areas such as kinetic design, i.e., where the movement of components in an assembly is an important functional and aesthetic product feature [4, 5]. Kinetic design is susceptible to significant changes to early design intent largely because of a lack of appropriate means to capture the qualitative specifications of movement, i.e., it is problematic to document the intent of 'sturdy elegant' motion when transitioning from conceptual to engineering design. This paper presents a method to bridge the decision making processes across early and late phase design areas. We do this by capturing the concrete late phase requirements in a scalable representation that is sufficiently unconstrained that early upstream decisions can incorporate and resolve. In particular, it is necessary to represent both the physics and the emotive features in the representation. This early resolution allows late phase requirements to be sufficiently incorporated to prevent later compromise of the difficult to capture design intent. As demonstration, bearing selection in kinetic design is taken as our case study to demonstrate how implementation decisions made in the late stages can heavily impact the early design intent. We thereby also show how capturing the myriad of alternative bearing selections early mitigates this effect.

Early design intent refers in this paper to the underlying goals beneath the specification of a design in the early stages of the NPD process where design concepts are initially formed, information is incomplete and representations are ambiguous [1, 2]. In some instances, maintaining early design intent can be achieved via technical documentation where materials and components can be unambiguously specified in engineering drawings. However, when a product is conceived with complex qualitative or sensorial criteria in mind, such as aesthetic features or for certain tactile or proprioceptive modalities, maintaining design intent from the early stages of product conceptualization to final specification remains an open challenge [3, 4]. This paper describes a proposed method to maintain design intent across design areas.

2 Background

Capturing the reasoning behind design components or attributes has been an active area of research [6]. Systems such as Design Rationale Capture illustrate the efforts to extract and use computer-interpretable rationale in engineering design teams [7]. Such systems typically consist of data capturing, representation and retrieval via coupled computer-aided design systems. Data capture consists of knowledge recording which involves encoding as much raw information as possible behind design decisions and representing it in a relevant schema, and design rationale construction which includes the processing and organization of such knowledge [8].

Thus, central to a design rationale system is the knowledge representation framework or schema. The work presented here develops an approach to capture early design intent in domains where decisions span across multiple knowledge schemas.

2.1 Design Decisions

New product development (NPD) is characterized by models including a variety of 'stage-gate' processes [1, 2]. NPD decisions are made *about* different aspects of a product across design stages. Moreover, *how* this reasoning is made differs significantly between stages. Early decisions are often partial and preliminary as they are made with incomplete and unstructured information and in the face of ambiguity and uncertainty. In later stages, as information and feedback about appearance and function is available from prototypes, decisions become more definite, comprehensive and in depth. Several studies have captured the differences in decision-making between product and engineering designers [9].

This work focuses on the transition between the concept design stage where initial decisions are usually made by industrial or product designers, and the early implementation stages of engineering design [2]. It is in this transition that early design intent is transferred across disciplinary areas in a process that is critical for developing the final product while maintaining its originally intended character. We view this handover as a creative process that remains largely unrecognized, poorly understood, insufficiently addressed in education and largely unsupported in professional practice.

The process of transferring design intent across stages can be a major source of conflict in teams where innovative products are developed starting with an original concept and finalizing with a comprehensive design specification. Design representations used by industrial design and engineering design professionals such as models and prototypes support different decision-making processes [10]. Industrial design models tend to be exploratory and built as block models or simple moving parts to capture appearance, use and ergonomic factors. Engineering design models comprise functional, assembly, and production models with limited or no reference to the intended appearance. Because each design area develops an array of representations to generate, develop and test their own priorities, it is apparent why conflicts may arise in the transfer of decisions.

2.2 Kinetic Design

To illustrate our conceptual approach, we select kinetic design as an exemplary area where design intent includes complex multi-sensorial factors. Kinetic design refers to the design of consumer products with movable components, particularly those where movement is a primary design feature. Examples range from doors to advanced robotic products. In such products, the "semantics of movement" or "language of motion" [4, 5] is envisioned -albeit challenging to articulate objectively- in the conceptual stage, and is difficult to translate into the selection of mechanisms that embody the aesthetic intention.

Beyond the particular case of kinetic products, this approach is applicable to the design of innovative products in general where the integration of desirability and feasibility criteria is important. In this work the decision making process for bearing selection is characterized ranging from coarse to fine criteria and addressing multiple requirements. In this way, we demonstrate an approach to balance early concept design criteria with engineering design criteria. The work presented here focuses on the engineering design process of defining rotational joints. We focus on the challenge to translate an early kinetic concept into a specification for fabrication. This process carefully integrates quantitative and qualitative criteria, i.e., a new product with a specific radial load capacity that conveys a sense of 'sturdy elegance'. Specialists tend to draw from a selection logic based on available data, previous implementation cases including the reasoning behind bearing selection under particular operating conditions and constraints. However, past experiences may reinforce the status quo and fail to support the embodiment of creative kinetic concepts. Moreover, when access to specialists is limited or when a novel product concept challenges the established know-how, a new type of support systems is required to maintain design intent in the transition from concepts to specification. We focus on bearings since they are widely used in kinetic consumer products and heavily influence the product semantics [4]. While physical attributes are customarily used in the selection process, our goal is to support early functional prototyping that integrates kinetic design concepts and functional requirements.

2.3 Design Space of Bearings

A bearing in a rotational joint is a part that connects structural components enabling the rotation in a single axis at a friction coefficient lower than that between the two materials. The design space includes: (a) plain bearings including sleeve bearings or bushings, dry rubbing bearings and semi-lubricated bearings; (b) fluid bearings or non-contact bearings including hydrodynamic and hydrostatic bearings; (c) rolling element bearings including spherical balls and rollers (d) thrust bearings; (e) flexure bearings; and (f) customized solutions such as air and magnetic bearings. Pressure and velocity (PV) metric is used to measure performance capability and determine the safe levels of load and rotational speed.

Customary requirements include type and direction of motion, temperature, vibration, humidity, dust, space constraints, maintenance, frequency of stop/start or change of rotating direction, start and running torque, life-span, axial and load-carrying capacity, noise levels, assembly tolerances, etc. Current selection techniques draw from such factors to provide guidelines such as "for low-speed applications and moderate loads, plain sleeve bearings with boundary lubrication provide reliable long-term service and are adequate alternative to rolling-element bearings". Such selection logic fails to account for decisions related to early concepts such as "what type of bearing produces a sturdy and elegant movement for this new product?". This lack of support complicates the integration of requirements across design areas in the earliest stage of prototyping. In addition to supporting the selection of an appropriate solution, the transfer of knowledge and expertise is of key importance in this work. The goal of the proposed system is thus to support the negotiation and learning processes by providing structured information that pertains to the integration of different types of requirements.

3 Bearing Selection Attributes

In order to efficiently test rotational joints in the early stage of functional prototyping, a broad understanding of what type of bearings are used in a range of conditions -and why- is required. Furthermore, given a set of kinetic and performance attributes, often times more than one type of bearing are applicable. Hence, it is important for designers to understand the tradeoffs and make well-informed decisions in the early stage of the prototyping process to minimize resources spent on iterations. This work seeks to support the selection process with a comprehensive view that allows designers to narrow down the range of products and study sample cases and the underlying reasoning. This way, designers can obtain focused knowledge in similar conditions and understand or challenge the standards.

3.1 Macroscopic Bearing Selection (MBS)

A focused study on the method of bearing selection was carried following an inductive research approach. The fundamental attributes are extracted through an inductive study of industry cases. Such macroscopic part selection (MBS) schema aims to narrow down the field of interest and to help define the problem according to the chosen attributes across design areas. The feasibility of MBS to be used in future KBS to search for solutions is demonstrated by applying these resulting decision trees in industry cases that are not included in the training set. A complete accuracy of the decision trees is not central to this approach since the goal of the MBS is to narrow down the product range for the prototyping stage, not to suggest a single final bearing type implementation.

Several dozen kinetic product design cases were collected from commercial catalogues, case studies, and academic and industry publications. The bearing types and the reasons for their selections were recorded. Then it was checked if these products existed with other types of bearings. If multiple types of bearings are used for the same product, the reason for selection ambiguity is investigated. In most cases, clear advantages and disadvantages exist for each bearing type. Consequently, extensive arguments written by the firms and the users about the strength of each product were studied and recorded. One example of such products are CPU fans, with continuous developments in both sleeve and ball bearing assemblies. A total of 54 cases are used to build the MBS, with the aim to maximize diversity of scales, speed, load and types of motion. The collection includes: vehicle wheels from RC toys, bicycles, trains and airplanes; fans of different sizes; crane pulley; door hinge; conveyor belt rollers; castor wheel; etc.

3.1.1 Attribute Selection: Functional and Kansei Requirements

Twelve engineering attributes were initially extracted from the literature for the MBS with the product dataset. These factors were further simplified in order to support early prototyping where not all details are known, and to facilitate design exploration. The more macroscopic the factors and simpler the categorization, the easier it is for the designers to utilize such tools earlier before having to develop a detailed understanding of the products' operation environments and states. The focused selection criteria and their value ranges are: (E1) Radial Load classified in: Light, Medium and Heavy; (E2) High Speed as True and False, with a border value of 200 RPM, distinguishing joints driven by motors or engines, and those actuated by hydraulics and human force; (E3) Oscillation as True or False to distinguish between oscillating and full rotation joints; (E4) Frequent Shock Load as True and False; (E5) Long Loaded Idle Time as True or False. True, if the joint is exposed to long inoperative status and False if the joint is regularly operated; (E6) High Contamination Environment as True or False; and (E7) Geometrical Constraints as True or False. True, if the joint has a geometrical constraint and False if minimizing the bearing size may provide a significant advantage. Attributes E1, E2 and E3 are numerically definable with clear cutoff values, whilst for the rest it is less trivial to set a cutoff value.

Five kansei attributes were taken from the literature to represent the qualitative requirements that describe the character of movement in kinetic products [11]. These attributes are defined as antonym pairs as is customarily in Semantic Differential (SD) evaluations: (K1) Dull-Exciting; (K2) Clumsy-Graceful; (K3) Unrefined-Elegant; (K4) Playful-Serious; and (K5) Delicate-Sturdy. The extraction of kansei values depends on the design goals defined by the team [11]. Special emphasis was put in evaluating the type of movement, rather than the overall product; namely, SD evaluations can contrast the artefacts "bicycles" versus "motorcycles", or a generic type, such as mountain bicycles vs. foldable bicycles, or a set of competing brands or models. Here the type of movement enacted by a rotational joint is evaluated, i.e., the semantic profile produced by bicycle handle brake levers is compared to the type of movement produced by bicycle pedals (Fig. 1).

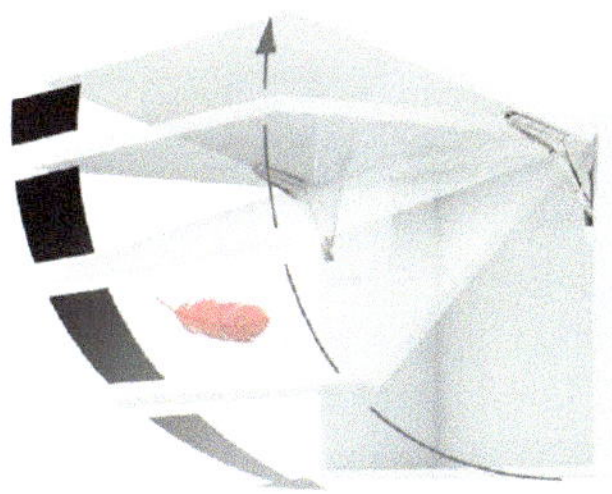

- This hinge damping system gently slows down the movement and closes the furniture door softly and quietly
- Smooth and soft closing action

Fig. 1 The 2014 hinge catalogue by Hafele™ refers to attributes 'smooth', 'soft', 'gentle'

3.1.2 Attribute Filtering and Assessment

The sets of engineering attributes were tested for their level of significance in influencing the bearing selection process. This preliminary analysis aimed to test the feasibility of building decision tree models and to screen out the non-significant attributes before validating the data with a panel of raters. The variables were first filled out by the authors and pre-tested using *RandomTree* classifier in WEKA [12]. For the engineering factors, the *RandomTree* analysis yielded only three mismatches—such 0.94 accuracy shows the feasibility of the engineering selection decision tree model and suggests that these attributes are good predictors of the bearing type selection. As the next step, the *SelectAttribute* function with *InfoGain* classifier and *Ranker* method in WEKA was used in order to narrow down the attributes to a few that had the most significant impact and still give a relatively accurate decision tree [12]. The most significant variables were identified as: (E1) Radial Load, (E2) High Speed, (E3) Oscillation, and (E4) Frequent Shock Load. These four engineering attributes were used for the assessment by raters. All five kansei attributes were used in the assessment with no preliminary filtering.

Seven design engineers were consulted for an inter-rater reliability study. These included three professors and four students enrolled in the MIT-SUTD engineering graduate program. The survey was conducted in two sets, each containing 27 products out of 54 and columns: Radial Load, High Speed, Oscillation, Frequent Shock Load and Level of Confidence. Each cell was rated from 1 to 6, 1 being 'Strongly Disagree' and 6 being 'Strongly Agree'. This scale was designed to avoid neutral values and to allow for the conversion of the numerical average into a binary of True and False, where average values between 4 and 6 were converted to True and values between 1 and 3 were converted to False. The column Level of Confidence represented the rater's degree of understanding of the particular product. Radial Load Class was not included in the survey as the weight ranges for each product is extracted from the literature.

Four product designers were consulted for the rating of the kansei attributes. The raters have more than 10 years of experience in product design. The kansei survey was conducted in a single set, with all raters evaluating the full product dataset applying the five attributes. The scoring system for this assessment ranged from 1 to 5, representing the corresponding term in the SD pairs, i.e., 1 for "Very Dull", 2 for

"Somewhat Dull", 3 for "Not applicable", 4 for "Somewhat Exciting" and 5 for "Very Exciting" in attribute K1. Neutral values are allowed since some kansei attributes may not be clearly applicable to certain products, either because the semantic character of the product dominates the characterization of its movement, or because the joint is usually blocked from view.

For both evaluations, the collected surveys were subjected to percentage overall agreement and the values for each column were calculated. Cohen's Kappa was included as a more robust measurement of the degree of agreement. The confidence level of the survey participants were used to spot items with exceptionally high disagreements or low average confidence level and to remove those outliers from the reliability and the rest of the decision tree study.

4 Decision Trees

The decision trees were built with WEKA using *RandomTree* classifier in the same manner as the preliminary analysis previously described. As an initial step, by using only the four most critical engineering attributes and the five kansei attributes, the system remains simple and becomes easily approachable to designers from different areas. The resultant decision trees and the pattern outliers were investigated as to why such combination of attributes prompted utilization of certain types of bearing and what unique conditions cause some products to fall outside the pattern, if any. Once the decision trees were built, the branches containing outliers were identified and the reasons for the discrepancies discussed. The trees were then updated by merging multiple branches or further developing branches in order to improve the accuracy of the rule base. In this paper, the focus is on the distinction between plain and rolling element bearings due to availability of information and sample products–future studies will analyze complementary criteria of scale, load and speed.

4.1 Engineering Criteria

Figure 2 shows the decision tree when considering the engineering requirements of a kinetic design using bearings. "Oscillation" has the largest effect as plain bearings dominate the area of oscillating joints—17 out of 20 cases. Examples include door hinges, radial dam gate trunnion bearing, rudder bearing in marine vessels, various construction machine joints and engine rocker arm bearing. However as the radial load, the rotating speed and the rotating angle increases, some transitions into other bearing types are registered.

As "Radial Load" increases, it presents a general shifting pattern from mostly plain bearing to mostly rolling element bearing utilization–except for lightweight products with very high RPM. This is due to the higher friction coefficient of the plain bearings that begin to substantially increase the force required to rotate, wear of the bearing and heat generation even at low speed with the significantly larger

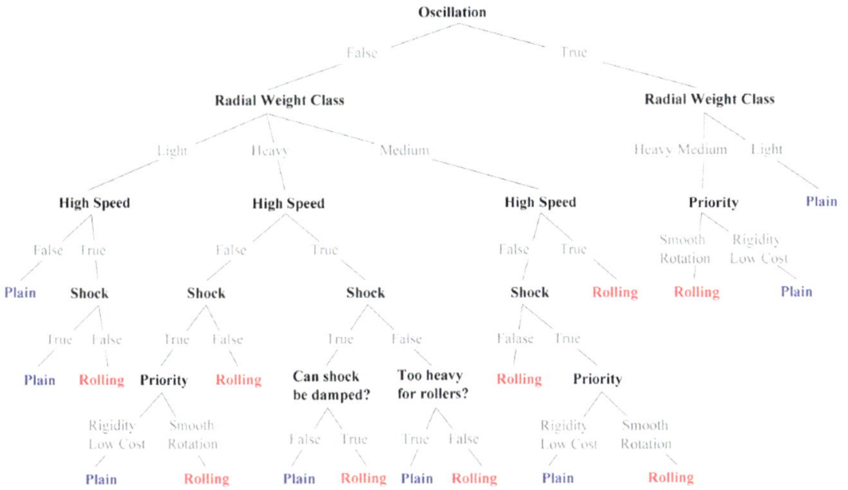

Fig. 2 Decision tree applying engineering attributes to the selection of bearing types

normal forces. "High Rotational Speed" is directly proportional to wear volume. Higher wear constants and lower hardness values of typical plain bearing materials naturally do not suit high RPM environments unless full hydrodynamic fluid film can be generated. Twelve out of 15 high speed full rotation cases have rolling element bearings. Examples of these products are ceiling fan, machine tools, induction motors and transmission shaft and gear bearings.

Lastly, higher "Frequent Shock Loads" can lead to quick premature failure of rolling element bearings. This attribute alone did not impact the bearing selection to a significant degree. Amongst all joints determined as frequent shock load exposed (32), fourteen are equipped with rolling element bearings while eighteen use plain bearings. Nevertheless, this attribute impacts the selection process when paired up with attributes such as low rotational speed or oscillatory movement. Five out of seven low speed, full rotational and frequently shock loaded joints were plain bearings while nine out of eleven oscillating and frequently shock loaded joints were plain bearings. Some examples include large construction equipment, swing bridge wheels and bike pedals. Under these circumstances, the lower rotational friction seems to be sacrificed for rigidity, durability and reliability just like in the case of radial dam trunnions.

4.2 Kansei Criteria

Figure 3 shows the decision tree considering the kansei requirements of a kinetic design using bearings. The kansei variables determined by the affinity clustering are termed K1–K4. Attribute K3 has the largest effect, with "Elegant" as the

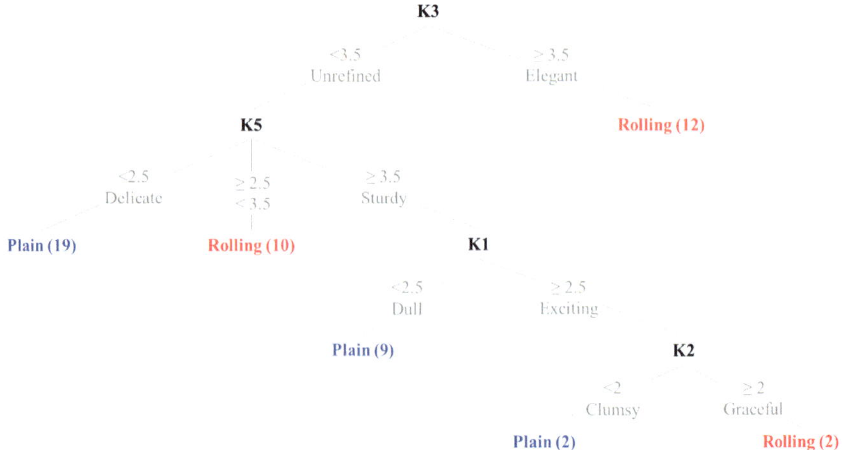

Fig. 3 Decision tree applying kansei attributes to the selection of bearing types

distinguishing factor for half of the 24 products using rolling element bearings. The set of products ascribed an elegant motion includes ceiling fan, Ferris wheel, paper mill roller, and train wheel axle.

Attribute K5 identifies three groups: products registered as having a "Delicate" movement separate 19 cases that use plain bearings, an intermediate group for which the raters agreed on neutral values for 10 rolling element bearings, and the remaining "Sturdy" products, all of which have plain bearings except for two products evaluated with high K1 and K2 values ("Exciting" and "Graceful"), which use rolling bearings. The two products with rolling bearings that combine such characteristics (Unrefined-Sturdy-Exciting-Graceful) are: wind turbine rotor and aircraft landing wheels. In comparison, the two products with plain bearings that differ only in the last attribute (Unrefined-Sturdy-Exciting-Clumsy) are: dump truck hydraulics joint and construction machine track chain.

Noticeably, attribute K4 ("Playful-Serious") does not have a significant impact, which is confirmed by this being the only attribute with a mode value = 3 (neutral) in the assessment by raters.

4.3 Decision Tree Testing

Beyond the engineering and kansei attributes considered here, other type of factors play a significant role in bearing selection, including market preference, availability, quality and cost trade-offs. In many cases, multiple bearing types may satisfy the requirements with not too significant cost, functionality and kansei differences. To help designers identify the trade-offs, the decision trees presented here are intended to support the transition from conceptual to engineering design. To validate this, a recent industry case led by the authors is presented. The kinetic design *Wave-Garden* initiated as a request to conceive, design and implement an interactive exhibition to embody principles of energy efficiency. The brief included requirements such as low maintenance, low power consumption, located outdoors in a tropical climate, and attractive appearance, among others. A space of 4 × 4 m was allocated. The conceptual design stage was led by a professional designer and supported by a group of first year undergraduate design students. The resulting design concept consisted of a modified version of a device demonstrated by J. Shive in his 1959 educational film "Similarities in Wave Behavior" [13]. The Shive wave machine sits on a tabletop and consists of an array of a few dozen small steel rods welded to a thin axle torsion bar that is perpendicular to the aligned rods. The torsion wire transmits energy from one rod to the next, while the high moment of inertia demonstrates wave transmission by ensuring the wave takes several seconds to traverse the entire series of rods. The design team chose the agile and smooth motion of the Shive wave machine as the main inspiration for *WaveGarden*. The concept was presented to the client as a four-meter long exhibition that could be actuated directly by hand with a gentle movement of the first rod. The client approved this design concept based on the appealing movement that would promote contemplation and trigger curiosity about the underlying science when displayed in a public area.

An example rendering of *WaveGarden* is shown in Fig. 4. The main challenge in the embodiment of *WaveGarden* was to achieve a trade-off between several parameters at the scale proposed for this kinetic design. A total of 60 thin-walled stainless steel pipes 1500 mm long were to rotate about the spine, and bearing selection became a critical step where engineering and kansei requirements needed to be balanced. It was decided that the pipes were to transmit the movement by

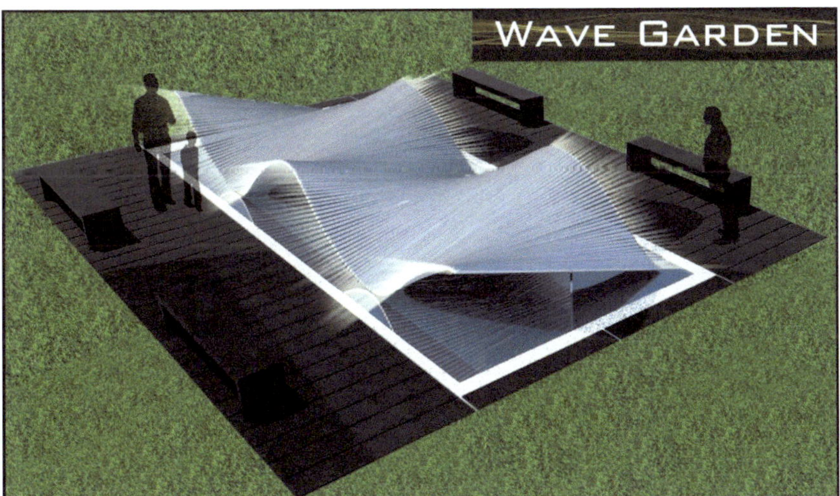

Fig. 4 Kinetic design project "WaveGarden"

connecting them with an elastic rope so that the rope tension could be used to calibrate the desired movement. A type of bearing was needed that would provide the minimum possible friction at this scale, as fluent movement gave this exhibition its unique character.

Based on the engineering decision tree, the decision to build initial prototypes was narrowed down to the use of plain bearings since the pipes oscillated and were less than 3 kg each (Fig. 2: "Oscillation" and "True and Radial Weight" Light). This type of decision aids save time and effort to narrow down the design choices and to avoid wasteful consideration and experimentation with other bearings, for example by assuming in this case that ball bearings would have a lower coefficient of friction. If ball bearings were used to prototype this design, the lubricant in the bearings would need to be removed in order to achieve a similar smooth rotation as plain ABS plastic sleeves due to the thickness of the grease, which of course would not be an appropriate solution for an outdoor exhibition.

In respect to the kansei decision tree, two values were selected to capture the intended type of motion of *WaveGarden*: Delicate and Graceful. Most low K5 values (Delicate) in Fig. 3 do point to plain bearings with a total of 14 of them also having high K2 values (Graceful). The norm in this subset is the use of plain bearings such as in trash bin rotating swing lid and windscreen wipers. The exceptions are products with high K3 values (Elegant) such as optical disk drive spindle which use rolling element bearings. In sum, *WaveGarden* fits well in the subset of "Oscillation" True, "True and Radial Weight" Light, low K5, and high K2. Other products with a light load, oscillate and denote delicate and graceful motion include: small walking robot leg joint, trash bin swing lid, and windscreen wiper joints. One key design consideration that could be misleading in the bearing selection process for *WaveGarden* is its atypical envelop size, as it is significantly larger than the size of the products in this category. However, the decision trees successfully narrow down the search and make the process manageable while maintaining the early design intent. This result was validated in the early prototyping phase. An early prototype was fabricated using roller bearings, and while it was known this selection would not be ideal for engineering reliability, it was surprising the compromise it made on the kinetic aesthetic. Dynamic misalignments were more obvious and the motion stalled out too quickly. Later prototypes with the plain bearing selection eliminated these issues and made for an improved dynamic motion.

5 Discussion

This paper offers a systematic approach to support design teams integrate quantitative and qualitative specifications maintaining early intent in the design and development of innovative kinetic products. Whilst maintaining design intent is important to ensure quality, support creativity and execute product strategy, it is often compromised in a chain of decisions, leading to innovativeness and implementation shortcomings. The approach is validated by applying the decision trees to a real industry case and demonstrating its supporting role to shortlist a manageable set of options for embodiment of kinetic designs. Future work will refine and complement the product datasets with other components, and will study the application of this macroscopic view in two complementary modes: "Maintain Intent" and "Derive Intent". The former consists of applying this approach once a conceptual design has been produced—as demonstrated in this paper. The latter consists of using this approach in creative ideation as an inspiration to imagine innovative kinetic design concepts based on juxtaposing sets of seemingly unrelated products to extract non-obvious applications of rotational joints.

References

1. Otto, K.N., Wood, K. L.: Product Design: Techniques in Reverse Engineering and New Product Development. Pearson, London (2003)
2. Krishnan, V., Ulrich, K.T.: Product development decisions: a review of the literature. Manage. Sci. **47**(1), 1–21 (2001)
3. Maura, M., Michele, G.: Aesthetic feature as a tool to preserve the design intent. In: International Conference on Engineering Design (ICED07), pp 399–400 (2007)
4. Hopson, B.: Kinetic design and the animation of products. Core77 (2009)
5. Parkes, A. Ishii, H.: Kinetic sketchup: motion prototyping in the tangible design process. In: Conference on Tangible and Embedded Interaction, pp 367–372 (2009)
6. MacLean, A., Young, R.M., Moran, T.P.: Design rationale: the argument behind the artefact. ACM SIGCHI Bull. **20**(I), 247–252 (1989)
7. Klein, M.: Capturing design rationale in concurrent engineering teams. Computer **26**(1), 39–47 (1993)
8. Regli, W.C., Hu, X., Atwood, M., Sun, W.: A survey of design rationale systems. Eng. Comput. **16**(3–4), 209–235 (2000)
9. Purcell, A.T., Gero, J.S.: Design and other types of fixation. Des. Stud. **17**(4), 363–383 (1996)
10. Pei, E., Campbell, I., Evans, M.: A taxonomic classification of visual design representations used by industrial designers and engineering designers. Des. J. **14**(1), 64–91 (2011)
11. Pérez Mata, M., Ahmed-Kristensen, S., Yanagisawa, H.: Perception of aesthetics in consumer products. In: International Conference on Engineering Design, pp 19–22 (2013)
12. Hall, M., Frank, E., Holmes, G., Pfahringer, B., Reutemann, P., Witten, I.H.: The WEKA data mining software: an update. ACM SIGKDD **11**(1), 10–18 (2009)
13. Shive, J.N.: Video: Similarities of Wave Behavior. AT&T Bell Labs, New Jersey (1959)

A Spatio-Temporal Network Representation for Manufacturing

Kumari Moothedath Chandran, Monto Mani and Amaresh Chakrabarti

Abstract Manufacturing systems are currently represented in the graphical space predominantly from the perspective of process duration and resource dependencies. Such representations of time are used for planning and scheduling purposes, with the aim of optimising the productivity and efficiency of the factory under consideration. Some examples of temporal representations are PERT, CPM, etc. In the case of globally distributed manufacturing systems, temporal representation alone fails to capture the whole network as location information is omitted. To capture the complexities of a globally distributed manufacturing network, a novel way of representing both space and time in the graphical space is required. This paper discusses a new representation mapping space and time, which captures location, time and duration information of a globally distributed manufacturing network. This representation aims to improve the understanding of process and resource dependencies across geographies, and has the potential for capturing global manufacturing network in varying levels of granularity.

Keywords Spatio-temporal representation · Manufacturing representation · Global manufacturing · Global supply chain

1 Global Manufacturing

Manufacturing industry of the 21st century is more geographically dispersed than ever before, with a supporting global supply chain network. With globalization of manufacturing industry, an original equipment manufacturer (OEM) source around 80 % of value of the final manufactured product from outside the company [1].

K.M. Chandran (✉) · A. Chakrabarti
Centre for Product Design and Manufacturing, Indian Institute of Science, Bangalore, India
e-mail: kumari@cpdm.iisc.ernet.in

M. Mani
Centre for Sustainable Technologies, Indian Institute of Science, Bangalore, India

A. Chakrabarti (ed.), *ICoRD'15 – Research into Design Across Boundaries Volume 2*,
Smart Innovation, Systems and Technologies 35, DOI 10.1007/978-81-322-2229-3_39

Large manufacturing plants have been replaced by networks of small to medium sized enterprises whose operations are integrated as part of an international vendor pyramid [1]. For example, manufacturing iPhones, designed and marketed by Apple, a California (USA) based company, involves nine companies, which are located in the PRC, the Republic of Korea, Japan, Germany, besides the USA [2]. Digital Equipment Corporation, once world's third largest vertically integrated computer company, had 12 plants in seven countries that focussed on a reduced set of core competencies [3]. Boeing sourced its major aerospace components such as wing and tailplane control surfaces from Sydney and Melbourne [1]. Globalization of manufacturing requires a highly coordinated flow of goods, information and cash within and across national boundaries [4], besides Intellectual Property (IP) protection, ensuring product reliability, quality and performance. International competitiveness has become paramount for companies wishing to participate in global manufacturing networks [1]. As competition is global and the complexity of environment within which companies operate is growing, managing an integrated international network has become an important challenge for manufacturing managers [5].

In networked manufacturing, Barteltte and Ghosal [6] recognize four types of relationships among subsidiary companies: physical goods, information, people, and financial resources [5, 6]. Global information and communication networks have allowed real-time operational monitoring, and information sharing through the supply chain [4]. Strategic initiatives such as just-in-time, just-in-time II, quick response, continuous replenishment, and collaborative planning, forecasting, and replenishment have supported global logistical arrangements [4].

Multinational manufacturing firms constantly test and question the design of their supply chains [3]. Manufacturing planners and decision makers need to compare alternative manufacturing networks and optimize supply chain networks. Several stochastic, deterministic and heuristic models exist for decision support in managing supply chains. Examples are Artzen et al.'s [3] Global Supply Chain Model (GSCM), Geoffrion and Graves [7] multi-commodity logistic network design model, Breitman and Lucas [8] framework for scenario description and analysis, and Cohen and Lee's [9] model for production scale economies. While all of these supply-chain manufacturing networks deal with time (delivery, commitment, supply) and space (where, how, transport) monitoring is still highly text-based and serves the purpose of communication (and record) only. A graphical spatio-temporal representation depicting global manufacturing networks and supply-chain alternatives are lacking in all these models. This paper presents a new concept to represent globally distributed manufacturing networks in a spatio-temporal manner, which can be used by decision makers and planners to choose, optimize and compare alternative supply chain networks for, e.g. sustainable manufacturing.

2 Graphical Representations in Manufacturing

Prior to developing a new graphical representation, a literature study on existing representations in manufacturing has been carried out, in order to understand how spatial distribution of manufacturing and its temporal nature are currently represented in the graphical space. A representation is defined by Winston [10 p. 21] as "…a set of conventions about how to describe a set of things. A description makes use of the conventions of a representation to describe some particular things" [10, 11].

There are various temporal representations used in describing aspects of manufacturing. Temporal points and intervals are the two most popular choices for temporal representation in manufacturing [12]. These are depicted using Activity-on-Node diagram (Fig. 1) and Gantt Charts. Activity-on-Arrow diagram (AOA) and Activity-on-Node diagram (AON) used in Program (or project) evaluation and review technique (PERT) and Critical Path Method (CPM) [13] respectively are also variants of temporal activity representation. These methods are used for project scheduling and monitoring, with critical path determining the shortest time to complete a project, particularly for large and complex construction projects. Gantt and Activity-on-Arrow are similar representations where the length of the line represents the duration of the activity. This aids in identifying parallel and concurrent activities, dependencies, and redundancies in estimating project completion times. Critical Chain Path Method (CCPM) by Eliyahu Goldratt, used for project scheduling in resource constrained situations, uses flowcharts to represent resource routing [14]. Petri-net representation of process model entities, proposed by Rudas and Horvath [15], are used for manufacturing process planning and production scheduling. Petri-nets comprise transitions, places and arcs to capture distributed systems, particularly those that are non-deterministic.

In production logistics, The Funnel Model (Fig. 2) and Throughput Diagram developed by Hans Kettner et al., are used for describing production processes [16]. Process chain diagrams or material flow diagrams are also used to map logistic operating curves to illustrate production networks (Fig. 3).

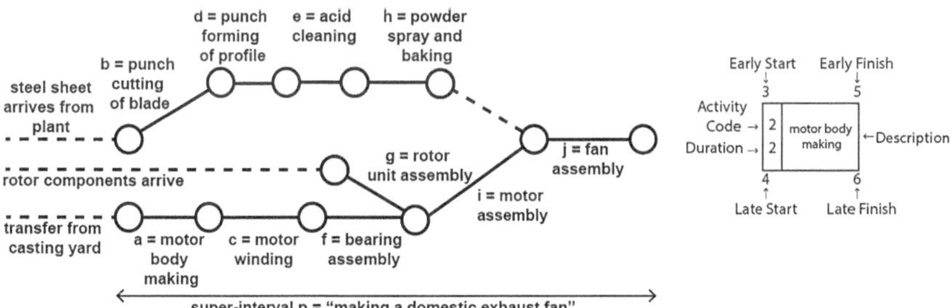

Fig. 1 Temporal point representation of domestic exhaust fan manufacturing using Activity-Node diagram (*left*) and node representation with time (*right*). Images adapted from [12, 13] for this example

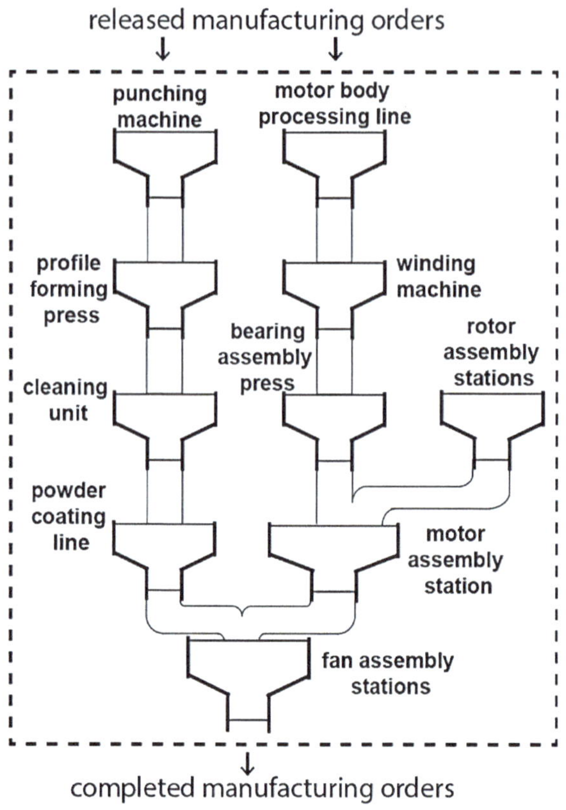

Fig. 2 The Funnel Model diagram of exhaust fan manufacturing. Image adapted from [16]

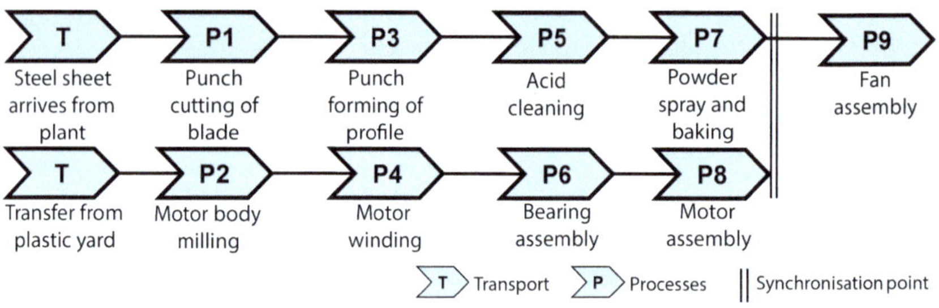

Fig. 3 Process chain diagram for exhaust fan manufacturing. Image adapted from [16]

In none of the above representations, spatial aspects of where the processes take place, are included. The spatial aspects are implied when 'transport' is added as a process (see process 'T' in Fig. 3) and transportation time is used in calculations to arrive at durations, and start and end times of the subsequent processes. In

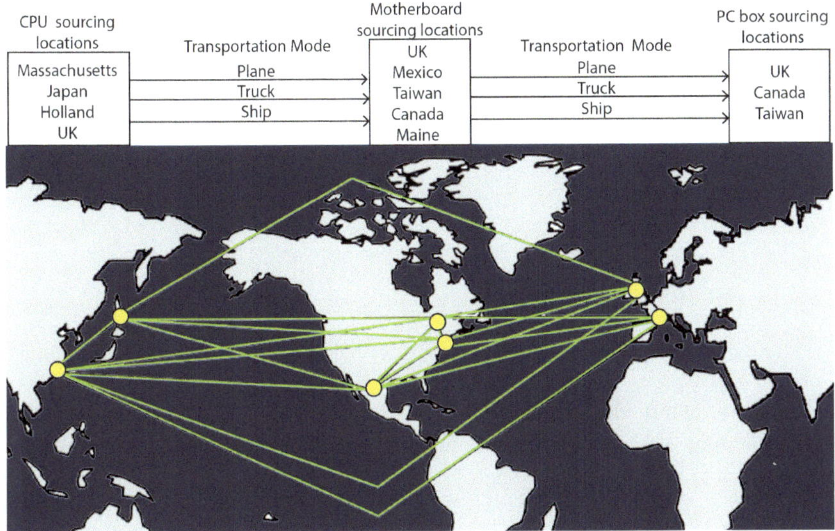

Fig. 4 Digital Equipment Corporation's global supply chain. Image adapted from [3]

manufacturing literature, a graphical representation of the 'spatial' aspect was found only as mapping of in-out flow of resources between plants plotted on world maps. An example of Digital Equipment Corporation's plants and subsidiaries, represented as a flow diagram plotted on the map, is shown in Fig. 4. Here the locations are entered as text in a flowchart below, to explain which component is manufactured where. The same locations may appear multiple times under each component. The flowchart suggests only the flow of components in effect, what goes into sub-assemblies and assemblies, with no temporal information. Another example of spatial illustration is shop-floor diagrams which show the process sequence and location of machines in a plant. Here, the spatial aspects are at a plant level, which do not capture the entire manufacturing network.

The temporal and spatial representations revised above are all flow-based, where the sequence of processes and the material flow are given importance. None of the representations capture the spatio-temporal aspect of manufacturing together in a single representation. Looking at domains other than manufacturing, for spatio-temporal representations, three dimensional time-geography representation for transportation networks depict space in a two dimensional plane, with time in an additional, orthogonal Z axis (Fig. 5). In several GIS representations, space is represented as a 2D map, and time or any other variable in the Z axis. Similar approaches for representing manufacturing networks will make it highly complex and difficult to comprehend and compare alternative networks for supply chain, even at a plant level.

In order to represent a global manufacturing network in a realistic manner while incorporating both spatial and temporal aspects, there is need for a representation

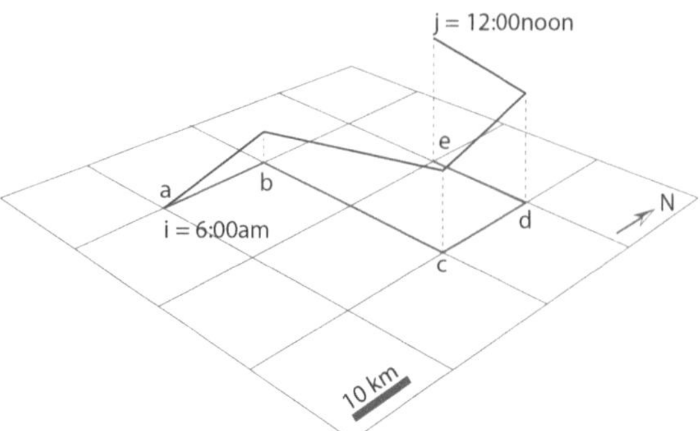

Fig. 5 Time-geography representation for a transportation activity. Image adapted from [17]

which can capture the essential aspects of this system. For this, a system level understanding of the network is required; this calls for a representation that can represent the complexities of such systems. There is need for a systems diagram which can capture the whole network 'as is', without losing much information at a micro or a macro level. The term 'system' is defined by Ludwig von Bertalanffy as "a set of elements standing in interrelation among themselves and with the environment" [18, 19]. In the systems thinking approach, according to Buchanan [18, 20], the emphasis is no longer on material systems, but on the human who experiences the system. He explains that "one of the most significant developments of system thinking is the recognition that human beings can never see or experience a system, yet we know that our lives are strongly influenced by systems and environments of our own making and by those that nature provides" [18, 20]. Buchanan argues that because humans cannot experience the totality of the system, they create symbols or representations that attempt to express the idea or thought that is the organizing principle [18, 20]. Examples of system diagrams are Harry Beck's Tube map for London Underground and Richard Wurman's Tokyo subway map.

As explained by Winston [10 p. 24], finding the appropriate representation is a major part of a problem-solving effort, and a good representation 'makes important things explicit' and exposes 'the natural constraints inherent in the problem' [10, 11]. The next section explains the new representation the authors propose, addressing the problems faced by existing representations.

3 Space-Time Network Representation for Manufacturing

The proposed Space-Time Network Representation for manufacturing networks aims to capture the entire network of parallel/simultaneous supply chain activities happening around the globe in a manufacturing endeavor in order to make the finished product and deliver to the end customer. This representation captures the spatial and temporal information of geographically distributed manufacturing activities, and generates the global network. In order to achieve that, information needed to convey the necessary spatial and temporal aspects have been selected, and details which could make the scenario complex and difficult to comprehend are omitted. A human-centered approach is taken to design this representation, to make manufacturing process drawings more intuitive, and easy to learn and draw, by stakeholders from plant engineers and supply-chain managers, to top-executives of a company.

The Space-Time Network representation plots space in one dimension and time in the second dimension. The spatial information needed in this representation is the set of geographic co-ordinate values (latitude and longitude) of plant or vendor locations. The temporal information needed for plotting are process durations, the start time of the day and location of the manufacturing plant in focus. Figure 6 explains the key notations of this representation.

Each location has a unique vertical line in the Space-Time Network representation. World Maps are used to project geo-locations to the space axis. The vertical lines are mapped following the rules listed below:

1. From West to East, the location coming first should be mapped as the first vertical line, starting from the left side of the diagram.
2. If rule one is not valid, from North to South, the location coming first is to be mapped in order from the left side of the diagram. Each location has a unique corresponding column in the process database used in this group.

A process is represented as a line segment between two points in the space-time diagram. Length of a line segment is the duration of the process represented.

In a factory, at every instant of time, many processes would go on at varying degrees of completion. Machines run continuously throughout the day (or night) producing outputs in batches. Similarly, at a given instant of time, production of parts, sub-assemblies and assemblies are all happening simultaneously. But, if we focus on a single lot (lot means a batch or a specific identified portion of a batch), the whole picture becomes clearer, and we should be able to trace the manufacturing network of that specific lot, forward and backward in time, across space. Hence, to plot time in our representation, a day and time of a process, and location i.e., a factory in focus, should be picked. At that snapshot of time at that location, one lot of a process chosen will be plotted and traced to generate the complete network.

Several notations are used for explaining incoming and outgoing material flows. The materials can be in the form of finished resources, end products, by-products, lubricants, effluents or packaging materials. These are explained as follows: Global Out or G_{out} denotes materials going out of a process from one location to another; Global In or G_{in} denotes materials coming into a process in a location from another location; Local Out or L_{out} denotes materials going out of a process in a location, but will not go outside that location; Local In or L_{in} denotes materials coming into a process in a location from the same location. These are represented by 45° line segments, with G_{out} denoted by right-outward line segment, G_{in} by left-inward line segment, L_{out} by left-outward, and L_{in} by right-inward line segments (see Fig. 7). Icons are used to represent men, machines, and transport involved along each process line segment (Fig. 6).

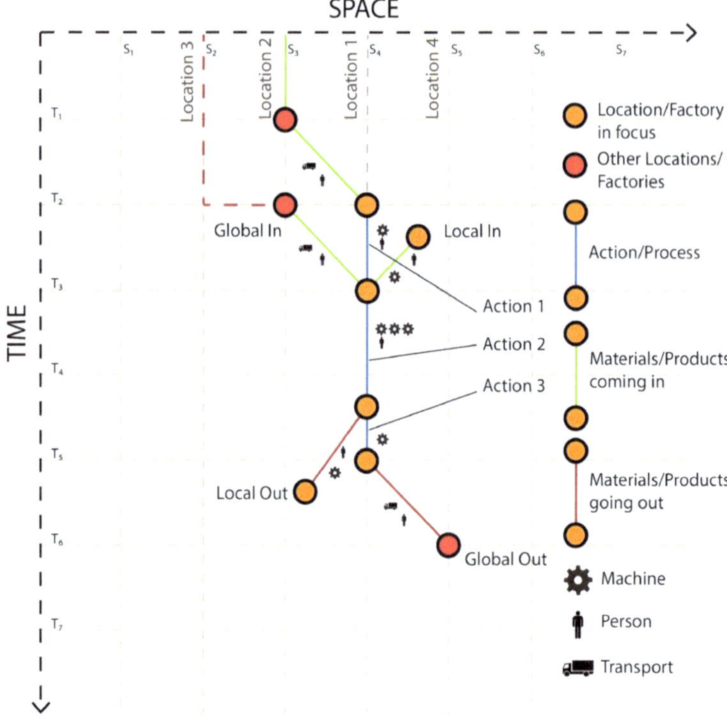

Fig. 6 Space-Time Network representation for manufacturing

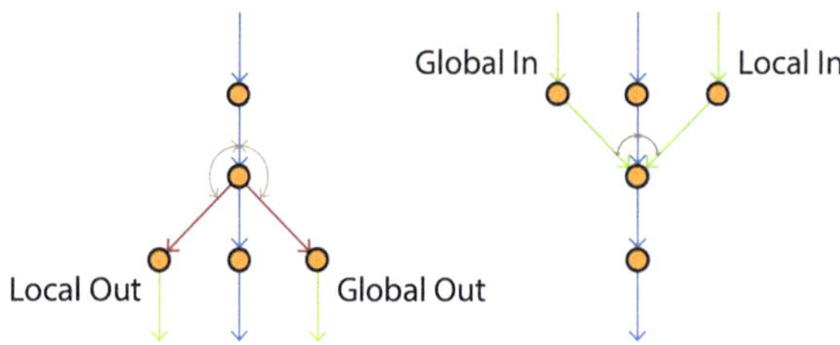

Fig. 7 Notations for Global and Local material flow

G_{out}, G_{in}, L_{out} and L_{in} act as the organizing elements of the space-time network representation. It connects the process chains of two or more locations using one or more of these notations to form a network. There are different possibilities of connections using these notations to generate a networked representation, as shown in Fig. 8.

An example of Coca Cola manufacturing process is used to explain the Space-Time Network Representation. The Coca Cola Company has a supply-chain network for beverage concentrate making plants, and bottling plants, where the drink is

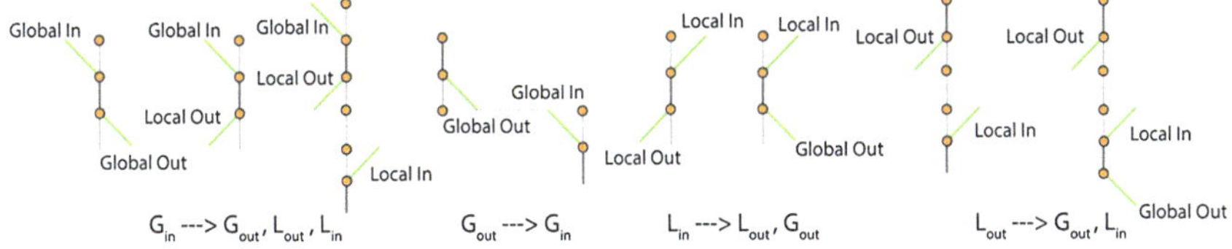

Fig. 8 Connecting networks using Global and Local In-Out notations

made from the concentrate and bottled to pet bottles, aluminum cans or glass bottles and packaged. The PET bottles, aluminum cans, glass bottles are manufactured and shipped to bottling plant by other vendors. The data available in open source for the Coca Cola manufacturing process used in this study was collected from the National Geographic Channel's television documentary on the company. Taking the bottling plant as our focus, the Space-Time Network is drawn for the activities taking place in this plant (See Fig. 9). A lot was traced using the process durations assumed. It is also possible to plot the activities of the PET bottle manufacturing plant to give the bigger picture of the company's manufacturing network. Figure 9 (left) shows the two networks together which gives insights into the dependencies of each process in one plant on processes happening in other plants, in a visual manner. The diagram clearly shows the relative temporal dependencies of a process over another process across space and time. For example, the filling process of a lot can start only after the PET bottle lot for this batch is manufactured and shipped, and join the process chain of bottling plant factory.

3.1 Advantages of Space-Time Network Representation

Unlike other spatio-temporal representations, as mentioned in Sect. 2, the proposed Space-Time Network method abstracts the two dimensional space into a single dimension. This helps in removing the complexity in visual representations so as to connect in and out flows of materials and processes across an otherwise two-dimensional spatial representation. Time is represented linearly in an axis going down, unlike in other space-time graphs where it is represented upwards. This helps in showing the processes which are completed above the ones which are going on at the moment, which is semantically a better way of presenting this information. If it were to be projected upwards, say in the coke example, PET bottle manufacturing process will be drawn below, and bottling process above. And one needs to look upwards to see the recent processes. Current conventions of writing, and drawing, for example, tree diagrams, are in a top-to-down manner. The left to right plotting of locations helps in identifying the relative positions of locations in the real world, even without a map. A location, plotted to the left of another, will be in the

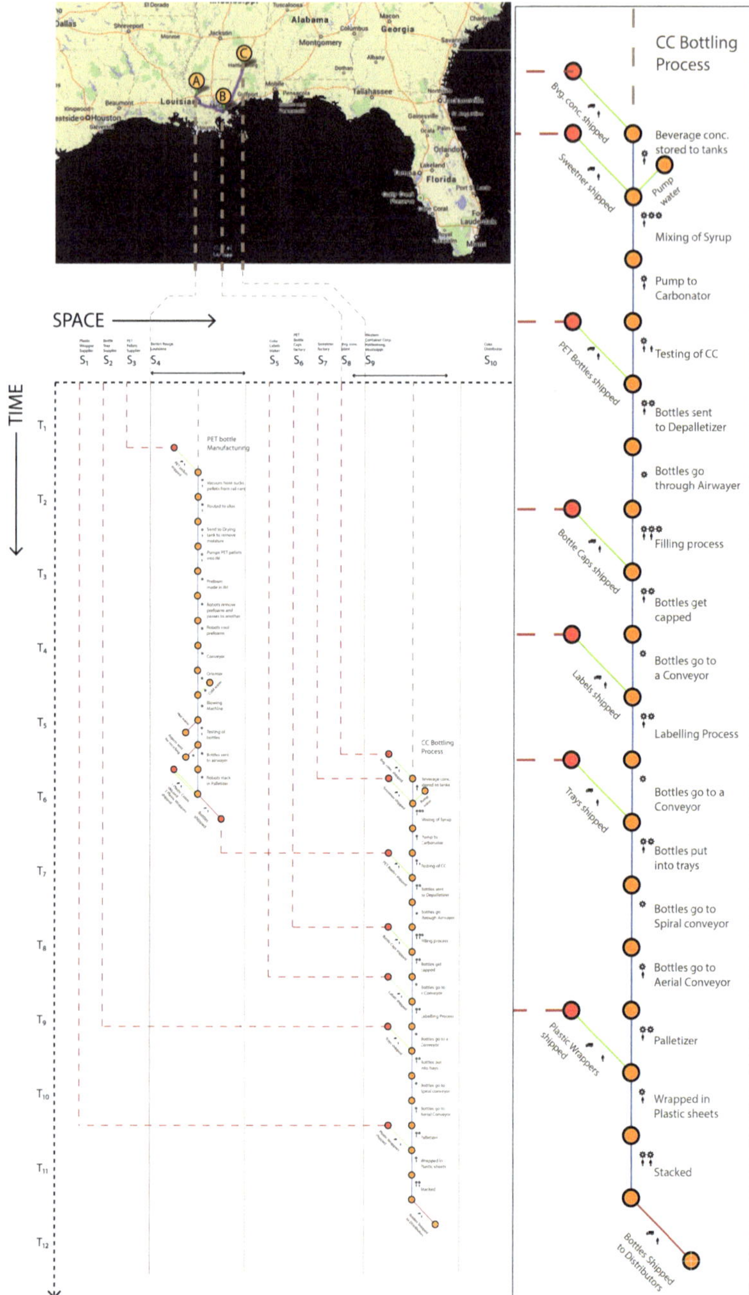

Fig. 9 Space-Time Network representation of PET bottle manufacturer and coke bottling plant (*left*). *Inset* bottling plant in focus (*right*)

geographic map, left of that location. The location-line, can also expand to form location slabs, if there are more information to be shown, without affecting the time axis (see S4 in Fig. 9). Time can be plotted using different units as demanded by the situation, either in seconds, minutes, hours, or days.

The model can capture the whole network as well as the partial network in multiple levels of hierarchy or granularity. If there is more information available for a network, more can be added in a modular manner without disturbing the rest of the diagram. This model can also be read in black and white, or in print medium, due to the use of unique notations for G_{out}, G_{in} L_{in}, and L_{out} for incoming and outgoing material flow. The model can also be easily drawn by hand by using these conventions for quick discussions. Vendors can also have their own space-time diagrams, and these could be put together to generate the complete network. Aggregation of vendor networks can be done easily in a hierarchical manner.

4 Conclusions

Diagrams and representations currently used within the manufacturing domain lacks the capability to capture aspects of the global manufacturing network. The Space-time network representation has the potential to fill this gap and become a tool to visualize, understand and comprehend complex manufacturing networks in a simple manner. There is need for a human-centred dimension in visual representations in manufacturing domain, which can be used by a spectrum of people in the industry, from vendors, engineers, managers, supply-chain planners and decision makers. The Space-time network representation is designed to give new insights into the manufacturing network. This method can also serve as a general set of rules or conventions to create such diagrams for a whole range of products, from consumer products, electronic products, automobiles and aerospace vehicles, to industrial machines and equipment. The method's capability, to be drawn by hand and its compatibility with print and digital versions, are of advantage in diverse manufacturing environments of different locations across the world, where access to computers is difficult. The representation was presented to experts from the industry. They were of the strong view that this form of representation will be useful in the actual planning, control and decision making of manufacturing projects. This representation is now being developed into a software tool for generating space-time network diagrams from datasets.

Acknowledgments The authors acknowledge the contribution of The Boeing Company for providing financial support under contract PC36018 at SID, IISc.

References

1. Adam, C., Robinson, P.: Manufacturing in the 21st century: an holistic future. Aust. Q. **66**(3), 17–36 (1994)
2. Xing, Y., Detert, N.: How the iPhone widens the United States trade deficit with the people's Republic of China. ADBI Working Paper 257. Asian Development Bank Institute, Tokyo (2010)
3. Arntzen, B.C., Brown, G.G., Harrison, T.P., Trafto, L.L.: Global supply chain management at digital equipment corporation. Interfaces **25**(1), 69–93 (1995)
4. Bowersox, D.J., Calantone, R.J.: Global logistics. J. Int. Mark. **6**(4), 83–93 (1998)
5. Vereecke, A., Van Dierdonck, R., De Meyer, A.: A typology of plants in global manufacturing networks. Manage. Sci. **52**(11), 1737–1750 (2006)
6. Bartlett, C.A., Ghoshal, S.: Managing Across Borders: The Transnational Solution. Hutchinson Business Books, Boston, MA (1989)
7. Geoffrion, A., Graves, G.: Multicommodity distribution system design by Benders decomposition. Manage. Sci. **29**(5), 822–844 (1974)
8. Breitman, R.L., Lucas, J.M.: PLANETS: a modeling system for business planning. Interfaces **17**(1), 94–106 (1987)
9. Cohen, M.A., Lee, H.L.: Manufacturing strategy concepts and methods. In: Kleindorfer, P.R. (eds.) The Management of Productivity and Technology in Manufacturing, pp. 153–188 (1985)
10. Winston, P.H.: Artificial intelligence, 2nd edn. Addison-Wesley, Reading, Massachusetts (1984)
11. Yuan, M., Mark, D.M., Egenhofer, M.J., Peuquet, D.J.: Extensions to geographic representations. In: A Research Agenda for Geographic Information Science, pp. 129–156. CRC, Boca Raton, FL (2004)
12. Parthasarathy, S., Kim, S.H.: Temporal reasoning for manufacturing: a hybrid representation and complexity management. Rob. Comput. Integr. Manuf. **6**(1), 67–81, ISSN 0736-5845 (1989)
13. Peurifoy, R., Schexnayder, C.: Construction Planning, Equipment, and Methods. McGraw-Hill, New York (2002)
14. Goldratt, E.M.: Production the TOC way. Revised Edition (2008)
15. Rudas, I.J., Horvath, L.: Modeling of manufacturing processes using a petri-net representation. Eng. Appl. Artif. Intell. **10**(3), 243–255 (1997)
16. Nyhuis, P., Wiendahl, H.P.: Fundamentals of Production Logistics: Theory, Tools and Applications. Springer (2009)
17. Miller, H.J., Bridwell, Scott A.: A field-based theory for time geography. Ann. Assoc. Am. Geogr. **99**(1), 49–75 (2009)
18. Jun, S., Kim, M., Lee, J.: The system diagrams: shifting perspectives. Des. Issues **27**(2) 72–89 (2011)
19. von Bertalanffy, L.: The history and status of general systems theory. Acad. Manag. J. **15**(4), 417 (1972)
20. Buchanan, R.: Design research and the new learning. Des. Issues **17**(4), 12 (2001)

Requirements Evolution: Understanding the Type of Changes in the Requirement Document of Novice Designers

Shraddha Joshi and Joshua D. Summers

Abstract This paper presents the findings from a case study conducted to investigate the evolution in the requirements documents of novice designers (students). The case study was conducted with four student teams working in parallel on senior design project. At the beginning of the semester, all four teams were given the design problem and set of initial requirements by the sponsor. For the purpose of this case study, data was collected from all four teams in the form of weekly requirements update. These requirements were then investigated to explore the evolution of requirements and identify the changes in individual requirements from initial weeks to final week. The findings from this case study suggest that the requirements document of novice designers changes in multiple ways. Finally, a set of recommendations and guidelines are developed to help the novice designers to maintain the requirements document and manage the changes in the requirements document.

Keywords Requirements evolution · Case study · Novice designer · Capstone design

1 Case Study Motivation

The goal of this research is to investigate the evolution in the requirements document of novice designers. In studying how requirements evolve from the start to the completion of a project, engineering educators can identify the types of changes that occur throughout the design project. This knowledge can then be used for developing tools and methods to support the novice engineers for both engineering design process and project management aspects.

S. Joshi
Mechanical Engineering, Carnegie Mellon University, Pittsburgh, USA

J.D. Summers (✉)
Mechanical Engineering, Clemson University, Clemson, USA
e-mail: jsummer@clemson.edu

A. Chakrabarti (ed.), *ICoRD'15 – Research into Design Across Boundaries Volume 2*, Smart Innovation, Systems and Technologies 35, DOI 10.1007/978-81-322-2229-3_40

In order to investigate the evolution of requirements, a case study is conducted with senior design students at Clemson University. Pattern matching is often used to study the analyzed data in case study research [1, 2] while counter-patterns are used to improve the qualitative objectivity [3]. Case study research has been used in the past to study novice engineers through capstone design projects at Clemson University [4–7].

The anticipated pattern for this study is that the requirements document of novice designers (students) will evolve in multiple ways; that there is no dominant mechanism for requirements change within a capstone project. This is based on the assumption that the students get feedback from the advisory committee throughout the project, thereby necessitating changes to the requirements. Moreover, these changes essentially illustrate the growth in problem understanding as the problem and solution co-evolve. Alternately, the counter pattern is that the requirements document will not change. This is based on the assumption that although the students receive weekly feedback, it is not specifically focused on the requirements document.

Capstone design class at Clemson University is a three credit, 15 week long class that falls at the end of the undergraduate curriculum and essentially serves as an exit exam for the mechanical engineering students. Student teams are created based on a balancing of expertise and work experience. Four teams of students are assigned to the same industrial sponsored project to work in parallel. The sponsored projects are funded to support prototyping, while serving as an incentive for the customer to be responsive to the team. The teams meet weekly with an advisory committee, consisting of graduate and faculty advisors. These meetings are design reviews where the teams get feedback on the project, but without direct management by the advisors. This course has been thus structured for roughly three decades, with many research studies derived from this course (e.g., [8]).

The case study to investigate the evolution of requirements in the requirements document of novice designers was conducted on one of the capstone projects from Spring 2011. The project was sponsored by Parker Hannifin and the problem given was:

Design and build a system to automatically splice seals.

For this project four teams of four students were given a presentation explaining the problem and sponsor requirements. Each team worked independently to further define and elicit requirements while developing solutions. The project is representative of a typical capstone design project at Clemson University, considering factors such as the complexity of the problem, the formation of the teams, the composition of the advisory committee, and the execution of the process.

As the goal of this study is to investigate the evolution of the requirements, the data collection entailed collecting weekly requirement documents from all four teams. In order to facilitate the data collection, a requirements update sheet was created as shown in Table 1.

This sheet captures details such as requirement, whether it was a constraint or criteria, the source of the requirement, justification for having the requirement, the

Table 1 Snap shot of requirements update sheet

Requirement number	1
Requirement	Grasp extruded circular cross-sections that range in thickness from 0.070″ to 0.250″ and diameters from 2″ to 2′
Type	Constraint
Source	Sponsor
Justification for requirement	Requested by sponsor
Date of elicitation	20-Jan
Target value	Described in requirement
Verification method	Calculations or actual results
Update	n/a
Reason for update	n/a
Description of change	n/a
Comments	

target value, verification method, and any updates. All four teams were asked to update the requirements and submit it every week. This requirements documentation tool is similar to that which is taught in the preceding course and is common in design textbooks [9].

However, submitting the weekly documents was not a course requirement, thus most teams did not submit the requirements every week. The frequency of submission varied between the teams and, thus, the comparisons presented in this paper are made between initial weeks and final weeks. As with best practice in case study research, the context of study was intentionally not altered. It may be noted that initial week may refer to either week one, two or three. While the self-reporting of the requirements documents were not sufficient to ensure weekly change evolution, the study presented here is sufficient to provide preliminary evidence of the types of changes that occur within capstone projects. To this point, no other studies have been found to have explored this aspect of requirements in engineering design education.

2 Requirements Tracing Protocol

Three evolution aspects are investigated: (1) the number requirement additions, deletions, modifications, and no changes, (2) the changes in the requirement elements (system, necessity, behavior, object and condition), and (3) the types of change per the taxonomy of [10]. Protocols were established to study each of these change types. It may be noted that since this is a preliminary study investigating change types in requirements of novice designers, the goal here is to identify whether or not the requirements documents of novice designers change. Therefore, the robustness of these protocols is not tested for this preliminary study. In order to

Number(state)	7(I)	7(F)	9(I)	9(F)	6(F)
Raw Requirement	Cool spliced rings	The design must cool the spliced rings in less than or equal to 3 minutes	Count the number of parts	The design must count the number of parts produced	The design must prevent breakdown in elastomer properties by ensuring that o-rings are not heated beyond 350 F
Subject		The design		The design	The design
Modal		must		must	must
Main Verb	Cool (T)	cool (T)	Count (T)	count (T)	prevent(T)
Direct Object (DO), NA, complement (CO)	spliced rings (DO)	the spliced rings(DO)	the number of parts (DO)	the number of parts produced (DO)	breakdown in elastomer properties (DO)
Adjunct		in less than, or equal to 3 minutes			by ensuring that o-rings are not heated, beyond 350F
Change Type	Introduction of new system Importance Specificity		Introduction of new system Importance		Addition
NL requirement element	System Necessity Condition		System Necessity		N/A

Fig. 1 Example requirements tracking sheet for team A

study this evolution, a requirement tracking sheet was created for each team and each requirement was traced as it evolved.

Figure 1 illustrates a snap-shot of requirements tracking sheet for Team A. The first row represents the requirement code and includes two elements, the requirement number and the state (I = initial weeks; F = final week). The mapping of the requirements from the initial document to the final document was done manually based on semantic content of the requirement. Some requirements were deleted and new requirements added. Thus the change type for these requirements was identified as 'addition' or 'deletion'. For example, in Fig. 1, there was no 6(I) requirement to correspond to the 6(F) requirement as this was only found in the final document. This requirement is designated as "addition" for change type. Similarly, if a requirement in the initial requirement document is not found in the final, then the change type is "deletion". If there was a change to the requirement, then the aspects of change are recorded. These change types are based on the work of [10].

The next step is to identify specific changes in requirement in terms of the elemental changes. The requirements are parsed into the elements: system (subject), necessity (modal), behavior/characteristic (main verb), and object and condition (complement or adjunct) [7, 11]. The elements that are different between the requirements are highlighted in Fig. 1. For instance, consider requirement 7 for Team A:

7(I): Cool the spliced ring.
7(F): The design must cool the spliced ring in less than or equal to 3 min.

In 7(F), the subject "design", modal "must", and adjuncts "in less than or equal to 3 min" are added to 7(I). These changes are highlighted (system, modality,

Table 2 Description of change types, adapted from [10]

Change type	Description (Identified as a change …)
Introduction of new system	In the system of the requirement. This could be addition of new component or system
Consistency	In vocabulary, units or terminology
Importance	In the necessity of requirement
Specificity	In the level of detail of a requirement
Application	In the application of part or system
Measurability/testing	In the measurability of a requirement
Withdrawal of system	In the system of the requirement. This could be a removal of system or component
Merging	When two or more requirements are merged into single requirement
Associated user	When the individuals associated with the requirement change
Splitting	When one compound requirement is split into two or more requirements
Scope change	Identified as change when the scope or focus of a requirement changes

adjunct). Identifying the requirement elements is based on the protocol of [11]. After identifying the change in the requirement elements, the next step is to identify the change types based on the taxonomy of change types established in [10]. The definitions and examples of each change type provided in the taxonomy are used to identify the change type for each requirement pair. Table 2 provides explanation of change types identified by examining the requirements for this study based on the types proposed by [12–14].

A requirement pair can have multiple change types, as seen in the changes for requirement 7(I)–7(F) with changes in the introduction of a new system, the definition of an importance for the requirement, and a change in the specificity of the requirement (Fig. 1). The change types for all requirement pairs for all teams are identified in this manner and a summary table created by counting the number of each change type for each team.

3 Findings from Requirements Tracing

3.1 Number of Additions, Deletions, Change and No Change

The first evolution type investigated in this case study is the high level modification of the requirement document. This includes changes such as addition of, deletion of, change to a requirement (Fig. 2). It may be noted here that, this is a preliminary study to investigate the change types. While only four teams were investigated, the goal was to identify whether or not any change types can be identified in the requirements documents of novice designers. Once, these patterns are identified,

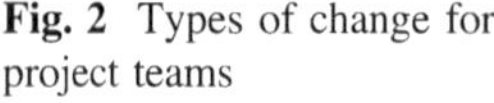

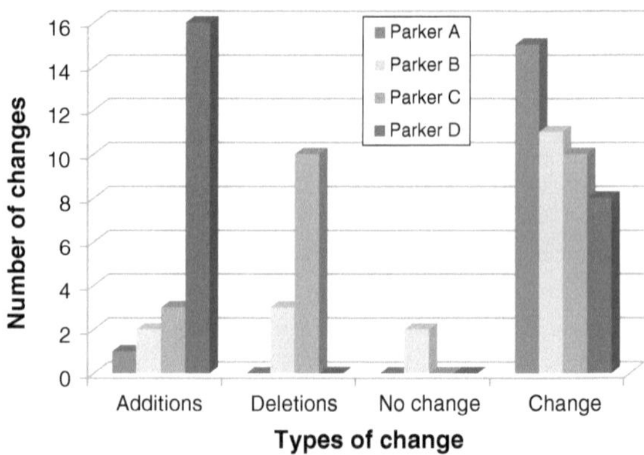

Fig. 2 Types of change for project teams

extensive studies with multiple teams and different projects will be conducted to further support the findings. However, these four teams are representative of the other senior design teams in terms of the complexity of project, execution of the design process and final deliverables. So the findings can be extended and the recommendations are valid for other senior design teams.

From Fig. 2, it is clear that each team had different types of changes to the requirement document, with the most common change being a detail change to an individual requirement. Team D had eight requirements in initial week and 24 requirements in final week resulting to the most number of additions. Team C, on the other hand, had 20 requirements in initial week and 14 requirements in final week resulting to the most number of deletions. Of the four teams, only team B has requirements with "no change". These findings suggest that requirement document of novice designers' change significantly from initial weeks to final weeks. Further, these changes are not just limited to addition of new requirements as the understanding of the problem grows but also include deletions and modifications of existing requirements.

3.2 Change in Natural Language Requirement Element

The second evolution type investigated in this case study is tracing the change in natural language requirement elements [10]. This accounts for changes in the system, necessity, behavior/characteristics, object or condition of a requirement. Figure 3 illustrates the overview of the changes in the natural language requirements elements.

From Fig. 3, it can be observed that all four teams had multiple changes in the natural language requirement elements. Most number of changes was observed in the 'system' element. The least number of changes were found in the 'object' element of the requirement sentences.

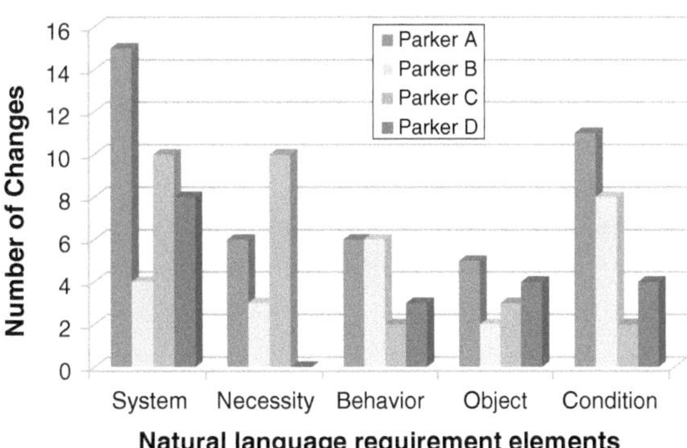

Fig. 3 Changes in requirement elements

Syntactically, 'system' is represented by the subject within a sentence. Thus, the system changes also represent the change in the subject of the requirement sentence. There were several types of system changes observed across the four teams such as introduction of new system, splitting, with drawl of a system and consistency.

The element of 'necessity' within a requirement statement is syntactically represented by a modal and it shows the importance of the requirement [10]. Necessity element of requirement changed for all teams except team D. Investigating the requirements tracking sheet for all four teams, it was found that for most teams, the change in 'necessity' stemmed from addition of a modal such as 'must' or 'should' in the final week requirements document.

Next, the element of 'behavior' within a requirement represents the function of the system or component being designed and is syntactically represented by a verb [10]. Most changes in the 'behavior' stemmed from either change in the function of the system (change type—application—for example—must "remove" excess adhesive and must "minimize" excess adhesive) or inconsistency in the vocabulary used to describe the function of the system (change type-consistency—for example —must "provide" economic advantage and must "present" economic advantage). Apart from these two, the behavior changes also resulted from merging of two requirements, splitting of a requirement into multiple requirements and change in the scope or measurability of requirement.

The next element is 'object' and it represents what the system is affecting [10]. The changes in the 'object' of a requirement stemmed from change in the vocabulary within requirements. So for instance, one of the requirement for team D changed from "*the system must count completed spliced parts*" to "*The final design solution must count spliced O-rings*". Here the object changed from "spliced parts" to "spliced O-rings", which is essentially inconsistency in the vocabulary as the completed spliced parts are the O-rings.

After object, the next natural language element of a requirement is 'condition'. The changes in the 'condition' stemmed from change in the vocabulary or change in the specific details of the requirement. For example, condition for requirement-2 for team B changed from "ranging from 2 inch to 2 feet in ring diameter" in initial week to "ranging from 2 inch to 24 inch in ring diameter" in final week. Here the unit changed from 2 feet to 24 in. Other type of condition changes resulted from change in measurability, application, change in associated user, change of scope or splitting of a requirement.

Thus, this section describes the changes in the natural language elements of a requirement. Each of these elements is critical within a requirement and any changes must be documented with the justification for the change. Next, Sect. 3.3 describes the changes in the requirements document as per the change taxonomy.

3.3 Change Type as Per Change Taxonomy

The next evolution type investigated within the requirement documents was to trace the 'type' of changes as per the change taxonomy established in [10]. It may be noted that while the taxonomy captured all the change types occurring in student requirement document, not all the change types described in the taxonomy were found in the requirement documents. Example of some of these change types include replacement of system, updating, incorrect raw data interpretation and correcting among others. Figure 4 illustrates the summary of change types for all four teams.

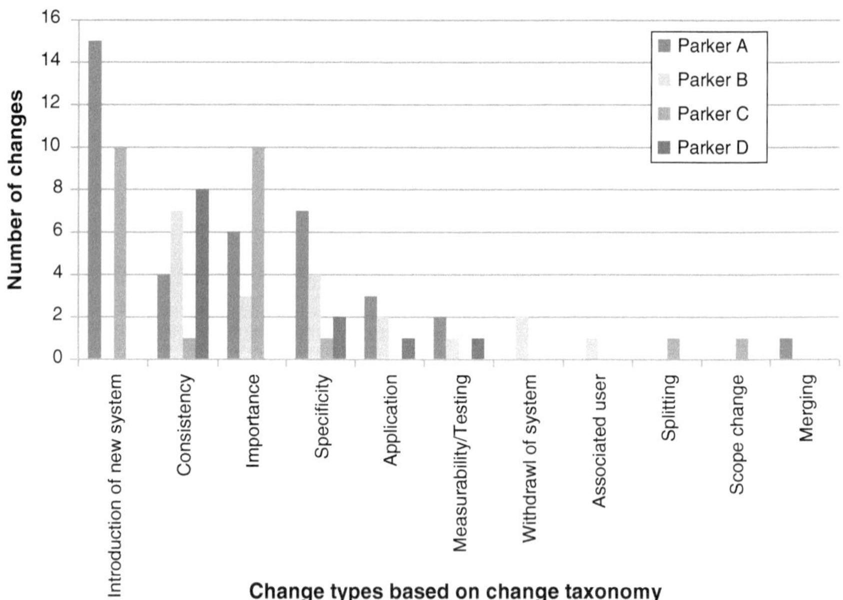

Fig. 4 Change type occurrences

It can be observed from Fig. 4 that all teams had multiple types of changes in the requirements, with the introduction of a new system being the most frequent change type. This change was observed in teams A and C as a result of addition of the subject in the requirements for final week. Other change types that are more frequently observed across all four teams include consistency, importance and specificity. These changes stemmed from change in vocabulary or units, addition or deletion of a modal and change in the specific details of requirement respectively.

The change types that are less frequent include application, measurability/testing and withdrawal of system, while the change types associated user, splitting, scope change and merging only occurred once. Further, the changes that are less frequent were mostly accompanied by other change types that are more frequent. So for instance, splitting and merging occurred once, but in each case they were accompanied by a new system.

Some of these changes can be more critical than others and if not properly document can lead to information loss. For instance the change types such as splitting or merging involve either separating a compound requirement or uniting two requirements into one. This may not always lead to major change in information as it is essentially re-writing requirements in different form. For example the requirements "*(the system) Must handle organic and silicone adhesives*" and "*(the system must) apply adhesive to ends of extrusion*" were merged into one requirement "*The design must have the capability to apply either organic or silicone adhesive to an extrusion end*" which is essentially conveying the same information.

On the other hand the change types such as introduction of new system, importance, specificity and consistency which can alter the purpose of a requirement can prove to be more critical making the documentation of these changes more important. For example for team C, the requirement in initial document changed from "*Apply* ***controlled amount of*** *organic or silicone adhesive to extrusion ends*" to "*The system must* ***apply 1 drop per square inch*** *of organic or silicone adhesive to extrusion ends*" in the final document. Here the team added specific detail about the "controlled amount" and thus this change is specificity change. Adding information about how much adhesive enhanced the purpose of this requirement making this change more critical compared to just merging or splitting a requirement. It is interesting to note that the changes which are more critical are also more frequently observed in requirement documents of the teams.

4 Conclusion and Recommendations

It was observed that requirements documents have multiple additions, deletions, and changes from initial weeks to final week, though the novice designers rarely states the justifications for the changes. It is interesting to note that other documents such as weekly summary, presentations, mid-term and final reports that the students are required to submit as a part of project deliverable do not mandate providing updates on requirements. None of these documents require the students to provide

Table 3 Recommendations for faculty

	Recommendation for faculty	Benefit	Implementation
1.	Mandate novice designers to maintain and submit weekly requirements update	This will help novice designers to manage and track the changes in the requirements document	Faculty may require the students to make weekly submission of requirements update sheet illustrated in Table 1 as a starting point
2.	Educate novice designers to identify high level document changes such as additions, deletions and change and be able to justify them	This will help novice designers to prevent accidental additions or deletions of requirements	Faculty may use in protocol 1 discussed in this paper to educate students about high level requirement document changes
3.	Educate novice designers to identify changes in requirement elements and be able to justify the changes	This will help prevention of potential information loss that could jeopardize the successful completion of the project	Faculty may use protocols 2 and 3 discussed in this paper to educate novice designers to identify changes in requirement elements and changes based on taxonomy

justification for the changes in requirements document. If there is no justification for these changes, valuable information could be lost. Further, the novice designers are not able to trace if a requirement was accidently deleted or added if they do have the documented justification for additions and deletions of requirements. Next, investigating the evolution in the requirement elements, it is shown that each type of element can change throughout the project, yet again the students rarely documented the justification for the change. A survey of design textbooks was conducted to investigate the tools pertaining to requirements [9] and none of the textbooks describe tools for managing the requirement changes. With the lack of tools or methods to track requirement changes, the novice designer cannot track down if a change was accidently introduced in a requirement. Finally, the evolution in the requirements document in terms of changes as per change taxonomy was investigated. Again, if the students are not taught to identify these change types, the criticalities associated with the change types and appropriate tools or methods to document and track these changes; it can lead to potential loss of valuable information and jeopardize the successful completion of the projects.

Several recommendations are made based on the study presented above (Table 3). These recommendations may serve as guidelines for faculty helping novice designers manage changes in the requirements documents.

References

1. Yin, R.: Case Study Research: Design and Methods. Sage, Thousand Oaks, CA (2003)
2. Tellis, W.: Application of a case study methodology. Qual. Rep. **3**(3), 1–17 (1997)
3. Teegavarapu, S., Summers, J.D., Mocko, G.M.: Case study method for design research: a justification. In: International Design Engineering Technical Conferences and Computers and Information in Engineering Conference, p. DTM–49980. ASME, Brooklyn (2008)
4. Joshi, S., Summers, J.D.: Investigating information loss in collaborative design: a case study with capstone design project. In: National Capstone Conference 2010, p. 51. Boulder, (2010)
5. Joshi, S., Morkos, B., Summers, J.D.: Requirements analysis: case study with capstone design project. In: National Capstone Conference 2012, p. 67. Champaign-Urbana, IL (2012)
6. Maier, J.R.A., Troy, T., Johnston, P.J., Bobba, V., Summers, J.D.: Case study research using senior design projects: an example application. J. Mech. Des. **132**(11), 111011 (2010)
7. Joshi, S., Summers, J.D., Morkos, B.W.: Requirements evolution: understanding the type of changes in requirement documents of novice designers. In: International Design Engineering Technical Conferences and Computers and Information in Engineering Conference, pp. DETC2014–35023. ASME, Buffalo (submitted) (2014)
8. Morkos, B., Summers, J.D.: Implementing design tools in capstone design projects: requirements elicitation through use of personas. In: National Capstone Conference 2010, p. 50. Boulder (2010)
9. Joshi, S., Morkos, B., Shankar, P., Summers, J.D.: Requirements in engineering design: what are we teaching? In: Horvath, I. (ed.) Tools and Methods for Competitive Engineering, p. 38. Karlsruhe (2012)
10. Morkos, B., Joshi, S., Summers, J. D.: Representation: formal development and computational recognition of localized requirement change types. In: International Design Engineering Technical Conferences and Computers and Information in Engineering Conference, pp. DETC2012–71417. Chicago (2012)
11. Lamar, C., Mocko, G.M.: Linguistic analysis of natural language engineering requirement statements. In: Horvath, I. (ed.) Tools and Methods for Competitive Engineering, pp. 97–111. Ancona (2010)
12. Rolland, C., Salinesi, C., Etien, A.: Eliciting gaps in requirements change. Requir. Eng. **9**(1), 1–15 (2004)
13. Stark, G.E., Oman, P., Skillicorn, A., Ameele, A.: An examination of the effects of requirements changes on software maintenance releases. J. Softw. Maint. **11**(5), 293–309 (1999)
14. Lin, L., Poore, J.H.: Pushing requirements changes through to changes in specifications. Front. Comput. Sci. China **2**(4), 331–343 (2008)

Integration of Strategic Flexibility into the Platform Development Process

Fatos Elezi, Rita Tschaut, Wolfgang Bauer, Nepomuk Chucholowski and Maik Maurer

Abstract Platform systems are used for obtaining and sustaining competitive advantage, as they make possible derivation of a variant-rich product portfolio and at the same time provide cost advantages in development, production and assembly. In this paper, a methodology for conception of a platform system that is robust to internal and external dynamic changes, and which at the same time supports flexibility with respect to these dynamic influences is proposed. The proposed methodology is evaluated in a real industrial setting and has shown much better results as an unstructured method for incorporation of planed flexibility. In addition, the methodology is to date incorporated into the platform design process of the industrial partner and is used on day-to-day basis.

Keywords Strategic flexibility · Product flexibility · Platform system · Modularity

1 Introduction

Manufacturing firms are operating in an environment where market power has shifted from producers to consumers. This is due to globalized markets, which have fractured from mass markets into heterogeneous niche markets [1]. The fractured market requires high quality, favorable, customized products in shorter product life cycles [2] while competitors are introducing new products in shorter intervals. As a consequence, typical market conditions have become extremely dynamic and turbulent. Companies have to respond quickly to these changing market demands for providing a versatile product portfolio for rapidly segmenting markets [3]. In order to do so, companies are seeking to incorporate strategic flexibility as an approach to deal with these dynamic and turbulent changes [4]. Strategic flexibility refers to the ability of companies to plan, adapt and respond to external changes. The first author

F. Elezi (✉) · R. Tschaut · W. Bauer · N. Chucholowski · M. Maurer
Institute of Product Development, Technische Universität München, Munich, Germany
e-mail: elezi@pe.mw.tum.de

A. Chakrabarti (ed.), *ICoRD'15 – Research into Design Across Boundaries Volume 2*,
Smart Innovation, Systems and Technologies 35, DOI 10.1007/978-81-322-2229-3_41

to extend the flexibility concept into strategy was Sanchez [5]. He states that strategic flexibility depends not only in the inherent flexibility of the firm's resources but also on firm's flexibility in applying those resources to different alternatives. This means that it is not enough just to have flexible resources but there need to be adequate processes that help operationalize these flexible resources. These processes, that according to Eisenhardt and Martin [6] are also called dynamic capabilities, are strategic decision-making, alliancing, and product development. In relation to product development, Buganza and Verganti [7] outline product flexibility and development process flexibility as two important perspectives of flexibility. In literature, however, the dynamic capabilities in relation to flexibility—strategic decision-making and product development [6]—are not treated in an integrative way, although in industry settings these two processes are highly interrelated and exactly in their interfaces most of the value in relation to product flexibility is generated. Strategic decision-making in relation to product flexibility is concerned more with deciding about the necessary functional flexibility in the products for achieving the sustainable competitive advantage, whereas product development in relation to flexibility is concerned about how to technically make possible this flexibility in an efficient and cost-sensitive manner. It is obvious that the stakeholders of these processes need to communicate in an efficient and pragmatic way so that the company incorporates fully the notion of product flexibility.

As a conclusion, the product flexibility can be implemented in an industrial setting only by having appropriate strategic decision-making processes and appropriate product development processes. In addition, these two approaches need to be synchronized and come together with a coherent result which is accepted and agreed upon in both levels—strategic and architectural (design) level. In the research project "SFB 768 transfer subproject", a methodology for integrating these two important processes for strategic flexibility is being developed. However, in this paper we focus only on the strategic decision-making in flexibility planning.

2 Theoretical Background

2.1 Modular Products and Platform Strategies

Modularization is described by Schuh [8] as a reduction of dependencies between product elements (modules) as well as a reduction of element interfaces. Therefore, modules can be understood as subsystems with different functions, but with standardized interfaces, that are both functionally and physically relatively independent of each other and allow diverse combinability [9–13].

According to the understanding of [12], a *platform system* constitutes of standardized and individualized elements. The standardized elements form a stable, uniform basis for all final products. The individualized elements use this uniform basis to form a number of variants of the product range. Accordingly, the modular product system within a platform strategy comprises a robust part and a flexible part, both clearly separated.

The term *platform strategy* describes the coordination of product portfolio- (internal) and market-oriented (external) strategic focus concerning product platforms with the objective of applying cost reduction potentials while offering a large and rapidly changeable product variety [12]. The platform strategy thus is a hybrid competitive strategy combining aspects of cost leadership- and differentiation strategies.

For the *implementation of a platform strategy* and its objectives modular product design is necessary. The cost-optimizing aspect of the platform strategy requires operative measures at the level of product architecture. The relatively large amount of standardized components within the platform system enable cost-effective production. The remaining flexible part serves the differentiating part of the strategy. Ley and Hofer [14] also point out that the platforms should be standardized across the entire product family and also should be stable in time. Besides product components also technologies, processes, knowledge and organizations are considered part of a platform system [15–17].

In summary, platform systems as a type of modular product systems provide two main benefits. For once, they allow quick response to dynamic markets by eased generation of new variants. Furthermore, cost efficiency can be raised due to realization of standardization and economies of scale in the stable platform components. To ensure long-term robustness of such a platform system, however, it is required to anticipate future changes relevant to platform system and in addition assess the required flexibility from the strategic perspective. Basic definitions for these notions are given in the following section.

2.2 Product Changeability, Anticipation and Planned Flexibility

Over time, i.e. during development phases, technologies applied in products evolve, market- and other external conditions vary, overall leading to altered requirements to a product [2]. To achieve long term value robustness for a platform system, it needs to have the ability to change, i.e. show *changeability* [18]. The more options for change a system has as a reaction to the effects of external change, the higher its changeability [18].

As a prerequisite to incorporating changeability in product system design, *anticipation* of future external and internal environmental changes needs to be performed. Anticipation is a future-oriented action, decision or behavior, which is based on predictions about the future [19]. According to Rosen [20]—founder of anticipation theory—an anticipatory system is defined as a system that contains predictive models of itself and/or the environment that allow the system state to change immediately on the basis of the predictions of the models. The focus of this work lies on anticipation of platform development activities.

For successful application of both anticipation and design for product changeability, these approaches need to be combined with the definition of *planned flexibility* according to [21]. Planned flexibility includes (1) anticipation of and (2) consideration of responsiveness to future changes in terms of flexibility of resources, flexible communication, parallel developments, redundancies and flexible production technologies, and modular product architecture [21]. Through a case study, Verganti [21] could prove that companies that anticipate in the early development stages and thereby plan responsiveness of the product system to later changes, generate shorter time to market and ensure higher product quality.

From the discussion above, it can be deduced that design for changeability is a form of implementing planned flexibility and thus must be a foundation for the development of platform systems. However, in existing procedures and methodologies suggested to support design of platform systems, neither planned flexibility, anticipation nor design for changeability in early development stages are explicitly included [22]. Therefore, a novel approach is required to incorporate planned flexibility into platform system conceptualization. The development of such a novel approach is the objective of this paper and is described in the next chapter.

3 Developed Methodology

3.1 Assumptions and Approach

As a basis for the presented approach it is assumed that incorporating dynamic, internal and external, predictable as well as unpredictable influence factors into conceptual design of platform systems increases value robustness of such systems during their whole lifecycle [23]. These dynamic influencing factors (DIF) thus represent the initiation of planning for flexibility in the presented approach. DIFs are part of superordinate trends. They originate in the designed system itself or in external environments such as the market. DIFs influence inputs and resources on which determining parameters of conceptual platform design is based. To enable transparent identification of said influence, the so-called change priority indicator (CPI) is proposed. The concept of CPIs is a variation of the FMEA method with the difference that for CPIs the degree of required product flexibility is assessed rather than risk of failure. Determination of DIFs and CPI is embedded in one of standard procedures for platform development [12] shown in Fig. 1 to incorporate planned flexibility into the development process of product platform systems.

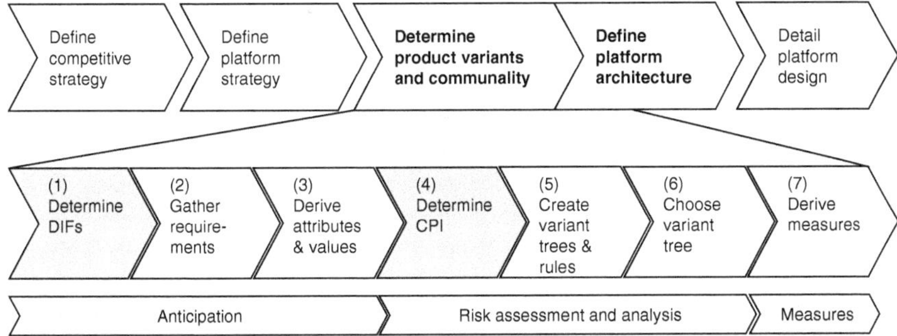

Fig. 1 Planned flexibility (*highlighted grey*) incorporated into standard procedure for platform system design

3.2 Methodology Steps

In the following, all seven suggested steps (see Fig. 1) of the methodology are briefly described regarding their purpose and content.

Step (1) is concerned with determining the DIFs for the investigated platform system based on the abovementioned assumption that dynamic influence factors are crucial for a platform system's economic robustness over time. It is suggested to collect DIFs based on e.g. marketing- and benchmarking data. The DIFs should be described concerning their dynamics, source, relevance and fields of impact.

Step (2) covers gathering of all internal and external requirements to the platform system. Based on the identified DIFs from step (1), additional future requirements or future changes to current requirements are to be anticipated in addition to common requirements analysis techniques.

In *step* (3), the platform system's products' attributes and their respective values are derived from the previously defined requirements. At this point, realized product variance and (part-/module-) communality of the platform system are determined.

Step (4) includes the CPI into the platform development process and thereby allows for quantitative assessment of required changeability. Subject to this assessment are the product attributes identified in step (3). Each attribute is rated in three categories on a defined scale from 1 to 10: probability of change (P), dynamics of change (D) and customer perception of change (C). These three factors are multiplied, resulting in the CPI as shown in Eq. 1.

$$\mathrm{CPI} = \mathrm{P} \cdot \mathrm{D} \cdot \mathrm{C} \qquad (1)$$

Differentiation into stable, invariant functions which represent the platform itself, and flexible product functions, later producing product variance, is

Table 1 CPI threshold values for stable and flexible platform components

Feature/function type	Factor values or	CPI value
Stable	C < 3	CPI ≤ 175
Flexible	P > 6, D > 8, C > 8	CPI > 175

realized by threshold values as listed in Table 1. These values are derived from scale definitions for the individual factors.

Step (5) is concerned with deriving alternative variant trees from the gathered product attributes, their necessary values and classification into stable and flexible components. Creating several alternatives ensures objective and transparent design decisions.

In *step* (6), the most suitable alternative created in the previous step is chosen in reference to the whole set of requirements, the determined CPI values and to costs for realization of the alternatives.

Step (7) concludes the herein presented procedure by deriving measures for further development. Management decisions, resource planning and design consequences are to be made referring to the CPI and thereby to risk assessment and anticipation of future developments. Categorization of DIFs concerning their fields of impact (cf. step 1) can be a useful source for decision making.

4 Case Study

The 2 steps for incorporating strategic flexibility into platform design processes described in the section above were applied in an industrial environment. At an international home appliances company, experts from departments of product management, product development, and design were engaged in moderated workshops to apply the method and subsequently evaluate applicability and quality of results.

4.1 Case Study Objective, Data Acquisition and Procedure

Initial identification of DIFs was performed with seven workshops, each involving two or three company experts. A list of known trends within the addressed industry served as a basis for DIF determination. Therefore, all categories of the context model [24] (politics, legislation, socioeconomics, technology, resources, knowledge, manpower and organization) had been considered at the levels of market, consumer, company, development system and development project. The list was filled with influencing factors utilizing the brainstorming technique. DIFs first were described by their time of initial occurrence, their dynamic characteristics and then

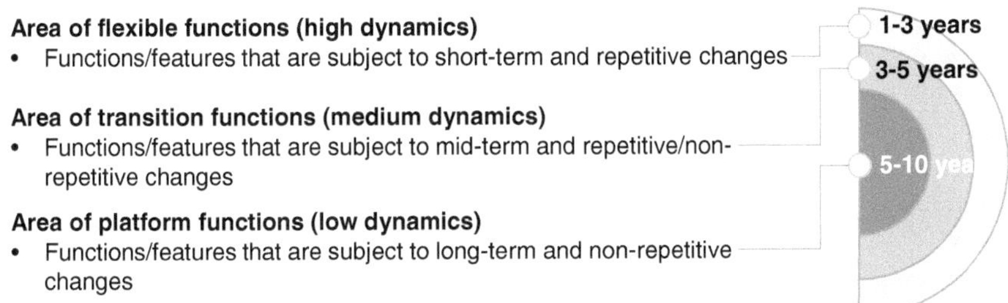

Fig. 2 The shell-model, *Source* adapted from industry partner's internal document

mapped to their fields of impact. This information served as input to the CPI workshops which represent the major focus of this case study.

Assessing the CPIs was undertaken within a workshop of seven product managers with experience ranging from 1 to 6 years. A second group of six specialists from comparable disciplines used the company's own, previously used shell-model (Fig. 2) to serve as a comparison group. Both groups used the same database of 41 consumer-relevant functions from eight strategic topics. Each function was separately discussed and rated during the workshops, both for the CPI-method and the shell-model. In addition, defining the CPIs was based on the previously collected DIFs as background information especially for assessment of the CPI-factors P and D (see Sect. 3.2, step 4).

To enhance comparability between the CPI and the shell-model approach, the CPI was normalized to CPI_N. It was decided by the workshop participants, that the areas for assessing flexibility are divided equivalent as in the shell-model (transition area was added in the middle of flexible and platform areas) and the CPI thresholds for these areas were set as following:

- Area of flexible functions: $70 \leq CPI_N < 100$.
- Area of transition functions: $40 \leq CPI_N < 70$
- Area of platform functions: $CPI_N < 40$

4.2 Results

In the initial DIF workshops, 233 different influencing factors were identified. From this number, 45 % of the DIFs could be characterized in terms of their dynamic properties. The time of initial occurrence could be determined for 25 % of DIFs. Mapping the DIFs to their respective fields of impact showed that about two thirds of all DIFs affect external fields such as market, customer requirements, technology and product definition, while one third impact internal topics such as finance, organization, production and strategy. The definition of DIF fields of impact was rated as adequate by seven of the eleven participants and the mapping results were

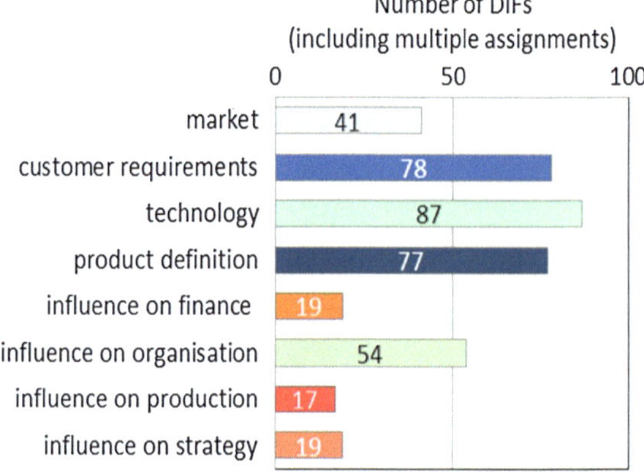

Fig. 3 Distribution of DIFs by fields of impact

rated as good. Figure 3 shows the defined fields of impact and the number of assigned DIFs.

In the CPI-assessment-workshops, a total of 41 product functions were considered. As a result of the CPI-method, 17 of those functions were assessed as platform functions whereas only six were rated as such using the conventional shell-model-method. The highest fraction of functions following the shell-model-method is the area of transition functions with 51 % (21 functions), compared to a fraction of 36.5 % (15 functions) resulting from the CPI-method. The area of flexible functions contained 14 functions using the shell-model method and nine using the CPI-method. Overall, there is an increase in platform functions and a decrease in flexible- and transition functions by applying the CPI-methodology instead of the conventional shell-model. The results of the CPI- and shell-model-workshop are presented in Fig. 4.

All functions rated as platform functions using the shell-model were ranked equivalently using the CPI-method. This consistency serves as an indicator for both the reliability of results from the CPI-method and the absence of a systematic error in the definition of the area of platform functions. Divergence of the results comparing both approaches can be explained as an effect of inadequate feature discussion as part of the shell-model-method and lack of knowledge about DIFs related to functions. Namely, when shell-model-method was applied, the workshop participants tended to be subject of group thinking. This can be observed in the

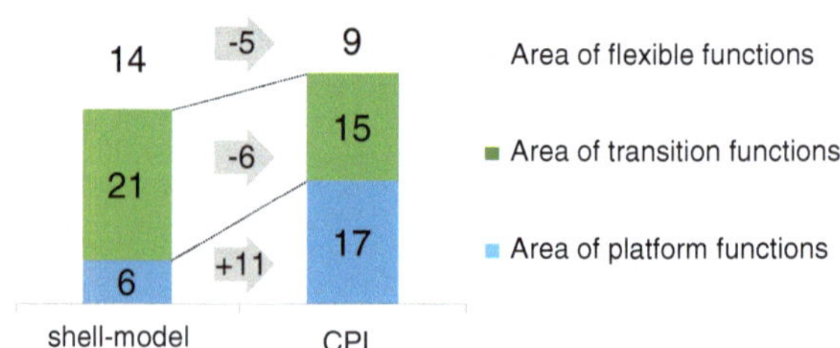

Fig. 4 Results of shell model and CPI—workshops

functions that belong to a strategic topic (for example, freshness functions were clustered together instead of discussing separately each of them). As a result, all functions within one function group (strategic topic) were rated equally when using the shell-model as opposed to more differentiated ratings for each function resulting from the CPI-method. This is observed especially in the area of platform functions, where the higher amount of platform functions as a result of the CPI-method supports the notion of developing a modular system that includes a robust platform.

4.3 Evaluation

Evaluation was conducted using standardized questionnaires with additional open questions. The rating scale ranged from 1 (very good) to 6 (unsatisfactory). Seven workshop participants were engaged in answering the questionnaire immediately after the workshops.

The results from the questionnaire showed that the participants were satisfied with the results of both the CPI- and the shell-method. The overall rating of the CPI-method in the categories consumer orientation, trust in results, definition of thresholds for the functions, and differentiation of functional areas was satisfactory (3) to sufficient (4) but inferior to the shell-model-method. Reasons might be time pressure during the CPI-workshop and uncertainty in handling a new method. Some participants conceived a lack of knowledge in the field of operational marketing as a potentially profound input concerning consumer orientation as beneficial to CPI-method. The participants of the shell-model-workshop on the contrary missed a differentiation between the importance of a function and the dynamics of a function. This shows that whereas the shell-model was conceived as more trustworthy procedure, its output lacks the clear input definitions and differentiation in terms of discussion (as can be taken from the previously compared results and from observing workshop discussions). Also, the quality of result documentation and thus transparency within the CPI-method was rated as very good and thereby superior to the shell-model-method.

Furthermore, participants rated the support offered by DIFs, their quantity and thematic mapping to the functions as good (2) in the workshop. This indicates that DIFs were used as background information for determining CPIs.

In general, the concept of using the factors P, D and C for determination of the CPI instead of using the simpler shell-model showed to be suitable for the industrial setting. This can be taken from replies of shell-model-workshop participants to the question which criteria they would subsequently use to determine necessary changeability. All of the factors named in the answers are covered by the suggested approach of determining feature CPIs based on initially identified DIFs.

5 Conclusions

The methodology presented in this paper is anchored between platform strategy definition and platform architecture design and enables systematic identification of the product variant structure under consideration of future developments.

Key elements of this methodology are two new steps: the method of determining dynamic influencing factors (DIFs) that supports the identification of dynamic influences acting on the platform system. To describe this level of awareness, the frequency of occurrence as well as the dynamic behavior of the dynamic influence factors are anticipated. Along with the importance of changes for the consumer, the so-called change priority indicator (CPI) method is used in order to quantify the effect of these influence factors on platform system. CPI is used as a quantitative measure of the necessary capacity for platform system flexibility and robust trade-off.

At the end of this paper we presented the case study, where it could be shown that the proposed methodology provides a concrete support for development of a platform system in an industrial setting. The methodology could be evaluated by the means of a survey among participants confirming successful integration of the proposed flexibility in a real platform system design process by adding transparency and objectivity to the strategic aspects of planning flexibility. It is worth noting that strategic flexibility alone may not be sufficient to address the needs for change caused by external and internal factors. The platform system itself may have to be updated or replaced by a new one, since there is always possibility that unanticipated influences occur. This means that for having always up-to-date platform system, the organization needs to embed a controller mechanisms that continuously monitors the "health" of the platform system. Therefore, as part of a SFB 768—transfer project 1, a lifecycle platform management controller that in a continuous manner monitors the platform system and has all the processes needed for platform-related decision-making is being developed.

Acknowledgments The authors would like to thank German Research Foundation (Deutsche Forschungsgemeinschaft—DFG) for funding this project as part of the collaborative research centre "Sonderforschungsbereich 768—Managing cycles in innovation process—Integrated development of product-service-systems based on technical products".

References

1. Fixson, S.K.: Product architecture assessment: a tool to link product, process, and supply chain design decisions. J. Oper. Manage. **23**(3), 345–369 (2005)
2. Fricke, E., Schulz, A.P.: Design for changeability (DfC): principles to enable changes in systems throughout their entire lifecycle. Syst. Eng. **8**(4) 2005
3. Sanchez, R.: Creating modular platforms for strategic flexibility. Des. Manage. Rev. **15**(1), 58–67 (2004)
4. Cingöz, A., Akdoğan, A.A.: Strategic flexibility, environmental dynamism, and innovation performance: an empirical study. Procedia-Soc. Behav. Sci. **99**, 582–589 (2013)

5. Sanchez, R.: Strategic flexibility in product competition. Strateg. Manage. J. **16**(S1), 135–159 (1995)
6. Eisenhardt, K.M., Martin, J.A.: Dynamic capabilities: what are they? Strateg. Manage. J. **21** (10–11), 1105–1121 (2000)
7. Buganza, T., Verganti, R.: Life-cycle flexibility: how to measure and improve the innovative capability in turbulent environments*. J. Prod. Innov. Manage. **23**(5), 393–407 (2006)
8. Schuh, G.: Produktkomplexität managen: strategien, methoden, tools. 2., überarb. und erw. Aufl. ed. München [u.a.]: Hanser. XIV, 326 S (2005)
9. Franke, H.-J.: Variantenmanagement in der Einzel-und Kleinserienfertigung. München. Hanser, Wien XXIII, 240 S (2002)
10. Göpfert, J., Steinbrecher, M.: Modulare Produktentwicklung leistet mehr. Harvard Business Manager **3**, 20–30 (2000)
11. Schuh, G.: Gestaltung und Bewertung von Produktvarianten: Ein Beitrag zur systematischen Planung von Serienprodukten, vol. Reihe 2, Fertigungstechnik. VDI-Verl, Düsseldorf VI, 170 S:Ill (1989)
12. Kraus, P.K.: Plattformstrategien—Realisierung einer varianz-und kostenoptimierten Wertschöpfung (2005)
13. Ulrich, K.: The role of product architecture in the manufacturing firm. Res. Policy **24**, 419–440 (1995)
14. Ley, W., Hofer, A.: Produktplattformen. IO Manage. **7**(8), 56–60 (1999)
15. Meyer, M.H., Lehnerd, A.P.: The power of product platforms: building value and cost leadership. Free Press, New York XIV, 267 S (1997) (u.a.)
16. Robertson, D., Ulrich, K.: Planning for product platforms. Sloan Manage. Rev., 19 (1998)
17. Sawhney, M.: Leveraged high-variety strategies: from portfolio thinking to platform thinking. J. Acad. Mark. Sci. **26**(1), 54–61 (1998)
18. Ross, A.M., Rhodes, D.H., Hastings, D.E.: Defining changeability: reconciling flexibility, adaptability, scalability, modifiability, and robustness for maintaining system lifecycle value. Syst. Eng. **11**(3), 246–262 (2008)
19. Rhodes, D.H., Ross, A.M.: Anticipatory capacity: leveraging model-based approaches to design systems for dynamic futures. IEEE International Conference on Model-Based Systems Engineering, pp. 46–51 (2009)
20. Rosen, R.: Anticipatory Systems. 2nd edn, (1985). Springer, New York (2012) (u.a.)
21. Verganti, R.: Planned flexibility: linking anticipation and reaction in product development projects. J. Prod. Innov. Manage. **16**(4), 363–376 (1999)
22. Schuh, G., Lenders, M., Bender, D.: Szenariorobuste Produktarchitekturen, in 5. Symposium für Vorausschau und Technologieplanung, pp. 99–119. Heinz Nixdorf Institut, Paderborn (2009)
23. Bauer, W., Elezi, F., Maurer, M.: An approach for cycle-robust platform design. In: DS 75-4: Proceedings of the 19th International Conference on Engineering Design (ICED13), Design for Harmonies, vol. 4. Product, Service and Systems Design, Seoul, Korea (2013) 19–22 Aug 2013
24. Langer, S., et al.: Development of an explanatory model of cycles within development processes by integrating process and context perspective. In: IEEE Industrial Engineering and Engineering Management (IEEM) (2010)

Research on Development of Liquid Composite Molding Parts: Situation and Framework

Paul Bockelmann, Klaus Drechsler and Amaresh Chakrabarti

Abstract Low weight and good mechanical performance make composite plastics the potential material of choice for a plethora of applications. Increasingly, liquid composite molding (LCM) is being applied as a means for industrialized production. While technological basics of composite design and manufacture seem to be sufficiently well understood, it is not clear why LCM part development often fails to fully exploit the material's potentials or leads to poor manufacturability of the parts. To shed more light on this question, we conducted case studies in the course of 25 interviews with members of LCM development teams. We propose that the context of exploration of LCM technology strongly influences the project variables, development strategy, procedure of development tasks and manufacturability. On one hand, the resulting project variables severely limit the scope for fiber-fair design; on the other hand, development strategies and procedures are mainly based on a product-oriented approach.

Keywords Carbon composites · Liquid composite molding · Engineering design · Contextual framework · Technological exploration

1 Introduction

High stiffness and strength, load-fair product design, low weight, beneficial damping characteristics—these and many more properties render carbon fiber-reinforced plastics (CFRP) as a preferred material choice for a variety of applications. Recent years have seen enormous efforts to introduce structural composite materials—such as continuously reinforced plastics—to high-volume applications.

P. Bockelmann (✉) · K. Drechsler
Institute for Carbon Composites, Technische Universität München, Garching, Germany
e-mail: bockelmann@lcc.mw.tum.de

A. Chakrabarti
Innovation, Design Study and Sustainability Laboratory, Centre for Product Design and Manufacturing, Indian Institute of Science, Bangalore, India

A. Chakrabarti (ed.), *ICoRD'15 – Research into Design Across Boundaries Volume 2*, Smart Innovation, Systems and Technologies 35, DOI 10.1007/978-81-322-2229-3_42

In this regard, the German automotive industry has been a pioneer, most notably BMW that now sells the first mass-produced electric vehicle which features a so-called "life module" made from carbon fiber.

Manufacturing of carbon fiber parts in industrial volumes requires highly automated and robust production processes. According to many e.g. [1], the key to producing structural components in a process that meets industrial standards is liquid composite molding (LCM). The technology is based on separation of processing of fibers and matrix: Fibers are aligned and stabilized in a so-called "preforming" step according to their intended location and orientation in the final part, and liquid resin is subsequently transferred into the fibrous structure before curing. Resin transfer molding (RTM) is representative for LCM processes favored for industrial application. Industrial use is mostly limited to high pressure RTM which is almost completely automated. However, high process costs and long production cycles still demand an optimization of production, shorter value chains and CFRP-fair part design [1].

Part design for LCM often fails both—tapping the full potential of the composite material and allowing for a good manufacturability of parts. Pure material substitution without geometry adaptation, which only draws on the benefit of a lower density of composites compared to metals, is commonly known as "Black Metal Design" [2]. One critical aspect in composites is the strong interdependence among part geometry, material selection and process definition, far more than in design and manufacture of metallic parts. Despite previous work on dealing with intertwined parameters, e.g. [3], it is unclear why the design problems still occur.

Our study uses the perspective of design methodology to understand why composite design of LCM parts often lack material exploitation and manufacturability contrary to the state of available knowledge. The material exploitation aspect refers to an overdimensioning of parts by not adjusting the fiber direction according to the loads. The manufacturability aspect is inhibited by unsuitable geometrical features and results in flaws of fiber structure and composite laminate. Based on preliminary interviews, we assume that the problems described emerge from the context of industry. We draw on Maffin's contextual framework for design [4] as a basis for the study since its construct elements allow us to consider links between different levels of factors—the industrial situation as a whole as well as the work of single engineers. In this course, we choose teams within firms as the unit of our analyses and investigate data from 25 interviews of team members. Through the case studies, we provide a qualitative description of the mechanisms at work and conclude with the assembly of a detailed contextual framework for future research.

2 Literature Review

2.1 Development of Liquid Composite Molding Parts

Holistic part development must always consider three aspects: Selection of the material from which a part should be made, design of the part geometry and determination of the production process that can reliably convert raw material into the final part shape. Considering part development for LCM, process selection is already determined. However, since material to be used is not merely selected from possible materials but is generated in the molding process itself, the process parameters, e.g. curing cycle, determine the properties of the resulting composite. Additionally, the production process is strongly influenced by part geometry due to fiber orientations and resin flow paths. Thus, the nature of the problem is not solely selection but interdepending adjustment of geometrical design, process parameters and material properties.

From the process point of view, this connection among geometry, process and material within LCM processes was subject to many experimental investigations, e.g. [5, 6]. Equally well established are the fundamentals of structural design with composites. Exploitation of the mechanical potential of a fiber material that is based on anisotropy of the fiber's mechanical properties has been subject of various publications, including [7].

Regarding the methodological aspects of development, material and process selection within the scope of engineering problems is widely discussed. Ashby et al. [8] give an overview on selection strategies that lead to a set of possible materials through evaluation based on requirements. No strategy explicitly supports the possibility to adapt part design, and works only if interference of all aspects is completely determined.

Methodology-focused research on composite part development with regard to the actual manufacturing technology is very scarce. There is a consensus that the principles of concurrent engineering apply to the development of composite parts. In 1990, Gandhi et al. [9] proposed a manufacturing-process-driven design methodology for composite components. This methodology starts—after establishment of specifications—with parallel material selection and part geometry. Considerations of manufacturing constraints iteratively follow thereafter. In the same decade, Potter [4] proposes a refinement of design strategies by concurrency, iterative nature and abstraction of the design problem through a procedure of connected loops. He discusses departmental structures that could hinder consistent cost monitoring, and states that integration of all needed disciplines—most notably manufacturing—is a prerequisite to avoid quality, yield and cost problems in production.

2.2 Initial Theoretical Framework

The current state of technological knowledge seems to be sufficient to reliably establish economical LCM products of high quality while tapping the full mechanical potential of fiber reinforcement. However, within the context of LCM-processing, the realization process of products often obstructs successful results. It is not clear, why low material exploitation, problems in fiber impregnation and uneconomic products are still often characteristic for the design of LCM

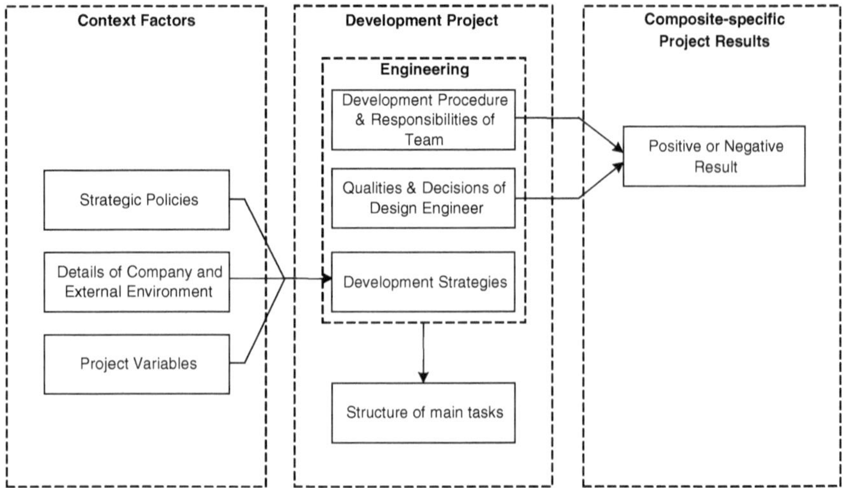

Fig. 1 Initial theoretical framework based on the contextual framework proposed by [5]

composites. From preliminary interviews, we assume that the context of LCM development projects plays a major role on the projects' outcomes.

Figure 1 shows the initial theoretical framework for the study as compiled from relevant literature. We apply Maffin's contextual framework to the situation within development projects for LCM parts and focus on the connection of contextual factors to the development strategies applied within the project. For judging the type of strategy, we draw upon Maffin's distinction between product-oriented and process oriented strategy. Process-orientation is indicated by problem analysis through abstraction, systematic concretization, generation of multiple solutions and progressive evaluation. On the other hand, assuming a solution due to lack of scientific knowledge and general use of rules of thumb are evidence for product-orientation [5]. Following the indication of the importance of team work by Potter, we consider two aspects regarding its influences on project results. First, the interaction of members within the development team and second, the qualities of the team's design engineer as the one being classically responsible for the main tasks. Drawing upon the scheme's iterative nature and concurrency of these three working steps, the structure of main tasks itself is also investigated.

3 Method and Sampling

In order to investigate the questions why design of LCM parts often lack material exploitation and manufacturability contrary to the state of available knowledge, we applied qualitative case study research [10, 11]. The links between context, methodological aspects of LCM development projects and their results have

Table 1 Overview of respondents that were interviewed in the course of the study

Category		Institution													Sum
		A	B	C	D	E	F	G	H	I	J	K	L	M	
Industry/ Institution	University			X/ X[a]											2
	Aerospace	X			X		X				X				4
	Automotive	X	X			X		X	X	X				X	7
	Machinery		X			X									2
Firm type	OEM		X			X	X	X			X	X		X	7
	Supplier	X	X		X	X	X		X	X		X	X	X	10
Region	Germany	X		X		X	X	X			X	X		X	8
	Austria		X		X				X	X					4
	Italy												X		1
	United Kingdom			X											1
Respondent	Material Eng.							1							1
	Process Eng.			1	1		1			1[b]		1			5
	Design Eng.					1				1[b]		1		1	4
	Simulation Eng.					1					1				2
	Designer											1	1		2
	R&D Leader	1	1	1	2	2		1	1	1		1			11
	Total	1	1	2	3	4	1	2	1	3	1	4	1	1	25

[a] Interview with independent scientists from a university in Germany and a university in the UK. The places of affiliation are not uniquely active in one industrial field

[b] Two persons attended one group telephone interview, therefore the total sum of interviewees is 25

hitherto not been addressed by research. Hence, the aim of this study is to build a theory based on empirical evidence for which case study research is especially appropriate [10]. For the investigation of procedures, strategies and responsibilities within development teams, it is argued that cases on project-level within single firms are best suitable. Firms were selected by theoretical sampling regarding three major aspects [11]. First, the firms had to be a user of LCM technology for serial production of composite parts. Second, they were part of one of the most relevant industries for composite growth, automotive, machinery and aerospace, and were categorized as either original equipment manufacturer (OEM) or supplier [12]. Third, the firms were located in Europe due to its significance for these industries [12–14]. Scientists in public institutions were added to the sample. Wherever possible, multiple team members from different disciplines of one firm were interviewed in order to take a broad range of perspectives into account and allow for triangulation of findings [11]. The sample of respondents is shown in Table 1.

The interviews were conducted using a general interview guide that was adapted to each respondent based on information available on the respondent. An interview typically took 40–60 min and was recorded (with the following exception: Preliminary interviews with Firm A, R&D Leader, Firm B, R&D Leader and Institution C, process engineering were not recorded. During interviews with Firm E, simulation engineering and R&D project leader only notes were taken due to confidentiality concerns). Relevant remarks and ideas were noted. The recordings were transcribed using Idea-Transcribe tool [15]; subsequent content analysis of transcriptions and notes was performed with the software Open Code. Following Eisenhardt [10], relevant statements were summarized in write-ups followed by a within-case analysis and a subsequent cross-case pattern search. Insights from emerging patterns were used to refine the constructs of the framework and understand the interviews [11].

4 Results

The results of our study concerning the main contextual influences on development strategies, the structure of main tasks and the influences regarding outcomes of LCM projects are presented. The findings regarding the current situation of the entire industry and detail the consequences characteristic for LCM development are discussed first.

4.1 Details of Companies and External Environment: Technological Exploration of LCM Processes

The interview results confirm our assumption that the technological context of LCM projects plays a major role on how well their outcome is. Currently, industrial use of LCM technology is undergoing a process of technological exploration, which is associated with terms like "search, variation, risk taking, experimentation, play, etc." [16]. A design engineer from Firm H reflects this by stating "Composites are a new area. That means you are […] developing business opportunities and eventually also parts that probably are not economical in the first moment. But generating knowledge and earning money with that one day—that is the motivation." This technological exploration leads to the fact that often "there is not a single part that is based on […] the technology, the insights of the previous project." (Firm H, R&D Project Leader). Furthermore, this learning process relates to product and process alike and amplifies the already strong concurrency. Both have to be advanced simultaneously in order to cater to the assumed requirements. The focus of technical learning, however, has to be differentiated: While firms operating in the automotive sector are mostly concerned with expanding the production throughput, those in the aerospace sector emphasize on the prediction of quality, e.g. in simulation, as it is currently the standard in prepreg technology. "Talking high volumes in composites means talking high uncertainties and high risks." As an R&D Leader from Firm I indicates, the first motive is particularly critical regarding process robustness. OEM-supplier partnerships, particularly in the automotive sector, are strongly shaped by the situation of technical learning. Collaboration is often complicated by unclear or inappropriate requirement formulation as well as a lack of explicit integration of the supplier into part development. In consequence, product features and geometries derived from traditional technologies pose difficulties in fiber alignment (preforming) as well as in resin flow processing during LCM.

4.2 Project Variables

The main project variables identified are installation space, LCM-fairness of quality specifications, industrial design of the entire product, and critical requirements. Each stems from technological exploration, the type of product or the nature of composite materials. Although different initial situations or demands for projects exist, the variables presented in the following were characteristic for the development projects of LCM parts.

The variable installation space originates from product exploration. As OEMs are learning to apply the materials within their products, they strive to minimize financial or technological risks. Most companies opt for picking single components from their products for material substitution to mitigate such risks. Since these parts typically stem from an assembly, redesign of these components is severely limited by the component's original installation space that is tailored to metal design but not composites.

Clear, unambiguous and complete product specifications are a prerequisite for targeted development. Beyond that, regarding composites R&D leader of Firm D observes, "To see as early as possible which defects are acceptable for the part or the product, even though they are defects, was a key factor for success." Defining properties of the product, such as porosities or deviations of the fiber orientation that are appropriate for liquid composite molding is pivotal for economical process adjustment. Since many firms are not familiar with development of products made from composites, they often adopt quality specifications of laminates produced with the more commonly used prepreg technology. These specifications prove to be ambitious and non-practical for LCM, due to its preforming and impregnation mechanisms.

Consumer products are often subject to intense design activities, e.g. in automotive industry. In that respect, the form of a product is intertwined with its concept and functionality [17]. In the automotive sector, for example, "most often, an application vehicle exists, which has a specific geometry, which is, I assume, essentially determined by the interior and exterior design", expresses the material engineer of Firm G. However, "departing from that overall design, often the wrong approach is chosen […]. This is reflected in the design of single elements, which [might] not be fiber-fair." (Firm H, R&D Leader). In addition, the material itself is emotionally charged as expressed by the designer of Firm K which alone may lead to a form language inappropriate for composites materials.

Single product requirements often seem to govern development of the associated composites production. These have to be carefully monitored throughout development, since "all can be somehow killer criteria" (Firm E, Design Engineer). The critical requirements named were part-weight in connection with costs, installation-space, number of units, information on environmental media and assembly/joining information.

4.3 Development Strategies

Since firms are only learning to process and apply newly developed LCM materials, a product-oriented strategy is common, as a material engineer of Firm G illustrates: "You know the part geometry; you know the boundary conditions […]. But currently, you don't have an exact idea of how thick such a laminate should be, how the fiber orientation should be. For example, [we] started with […] a 2 mm wall thickness and a quasi-isotropic lay-up […]. Then you produce parts and see how they perform […]. You take this in turn to optimize the part geometry or the layup." A simulation engineer of Firm J even claims that "you just start designing randomly". It becomes apparent that especially in preform layout, approximations and rules of thumb are used to generate quick solutions and subsequently refine them by experiment. Since in most cases a part's rough geometry or a narrow installation space are predefined, only wall thicknesses and fiber architecture can be determined, which reduces the possibility for fiber-fair design. With known material combinations, analytical load calculations via the rule-of-mixture [7] often results in a quasi-isotropic fiber layup which mimics the isotropy of homogeneous materials. Consequently, the part is able to sustain loads in directions in which no loads are to be expected during regular application, which impairs material exploitation.

However, a scheme of systematic trials for material characterization and process validation which is to be judged as a process-oriented approach was repeatedly reported. Typically several possible material concepts are tested and specific trials are often performed at abstracted geometries or critical details of the actual part as the simulation engineer of Firm J reports: "Intermediate steps certainly include sub-prototypes or coupon tests of single, critical elements. If you have big, complex parts it makes sense to manufacture and test prototypes of details."

4.4 Structure of Main Tasks

We could not find evidence for a predominant sequence of all three elements of the main tasks. However, as previously described, geometry, mostly through installation space, is usually defined first and dominates development. Subsequently, all tasks are considered simultaneously in a stepwise manner from rough to detailed determination.

Materials are usually chosen from a database of known material combinations which also contain appropriate processing specifications. If new materials are applied, basic characterization has to be carried out first. In aerospace however, the OEM typically prescribes which materials are to be used since only a few are officially certified.

The production process is usually determined based on the rough part geometry and the requested number of units. Further definition of processing steps, critical for serial production, usually alternates with specific production trials with selected materials.

4.5 Team Interaction and Quality of the Design Engineer

Team interaction in development was often declared by the respondents as one of the most important influences on a project's success. General aspects such as communication, decision making and coordination of responsibilities—characteristics of integrative development as reported elsewhere [17]—were seen to play significant roles. However, team members admitted that collaboration between product experts and experts for materials/processes is insufficient due to project or organizational structure.

Technological exploration and its effect on project variables and development strategies leave only a small scope for establishing a fiber-fair design of the part by design engineers. Three main qualities of design engineers of LCM parts that strongly influence the quality of design were identified with a strong emphasis on understanding of production technology. In order to judge the producibility of a part, a design engineer must understand the LCM production technology, most notably effects of fiber drape and resin flow processing. This should result in advantageous feature design, summarized in material and process-driven geometrical criteria: no sharp edges, wide radii as well as avoiding undercuts, beadings, notches, holes, mountings and strong double curvatures. Structural comprehension relates to the beneficial application of anisotropic material in order to achieve load-oriented stiffness and strength allocation within the composite part.

5 Conclusions

With our study, we hope to contribute to a better understanding of the contextual factors that influence the development of LCM parts and to inspire further research in the field. Technological learning seems to be a major contextual driver for development projects. It involves concurrent part and process development to satisfy requirements for new applications and bears considerable financial and

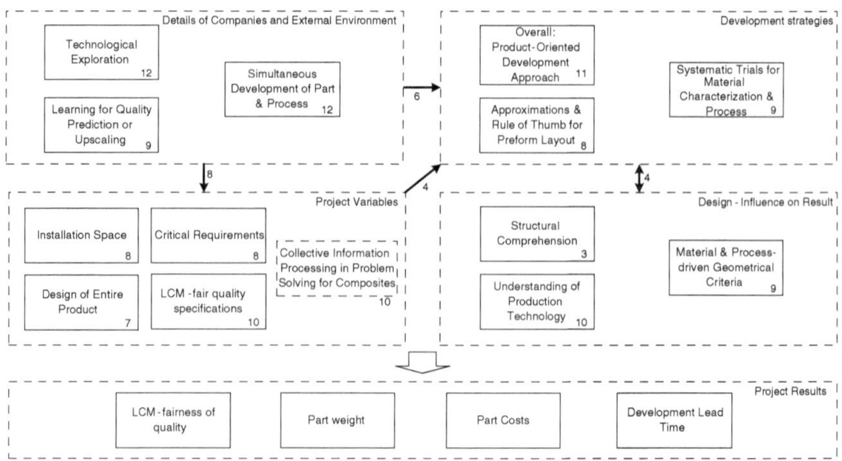

Fig. 2 Contextual framework for the development of liquid composite molding parts

technological risks. Firms try to mitigate these risks by picking preexisting metal components for replacement, thus imposing severe geometrical restrictions that result in a narrow scope for fiber-fair part design. The contextual framework for LCM development is summarized in Fig. 2 with quantitative references to explicit statements of the respondents. It incorporates our findings in the form of elements allocated to the construct elements of the initial framework in Fig. 1. Its arrows underscore the influences between elements.

To reduce adverse effects of the current situation in part design, we propose that OEMs recognize the importance of concurrent product and process development for LCM parts by integrating suppliers closer into development. Supplying firms should foster systematic process development and strive for more responsibilities in product development.

This study is limited to providing an overall picture of the current situation of development. No detailed investigation of relations between framework constructs has been carried out explicitly drawing upon the perspectives of certain domains of research, e.g. technological exploration. This bears possibilities for future research on single links to gain detailed information on the construct interdependencies.

An important question to ask is how development may be supported best. In order to answer this question, we believe that a better understanding of the situation must be gained first. We propose two non-exclusive approaches. One is to investigate details of lean development which, among other, fosters supplier integration [18]. A second approach is to gain further insights into organizational learning of respective firms.

Acknowledgments The authors are grateful to the Bavarian Research Foundation (Bayerische Forschungsstiftung) for funding this study under the grant PIZ-201-13.

References

1. Drechsler, K.: CFK-Technologie im Automobilbau: Was man von anderen Märkten lernen kann. C.C.e.V. Automotive Symposium. Neckarsulm, June 2010
2. Tsai, S.W., Patterson, J.M., Pérez, J.L., Donaldson, S.L.: Analytical qualification of aircraft structures. In: AGARD Report No. 772, Bd. **772**, 7.1–7.6
3. Potter, K.: Design of composite products—a personal viewpoint. Compos. Manuf. **3**(3), 173–182 (1992)
4. Maffin, David: Engineering design models: context, theory and practice. J. Eng. Des. **9**(4), 315–327 (1998)
5. Bickerton, S., Sozer, E.M., Graham, P.J., Advani, S.G.: Fabric structure and mold curvature effects on preform permeability and mold filling in the RTM process. Part I. Experiments. Compos. Part A **31** (5), 423–438 (2000)
6. Lundström, T.S., Gebart, B.R: Influence from process parameters on void formation in resin transfer molding. Polym. Compos. **15** (1), 25–33 (1994)
7. Schürmann, H: Konstruieren mit Faser-Kunststoff-Verbunden (2007)
8. Ashby, M.F., Bréchet, Y.J.M., Cebon, D., Salvo, L.: Selection strategies for materials and processes. Mater. Des. **25**(1), 51–67 (2004)
9. Gandhi, M.V., Thompson, B.S., Fischer, F.: Manufacturing-process-driven design methodologies for components fabricated in composite materials. Compos. Manuf. **1**(1), 32–40 (1990)
10. Eisenhardt, K.M.: Building theories from case study research. Acad. Manag. Rev. **14**(4), 532–550 (1989)
11. Patton, M.Q.: Qualitative research and evaluation methods. 3. Aufl. Sage Publications, Thousand Oaks (2002)
12. Lässig, R., Eisenhut, M., Mathias, A., Schulte, R., Peters, F., Kühmann, T.: Serienproduktion von hochfesten Faserverbundbauteilen. Perspektiven für den deutschen Maschinen- und Anlagenbau. Hg. v. Roland Berger Strategy Consultants (2012)
13. Heuss, R., Müller, N., van Sintern, W., Starke, A., Tschiesner, A.: Light weight, heavy impact. Hg. v. McKinsey. McKinsey (2012)
14. Jahn, B., Witten, E.: Composites market report 2013. Market developments, trends, challenges and opportunities. Hg. v. C.C.e.V. und AVK (2013)
15. Sarkar, P., Chakrabarti, A.: A support for protocol analysis for design research. Des. Issues **29** (4), 70–81 (2013)
16. March, J.G.: Exploration and exploitation in organizational learning. Organ. Sci. **2**(1), 71–87 (1991)
17. Ehrlenspiel, K.: Integrierte Produktentwicklung. Denkabläufe, Methodeneinsatz, Zusammenarbeit. 4. Aufl. München, Wien: Hanser (2009)
18. Hoppmann, J., Rebentisch, E., Dombrowski, U., Zahn, T.: A framework for organizing lean product development. Eng. Manag. J. **23**(1), 3–15 (2011)

Knowledge Sharing in Design Based on Product Lifecycle Management System

Pierre-Emmanuel Arduin, Julien Le Duigou, Marie-Hélène Abel and Benoît Eynard

Abstract Working together within an extended enterprise is not as natural for people as working alone to pursue their own objectives. It has been observed that the specification and development of information systems often strengthens this preference for working alone: interoperability as much as interpretation variance restrain the ability of people and systems to interact and to work together within an extended enterprise. This paper introduces an approach in order to highlight how product lifecycle management (PLM) systems could be extended so that not only data and information are shared in design, but also knowledge, meaning and individual cognition. This approach will be practically studied through the use of the MEMORAe approach at the end of this paper.

Keywords Product lifecycle management · Knowledge management · Ontology · Design · Tacit knowledge

1 Introduction

Product lifecycle management (PLM) aims at an integrated management of all product-related information and processes through the entire lifecycle for [1, 2]. Within an extended enterprise, i.e. a network of firms collaborating in a project to achieve a common goal [3], systems try to interact in order to share data and information throughout a product lifecycle. This try has to be successful in order to ensure performance. Beyond this computing point of view, it can be considered that

P.-E. Arduin (✉) · M.-H. Abel
Heudiasyc UMR CNR 7253, Université de Technologie de Compiègne, CS 60319, 60203 Compiègne Cedex, France
e-mail: pierre-emmanuel.arduin@utc.fr

P.-E. Arduin · J. Le Duigou · B. Eynard
Roberval UMR 7337, Université de Technologie de Compiègne, CS 60319, 60203 Compiègne Cedex, France

A. Chakrabarti (ed.), *ICoRD'15 – Research into Design Across Boundaries Volume 2*, Smart Innovation, Systems and Technologies 35, DOI 10.1007/978-81-322-2229-3_43

workers may be spread all over the world and that is the reason why the same information on the same product within the same extended enterprise may lead to different interpretations. Knowledge and particularly tacit knowledge as introduced by [4] have to be considered and shared during design.

The aim of this paper is to introduce an approach relying on the MEMORAe approach (see [5]) in order to share knowledge and particularly individual cognition through PLM systems. The design of such systems should be thought as closer as possible from the needs of final users. Ethnographic workplace study, as introduced in [6], allows for example to understand human behavior and to consider tacit knowledge during specification, development and implementation of information systems as discussed in [7]. Considering tacit knowledge may indeed improve collaborative work, as discussed in [8].

After remembering the considered vision of knowledge in the organization, PLM and its objectives are presented. Then the MEMORAe approach is introduced, as well as its use within PLM systems in order to share knowledge in design. This work is a work in progress and its added values as well as its perspectives will finally be discussed at the end of this paper.

2 Background Theory and Assumptions

Relying on the assumption that several people may interpret differently the same information, this work proposes to focus on knowledge as an individual interpretation and to share it through PLM systems. The proposed approach ensures that knowledge is shared by such systems through the use of the MEMORAe approach. So that not only data and information are shared throughout a product lifecycle, but also knowledge and particularly individual tacit knowledge.

The considered vision of knowledge in the organization is introduced in the first part of this section. A few models of PLM are then presented in the second part of this section.

2.1 A Vision of Knowledge in the Organization

As the authors of this paper, we have got tacit knowledge, i.e. an individual cognitive construction, that we have structured into information during a process of sense-giving. As the readers of this paper, you have interpreted this information perceiving forms and colors, absorbed words, data, during a process of sense-reading, possibly creating new tacit knowledge for you (see Fig. 1). Sense-giving and sense-reading processes are defined in [9] as follows: “Both the way we endow our own utterance with meaning and our attribution of meaning to the utterances of others are acts of tacit knowing. They represent sense-giving and sense-reading within the structure of tacit knowing” [9, p. 301].

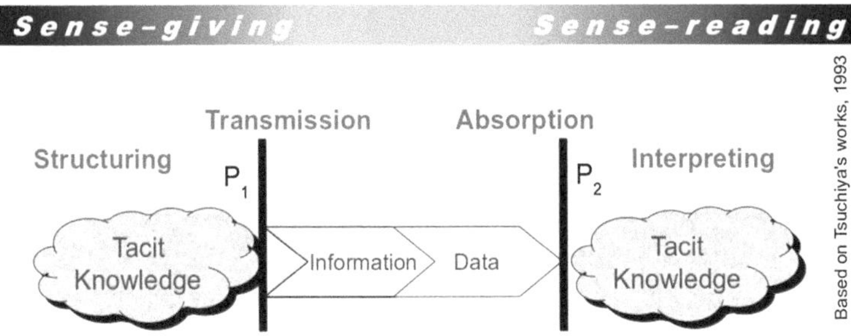

Fig. 1 Tacit knowledge transfer

Information is continuously interpreted during sense-reading processes. Within extended enterprises, information can be transmitted by speaking, writing or acting, and more generally, by information systems. Knowledge can then be:

- *explicited*, i.e. it has been made explicit by someone within a certain context, it is socially constructed and can be supported by information technology. Individuals, as well as computers are "information processing systems" as said by [10] (p. 9),
- *tacit*, it is not always articulated and cannot always be articulated, relying on [4] notably: "we can know more than we can tell".

So that explicited knowledge is tacit knowledge made explicit by someone and information source of tacit knowledge for someone. It is "what we know and can tell" answering to [4] quoted above. The term "explicit knowledge" is often used (by [11, 12] notably), whereas it does not reflect the dynamic of the explicitation of tacit knowledge by someone as well as the term of "explicited knowledge". That is the reason why we prefer to use this expression, which clearly shows how every information can be seen as an individual cognitive construction within an organization.

Authors as [13] consider that whereas explicited knowledge can be regarded as a resource necessary to carry out processes, tacit knowledge is something that ensures their efficiency: routines, non-written procedures, skills, crafts, job secrets, etc. That is notably the reason why PLM systems should share knowledge, and particularly individual tacit knowledge throughout a product lifecycle. As remarked by [14] (p. 101), knowledge transfer is not reduced to the transmission of information: "Transfer = Transmission + Absorption (and Use)". We interpret information and create a piece of tacit knowledge, which has a meaning for us. Within an extended enterprise, there can be someone who has received the same information and, interpreting it, has created a piece of tacit knowledge, which has a meaning for him/her. Tacit knowledge is the result of the interpretation by someone of information [15], it is an individual cognitive construction, which can differ from one person to another [16].

Design based on PLM systems should integrate the possibility of meaning variance, that is notably the reason why the proposed approach has been developed and is introduced in this paper. Now a few models of PLM models are presented. Being a work in progress, this work will of course be refined in the future with the use of others models.

2.2 Modeling Product Lifecycle Management

The Product lifecycle management (PLM) is defined by [1, 2] as a systematic concept for the integrated management of all product-related information and processes through the entire lifecycle, from the initial idea to end-of-life. PLM capabilities can be used in order to facilitate capitalizing on knowledge, in the sense introduced by [17]. Indeed PLM focuses on the capture of information source of knowledge from various enterprise information systems. For [18]: "the need of different knowledge categories covering, on the one hand, product and process characteristics and, on the other hand, the expert viewpoint and all links between product-process properties and performance indicators [is] one of the major challenges" [18, p. 453]. Knowledge management (KM) and information systems are key points in order to tackle this challenge. The first considers notably tacit knowledge, which is inherent to individuals and to their cognitive interpretations, whereas the second points out explicited knowledge, which can be stored, processed, and disseminated within information systems. Liao [19] Strengthen this position when they argue that PLM approaches can improve product and process knowledge consideration from several applications. Within a particular context, a PLM approach may be used to generate, to organize, to share and to reuse relevant knowledge according to projects' goals.

So that even if several systems have been developed with the aim to support modeling, storage and reuse of explicited knowledge [19, 20], PLM systems have good integration properties within extended enterprises, and this is one of their major strengths.

For product development process, two main categories of knowledge are currently distinguished according to [18]: product engineering knowledge and manufacturing process knowledge. Within extended enterprises, experts use different documents and information systems components in order to structure and to share this knowledge. While the first can be handled based on systems such as computer-aided design (CAD), the second is included in standards such as process scheduling, resources allocation, operations planning, etc. Nevertheless, product definition may change during its lifecycle, that is the reason why authors as [21] consider that knowledge management is an important challenge in manufacturing. For [22], lifecycle knowledge is the knowledge which has been created or used by processes throughout the product's lifecycle. Even if information technology support for PLM results from the integration between heterogeneous systems [23], the consideration devoted to knowledge and particularly to tacit knowledge is extremely weak, not only in the literature, but also within industrial fields.

The purpose of this paper is to highlight how knowledge, whether tacit or made explicit by someone, may be shared in design through a PLM system.

3 Methodology Proposal

People and information technology work together throughout a product lifecycle in extended enterprises. The aim of our approach is to improve their abilities to work together in order to catalyze knowledge sharing in design.

The first part of this section introduces the MEMORAe approach, an approach and a platform facilitating knowledge sharing through the collaborative development of ontologies. The second part of this section highlights how the MEMORAe approach may lead to a better integration and management of knowledge within PLM systems. Finally, the last part of this section discusses the added values and the perspectives of this work, which is a work in progress.

3.1 The MEMORAe Approach

In [5] the MEMORAe approach is presented as aiming to offer an alternative to the loss of competencies and knowledge in an organization. A competency is considered as a way to put into practice some knowledge within a specific context [5]. The MEMORAe approach has been developed relying on the concept of learning organizational memory.

Knowledge obsolescence and loss of know-how are two key points observed within industrial fields that leaded to propose the MEMORAe approach, which offers an ontology-based learning organizational memory. As presented in the first section of this article, transmitting information is not sufficient to share knowledge [14], due to the existence of individual cognition in sense-giving and sense-reading processes [9]. So that sharing, disseminating and receiving information is not sufficient to integrate, to manage, and to share knowledge. Nevertheless, collaboratively developing ontologies leads stakeholders to produce a shared understanding, so that knowledge, interpretation, and meaning are shared. That is notably the reason why the MEMORAe approach has been chosen for this work: it allows stakeholders, regardless where they are, to collaboratively produce a shared understanding through the collaborative development of ontologies. The reader interested in other approaches may have a look to [24].

Ontologies reflect a shared world view and for [25] they can support communication and knowledge sharing through a community of practice. Wenger [26] presented what kind of practices are involved within a community of practice: “Such a concept of practice includes both the explicit and the tacit. It includes what is said and what is left unsaid; what is represented and what is assumed.” [26, p. 47]. With Web technology, communities become present online, and for [17]

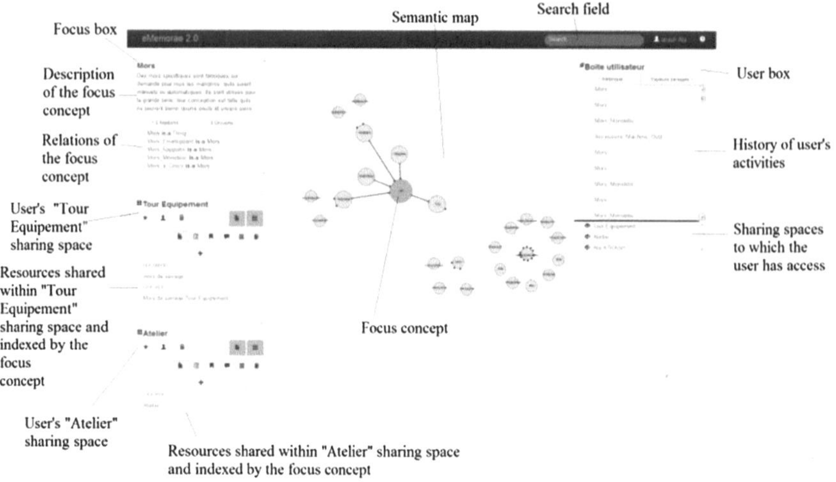

Fig. 2 The platform of the MEMORAe approach

an online community consists of people, a shared purpose, protocols and rules that guide interaction and computer systems.

The platform of the MEMORAe approach (see http://www.hds.utc.fr/memorae/ and Fig. 2) is a prototype showing the ways to help users to collaboratively develop ontologies. Entry points at the left of the screen provide direct access to a concept, whereas a history of the navigation is shown on the right of the screen. The part of the ontology describing current resources appears in the center of the screen and is framed by a short description of the current concept and by a list of shared resources related to the current concept at the left of the screen. Within different spaces (for example "Tour Equipment" in Fig. 2), individuals may have access to different resources, may be authorized to interact with different stakeholders (by chatting or through a wiki for example), and may add new concepts in the ontology or see who added them. By discovering new concepts, users can ensure that they interpret them correctly by browsing the entire screen of the platform. The design of such environment has been thought in order to facilitate individual cognition and collaborative understanding.

The use of this prototype has been tested notably by several students during courses and it has been observed that they appreciated it and shared knowledge in majority. Observed students learned rapidly studied concepts and they all understood the same thing, whereas some of the studied concepts usually lead to different interpretations. They were all interacting through the platform in order to develop an ontology of the studied concept. That leaded them to collaboratively produce a shared understanding, to share knowledge and meaning. These early experiments strengthen our proposition that developing ontologies *together* is a good way to produce a shared understanding between stakeholders.

The use of the platform of the MEMORAe approach within extended enterprises can be seen as a challenge because of the huge amount of information source of knowledge to represent. Nevertheless the pragmatic approach of PLM can lead not only to perform the implementation of such platforms within extended enterprises, but also to improve knowledge integration and management within PLM systems in design, which is the purpose of this paper.

3.2 Knowledge Integration Within PLM Systems Based on the MEMORAe Approach

Collaborative tools such as Computer-Supported Cooperative Work (CSCW) result of information technology whose implementation has been polarized around the individual user for [27]. Individuals are users and they are individually using an information system. Arduin et al. [7] notice that neglecting social activity leads to "meaningless conclusions", and [6] insists when she reminds that knowledge is not only based on the group but is also tacit, embodied in individual cognition: "We believe that there is yet another dimension that needs to be explored and that is the knowledge that is not only group-based but also tacit, implicit, embodied, and not articulated" [6, p. 18].

So that sharing knowledge, and particularly through PLM systems, needs to consider not only communities sharing knowledge, but also individual cognition. Let us remind that according to [14], knowledge is shared when information is transmitted, when it is absorbed, i.e. interpreted into tacit knowledge through individual cognition, and when this knowledge is used, knowledge being linked to the action: "Transfer = Transmission + Absorption (and Use)" [14, p. 101].

Throughout a product lifecycle, knowledge is created and used by several actors, within several information systems and in every corner of the globe in an extended enterprise. The platform of the MEMORAe approach allows individuals to collaboratively develop ontologies and to produce a shared understanding. So that knowledge is shared, even if it cannot be made explicit. Such platform relies on standards of the Web semantic so that our proposition is not reduced to that platform.

In the MEMORAe approach, institutional or dynamic groups can be created:

- institutional groups, which are created and leaded by a manager and include two or more collaborators,
- dynamic groups, which are created by any member of the organization without requiring a validation from the hierarchy. It can be a community of interest around a specific topic, person or both.

So that the proposed approach offers a shared environment allowing people, regardless where they are and regardless who they are, the means to produce a shared understanding and to share knowledge in design through PLM systems.

They may actively participate in communities by creating them, by managing them, or by inviting relevant persons in them. This shared environment has been observed as being the starting point of knowledge sharing and its integration by stakeholders. The proposed approach facilitates collaborative work in the sense that it gives the means to disseminate information and to ensure its interpretation within identified groups of characterized individuals: they all give the same meaning to the same information. Not only information is disseminated, but also knowledge is shared in design based on PLM systems with the proposed approach.

3.3 Discussing the Added Value and the Perspectives of This Work

As pointed out in the beginning of this paper, this is a work in progress which will be of course refined in future works. More than the statement of ideas such as the merging of different research fields, this study must now rely on industrial field-works in order to identify the conditions so that our approach will be useful throughout a product life cycle.

A product life cycle involves several actors using several information systems in every corner of the globe in an extended enterprise. Different entry points exist for information source of knowledge. Our approach aims at giving stakeholders the means to produce a shared understanding in order to share knowledge in design. To do so they may collaboratively develop ontologies with the platform of the MEMORAe approach. This idea of shared understanding in design has already been discussed notably by [28].

The MEMORAe approach allows actors to express what they think about given subjects and to create their own subjects. This ensures that everyone can access to every element correlated with the product throughout its life cycle. For the moment it is not possible to show an industrial case study validating or invalidating the proposed approach. This will be developed in future works: we are notably focusing on studying the ease of implementation, ease of training, ease of use, and scalability of such approach in order to understand what could be done to improve knowledge sharing in design based on PLM systems. Such indicators will notably allow the comparison of the MEMORAe approach with alternative approaches, as those presented in [16].

The MEMORAe approach gives the means to involve the stakeholders in knowledge creation, knowledge sharing and knowledge use and re-use. Created ontologies may be evaluated as suggested by [29], even if this work focuses less on the development of normalized ontologies than on the production of a shared understanding through the collaborative development of ontologies.

4 Conclusions and Perspectives

This paper has presented a work in progress in order to share knowledge in design by extending PLM systems through the use of the MEMORAe approach. The stakeholders are given the means to create communities, to manage them, or to invite relevant persons in them. Moreover, they become involved in the processes of developing ontologies together and it has been observed that it leaded them to collaboratively produce a shared understanding so that knowledge, meaning and interpretation are shared.

After having summarized the considered vision of knowledge in the organization relying particularly on sense-giving and sense-reading processes introduced by [9], PLM principles are presented, the integrated management of all product-related information and processes through the entire lifecycle for [1, 2]. Then the MEMORAe approach from [5] is presented and its platform is introduced. Finally the added values as well as the perspectives of this work are discussed.

The proposed approach focuses on a vision of knowledge in the organization, which considers knowledge as the result of the interpretation by someone of information through individual cognition. The proposed approach focuses also on a strength of PLM systems, which are integrated through the whole product life cycle. The proposed approach finally takes into consideration a strength of the MEMORAe approach, which gives the means to individuals to collaboratively develop ontologies, producing then a shared understanding. It is proposed to merge these different research areas in order to ensure knowledge sharing in design based on PLM systems. The proposal leads to focus on the impact of individual cognition in design within extended enterprises. Next steps of the study will concern the analysis of industrial applications, the implementing conditions and their limitations. They will be presented in future works.

Acknowledgments This work was carried out and funded in the framework of the Labex MS2T. It was supported by the French Government, through the program "Investments for the future" managed by the National Agency for Research (Reference ANR-11-IDEX-0004-02).

References

1. Saaksvuori, A., Immonen, A.: Product Lifecycle Management, 3rd edn. Springer, Berlin (2006)
2. Terzi, S., Bouras, A., Dutta, D., Garetti, M., Kiritsis, D.: Product lifecycle management—from its history to its new role. Int. J. Prod. Lifecycle Manag. **4**, 360–389 (2010)
3. Ross, J.W., Weill, P., Robertson, D.: Enterprise Architecture as Strategy: Creating a Foundation For Business Execution. Harvard Business School Press, Cambridge (2006)
4. Polanyi, M.: Personal Knowledge: Towards a Post Critical Philosophy. Routledge, London (1958)
5. Abel, M.-H.: Competencies management and learning organizational memory. J. Knowl Manage. Spec. Issue Competencies manage **12**, 15–30 (2008)

6. Jordan, B.: Ethnographic workplace studies and computer supported co-operative work. In: Shapiro, D., Tauber, M., Traunmüller, R. (eds.) The Design of Computer-Supported Cooperative Work and Groupware Systems, pp. 17–42. North Holland/Elsevier Science, Amsterdam (1996)
7. Arduin, P.-E., Grundstein, M., Rosenthal-Sabroux, C.: Considering Tacit Knowledge when Bridging Knowledge Management and Information System for Collaborative Decision Making, pp. 131–158. Chapter in Information Systems for Knowledge Management, Wiley-ISTE (2014)
8. Arduin, P.-E., Rosenthal-Sabroux, C., Grundstein, M.: Considering Tacit Knowledge When Bridging Knowledge Management and Collaborative Decision Making. In: Proceedings of the EWG-DSS Liverpool-2012 Workshop on Decision Systems, pp. 62–70. Liverpool, UK, 12–13 April (2012)
9. Polanyi, M.: Sense-giving and sense-reading. Philosophy J. Roy. Institute Philosophy **42**, 301–323 (1967)
10. Hornung, B.R.: Constructing Sociology From First Order Cybernetics: Basic Concepts For a Sociocybernetic Analysis of Information Society. In: Proceedings of the 4th Conference of Sociocybernetics, Corfu, Greece (2009)
11. Nonaka, I., Konno, N.: The Concept of "Ba": Building a Foundation for Knowledge Creation, vol. 40. California Management Review (1998)
12. Nonaka, I., Takeuchi, H.: The Knowledge-Creating Company: How Japanese Companies Create the Dynamics of Innovation. Oxford University Press, Oxford (1995)
13. Nelson, R.R., Winter, S.G.: An Evolutionary Theory of Economic Change. Harvard University Press, Cambridge (1982)
14. Davenport, T., Prusak, L.: Working Knowledge: How Organizations Manage What They Know. Harvard University Press, Cambridge (1998)
15. Tsuchiya, S.: Improving Knowledge Creationability Through Organizational Learning. In: ISMICK 1993: Proceedings of the International Symposium on the Management of Industrial and Corporate Knowledge, pp. 87–95 (1993)
16. Arduin, P.-E.: On the use of cognitive maps to identify meaning variance. Lecture Notes in Business Information Processing **180**, 73–80 (2014)
17. Grundstein, M.: From capitalizing on company's knowledge to knowledge management. Knowl. Manage. Classic Contemp. Works **12**, 261–287 (2000)
18. Bosch-Mauchand, M., Belkadi, F., Bricogne, M., Eynard, B.: Knowledge-based assessment of manufacturing process performance: integration of product lifecycle management and value-chain simulation approaches. Int. J. Comput. Integr. Manuf. **26**, 453–473 (2013)
19. Liao, S.: Knowledge management technologies and applications literature review from 1995 to 2002. Expert Syst. Appl. **25**, 155–164 (2003)
20. Virtanen, P., Helander, N.: Knowledge management. InTech (2010)
21. Rentzsch, W., Shercliff, H., Brechet, Y., Legras, L.: Knowledge management for materials processing. Adv. Eng. Mater. **7**, 30–39 (2005)
22. Ameri, F., Dutta, D.: Product lifecycle management: closing the knowledge loops. Comput. Aided Des. Appl. **2**, 577–590 (2005)
23. Schuh, G., Rozenfeld, H., Assmus, D., Zancul, E.: Process oriented framework to support PLM implementation. Comput. Ind. **59**, 210–218 (2008)
24. Studer, R., Fensel, D., Decker, S., Benjamins, V.R.: Knowledge engineering: survey and future directions. In: XPS-99 Knowledge-Based Systems, Lecture Notes in Computer Science, vol. 1570, pp. 1–23 (1999)
25. Domingue, J., Motta, E., Buckingham, S., Vargas-Vera, M., Kalfoglou, Y., Farnes, N.: Supporting Ontology Driven Document Enrichment Within Communities of Practice. In: Proceedings of K-Cap'01, pp. 30–37. Victoria, 22–23 Oct 2001
26. Wenger, E.: Communities of Practice: Learning, Meaning, and Identity. Cambridge University Press, Cambridge (1998)
27. Schmidt, K., Bannon, L.: Taking CSCW seriously: supporting articulation work. Comput. Support. Coop. Work **1**, 7–40 (1992)

28. Pavković, N., Štorga, M., Bojčetić, N., Marjanović, D.: Facilitating design communication through engineering information traceability. Artif. Intell. Eng. Des. Anal. Manuf. **27**, 105–119 (2013)
29. Štorga, M., Andreasen, M.M., Marjanović, D.: The design ontology: foundation for the design knowledge exchange and management. J. Eng. Des., vol. 21 (2010)

Part VI
Applications in Practice (Automotive, Aerospace, Biomedical Devices, MEMS, etc.)

Development of a Cellular Lightweight Cement (CLC) Block Cutting Machine for Small Scale Industries

Vikas Patil, Aniket Patil, Bhagyesh Deshmukh and Roohshad Mistry

Abstract Cellular lightweight Cement block is one of the alternatives for conventional clay bricks. Currently such blocks are manufactured by small scale industries at different places. The way of manufacturing is conventional that requires more labors as well as time. These blocks are casted in molds having sub compartments of required block size. Setup work required for setting of the mold is labor and time consuming. To cater the cost effective needs of industry it is required to reduce labor work by using machines. Therefore a CLC block cutting machine is introduced which eliminates the background work required for setting of mold and disassemble it.

Keywords CLC block cutting machine · Cost effectiveness · Setup work

1 Introduction

It is the need of industry to reduce the labor work, time required for production and production cost. In large scale industry it is possible due to potential use of automatic machines (high capital investment) however this may not possible for small scale industry.

Hence there is a need to find the solution for reducing labor, process time and expenditure. CLC block manufacturing is a small scale industry facing such problem. In this industry the time and labor required for making the molds for casting of the block is more. It is hence required to reduce these parameters in order to increase the rate of production. CLC block cutting machine is specially designed for small scale industries.

V. Patil (✉) · A. Patil · B. Deshmukh · R. Mistry
Department of Mechanical Engineering, Walchand Institute of Technology, Solapur, Maharashtra, India
e-mail: patilvikas492@gmail.com

A. Chakrabarti (ed.), *ICoRD'15 – Research into Design Across Boundaries Volume 2*, Smart Innovation, Systems and Technologies 35, DOI 10.1007/978-81-322-2229-3_44

The process carried out with the help of developed machine, provides predetermined size of the block and a typical surface texture. The machine is optimized with proven mechanism which helps to carry out the process efficiently. The modeling and simulation of the machine are undertaken in order to obtain a motion synthesis and the results obtained confirmed the correctness of the mechanism.

2 Literature Review

At present state [1] the mold used to produce blocks is of two L-Shaped plates which are bolted together. Plates of size $2' \times 8''$ are placed on equal distance in mold to make the bricks of required size. Each block requires plate on all side for maintaining corner and surface finish. Oil spraying is required to avoid adhesion between block and plate. The slurry is poured into molds of size $8' \times 2'$ sequentially in each compartment. The molds are opened after 48 h by removing fasteners. Plates are then removed one by one and each block is placed outside the processing unit in curing area.

There is no such data available/communicated by researchers regarding this type of machine. The machines available are manufactured by local fabricators and are not well engineered. The large scale industry uses highly automated machines. Some of the companies like Nanolite Infratech Pvt. Ltd. [2], THT Vietnam [3], Maruti Hydraulics Pvt. Ltd. [4] and Balaji Construction Machine [5] are producing such machines. The machines are made considering large scale production rate. Setups required for such machines are fully automated consisting of guide rails, hoists, etc which eventually results in higher cost. Many companies from Hyderabad (India), Nashik (India) design machines required for small scale industry. The variants are available in semiautomatic and manual. Though these machines simplify the work of CLC block cutting, the blocks are required to move towards the machine bed for cutting. This can lead it towards the damage of blocks. Therefore due care is required while moving the blocks. The movement of blocks is carried twice that increases the labor work and requires time for handling. The weight of the machine and occupied space is larger. Some machines involve tilted frame for cutting that causes balancing problem in setup. Surface finish obtained by these machines is unable to hold the plastering of Plaster of Paris (POP).

Though the literature mentions various machines, a scope remains to make them user friendly and reliable particularly for small scale industry. Apart from that, this industry is facing problem like availability of skilled labor. Hence it is difficult to assign more labors for the process to minimize the productive time as well as to achieve good quality of product. This raises the intuition of designer to develop user friendly and reliable machine for cutting of CLC block. The design methodology, design built-up and synthesis for the proposed machine is based on the concepts [6] produced by researchers.

3 Design Methodology

3.1 Possible Solutions

After many iteration following possible solutions are obtained according to need. Figure 1 shows the first idea in which the cutting of block is carried out by impact of thin plate in cast. The plates are placed on equal distance in a frame. This frame is released from some height thus this fall on mold and shears it off into required size. Mechanism of this machine is simple with less number of moving parts. However limitation with this design is the damaging of edges and corners of the block during shearing. This cannot be ignored for use of block in construction. The impact affects the constructed ground base of factory. The vibration produced due to impact cause cracks in block. Figure 2 gives the idea of cutting by continues wire. In this design cutting is carried out by passing continues wire on two pulleys which are placed in a frame. Feed is given by moving this frame downward. Wires rotate continuously with pulley and cause cutting action. Direction of motion of wire is opposite on upper and lower side of pulley. This causes uneven surface lines and also damages the sides of block.

The idea of reciprocating frame is shown in Fig. 3. In this design a sub frame supported by main frame is made to reciprocate by crank and connecting rod mechanism. Cutting wires are placed in sub frame. Reciprocation of this sub frame causes the cutting action. Feed is given by main frame which slides vertically on main structure. Numbers of reciprocating parts are more in this design which produces high vibrations. This may causes crack in block also structural rigidity

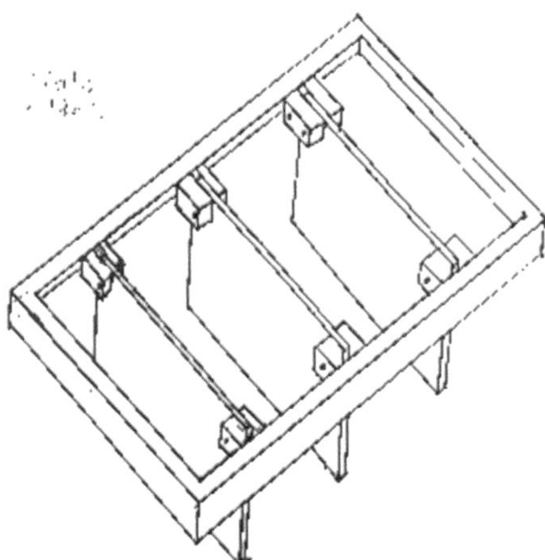

Fig. 1 Cutting by shearing

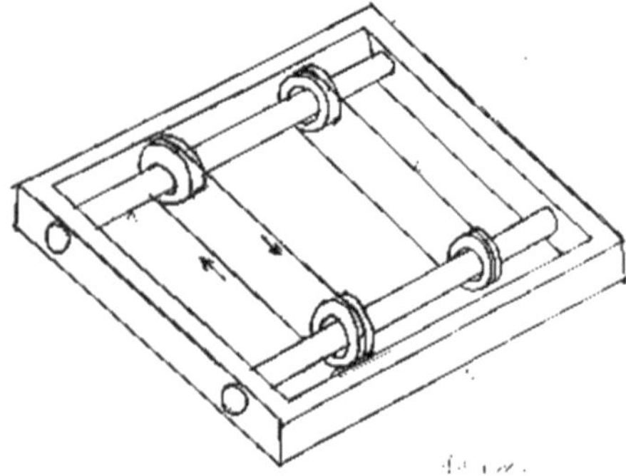

Fig. 2 Continues rotating wire

required is higher. Figure 4 shows the idea of dual fork design. This design consists of four forks for each wire.

Two of them are driving placed on upper side and two are idler placed on lower side of the frame. Driver fork is made to reciprocate by crank and connecting rod mechanism. This reciprocation causes the wire to reciprocate and results in cutting of block. Moving parts are more in number and because of their inertia vibrations are produced. Weight of frame is more which require high power for lifting. Height of machine is required to be kept more because of mechanism. Many number of parts lead to frequent maintenance and lubrication.

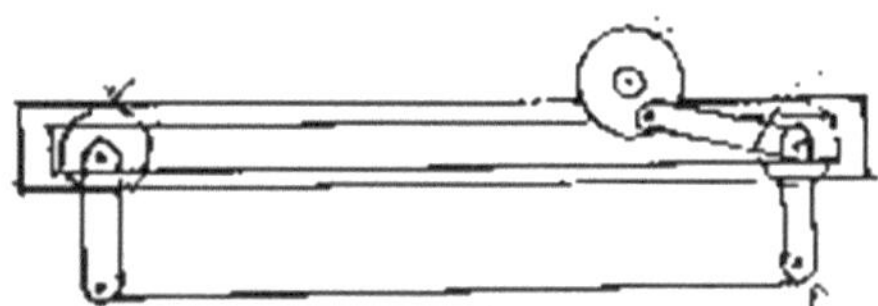

Fig. 3 Reciprocating frame

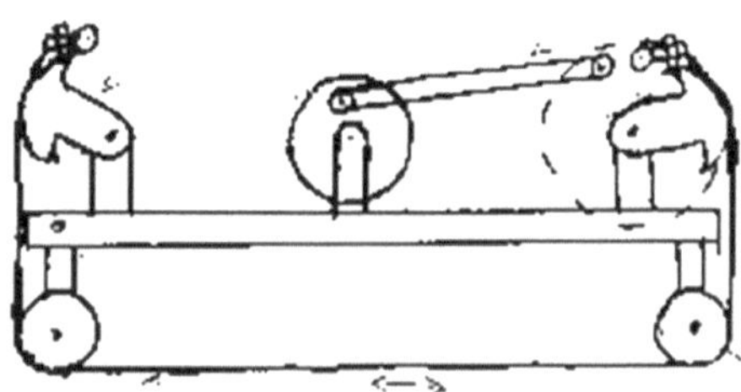

Fig. 4 Dual fork

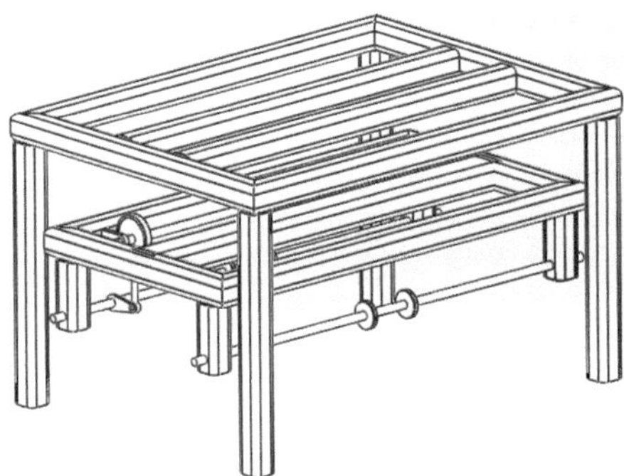

Fig. 5 Primary model of reciprocating wire mechanism

Idea of reciprocating wire mechanism is shown in Fig. 5. This mechanism consists of wire held on two pulleys placed on opposite shafts. The shaft is made to reciprocate by crank and connecting rod mechanism. Driver shaft is connected by two arms and a link. Wires reciprocate with pulleys and cause the cutting action. Feed is given by moving the frame in downward direction. Whole mechanism is supported by a single frame. Numbers of moving parts are lower which result in less vibration. The mechanism is simple and easy for maintenance and space requirement is less.

The mechanism shown in Fig. 5 is found to be the best possible solution as it fulfills the requirement about the machine. It is reliable for use and user friendly (Fig. 6).

Fig. 6 Prototype

3.2 Working of the Machine

Developed machine contains three frames viz. main frame (a) sub frame (b) and cutter frame (c). Sub frame and cutter frame are attached together with nut-bolts. The inner frames are provided with guide wheels (f) and main frame with guide ways. Both these frame moves upward and downward inside the main frame. Two electric motors are used as prime movers. One motor (d) placed is on main frame to control the motion of sub frame and another (e) is to give reciprocating motion to wire.

The motor (e) on sub frame rotates the pulley (g). This pulley placed on one end of the shaft with the crank (h) on the other end. As pulley and crank are on single shaft the crank also rotates with pulley. An arm (l) is attached on one end of shaft. This arm is connected to crank (h) by a connecting rod (k). The cutting frame (c) contains two shafts on which pulleys (i) are placed at required distance. The wires (j) are held by holding fasteners provided with adjuster for maintaining tension in wire. The rotary motion of the crank (h) is now converted to reciprocating motion of the arm (l). Ultimately the shaft, pulley, wire reciprocate. This causes cutting action on block (m). Feed given to the machine is controlled by motor (d). For easy movement of the machine in the factory premises caster wheels (n) are provided to all four legs (Figs. 7 and 8).

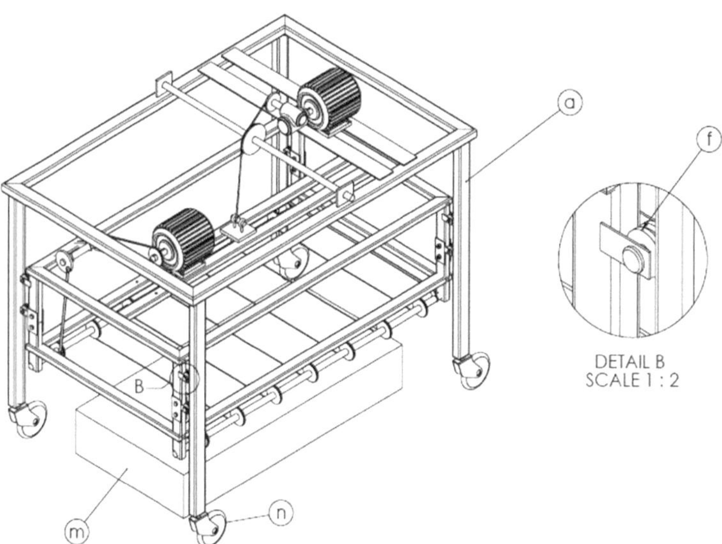

Fig. 7 Assembly view

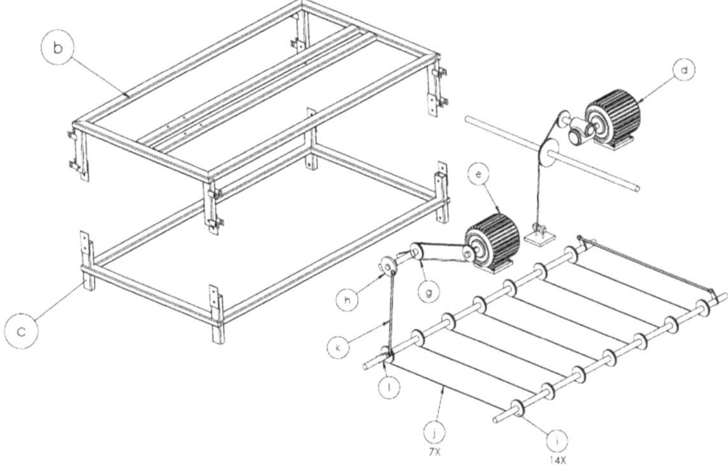

Fig. 8 Exploded view (without Main Frame-a)

3.3 Machine Built-Up

The built up of the machine is carried out by considering different concepts like design for assembly (DFA), design for reliability, design for disassembly (DFD) [6]. Application of these concept lead to design machine for full satisfaction of the industry. Some of the important points considered while designing this machine are as below

- Compact design
- Modular design
- Surface texture

For designing of system, proper care is taken regarding shape, size, maintainability, ergonomics etc. Basic dimensions of machine are based on size of mould, keeping minimum safe distance from block as it is brittle and can get damaged during operation. Height of machine satisfies basic needs of operation and also helps ergonomically easy to operate. Minimum numbers of parts are used to achieve simplicity in design which ensures low maintenance. Consideration of this point results in developing compact design of the machine. As the design of machine is compact it can be hovered easily in factory and avoid handling of the blocks.

The product (CLC Blocks) are required in three different dimensions i.e. $2' \times 4'' \times 8''$, $2' \times 6'' \times 8''$, $2' \times 8'' \times 8''$. As per demand, the supplier needs to produce different size blocks thereby the need of the modular design is raised. The proposed design can be used for all three types of variants of blocks. The cutting frame can be replaced by other cutting frame with different configuration to obtain required sized product. Instead of making adjustments in single unit, cutting frame of three configurations is provided. This system insures simplicity of machine. User can easily shuffle without any skillful operation training. A notable added feature delivered by this machine is surface texture. The texture got after cutting is uniform and improves aesthetics of block. This gives added advantage of good adhesion between POP and brick. Because of the texture brick can be used for eco building. There is no need of plastering if brick is used for an application such as construction of fencing walls. This can help in saving the construction material.

4 Discussion

The design of machine is on primary level with consideration of factory constraints, needs and other factors. This satisfies basic need about the machine for reliability and user friendliness. It also gives benefit of surface texture, modularity, etc. The design creates perfect base for further development in such type of machines. A flow chart shown in Fig. 9 mentions guidelines about the steps involved in design of machine. The design starts from the CLC block on which cutting is to be carried out. The block shows different properties for different time of curing that affect selection of cutting wire. Tension on wire is considered while designing the shaft and pulley. The pulley, shaft, wire and arm are included in cutting frame. Connecting rod remains separate and links the power from main frame to cutting frame by crank. Main frame involves a motor, pulley, shaft and crank.

As machine is specially designed for the small scale industry the cost is estimated around INR 51,000 ($842).

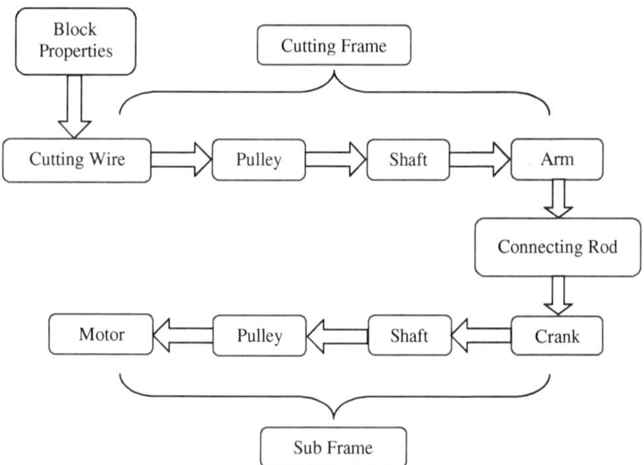

Fig. 9 Guide lines for designing of machine

To improve any system for getting better performance, efficiency and accuracy there is always need of some subsystem. The following systems can enhance the workability and efficiency of the machine.

- Continuity sensor for identifying failure of the wire.
- Block condition identifier.
- Proximity sensor for machine setting.
- Powered wheel.
- Remote controlled operating.
- Wire tension indicator.
- Overhead wire connections.

5 Result

Table 1 gives the information about the results obtained from the testing of machine against the traditional method of block manufacturing. These results are calibrated on the basis of prototype outputs. Total saving in time behind each mold is 42 min.

The surface textures obtained by cutting of the block at different time intervals from pouring the mold are illustrated in Figs. 10, 11 and 12.

Table 1 Output result

Operation	Time required (min)		Saving in time (in min)
	Conventional method	Developed machine	
Inserting of plates in mold and oil spraying	20	5	15
Disassemble the mold	30	3	27

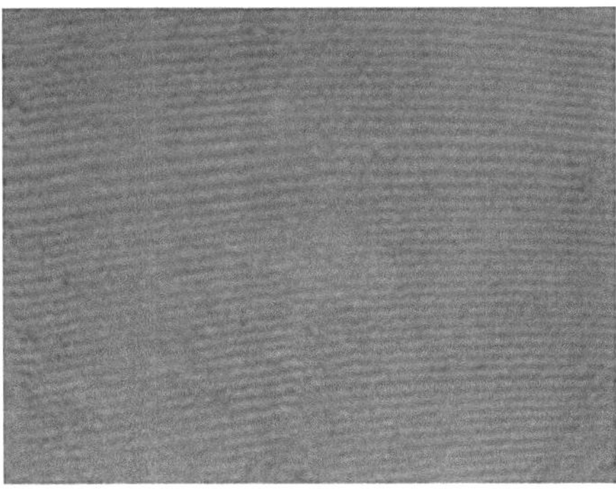

Fig. 10 Surface texture on block after 6 h from pouring

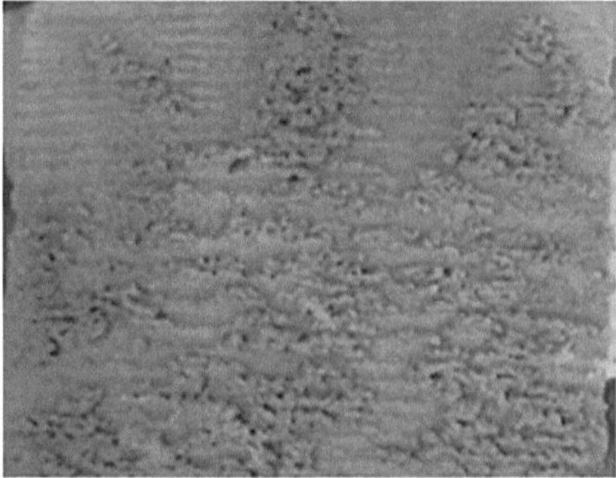

Fig. 11 Surface texture on block after 12 h from pouring

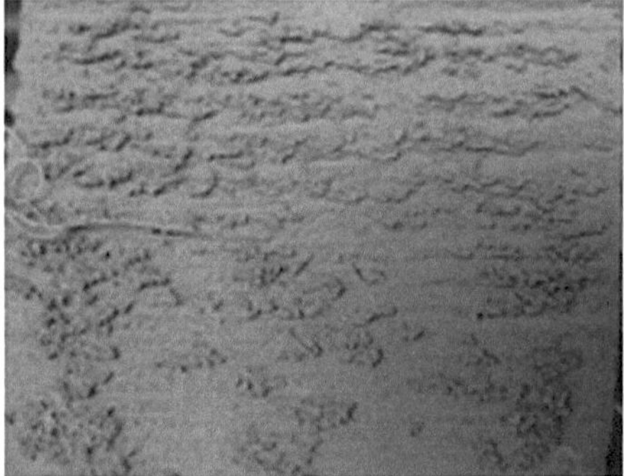

Fig. 12 Surface texture on block after 24 h from pouring

6 Conclusion

The results obtained from prototype indicate that the time required for the cycle is significantly reduced to 8 min in comparison to earlier method (50 min). The machine is hence a potential alternative/solution to the problems discussed in current practices. Use of modular concept extends the utility of the machine to be used for varied block sizes. This machine overcomes the drawbacks of the present available machine thereby satisfying the need of a small scale CLC block manufacturing plant regarding cutting of the block.

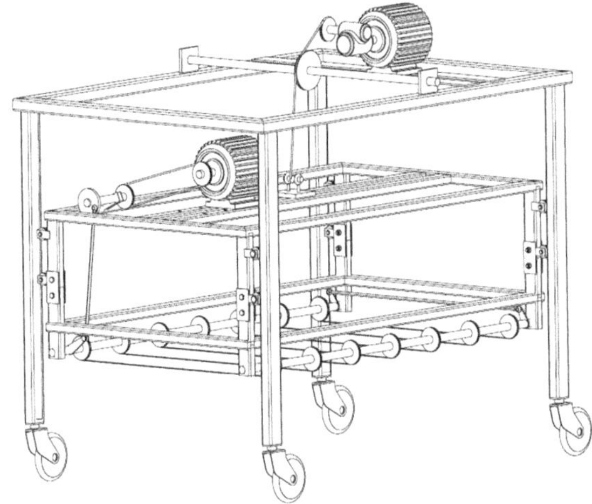

Fig. 13 CAD of final model

7 Cad Model of Final Design

See Fig. 13.

References

1. Krishna Bhavani Siram, K.: Cellular Light-Weight Concrete Blocks as a Replacement of Burnt Clay Bricks. Int. J. Eng. Adv. Technol (IJEAT) **2**(2), (2012). (ISSN: 2249–8958)
2. Nanolite Infratech Pvt. Ltd., Hyderabad (India). http://www.nanoliteinfratech.com
3. THT International Co. Ltd., Vietnam. http://www.thtvietnam.com
4. Maruti Hydraulics Pvt. Ltd., Nashik, Maharashtra (India). http://www.marutihydraulics.com
5. Balaji Construction Machine, Satara (India). http://www.aacconcrete.net
6. Roozenburg, N.F.M. In: Chakrabarti, A. (ed.) Engineering Design Synthesis. Springer, London (2002)

Novel Design Features for Matched Die Moulding for Bio-composites

D. Saravana Bavan, P. Kamalbabu and G.C. Mohan Kumar

Abstract The present work is focused in designing and fabricating a matched-die compression mould for a bio-composite material, suitable for testing mechanical properties (tensile and flexural). The mould is designed as per the guidelines and is fabricated. The bio-composite material is tested for mechanical properties and the obtained results were appreciable. The fabricated matched-die mould suits for bio-composites, wood composites and polymeric materials. Matched-die mould design features and compatibility of the design are carried out to make it simple and easier.

Keywords Bio-composite · Polymers · Matched-die · Design features

1 Introduction

Developing a bio-composite material from the existing natural fibres is increasing daily due to the added advantages such as low cost, low density, recyclability, biodegradability, reduced energy consumption, toughness and specific strength [1–4]. The properties of natural fibres depend on various factors such as chemical composition, location of growth, source of the plant part, environmental factors and climatic conditions. Bio-composites find applications in automotive sector, building, construction, packaging and textiles. They have the ability to replace the traditional fibres in composite sectors [2, 5–7].

Bio-composite material can be processed by different methods such as hand layup, injection moulding, bulk moulding, compression moulding, resin transfer moulding and vacuum bagging [8–11]. Among that, compression moulding satisfies the production requirements such as high rate of production, precise fiber

D.S. Bavan (✉) · P. Kamalbabu · G.C. Mohan Kumar
Department of Mechanical Engineering, National Institute of Technology,
Mangalore, Karnataka, India
e-mail: saranbav@gmail.com

A. Chakrabarti (ed.), *ICoRD'15 – Research into Design Across Boundaries Volume 2*,
Smart Innovation, Systems and Technologies 35, DOI 10.1007/978-81-322-2229-3_45

control, and high degrees of part integration. Design and manufacturing a suitable mould is more important and they play a vital role in the production cycle for making products at a faster rate. The cost and time taken for production of these bio-composites materials is purely based on the quality, material and lead-time of the mould. The successful use of bio-composites material in structural application is because of rapid and low cost production with included material characteristics.

In the present work, natural fibres with thermoset polymer of epoxy are considered for making a bio-composite material. A matched-die compression moulding is considered for processing a bio-composite material. The design features for the obtained material is carried out and fabrication of the mould is made according to the available features of the design. The obtained part material is readily suitable for testing mechanical properties such as tensile and flexural test.

1.1 Bio-composites

Bio-composites consist of synthetic polymers/bio polymers as the matrix material and natural fibres as the fiber material and these materials have one or more phases from biological sources. The matrix material usually the polymers (vegetable oil, starch) which holds the fibres and maintains preferred direction of fibres. It also helps in transferring the applied load to the reinforcement material and protects the fibres [4, 12]. The reinforcement are non continuous and one or more dispersed phases. The reinforcement is the fibres, which helps in giving strength and stiffness to the structure of the composite. These bio-composite materials are light weight having good stiffness and high strength to weight ratio [12]. Some of the problems associated with bio-composites are low thermal stability, hydrophilic nature of fiber surface that leads to swelling of composites, and moisture content that decreases the properties of the composite material [4, 7].

Natural plant based composites are mainly composed of cellulose, hemi cellulose, lignin, pectin and non structural cell wall components (waxes). Cellulose fibres show high tensile strength and modulus, whereas hemi cellulose acts as a bridge between cellulose fibres and lignin. Lignin holds the fibres and act as a stiffening agent for cellulose in the fibres [1, 13–15]. The mechanical properties of these fibres are competitive to synthetic fibres [2, 16]. Fibre geometry, morphology nature, fibre extraction and aspect ratio are some of the factors that influence the performance of bio-composites [12].

1.1.1 Classification of Bio-composites

Bio-composites or Natural fiber composites consists fibres from plant, animal and mineral sources [2–4]. Plant fibres are the most important fibres and they are further classified based upon the part of growth such as bast fibres (jute, flax, hemp, ramie and kenaf); seed fibres (kapok, cotton, milk weed); fruit fibres (coir, oil palm); stalk fibres (rice, wheat, barley, maize and oat) and grass fibres (bamboo and canary). Animal fibres consist of proteins e.g. silk, wool, feathers, fur, human hair, goat, camel hair and mohair. Mineral fibres are in small amount and produce carcinogenic effect. e.g. Asbestos. fibrous brucite and wollastonite.

Polymer matrix can be divided into fully degradable and partly degradable [13, 14, 17]. Fully degradable are further divided into biodegradable-natural based polymer e.g. polylactide (PLA), starch, cellulose, polyhydroxy alkanoate (PHAs) and biodegradable petroleum based polymer e.g. Aliphatic polyester, poly ester amide, poly vinyl alcohol. Partly degradable can be divided into non degradable-petroleum based polymer. e.g. Polyester, polypropylene, polyethylene.

2 Compression Moulding for Bio-composites

Bio-composites material processed by compression moulding using thermoset polymer of soya bean oil and flax plant fibres was carried out [18] and the prepared composites were tested with flexural and impact testing. Compression resin transfer moulding which combines features of compression moulding with resin transfer moulding infusion was studied [11, 19]. Natural hybrid composites of banana and sisal fibres reinforced with polyester resin were fabricated by compression moulding and the mechanical properties such as flexural, tensile and impact were studied [9]. Static properties on tensile, flexural and dynamical properties of compressed mould with vinyl ester resin composites were analysed [8]. Composites of high density polyethylene filled with bio inert and bio active ceramics fabricated by compression moulding was examined [20]. Bio-composites of grass fibres were fabricated by different process such as injection moulding, sheet moulding compounding and compression moulding and they revealed that compression moulding composites showed high impact strength than that of injection moulded samples [21].

2.1 Manufacturing Factors for Compression Moulding Process

Matched die compression mould are fabricated according to the available standard design and calculations. Compression moulding is classified into three types such as positive, semi positive and flash type. The important factors of compression moulding process consist of four aspects [22]. First is to find the required amount of material to fill the cavity when it is closed; Secondly to determine the time taken to heat the blank material to attain the processing temperature; Thirdly in selecting the suitable mould temperature and purely selecting the polymer and fiber material (for a bio-composite material the mould temperature is around 80–120 °C) and lastly to calculate the force required to fill the mould and normally the moulding pressure for polymer based bio composite materials is 250–300 psi. If the pressure applied is too fast then the material will squirt, and if the applied pressure is too slow then the material will not fill properly. Hence pressure applied should be evenly based on the cure time of the polymer material and viscosity rise. Additional factors in developing design features for matched-die moulding are gel time (mould filling should be complete before the resin gels) and demoulding time (time to attain a sufficient degree of cure to de mould the part) [10].

2.2 Mould Part Design Guidelines

During composite processing, cavity, core and vent are in contact with the part material and they are at high temperature and pressure to form the required shapes. Therefore these mould materials must have sufficient rigidity, hardness, surface roughness, abrasion resistance and corrosion resistance. Basic mould design guidelines in compression moulding includes direction of mould, closure and location of parting surface, part must be easily ejected from the mould after solidification, avoid undercuts, thick walls and sharp corners, design the parts with uniform cross section and number of gates to allow the flow of material in the mould uniformly and evenly.

2.3 Factors Considered for Working Design

Dimension designs for moulded parts are based on the dimension of mould which is in direct contact with the part material. Accurate dimension of cavity and core influences the part dimensions [23]. Three factors are generally considered for dimension design of part material.

- Part Shrinkage: Normally plastic parts will expand or shrink based on its processing temperature and the cavity will not fill thoroughly. Average shrinkage percentage considered for thermo set material is around 2 %.
- Tolerance level for cavity and core: Tolerance for the cavity and core material has great influence in the dimensions of part material. Tolerance of the part material is taken as manufacturing tolerance (±1 mm for length wise and ±4 mm for width wise).
- Abrasion loss in cavity and core: Abrasion and restoration factors make the cavity dimension get larger and core dimension get smaller.

3 Design for Working Dimension

3.1 Calculation for Cavity Dimension

Cavity dimension influences more in overall dimension of the bio-composite material [23] and these dimensions are calculated as shown in Eqs. (1)–(3). While designing, lower limit is taken as containment dimension and upper deviation as dimension tolerance. Integrated embedded cavity and detailed dimension of the mould is shown in Fig. 1.

$$\text{Length of cavity, } L_L = [L_{PL}(1 + K) - (3/4)\Delta_L]^{+\delta} \tag{1}$$

$$\text{Width of cavity, } L_W = [L_{PW}(1 + K) - (3/4)\Delta_W]^{+\delta} \tag{2}$$

$$\text{Depth of cavity, } H = [H_P(1 + K) - (2/3)\Delta_D]^{+\delta} \tag{3}$$

where δ is manufacturing tolerance of mould (1/3–1/6 of the dimension tolerance of part material), L_{PL} is the nominal dimension (length) of the part material, K is the average shrinkage of polymer material, Δ_L is the dimension tolerance of part material (length), L_{PW} is the nominal dimension (width) of the part material, Δ_W is the dimension tolerance of part material (width), H_p is the nominal dimension in the altitude direction of part material, Δ_D is dimension tolerance of part material (depth).

3.2 Calculation for Core Depth Dimension

Core depth dimension influences more in the part thickness for a bio-composite material. Working dimensions are kind of containment dimension of core which are gradually decreased because of abrasion in core. While designing, upper limit is taken as containment dimension and lower deviation as dimension tolerance. Core depth dimension is shown in Eq. (4).

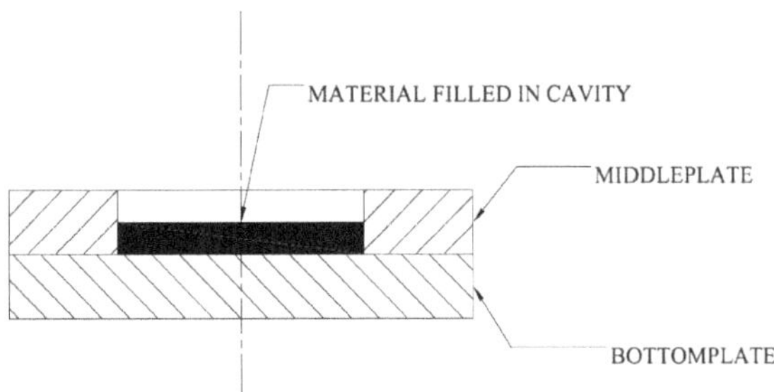

Fig. 1 Integral embedded cavity

$$\text{Depth dimension of core, } h = [h_P(1 + K) + (2/3)\Delta_D]_{-\delta} \tag{4}$$

where h_P is the nominal dimension in the altitude direction of core material.

3.3 Mould Design Calculation for Bio-composites

ASTM standards D790 for plastic material for flexural test include size of 127 mm length, 12.7 mm in width and thickness of 3.2 mm. These dimensions are considered for compression moulding process and cavity design calculation is shown below. Epoxy shrinkage allowance chosen value is 2 %. Calculations for cavity and core dimensions are shown in Eqs. (5)–(8).

3.3.1 Calculation for Cavity Dimension (Flexural Test)

Length dimension of cavity

$$\begin{aligned} L_L &= [127 \times (1 + 0.02) - (3/4) \times 2]^{+(1/6\times2)} \\ L_L &= 128.0\ \text{mm (length wise, approximate value)} \end{aligned} \tag{5}$$

Width dimension of cavity

$$\begin{aligned} L_W &= [12.7 \times (1 + 0.02) - (3/4) \times 0.4]^{+(1/6\times0.4)} \\ L_W &= 12.7\ \text{mm (width wise)} \end{aligned} \tag{6}$$

Depth dimension of cavity

$$\begin{aligned} H &= [9.2 \times (1 + 0.02) - (2/3) \times 0.4]^{+(1/6\times0.4)} \\ H &= 9.1\ \text{mm} \end{aligned} \tag{7}$$

3.3.2 Calculation for Core Dimension (Flexural Test)

Depth dimension of core

$$\begin{aligned} h &= [5.5 \times (1 + 0.02) + (2/3) \times 0.4]_{-(1/6\times0.4)} \\ h &= 5.9\ \text{mm} \end{aligned} \tag{8}$$

Detailed dimension of the mould is shown in Fig. 2. Similarly tensile specimen for the bio-composite material as per ASTM standard D638 are fabricated and used for testing. Exploded view is shown in Figs. 3 and 4.

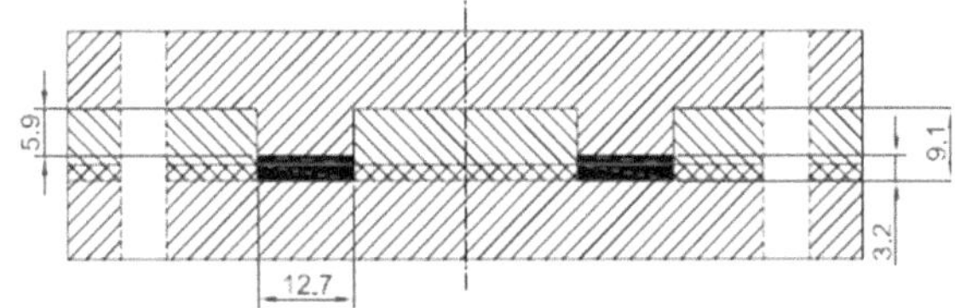

Fig. 2 Detailed dimension of the mould

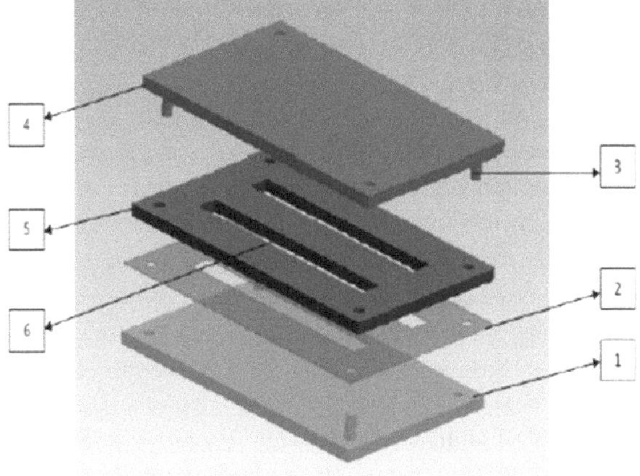

Fig. 3 Exploded view of compression mould for flexural specimen. *1* bottom plate, *2* teflongasket, *3* stud, *4* top plate, *5* middle plate, *6* cavity area to fill material

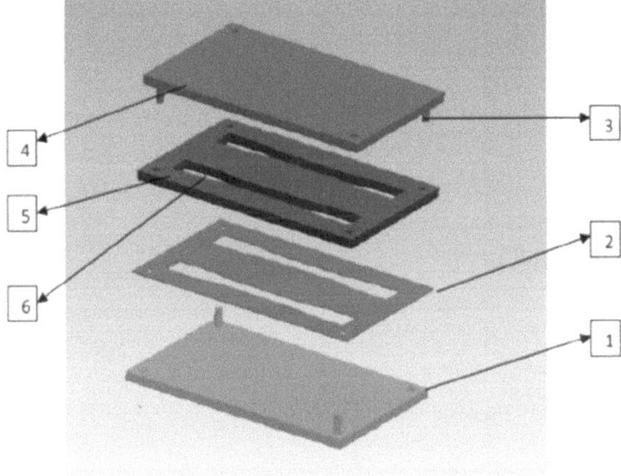

Fig. 4 Exploded view of compression mould for tensile specimen. *1* bottom plate, *2* teflongasket, *3* stud, *4* top plate, *5* middle plate, *6* cavity area to fill material

4 Results and Discussions

For bio-composite materials, the important criteria in compression mould are in finding more required amount of material, selecting the right type of mould, choosing mould temperature and the correct amount of applied pressure. While designing the mould, factors to be considered are material shrinkage and tolerance level. Among them mould dimensions has great influence in overall dimension for a bio-composite material. Factors such as selecting the suitable mould material, parts like integral embedded type cavity and core and electric heating coil will increase the production rate and dimensional stability for a bio-composite material. Processing parameters for thermoset epoxy polymers are shown in Table 1. The Table 2 shows the control level of process parameters and the problems occurred in compression moulding. The process parameters are mould temperature, material pre-heat temperature, applied pressure and cure time. Figures 5 and 6 shows various stages in obtaining a bio-composite material processed through matched-die compression mould for flexural and tensile specimen.

Bio-composite materials (10 % fibre content) with use of matrix of thermoset polymers are tested (six samples each) for tensile and flexural test with cross head speed of 2 mm/min and the results are plotted as shown in Fig. 7. The obtained best values for tensile test was 18.21 MPa (maximum tensile stress) and 1.71 GPa (Young Modulus) where as for flexural test 25.69 MPa (maximum flexural stress) and 1.27 GPa (Flexural Modulus) with maximum deflection of 6.89 mm. If the specimen dimensions were changed it would lead to improper values and the stress transfer between fiber and matrix will keep on changing. Some procedural modifications such as conditioning of samples, maintaining the exact and same gauge length, relative humidity as per standards, low cross head speed, controlled atmosphere testing and maintaining the same thickness throughout the samples were carried out, so that the effect of size, shape temperature and stress concentration will not affect the sample value and if so it does, it just play a minor role. These tested samples were suited for structural applications and bio-material applications. The mould is suited for bio-composites, wood-composites and polymer composites.

Table 1 Processing parameters for thermoset (epoxy) polymers

Press temperature (°C)	Pressure (kg/cm^2)	Press time (min)
80–120	3–30	30–45

Table 2 Control levels of process parameters processed through compression moulding

Process parameters	Mould temp	Material pre-heat temp	Applied pressure	Cure time
Problems occurred				
Flash excessive of material	✓	✓	✓	–
Mould staining	✓	✓	–	–
Non fills	×	✓	✓	✓
Part shrinkage	×	–	–	×
Sticking in mould	✓	–	–	✓

(✓ = increase; × = decrease)

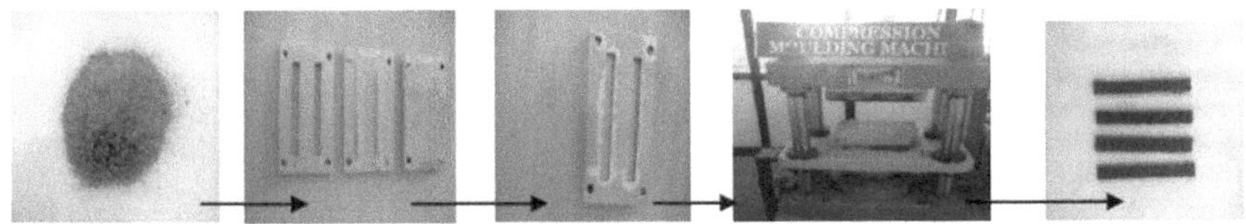

Fig. 5 Various stages in obtaining a bio-composite material processed through matched die compression mould for flexural specimen

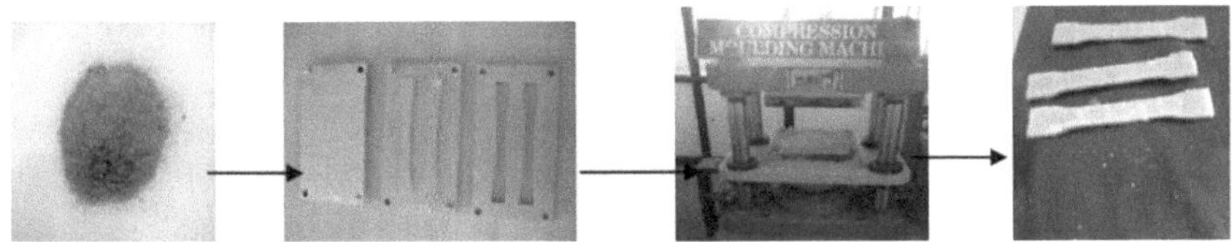

Fig. 6 Various stages in obtaining a bio-composite material processed through matched die compression mould for tensile specimen

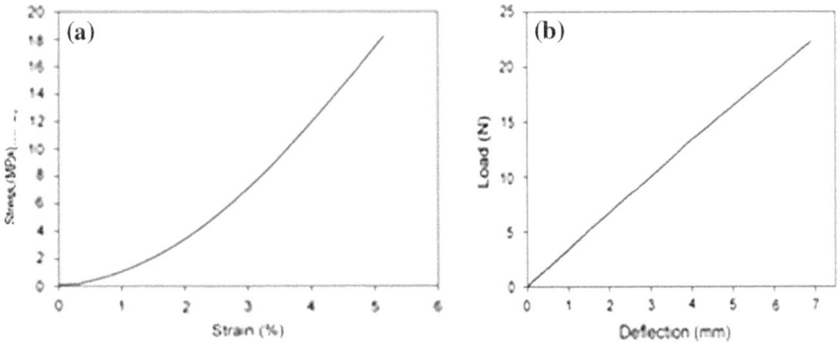

Fig. 7 10 % fiber content bio-composite material (natural fibres) **a** Tensile test. **b** Flexural test

4.1 Pros and Cons for the Obtained Mould with Use of Limitation

The prepared mould is focussed in generating more readymade tensile and flexural samples at a faster rate with easier technique and can be directly used for mechanical testing. The goal is to make innovative ideas with minimum cost and maximum flexibility for the product. The particular challenges for the mould are they suit only for D790 and D638 Type-1 rigid material and not for other standards. It is not suitable for laminated, woven fabric, more thickness kind of material and polymer matrix composite materials reinforced with high modulus fibres. The mould doesn't suit for non-rigid plastics and for high modulus composites. Getting thermoplastic type of material was easier rather than getting thermoset material. In thermo set kind of bio-composites, getting even thickness was difficult because of viscous of polymers.

5 Conclusions

Depletion of petroleum resources and global warming are making researchers to find an alternate solution to the existing problems. In that sense, bio-composite material can lead the path in materials and composites because they are green materials with less emissions and carbon neutral. Biodegradability and low cost finds these materials more economical that suits for composites sector. For thermoset based bio-composite material, compression moulding is the hopeful methodology compared to other existing methods. Designing the suitable mould with the right type of parts and the material dimension gives the better quality and also reduces the production time. The designed and fabricated flexural and tensile mould suits for bio-composite and polymer composite materials. The mould is prepared as per ASTM standards and can be used directly for testing mechanical properties such as tensile tests and flexural tests.

Acknowledgments The authors sincerely thank Government Tool Room and Training Centre (GT&TC), Baikampady, Mangalore for giving suggestions and in fabricating the mould.

References

1. John, M.J., Thomas, S.: Biofibres and Bio-composites. Carbohydr. Polym. **71**(3), 343–364 (2008)
2. Joshi, S.V., Drzal, L.T., Mohanty, A.K., Arora, S.: Are natural fiber composites environmentally superior to glass fiber reinforced composites? Compos. A **35**, 371–376 (2004)
3. Mohanty, A.K., Misra, M., Hinrichsen, G.: Biofibres, biodegradable polymers and bio-composites: an overview. Macromol. Mater. Eng. **276**(277), 1–24 (2001)

4. Mohanty, A.K., Misra, M., Drzal, L.T.: Natural Fibres, Biopolymers, and Biocomposites. CRC Press, Taylor & Francis Group, Boca Raton FL (2005)
5. Dweib, M.A., Hu, B., O'Donnell, A., Shenton, H.W., Wool, R.P.: All natural composite sandwich beams for structural applications. Compos. Struct. **63**, 147–157 (2004)
6. Marsh, G.: Next step for automotive materials. Materials today, (2003)
7. O'Donnel, L.,A., Dweib, M.A., Wool, R.P.: Natural fiber composites with plant oil-based resin. Compos. Sci. Technol. **64**, 1135–1145 (2004)
8. Enos, J.H., Erratt, R.L., Francis, E., Thomas, R.E.: Structural performance of vinyl ester resin compression moulded high strength composites. Polym. Compos. **2**(2), 53–57 (1981)
9. Idicula, M., Sreekumar, P.A., Joseph, K., Thomas, S.: Natural fiber hybrid composites-a comparison between compression moulding and resin transfer moulding. Polym. Compos. **30**(10), 1417–1425 (2009)
10. Lasell, D.M.: Part and mould design guidelines for the high volume compression moulding of carbon fiber reinforced epoxy. Think Compos. LLC Palm Harbor, FL (2011)
11. Merotte, J., Simacek, P., Advani, S.G.: Resin flow analysis with fiber preform deformation in through thickness direction during compression resin transfer moulding. Compos. A **41**, 881–887 (2010)
12. Hull, D., Clyne, T.W.: An Introduction to Composite Materials. Cambridge University Press, Cambridge (1996)
13. Wallenberger, F.T., Weston, N.: Natural fibres, Plastics and Composites. Kluwer Academic Publishers, Netherlands (2003)
14. Wool, R., Sun, X.S.: Bio-Based Polymers and Composites. Elsevier, Amsterdam (2005)
15. Young, R.A., Rowell, R.M.: Cellulose Structure, modification and hydrolysis. Wiley, New York (1986)
16. Fowler, P.A., Hughes, J.M., Elias, R.M.: Review biocomposites: technology, environmental credentials and market forces. J. Sci. Food Agric. **86**, 1781–1789 (2006)
17. Velde, K.V., Kiekens, P.: Biopolymers: overview of several properties and consequences on their applications. Polym. Test. **21**, 433–442 (2002)
18. Adekunle, K., Akesson, D., Skrifvars, M.: Biobased composites prepared by compression moulding with a novel thermoset resin from soybean oil and natural-fiber reinforcement. J. Appl. Polym. Sci. **116**(3), 1759–1765 (2010)
19. Bhat, P., Merotte, J., Simacek, P., Advani, S.G.: Process analysis of compression resin transfer moulding. Compos. A **40**, 431–441 (2009)
20. Tripathi, G., Dubey, A.K., Basu, B.: Evaluation of physico-mechanical properties and in vitro biocompatibility of compression moulded HDPE based bio composites with HA/Al_2O_3 ceramic fillers and titanate coupling agents. J. Appl. Polym. Sci. **124**(2), 3051–3063 (2012)
21. Liu, W., Thayer, K., Misra, M., Drzal, L.T., Mohanty, A.K.: Processing and physical properties of native grass-reinforced biocomposites. Polym. Eng. Sci. **47**(7), 969–976 (2007)
22. Altan, T., Lilly, B., Yen, Y.C.: Manufacturing of dies and moulds. CIRP Ann.—Manuf. Technol. **50**(2), 404–422 (2001)
23. http://www.plenco.com/plenco_processing_guide/

Design Methodology and Dynamic Simulation of Fixed Displacement Swash Plate Compressor

Raushan Kumar Jha, Selvaraji Muthu and S. Sivakumar

Abstract Swash plate compressor is widely used in automotive air conditioning industry for its light weight, small size and compact structure. In this article the detailed simulation of swash plate compressor with the five double acting pistons is performed to predict the dynamic motion. The pressure and force calculation are performed to arrive the radial and thrust loads and its trends. The variation of different parameters (valve losses, frictional losses and resistance torque) with rotation angle is simulated and the trend is plotted. The proportion of each factors contributing to total power are shown. Resultant shaft torque fluctuation is predicted. The power consumption by the compressor is calculated at different speeds and validated with the test data.

Keywords Swash plate · Dynamic motion · Resistance torque · Radial and thrust loads

Nomenclature:

R_p	Pitch circle radius (m)
θ	Shaft rotation angle (°)
F_g	Gas force (N)
F_I	Inertia force (N)
P_d	Discharge pressure (N/m^2)
V_c	Clearance volume (m^3)
S	Stroke (m)
ORR	Oil Retention Rate (%)
α	Swash angle (°)
ω	Angular velocity (rad/s)
F_f	Friction force (N)
Ps	Suction pressure (N/m^2)
Ap	Area of piston (m^2)

R.K. Jha · S. Muthu (✉) · S. Sivakumar
SUBROS Ltd, C51, Phase 2, Noida 201304, Uttar Pradesh, India
e-mail: selvaraji.muthu@subros.com

A. Chakrabarti (ed.), *ICoRD'15 – Research into Design Across Boundaries Volume 2*, Smart Innovation, Systems and Technologies 35, DOI 10.1007/978-81-322-2229-3_46

Vs	Swept volume (m^3)
ac	Clearance volume ratio (%)
OCR	Oil Circulation ratio (%)

1 Introduction

The most popular compressor types used in the mobile air-conditioning applications are swash plate compressors. The reciprocating motion of piston is parallel to the axis of the rotating shaft, so is called axial piston compressor. Fixed displacement and variable displacement swash plate compressors are the two generations of compressor family apart from rotary compressors like scroll and vane compressor.

The fixed displacement compressor with double headed pistons has swash plate mechanism with a fixed angle that drives the pistons to and fro, which is responsible for compressing the refrigerant gas in one cylinder and simultaneously suction on opposite cylinder as it drawn through the compressor. The driving mechanism for the compressor has a rotating pulley mounted on the front cover of compressor and a belt is mounted over this pulley, which rotates at a speed proportional to the engine crankshaft rotation, as shown in Fig. 1.

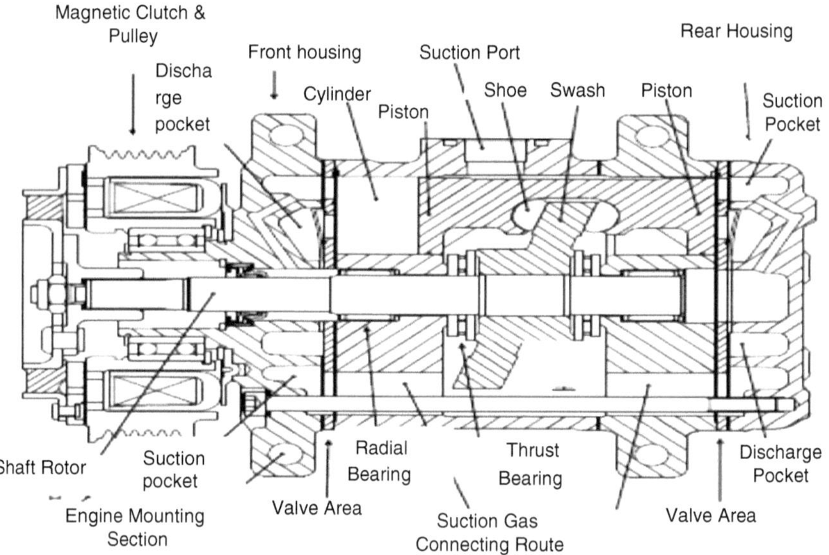

Fig. 1 Sectional view of swash plate compressor [8]

Swash plate is press fitted into shaft, which is supported by front and rear cylinder block with radial bearings and thrust bearings on either side to take care of axial and radial load generated by the compression of refrigerant gas respectively.

2 Working Principle

The rotating motion of swash plate is converted to reciprocating motion of piston by a pair of shoe, which wobbles between swash plate and piston. The flat surface of shoe is in contact with swash plate and convex portion in contact with concave pocket of piston. Reed Suction and discharge valves of cantilever type are located on valve plate, operates based on pressure difference between swash chamber inside the compressor and suction and discharge chamber in the housings respectively.

Below and Miloslavich [1] investigated the dynamics of the swash plate and developed the mathematical relationships necessary for evaluating stresses, bearing loads, and for sizing and locating counterweights for a proper running balance. Tani et al. [2] demonstrated the features of the swash plate type compressor with respect to the degree of influence on the volumetric efficiency of the compressor according to the internal temperature distribution. Liu and Hui [3] simulated the dynamics of four cylinder wobble plate compressor using program in MATLAB to come up with motion curve of the piston along with the piston forces and resistance moment. The stress distribution of piston at various rotational angles is analyzed by Wu et al. [4] using commercial finite element software. It is shown that to increase the fillet size could decrease stress concentration effectively. Xu [5] described the Stress Analysis of the swash plate and spindle of the swash plate compressor with theoretical model along with influence of radial and axial clearance management in the assembly.

Xu [6] analyzed the lubrication issues in swash plate type refrigerant compressor between piston and cylinders. Shu and Zhu [7] calculated the balancing performance of swash plate compressors by visualization and representation of forces using MATLAB to quickly find the main source of vibration and noise.

3 Swash Plate-Piston Motion Simulation

3.1 Piston Motion Equations

The displacement of double headed piston, velocity and acceleration are as per Eqs. (1)–(4).

$$\text{Piston displacement, } x = R_p \times (1 - \cos(\beta + \theta)) \times \tan\alpha \quad (1)$$

$$\text{Stroke (Max displacement)}, S = 2 \times R_p \times \tan \propto \quad (2)$$

$$\text{Piston velocity}, v = -R_p \times \omega \times \sin(\beta + \theta) \times \tan \alpha \quad (3)$$

$$\text{Piston acceleration}, a = R_p \times \omega^2 \times \cos(\beta + \theta) \times \tan \alpha \quad (4)$$

4 Dynamic Analysis

Compressor working is divided into expansion, suction, compression and discharge process respectively. During expansion process, the gas re-expansion takes place due to dead end clearance and till the suction pressure so to inhale the gas in suction process. The gas is sucked at suction pressure till BDC (Bottom Dead Centre). After reaching to BDC, the piston is moving towards TDC (Top Dead Centre) to compress the refrigerant gas.

4.1 Piston Force

The total piston (F_p) force includes gas force (F_g), reciprocating inertia force (F_I) and the friction force (F_f) between cylinder wall and the piston, as shown Fig. 2. Pressure drop due to valve lift and suction and discharge losses are considered in this study.

$$F_p = F_g + F_f + F_I \quad (5)$$

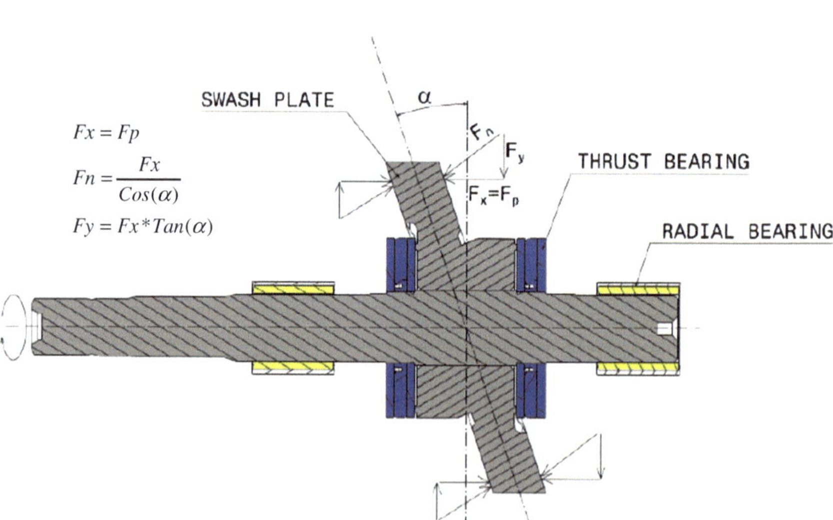

Fig. 2 Schematic view of swash plate subassembly

4.2 Gas Force Equation

The gas force acting on the piston is the product of resultant gas pressure on both sides of the piston and the area. The piston force direction towards compression is defined as positive.

$$F_g = [P - P_s] \times A_p \tag{6}$$

In the discharge process, the net pressure drop is the sum of pressure losses due to valve lift, orifice effect in port and circumferential flow. The compression cycle is considered poly-tropic.

4.2.1 Re-expansion Process

$$P = \left(\frac{V_c}{V}\right)^m \times P_d = \left[\frac{S_o}{S_o + x}\right]^m \times P_d \quad 0 < \theta < \theta_1 \tag{7}$$

The equivalent dead end clearance (S_o) can be calculated as $S_o = V_c \times S$ where V_c is between 0.7 and 1.5 %

Re-expansion process takes place till the shaft rotation angle θ say (θ_1), until chamber pressure is equal to the suction pressure.

4.2.2 Suction Process

$$P = P_s \text{ when } \theta_1 < \theta < \pi \tag{8}$$

4.2.3 Compression Process

$$P = \left(\frac{V_s}{V}\right)^m \times P_s = \left[\frac{S_o + S}{S_o + x}\right]^m \times P_s \quad \pi < \theta < \theta_2 \tag{9}$$

Compression process takes place till the shaft rotation angle θ say (θ_2), until chamber pressure is equal to the discharge pressure.

4.2.4 Discharge Process

$$P = P_d \text{ when } \theta_2 < \theta < 2\pi \tag{10}$$

4.3 Friction Force

Friction force acting between the piston OD and cylinder bore is assumed constant throughout the process. The friction is negative in the suction and re-expansion process and positive in the compression and discharge process. Empirical correlation for the friction force is as follows [3]:

$$F_f = \frac{A_p}{16} \times [P_s + P_d] \times \left[\frac{1}{\eta_m} - 1\right] \tag{11}$$

where, η_m is the mechanical efficiency of the compressor.

4.4 Total Resistance Moment (Torque)

The resultant torque is the sum of the resistance moment by each cylinder.

$$M = \sum M_i, \; where\; i = [1, n], n = No\; of\; pistons \tag{12}$$

$$M_i = F_p \times R_p \times \tan\propto \times \sin(\beta + \theta) \tag{13}$$

$$F_y = F_p \times \tan\propto = P_s \times A_p \times \tan\propto \tag{14}$$

where β is phase angle $= \frac{360}{n}$.

5 Results and Discussions

Figure 3 shows the displacement of piston at various angle of shaft rotation. Piston 1, which is at TDC position is the beginning of Re-expansion and then suction process, by which the gas volume increases, attains maximum displacement (stroke) at 180°. The displacement decreases after 180° which indicates compression and discharge process up to 360°.

From Figs. 4 and 6, the cylinder gas pressure and piston force at various angle of shaft rotation are shown. Piston 1, at TDC position travel towards BDC is taken as reference. During the Re-expansion process, the gas volume in the clearance space

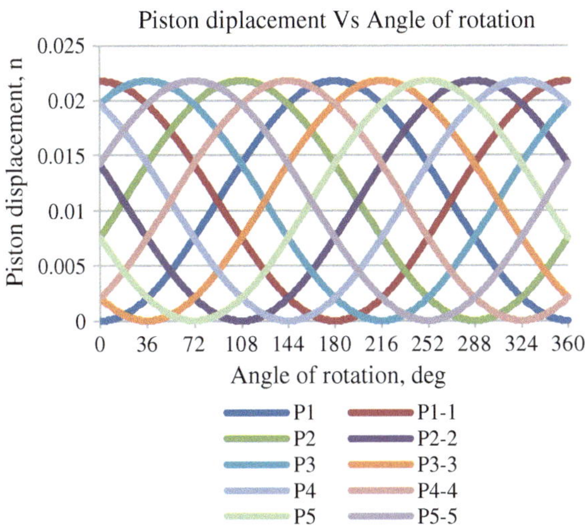

Fig. 3 Displacement versus angle of rotation

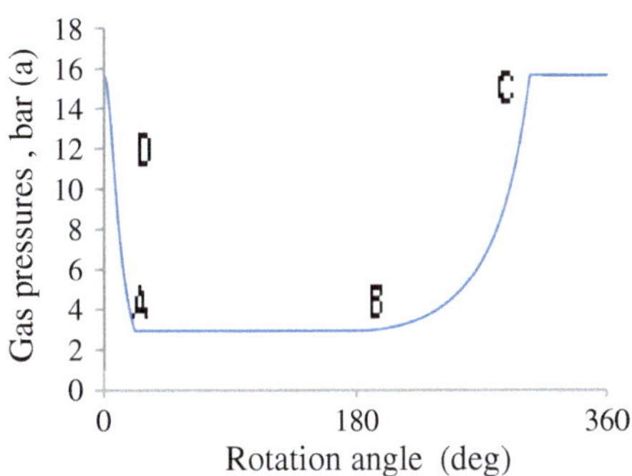

Fig. 4 Gas pressure

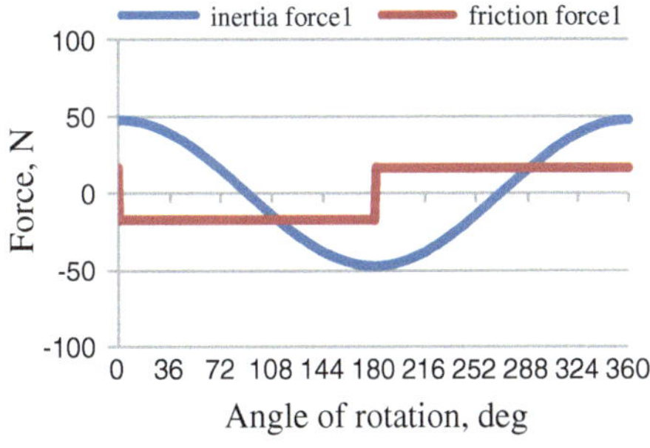

Fig. 5 Inertia and friction force

is expanded up to 21° of shaft rotation. The suction process of piston 1 is complete at shaft rotation angle of 180° i.e. the suction valve closes which is represented by B. The refrigerant gas is compressed to discharge pressure which follows the

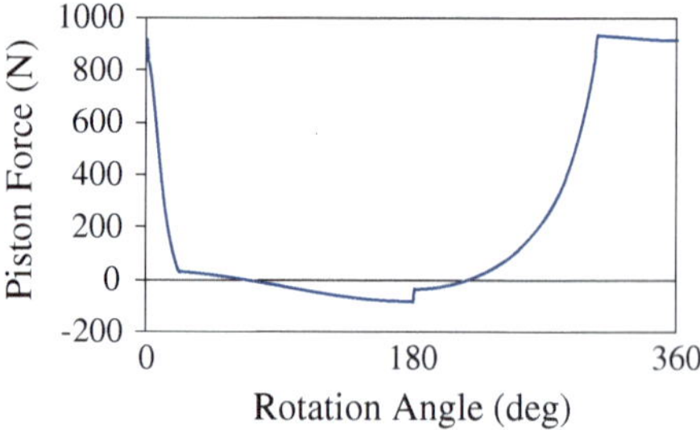

Fig. 6 Piston force

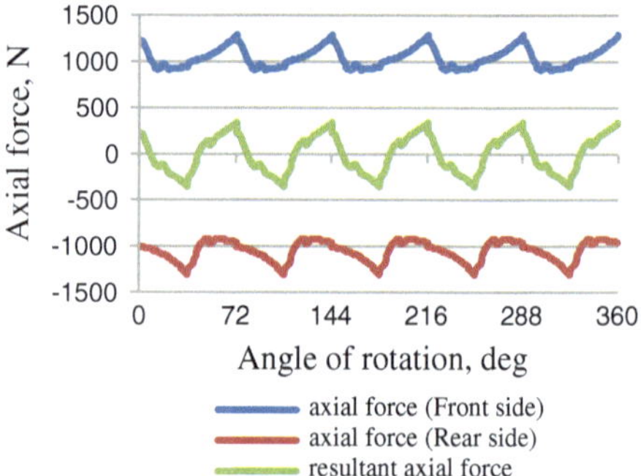

Fig. 7 Axial force

polytrophic relation up to C. At point C, i.e. at shaft rotation angle of 305°, the discharge reed valve opens the changes of the reciprocating inertia and friction force of piston 1 against angle of rotation are shown in Fig. 5. The friction is constant throughout the cycle, where it is negative in the re-expansion and suction process and positive in the compression and exhaust process. When the piston reaches the maximum and minimum displacement, the reciprocating inertia force reaches the maximum and at shaft rotation of 90° and 270°, rotation.

In Figs. 7 and 8, the variation of radial and axial forces on front and Rear sides of the piston are plotted. Due to symmetrical reciprocating action of double headed piston, the axial and radial forces are balanced. Therefore the resultant axial and radial forces are minimum, which are taken care by the thrust and radial bearings mounted on rotating shaft. This is the one of the major outcome of this simulation. In Fig. 9, the variations of all the piston moment with rotation angle are plotted. There are 10 peaks and 10 valleys corresponding to number of cylinders, which is one of the design parameter for magnetic clutch. The peak values are considered for determining the factor of safety for design of magnetic clutch slip torque. In Fig. 10, the magnitude and direction of the piston force is varying with phase angle of 36°. In Fig. 11, the resultant shaft torque is plotted.

In Figs. 12 and 13, the effects of rotation speed and discharge port size on the torque are shown. The torque fluctuation is increasing with speed and following the

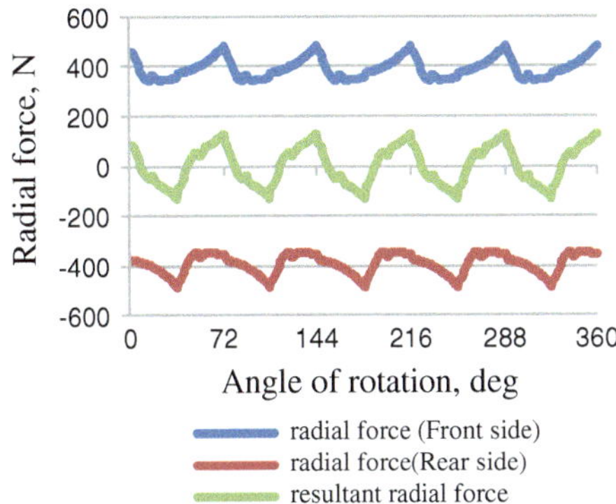

Fig. 8 Radial force

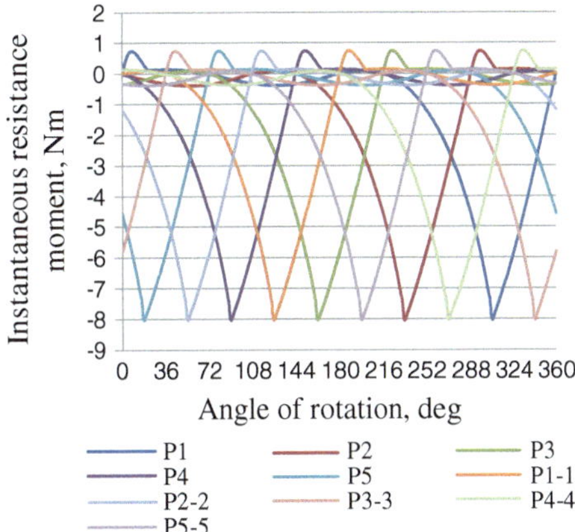

Fig. 9 Piston moments

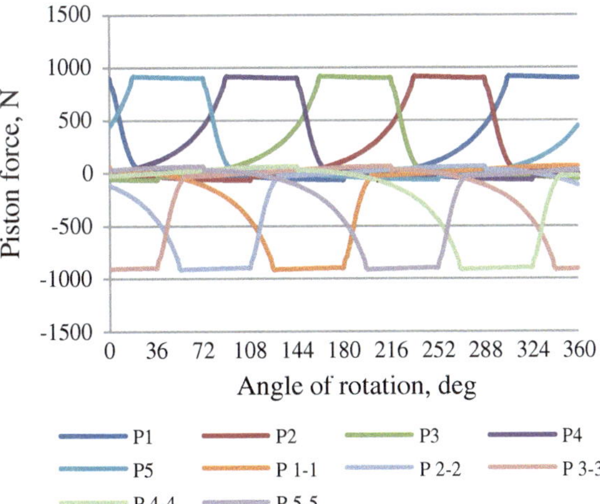

Fig. 10 Piston force

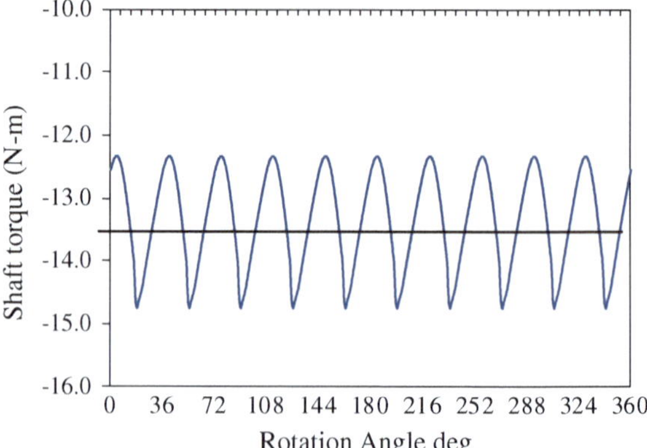

Fig. 11 Shaft torque

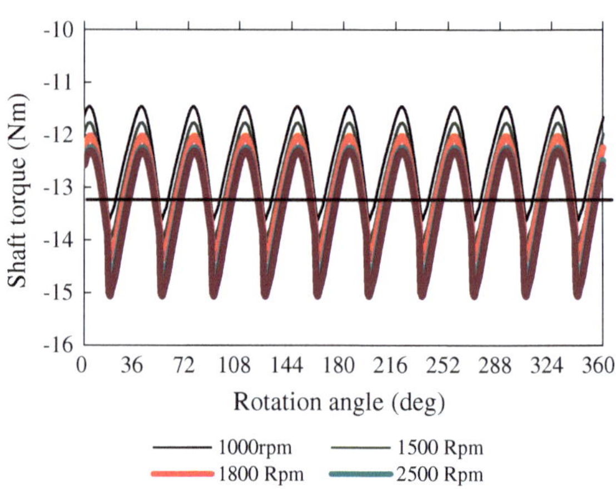

Fig. 12 Shaft torque at various speed

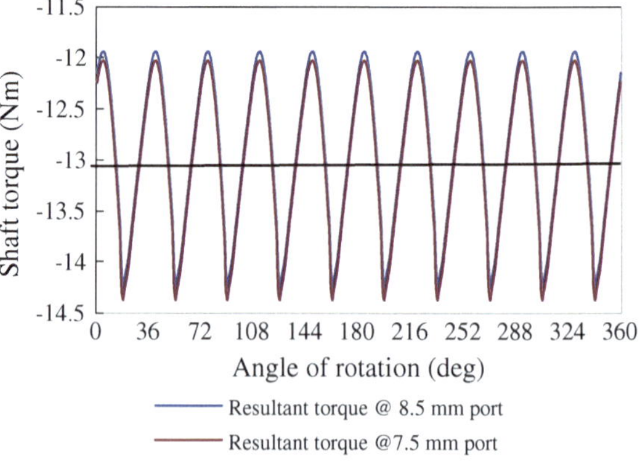

Fig. 13 Effect of port size

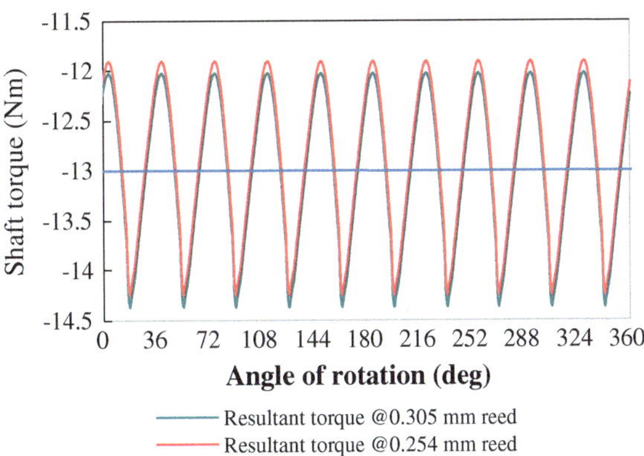

Fig. 14 Effect of reed thickness

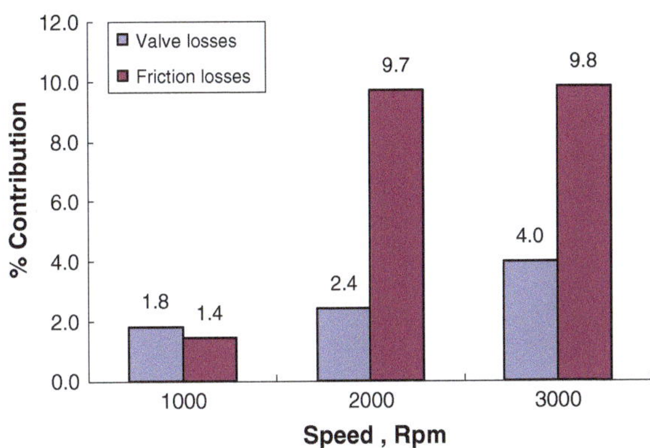

Fig. 15 Factors contributing to power consumption (%)

same trend, which contributes for the noise produced from compressor. The higher port size corresponds to low pressure loss, which has effect on less power consumption.

In Fig. 14, the effect of discharge reed thickness on torque is discussed. The optimum thickness of reed valve reduces the work done in lifting the reed, which contributes to less pressure loss and the pressure fluctuation in the compressor. In Fig. 15, the factors responsible for power losses are shown at various shaft speeds.

6 Validation

The experimental validation is done at standard operating condition (Pd = 15.7 Bar, Ps = 2.96 Bar, SC = 5 K, SH = 10 K) in component calorimeter. The torque is captured using torque transducer.

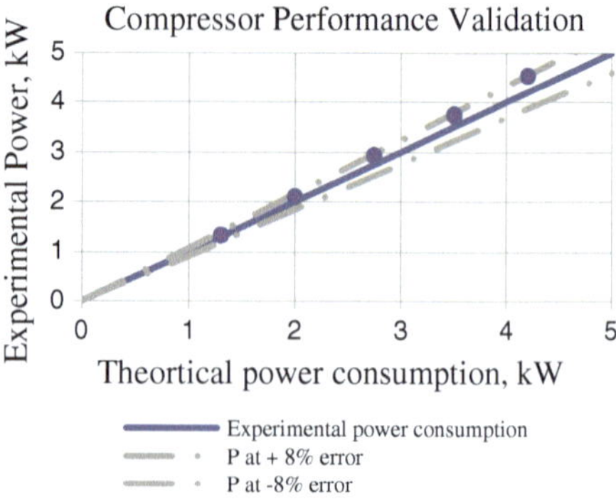

Fig. 16 Compressor performance validation

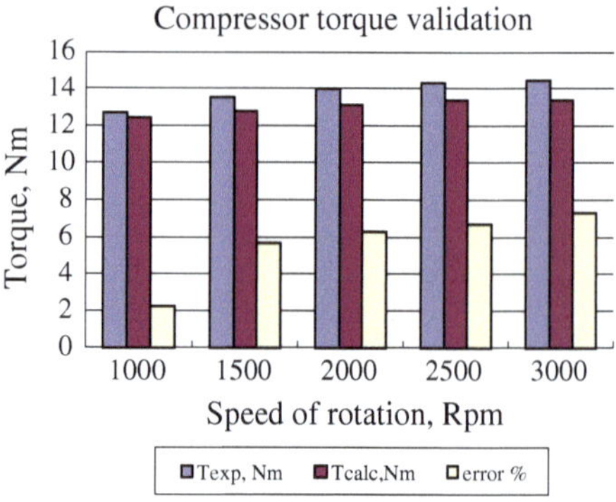

Fig. 17 Experimental versus calculated torque

Figure 16 shows the calculated versus experimental power consumption, in which the calculated power is always higher due to viscous effect, oil churning inside compressor, OCR and ORR are not considered in the calculation. Figure 17 shows the calculated torque versus experimental torque. The % error in calculated torque even for higher speed of 3,000 Rpm is within 7 % for 1,000–3,000 Rpm, which is considered to be acceptable value at design conditions.

7 Conclusion and Scope of Work

1. The dynamic motion of fixed displacement swash plate compressor with shaft rotational angle is mathematically simulated, axial and radial forces, shaft torque at various speeds are predicted.
2. At higher speed, frictional force and flow losses increase and hence mechanical efficiency is reduced and the power consumption is more than the proportionate.
3. Design and selection of magnetic clutch requires the peak torque (periodic repetition), which is predicted in this work.

4. Port optimization linked with resultant shaft torque is further explored in this work, by which the torque fluctuation is reduced by 5 %.
5. The design and selection of reed thickness is assessed in terms of pressure drop and hence the power loss.
6. The work can be further extended for design and selection of radial and thrust bearings and shaft.

References

1. Below, J.F., Miloslavich, D.A.: Dynamics of the Swash Plate Mechanism. International Compressor Engineering Conference (1984)
2. Tani, S., Kanemoto, K., Kuwahara, K.: Swash Plate Type Car Cooler Compressors. Hitachi Mechanical Engineering Research Laboratory, Hitachi review, vol. 53(5) (1971)
3. Liu, Y., Hui, G.: Motion simulation and dynamic analysis of a four-cylinder wobble-plate compressor. In: Proceedings of 2012 International Conference on Mechanical Engineering and Material Science (MEMS 2012), Paper 437 (2012)
4. Hou, J., Wu, J., Sun, C., Huang, L., Tang, T.: Fracture Analysis on Piston of Swash plate Compressor for Automotive air conditioner (Chinese). Chongqing Jianshe Automotive Air Conditioning Co. Ltd., Chongqing
5. Xu, Z.: Stress Analysis of the Swash Plate and the Spindle of the Swash Plate Compressor (Chinese), Shougang Air Conditioner Factory, vol. 22, No. 2, Fluid Machinery (1994)
6. Xu, Z.: Swash Plate Refrigeration Compressor Piston Dynamic Load Hydrodynamic Lubrication (Chinese), Fluid Engineering, vol. 8 (1990)
7. Shu, A., Zhu, P.: Matlab in Balancing Performance Analysis of Swash Plate Compressor (Chinese), Wuhan Institute of Mechanical Engineering, Hubei (430073)
8. Refrigerant compressor (Japanese), Japan Society of Refrigerating and Air Conditioning Engineers. (2013) (ISBN 978-4-88967-119-3)

Compliance Monitored Clubfoot Brace "PADMAPADA"©

Pradyumna K. Kammardi, Kaushik D. Sondur, N.S. Dinesh, Chalapathi Rao Nori, Vrisha Madhuri and Sanjay Chilbule

Abstract Padmapada is a compliance monitoring brace for clubfoot deformity used to provide corrective forces to the manipulated foot. The brace provides freedom to move the lower limbs of the child keeping the foot in the desired plantarflex position. The sensors placed in the brace insole detect the presence of the foot and acquires data periodically. The downloaded data is also stored for future use on the computer which can be accessed by the doctor for comparative analysis and to monitor progress during the course of the treatment.

Keywords Clubfoot · Brace · Compliance monitoring

1 Introduction

Clubfoot is a common disorder affecting the musculoskeletal system. It is also a congenital disorder, i.e., it is present at child's birth. It affects 2.5 in 1,000 live births. The disorder can be unilateral or bilateral. The affected foot is turned inwards and plantarwards. This disorder is known to exist for several centuries and the causes of the deformity are known. Figure 1 shows a child with the deformity in right leg.

The success of the treatment depends on the treatment method and the skill of the orthopedist. Treatment of the disorder in early stages is almost always conservative. Manipulation is done using plaster cast in cases where the deformity is treated early. In such cases, the doctors monitor the response of the deformed foot at regular intervals and change the cast as and when it is deemed necessary. The manipulation is repeated till the correction is achieved. Figure 2 illustrates the

P.K. Kammardi · K.D. Sondur · N.S. Dinesh (✉) · C.R. Nori
DESE, IISc, Bangalore, India
e-mail: dinesh@dese.iisc.ernet.in

V. Madhuri · S. Chilbule
Unit of Paediatric Orthopaedics, Christian Medical College, Vellore, India

A. Chakrabarti (ed.), *ICoRD'15 – Research into Design Across Boundaries Volume 2*, Smart Innovation, Systems and Technologies 35, DOI 10.1007/978-81-322-2229-3_47

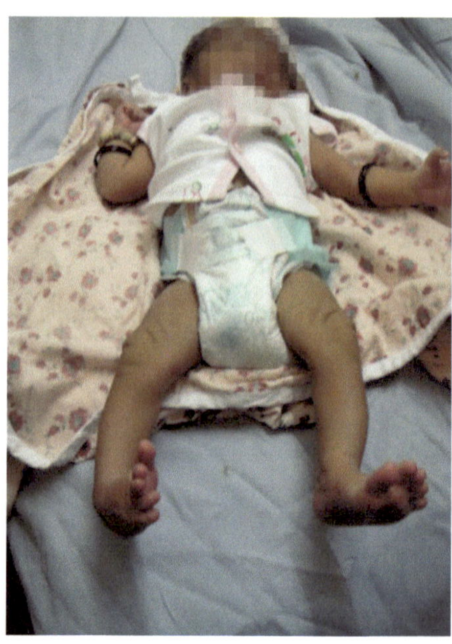

Fig. 1 Child with clubfoot in right leg

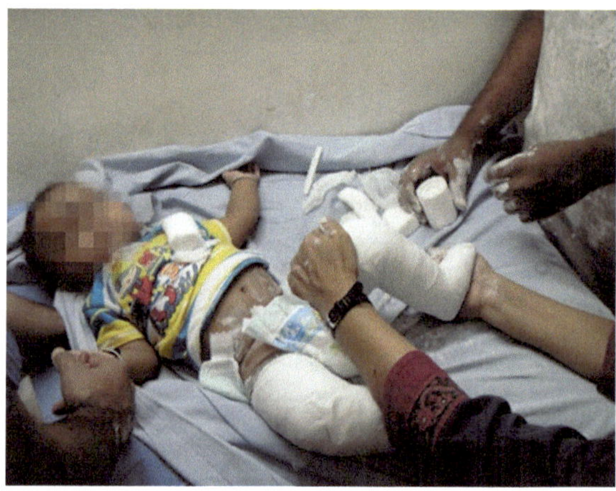

Fig. 2 Initial manipulation process involving plaster cast

typical manipulation process using plaster cast. The other way of treating deformity is through surgical procedure depending upon the nature of deformity.

Irrespective of the choice of the initial treatment procedure, clubfoot is one of the few common yet complex conditions which require regular medical monitoring for a successful treatment and to avoid reoccurrence during early childhood. The use of braces for 3–5 years is critical in the successful treatment of the deformed foot irrespective of manipulation process i.e. by using plaster cast or by surgery. The effectiveness of the treatment depends crucially on the duration for which the brace is worn by the child. On average, affected children are expected to wear the brace full time for a minimum period of 3 months. Beyond this, the child may have to wear the brace for up to 5 years. During this period, the wearing period is for 14 h. The details of the treatment procedures and some of the early versions of the brace used in treatment are discussed in [1, 2].

1.1 Problems

It is difficult to assess the progress of the treatment for this disorder and to predict the response for a treatment method. The deformity can also recur during treatment. This can be due to improper fitting of the shoes. However, it is noticed that one of the main causes for recurrence is the removal of the brace and non-compliance in using the brace for the recommended duration of time each day. The correction forces applied on the affected foot can create discomfort for the child and the brace is removed by the members of the family to pacify the child when it cries. As a result of this, the doctors find it difficult to assess the reasons for the recurrence since either of the two mentioned causes can affect the treatment.

Large number of children is affected by clubfoot worldwide; many types of braces are available in the market [2]. All Manipulative techniques including Ponseti's Technique require some form of bracing after correction. Dennis Browne Brace [2, 3] is one such brace used for this purpose. It has a cross bar between a pair of footplates or shoes. The external rotation of the affected foot is fixed to 10° in the initial version and 50°–70° in the modified brace for the affected and 30°–40° in the unaffected foot. The single bar connecting the shoes provides necessary dorsiflexed abduction but does not allow the child to crawl on its knees because moving lower limbs independent of the other is not possible.

Dobb's Dynamic Clubfoot Brace [2, 5], Markell brace [2, 4] unlike the Dennis Browne Brace utilizes a dynamic bar that allows the child to move both legs independently but due to the design dorsiflexion is difficult to achieve.

2 Objective

The objectives set for this clubfoot brace (Padmapada) are:

1. A wearable shoe which is well constructed and fixed with the desired external rotation.
2. Shoe of different sizes and rods of different lengths matching to the child's growth must be available.
3. A significant freedom of movement allowed for the lower limbs so that child will be able to crawl and roll over sideways without affecting the feet under abduction force of the brace.

4. The child should be able to stand with some support with the braces on.
5. The material used for the brace must withstand the kicking movements.
6. Implement a low cost and low power consuming electronic module with sensors for monitoring the brace for compliance and wearing.

3 Brace Construction

Material used here to build the brace is Aluminum alloy 5052. Aluminum alloy 5052 in H32 temper has very good corrosion resistance to marine and industrial atmosphere. Using this material the overall weight of the brace is reduced yet ensuring sufficient strength and rigidity. It does not suffer elongation or deformation under high performance conditions.

For a proper correction the planes of the feet must be kept parallel to each other under all feet positions. This is achieved in Padmapada with a four bar linkage mechanism. This is a planar linkage with the property that its entire links move in parallel planes [6]. Figure 3 shows the construction of the brace with four bar mechanism.

Shoes are mounted on the extreme ends of the linkage on the Foot plates. This design provides the child to move the leg independently keeping both the feet parallel to each other. This helps the child to crawl on its knees while the abduction force is still being applied on the corrected foot. To make sure that the abduction force is applied, the foot is placed at the correct position inside the shoe and a lace is tied firmly such that the foot touches the insole with a maximum area of contact. To verify whether the foot is properly positioned, a hole is provided near the heel of the shoe through which observation can be made, shown in Fig. 7a.

Shoes mounted on the extreme ends are at an angle of 5° upwards to correct the equinus by dorsiflexing the foot. Even after the surgery or casting, the foot tends to grow inwards due to the differential growth of the tendons. To restrict this, the foot is held up (dorsiflex position). This is done by elevating the shoes itself while

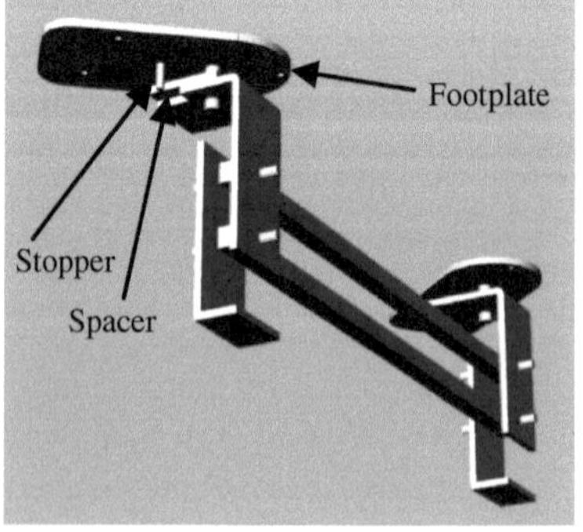

Fig. 3 Four bar linkage mechanism in the Padmapada

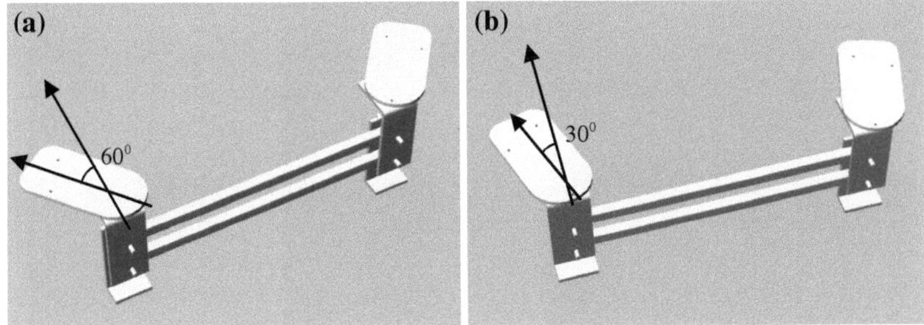

Fig. 4 Both right and left footplates resticted inwards to 60° in (**a**) and 30° in (**b**)

mounting it on the brace. By adding a small spacer at the front, below the shoe keeps it at the desired elevated dorsiflexed position. This holds good even when the child is rotating its feet or while crawling or just wearing the brace. The elevated dorsiflexed position is not implemented in the braces that are currently commercially available. Dobbs brace though it helps in crawling, at certain positions the feet are pantarflexed which is detrimental to the corrective process. In Padmapada the rotation of the affected foot inwards can be restricted to 60° (Fig. 4a) and that for the unaffected foot to 30° (Fig. 4b). A stopper is provided at the base of the foot plate which restricts the inward movement. The angle of restriction for the abnormal foot can be increased with an additional spacer shown in Fig. 3. The length of the bars connecting the shoes is equal to child's shoulder length which can be adjusted as per requirement. Different size shoe can be mounted on the footplate according to the child's feet size over the period of the child's growth. The brace is powder coated to have an attractive and durable surface finish. It protects the material from corrosion.

4 Compliance Monitoring

Figure 5 shows the functional block diagram of the electronic module which monitors the compliance.

Compliance monitoring involves detection of presence of foot and detection of proper contact of the foot with the shoe insole. For this purpose capacitive based sensors are used. Three main areas are recognized for the effective sensing: the heel area; the big toe (first Meta tarsal); the small toe (fifth Meta tarsal) areas (shown in Fig. 6). The reason for choosing these areas is that in a normal foot they touch the shoe insole completely. The capacitive sensors are designed to sense the foot in the presence of insole which comes over the sensor and the socks the child may wear. The sensor output varies with environmental conditions and the condition of the socks and the insole. The system after calibration of the sensors can detect the presence of foot under variable conditions.

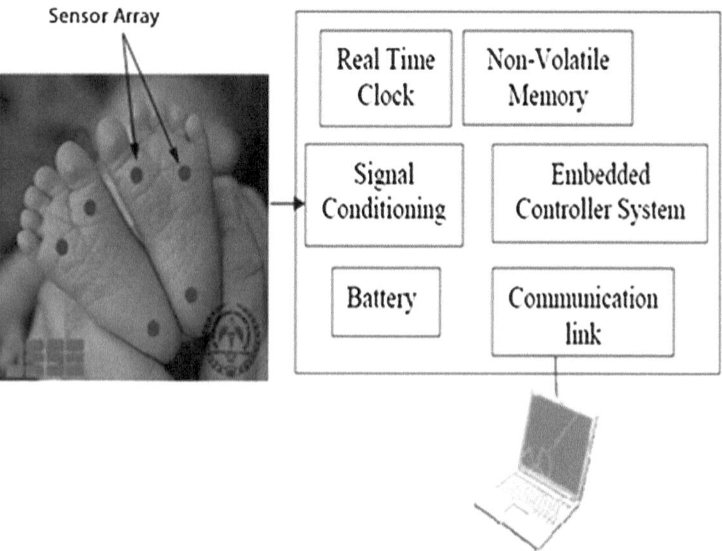

Fig. 5 Block diagram of the electronic module

Fig. 6 Three sensing regions in the insole

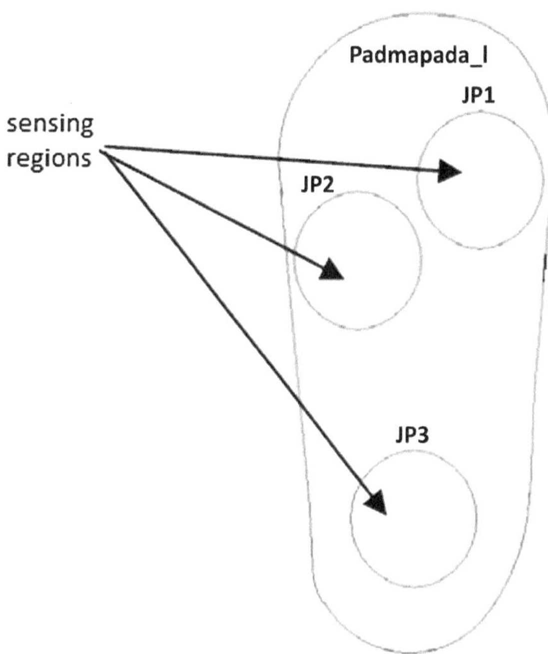

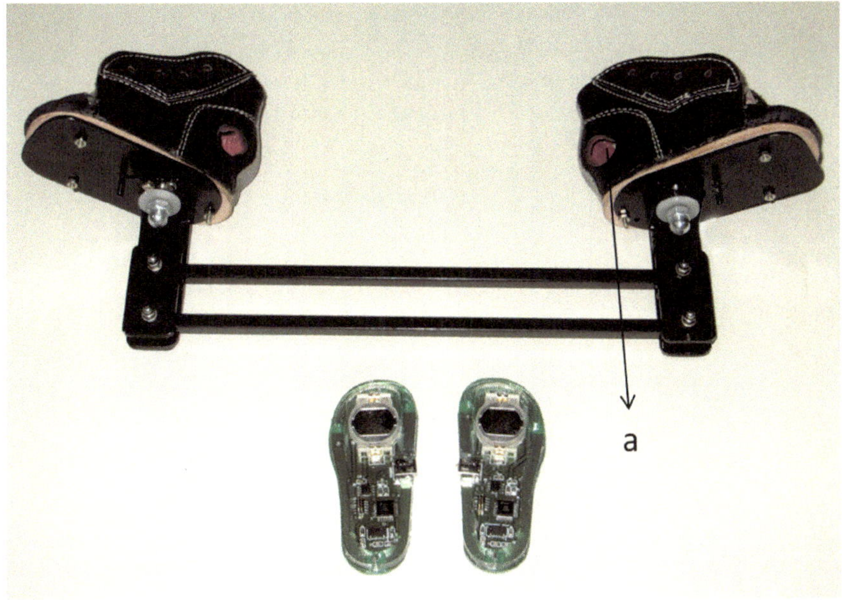

Fig. 7 Complete brace and electronic module (*a*) hole provided to check whether the shoe is worn properly

Separate foot-shaped sensor boards with data acquisition electronics are placed inside each shoe (shown in Fig. 7). The sensors are arranged to stay below the insoles for sensing the foot position. The electronic module is designed to consume low power. The system is kept in sleep mode to conserve battery energy when data is not acquired. It is woken up by a timer—Real Time Clock, for every hour to acquire sensor data. These data are stored in a non-volatile memory. It can store up to 12 months of data. The electronic modules are powered with a 3 V–240 mAh (CR2032) Lithium Ion coin cells which can power the system for one year.

This arrangement makes the electronics modular and avoids requirement of any interconnections between the shoes. This feature also provides for use of only one electronic module in case the child has only one deformed foot and that has to be monitored. This will reduce the cost of the brace.

The information logged by the electronic module can be downloaded to PC for analysis. A graphical user interface (GUI) has been developed to indicate wearing pattern on hourly basis for all the days as shown in the Fig. 8 with vertical axis showing the hours for which the brace was worn and the horizontal axis date. There is also a provision to see the response of the individual sensors separately which gives the information about whether that part of the foot is making a proper contact with the insole. Figure 9 shows a graph depicting the sensors being on or off at a given hour during the period of data acquisition. This information is useful for tracking the correct growth of the manipulated feet and take remedial measures if required. The variations in data can be seen either due to the child not wearing the

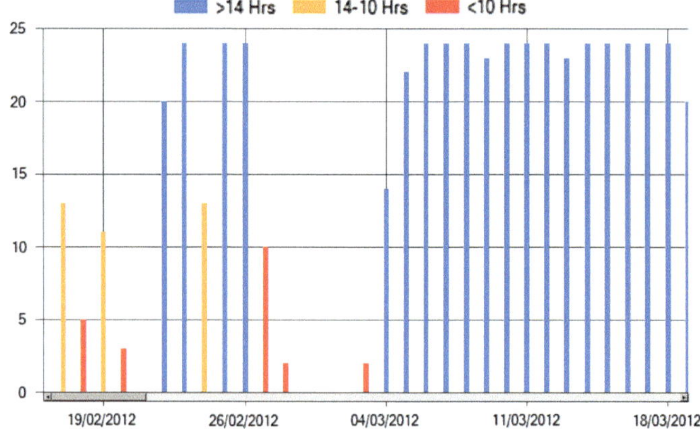

Fig. 8 GUI—graph depicting total hours worn along *vertical* axis and days on *horizontal* axis

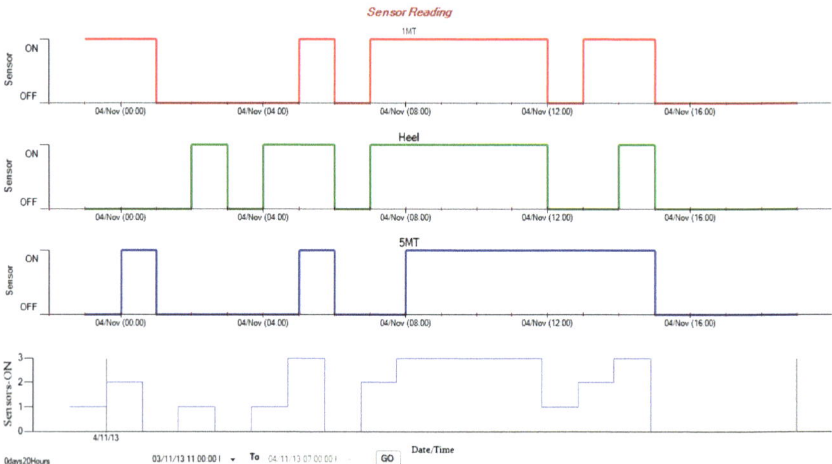

Fig. 9 GUI—graph showing individual sensor data

brace where none of the sensors would be on or one or more sensors are off continuously indicating an area of the foot not being sensed.

The electronic module is designed and packaged to withstand vibrations due to child's kicking actions. Also the packing prevents the electronic module from getting affected when the shoe gets wet. The growth of the child is quite fast and the foot size varies significantly. This necessitates resizing of the electronic module to accommodate varying size of the foot. The standard shoe sizes range from size 5 to 15. In order to reduce inventory the electronic module is designed to match to two

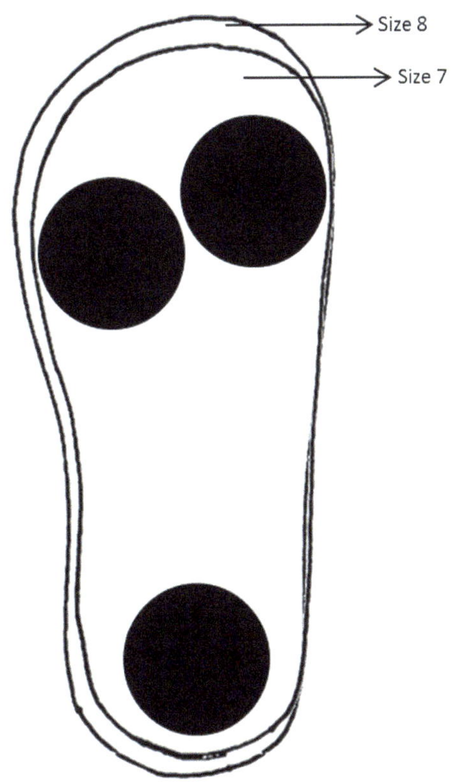

Fig. 10 Placement of sensors according to standard size 7 and 8

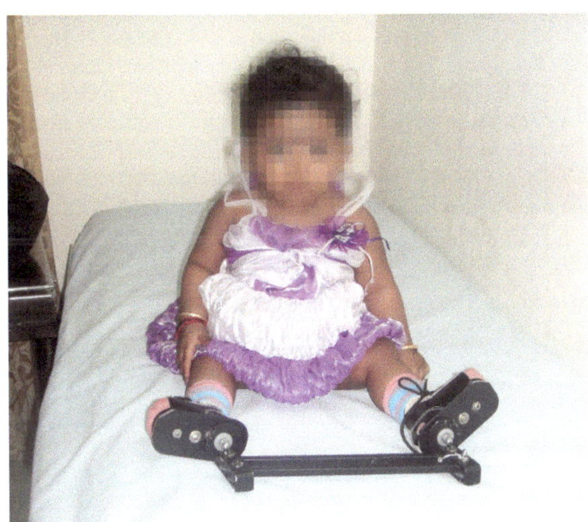

Fig. 11 Child wearing Padmapada

consecutive shoe sizes. This is achieved through scaling of the sensor boards for the required sizes. Figure 10 shows the placement of sensors matching shoe sizes 7 and 8. The report on standard foot sizes is discussed in [7].

5 Conclusions

Padmapada has gone through successful initial trials and is now going through clinical trials at Christian Medical College Hospital (CMC), Vellore. This is a collaborative research between IISc and CMC Vellore. Figure 11 shows a picture of the patient wearing padmapada at field trials. We now have an abduction brace which is light weight and allows 4° of freedom for crawling but avoids unfavorable inward position for the foot. The capacitance sensor embedded in the insole assesses the effectiveness of the brace.

Acknowledgments We would like to express our gratitude to Department of Science and Technology, Government of India for providing us with financial assistance for this project. We sincerely thank the staff members from Department of Systems Engineering, IISc who helped making the prototypes of the brace.

References

1. Ponseti, I.V.: Current concepts review—treatment for congenital clubfoot. J. Bone Joint Surg. **74-A**(3) (1992)
2. Desai, L., Oprescu, F., DiMeo, A., Morcuende, J.A.: Bracing in the treatment of children with clubfoot: past, present, and future. Iowa Orthop. J. **30**, 15–23 (2010)
3. Yamamoto, H., Furuya, K.: Treatment of congenital clubfoot with a modified Denis Browne splint. J. Bone Joint Surg. **72-B**(3), 460–463 (1990)
4. Garg, S., Porter, K.: Improved bracing compliance in children with clubfeet using a dynamic orthosis. J. Child. Orthop. **3**(4), 271–276 (2009)
5. Dobbs, M.B., Gurnett, C.A.: Update on clubfoot: etiology and treatment. Clin. Orthopaed. Relat. Res. **467**(5), 1146–1153 (2009)
6. McCarthy, J.M., Soh, G.S.: Geometric design of linkages, 2nd edn. Springer, Berlin (2011)
7. Steenbeek, H.M., David, O.C.: Steenbeek brace for clubfoot, 2nd edn. Global Help Health education using Low-cost Publications (2009)

Lumped Parameter Modeling of Vibratory Soil Compactor

Gomathinayagam Arumugam, G.S. Narayana, S. Babu and S. Mohamed Ebrahim

Abstract The mathematical modeling of soil–drum interaction in a vibratory soil compactor is a complex process. In this paper an effort is made to capture the behavior of soil–drum interaction through lumped parameter modeling. The amplitude of the structures are evaluated for static and dynamic conditions using the mathematical models and compared with the experimental results. The different modeling approaches are considered and the results reveal that the properties of the soil as input play a vital role in the accuracy of the modeling.

Keywords Vibration · Compactor · Modeling

1 Introduction

Vibratory compactors are primarily used for compaction tasks in earthwork and road building. Soil compactors are commonly used to transform loosely placed granular and mildly cohesive soils into densely packed load bearing earth structures. They use combined static and dynamic forces (weight and vibrations).

Figure 1 shows a typical vibratory soil compactor. It uses hydrostatic power transmission system. The rotation of the drum is generated with built-in hydro motor, but vibrations are produced with mechanical components. Those are shaft and weights eccentrically mounted on it. Hydro motor achieves various frequencies and amplitudes by regulating speed with variable pump and by moving weights. Thereby, the mass of the vibratory drum and weights cause centrifugal force required for soil compaction.

Previous works [1–3] showed that the mathematical modeling include the soil, drum and the frame. They considered the components above the drum as a single mass, called frame. They did not consider the canopy or operator platform as a

G. Arumugam (✉) · G.S. Narayana · S. Babu · S.M. Ebrahim
Product Development Center, Larsen and Toubro Limited, Coimbatore, India
e-mail: agnayak@yahoo.com

A. Chakrabarti (ed.), *ICoRD'15 – Research into Design Across Boundaries Volume 2*, Smart Innovation, Systems and Technologies 35, DOI 10.1007/978-81-322-2229-3_48

Fig. 1 Vibratory soil compactor

separate entity. They used 2, 3 and 4 DOF systems for modeling soil–drum interaction. They also used contact/loss of contact and linear/non-linear models for the same.

This paper explores the capability of 3 and 4 DOF lumped parameter models with linear elements to reproduce the vibratory compactor behaviour observed experimentally. The various model approaches are investigated and compared with the experimental data. The paper also addresses the role of soil properties on the model results.

2 Mathematical Modeling

The general framework for the roller soil model is shown in Fig. 2 where both the roller and soil are modeled with mass-spring components. While x and y motion is kinematically possible, only z motion (vertical) is considered here because it directly contributes to the determination of soil stiffness.

The drum mass includes the eccentric assembly while the frame mass is determined by subtracting the drum weight from the total static weight measured under the drum. One set of rubber mounts provide isolation between the drum and the frame while another set provides isolation between the frame and the canopy.

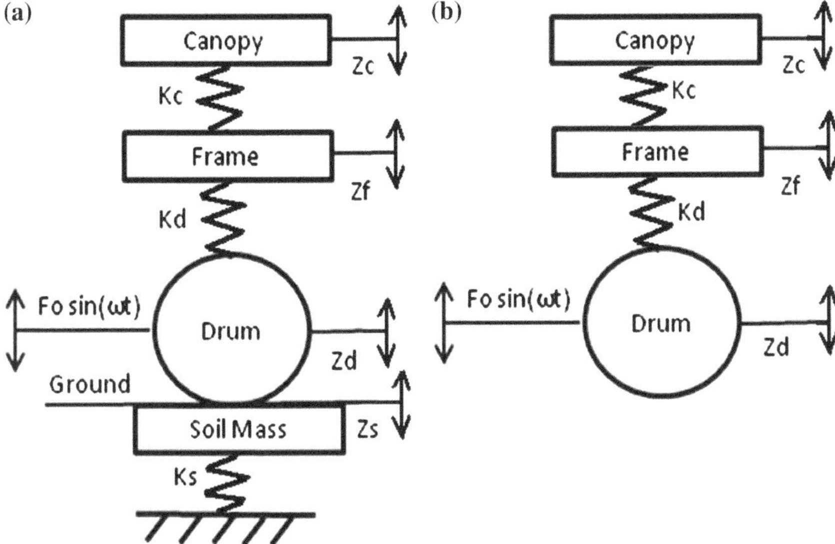

Fig. 2 Mathematical models of soil compactor. **a** Contact model. **b** Loss of contact model

The soil model employed here (mass-spring) is a simple analog that has been proven effective in modeling surface vibration of foundations resting on homogeneous and layered elastic half spaces [4–7]. During coupled drum/soil vibration, a portion of the involved soil is in synchronous motion with the drum. The amount of added or apparent soil mass has been debated throughout the foundation vibration literature with no clear consensus. The inclusion of an apparent soil mass equal to 20 % of the drum mass provided a good amplitude match between the models and the experimental data [3].

During compaction, the vertical excitation force can exceed the static weight of the roller at the drum location and cause drum/soil decoupling during a portion of each cycle. Only contact and partial loss of contact operation is considered during modeling.

2.1 Contact Model

During compaction, the drum and soil motion are coupled together and the compactor is modeled as shown in Fig. 2a. The coupled drum/soil motion is described as Z_d while the frame and canopy displacements as Z_f and Z_c respectively. Soil stiffness (k_s) and the stiffness of mounts (k_d) are considered. The equations of motion during contact are given below:

$$(m_d + m_s)\ddot{z}_d + k_s z_d + k_d(z_d - z_f) = F_0 \sin \omega t \tag{1}$$

$$m_f \ddot{z}_f - k_d(z_d - z_f) + k_c(z_f - z_c) = 0 \tag{2}$$

$$m_c \ddot{z}_c - k_c(z_f - z_c) = 0 \tag{3}$$

2.2 Loss of Contact Model

In this model, the drum is decoupled from the soil as in Fig. 2b. The drum displacement Z_d varies with respect to eccentric force (F_0). The frame and canopy displacements are described as Z_f and Z_c respectively. Equations (4)–(6) describe the motion during loss of contact.

$$m_d \ddot{z}_d + k_d(z_d - z_f) = F_0 \sin \omega t \tag{4}$$

$$m_f \ddot{z}_f - k_d(z_d - z_f) + k_c(z_f - z_c) = 0 \tag{5}$$

$$m_c \ddot{z}_c - k_c(z_f - z_c) = 0 \tag{6}$$

2.3 Input Parameters for Model

A vibratory soil compactor of 20 ton is considered for the study. The input parameters for modeling approach are given in Table 1.

Table 1 Input parameters of 20T compactor

Parameter	Value	Unit
Mass of the drum, m_d	5,877	kg
Mass of the frame, m_f	7,200	kg
Mass of the canopy, m_c	512	kg
Stiffness of the soil, k_s [3]	72	MN/m
Exciting frequencies, f	28 and 33	Hz
Corresponding eccentric forces, F_o	340 and 260	kN
Circular frequency, ω	2πf	Rad/s
Stiffness of the drum mounts, k_d	5.83	MN/m
Stiffness of the canopy mounts, k_c	1.04	MN/m

2.4 Solution to the Equations

Substituting the following terms in the Eqs. (1)–(6) will make them simple and they can be solved using Crammer's rule.

$$z = Z \sin \omega t$$
$$\ddot{z} = -\omega^2 Z \sin \omega t$$

After simplifications, the Eqs. (1)–(3) as well as (4)–(6) are solved using MathCAD software for the given input parameters.

3 Experimental Work

The experimental work is carried out and vibration levels (acceleration) are measured at the drum, frame and canopy using tri-axial accelerometers and eDAQ data acquisition system as shown in Fig. 3. The experimental data are post processed

Fig. 3 Experimental set up

using Glyphworks software. In case of canopy, the accelerations are measured at four places in and the average value is considered for comparison.

4 Results and Discussions

Model results are presented together with the experimental data to illustrate the efficacy of the lumped parameter modeling.

4.1 Static Jacked Condition

Before comparing and analyzing the mathematical and experimental models, the design value of drum amplitude for the given frequency and centrifugal force is calculated for the static jacked condition. Then the compactor is jacked at four supports and drum is put on vibration mode. The frequency and amplitude are measured using accelerometer. It is found that a very good correlation exists between two methods as in Table 2.

4.2 Dynamic Working Condition

The displacement (Z), velocity (V) and acceleration (A) for the Drum, frame and canopy for low frequency-high amplitude during contact in dynamic condition using analytical method for the input parameters specified in Table 2 are given in Fig. 4. The graph in the figure is for one cycle and it clearly depicts the phase differences between the displacement, velocity and acceleration.

The displacement at the drum, frame and canopy using analytical (contact and loss of contact) and experimental methods are shown in Fig. 5. The experimental values for drum are 73 and 80 % of analytical values for the contact and loss of contact models which is a good correlation.

Table 2 Comparison of analytical and experimental results for static jacked condition

Static jacked condition	Frequency in Hz		Amplitude in mm	
	Specification	Measured	Specification	Measured
Low frequency (LF)-High amplitude (HA)	28	28.6	2	2.04
High frequency (HF)-Low amplitude (LA)	33	33	1.1	1.11

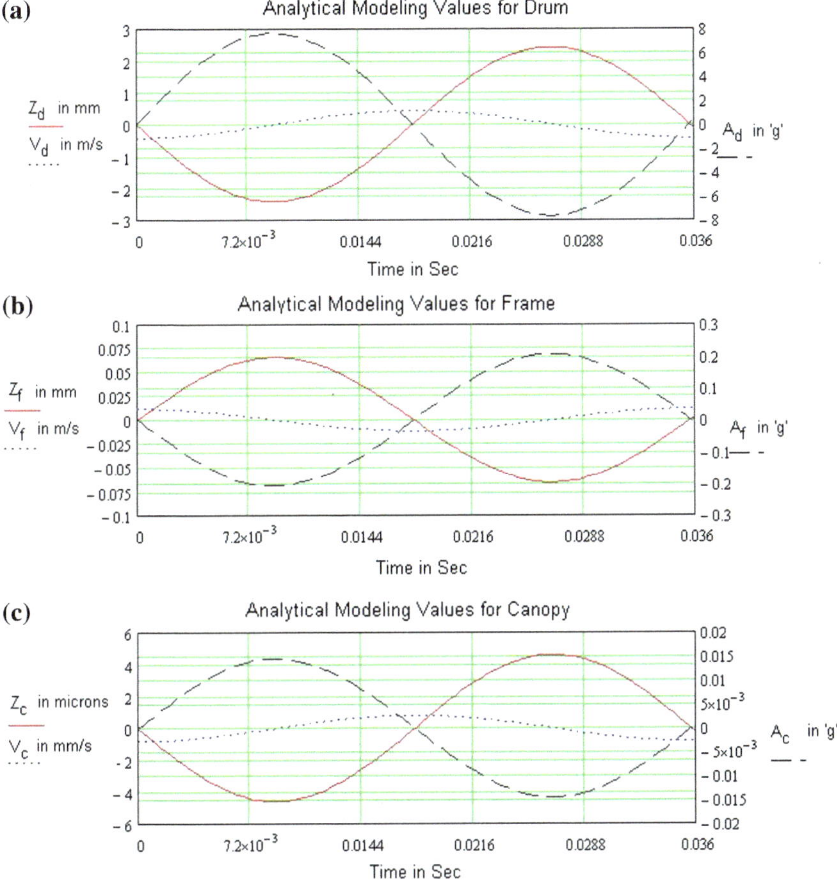

Fig. 4 Response of the structures (analytical-contact) for low frequency-high amplitude

In case of the frame and canopy, the deviations in results are high. The higher deviations are due to the following reasons:

- The amplitude of drum depends more on the soil properties for the dynamic condition (contact model).
- The frame amplitude depend on the soil properties, isolation mounts and its connection with the other structures.
- The amplitude of canopy depends on the soil properties, isolation mounts, connection with other structures, vibrations of other contributors like engine and the stiffness of tires.
- Here the tire stiffness, engine harmonics and connection with other structures have not been modeled.

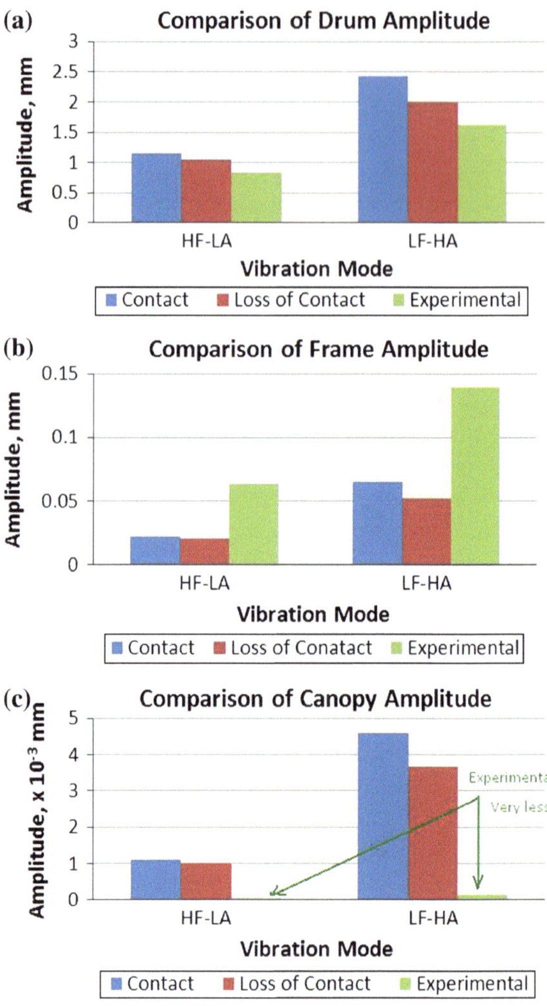

Fig. 5 Comparison of analytical and experimental results

4.3 Influence of Soil Stiffness

The soil property plays a major role in the amplitude of the structures. In this paper, only the stiffness of soil is considered and damping is not considered as it increases the complexity of the system. Figure 6 shows the influence of soil stiffness on the amplitude of the structures using analytical method for contact model.

The amplitude of structures increases with increase in soil stiffness. The soil stiffness at the test site where experimental work is carried out is not available. Hence if the exact soil stiffness is known, the amplitude of the structures can be compared against the theoretical values.

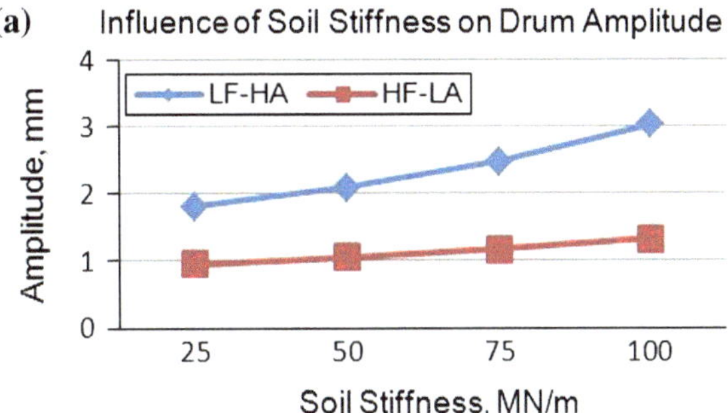

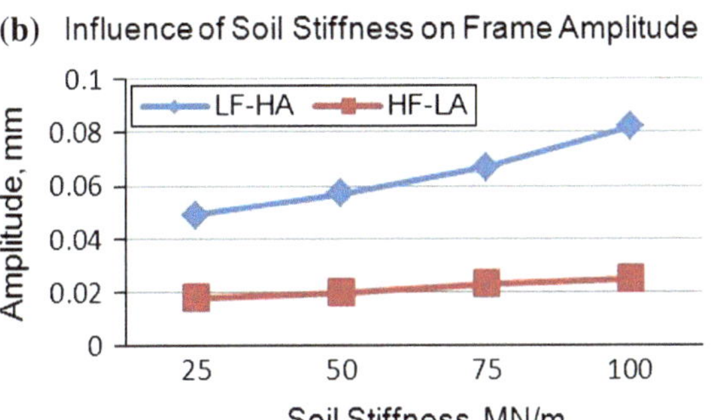

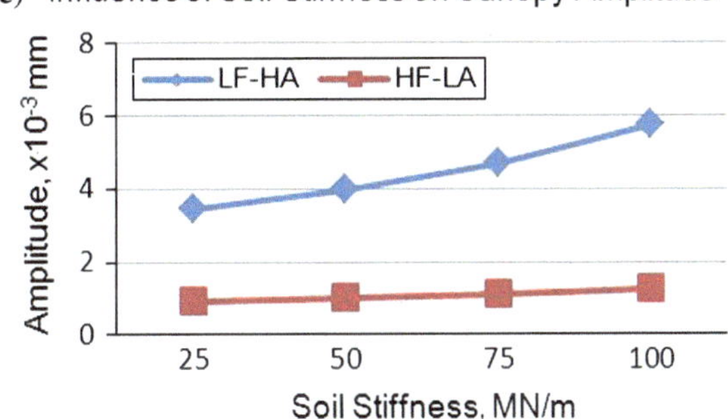

Fig. 6 Influence of soil stiffness on amplitude of structures

5 Conclusions

The capability of lumped parameter models with linear elements to reproduce the vibratory compactor behaviour is studied. The amplitude of the structures is evaluated using mathematical models and compared with the experimental results. A good correlation exists between the mathematical and experimental results for the static jacked condition. The deviation between the mathematical and experimental values in the dynamic condition is due to assumptions in modeling and soil properties. The results show that the properties of soil play a vital role in the results of mathematical model for dynamic condition.

Acknowledgments Authors wish to thank the management of Larsen and Toubro Limited for the use of their Soil Compactor 20T—Proto Machine data and granting permission to publish this article. They also thank their colleagues for their assistance in carrying out the experimental work.

References

1. Pietzsch, Dieter, Poppy, Wolfgang: Simulation of soil compaction with vibratory rollers. J. Terrramech. **29**(6), 585–597 (1992)
2. Siminiati, D., Hren, D.: Simulation on vibratory roller—soil interaction. J. Adv. Eng. **2**, 112–120 (2008). ISSN 1846-5900
3. van Susante, P.J., Mooney, M.A.: Capturing nonlinear vibratory roller compactor behaviour through lumped parameter modeling. J. Eng. Mech. **134**, 684–693 (2008) (ASCE, AUG)
4. Gazetas, G.: Analysis of machine foundations: state of the art. Int. J. Soil Dyn. Earthquake Eng. **2**(1), 2–42 (1983)
5. Lysmer, J., Richart, F.E.: Dynamic response of footings to vertical loadings. J. Soil Mech. Found. **92**(1), 65–91 (1966)
6. Baidya, D.K., Muralikrishna, G., Pradhan, P.K.: Investigation of foundation vibrations resting on a layered soil system. J. Geotechnol. Geoenviron. Eng. **132**(1), 116–123 (2006)
7. Wolf, J.P.: Foundation Vibration Analysis using Simple Physical Models. Prentice-Hall, Englewood Cliffs (1994)

Automated Puppetry—Robo-Puppet©

M.A. Aravind, N.S. Dinesh, Nori Chalapathi Rao and P. Ram Charan

Abstract Many types of puppet theatres and automation systems exist. The paper presents a fully automated marionette theatre. The system presented consists of robot manipulators each with sixteen degree freedom which can be adapted to different degree of freedom. The manipulators are linked through a WIFI network to a master controller (Linux PC). Software developed provides for the user interface at the front end and a back-end module which synchronizes motion data, audio and video streams in real-time. The theater provides for a puppet motion capture in three modes. The motion programming modes are—direct human body motion capture, motion programming through joystick and use of motion library modules. The user interface designed provides for an easy and efficient way for the puppeteer to record the motion in sync with the audio and video. The theater can support simultaneous operation of six marionettes with automated stage lighting and curtain (Patented as Robo Puppet).

Keywords Robo puppet · Manipulator · Marionette · Puppet theatre

1 Introduction

Puppetry is one of the most ancient forms of entertainment in the world. Over the centuries, puppetry has developed into a powerful media of communication. Traditionally, India has a rich heritage of puppetry. Puppeteers from different regions conduct their shows in their colloquial languages, limiting the effectiveness of communication to the audience who know the language. The objective of the present work is to develop an automated Puppet Theater system which derives

M.A. Aravind (✉) · N.S. Dinesh · N.C. Rao · P.R. Charan
DESE, Indian Institute of Science, Bangalore, India
e-mail: aravindma1990@gmail.com

A. Chakrabarti (ed.), *ICoRD'15 – Research into Design Across Boundaries Volume 2*,
Smart Innovation, Systems and Technologies 35, DOI 10.1007/978-81-322-2229-3_49

benefits from modern technology. A conscious effort has been put to realize a system which adapts to conventional puppets without modifying the traditional construction of the puppets.

There are many forms of puppets and they differ in their construction and manipulation techniques. Most types of puppets in use in India today fall into five broad categories [1]: hand puppets, rod puppets, marionettes, rod-string puppets, and shadow puppets. We have chosen stringed puppet (Marionette) for our study. From literature we find that Marionettes have strings ranging from 2 to 120 depending upon the complexity of motion required. It is observed that most of the puppet manipulations can be achieved with 10 strings and the puppets with more strings are considered special puppets and may need highly skilled puppeteer to manipulate them. Therefore we have presently considered a marionette with up to 12 strings for the automation. A robot manipulator with a provision to manipulate up to 16 strings has been developed to manipulate the puppet. The puppet theater developed by us is able to accommodate 5 robots which can simultaneously manipulate 5 puppets, playing the part of 5 puppeteers. The system has provision for the automated stage lighting, curtain controls, video effects and possible special effects.

The motions of the puppets are precise sequence of choreographed motions as decided by the puppeteer and importantly all the motions have to be in synchronization with the audio/video used in the show. This calls for a real-time audio/video and motion synchronization capability for the electronic system used for the control of the entire show. Effective programming of the puppet theater is the key to the success of the technology envisaged.

Many ways of motion capturing for puppetry has been explored earlier [2, 3–4]. Most of them use elaborate methods of using markers and multiple angle capture cameras. The proposed technology provides for cost effective means of automation and helps create sophisticated shows with puppets built with traditional technology. The developed system may emerge as a new form of art for the artists to exploit.

2 Puppet Construction

The puppet theatre was constructed for marionettes. These marionettes were constructed out of papier-mâché so that they are of light weight. The typical weight of the constructed marionette was around 200 g. This was an important parameter to decide the specifications of the actuators and construction of the manipulator. The dimensions of the puppet were kept to the general standards of a marionette puppet. Figure 1 shows the puppet construction.

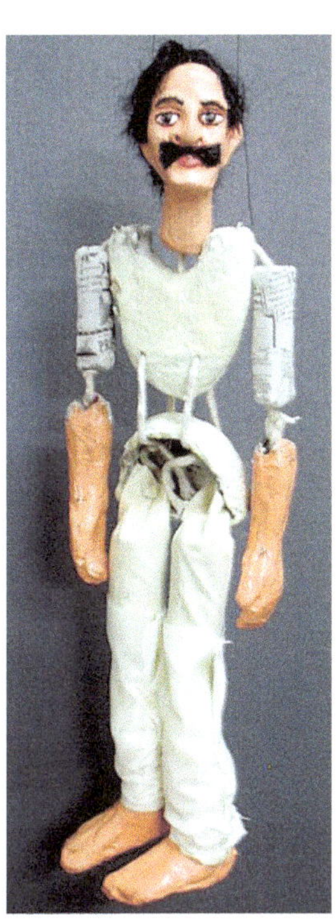

Fig. 1 Puppet stage with reference

3 Manipulator Design

The puppet manipulator was designed keeping in mind the weight and speed consideration of the marionette. A typical puppet theatre can house around 4–5 puppets at a time. The frame of reference of the whole puppet systems is as shown in Fig. 2.

Movements of different limbs of puppet:

The puppet is given 9° decoupled of motion. The entire puppet can move in the X direction from one end of the stage to the other. The wrist is given two degrees motion in X and Y as shown in Fig. 3. While only two degree freedom is given to the wrist depending upon the position of the Y axis, the loose passive joints of the arm will take different positions due to gravity and give the illusion of more than two axes motion. The motion of the wrist is the largest in the puppet. For the puppet designed by us it is 30 cm from end to end in Y axis. The other motions that were given are the following: The leg is given one degree motion along Y direction. The entire body is movable in Y direction and rotatable around puppets vertical axis. The actuation is done by DC servo motors. These servo motors were housed in a manipulator along with an assembly of gears and pulleys to bring about the various effects of the system. Servo motor specifications were chosen depending on the load characteristics of the limbs of the puppet. Other similar designs existed like [5, 6], but these systems have a static manipulator with pulleys and gears. In our system

Fig. 2 Puppet stage with reference

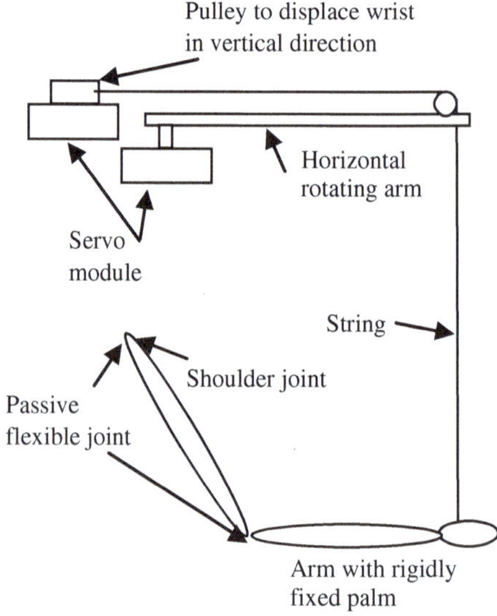

Fig. 3 Arrangement for hand motion with two axes decoupled motion

the whole manipulator could rotate about an axis as well as laterally on a rail mechanism. This gave two extra degrees of freedom compared to other designs.

The commercially available servos generate a maximum of 180° of total motion. Therefore we used gear train in order to achieve small size of the pulley and get amplified motion (1.5 turns). In our prototype system, the puppets weighed 200 g with each limb weighing roughly 10 g. We could actuate the arm and legs with a

Fig. 4 Robot manipulator

required speed of 50 cm/s with servos which provided 34 N-cm torque. The entire puppet body was actuated with a servo of 100 N-cm at 16 cm/s speed. The manipulator construction is shown in Fig. 4.

4 Controller for the Manipulator

The manipulator controller is a system which receives motion command stream from a master controller and controls the servo motors accordingly. In literature [7] we find some controllers are built with wireless trans-receivers working in master slave mode. But in the present work (Fig. 5) the master controller and many manipulator controllers are networked using WIFI link. This provides for real-time streaming of motion, audio and video on a single network. The master controller is a Linux PC which synchronizes all the three streams within 30 ms. The synchronizing is done at

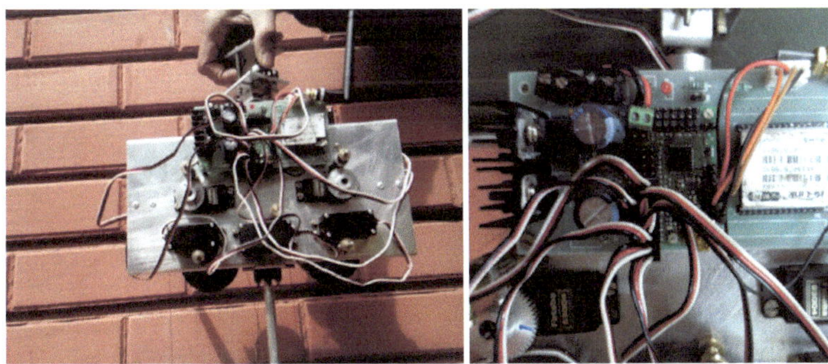

Fig. 5 Controller for manipulator

the video frame rate of 30 FPS. The manipulator controller is an ARM cortex based system which can control up to 16 servo motors. There is provision to monitor the health of each controller in real time and transmission error detection and correction.

There is a stage management module (SMM) which is also WIFI linked to the master controller. The SMM is used for the stage lighting, curtain control and special effects. The SMM works like a Programmable logic controller which works in synch with the puppet motion.

5 Software Design

The software has following two main modules: Front-end module and the back-end module.

The front end module forms the Graphic User Interface (GUI) which interfaces the user to the back end functions for the real time controlling of the puppets.

5.1 GUI Design

A GUI based program was created to interface to all the puppets. Five main modes defined the whole system: Test mode, Record mode, Playback mode, Edit mode, Real-Time mode.

5.1.1 Test Mode

This mode is designed to test all the functionalities of the puppet (Fig. 6). This is also used to do any calibrations that might be necessary for the initial setup of the puppet. Some pre-defined motion patterns could be played through selected axis for testing. Provision is made for testing and calibrating joysticks and Microsoft Kinect which form input devices for motion data.

5.1.2 Record Mode

Record mode is used to generate motion and stage management data in sync with recorded audio/video stream. Three types of inputs were used for capturing motion data.

Microsoft Kinect

Kinect is a 3D motion capture gaming device. Many open source tools allow the developer to access the 3D view of the Kinect. Human motion can be detected

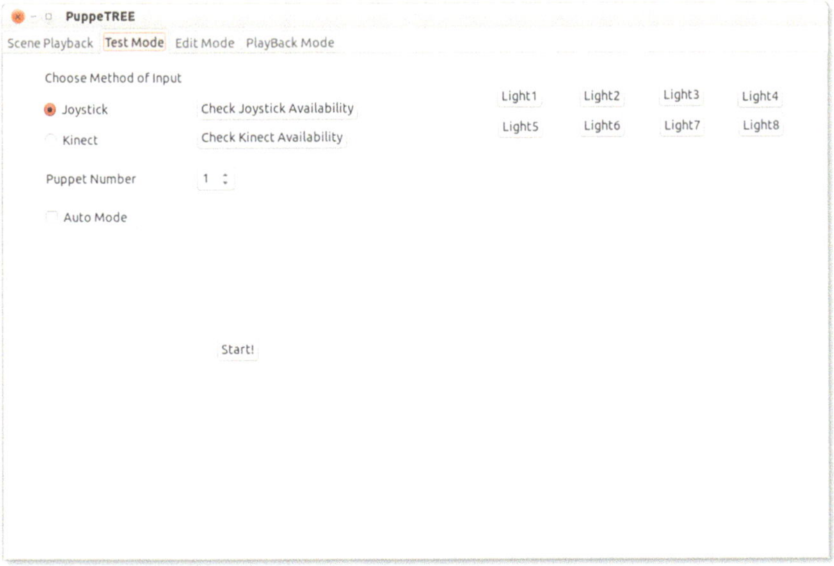

Fig. 6 Test mode

through some image processing and various points can be accessed on the skeletal frame of the human. Motion capturing techniques were used to extract the user skeleton from the images and apply the required positions onto the puppet. These points were scaled and mapped to the joints of the puppet respectively. Thus generated motion data is transmitted in real time to various puppets. During our studies, it was found that the motion data had noise. Averaging techniques were used on the values sent, to give a smooth transition between values as shown in Fig. 7. This reduced a lot of jitters that came from image processing of the user's gestures from Kinect. Also, the system achieved minimum of one frame delay from the human body motion to the puppet motion.

It was required to scale the human gestures to the size of a puppet. In the case of Kinect, the size of the body captured on the camera depends on how far the user is from it. Therefore, direct mapping of limb positions cannot be done in the Cartesian co-ordinates. The problem was overcome by mapping the angles between the limbs to the puppet's respective limbs. With this, the motion data is made insensitive to the position of the user within play area of the camera.

Joysticks

Two 2-axis joysticks can be interfaced to this system. Provision is made in GUI to map virtually, the joystick axes to any limbs of the puppet as the user requires.

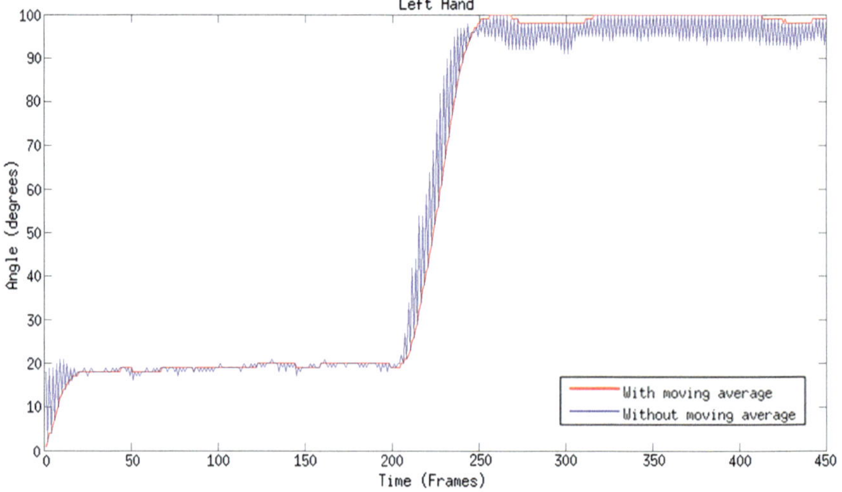

Fig. 7 Moving average for left hand (angles vs. time (frames))

Motion Library Modules

In order to simplify motion programming, pre-defined motion library files were created for general moves like walking, lifting and waving hands, sitting down or standing up etc. These can be directly used for the motion programming. The motion file is created compatible to Microsoft excel format which provides for easy file manipulation.

Playback Mode

This mode provides an interface for the user to playback an earlier recording (Fig. 8). The interface provides for selecting the start and end time as well as the number of puppets to be played simultaneously.

5.1.3 Edit Mode

This mode allows the user to edit the pre-recorded motion file (Fig. 9). Provisions have been made to copy motions of one puppet to another puppet. User can cut and copy motion from one part of the play to any other puppet in any other part of the play. User can also mirror this motion while copying with the enable of a button. This brings about complete sync between puppets. This provision helps puppeteer easily sync multiple puppet motions which otherwise needs highly skilled puppeteers to achieve the same in manual puppetry. This feature of cutting copying the motion and mirroring of motion help create dance sequences effectively and easily.

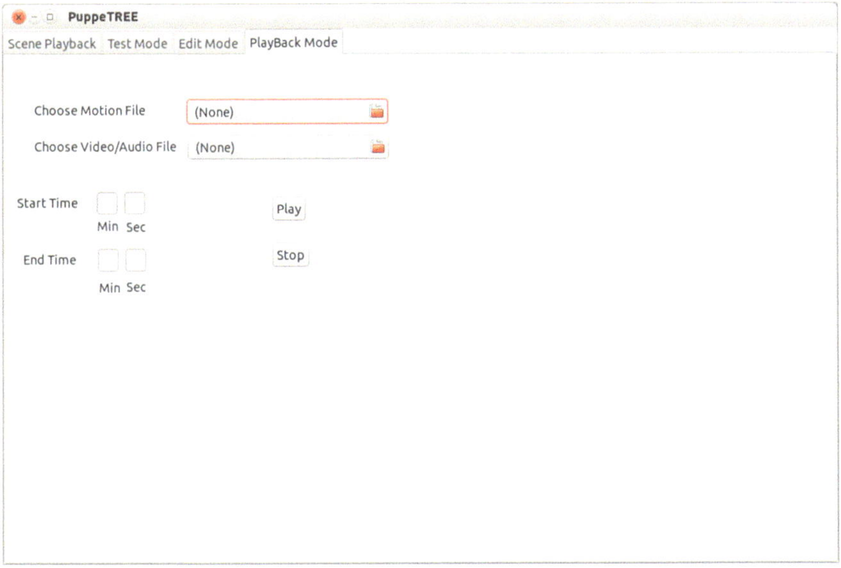

Fig. 8 Playback mode

5.1.4 Real-Time Mode

In this mode, four puppeteers can manipulate four different puppets in real time simultaneously using Kinect. An audio/video could be played on the screen behind the puppet. This enables the user to interact directly with the audience and respond to them. This is a novel approach to interactive puppetry.

5.2 Synchronization of Motion Data with Audio/Video

In order to synchronize the motion to the audio/video, it was required to transmit one frame of motion data, which consists of data for all axes, within one frame of the video. We have designed the system for 30 FPS video stream where, duration of one frame is 33 ms.

A new media player was designed for this purpose. The media player synced a motion frame with that of an audio/video frame. For this purpose, a file containing the motion data was created. When the video codec set up the next frame on the video buffer, the corresponding motion data for that frame was extracted from the motion data file and placed on the data buffer connected to the WIFI protocol. This would then transmit the data to the controller on the manipulator just as the video frame would come up on the screen. The audio was synced to this video frame just like any other media player.

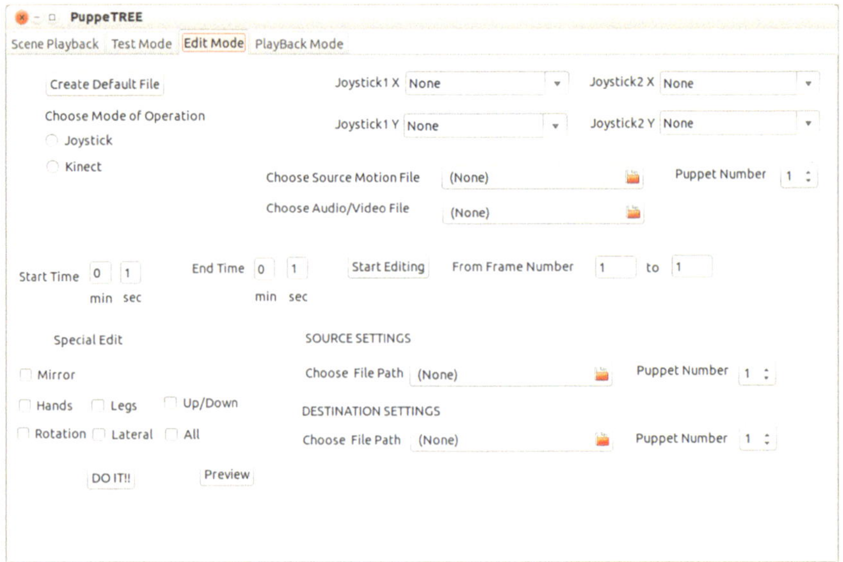

Fig. 9 Edit mode

The setting up of the video and audio frame took an overhead of 15 ms on the software. The WIFI module had a round trip ping time of around 6 ms for sending one frame of motion data for 5 puppets. This still provides for an extra buffer time of around 8 ms which was reserved for re-transmissions, health monitoring of manipulators and future use.

Provisions were also made to fast forward and rewind the audio/video/motion data. This way the play could be reset to start from any position with respect to the video. This enabled the user to fine tune a particular move by quickly replaying the same sequence on the press of a button. A resolution of 30 ms was achieved for editing the motion file.

6 System Integration

The Robo puppet system architecture is shown in the block diagram in Fig. 10. It consists of a master controller, a Linux machine which interacts with several slave puppet manipulators. The communication happens through Wi-Fi. Each puppet manipulator consists of electromechanical actuators with ARM processor based controller. Each slave has a unique ID through which it will communicate with the master. The entire communication between master and slaves takes care of real-time synchronization of motion and audio/video streams. The master controller also gets inputs from Kinect, joysticks and keyboard which are used for programming the puppets.

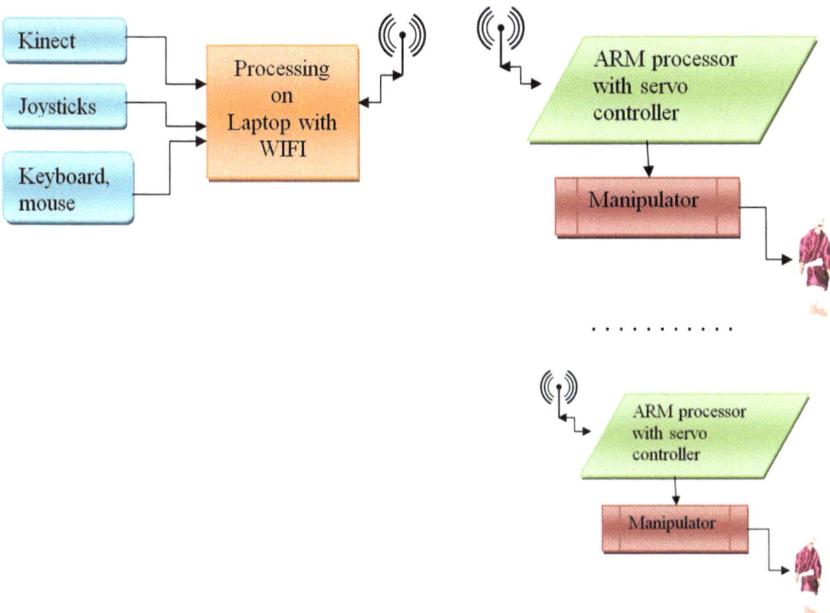

Fig. 10 System Integration Flowchart

7 Conclusion

This work presents a technology which automates Marionette Theater. Papier-mâché puppets were constructed. A robot manipulator was designed based on the requirements for effective marionette manipulations. An automated puppet theatre was built with provision to play 5 puppets simultaneously with programmable stage management. A user friendly GUI was created to capture, edit, record and playback motion data in sync with audio/video file. Some of the highlights of this technology are:

- The puppet show can be dubbed into any language making it transcend language barrier. The audio track is made up of key frames, which are synchronized with the video and the motion streams. A provision has been made to substitute the existing audio track with a new audio sound track of different language. This enables the puppeteer to present the show in any language.
- It does not need highly skilled people to choreograph the puppets.
  As mentioned in Sect. 5.1.2, the system has been designed to program the puppet in three different modes—Keyboard, body motion, pre-defined motion libraries. In conventional puppetry, puppeteer will never get to see his own actions through the marionettes, whereas in the proposed system, the puppeteer gets direct visual feedback while programming through any of the above modes. This enables even a novice puppeteer to choreograph the puppets reasonably well. Certainly an experienced puppeteer can make his shows more effective.

- Provisions of cut and paste and mirroring will help a person to program the entire show single handedly.
  A character in a show may have a distinct style of walking which can be programmed once using any of the above modes. The motion thus created, can be directly used anywhere within the show without requiring programming. Mirroring the motion sequence will add to the aesthetics of dance sequences or other general choreography. This feature provides for ease of programming and creation of entire show single handedly.
- Puppet show gives consistent performance every time.
  An effective control system and mechanical design ensures that, once choreographed, the system will perform consistently in each show as against a conventional marionette show where the outcome is dependent on the puppeteers experience and human error. This enables us to archive entire puppet shows which can be recreated anytime at will.
- Traditional marionettes can be directly used without modifications.
  Unlike other puppet systems like the Animatronics where the actuators and the electronics are integrated inside the puppet, the present system has been designed consciously to work with conventional marionettes which are built by traditional techniques. Therefore, this technology can be used by traditional puppeteers without any technical knowledge of robotics or electronics.

References

1. Currell, D.: Puppets and Puppet Theatre. Crowood Press, Wiltshire, UK (1999)
2. Moeslund, T., Granum, E.: A survey of computer vision-based human motion capture. Comput. Vis. Image Underst. **81**, 231–268 (2001)
3. Nguyen, K.D., Chen, I.M., Yeo, S.H., Duh, B.L.: Motion control of a robotic puppet through a hybrid motion capture device. In: IEEE Conference on Automation Science and Engineering Scottsdale, pp. 754–758 (2007)
4. Yamane, K., Hodgins, J.K., Brown, H.B.: Controlling a marionette with human motion capture data. In: IEEE International Conference on Robotics and Automation, pp. 3834–3841 (2003)
5. Chen, I.M., Xing, S., Yeo, S.H.: Robotic marionette system: from mechatronic design to manipulation. IEEE International Conference on Robotics and Biomimetics, pp. 228–233 (2005)
6. Hu, J.S., Wang, J.J., Sun, G.C.: Self-balancing control and manipulation of a glove puppet robot on a two-wheel mobile platform. In: International Conference on Intelligent Robots and Systems, pp. 424–425 (2009)
7. Setyawan, L., Kesaulja, L.M., Darma, J., Kuantama, E.: Hand-puppet Robot Control. IEEE (2011)

Math Based Model for Quick Estimation of Heat Generated in an Automotive Li–Ion Battery Pack at Various Operating Conditions

Vikrant Singh, Chiru Venkat Reddy, Justin R. McDade and Rashed S. Rabaa

Abstract With an emphasis on cutting down usage of fossil fuels and creating cleaner transport solutions, many automotive organizations have embraced Li–ion based battery technology to power their next generation hybrid and electric vehicles. The life and performance of the Li–ion battery packs, used on such vehicles, depends greatly upon the temperature of operation. An accurate estimation of the heat generated within a battery pack is essential in terms of sizing and designing the on–board coolant system. This paper demonstrates a simple approach for quick estimation of the battery pack temperature rise under different operating conditions. A Saber–Simulink co–simulation modeling capability has been developed to virtually represent the thermal and electrical components of a typical battery pack. Cell heat source has been evaluated using a reduced order model based on Newman's 1–D approach. The cell model for heat generation has been integrated with a virtual battery pack electrical schematic in Saber™ and pack heat generation simulated with varying ambient and fluid flow conditions.

Keywords Li–ion battery · Thermal management · Reduced order model · EV/HEV

V. Singh (✉)
Battery CAE, General Motors, Bangalore, India
e-mail: vikrant.singh@gm.com

C.V. Reddy
Allegis Inc., Bangalore, India

J.R. McDade
Electrification CAE, General Motors, Warren, MI, USA

R.S. Rabaa
Electrical Battery Cell/Pack, Electrification CAE, General Motors, Warren, MI, USA

A. Chakrabarti (ed.), *ICoRD'15 – Research into Design Across Boundaries Volume 2*,
Smart Innovation, Systems and Technologies 35, DOI 10.1007/978-81-322-2229-3_50

1 Introduction

Major automotive manufacturers worldwide are adopting electric powertrains in order to cut down the dependence on fossil fuels for transportation. Li–ion batteries have emerged as a strong contender for on–board electrical energy storage on such vehicles. The energy density of Li–ion batteries is higher than that of lead–acid and Ni–MH batteries and as such enabling automotive manufacturers to produce vehicles powered exclusively with electrical energy. These Li–ion batteries produce heat during operation and they should be maintained within a temperature range of around 25–30 °C for optimum performance and life [1]. Hence, an onboard coolant system for battery thermal management is necessary. A single cell cannot provide all the energy required for transportation. Hence, multiple cells are assembled together to form a module and many such modules are assembled in series to form a battery pack. Reliable sizing of the coolant systems for a module necessitates a proper understanding of the cell level heat generation. Hardware testing and coolant sizing is an expensive exercise and hence virtual modeling could be relied upon to enable an understanding of the coolant requirements for the automotive pack under various operating conditions. In literature [2, 3], the electrochemical and heat generation behavior of a cell is often numerically modeled using Newman's 1–D model. Newman's framework is characterized by coupled heat and mass transport equations in addition to charge conservation equations. Heat is produced in the cell due to the electrochemical reactions and ohmic heating. An implementation of Newman's 1–D model is available on various commercial tools like BDS [4] and COMSOL [5]. However, the time required (CPU time) to run such simulations for a single cell is high as one needs to solve multiple, coupled partial differential equations. Therefore, simulating a module or a full pack is computationally expensive. Many researchers have attempted to solve this problem through simplifying the overall cell level model by including a less complex heat generation model [6, 7] or alternatively, to exclusively use an equivalent circuit [8] coupled with thermal modeling and/or CFD [9, 10]. Equivalent circuit based models are quick to run but often do not work well for low temperature cases where even a small change in temperature could impact cell internal resistance (and hence heat generation and voltage). There have been some attempts at using simplified physics models for simulations like BDS NTG model [4] and models with simplified Butler Volmer kinetics [11], but such models often rely on some fitting parameter evaluated through a series of physical tests.

In the current study, we extend an isothermal Reduced Order Model (ROM), developed earlier [12] to incorporate temperature dependency and model the cell heat generation at different ambient conditions. The cell physics based ROM is integrated with a module level electrical–thermal–model on Saber™ and electrical–thermal co–simulations are carried out to assess the battery module temperature under various cell loading conditions.

2 Mathematical Modeling

Figure 1 shows the schematic of Newman's math model (1–D approach) and the basic philosophy of ROM. Newman's 1–D model is expensive to solve computationally as one has to deal with a number of control volumes for a set of coupled transport equations. Li^+ ion transport in the solid and the liquid phase (Fickian diffusion) is solved for along with the heat generation equations and Ohm's law (charge conservation). The reversible reaction kinetics is defined by the Butler Volmer equation. Upon introducing simplifications like the uniform reaction rate approximation and a profile based concentration approximation, the complex system of partial differential equations (PDE) can be reduced to a system of ordinary differential equations (ODE) with some algebraic expressions [12]. The solid phase transport is addressed in anode and cathode by the single particle model [13, 14]. Thus, starting from a set of PDEs, one has to solve for a set of ODEs and algebraic expressions to solve for the dynamics of the cell. In the current paper, an isothermal ROM model developed earlier is extended to incorporate temperature dependency and account for cell heat generation and dissipation in a module.

2.1 ROM Formulation—Cell Physics

Details on the derivation of ROM are available elsewhere [12] and the final equations are provided below:

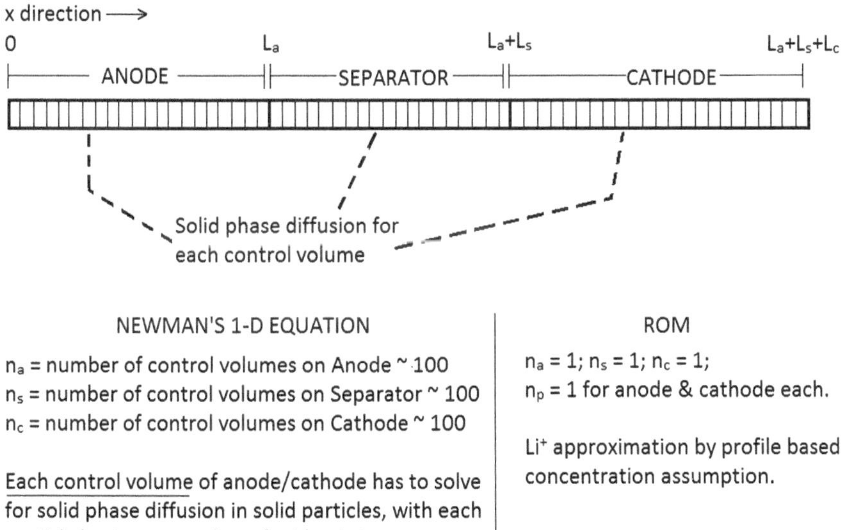

Fig. 1 Numerical framework of the reduced order model (ROM) and Newman's 1D model

Transport in liquid phase (2 ODE):

$$A\frac{dq_{in}}{dt} + B\frac{dq_{ip}}{dt} = -q_{in} + (1 - t_+)a_n l_n \langle j_n \rangle \tag{1a}$$

$$C\frac{dq_{in}}{dt} + D\frac{dq_{ip}}{dt} = q_{ip} + (1 - t_+)a_p l_p \langle j_p \rangle \tag{1b}$$

where, q_{in} and q_{ip} are the interfacial flux of Li^+ ions at the anode/separator and cathode/separator boundary respectively. *a, l* and *j* denote the active area per unit volume, length scale and reaction rate at anode and/or cathode, with suffix *n* and *p* denoting anode and cathode respectively. *A, B, C, D* are variables in terms of local cell properties and/or dimensions and t_+ denotes the transport number.

Transport in solid phase (4 ODE): A quartic profile for Li^+ concentration within the solid phase was assumed, as in [13] and the volume averaged transport equations for the same may be represented as:

$$\frac{dc_{avg,n}}{dt} = -\frac{3\langle j_n \rangle}{R_n}; \quad \frac{dc_{avg,p}}{dt} = -\frac{3\langle j_p \rangle}{R_p} \tag{2a}$$

$$\frac{dc_{r,n}}{dt} + \frac{30D_n c_{r,n}}{R_n^2} = -\frac{45\langle j_n \rangle}{2R_n^2}; \quad \frac{dc_{r,p}}{dt} + \frac{30D_p c_{r,p}}{R_p^2} = -\frac{45\langle j_p \rangle}{2R_p^2} \tag{2b}$$

where, $c_{avg,n/p}$ and $c_{r,n/p}$ denote the Li^+ average concentration and radial gradient of Li^+ concentration at the anode (*n*) or cathode (*p*) respectively.

The above Eqs. 1a, 1b and 2a, 2b are linked through the reaction rate term and the Butler Volmer Kinetics expression [3, 10–14].

2.2 Modeling Cell Heat Generation

Source of heat generated in a Li–ion cell may be electrochemical or ohmic in nature. Equation 3 gives the total heat generated in a cell, where Q_o denotes the ohmic heat and Q_e denotes the electrochemical heat generated in the cell.

$$\dot{Q} = \dot{Q}_o + \dot{Q}_e = I^2R + I\left(V - U - T\frac{\partial U}{\partial T}\right) \tag{3}$$

$$m_{cell}Cp_{cell}\frac{dT_{cell}}{dt} = \dot{Q} - heat\ dissipated \tag{4}$$

where *V* is the cell voltage, *U* the Open Circuit Voltage (OCV), *T* the temperature, *I* the current, *R* the resistance of metal current collectors and m_{cell}, Cp_{cell} as the mass and heat capacity of the cell. The ohmic heat is calculated for Copper as well as

Aluminum current collectors. The entropy term TdU/dt has been neglected in the current study. Eqs. 1a–3 define the non–isothermal ROM for the estimation of heat generation in a cell. The transport properties in a cell are impacted by any change in the cell temperature. The non–isothermal ROM cell model was validated with a Full Order Model (FOM) at different operating conditions.

2.3 Modeling Battery Module—Heat Dissipation

Design of a module determines the heat dissipation from a cell during operation. A typical battery module is built by connecting cells in series/parallel, as shown schematically in Fig. 2. Two cells and a cooling plate form the repeating unit for the module. The cooling plate consists of micro–channels through which the coolant liquid flows and removes the heat generated in the cell. The virtual battery module was created in Saber™. The electrical schematic of the whole module was combined with the thermal model for the module. The thermal model included RC circuits for heat transfer modeling through various components of the module (like cell to cooling plate and heat transfer to ambient). The RC thermal network model was verified with a detailed CFD-thermal model (details of verification are not addressed here). The cell dynamics (voltage, temperature, State of Charge), described by the ROM, were implemented in the same tool that also provides heat generation in the cell i.e. Simulink. A co–simulation of Saber–Simulink was set up to carry out coupled electrical thermal simulations of the whole module (assumed to be composed of 32 cells).

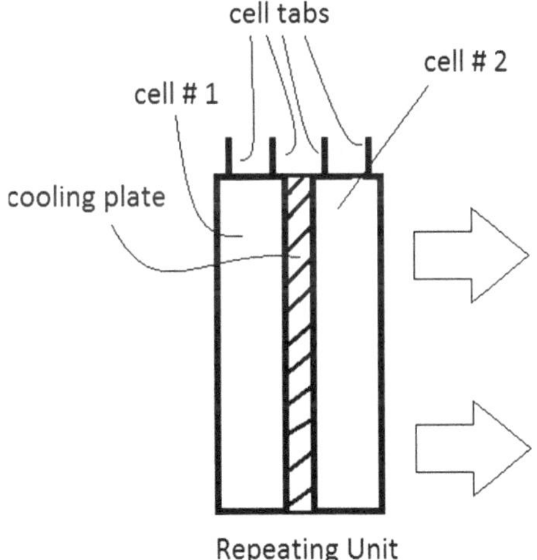

Fig. 2 Schematic of the repeating unit of cells to form a battery module

3 Results and Discussion

3.1 Cell Model (ROM) Verification with a Full Order Model (FOM)

Properties of cell assumed for current simulation are provided in Table 1. ROM was verified against the full order model (FOM) for a single cell. Figure 3a–c shows the plot of cell voltage and cell temperature with time, at different ambient conditions.

Figure 3a shows the plot for discharge of a cell at 1 C rate and 25 °C ambient temperature. It can be observed from the figure that the match between ROM and FOM for cell voltage is good (r.m.s. error <2 % = |1–ROM voltage/FOM voltage|). The temperature prediction from ROM mimics the temperature prediction from FOM well too. The peak in the temperature profile is not accurately captured but the difference is small (~ 1 °C) and could be termed as acceptable. Figure 3b shows a similar plot for cell discharge at 1 C rate at 40 °C ambient temperature. Again, the ROM matches very well with the FOM, for both temperature and voltage profiles. Figure 3c shows the discharge profile at 0.5 C rate for 0 °C ambient temperature case. The match between ROM and FOM is again found to be reasonable. The peak temperature for 0.5 C rate is under-predicted in ROM by 0.75 °C, which may be termed as acceptable. We may thus conclude from Fig. 3a–c that ROM captures the cell physics (voltage, temperature) well at different ambient conditions.

Table 1 Properties of cell used in the simulation

Property	Anode	Separator	Cathode
Thickness	74 μm	25 μm	70 μm
Current collector thickness	10 μm	–	20 μm
Porosities	0.27	0.41	0.22
Bruggeman coefficient	1.5	1.5	1.5
Particle size (radius)	7.5 μm	–	6 μm
Density of solid (gm/cc)	5.16 gm/cc	–	2.26 gm/cc
Solid surface area	0.215056 m^2/gm	–	0.126831 m^2/gm
Exchange rate constant (at 25 °C)	0.169583 mA/cm^2	–	0.176159 mA/cm^2
Kinetic rate E_a	20,000 J/mol	–	58,000 J/mol
Solid diffusion coeff. (at 25 °C)	3.93049e^{-10} cm^2/s	–	5.03236e^{-10} cm^2/s
Solid diffusion E_a	35,000 J/mol	–	29,000 J/mol
Transport number	0.4	0.4	0.4
Material maximum capacity	372 mAhr/gm	–	148 mAhr/gm
External cell area (cm^2): 424.089			
Cell heat capacity (J/g-K): 1.062			

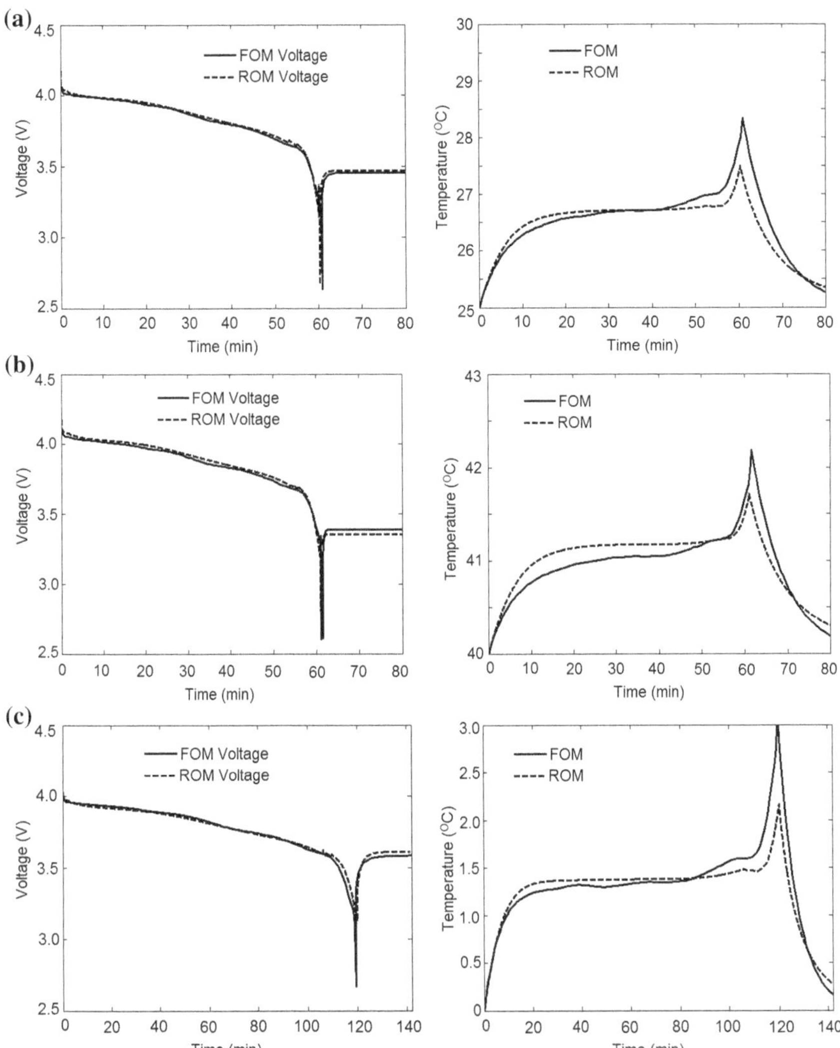

Fig. 3 **a** Voltage and temperature profile for 1 C discharge at 25 °C ambient. **b** Voltage and temperature profile for 1 C discharge at 40 °C ambient. **c** Voltage and temperature profile for 0.5 C discharge at 0 °C ambient

3.2 Module Temperature Simulations

The cell model developed above was integrated with electrical schematic of a module and was used to simulate temperature profiles at the center and at the end of the module. The module thermal circuit was developed as described in Sect. 2.3. The non–isothermal ROM model in Simulink was imported into Saber^{TM} where a

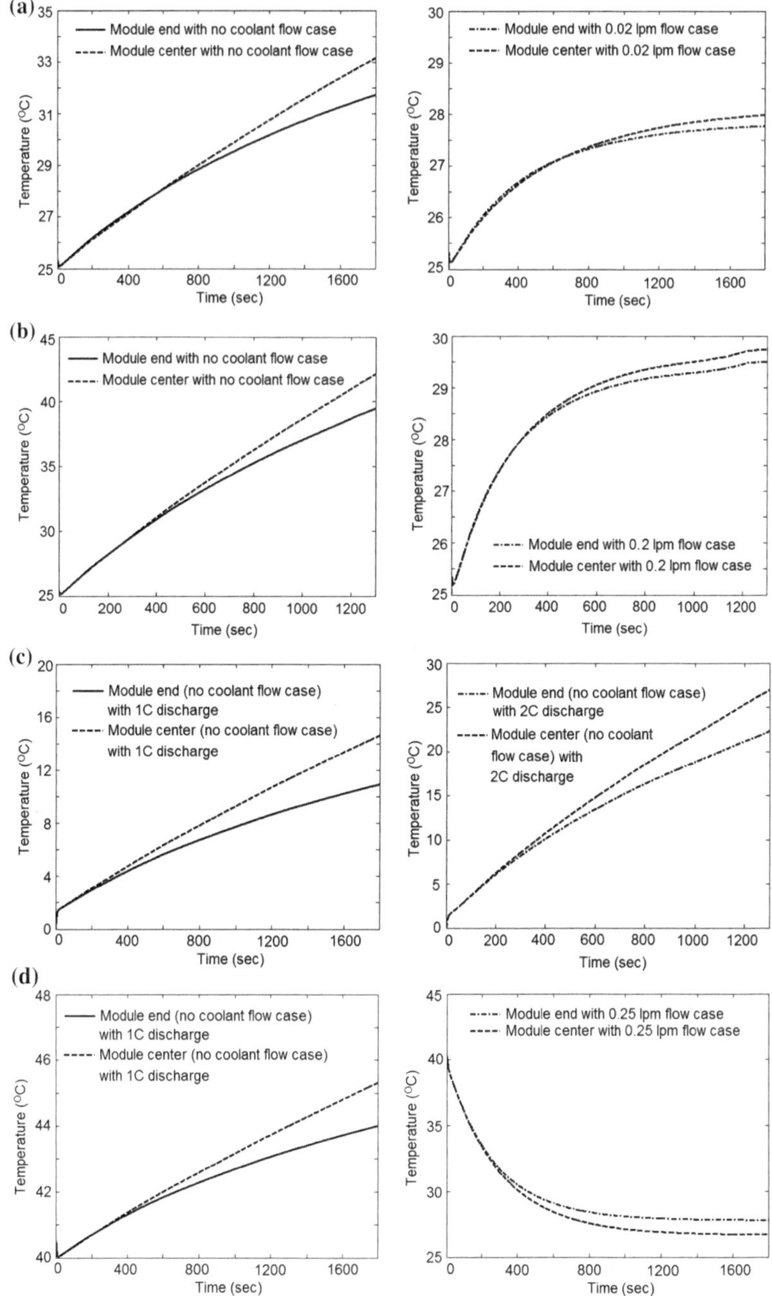
(a)
Module end with no coolant flow case
Module center with no coolant flow case
Temperature (°C)
Time (sec)
Module end with 0.02 lpm flow case
Module center with 0.02 lpm flow case
(b)
Module end with no coolant flow case
Module center with no coolant flow case
Module end with 0.2 lpm flow case
Module center with 0.2 lpm flow case
(c)
Module end (no coolant flow case) with 1C discharge
Module center (no coolant flow case) with 1C discharge
Module end (no coolant flow case) with 2C discharge
Module center (no coolant flow case) with 2C discharge
(d)
Module end (no coolant flow case) with 1C discharge
Module center (no coolant flow case) with 1C discharge
Module end with 0.25 lpm flow case
Module center with 0.25 lpm flow case

◀**Fig. 4** **a** Temperature at center and end of module, with and without coolant flow for 25 °C ambient and 1 C rate discharge at cell level. **b** Temperature at center and end of module, with and without coolant flow for 25 °C ambient and 2 C rate discharge at cell level. **c** Temperature at center and end of module, without coolant flow for 0 °C ambient; 1 and 2 C rate discharge at cell level. **d** Temperature at center and end of module, with and without coolant flow for 40 °C ambient and 1 C rate discharge at cell level

RC circuit based thermal model was developed to estimate heat transfer from a cell in the module to its surroundings. The coolant used in all simulations was water with an initial temperature of 25 °C. Co–simulations were then carried out for assessing coolant flow–rate requirements (constant flow–rate assumed) for keeping the module between 25–30 °C during operation.

Figure 4a–d shows a variation in temperature at the module center and at the module ends for different operating conditions. Figure 4a, b shows that the temperature of the module may rise as high as 43 °C (during 2 C discharge rate) if no cooling mechanism is present (no coolant flow case). Also, as was expected, the module center temperature was higher compared to the module end temperature (higher heat loss at the ends of the module). With coolant flow, the temperature of the battery module was contained to a value below 30 °C. As can be observed from Fig. 4a, a low coolant flow rate of 0.02 lpm was required to cool the battery pack to a value below 30 °C. Again, the module end temperature was lower than the module center temperature and both were at a value below 30 °C. Similarly, with 2 C rate discharge at 25 °C, 0.2 lpm of coolant flow–rate was found (see Fig. 4b) to be sufficient to keep the maximum module temperature below 30 °C. Figure 4c shows the module operation at low temperature (0 °C ambient temperature). It was found that at 1 C as well as 2 C discharge rates, the module maximum temperature never crossed a value of 30 °C, even in the absence of coolant flow. Figure 4d shows the module temperature, with and without coolant flow, at 40 °C ambient temperature conditions. As was expected, in hot weathers (40 °C and higher), the coolant flow-rate requirements are higher as compared to ambient conditions of around 25 °C.

4 Conclusion

A non–isothermal physics based cell model has been developed and coupled with Saber™ to carry out fast electrical–thermal simulations for assessing cooling requirements for a battery module. The model is intended to provide results such that design engineers can quickly iterate through multiple designs. It provides a good platform for integrating multi-physics modeling whereby physics based cell level models can work, relatively inexpensively, in tandem with detailed electrical and thermal models. Such flexibility leads to better decision making using CAE tools. The cell level model was validated with a detailed electrochemistry model

(Full Order Model FOM) for assessing heat generation during cell operation. A detailed electrical-thermal schematic for the module was then developed in Saber^{TM} and the cell non–isothermal ROM integrated for co–simulation capability. The simulation tool developed is able to simulate battery module temperature profile variations at different ambient and operating conditions. The simulations show that more coolant is required as the ambient temperature rises. Higher C rates or faster discharge also leads to faster rise in module temperature and hence more coolant requirements. This tool can enable a quicker decision making during virtual proto–typing stage of the product and enable faster product development at a reduced (hardware, testing) cost.

References

1. Paseran A.: Thermal management studies and modeling. In: Annual Merit Review Meeting, US DoE Hydrogen program and Vehicle Technologies program, Arlington, Virginia, USA 18–22 May 2009
2. Bernardi, D., Pawlikowski, E., Newman, J.: A general energy balance for battery systems. J. Electrochem. Soc. **132**(1), 5–11 (1985)
3. Rao, L., Newman, J.: Heat generation rate and general energy balance for insertion battery systems. J. Electrochem. Soc. **144**(8), 2697–2704 (1997)
4. Battery Design Studio, CD–adapco. www.cd-adapco.com/products/battery-design-studio
5. COMSOL: http://www.comsol.co.in/
6. Chen, Y., Evans, J.W.: 3D thermal modeling of lithium polymer batteries under galvanostatic discharge and dynamic power profile. J. Electrochem. Soc. **144**(11), 2947–2955 (1994)
7. Paseran, A., Vlahinos, A., Stuart, T.: Cooling and preheating of batteries in hybrid electric vehicles. In: 6th ASME–JSME Thermal Engineering joint conference, USA, 16–20 Mar 2003
8. Hu, Y., Yurkovich, S., Guezennec, Y., Yurkovich, B.J.: Electro-thermal battery model identification for automotive applications. J. Power Sources **196**, 449–457 (2011)
9. Khateeb, S.A., Amiruddin, S., Farid, M., Selman, J.R., Al-Hallaj, S.: Thermal management of lithium ion battery with phase change material for electric scooters. J. Power Sources **142**, 345–353 (2005)
10. Chen, S.C., Wan, C.C., Wang, Y.Y.: Thermal analysis of lithium ion batteries. J. Power Sources **140**, 111–124 (2005)
11. Kim, U.S., Shin, C.B., Kim C.S.: Modeling for scale up of a lithium ion polymer battery. J. Power Sources **149**, 841–846 (2009)
12. Vadivelu, S.: Reduced order model for a lithium ion cell with uniform reaction rate approximation. J. Power Sources **222**, 426–441 (2013)
13. Subramanian, V., Diwakar, V.D., Tapriyal, D.: Efficient macro-micro scale coupled modeling of batteries. J. Electrochem. Soc. **152**(10), A2002–A2008 (2005)
14. Zhang, D., Popov, B.N., White, R.E.: Modeling lithium intercalation of a single spinel particle under potentiodynamic control. J. Electrochem. Soc. **147**(3), 831–838 (2000)

Ideal Workstation Design of Driver's Cab for New EMU Rake of Mumbai Local

Amar Kundu, Kiran Gangadharan, Nishant Sharma and Gaur G. Ray

Abstract The Mumbai Suburban Railway is the first rail system in India which began services in Mumbai in 1867 transports today about 6.3 million passengers daily and has the highest passenger density in the world. Considering the age old existing workstation design a new ergonomics based design approach is undertaken to redesign the workstation with advanced instrumentation. In present situation drivers have difficulty in external visibility along with associated musculo-skeletal and psychophysical stresses as obtained from direct discussions and questionnaire feedback. A user centered approach on Drivers behaviour and understanding the issues while driving were made on board by personal interview and observation method. Based on collected anthropometric data a 3D model of the chair and the workstation were made by using solidworks CAD software. The CAD drawing was then converted to a full scale prototype for simulation study. Special care was taken for rearrangement of different control panels according to drivers' suitability. Driving comfort was also taken into consideration while designing the ideal workstation.

Keywords Workstation design · EMU rake · Ergonomics · Motorman · Mumbai suburban railway

1 Introduction

Mumbai Suburban Railway is the first rail system in India which began services in Mumbai in 1867. Today it is spread over 465 km, plying with 2,342 number of trains and carrying more than 7.24 million passengers daily having highest passenger density in the world. While transfer of daily passengers is of prime concern, their lives are in the hands of the motormen who drive the suburban trains

A. Kundu (✉) · K. Gangadharan · N. Sharma · G.G. Ray
Indian Institute of Technology Bombay, Mumbai, India
e-mail: kundu.amar2010@gmail.com

A. Chakrabarti (ed.), *ICoRD'15 – Research into Design Across Boundaries Volume 2*, Smart Innovation, Systems and Technologies 35, DOI 10.1007/978-81-322-2229-3_51

or Electro Motive Units (EMU) coaches. With high mental load in terms of trespassing, daily passengers' issues, management issues, it is really a miracle to see that these trains are running on time and safely. The high level of Physical, biomechanical, psychological and environmental loads the motormen (drivers) are exposed to while operating the age old motorman's cabin is one of the major concern in stress accumulation. With the advancement of technology, and demand for better quality of service, it is now imperative to relook at the motorman's cabin and redesign it considering different issues like motorman's operational ergonomics, work posture, remote communication, visual limitations, and fatigue accumulation and so on. The Mumbai Railway Vikas Corporation Ltd. (MRVCL), a public sector undertaking of Govt. of India, under the Ministry of Railways, is responsible to execute the New Generation Mumbai Urban Transport Trains as sanctioned by Ministry of Railways. Out of multifaceted approaches, one of the major approach was redesigning the Motorman's cabin with integration of state of the art technology and ergonomics.

2 Methodology

To begin with, the user study, activity analysis and problem identification were carried by using personal questionnaire and observation method. Indian Anthropometric Data were considered for redesign purpose based on available published data. Based on the principle of "Design for all", 5th and 95th percentile anthropometric were considered which covers 90 % of the population. Statistical methods were used to obtain 5th, 50th and 95th percentile value of each anthropometric dimension. 2D Manikins were developed and used by applying those anthropometric data.

3 Initial Observations

To understand the issues while driving the EMU racks, 12 numbers of experienced (more than 5 year) motormen were observed while driving the train to and fro from one end to the other end of their travel rout. Each rout was comprised of about 2–2.5 h journey. Video photographs were recorded and a questionnaire was introduced to obtain the issues of dissatisfaction/discomfort felt by them. Motormen were also allowed to give their suggestions for improvising the driving system. The study was conducted in Mumbai and all the motormen were responsible in driving the Mumbai Suburban Electro Motive Unit (EMU) trains. The primary controls (as per frequent usability) with existing workstation are shown in (Fig. 1).

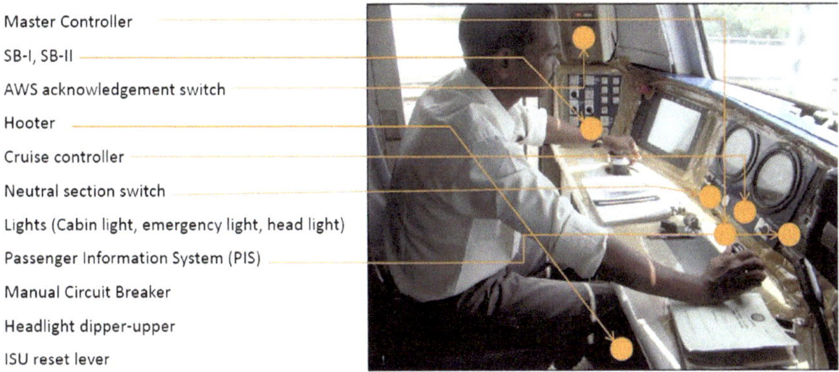

Fig. 1 Position of most frequently used controls in existing EMU local train

4 Problem Identification

During driving motorman faces multiple problems related to workstation layout. The main control is placed in the front work table. Motorman needs to hold/operate this control for long time. This activity demands a sustained bending posture that creates musculoskeletal disorder (MSD) in lower back, shoulder, hand, wrist etc. To reduce the stress accumulated on the right hand in the present absence of hand rest situation they use log book to support hands. But it does not fulfill actual armrest criteria. At the time of using emergency brake an unnatural wrist bending is observed causing accumulation of pain in wrist. Due to the low height and depth of the table top, motorman could not acquire suitable leg space room. For safe driving, certain minimum visibility is essential (as per UIC code). But due to improper desk height, chair height and window dimensions, shorter motormen could not have the required visibility. The controls are arranged erratically without following any suitable order. Some controls (Cabin light, fan etc.) are placed on right side of body though right hand is always engaged with Dean Man's Handle (DMH) operation. Some frequently usable controls (SB-I, SB-II) are placed out of primary hand reach area demanding motormen to raise their left hand above shoulder level repeatedly. Due to inappropriate storage space motormen put their emergency tool kit, torch, water bottle etc. in unorganized manner as per available space here and there. Causing accumulation of pain in wrist.

5 Workspace Analysis for Conceptual Model

According to the UIC code, Motormen need to see the lower level of signal at a distance of 15 m and upper level of signal at a 10 m distance from the front buffer while in sitting posture. But as per feedback from the motormen they want maximum visibility possible. Trespassing and passengers crossing the train line are the underline issues specially for the Mumbai track. To know the environmental situation, motorman wants extended visibility through windscreen. In order to drive the train safely and effectively, the motorman must be able to see all the displays, indicators and signals. He should be able to move around quickly to operate the controls, and should be in a position to communicate with the co-motorman or station master as and when required. The arrangement of controls and switches should, therefore, be based on distance travel by hands and other body component, frequency of use, easy visibility and accessibility. Controls and displays should be grouped according to functional classification representing sequence of use. Conceptual model is designed with all basic principles and design constraint to overcome the existing problem faced by the motormen. For placement of controls, display and indicators, four separate aspects were considered.

1. The position of a control with respect to another control
2. The position of display with respect to another display
3. The position of a control with respect to display, and vice versa and
4. The shape and dimension of the workspace from the view point of visibility, and reach ability.

The essential features that should be considered for conceptual model are mentioned below:

- In an ideal design the right hand should be used only for the master control and the left hand for all other controls and switches. Master control should be placed in the right side of motorman and there should be proper armrest.
- Monitoring displays should be shifted to the front side.
- Frequently used switches needs to be located in the effective reach zone.
- Emergency brake will be placed on the left side of the motorman. It should be within easy reach zone for operation in emergency situation. It can also help for resting left hand.
- If some controls are inevitable for the right side, they should be used only when the train is at a halt/stop.
- Switches and knobs should be designed/placed based on activity analysis.
- There should be some mechanism to prevent the loose papers from flying off due the high air flow inside the cabin.
- The journey card/caution note, log book and personal log book should be kept firm on the workspace and in readable distance.
- There should be sufficient leg room under the table to keep legs and operate the hooter knob easily.
- There should be some definite storage space for tool kit, torch and water bottle.
- Motormen frequently watch speedometer as they need to stop the Train at a very specific position on each station. So, they preferred speedometer to be in the primary visual area on the workstation.
- In the present scenario main control was on the table whose depth was around 10 cm which occupied the leg room space and led to uncomfortable sitting posture. It also reduces visibility, since height of the table was more. Operating the

master control from that position caused stretched shoulder, twisted wrist. Since master control is being used for full time, it should be positioned in such a way that during operation it provides a very comfortable body posture throughout.

- Although the operators visual field is quite wide, his accuracy and speed of response depends largely on the position in the visual field at which stimulus occurred. The fastest time to response in the visual field falls in the area covered by about 8° cone above and below the central line of sight, up to 40° to the right and left of the midpoint. So it finds its place just beside primary vision cone.

On the basis of above mentioned problems and recommendations, a conceptual work station was developed by using basic ergonomics understanding. The work station was divided in prominent two parts, primary and secondary reach areas. The frequently usable controls (SB-I, SB-II, Cruise controller, Neutral section switch etc.) are placed within primary area. The other controls are placed in secondary area.

The main control is placed just right side of the body in front of working table with appropriate height (seating elbow height). It will prevent sustained bending posture. One adjustable armrest/hand support was also attached with the workstation. In front of main control, a control panel was made for cabin light, fan etc. On left hand side another control panel was created for other frequently used controls/ switches. The communication related controls (microphone, mike etc.) are placed

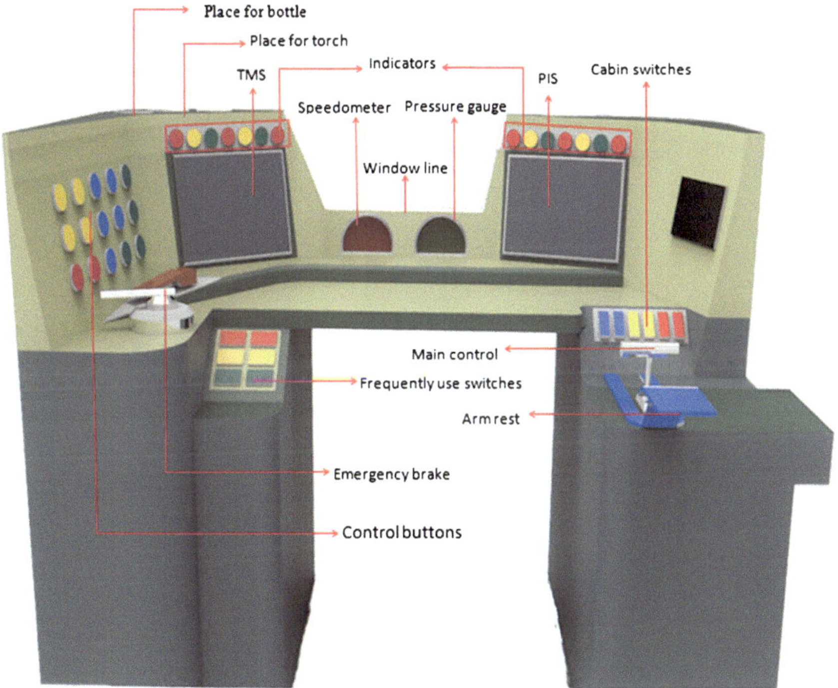

Fig. 2 Conceptual model for EMU local

on right side panel as they will be used only when the train is in halt mode. But the telephone handle is placed on left side secondary accessible area.

Speedometer and pressure gauges are placed just in front of driver's eye demanding frequent visual monitoring. Truck Management System (TMS) and Passenger Inquiry System (PIS) are placed left and right corner of motormen respectively. All the indicators are placed above the TMS and PIS. No elements were put on the vertical plane of the workstation to achieve maximum "V" shaped visual clearance through the front windscreen. Emergency off button is placed in front panel between speedometer and pressure gauges within both hand reach and also little away from primary reachable area to prevent unconsciously activation. A dedicated storage space is given just below the main control to keep essential luggage within easy arm reach. Based on the CAD model of existing nose-cone, as received from MRVCL, the conceptual workstation was developed by using SolidWorks 3D software (Fig. 2).

6 Ergonomic Analysis of Conceptual Model

6.1 Seat Height Analysis

The conceptual 3D model was converted in 2D line diagram for ergonomic analysis (Fig. 3) by using anthropometric data pack derived from actual motorman.

Based on UIC recommendation, sufficient space and clearance for 95th percentile motorman was kept between the seat backrest and the cabin wall. According to anthropometric data, the ideal working table height should be 755 mm and main control height should be of 747 mm. This helps to obtain 2.8 m visibility fulfilling motormen's requirement. The available floor length of 1,740 mm (330 mm more than existing available floor length) was achieved when 95th percentile motorman

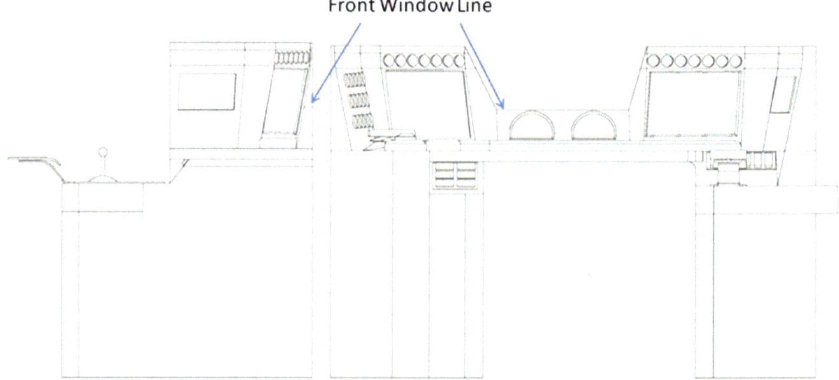

Fig. 3 Line diagram of conceptual model. (**a**) Side view, (**b**) front view

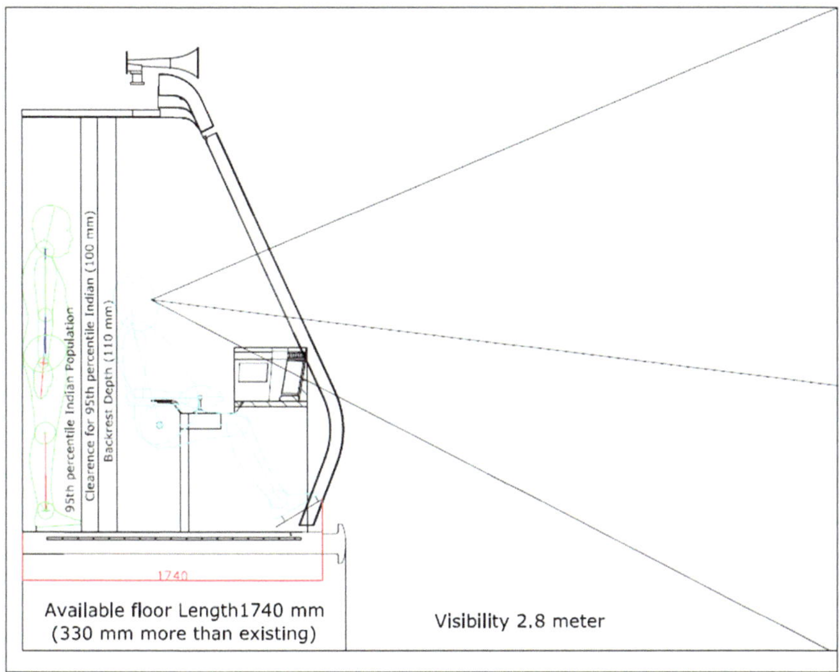

Fig. 4 Dimensional analysis for 95th percentile motorman

manikin was placed in comfortable sitting posture by acquiring knee and backrest angles 110° and 105° respectively as shown in Fig. 4.

The workstation designed for 95th percentile was then assessed for 5th percentile motorman. 5th percentile motorman required more hand rotation to hold main control. They could obtain only 3.8 m visibility, which did not fulfill motorman's requirement. To get minimum 3 m visibility, either the window height had to decrease by 62 mm below the window line or increase seat height as well as footrest height as required (Fig. 5).

The main control height (654 mm) was then fixed according to 5th percentile motorman and checked by using 95th percentile motorman (Fig. 6).

The other relevant dimensions used for designing the workstation are given in Table 1.

6.2 Arm Reach Analysis

This analysis was done to visualise the region covered by each hand and also the common region between them. To validate the primary and secondary reachable area, both normal and maximum hand arm reach analysis done by using manikin.

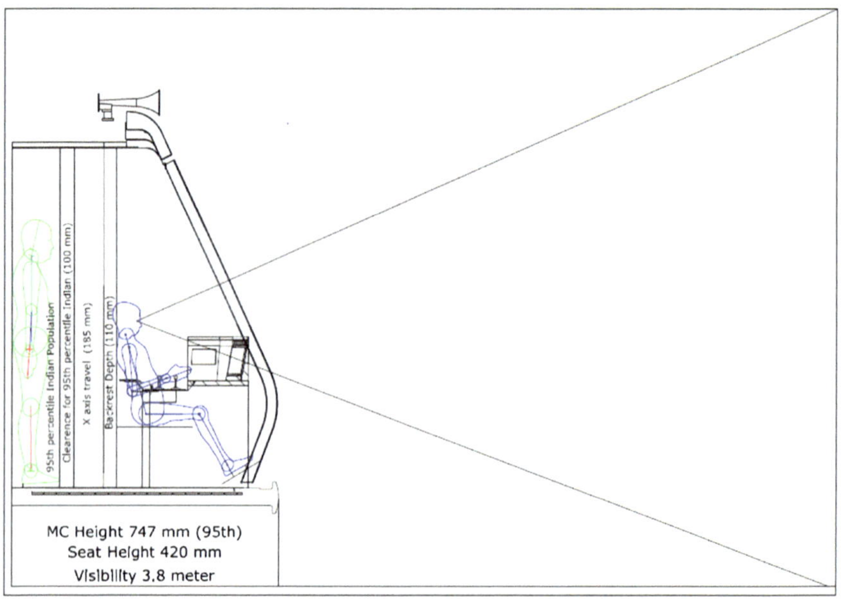

Fig. 5 Dimensional analysis for 5th percentile motorman

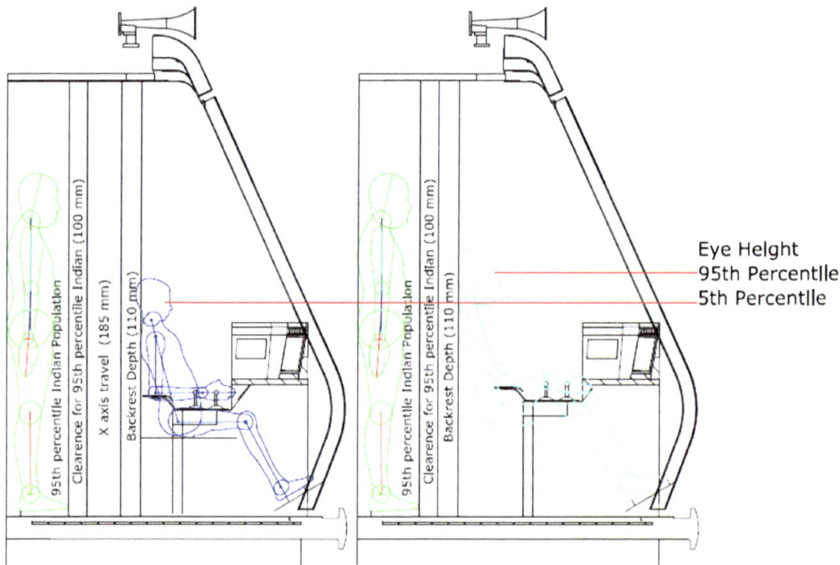

Fig. 6 Dimensional analysis (main control height) for 5th and 95th percentile motorman

Table 1 The relevant dimensions used for ideal workstation of EMU local

Dimension	5th percentile	95th percentile	Adjustable
Working table Ht	755 mm	755 mm	Nil
Main control Ht	706 mm	706 mm	Nil
Seat height	420 mm	470 mm	50 mm
Seat movement along X axis	625 mm	440 mm	185 mm
Footrest Ht	30 mm	30 mm	Nil
Footrest angle	30°	30°	Nil
Visibility	3.8 m	2.8 m	Nil

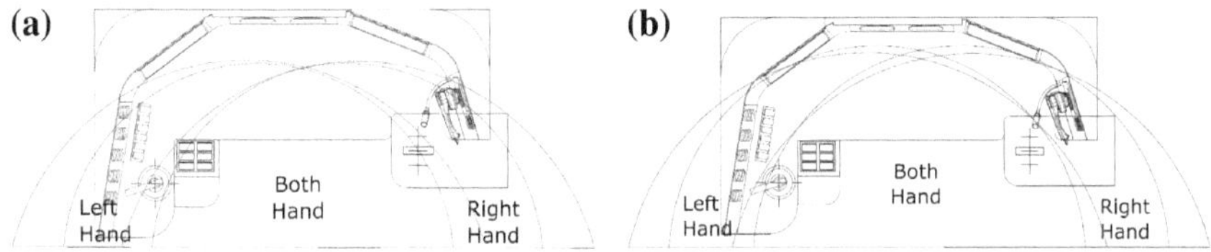

Fig. 7 Arm reach analysis for 95th percentile (*red*) and 5th percentile (*Black*). (**a**) Normal arm reach, (**b**) maximum arm reach

This analysis helped to define the table width and understand the position for switches, indicators master control, emergency break, phone, mike, track management system (TMS), passenger inquiry system (PIS) and emergency switch (Fig. 7).

After Dimensional analysis, the conceptual model was converted to a full scale (1:1) prototype model for actual simulation by drivers/motormen.

7 Final Model

In the final model all control switches were arrange as per their priority in operation. Altogether 31 different control operations were identified which were achieved through 10 rotary knobs, and 21 push buttons. The panel also had to house 10 different indicators for indicating the status of different operational situations. Figure 8 denotes the control panel design highlighting the position of each control and display. Based on activity analysis all controls were divided in two parts based on their frequency of usage. The frequently used knobs were positioned towards proximity of the body. Others were placed on the panel based on frequency and reach criteria. By following the group theory, all rotary switches were placed on the left side of panel and indicators on the right side of panel. Press buttons were placed between the rotary switches and as per their relations with the indicators. Same type control button and indicator were placed in same alignment and in same

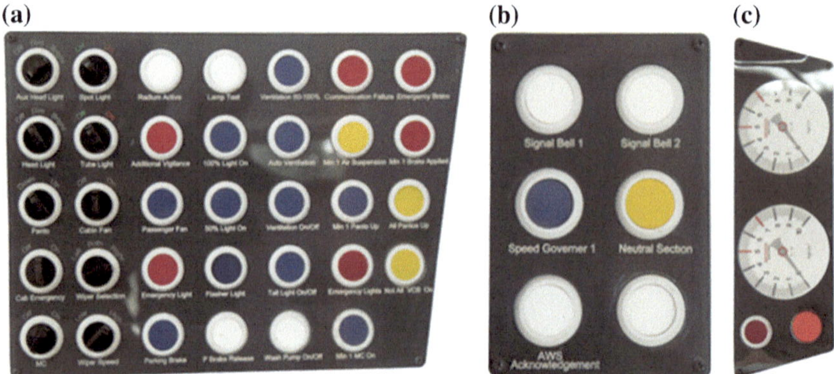

Fig. 8 Layout of control panel. (**a**) Less frequently used controls (*left*), (**b**) frequently used controls (*middle*), (**c**) Emergency off button, speedometer and pressure gauges

Fig. 9 Denotes the overall layout of the ideal workstation which is under evaluation of the motormen for freezing the design

group. Emergency off indicator and switch were placed in the front panel beside the pressure gauges which should be operation by both hands in emergency. Audio-visual indicator and Public address (Microphone on/off) switch were placed in the right hand side panel beside the microphone. Track Management System (TMS) and Passenger Information System (PIS) displays were installed in their existing position. Telephone handle was also shifted to right side panel with other communication device.

To maintain driver's body balance, emergency brake handle height was aligned with main control (DMH) height. A knob type handle was designed for emergency brake to prevent the wrist bending. A provision was made for storage space of torch and water bottle on the left hand side below the frequently used control buttons of the workstation considering easy reach. The final model also was converted to a full scale prototype (non functional) for simulated study and testing by people with different body dimensions. It is now being awaited for motormen's feedback (Fig. 9).

Acknowledgments Authors convey sincere thanks to the motormen community for their active participation throughout the study. Heartfelt thanks to MRVCL for sponsoring the project. Thanks are also rendered to ICF and Bombardier Transport for their support. Special thanks to ARAI Pune, for their support in determining the 3D anthropometric dimensions of the motormen.

References

1. Panero, J., Zelink, M.: Human Dimensions and Interior Space: A source Book of Design Reference Standards Hardcover. Random House LLC, New York (1979)
2. Code, U.I.C.: Layout of Driver's Cabs in Locomotives, Railcars, Multiple Unit Trains and Driving Trailers, 4th edn (2002)
3. Chakrabarti, D.: Indian Anthropometric Dimensions for Ergonomic Design Practice. National Institute of Design, Ahmedabad (1997)
4. Anthropometric and Strength Data of Indian Agricultural workers for Farm Equipment Design. Central Institute of Agricultural Engineering Bhopal, India (1985)
5. Sanders, M.S., McCormick, E.J.: Human Factors in Engineering and Design, 6th edn. McGraw-Hill International Editions, New York (1987)

Characteristics of Jewellery Design: An Initial Review

Noor Adila Mohd Rajili, Elin Olander and Anders Warell

Abstract Previous studies in jewellery design have primarily focused on the techniques of the jewellery design process, such as the use of tools including computer aided design, rapid prototyping and other design technology. However, limited attention has been devoted to understanding the jewellery design practice itself, including methodologies and rationale of the work. Therefore, the focus of this position paper is to understand the nature of the practice, how designers reason about jewellery making, and how they create the knowledge they need in the design process. A literature review on the topic revealed that the characteristics of jewellery design can be described by three aspects; the practice of the jewellery designer, characteristics and rationale of jewellery designers, and knowledge creation in jewellery design. In conclusion, jewellery design can be understood as an intuitive practice with an artistically driven standpoint and experimentally based approach, focusing on resolving design challenges related to the object itself.

Keywords Jewellery · Jewellery practice · Design practice · Knowledge

1 Introduction

Jewellery design practice aims at creating ornamental objects, art forms or jewellery objects which "enhance and decorate the wearer" [1], using a variety of materials such as precious or semi-precious metals and gemstones. Through the centuries, jewellery design has evolved and became a discipline of its own. The development has changed the way jewellery making is practiced today. Conceptual, historical and material aspects have influenced and shaped the practice of jewellery into a

N.A.M. Rajili (✉) · E. Olander · A. Warell
Department of Design Sciences, Division of Industrial Design,
Lund University, Lund, Sweden
e-mail: adila.mohd_rajili@design.lth.se

A. Chakrabarti (ed.), *ICoRD'15 – Research into Design Across Boundaries Volume 2*,
Smart Innovation, Systems and Technologies 35, DOI 10.1007/978-81-322-2229-3_52

field in which designers communicate their ideas and convey their self-image and identity through their work.

The practice of jewellery employs approaches based on crafts and arts, such as applied art, decorative art and fine art [2]. According to MacDonald [3], 'craft' originally implied something made by hand, however in the nineteenth century the term 'manufacturing' was introduced, as craft-based products became increasingly industrially produced.

Historically, jewellery making and design inspiration were derived from the history of art. From the earliest times people made jewellery because they were passionate about "materials, the challenge of constructing pieces with inevitable problem-solving, a strong desire in creating own designs and expressing ideas" [4]. The professional skills and traditions of jewellery making are well known in traditional and contemporary jewellery, however, how designers' reason and develop the knowledge needed for their practice has been very sparsely explored and thus not properly understood. In particular, there is a need to deepen the understanding of how knowledge dissemination from theory to practice in jewellery design occurs in order to increase the understanding of the discipline, and how jewellery practitioners understand their practice, develop and communicate it to others such as customers or students within jewellery education. In this context, knowledge dissemination is related to the transfer and sharing of knowledge as suggested by Gagnon [5]. Hutchinson and Huberman [6] note that knowledge dissemination has a "different meaning to different people", and has conceptual (as learning or the possession of new perspectives or approaches) and instrumental (the improvement of the practice or new practice) components. While, theory in jewellery design can be defined as a basic principle of designers' expression which involved with both design and technical aspects in relation to the making of jewellery such as history, composition, movement, forms, techniques, tools and materials. In this context, from the researcher point of view, understanding the theory is fundamental for jewellery designers in order to understand their processes and jewellery products. Various developments of the profession, such as the introduction of digital tools and contemporary methods in jewellery design, have motivated the authors to study the characteristics of the discipline. However, research in jewellery design rarely focuses on the characteristics of the practice, i.e., how jewellery designers engage with the processes of design, such as how they argue and reason about the motivations and rationale of their work, how this knowledge can be characterized, and how it may contribute to developing the jewellery design profession and education to meet contemporary and future needs.

In order to investigate the above issues, a literature study has been undertaken in which literature relating to jewellery design, jewellery practice and design practice in general was reviewed. The relationship between jewellery design practice and other design practices with respect to processes and thinking is particularly in focus, related to the three research questions of this work: (1) What practices do jewellery designers employ in their work? (2) How do jewellery designers reason about their work and describe the rationale for their practice? (3) How do jewellery designers create the knowledge needed for their design activities?

2 The Practice of Jewellery Designers

Jewellery design is a specialised field of knowledge and practice, which intimately involves processes of conceptualising and making jewellery products. This field engages designers with a highly developed artisan skills and requires knowledge of basic metallurgy and gemstones in which particular tools, techniques, and methods are used, including the knowledge of visual language. Jewellery practitioners make use of a range of principles in their work. In this context, 'principles' refers to the activities of designing and making jewellery, involving stages of finding inspiration, exploring and generating ideas, and evaluating and translating ideas [7]. In relation to the stage of translating ideas from sketch to product, decisions are made concerning materials, dimensions, colours, flexibility, wearability, weight, cost and the use of techniques and tools. McGrath [7] states that jewellery designers work in a variety of different ways by applying a range of jewellery techniques such as cutting and piercing, construction, forming, carving and casting, mechanisms, colouring and texturing, stone setting, stringing beads and outworking, as described by Young [8]. The variety of techniques indicates that the process of creating jewellery would exhibit a range of methodological approaches in actual design work. However, previous studies on this topic have not been found as part of this research.

The practice of jewellery design has similarities with other design disciplines such as ceramics and furniture design. For example, Cooper and Press [9] state that in ceramics, furniture and jewellery design, the process of designing and making are one and the same process. This is in contrast to other design disciplines, where they are often separated. This position is also taken by Mäkinen [10], claiming that the phases of design and production in her context of study of small firms are united, as they are part of an integrated craft production process. It should be noted that until a few decades ago, jewellery was understood as craft discipline [2], similar to ceramics, glass, textiles and furniture. However, through industrial development, these fields are now recognized as design disciplines. For example, Nimkulrat [11] states that in Finland, the term 'design' was adopted by the craft disciplines of ceramics and textiles as a result of the development of industrial production in the 1950s. As previously stated, Greenhalgh [2] proposed that "ceramics, furniture making, glassmaking, jewellery, metalsmithing, painting, sculpture and tapestry" can be described as belonging to the same genre, why it may be argued that jewellery making can be regarded as a design discipline.

Jewellery practice may also be characterised as positioned between craft and design-based approaches. Shaobin [12] goes one step further, claiming that jewellery belongs to the category of industrial design. To support her claims, she notes that jewellery students employ the principles of jewellery design and industrial design as a procedure in jewellery making, by illustrating how both types of working have been applied in jewellery design education. For example, jewellery students need "to make an overall plan and to consider a factor of sense, inspiration, creative idea, modelling, colour and material, texture, shape, technology, function and trend with tendency" [12]. Shaobin's [12] work, however, lacks the explicit explanation on the integration of industrial design activities with the principles of jewellery design in the jewellery making processes with respect to the above examples. It can be argued that if jewellery design students are taught using industrial design methods and tools, this will influence how they work and may not be a sign that the principles of the two disciplines have anything natural in common. Therefore, this is need to be studied further.

3 Characteristics and Rationale of Jewellery Design Processes

Jewellery designers use an intuitive approach and experimental practice in their work. Pereira and Tschimmel [13], Amadue [14] and Wallace et al. [15] show that intuition is presented in the jewellery process. However, they claim that intuition in a very different way for instances; Amadeu [14] states that intuition is possibly involved in the early stages of the creative process, whereas Wallace et al. [15] suggests that intuition within idea development arises through an emphatic and iterative process. In their studies, Pereira and Tschimmel [13], Amadue [14] and Wallace et al. [15] agreed that intuition is present in the jewellery process, however they claimed that intuition may be related to different aspects of their work.

Furthermore the literature study revealed that jewellery is an experimental practice [13, 14]. Studies suggest that since the 1960s, the contemporary jewellery movement, together with the new ceramics, fibre art, studio glass and book arts have grown as experimental practices [2]. This can be seen in jewellery field in which jewellery practitioners also often integrate “intuitive and experimental practices” during their creation processes [13, 14].

Wuytens and Willems [16] propose a conceptual framework describing variation within design processes of studio jewellers. The framework describes three basic design process; ‘design parameter reflection’ (imposing constraints on a design by looking backward to the set of constraints already imposed and by looking forward to estimate the design qualities based on heuristics), ‘design cluster reflection’ (imposing a hierarchically structured order onto the design) and backtracking (reframing the problem by considering previously imposed constraints or hierarchically structured). According to Wuytens and Willems, the way they describe the design process is useful in educational situations, innovation and design strategies. Wuytens and Willems found three sources of diversity based on their conceptual model. The diversity can be distinguished by designers’ reflection in the use of ‘different orderings of design clusters’, ‘the proportion of design parameters constrained by thinking and doing’ and the experience of the designers. Wuytens and Willems’s [16] work is a good beginning for understanding the jewellery designers’ working processes as showed in the diversity of their design processes.

A majority of the existing research in the field of jewellery focuses on understanding how certain aspects of jewellery design are supported by methods and tools, such as form specification using Computer Aided Design (CAD) software tools [17, 18], and rapid prototyping (RP) [19]. Although this research has an impact on understanding a specific aspect of jewellery design practice, it focuses on the use of computer technology in jewellery processes, i.e. the techniques of jewellery design. As technology has developed in society at large, there has also been a shift in how methods and tools are applied in the design process; i.e. the methodology of the process. This development has changed the way designers work, i.e. the processes, in which techniques and methods are employed. As stated before, in contrast to e.g. industrial designers, jewellery designers often integrate the conceptual and making phases in their design process [10]. One reason of this is that jewellery practitioners still employ craft based methods as their main way of working. According to Mäkinen [10], computer technology is not a suitable tool for these designers, who work on a small-scale basis. However this is not valid for large design organisations, which depend on computer technology for their mass production of jewellery pieces [10].

4 Knowledge Creation in Jewellery Design

The third research question relates to how jewellery designers create the knowledge they need in their work. In this context, the knowledge refers to how jewellery designers understand and make decisions regarding activities, strategies and tools, relating to e.g. materials, processes, meaning, form, and market. The literature review revealed no earlier studies on this topic.

Schön [20] describes reflection-in-action is the way designers reflect in both situation; during and after the designing process. Schön [20] introduced the term reflection-in-action, describing a strategy that professional practitioners use in their everyday practices, while working under situations of complexity, uncertainty, uniqueness, and value conflict. According to Schön [20], practitioners participate in reflective practices with others on their knowing and reflecting-in-action, which allows them to reconstruct their theories of action making their action-strategies able to be explicitly formulated. In relation to the designing and making process of jewellery, designers need to reflect on their actions, for example; how jewellery designers converse and reflect during the process of designing and making jewellery and how they make sense of it are important to enhance understanding.

Another study that related to this knowledge creation process [21] which also correspond to the level of cognitive processing information. This study shows that four aspects such as physical, perceptual, functional and conceptual were cognitively interacted in the processing information in the design process at sketches stage by practicing architect. In relation to this study, the four aspects would be also integrated on how jewellery designers create the knowledge they need their activities.

5 Discussion

As the focus of this paper is to review the literature on jewellery design and position the research in relation to other work, aspects related to how jewellery designers' reason about their practice and the characteristics of jewellery design has been the focus of this paper.

However, there is also a need for empirical evidence on how designers work in practice. Therefore, an empirical study has been conducted to investigate the research questions by interviewing jewellery designers about their practice. The empirical study contributes to a deeper understanding of how designers describe and make sense of their practice.

As previously stated, jewellery design as a discipline is derived from craft-based activities. The nature of working with jewellery is dominated by an intuitively driven approach to design [13–15]. Therefore, in order to develop the practice of jewellery design and to strengthen the disciplinary status of jewellery design, further research relating to thinking and reasoning, for instance focusing on structured ways of working and being explicit and conscious about the practice, is necessary. It is important that jewellery designers develop the conceptual understanding of their discipline as have other design disciplines. Developing a shared terminology will also benefit to develop the identity of the profession and will position jewellery design in the community of design practice.

This study is significant in order to motivate the practice to integrate intuitive and experimental processes in a 'thought process' [22] of jewellery design. Lawson [22] states that a 'thought process is required to identify and understand design problems and create design solutions' for designers' outcome and competency. For example, Lawson [22] claims that there are different modes of thinking; 'thinking', which is about concentration or paying attention in the process of designing; 'imaginative thinking, which refers to what is possible but not actual; and 'reasoning', which means "to control the direction of thought, reflective thought and problem-solving" [22]. It is likely that these modes of thinking also apply to jewellery design processes.

In order to further investigate the third research question, a situated case-based research study using e.g. participatory observation is under consideration. This study will explore how jewellery designers working situation, not just by talking about their work.

In this paper, it is proposed that the practice of jewellery is dominated by intuitively and experimentally based approach to design, focusing on resolving design challenges related to the object itself. Concerning this issue, one might ask how the practice of designing jewellery may change if designers focused on questioning the status quo of jewellery and the wearer, and what the implications of such a change may be. As a future direction of the research, a further examination of this study on how jewellery designers make sense of their own profession will be carried out.

Acknowledgments I would like to acknowledge the financial support of the *Ministry of Higher Education of Malaysia* and *Universiti Teknologi MARA* for the duration of my PhD Studies.

References

1. Galton, E.: Basics Fashion Design 10: Jewellery Design: From Fashion to Fine Jewellery. AVA Academia (2012)
2. Greenhalgh, P. (ed.): The Persistence of Craft: The Applied Arts Today. Rutgers University Press, Jersey (2002)
3. MacDonald, J.: Concepts of Craft, Exploring Visual Culture: Definitions, Concepts, Contexts. Chapter 3, 34 (2005)
4. Eddy, M.: Jewelry making: a brief history. In: Young, A. (ed.) The Workbench Guide to Jewelry Techniques, pp. 12. Thames & Hudson, London (2010)
5. Gagnon, M.L.: Moving knowledge to action through dissemination and exchange. J. Clin. Epidemiol. **64**(1), 25–31 (2011)
6. Hutchinson, J.R., Huberman, M.: Knowledge dissemination and use in science and mathematics education: a literature review. J. Sci. Educ. Technol. **3**(1), 27–47 (1994)
7. McGrath, J.: The Complete Jewellery Making Course; Principles, Practice and Techniques: A Beginner's Course for Aspiring Jewellery Makers, p. 1 (2007)
8. Young, A.: The Workbench Guide to Jewelry Techniques. Thames & Hudson, London (2010)
9. Cooper, R., Press, M.: The Design Agenda: A Guide to Successful Design Management. Wiley, Ney York (1995)
10. Mäkinen, H.: Product Design as a Core Competence in a Design-Oriented Industry. Adv. Appl. Bus. Strat. **9**, 103–126 (2005)
11. Nimkulrat, N.: Hands-on intellect: Integrating craft practice into design research. Int. J. Des. **6** (3), 2012 (2012)
12. Shaobin, L.: Design elements in jewelry design, computer-aided industrial design and conceptual design (CAIDCD). In: 2010 IEEE 11th International Conference 2010, vol. 1–2, pp. 257–259, Conference Paper (2010)
13. Pereira, Á., Tschimmel, K.: The design of narrative jewelry as a perception-in-action process. In: Proceedings of the 2nd International Conference on Design Creativity, Glasgow: The Design Society, vol. 97, p. 106 (2012)
14. Amadeu, F.: Creativity and emerging knowledge: intuitive practice in design and crafts. Trans-techresearch.net (2012)
15. Wallace, J., Dearden, A., Fisher, T.: The significant other: the value of jewellery in the conception, design and experience of body focused digital devices. AI Soc. **22**(1), 53–62 (2007)
16. Wuytens, K., Willems, B.: Diversity in the design processes of studio jewellers. EKSIG: Experimental knowledge, method & methodology. Availabe at: http://www.academia.edu/885664/Diversity_in_the_design_processes_of_studio_jewellers (2009)
17. Molinari, L.C., Megazzini, M.C., Bemporad, E.: The role of CAD/CAM in the modern jewellery business. Gold Technol. **23**, 3–7 (1998)
18. Wannarumon, S., Bohez, E.L.: Rapid prototyping and tooling technology in jewelry CAD. Comput.-Aided Des. Appl. **1**(1–4), 569–575 (2004)
19. Hohkraut, U.: Rapid prototyping and jewelry design. In: Annals of DAAAM and Proceedings (2010)
20. Schön, D.A.: The Reflective Practitioner: How Professionals Think in Action, vol. 5126. Basic books, New York (1983)
21. Suwa, M., Purcell, T., Gero, J.: Macroscopic analysis of design processes based on a scheme for coding designers' cognitive actions. Des. Stud. **19**(4), 455–483 (1998)
22. Lawson, B.: How Designers Think: The Design Process Demystified. Routledge, London (1997)

QFD Based Methodology to Decide upon Contextually Appropriate Solution Category with Specific Reference to Pineapple Peeling Equipment Design

Prakash Kumar and Debkumar Chakrabarti

Abstract Framing product design specifications is an important part of design process. The decisions taken at this stage affect all the downstream phases, right from conceptualization, embodiment and prototyping to manufacturing and dissemination. In engineering design, to determine priorities of different engineering specifications of product, Quality Function Deployment (QFD) method is used. Similar approach was used for deciding upon some design specifications as well. The present study elaborates on a QFD based methodology which helped in deciding an important specification, in the context of pineapple peeling equipment design. Here, a pineapple peeling solution was to be designed for small fruit processing units. During study of the existing solutions, it was found that there were broadly four categories of solutions and final solution was likely to belong to one of them. Hence, specifying the appropriate solution category, prior to conceptualization, was thought to be reducing both, time and effort. The paper, based in this backdrop, presents a QFD based approach used for the peeling equipment design. At first, priority of different user requirement was determined. Then, the relation between these requirements and different solution categories was established. These inputs were, then, put through QFD based process that generated scores indicating an order of appropriateness of solution categories to meet larger users' priorities. The order of appropriateness, obtained through the methodology, was evaluated through users' responses. Through this case study, the paper attempts to show proof of concept of the adopted methodology supporting adoption of similar approach for design of other fruit processing equipment also advocates adoption of specific design methods in new design contexts.

Keywords Equipment design · Product design specification · Key requirements · User priority · Category-requirement relationship · Solution category

P. Kumar (✉)
Department Art, Design and Performing Arts, Shiv Nadar University, Greater Noida 201314, Uttar Pradesh, India
e-mail: prakash.iitg@gmail.com

D. Chakrabarti
Department of Design, Indian Institute of Technology Guwahati, Guwahati, India

A. Chakrabarti (ed.), *ICoRD'15 – Research into Design Across Boundaries Volume 2*, Smart Innovation, Systems and Technologies 35, DOI 10.1007/978-81-322-2229-3_53

1 Introduction

Product Design specifications are like guidelines that intend to ensure that the subsequent design and development of a product meets the needs of the user [1]. Product design process has several phases of divergence and convergence [2] which gradually narrows down towards a final solution (Fig. 1).

During the process, there is a divergence to get information regarding the various aspects of the problem, existing solutions and practices and their respective limitations followed by convergence to define problem areas that were going to be addressed by putting down the product design specifications. This gives appropriate direction to diverge for ideas and concepts for more effective solutions.

In the present study, the ergonomic issues related to pineapple peeling task were identified and was decided to be addressed through equipment design that would make pineapple peeling task easier, effective and faster. To achieve this, a generic methodology was followed which has been shown in Fig. 2.

During problem analysis, various aspects of the problem was tried to be understood with a holistic approach leading to design specifications (Fig. 3). Study of the concerned system, for which solution was meant, was carried out from different perspectives with a holistic approach and different key requirements of stakeholders were identified. During study of existing solutions and practices, it was found that the solutions could be classified into four broader categories i.e. fully automated solutions, semi-automated solutions, hand powered solutions and small hand tool like solutions. Besides these studies, structure and properties of pineapple fruit was also analyzed. These studies from various perspectives contributed to knowledge and experience that forms basis for the specifications constituting design brief that gives thoughts and directions for conceptualization [3].

During specification framing for pineapple peeling machine, an important decision, to be taken, was regarding the broader category of solution that would be suited to larger section of the user population. In design, there are many decisions based on intuition and in present context, this was one such decision. This decision depends upon a number of requirements and their respective priority for the users/ buyers varying from unit to unit. Hence, to take a proper account of the requirements without missing any important information, a systematic approach was adopted for deciding upon the most appropriate solution category adopting an

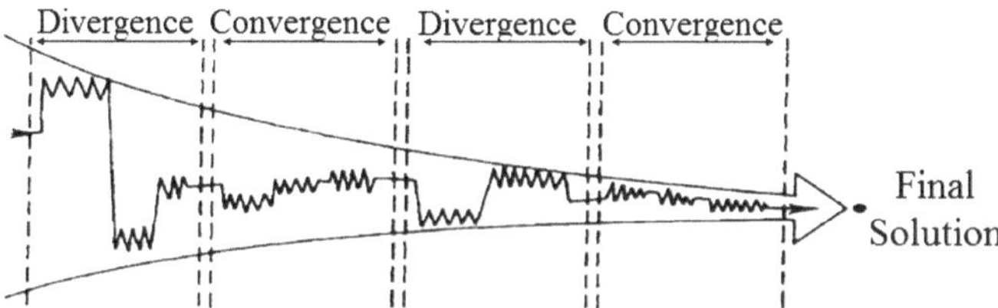

Fig. 1 Divergence and convergence during design process. *Source engineering design methods—strategies for product design*

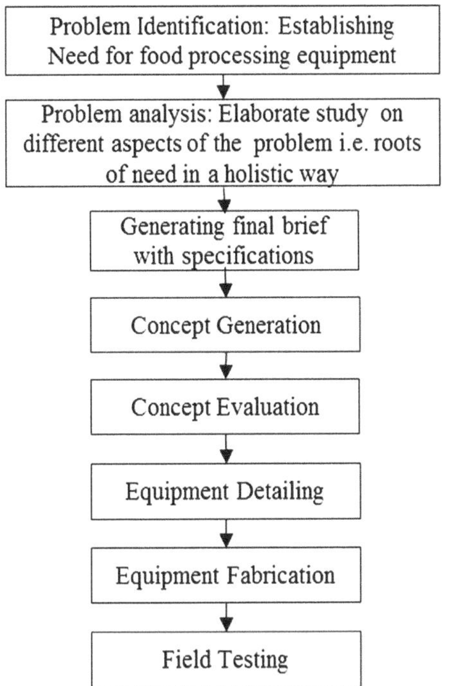

Fig. 2 Methodology adopted for pineapple peeling equipment design

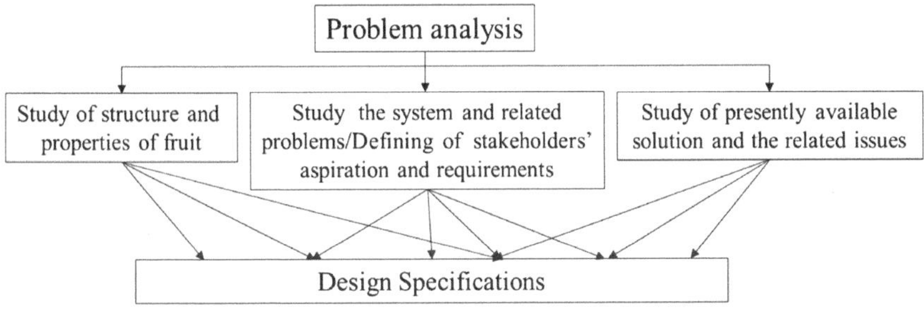

Fig. 3 Holistic problem study during problem analysis phase leading to design specifications

approach based on Quality Function Deployment (QFD) method, used for prioritizing engineering specifications of products in engineering design [4]. The paper, explains the process of deciding the solution category suitable for larger user population using this QFD based approach and secondly, it describes the validation of the approach by testing its outcome through stakeholders' responses. Finally, the paper reflects upon the deviation of users' preference findings from outcome of the methodology and lays down future improvement and scopes of use for the approach.

2 Method

This study adopted a QFD based approach to prioritize the appropriateness of the solution categories that can indicate the broader category of solution that is more suitable to the larger section of the users/ buyers. For this, users' priorities of key requirements and the broader categories of available solutions were required as inputs (Fig. 4). The method generated scores that indicated suitability of different solution categories for given context, as the outcome.

The users, considered here, were the entrepreneurs who were the actual purchaser of equipment. As per identified key requirements, the solution was supposed to have:

- Lower cost
- Lower wastage of raw material
- Better Portability
- Higher productivity
- Better peeling effectiveness (different sizes)
- Uniform quality of output
- Lesser repair and maintenance
- Higher aesthetic appeal
- Power-independent operation
- Less Space occupying
- High Safety of workers (accident/injuries)
- Quicker Adaptability of workers (familiarity to technology)
- Better comfort of workers
- Multipurpose utility

The user priority for all these key requirements was assessed through their ratings and the mean values were considered for further calculations. Also the relationship between these requirements and the different categories of solution was ascertained through mean score obtained from the ratings by the senior PhD researcher in design from backgrounds of engineering, ergonomics and design. Eventually, the evaluation was done by briefing users about the four categories of solution and asking them about category of solution, they felt, would be more preferable to their context of use. They were also told to arrange all the solution categories in their order of preference in descending order.

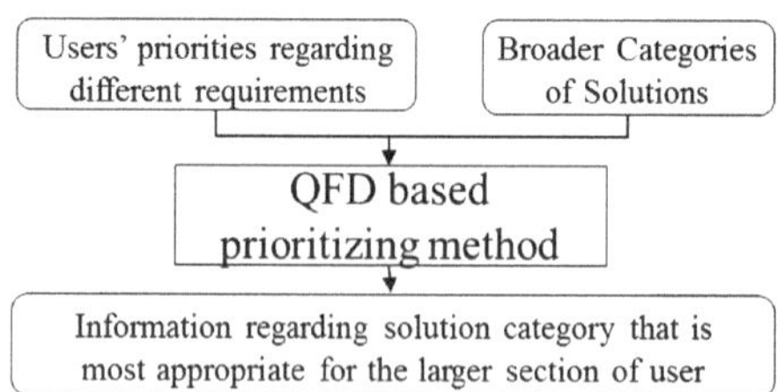

Fig. 4 QFD based approach for deciding most appropriate solution category

2.1 Determining Users' Priorities for the Key Requirements Using Questionnaire

To know how prior the various requirements are for the entrepreneurs, they were asked to rate various requirements as per their priorities using questionnaire. The questions were put in an indirect way for true understanding of their priorities. Total of 28 questions were put in the questionnaire with some of the questions being reverse coded. All the 14 key requirements were covered in this questionnaire and 1–7 scale was used for rating. The questionnaire was administered to 17 entrepreneurs and the mean values of their ratings were considered for further calculations.

2.2 Establishing the Category-Requirement Relationship

Each broader category of solutions was related to a number of requirements directly or indirectly influencing entrepreneurs' purchasing preference of new equipment. The key requirements which have been identified earlier during system study were analyzed and rated for their relationships with different categories of solutions. To avoid biasness, the rating was done by 11 PhD researchers in Design with different fields of interest i.e. mechanical engineering, design, agriculture engineering, ergonomics and sustainability. The ratings were done on 1–7 rating scale "7" being most positive relation and "1" being least positive relation. All these ratings were analyzed and mean values were considered for further calculations.

2.3 Incorporation of Ratings for Determining Most Suitable Category of Solution

In the proposed QFD based approach, users' priority ratings and category-requirement relationship score helped to establish an overall importance of each solution category. In present context, users' priorities, for key requirements, are arranged in different rows and different solution categories in different columns from left to right. Category-requirement relationship score were multiplied with the corresponding priority scores. Finally, all these individual scores for each category were added generating final category score. The final score of each category indicated its relative importance with respect to other solution categories (Table 1).

The individual mean priority scores were put in terms of percentage and the category-requirement relationship scores was determined from 1 to 7 points scale of rating.

Table 1 Table representing various components of the QFD based method

Effectiveness with which the solution categories fulfils the requirements (rank 1–7 scale) Priority for each requirement (in % age)	C1	C2		Cn
P1	P1 * RC11	P1 * RC12	————	P1 * RC1n
Pn	Pn * RCn1	Pn * RCn2	————	Pn * RCnn
Total score	∑Pn * RCn1	∑Pn * RCn2		∑Pn * RCnn

The final score for each category can be calculated using the following numerical formula:

$$\text{Final score for a category} = \sum (\text{Pn x RCnn})$$

where

Pn priority of nth requirement among all the requirements
Cn nth category of solution
RCnn Rating for relationship of nth category with nth requirements

2.4 Evaluating Outcome of the Proposed Approach

To check validity of the approach, the entrepreneurs were explained about the different category of solutions and they were asked to rate them for appropriateness as per their context of use. The process was divided in two parts. On one hand, the entrepreneurs were asked to directly arrange the categories in descending order as per appropriateness to the context of use. On the other, the solution categories were rated for contextual appropriateness on 1–7 scale, “1” being least appropriate and “7” being most appropriate. The survey was done over the same 17 entrepreneurs. The survey was conducted prior to discussing and disclosing the outcome obtained from the proposed approach. All the above data were analyzed using SPSS 20.0.

3 Results

3.1 Users' Priorities for the Key Requirements

Among the various key requirements, the user priority was found to be the greatest for "Higher safety of the workers" with a mean score value of 6.17 of 7. This was followed by the priority for "Better repair and maintenance", "Higher productivity" and "Wastage minimization" with a mean score of 5.47, 5.29 and 5.29, respectively. The priority for "Peeling effectiveness" and "Peeling quality" both had a mean score of 4.09. This relatively lesser priority for the two can be understood in the light that most of the units do not have canning facilities and generally they prepare juice based product for which they only required roughly and unevenly peeled pineapples that would be crushed and filtered to extract juice for further preparations. Hence the priority scores for the two requirements are, comparatively, lower (Table 2).

3.2 Category-Requirement Relationship Scores

In rating for the category-requirement relationship, "Lower cost", majority of the respondents were of the view that hand operated tools would be cheaper than other categories of solutions but since the productivity of such solutions is lowest and hence they have highest running cost per unit. Hence considering both the aspect,

Table 2 User priority for different key requirements (n = 17)

S. no.	Users' requirement parameters	Average priority score out of 7	Priority *w.r.t* other parameters in %
1	Lower cost	4.53	7.24
2	Lower wastage of raw material	5.29	8.45
3	Better portability	4.97	7.94
4	Higher productivity	5.29	8.45
5	Better peeling effectiveness (different sizes)	4.09	6.54
6	Uniform quality of output	4.09	6.54
7	Lesser repair and maintenance	5.47	8.75
8	Higher aesthetic appeal	1.79	2.86
9	Power-independent operation	4.47	7.15
10	Less space occupying	4.18	6.68
11	High safety of workers (accident/injuries)	6.12	9.78
12	Quicker adaptability of workers (familiarity to technology)	4.09	6.54
13	Better comfort of workers	3.97	6.35
14	Multipurpose utility	4.21	6.73

Table 3 Category-requirement relationship score (n = 11)

S. no.	Categories of food processing machine / Users' requirement criteria	Hand operated small tools	Hand powered machines	Semi-Automated Machines	Fully automated Machines
1	Lower cost	3.91	5.00	3.82	1.73
2	Lower wastage of raw material	5.55	4.73	4.64	4.36
3	Better Portability	6.45	5.27	4.00	2.00
4	Higher productivity	1.73	4.90	5.55	7.00
5	Better peeling effectiveness (different sizes)	5.82	4.45	3.82	3.18
6	Uniform quality of output	3.00	5.00	5.73	6.18
7	Lesser repair and maintenance	6.64	5.09	3.45	2.00
8	Higher aesthetic appeal	3.45	3.82	4.64	4.18
9	Power-independent operation	6.73	6.09	2.36	1.27
10	Less space occupying	6.64	5.55	3.82	1.64
11	High Safety of workers	2.64	4.27	5.00	5.91
12	Quicker adaptability of workers (familiarity to technology)	5.91	5.55	5.00	3.36
13	Better comfort of workers	2.45	4.45	5.64	6.64
14	Multipurpose utility	5.73	3.27	3.00	2.18

Hand powered solution was rated for greater association. For "Lesser wastage", "Hand tool category of solution" was rated for largest association (m = 5.55) as manual peeling was perceived to have greater flexibility in peeling different shapes and sizes without much wastage. Similarly for "Better portability", "Better peeling effectiveness", "Lesser repair and maintenance", "Better adaptability" and "Multipurpose use", "Hand tool category of solution" was rate was better than other categories. For "Better safety", "Fully automated solution category" was rated highest (m = 5.91) as for this category, human interference is least. For "power independent operation", both "Hand tool category of solution" and "Human powered solution category" rating were higher as for both these categories power is not required (Table 3).

3.3 Overall Scores Obtained by Different Solution Categories

The overall scores obtained by different categories of solution considering users' priorities for different key requirements using the QFD based were tabulated in Table 4. "Human powered solution category" got the highest score (482.62)

Table 4 Final score obtained for each solution category

Sr. no.	Category of fruit processing machine	Final scores
1	Hand powered solutions	485.62
2	Small hand tool like solutions	477.75
3	Semi- automated solutions	431.43
4	Fully automated solutions	371.71

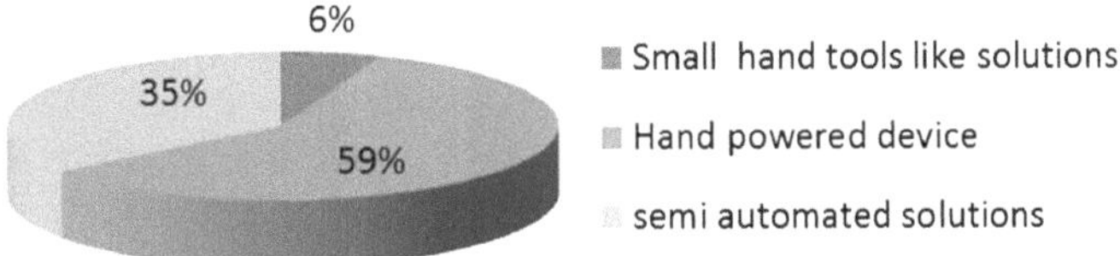

Fig. 5 First preference of the users (n = 17)

indicating that it is more suitable for the larger user population followed by "Hand tool like solution category" (477.75) followed by "Semi-automated solution category" (431.43).

3.4 Evaluation Results Regarding Outcomes of QFD Based Method

For preference of different categories of solution, "Hand powered solution category" was the first preference for 59 % of the respondents. This was followed by "Semi-automated solution category" (first preference of 35 % respondents), followed by "Small Hand tool like solution category" (first preference of 6 % respondents), whereas there were no first preferences for "Fully automated solution category" (Fig. 5).

The respondents were also asked to rate the suitability of different solution categories in their context of use. Mean values of the ratings were calculated for further consideration (Table 5). The rating was found highest for the "Hand powered solution category" (5.25) followed by "Semi-automated Solution category" (4.57) and "Fully automated solution category" (1.49).

Table 5 Mean preference scores for each category (n = 17)

Category	Mean
Small hand tools like solution	4.10
Hand powered solutions	5.25
Semi-automated solutions	4.57
Fully automated solutions	1.49

4 Discussion

The present study describes QFD based approach to decide upon the broader solution category for a given context, presently for pineapple peeling equipment design that is more suited to the larger user population, helping in coming up with a more appropriate design solution. The decision regarding selection of broader category of solution is generally intuitive and based on the various contextual factors. The present approach goes a step further and tries to consider different key requirements in a systematic way so as to not to skip any important factors and based on it, indicates the broader solution category that is most suitable for the larger user population. The study shows that categories identified through the method as most and least appropriate were, confirmed to be true in the user survey. But there were differences in the ranking order of "Hand tool like solution" and "Semi-automated solution". The methodical outcome predicted "Hand tool like solution" at 2nd position and the later at 3rd. Whereas, the user survey indicates the reverse order of these two categories.

From equipment's perspective, this difference in order may be due to the reason that the score, indicating suitability of the category, takes into account many requirements. Many of those with comparatively low priorities have exceptionally high positive association with a solution category pushing their score high, making it more suitable than other. Form users' preference perspective, it might be understood as the fact that users are fairly more critical about different factors while making the first preference. But, if given the option for second preference, they tend to opt for a more ideal solution even at the cost of skipping certain important factors. The above reasons, largely, explains why there was a difference in ranking as suggested by the methodology and as preferred by the users. The approach can be further improved upon for designing other food processing equipment and may prove helpful especially to the novice designer and give them a better perspective of looking at and analyzing the problems. The approach also advocates use of specific design methods in different design contexts.

References

1. O. University, Manufacture Materials Design Manufacturing for Engineers and Managers, p. 10 (2001)
2. Cross, N.: What Is design Strategy? In: Engineering Design Methods Strategies for Product Design, 3rd edn, p. 187 (2005)
3. Iino, K., Hatamura, Y.: Decision-Making in Engineering Design: Theory and Practice, p. 279. Springer, Berlin (2006)
4. Pahl, G., Beitz, W., Feldhusen, J., Grote, K.-H.: Engineering Design: A Systematic Approach, p. 638. Springer, Berlin (2007)

Role of Colour and Form in Product Choice and Variation of Preferences Across Product Categories: A Review

Swathi Matta Reddy, Anirban Chowdhury, Debkumar Charkrabarti and Sougata Karmakar

Abstract Among various physical attributes of a product colour and form capture maximum visual attention and thus play significant role in consumers' product purchase decisions. Since consumers buy different products for different use, they perceive product's physical attributes differently and their aesthetic appraisal differs depending on the product category. In present paper, following extensive literature review, it has been observed that there is rarely reported empirical evidences regarding how people make choice of colour and form for a particular product and how consumers' preference of colour and form varies from one product category to another. Empirical research involving advanced techniques like eye-tracking to understand consumers' visual behavior is urgently needed in this unexplored area to bring out quantitative data which can augment valid results. The upshot data produced regarding the preferences of colour and form among different categories would be useful to product designers and marketing managers to formulate design strategies.

Keywords Product design · Visual perception · Colour · Form · Market research

1 Introduction

There is a level of exhaustion reached by the technology in the present market scenario and every brand provides almost similar services in terms of quality, features, cost etc. leaving the consumers in a dilemma of which one to choose [1]. In such a scenario, companies are preparing themselves to manufacture/produce aesthetically appealing products to grab more buyers. Therefore, importance of the aesthetic facets of a product (e.g. shape, color, texture, etc.) as a determinant of consumer behavior is getting momentum day by day. In today's highly competitive

S.M. Reddy · A. Chowdhury · D. Charkrabarti · S. Karmakar (✉)
Department of Design, Indian Institute of Technology Guwahati, Guwahati, India
e-mail: karmakar.sougata@gmail.com

A. Chakrabarti (ed.), *ICoRD'15 – Research into Design Across Boundaries Volume 2*, Smart Innovation, Systems and Technologies 35, DOI 10.1007/978-81-322-2229-3_54

market, visual attractiveness of products serves as a marketing tool. Research says that among the widespread values, aesthetics has exceptional relevance to consumer behavior in three ways, e.g. aesthetics is naturally appreciated [2], has evolutionary benefits [3], and is relevant in a product context [4]. It is found that unlike the other collective values identified, such as social, political, theoretical, economic, and religious; aesthetics is the foremost one with a direct expression in consumer products [4].

People feel attraction towards different physical attributes of products in different ways. Visual appearance of products plays a large role in decisive consumer response [5]. Human perception is dominated by vision and the attractiveness of products is immensely rooted in visual perception [6]. If it is true that "seeing is believing" and "believing is buying", it is necessary to check on the consumers eye-on products [7]. The focus is on visual sense because it is the most influential and dominant one in perceiving things [8]. Visual attention is influenced by the physical properties of product; and brands having products with colour and form options are more competitive compared to other brands. Therefore it can be stated that the physical attributes of a product play important role while a consumer is choosing a product.

Interpreting consumer's product selection process is an area of interest for developing marketing strategy. In marketing, advertising and applied academic research, colour is considered to play a vital role because colour and product associations have an influence on consumer's intent to purchase [9]. Form is another important physical attribute. Form of a product relates to the consumers' psychological and behavioral responses; it creates a primary impression and generates inferences concerning other product attributes [10]. Colours and shapes convey about 80 % of all visual communication [11]. Therefore, in present paper, our focus is on these two attributes of the product.

Reviewing available literatures it is found that there is reported research on individual/single product category concerning colour, form and other product attributes but there are no empirical studies which indicate how consumers' preference towards colour and form varies across different product categories. It can be interesting to study colour and form preference of consumers across different product categories where the functionality product changes. The aim of the current paper is to present a thorough literature study with intention of understanding of colour and form preference of consumers across various product categories.

2 Physical Attributes of Products and Their Role in Product Choice

Malcolm Gladwell put forward that the subconscious mind takes a decision within a few seconds when a product is presented [12]. A baseline impression is created in mind of consumer by the appearance of product even before the consumer gets a chance to evaluate the product features, function etc. Consumer behavior towards product results in cognition and affect, which are followed by behavior [10]. Visual attention is captured by the physical attributes of the product: colour, form, texture, graphics, typography, detailing, [13, 14] product display environment (background, illumination). There are three key characteristics of product to illustrate the overall physical description and contours of objects—form, surface (including colour) and details (refers to grooves, buttons, grills, and other elements of the object) [15]. Unlike before, the market now provides a range of colours and styles and variations in other physical attributes in consumer products such as clothes, laptops, mobile phones, kitchen appliances like mixers etc. which influence the consumer buying behavior. Most of the interaction with objects is limited to a physical engagement with surfaces or the outer cover of the objects [15]. Visual appearance of the product triggers instant effect on consumers and is influenced by physical attributes of a product. It is found that appearance has an impact on product preference [16]. In their study, [17] showed that visual appearance of a product can control consumer product appraisals and choice making in several ways. The initial visual impact that a product makes on a consumer is a major factor when making purchasing decisions [18]. If various color, form or texture or the combinations of these and various other physical attributes of products are presented to the consumers at the market while they are making a purchase, they are bound to be attracted to specific ones of their personal choice and it is rare to let go off something a person has peculiarly chosen among many, thus starts creating an attachment. It is found that it is aesthetics that distinguishes products and makes the difference between the success or failure of a product [19]. Consumers would like to express themselves to others through purchasing products of a kind [20, 21] and such a communication is easiest through the physical attributes of the product or visual aesthetics [4]. Today, 70 % of the consumer choices are made at the shelf for buying, 85 % of the products are chose without picking up the competitive item and 90 % of the choices are made only by seeing the front face of the product [22]. Therefore, it can be stated that physical attributes of a product have a profound influence in choice making while purchasing a product.

3 Role of Colour in Product Selection

Decision to buy a product or not, can largely be influenced by its colour and it can dominate over other factors which lead towards product choice. Colour is one of the prominent aspects of visual aesthetics of a product. Colour has an effect on how a consumer perceives a product and can influence the purchase decision [23, 24]. Colour is an extremely significant product attribute because it segregates similar kind of products and it plays a major role in making buying decisions of consumers for different products. It is said that colour acts as a visual stimulus that attracts consumers to touch and feel the product and also to an extent convinces people to buy. Colour connects the viewers with objects more quickly than any other

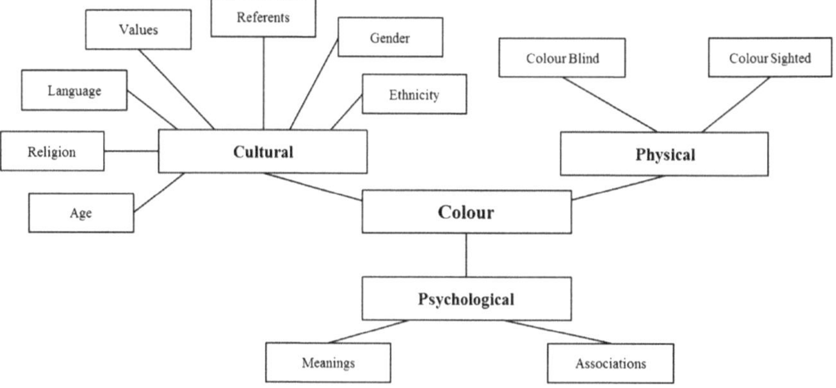

Fig. 1 Elements of colour perception. *Source* This is adapted from [30]

identifying characteristic [25]. The influence of colour is very strong on visual aspects, appreciation, responsiveness and identification of objects [26]. It is found that the colour of a product affects consumer product choice intensely [23] and can create an image about products that can be good or bad [27, 28]. Research says that colour is capable of affecting consumer's perception of products, pull attention to and generate purchase decisions [24, 29]. Human sub-consciousness establishes an impression instantly when an object is seen and colour is a huge part of that impression. The corporate marketing communications strategy has colour as important integral element and is considered an essential tool for shaping the consumer's responses [23]. Colour impacts consumer perception and behaviours and helps corporations' position or make a distinction from the competition [27, 30]. Research says that an average consumer's judgment to purchase a product is about 60 %, on its color [31]. Colour perception of colour consists of three elements: Cultural, psychological and physical as shown in Fig. 1 [30].

4 Role of Form in Product Selection

Consumers are attracted to product form that display its functionalism (form follows function), semantics (form follows meaning) [32] and pleasure [33]. Form of a product relates to behavioural responses and consumers psychology. The design or form of a product can contribute to its success in numerous ways. In clustered markets, form contributes to gain customer notice and is an important means of communicating information to customers. An initial impression is created by the product form and it generates inferences regarding other product attributes [10]. It is stated that "The reality of product design is that both product form and consumer response are determined by a vast array of factors and that these factors interact with one another in a complex and unpredictable way" [34]. Product form

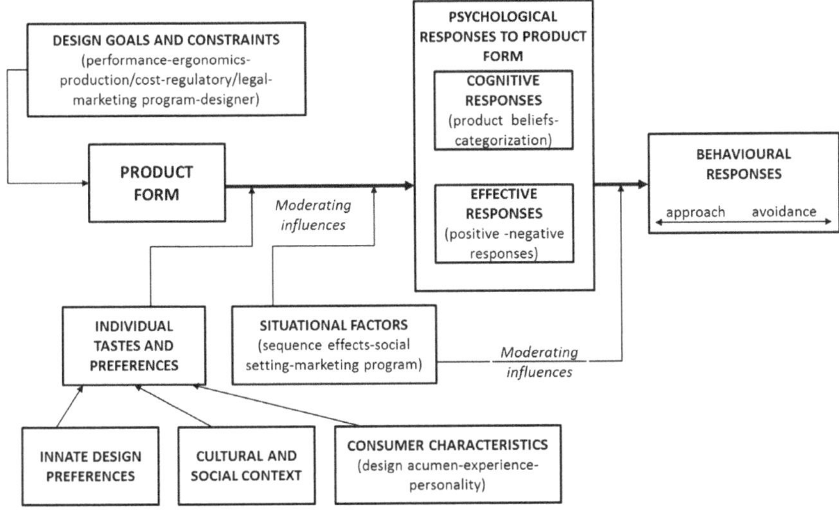

Fig. 2 A model of consumer responses to product form. *Source* This source is adapted from [35]

sometimes portrays a metaphor in order to give product experiential properties and they can be very dominating [33]. There are forms that have become iconic in many product areas. There are particular forms that people of certain cultures have got associations with particular products. Changing such icons can be either disadvantageous or advantageous depending on the attitudes of people at whom the product is intended. Figure 2 shows a systematic approach model of consumer responses to product form. It is drawn from number of disciplines including engineering, art psychology and ethology, and also relevant work from marketing and consumer behaviour [35].

5 Colour and Form Preference Across Different Products Categories

The first thing which human being registers during assessing anything is colour and it is always obvious to make associations with certain colours. Earlier, consumer durables were only white and were referred as 'white products' but now colour choice is a demand from consumers and colour has become the 'hot' issue among the manufacturers [33]. Many investigations in consumer psychology on colour have concentrated on specific dimensions of colour and its effect on consumer responses [36]. Bellizzi and Hite [37] found that red caused more arousal than blue and conversely, products when presented with blue coloured background were liked more than products presented with red backgrounds [37]. However, these studies did not reveal its application in any specific product category and also such finding cannot be generalized across all product categories. For example, consumers may prefer cheerful colours for mobile phones but their colour choice may change to neutral colours like black or white when selecting home appliances like washing machines, TV etc. This difference in colour preference could be due to specific use of the product and the environment in which it is being used.

Form of a product is accepted differently by different people depending on their mind-set. For example, Juicy Salif Lemon Juicer by Philippe Starke is an extreme design that attracts attention. While there are buyers who may enjoy the creativity and humor of the product, there must also be consumers who may feel unsure about buying such a product which is far from the conventional products. Swatch International launched a variety of uncommon product forms and could efficaciously stand out in the established market of wrist watches [38].

From literature, it is found that there are semantic differential studies concerning product form perception where researchers attempted to examine the relationship between the participants' evaluation of telephone samples and form design elements [39]. Chuang et al. [40] tested the relationship between consumer preference perception of mobile phones and elements of form design [40]. Three parallel studies were done to uncover fundamental dimensions of affective responses to product shapes where the structure of product shapes and affective responses of three product categories- automobile, sofa, and kettle were investigated [41]. During study of product form classification of mobile phones, Chang and Wu [42] aimed to extract the participant's mental structure in order to discover the principal factors of product form classification [42]. In a study to examine consumer preference for colour combinations in sports shoes, it was found that people like to combine colours which are comparatively close and also that people emphasize on hue and saturation and focus less on lightness [43]. Augustin et al. (2012) conducted studies on different product categories concerning physical appearance related to terminology for description and measurement of aesthetic impressions of products [44]. In another similar study, researchers worked on relationship between aesthetics and usability with washing machines and digital cameras (as stimuli), and they investigated the effects of novelty in product appearance by changing colour and form of products. Here, they used black colour for projecting the novel look in case of washing machines and the shape of digital camera was manipulated and deviated from typical shape to present in high novelty conditions [45]. A related study to check the effect of design newness on consumer response was performed where hairdryer was included with washing machines and digital cameras in the stimulus material in order to increase the generalizability of results [46].

From all the above mentioned studies involving various product categories, it is evident that consumers can have different responses to product colour and product form with variation of product categories.

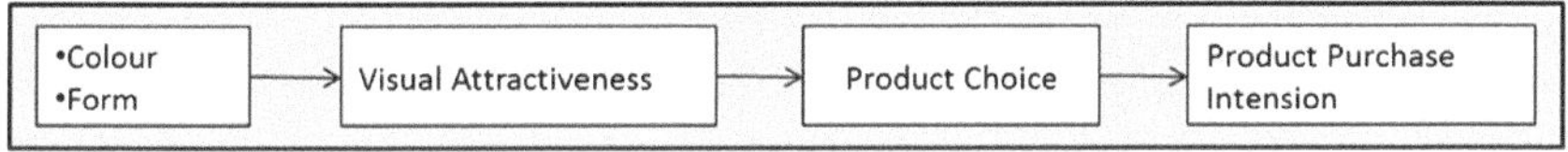

Fig. 3 Product selection process by consumers

6 Summary and Conclusion

Fenko et al. [47] found that vision is the most vital sense in the product-buying experience [47]. Therefore, understanding the visual interest of consumers is very important. Visual interest is influenced by the product's physical attributes or physical properties as discussed in the previous sections. The product selection process can be analyzed in a step by step process (Fig. 3).

Blijlevens et al. [48] described a two-step model of product appearance perception. They mentioned that product's physical properties like colour (hue, saturation, combinations), shape (round, rectangular), texture, surface, size, contrast, symmetry, orientation, weight etc. create a perception of modernity, simplicity, unity etc. [48]. Thus, it can be assumed that the physical properties elicit certain appearance attributes which are contextual depending on the category of the product. Since we buy different products for different use, we perceive them differently and our aesthetic appraisal differs depending on the product category. Therefore, preference of colour and form among different product categories may differ. It is found that there is rarely reported literature regarding how people make choice of colour and form while planning to purchase product. Information is also lacking concerning how consumers' preference of colour and form varies from one product category to another.

Preference towards products' physical attributes can be studied and analyzed using subjective evaluation through psychological questionnaires as well as objective measurement using various advanced techniques (e.g. eye-tracking, electro-encephalography, nuclear magnetic resonance imaging etc.). In questionnaires it is important to construct a suitable measurement scale aiming at the preference study which is associated to visual attractiveness, likeliness for colour and form, product purchase intensions, etc. Researchers have often used seven-point rating scales for similar studies [45, 49 and 50].

Studying visual behavior of consumers during product selection process is very essential in marketing research as colours and shapes convey about 80 % of all visual information [11] as stated earlier in 'Introduction' section. Thus, among various objective evaluations for product choice, now-a-days 'eye tracking' techniques are being widely used by marketing researchers [51]. This technique is used since it is capable of capturing visual attention and visual scanning behavior of consumers on products by recording eye-movements [49, 52–56]. Eye tracking variables like fixation duration and fixation counts are commonly used metrics for evaluation of area of interest (AOI) and perception of viewers. Fixation duration is the amount of time spent in viewing a particular AOI and is shown in milliseconds and Fixation count is the number of considerable eye fixations a viewer shows in a defined AOI. The inference from these metrics is that, the fixation duration and fixation time of an AOI is more if the viewer is interested in a particular area and vice versa. Heat maps and scan paths represent the AOI in a pictorial way. Scan paths show the order of fixations from the time of start till the end and is represented by connecting lines and circles, also the diameter of the circle increases with increase in duration time. Non-intrusive methods for eye detection and tracking like eye-tracking are useful for many applications of vision-based evaluations [57]. Authors of the present paper feel that eye tracking based studies to observe preference of colour and form across different product categories can bring out quantitative data that can augment valid results.

Understanding the perceptions and preferences of consumers has become increasingly important for the design studios, design schools, marketers and there is a need to conduct studies on the preference of colour and form by consumers during their purchase. The preference of colour and form among different product categories is an area unexplored and there is research scope in this direction. The upshot data produced regarding the preferences of colour and form among different categories would be useful to product designers and marketing managers to formulate design strategies. A reliable marketing database on consumer preferences can be created that can serve as a ready manual providing the design methodology to conduct similar colour and form studies.

References

1. Veryzer, R.W.: The place of product design and aesthetics in consumer research. In: Kardes, F.R., Provo, M.S. (eds.) NA—Advances in Consumer Research, vol. 22, pp. 641–645. Association for Consumer Research, UT (1995)
2. Langlois, J.H., Ritter, J.M., Roggman, L.A., Vaughn, L.S.: Facial diversity and infant preferences for attractive faces. Dev. Psychol. **27**(1), 79–84 (1991)
3. Dutton, D.: Aesthetics and evolutionary psychology. In: Levinson, J. (ed.) The Oxford Handbook of Aesthetics. Oxford University Press, New York (2003)
4. Townsend, C., Sood, S.: Self-affirmation through the choice of highly aesthetic products. J. Consum. Res. **39**(2), 415–428 (2012)
5. Crilly, N., Moultrie, J., Clarkson, P.J.: Seeing things: consumer response to the visual domain in product design. Des. Stud. **25**(6), 547–577 (2004)
6. Baxter, M.: Product Design: Practical methods for the systematic development of new products, p. 58. Chapman & Hall, London (1995)
7. Wedel, M., Pieters, R.: Eye tracking for visual marketing. Found. Trends. Mark. **1**(4), 231–320 (2006)
8. Bundesen, C., Habekost, T.: Principles of Visual Attention: Linking Mind and Brain. Oxford University Press, Oxford (2008)
9. Vantturley, V.L.: The influence of packaging colour on consumer purchase intent: The influence of colour at the point of purchase, p. 36. ProQuest, Michigan State University, USA (2007)
10. Bloch, P.H.: Seeing the ideal form: Product design and consumer response. J. Mark. **59**(3), 16–29 (1995)

11. LaCroix, J.: In the blink of an eye: the three factors that determine whether your packaging will do its sales job. Mark. 103(24), pp. 29–37 (1998)
12. Gladwell, M.: Blink: The Power of Thinking Without Thinking. Penguin Publishers, UK (2005)
13. Hannah, G.G.: Elements of Design. Princeton Architectural Press, New York (2002)
14. Sausmarez, M.D.: Basic Design: The Dynamics of Visual Form. Herbert, London (1983)
15. Boradkar, P.: Designing Things: A Critical Introduction to the Culture of Objects, p. 135. Berg, USA (2010)
16. Yamamoto, M., Lambert, D.R.: The impact of product aesthetics on the evaluation of industrial products. J. Prod. Innov. Manage **11**(4), 309–324 (1994)
17. Creusen, M.E.H., Schoormans, J.P.L.: The different roles of product appearance in consumer choice. J. Prod. Innov. Manage **22**, 63–81 (2005)
18. Bruseberg, A., McDonagh, D.: Product handling and visual evaluation supporting new product development. In: McCabe, P.T. (ed.) Contemporary Ergonomics 2002, pp. 303–308. Taylor & Francis, London (2002). ISBN 0-415-27734-5
19. Duncum, P.: Aesthetics, popular visual culture, and designer capitalism. Int. J. Art. Des. Edu. **26**(3), 285–295 (2007)
20. Belk, R.W.: Possessions and the extended self. J. Consum. Res. **15**, 139–167 (1988)
21. Kleine, R.E., Kleine, S.S., Kernan, J.B.: Mundane consumption and the self: a social identity perspective. J. Consum. Psychol. **2**(3), 209–235 (1993)
22. Clement, J.: Visual influence on in-store buying decisions: an eye track experiment on the visual influence of packaging design. J. Mark. Manag. **23**(9), 917–928 (2007)
23. Clarke, I., Honeycutt, E.D.: Color usage in international business-to business print advertising. Ind. Mark. Manage. **29**, 255–261 (2000)
24. Sable, P., Akcay, O.: Color: cross cultural marketing perspectives as to what governs our response to it. Proc. ASBBS **17**(1), 950–954 (2000)
25. Akcay, O., Dalgin, H., Bhatnagar, S.: Perception of color in product choice among college students: a cross-national analysis of USA, India, China and Turkey. Int. J. Bus. Soc. Sci. **2** (21), 42–48 (2011)
26. Wichmann, F.A., Sharpe, L.T., Gegenfurtner, K.R.: Learning, memory and cognition. J. Exp. Psychol. **23**(3), 509–520 (2002)
27. Grossman, R.P., Wisenblit, J.Z.: What we know about consumers' color choices. J. Mark. Pract. **5**(3), 78–90 (1999)
28. Singh, S.: Impact of color on marketing. Manag. Decis. **44**(6), 783–789 (2006)
29. Kerfoot, S., Davies, B., Ward, P.: Visual merchandising and the creation of discernible retail brand. Int. J. Retail. Distrib. Manag. **31**(3), 143–152 (2003)
30. Aslam, M.: Are you selling the right colour? a cross-cultural review of colour as a marketing cue. J. Mark. Commun. **12**(1), 15–30 (2006)
31. Wagner, C.: The Wagner Color Response Report, pp 29–95. The Wagner Institute for Color Research (Revised Edition) (1988)
32. Krippendorff, K.: On the essential contexts of artifacts or on the proposition that design is making sense (of things). In: Margolin, V., Buchanan, R. (eds.) The Idea of Design, pp. 156–184. MIT Press, Cambridge (1995). (Second Printing)
33. Jordan, P.W.: Designing Pleasurable Products: An Introduction to the New Human Factors. CRC Press, Boca Raton (2000)
34. Crilly, N., Moultrie, J., Clarkson, P.J.: Shaping things: intended consumer response and the other determinants of product form. Des. Stud. **30**(3), 224–254 (2009)
35. Bitner, M.J.: Servicescapes: The impact of physical surroundings on customers and employees. J. Mark. **47**, 69–81 (1992)
36. Gorn, G., Chattopadhyay, A., Yi, T., Dahl, D.: Effects of color as an executional cue in advertising: They're in the shade. Manage. Sci. **43**(10), 1387–1400 (1997)
37. Bellizzi, J., Hite, R.E.: Environmental color, consumer feelings, and purchase likelihood. Psychol. Mark. **9**(5), 347–363 (1992)
38. Hollins, B., Pugh, S.: Successful Product Design. Butterworths, London (1990)

39. Hsu, S.H., Chuang, M.C., Chang, C.C.: A semantic differential study of designers' and users' product form perception. Int. J. Ind. Ergon. **25**, 375–391 (2000)
40. Chuang, M.C., Chang, C.C., Hsu, S.H.: Perceptual factors underlying user preferences toward product form of mobile phones. Int. J. Ind. Ergon. **27**, 247–258 (2001)
41. Hsiao, K.A., Chen, L.L.: Fundamental dimensions of affective responses to product shapes. Int. J. Ind. Ergon. **36**, 553–564 (2006)
42. Chang, C.C., Wu, J.C.: The underlying factors dominating categorical perception of product form of mobile phones. Int. J. Ind. Ergon. **39**, 667–680 (2009)
43. Deng, X., Hui, S.K., Hutchinson, J.W.: Consumer preferences for color combinations: An empirical analysis of similarity-based color relationships. J. Consum. Psychol. **20**, 476–484 (2010)
44. Augustin, M.D., Wagemans, J., Carbon, C.C.: All is beautiful? generality vs. specificity of word usage in visual aesthetics. Acta Psychol. **139**(1), 187–201 (2012)
45. Mugge, R., Schoormans, J.P.L.: Product design and apparent usability. The influence of novelty in product appearance. Appl. Ergonomics **43**(6), 1081–1088 (2012)
46. Mugge, R., Dahl, D.W.: Seeking the ideal level of design newness: consumer response to radical and incremental product design, Product Development & Management Association. J. Prod. Innov. Manag. **30**(S1), 34–47 (2013)
47. Fenko, A., Schifferstein, H.N.J., Hekkert, P.: Shifts in sensory dominance between various stages of user–product interactions. Appl. Ergonomics **41**, 34–40 (2010)
48. Blijlevens, J., Creusen, M., Schoormans, J.: How consumers perceive product appearance: the identification of three product appearance attributes. Int. J. Des. **3**(3), 27–35 (2009)
49. Carbon, C.C., Talker, C.: Evaluating design perception of 36 car interiors. Unpublished Research Report. Faculty of Psychology, University of Vienna (2006)
50. Carbon, C.C.: The cycle of preference: Long-term dynamics of aesthetic appreciation. Acta. Psychol. **134**(2), 233–244 (2010)
51. Duchowski, A.T.: Eye Tracking Methodology: Theory and Practice. Springer, Berlin (2007)
52. Kukkonen, S.: Exploring eye tracking in design evaluation, joining forces. University of Art and Design, Helsinki (2005)
53. Marshall, S.P.: What the eye reveals: measuring the cognitive workload of teams. In: Duffy, V. G. (ed.) Digital Human Modeling, HCII 2009, pp. 265–274. Springer, Berlin (2009)
54. Zambarbieri, D., Carniglia, E., Robino, C.: Eye tracking analysis in reading online newspapers. J. Eye Mov. Res. **2**(4), 1–8 (2008)
55. Tzanidou, E.: Eye tracking as a complimentary usability evaluation technique for E-commerce sites, Department of Computing. The Open University, Milton Keynes (2006)
56. Tonkin, C., Ouzts, A.D., Duchowski, A.D.: Eye tracking within the packaging design workflow: Interaction with physical and virtual shelves. J. Comput. Appli. Soc. behav. Sci. (2011)
57. Gulliver, S.R., Ghinea, G.: "Region of interest displays: addressing a perceptual problem?" In: IEEE 6th International Symposium on Multimedia Software Engineering (ISMSE'04), [ISBN: 0-7695-2217-3/04]. Department of Information Systems and Computing, Brunel University, Uxbridge, pp. 2–9, 13–15 Dec (2004)

What Do You See? Research on Visual Communication Design to Promote Positive Change for Unorganized Workers in Karnataka, India

Sabina von Kessel

Abstract The principal objective of this paper is to indicate the role of inclusive visual communication design in development. This case study shows the research on semiotics with Unorganized Workers in a rural context in Karnataka. It demonstrates how images have to be specific to its cultural context to create awareness and stimulate positive change among mostly illiterate participants in a rural context. It demonstrates the learnings: a strategic approach of communication and the specific design and use of visual media as a tool to create awareness and stimulate positive change. These examples are relevant for social designers and people who deal with complex social problems and communication4development in a joint project involving several stakeholders and government agencies.

Keywords Visual communication design · Visual communication4development · Behaviour change communication · Unorganized workers · Social security schemes

1 Social Security and the Information Gap

An improved social security system is crucial to alleviating poverty and to advancing inclusive economic growth. Many social security programmes in India, such as health insurance and old-age pensions, are limited to workers in the organised sector. Yet more than 94 % of the country's workforce is engaged in the unorganised sector, in small-scale farming, and in small businesses. Breman [1] states, that "the informal economy consists of all activity generating work and employment that is not registered and administered by public regulation" with no protection against arbitrary dismissal, accidents and other risks. Thus unorganized workers are extremely vulnerable.

S. von Kessel (✉)
Transcultural Design Consultancy International, Bangalore, India
e-mail: Sabinavkessel@gmail.com

A. Chakrabarti (ed.), *ICoRD'15 – Research into Design Across Boundaries Volume 2*, Smart Innovation, Systems and Technologies 35, DOI 10.1007/978-81-322-2229-3_55

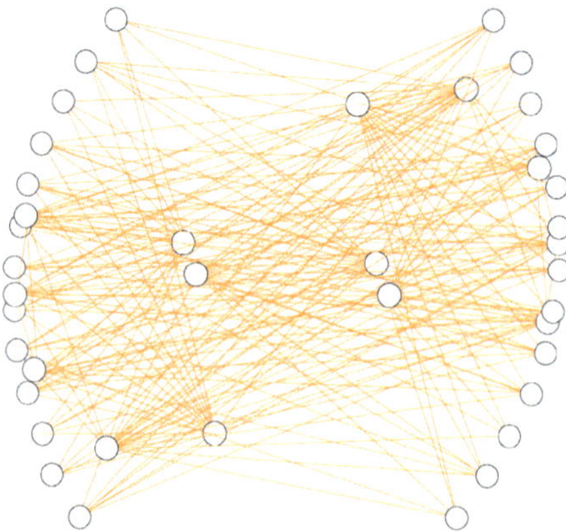

Fig. 1 A multiplicity of schemes both at the central and state levels aiming at social protection of the underprivileged have been framed at random at various points of time, which makes it in general difficult to understand who can benefit from which scheme. Design: Sabina von Kessel

The Indian government has set up at various points of time a multiplicity of schemes which aim to ensure that the vulnerable sections of the population have access to basic entitlements, such as food and nutrition, housing and health, education and employment, etc.

The problem is that there is no overall conformity and design of those schemes and a lack of consistent policy and the coverage of unorganized workers under the schemes has been minuscule. Unorganized workers have generally no information about their entitlements and therefore no access to pensions, healthcare or other forms of social security. For Government stakeholders it is also difficult to understand who can benefit from which scheme (Fig. 1).

In Karnataka, information about and access to these schemes has been made easier by the Cooperation between the Department of Labour in Karnataka, in collaboration with Deutsche Gesellschaft für Internationale Zusammenarbeit (GIZ).

Since 2008 unorganized workers have access to efficient and improved social security services via newly established worker facilitation centres (WFCs) in five districts in Karnataka. Community facilitators (CF) have surveyed and identified beneficiaries and have offered easier access to social security schemes through a Single-Window-Service (SWS) in a pilot project, which has been taken over by the Department of Labour, Government of Karnataka in recent month.

In spite of the efforts made by the GIZ, there is still a lack of information and awareness about Social Security Schemes within the society in general and the Unorganized Workers and Government stakeholders in particular.

None of the Unorganized Workers, we interviewed at random in different districts, had heard of any social security schemes or knew that they were entitled to benefit from various schemes. On the other hand, all of the beneficiaries who had obtained their “cards” (benefits), got these only through Interpersonal Communication with the Community Facilitator in the Single-Window-Service Centres, established by the GIZ. There were no other forms of communication, which reached the participants other than interpersonal communication. Visual communication was either missing, did not reach the participants or simply did not use a visual language which communicated successfully with participants, who in their majority are illiterate.

1.1 Every Picture Tells a Story

This chapter reviews the key questions of visual communication design and visual semiotics in relation to the participants and the cultural context.

Using visual communication to stimulate positive change and promote development goals has a long tradition. UNICEF states that C4D, "Communication for development goes beyond providing information. C4D involves understanding people, their beliefs and values, the social and cultural norms that shape their lives. It involves engaging communities and listening to adults and children as they identify problems, propose solutions and act upon them. Communication for development is seen as a two-way process for sharing ideas and knowledge using a range of communication tools and approaches that empower individuals and communities to take actions to improve their lives" [2].

Though the statistics clearly show that illiteracy is a significant global and local issue and even today among ten in India about four are illiterate, very little has been done to tackle the challenge of communicating to illiterate people in general and, more specifically to an illiterate rural population, to which Unorganized Workers in their majority belong. The repercussions of an inability to communicate to and with them are demonstrated here (Fig. 2).

There has been very little research on and how participants in rural environments in India read and understand visuals, which are mostly created by design offices in cities for participants but not in communication with them. Visual images, like all

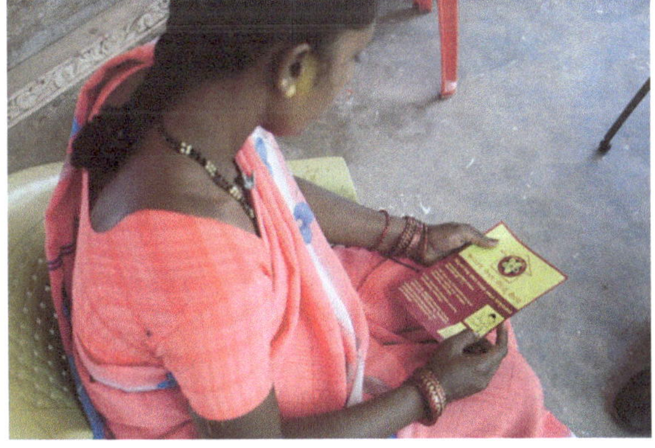

Fig. 2 This illiterate unorganized worker in Hoskote (rural Bangalore) is not able to understand the brochure, provided by GIZ/ DoI. Photo Sabina von Kessel

Fig. 3 Painted agricultural trailers from Rajasthan (*left*) and Karnataka (*right*) show a significant different visual language. Photos Sabina von Kessel

representations, are never innocent or neutral reflections of reality…"they represent for us: that is, they offer not a mirror of the world but an interpretation of it" [3].

Our interpretation of symbols and graphic language depends greatly on our cultural language. There is very little uniformity between different cultures. Evamy [4] states that "it is a great misconception to assume that graphic information becomes magically accessible to everyone, everywhere when you remove words" (Fig. 3).

Visual communication in India is unique in the fact that it reflects its cultural context and social patterns extremely accurately. A vernacular visual language in signs, graphic, photographic and audio visual information depicts wealth, caste, class, religion, ethnicity on billboards, wall paintings and signs in public places, TV shows and film posters.

These images are successful because they reflect existing cultural norms and attitudes prevalent in Indian society.

As Art historian Keith Moxey argues: "Semiotics makes us aware that the cultural values with which we make sense of the world are a tissue of conventions that have been handed down from generation to generation by the members of the culture of which we are a part. It reminds us that there is nothing 'natural' about our values; they are social constructs that not only vary enormously in the course of time but differ radically from culture to culture. It reminds us that there is nothing 'natural' about our values: they are social constructs that not only vary enormously in the course of time but differ radically from culture to culture" [5].

1.2 Visual Literacy and the Process of Sensemaking

This chapter looks at the 'visual' age and the meaning of 'visual literacy'.

Humans have always created images, in fact for a much longer period of time than they were able to read and write. We live in a "visual" age, in an increasingly visual world, in which we are constantly looking at screens of different sizes with superior resolutions at handheld devices. We also live in a world with more data available to us than ever before. IBM says that 90 % of the world's data was produced only in the last 2 years and that we produce 2.5 quintillion bytes of data daily.

After centuries of the predominance of the alphabet, the contemporary culture relies not so much on thinking and communication, but on image modes. Nowadays the ability to read and write is dominated by the ability to decode images as sources of information.

In a growing world of visualizations, our brains understand images faster than they do text.

The use of images as sources of information and learning demand new abilities in the global world. These new abilities, coined Visual Literacy have transformed our ways to communicate and to decode communication. Children with access to digital technologies learn an image-based use of language and become visually literate at a very early age.

"Youths who spend a third of their waking hours in front of a screen are saturated with images. The ubiquity of images in young people's lives has transformed the way they learn and perceive the world" [6].

The majority of people in India has not yet got access to the internet and there is a huge digital divide between the "flat world, in which you can plug and play, collaborate and connect, if you are from Bangalore or Brooklyn, your life chances and opportunities hold more potential than ever before" [7] and the realities of Indian people (Fig. 4).

In India, internet penetration in the country is a mere 10.2 % (December 2011) [8], but lately being fueled by the ever growing popularity of smart phones.

In 2014 India will be holding the largest proportion of smartphones sold to new users. A smartphone explosion is expected with China (283m) followed by India (225m).

Fig. 4 A painter from Hoskote and a construction worker from Mangalore show their simple cellphones. The majority of unorganized workers in Karnataka use cellphones only to make phone calls. They hardly use other functions, like SMS. Photos, collage: Sabina von Kessel

Though India is becoming the second-largest country for smartphone use in the world, it seems that the majority of the population in India is until today not able to decode the abstract global sign system.

1.3 The Research: What Do You See?

> What is the use of a book' thought Alice, 'without pictures or conversations.' Lewis Carroll, Alice in Wonderland [9]

Planning and evaluation research should accompany every planning and design process for a communication campaign, since communication is "what arrives and causes an effect, not what you send out" [10].

The research undertaken by the author, with the help of a local interpreter, was qualitative. The primary research was conducted through interviews, with the help of a local interpreter, with Unorganized Workers and Self-Help-Groups (SHGs) as well as members of the Panchayat Rajs (Village Councils) and Community Facilitators in three different districts of Karnataka. The research was documented with audiovisual media.

We conducted in total 90 interviews in three different districts, Hoskote, Mangalore and Gulbarga with 61 Unorganized Workers, 15 Community Facilitators and 14 members of the Village Councils (Panchayat Raj). Questionnaires, as well as cards, on which pictograms or photos were printed, were shown to participants in isolation from the print media, brochures and books, they had originally been used in.

Parallel to the testing of the pictograms and photos, fresh pictograms, illustrations and a new logo were developed and tested (Fig. 5).

The purpose of the research in the field was to examine the visual perception (semiotics) of our stakeholders and their use of media and technology, in detail

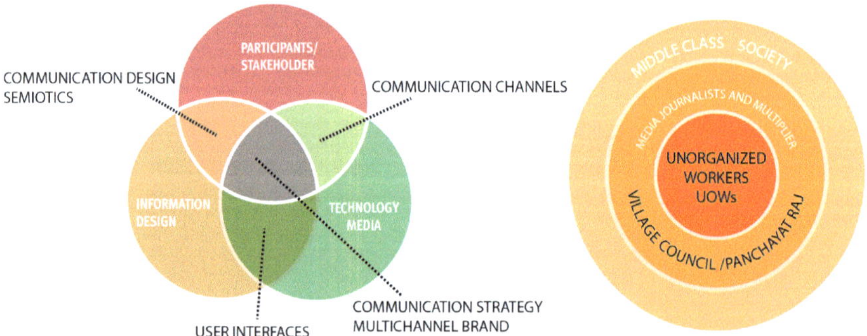

Fig. 5 Field research was conducted to understand and examine the perception of stakeholders (target groups), their use of communication channels and technology. The aim was to implement the learnings into a communication strategy and multichannel brand. Design: Sabina von Kessel

1. to examine and test the print material used so far
   - to study the visual language and cultural perceptions of the participants
2. to investigate the use of communication channels:
   - cellphones (types of cellphones and their use)
   - mass media/entertainment (which TV channels and which soaps are watched)
   - actors or celebrity spokespersons (known and liked)
3. to find out which consumer products/brands are commonly bought (to identify a brand for a public-private partnership)

In addition to the field research secondary research took the form of topic searches in the internet and academic texts, leading into commercial marketing, neuro-marketing and the emotional aspects of branding. The research was extended into communication campaigns and behaviour change theory and some well written papers from UNICEF led to learnings for this project.

1.3.1 Field Research and the Findings of the Survey

The examination of print material and images shows that unspecifically designed visuals do not communicate with/to participants (Fig. 6).

79 % of our interviewees own an average of 2 cellphones with basic technology per family, from those only 15 % use it to send, receive sms. 19 % use the cellphone to listen to radiostations. None of the participants had internet access.

80 % watch TV and from those 82 % prefer the channel Udaya, 39 % (mostly women) watch and like the Kannada TV Soap Bangara, 55 % like Shiv Raj Kumar and Puneet Raj Kumar, 51 % Upendra. 28 % watch Kannada movies in film theatres.

1.4 Learnings Overall

The research shows, that Unorganized Workers and participants in Panchayat Rajs (Village Councils) struggle to interpret icons, pictograms, illustrations, graphs and info graphics and other representations which are used globally. Therefore visuals used in communication campaigns which intend to persuade and change behaviour have to use a vernacular visual language or adaptations of the same. They have to be specific towards detail, a tool, the crop growing on the field or the footwear of a worker has to look right.

The research shows that colourful, flat depictions which do not use perspective and are large enough, work best. Many UOWs cannot afford to wear glasses, and cannot read or see smaller letters or illustrations. Printed material cannot be the only

Fig. 6 (*Clockwise*) Pictograms, photos and illustrations were shown in interviews with unorganized workers and members of Panchayat Rajs (village councils) in three different districts in Karnataka to evaluate if participants can derive meaning from the print material which was disseminated by community facilitators. Interviewer Johny Jaganath, Photos: Sabina von Kessel. (*Left*) This leaflet was disseminated in thousands into the households of unorganized workers. It used inconsistently designed cliparts, randomly taken from the internet to communicate social security schemes and single window service to rural, illiterate stakeholders. The logo (*on top*) is too complex, uses unspecific cliparts and does not fulfill technical requirement, eg. cannot be recognized in small scale. Many people in rural areas simply do not wear glasses and cannot read or see smaller letters or illustrations. Design: GIZ, DOL Karnataka

and first choice, since it is often unpractical for people who do not have bookshelfs and is often used best to light the cooking fire in the evening.

Print media in form of brochures and posters can only be one, out of a multiplicities of media, which tries to persuade or influence behaviour.

Communication4Development needs to be strategic and use multi channels to reach the participants. *Only strategic communication works to influence behaviour. One single media does not move anybody.*

1.5 Put into Practise

The illustrations we developed during the research process went through a cycle of testing and redesigning multiple times before they "worked". The participants were very specific towards details. We learnt that details like the right shape of tools or the look of the crops growing on the fields in the background of an image matter a lot to the understanding of the whole image. We have adapted the P-process [11] for our work. All of the developed material is reevaluated and if necessary redesigned and tested again, parallel to the processes of participation and capacity strengthening.

1.6 Consequences: Developing New Design Strategies

The learnings resulted in designing an integrated communication and PR plan, creating a new visual identity and brand, as well as producing new print media. TV spots and radio jingles as part of a brand campaign for Single Window Service will be broadcasted in local stations (Figs. 7, 8, 9 and 10).

Fig. 7 The idea for the new visual identity of the single-window-service stems from the sacred geometry of rangolis, which show a great mode of abstract recognition and memory. Design: Sabina von Kessel

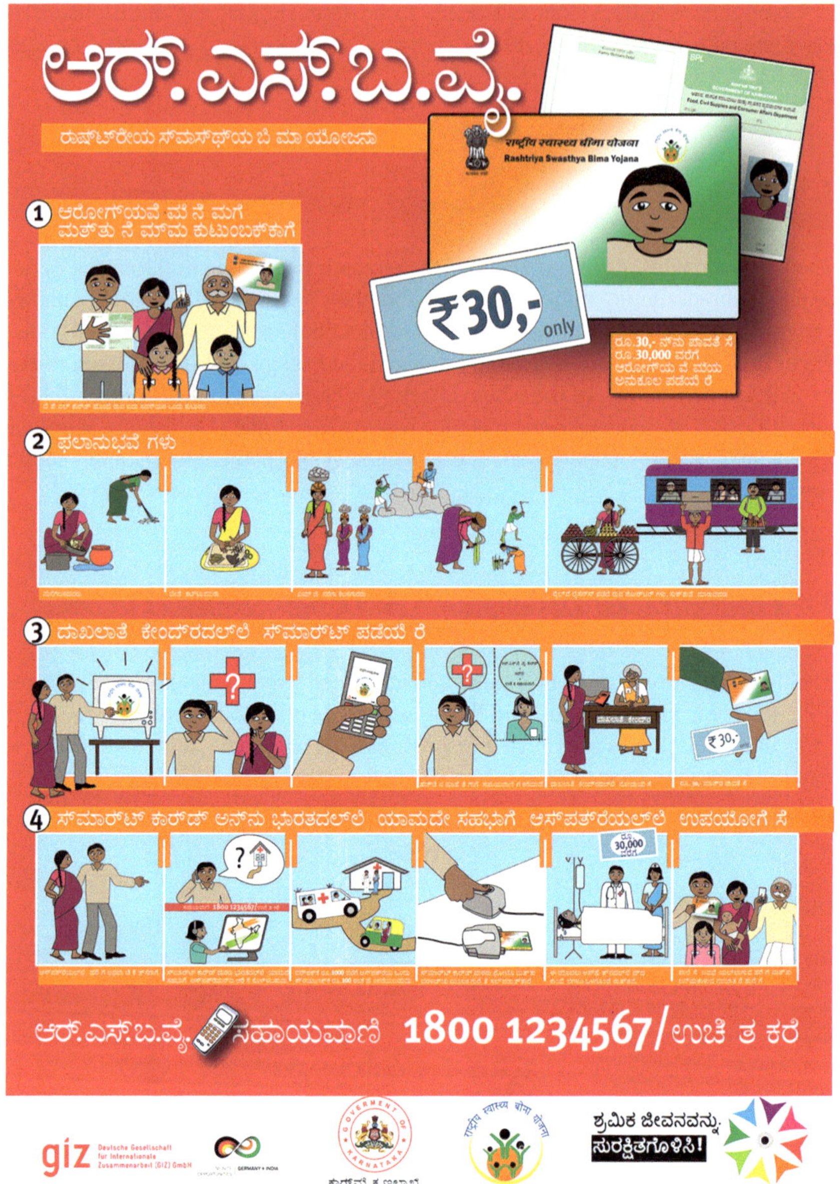

Fig. 8 Poster (A1) to promote RSBY, the national health insurance scheme for unorganized workers with illustrations which translate not only insurance benefits into a visual language, but also who the beneficiaries are, how they get access and how much they have to pay. Illustrations Juliane Denogent, design Sabina von Kessel

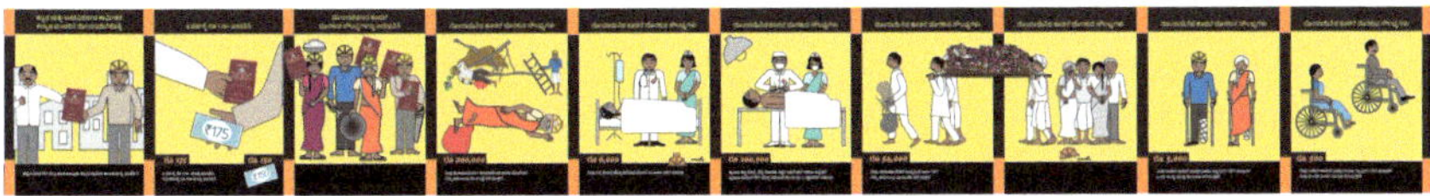

Fig. 9 The brochure for construction worker welfare board (*CWWB*), has, when folded, the size of a shirt pocket with illustrations which translate accessability and insurance benefits into a visual language. Illustrations Juliane Denogent, design Sabina von Kessel

Fig. 10 Scenes from the production of TV spots as part of the branding campaign to promote single window service for unorganized workers. Filmproduction Firiri films, photos Sabina von Kessel

References

1. Breman J (2013) At work in the informal economy of India: a perspective from the bottom up. Oxford University Press, Oxford 19 Mar 2013
2. www.unicef.org
3. Midalia S (1999) "Textualising gender", interpretations, 32 (1) quoted in Hurrell G 2001, "masculinities in the english classroom: fracturing stereotypes". English in Australia, No 131
4. Evamy M (2003) World without words, vol 24. Crown Publishing Group, New York
5. Moxey KPF (1994) The practice of theory: poststructuralism, cultural politics, and art history, vol 61. Cornell University Press, Ithaca
6. Learning in a visual Age. Paper by National Art Education Association, 1916 association drive, Reston VA, 20191. www.arteducators.org
7. Friedman TL (2005) The world is flat, a brief history of the twenty-first century, vol 63. Farrar, Straus & Giroux, Hardcover, New York
8. http://www.internetworldstats.com
9. http://www.cs.indiana.edu/metastuff/wonder/ch1.html
10. Macnamara J (2005) Jim Macnamara's public relations handbook, 5th ed. Archipelago Press, Sydney (Chapter 18, pp 243–312)
11. The P-process is based on work at Johns Hopkins Bloomberg School of Public Health

A Study on DSM Partitioning Through Case Study Approach

Purva Mujumdar, Soma Bhattacharya and J. Uma Maheswari

Abstract Planning for design process is often ignored and given least importance. Design phase of any project is associated with continuous and abundant information exchanges across the different teams. If these evolving design information exchanges are not perceived and planned during the early stages of a project, it results in adhoc information exchange, mistakes or rework. Over the past decade, researchers had identified DSM (Design Structure Matrix) as a potential tool to manage the information interdependencies and to evaluate the sequence of execution. DSM is a square matrix with the same elements (elements signify teams, components, activities or parameters) in the rows and columns and the off-diagonal cells represent the dependency relationship between those elements. The basic DSM operations such as partitioning and tearing enable the user to evaluate the feasible execution sequence. In reality, generating a basic DSM is tedious and time-consuming. Hence, an attempt was made in the present study to analyze if there exist any correlation among the various types of DSMs (team DSM to parameter DSM or activity DSM to team DSM, etc.) formed. To achieve this objective, data was collected for an offshore project through interactions and discussions with experts of the eight teams for several weeks. For the current study, four teams—mechanical, HVAC, piping and structure which interact frequently was considered. To ease the analysis, deliverables DSM and team DSM was only considered. Through partitioning process, the sequences of deliverables and teams were determined and other elements such as parameters or activities were ignored. It was observed that there exists no relation between the different DSMs formed.

Keywords Sequencing · Partitioning · DSM

P. Mujumdar (✉) · J. Uma Maheswari
Department of Civil Engineering, Indian Institute of Technology Delhi, New Delhi, India
e-mail: purvamujumdar@gmail.com

S. Bhattacharya
Samsung Heavy Industries, Noida, Uttar Pradesh, India

A. Chakrabarti (ed.), *ICoRD'15 – Research into Design Across Boundaries Volume 2*,
Smart Innovation, Systems and Technologies 35, DOI 10.1007/978-81-322-2229-3_56

1 Introduction

Engineering design is an ill-structured process of frequent information exchanges among various design disciplines. Further, design projects are executed by the design groups from varied geographical locations. Every design group carries out their design independently without having a holistic perspective of the entire project.

Typically, design activities share three types of information dependencies among them—dependent, independent and interdependent i.e. loops/cycles [1, 2]. Traditional network techniques such as Critical Path Method (CPM) and Program Evaluation and Review Technique (PERT) are not considered appropriate due to their inability to model activity interdependencies and feedback loops. Design Structure Matrix (DSM) is a suitable technique to model information interdependencies and to evaluate correct sequence of design activities [3–5]. Few applications of DSM have been attempted in construction industry [6–12]. Research attempt in mapping the relationship between the various DSM (such as parameters to teams or teams to activities, and so on) sequences were not investigated so far. This implies several sequences can be generated for the DSM types (teams, parameters or activities) and the relationship between team DSM and the parameter DSM or the activity DSM needs to be investigated.

Exact sequencing of design deliverables and the corresponding teams responsible for those deliverables is critical for timely completion of a project. In reality, collecting data for generating the DSMs with regard to teams, components, parameters, etc. from every design expert through discussions and experts is cumbersome. Hence, the objective of the present study is to investigate the existence of any relationship between the various DSMs (team DSM or parameter DSM or activity DSM) formed. By deriving the relationship, several abstract-level DSMs could be generated automatically when the primitive or lower-level DSM has been formed.

To achieve this objective, data was collected for an offshore project through interactions and discussions with experts of the eight teams for several weeks. For ease of analysis, four teams that interact more frequently (mechanical, HVAC, piping, structural) and two elements which was easy to identify and list (teams, deliverables) were considered in the present study. Several DSMs were formulated and partitioned for determining the sequence. Partitioning process is the process of identifying the feasible sequence of execution along with grouping cycles/loops which is termed as block formation. The data analysis revealed that it was difficult to map the above-mentioned relationship between the teams and deliverables DSMs formed.

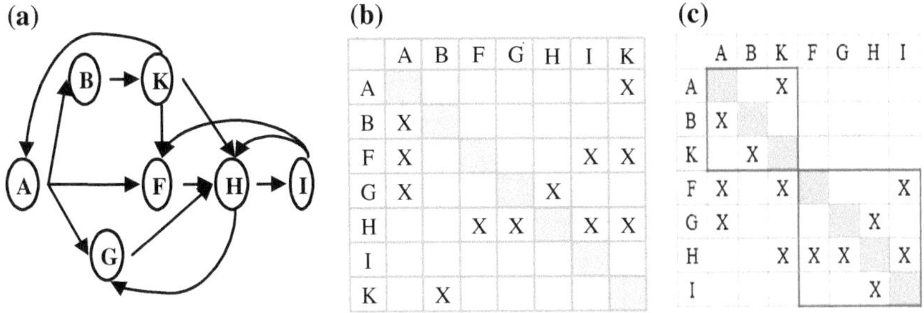

Fig. 1 Graphical and DSM representation. (a) Graphical Representation, (b) Activity DSM, (c) Partitioned DSM

2 Design Structure Matrix—An Overview

A basic activity DSM is a square matrix with the list of activities in the same order in the rows and columns in a matrix form [3–5]. To illustrate, a simple network of seven activities as shown in Fig. 1a is considered. From the network representation, it is difficult to identify the start or the end activities.

The DSM representation equivalent for this graphical network is as seen in Fig. 1b. The dependencies among the activities are represented as 'X' marks in the off-diagonal cells. For instance, the X mark in the first row implies that activity K gives the information to activity A. Diagonal cells do not represent any relationships and can be used for representing duration of those corresponding activity.

Partitioning is the process of rearranging the order of activities in an effort to bring the dependency relationships below the diagonal (dependent or independent) or close to the diagonal (interdependent) to form blocks as illustrated in Fig. 1c [3, 4, 13–15]. For this network two dependent blocks ABK and FGHI are identified using topological sorting-matrix multiplication-condensation process in cycles [9]. The algorithm used for partitioning process adopted from [9] is as shown in Fig. 2.

3 Case Study and Methodology

Design data was collected through discussions and interactions with the different experts for several weeks on a completed offshore Floating Production, Storing, and Offloading (FPSO) project which involves developing a floating offshore facility. Nine groups involved in this project were Process (PC), Mechanical (ME), HVAC (HV), Architecture (AR), Piping (PI), Structure (ST), Electrical (EL), Instrumentation (IN), Health Safety and Environment (HSE). Every design group has their own priority and perspective of the entire project as portrayed in Fig. 3. Currently, the teams interact and exchange information very often and the sequence of execution for the complete project is adhoc [16].

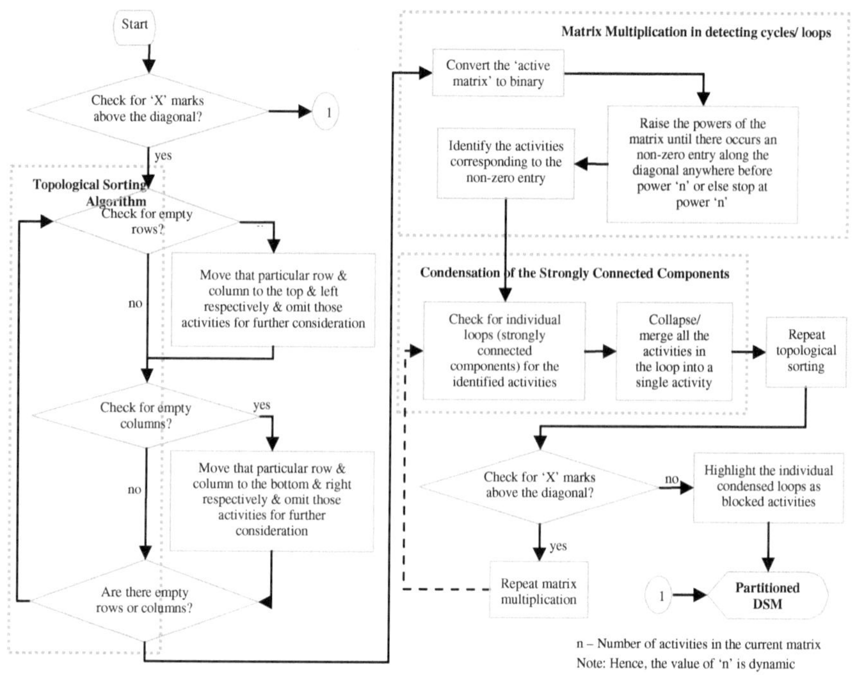

Fig. 2 Flowchart for partitioning algorithm [9]

A complete list of deliverables is identified when the scope of the project is defined through initial kick off meeting held with the lead engineers of each group. The actual design work is initiated once each of the nine individual teams finalizes their list of deliverables. Simultaneously, the Work-Breakdown Structure and the project organization chart is studied to identify the design teams and the corresponding deliverables. After sufficient design progress has been obtained, all the teams meet together and interact for exchanging and updating their information. In the present study, this data is captured in matrix form for the design deliverables and this was referred as Deliverable DSM (DDSM). The methodology adopted for the present study is shown in the Fig. 4.

In simple terms, the methodology of the entire study was primarily in comparing and observing the relationship between (1) individual team deliverable sequences to the project deliverable sequences and (2) project deliverable sequences and project team sequences. Generally, as stated in Fig. 3, different teams have varying viewpoints and hence generating a complete project DSM may not be feasible and this formed the reason for the first comparison. As stated earlier, if there exists a relationship between teams and parameters and the deliverables, when the larger deliverable DSMs are formed, activity DSM sequences or team DSM sequences should be generated. This formed the reason for the second comparison.

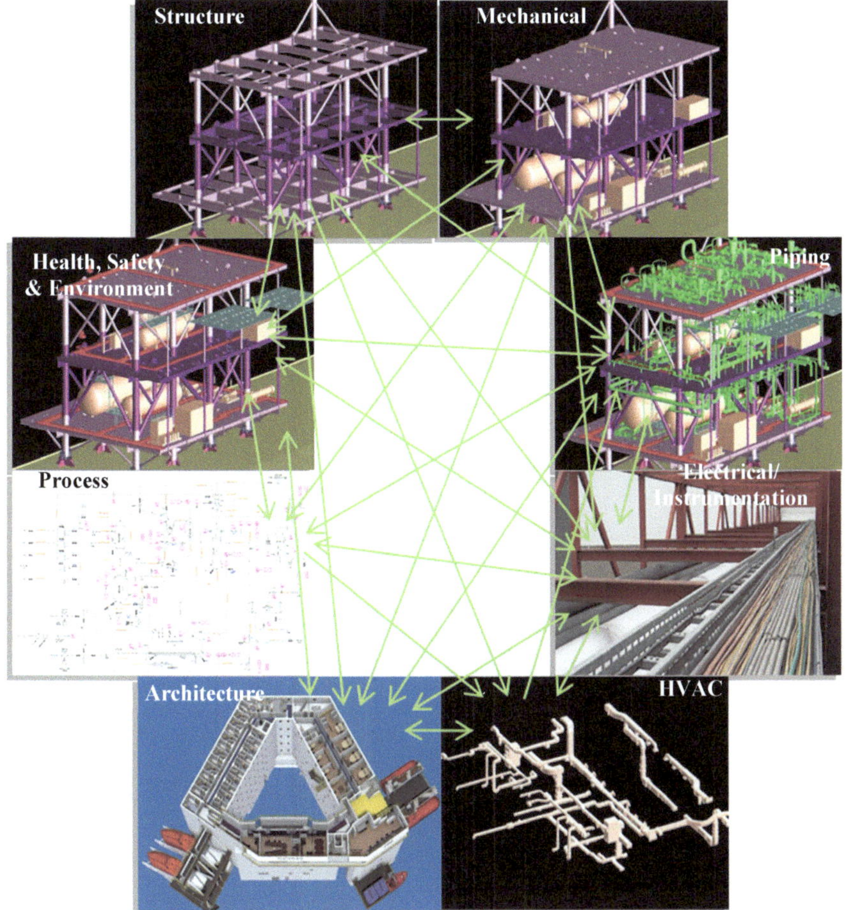

Fig. 3 Viewpoints of design domains and their interactions

4 Data Collection and Analysis

As stated earlier, only four teams (mechanical, HVAC, piping, structural) and two elements (teams and deliverables) were considered for the present analysis and the procedure as seen in Fig. 4 was used for generating the team DSMs and the deliverable DSMs. Initially, deliverables responsible for all the four identified teams were identified and the deliverable DSMs were formed individually. Once, the DDSMs are formed for every individual team, and then the DDSMs are partitioned as seen in the Figs. 5, 6, 7 and 8 using the procedure as illustrated in Fig. 2. The sequence of execution revealed the presence of independent and dependent activities and there were no interdependent activities.

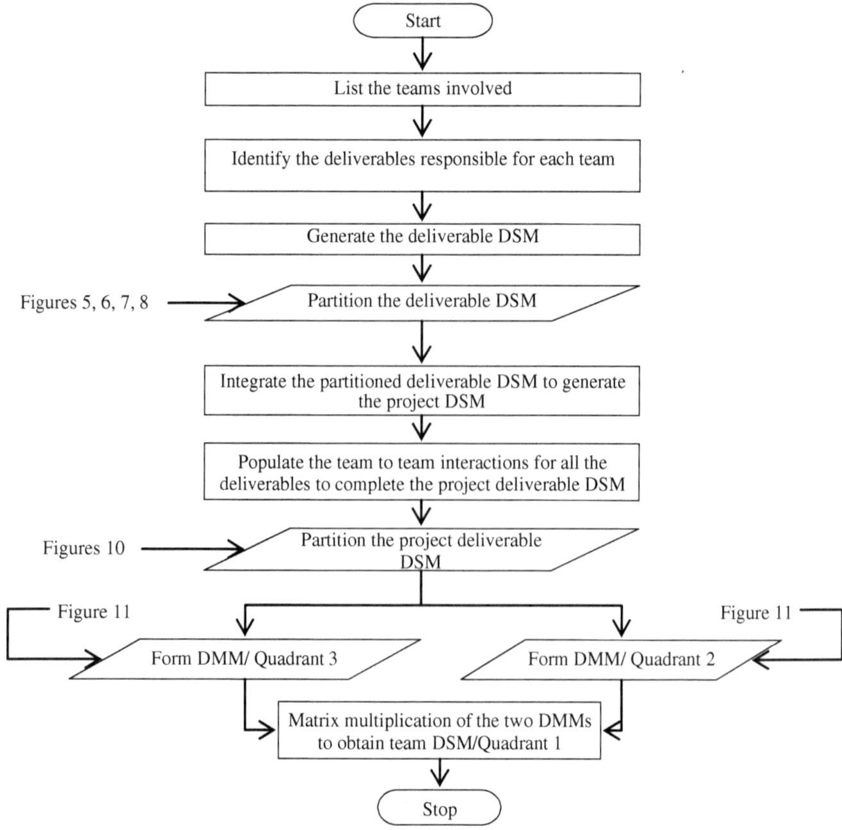

Fig. 4 Methodology adopted for the study

		ME1	ME3	ME4	ME2
Material Handling Philosophy	ME1	7			
Equipment Datasheet	ME3		30		
Package Datasheet	ME4			30	
Material Handling General Arrangement Drawing	ME2	X	X		3

Fig. 5 Partitioned deliverable DSM for mechanical team

		HV3	HV5	HV6	HV4	HV7	HV8
Design Basis for HVAC	HV3	7					
Ducting and Instrumentation Diagram	HV5	X	15				
HVAC Ducting Layout	HV6		X	10			
HVAC System Description and Control Philosophy	HV4	X			6		
HVAC Equipment Layout	HV7		X			5	
HVAC Penetration Drawing	HV8			X			3

Fig. 6 Partitioned deliverable DSM for HVAC team

		ST1	ST2	ST3	ST4	ST5	ST6
Standard Drawings	ST1	8					
Framing Plan Drawing	ST2		40				
Framing Plan Elevations	ST3			10			
Secondary Outfitting and Tertiary	ST4				20		
Weight Control Procedure	ST5					7	
Weight Control Report	ST6	X	X	X			3

Fig. 7 Partitioned deliverable DSM for structure team

		PI4	PI5	PI10	PI1	PI2	PI3	PI7	PI11	PI6	PI8	PI9
Equipment Layout	PI4	50										
Piping General Arrangement Drawing	PI5		100									
Stress Analysis Philosophy	PI10			7								
Piping Material Specification	PI1				10							
Valve Material Specification	PI2					7						
Overall Layout	PI3						50					
Piping Detail Standard	PI7							7				
Insulation Philosophy	PI11								5			
Pipe Support Layout	PI6		X							2		
Nozzle Orientation Drawing	PI8	X	X								30	
Critical Line List	PI9			X								4

Fig. 8 Partitioned deliverable DSM for piping team

As stated in Fig. 4, the project deliverable DSM was formed by integrating all the four teams' individual deliverable DSMs as illustrated in Fig. 9. Then, the team to team interaction of all the deliverables was captured to populate the project DDSM. The partitioned DDSM (using the same methodology as stated in Fig. 2) for the entire project is seen in Fig. 10. Unlike the sequences of the individual team's deliverables, the partitioned DDSM for the entire project revealed a bigger block as seen in Fig. 10.

To determine whether there exists any relationship between the various DSMs, the team DSM was not formed in isolation. Figure 11 shows the methodology for generating the team DSM. As a first step, the deliverables DSM obtained using Fig. 10 is retained in the 4th quadrant as presented in figure. The corresponding teams that give the input to these deliverables are captured in the 2nd quadrant (team-deliverables DMM). Similarly, the corresponding teams that receive the input from these deliverables are plotted in the 3rd quadrant. DMM is the Domain Mapping Matrix which captures the relationship between two different elements [13, 17, 18]. As elaborated in the Fig. 4, the team DSM is formed by multiplying the two matrices available in the 2nd and 3rd quadrant. Figure 12 represents the partitioned DSM of the teams (using the same methodology as stated in Fig. 2).

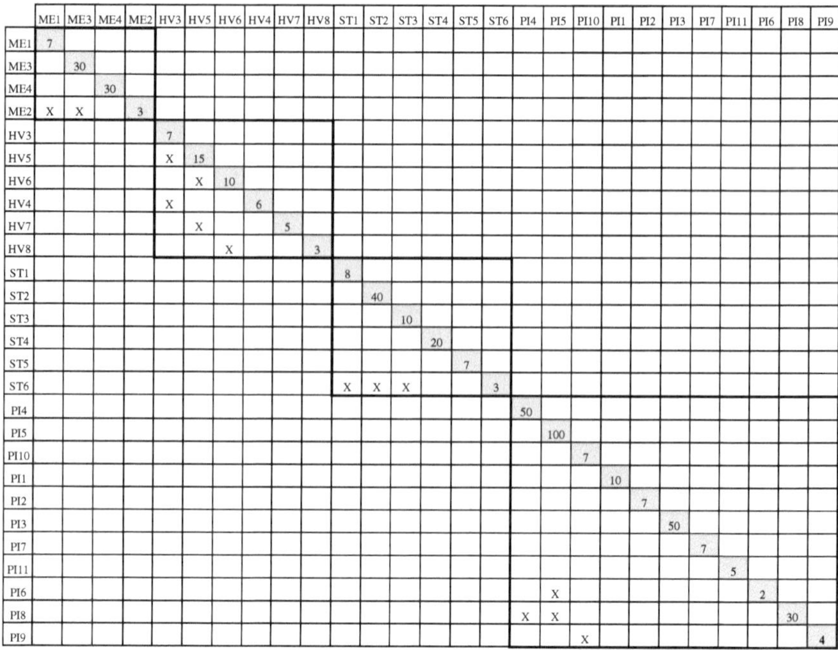

	ME1	ME3	ME4	ME2	HV3	HV5	HV6	HV4	HV7	HV8	ST1	ST2	ST3	ST4	ST5	ST6	PI4	PI5	PI10	PI1	PI2	PI3	PI7	PI11	PI6	PI8	PI9
ME1	7																										
ME3		30																									
ME4			30																								
ME2	X	X		3																							
HV3					7																						
HV5					X	15																					
HV6						X	10																				
HV4					X			6																			
HV7						X			5																		
HV8							X			3																	
ST1											8																
ST2												40															
ST3													10														
ST4														20													
ST5															7												
ST6											X	X	X			3											
PI4																	50										
PI5																		100									
PI10																			7								
PI1																				10							
PI2																					7						
PI3																						50					
PI7																							7				
PI11																								5			
PI6																		X							2		
PI8																	X	X								30	
PI9																			X								4

Fig. 9 Integrated individual DDSM for the four teams

	ME1	PI1	PI2	PI3	HV3	HV5	ME3	ME4	HV6	HV8	PI4	PI5	PI8	ST1	ST2	ST3	ST4	ST5	PI10	ME2	HV7	PI6	PI7	PI9	PI11	ST6	HV4
ME1	7																										
PI1		10																									
PI2			7																								
PI3				50																							
HV3				X	7																						
HV5					X	15																					
ME3		X		X			30				X	X	X				X										
ME4		X		X				30			X		X				X										
HV6				X		X			10		X	X			X	X											
HV8				X					X	3	X	X			X	X	X										
PI4							X	X			50			X	X	X	X										
PI5							X	X				100		X	X	X	X	X									
PI8							X				X	X	30		X	X	X										
ST1											X			8													
ST2				X			X			X	X				40												
ST3				X			X				X					10											
ST4				X			X				X	X					20										
ST5							X					X						7									
PI10							X							X					7								
ME2	X			X			X				X	X			X	X	X			3							
HV7				X		X					X	X			X	X					5						
PI6												X			X	X	X					2					
PI7														X									7				
PI9															X	X	X		X					4			
PI11							X											X							5		
ST6							X				X	X		X	X	X										3	
HV4					X																						6

Fig. 10 Partitioned DDSM for the project

Team DSM (Quadrant 1)	Team-Deliverables DMM (Quadrant 3)
Team-Deliverables DMM (Quadrant 2)	Deliverables DSM (Quadrant 4)

Fig. 11 Methodology for generating team DSM

	ME	PI	ST	HV
ME		X	X	
PI	X		X	
ST	X	X		
HV	X	X	X	

Fig. 12 Partitioned team DSM

5 Discussions and Observations

At the end of applying the methodology (Fig. 4), three partitioned DSMs were obtained. As stated earlier, Figs. 5, 6, 7, 8 and 10 were considered for comparing/mapping the relationship between the sequences of the deliverables associated with individual teams and the project deliverables sequence as a whole. The sequence determined for the complete project consisting of the four teams (Fig. 10) cannot be used to determine the individual team sequences (Figs. 5, 6, 7 or 8) and vice versa. Subsequently, Figs. 10 and 12 were utilized for determining the relationship between the project deliverable sequences and the team sequences. The observation revealed any relationship to hardly exist between the various partitioned DSMs. Further, random analysis was done for the other teams and there was hardly any mapping revealed among those partitioned DSMs.

This study analyses the partitioning process of the various DSM types. From the case study data and analysis, it was revealed that there hardly exists any relationship among the different DSMs. There could be many reasons deviating the mild relationship that could be drawn among the different DSMs. Some of the reasons can be that the project was quite complicated with nine teams and more than 250 deliverables in total, identification of the dependency relationship was subjective, etc. A simple and ongoing project can be considered and if the design process is monitored continuously from the start of the project, dependency relationships could be captured with clarity and there could be a relationship existing under those circumstances. Research work in this direction with more rigorous methodology is already in progress.

Acknowledgments The first and third author would like to acknowledge DST (Department of Science and Technology, Government of India, New Delhi) for providing the financial support to execute the work in this area.

References

1. Eppinger, S.D.: Innovation at the speed of information. Harv. Bus. Rev. **79**(1), 149–158 (2001)
2. Mujumdar, P., Maheswari, J.U.: A design iteration framework for construction projects. In: Paper Presented at the RICS Cobra Conference (RICS 2013), 10–12 Sept 2013
3. Eppinger, S.D., Browning, T.R.: Design Structure Matrix Methods and Applications. The MIT Press, Cambridge (2012)
4. Steward, D.V.: The design structure system: a method for managing the design of complex systems. IEEE Trans. Eng. Manage. **28**, 71–74 (1981)
5. Yassine, A.A.: An introduction to modeling and analyzing complex product development processes using the design structure matrix method. Quaderni di Manage. (Ital. Manage. Rev.) **9**, 1–17 (2004)
6. Austin, S.A., Baldwin, A., Li, B., Waskett, P.: Analytical design planning technique (ADePT): a dependency structure matrix tool to schedule the building process. Constr. Manage. Econ. **18** (2), 173–182 (2000)
7. Browning, T.: Applying the design structure matrix to system decomposition and integration problems: a review and new directions. IEEE Trans. Eng. Manage. **48**(3), 292–306 (2001)
8. Fayez, M., Axelsson, P., Oloufa, A., Hosni, Y.: DSM versus CPM: issues for planning design and construction activities. In: Proceedings of the ASCE construction congress, 10–14 March 2003
9. Maheswari, J.U.: Modeling activity sequencing for construction projects using dependency structure matrix, PhD Thesis, Department of Civil Engineering, Indian Institute of Technology Madras, Tamil Nadu, India (2006)
10. Maheswari, J.U., Varghese, K.: Project scheduling using dependency structure matrix. Int. J. Project Manage. **23**, 223–230 (2005)
11. Maheswari, J.U., Varghese, K., Sridharan, T.: Application of dependency structure matrix for activity sequencing in concurrent engineering projects. J. Constr. Eng. Manage. **132**, 482–490 (2006)
12. Wang, W.C., Liu, J.J., Liao, T.S.: Modeling of design iterations through simulation. Autom. Constr. **15**, 589–603 (2006)
13. Lindemann, U.: The Design Structure Matrix. http://www.dsmweb.org/. Last visited 14 May 2014
14. Warfield, J.N.: Binary matrices in system modeling. IEEE Trans. Syst. Man Cybern. **3**, 441–449 (1973)
15. Yassine, A., Falkenburg, D., Chelst, K.: Engineering design management: an information structure approach. Int. J. Prod. Res. **37**(13), 2957–2975 (1999)
16. Mujumdar, P., Muraleedharan, P., Maheswari, J.U.: Structured methodology for applying multiple domain matrices (MDM) to construction projects. In: Paper presented at the 16th International Dependency Structure Modelling Conference (DSM 2014), 2–4 July 2014
17. Danilovic, M., Browning, T.R.: Managing complex product development projects with design structure matrices and domain mapping matrices. Int. J. Project Manage. **25**(3), 300–314 (2007)
18. Danilovic, M., Sandkull, B.: The use of dependence structure matrix and domain mapping matrix in managing uncertainty in multiple project situations. Int. J. Project Manage. **23**(3), 193–203 (2005)

A Paradigm-Shift towards User-Centred Empirical Methodology in Interactive Multimedia Communication

Manoj Majhi and Debkumar Chakrabarti

Abstract With multiple mediums of interaction available, it has become essential for a designer to be able to choose the medium that can be most effective for the chosen user. The popular mediums identified for the interactive study are special effects, Cel-animation, Puppetry, Video and Stop-motion animation. Case studies were made available to a group of 40 users, so as to be able to get an exposure of the available mediums of interactive multimedia communication. The findings from the study gave a surprising revelation of the preference of the users in choosing the medium for the empiricism with more than half the users preferring the stop-motion medium to be used to communicate. The study focuses on the user-centered empirical methodology to reveal the vein of the interactivity which was the user having a flexibility of using the expertise of using observation, being able to create which is an inventive process, an appreciation of the invention to interact by studying the inventive user-centered empirical methodology to make the design effective for its aesthetics, semiotics and semantics in design.

Keywords SFX · Cel-animation · Puppetry · Videography · Photography · Stop-motion animation

1 Section

The methodology followed to carry out the user-centered empiricism has been divided into the following

M. Majhi (✉) · D. Chakrabarti
Department of Design, Indian Institute of Technology Guwahati, Guwahati, India
e-mail: manojmaj@iitg.ernet.in

D. Chakrabarti
e-mail: dc@iitg.ernet.in

A. Chakrabarti (ed.), *ICoRD'15 – Research into Design Across Boundaries Volume 2*, Smart Innovation, Systems and Technologies 35, DOI 10.1007/978-81-322-2229-3_57

1.1 Review of Studies

A review to study of the user experience in the evaluation of the effectiveness of interactive multimedia in design was taken up in which it was observed that [1] there were 15 % of users who did things differently, and still being able to help deliver a solution effectively by the users who were empiric. The users were allowed to interact with multiple mediums to be able to effectively use multiple mediums to be able to communicate.

1.2 Alternative Design Approaches and Methods

As mentioned a design is a plan that could have a model of a particular method. Similarly some methods that have been used are the KISS {Keep It Simple Stupid} [2] principle that experiments to find solutions to unnecessary complications, TIM-TOWTDI {There is more than one way to do it} [3] a philosophy that uses multiple ways of doing things, use-centered design emphasises on the connected solutions and experiments with the use of content and not the end user, an element of interactivity can help the design become more effective.

1.3 User Centric Design Evaluation of the Effectiveness of Interactive Multimedia in Design

An user centered design's is am evaluation of the effectiveness of the interaction in systematic plan, study and design of the communication of the user as well as the medium of interaction, that borrows from each other the techniques such as computer graphics, operating systems, the encrypted program written for the generation of the content for the story line it is designed for is crucial for the effectiveness of final designs, analysis is essential to categorise the interactive complex empiricism in smaller sections such as to be able to effectively understand the medium such as qualitative, quantitative as well as an interactive.

1.3.1 Qualitative Instruments for Analysis

Qualitative research is a method of inquiry employed in many different disciplines, to gather an in-depth understanding of human behavior and the reasons that govern the behavior thus, smaller but focused samples are more often needed, rather than large samples. Qualitative methods produce information only on the particular cases studied, and any more general conclusions are only hypotheses (assumptions). Quantitative methods can be used to verify which of such hypotheses are true.

1.3.2 Quantitative Measurement Tools for Analysis

In the analysis provided by qualitative and quantitative tools and methods to measure and analyse the results achieved at the experiments, process, program and enterprise levels. Six Sigma has made its way into design disciplines, more often the need to know how to leverage Six Sigma within software process improvement initiatives. Result-Driven Measurement Result-driven measurement helps to identify and define software measures to support the experiments goals. Experiments run on data. It is used to manage experiment, make decisions, and guide improvement. But how reliable is the data collected? How good is the data? Are the right things done in measurement and analysis? Is it doing well? How good is the information generated? How to improve measurement processes and improve information quality?

1.3.3 Interactive Multimedia Methods in Design

In design the methods have a broad focus area i.e., Divergence that Explores many possibilities and constraints of received conditions by applying critical thinking through qualitative and quantitative research methods to create new understanding toward better design solutions using the available mediums of communication i.e., multimedia. Transformation which redefines specifications of design solutions which can lead to better guidelines for traditional and contemporary design activities (architecture, graphic, industrial, information, interaction, et al.) and/or multidisciplinary response and Convergence that prototypes possible scenarios for better design solutions that incrementally or significantly improve the originally inherited situation, Sustainability to manage the process of exploring, redefining and prototyping of design solutions continually over time and finally articulation of the visual relationship between the parts and the whole.

1.4 Planning the Design Experiments

Design of experiments or experimental design is the design of gathering information exercises where the variations are visible, if under complete control of the experimenter or not.

Step 1: Begin with a collection of subjects for example 10 people.
Step 2: Devise a method to randomise that is purely mechanical (e.g. flip a coin).
Step 3: Assign subjects with “Heads” to one group i.e., Control Group. Assign subjects with “Tails” to the other group.

1.4.1 Medium Based Studies

With the changing values and awareness among the diasporas it has become easier to deliver the intended message to the chosen user with ease, but the appropriate medium for the message is essential to be able to make it effective for the user, a study to identify the most empirically effective mediums of communication to deliver the message.

1.4.2 Selection of the Mediums

In the selection of a medium of interactive multimedia communication it becomes essential for an user to be able to be able to derive maximum effectiveness out of it so as to be able to communicate effectively, the mediums available are much more than that were available a few years back, today's user has assumed that information is everywhere one just needs to interact to be able to get it.

1.4.3 Analysis of Elements of Interactive Multimedia as a Tool

With various mediums of interactive communication available such as Animation a medium of communication as an educational tool for users to learn in a creative, exploring environment, working with practical animation techniques in small teams. Animation is creating life—bringing objects to life—telling a story about the world we all are part of. It is fun, exciting and fascinating, and the user will experience being absorbed in a world where time and place do not exist. It is like being in Wonderland, and it will be beneficial to all children to have this experience. Theories of learning and instructive will be practiced, where the used will carry out an experiment with the pedagogical possibilities of Animation as a learning tool. To give user an experience of culture and traditions of learning.

1.4.4 Stages of the Interactive Multimedia Story Generation Process of the Medium

Storytelling is imparting events in Multimedia using the mediums of words, images and sounds, etc., stories or narratives have been shared effectively in every culture as a means of entertainment, education, cultural preservation and in order to instill moral values. Crucial elements of stories and storytelling include plot, characters and narrative point of view using the stories. Using animation as a tool to encourage and develop Youngsters learning is not only fun but effective.

1.4.5 Interactive Multimedia Communication Tool Centric Study

Communication is the process of transferring information. Communication requires a sender, a message, and an intended recipient [4], though the receiver need not be aware or participating of the sender's intent to communicate at the time of communication, The communication theory has three levels of the problem as argued by the Shannon and Weaver's model i.e. the clarity of the message transmitted, a the technical design problem, the accuracy of the meaning conveyed i.e., the semantic problem and its effectiveness can the behavior be affected by the meaning received, the effectiveness problem.

1.4.6 Objectives of Interactive Multimedia Communication

Tool Centric Study "Multimedia" is a term frequently heard and discussed among educational technologists today."Multimedia is the exciting combination of computer hardware and software that allows you to integrate video, animation, audio, graphics, and test resources to develop effective presentations on an affordable desktop computer" [5]. Too many of us, especially in the field of design, are caught in this modern tug-of-war. Throughout the 1980s and 1990s, the multimedia concept took on a new meaning, as the capabilities of satellites, computers, audio and video converged to create new media with enormous potential. Together with advances in hardware and software, these technologies were able to provide enhanced learning facility and with attention to the specific needs of individual users. The fundamental objective of interactive multimedia material is not to replace the instructor but to change the interaction concept completely.

1.4.7 Planning the Methodology for the Interactive Multimedia Communication Study

Today's multimedia is a careful mixture of the combination of text, graphics, audio, animation, and video elements. When an user is allowed i.e., the viewer of a multimedia project, to control 'what' and 'when' and 'how' of the elements that are delivered and presented, it becomes interactive multimedia.

1.4.8 Instrument of Measurement for the Interactive Multimedia Communication Study

In this section, it has been argued that interactivity is not singled out to computers or networks and cannot be reserved solely for the discussion of so-called New Media. The term New Media is now not very effective and is misleading, as in the present scenario a latest technology can be new, and to take it ahead, new concepts need not be devised, even though the take the limelight.

1.4.9 Design of Questionnaire

to analyse the interactivity of Video + SFX, Puppetry/2D–animation, 3D/character/ physical FX, CGI animation, Cel-animation and Stop motion A questionnaire is a methodological set of questions for deriving information from the respondents. The main objective is to translate the researcher's information needs into a set of specific questions that respondents are willing and able to answer. While this may seem straightforward, questions may yield very different and unanticipated responses. For example, how would you answer the following question: "Which State is larger, Assam or Bengal?" where in there is ambiguity of if the answer is based on population or area? The mediums were categorised as the mediums of Video + SFX in which Bindas bol, Commonwealth games 2012, and Anti-smoking campaigns were analysed. The medium of Puppetry had three case studies Haddi raja, gali gali sim sim and Steam boat willie, the medium of 3D/character/physical FX had Discrimination awareness, Mobile service promotional and National integration, the CGI animation had Hanuman, Anti-war campaign and smurfs campaign. The medium of cel- animation had Discrimination, Ek anek and chotta bheem. And based upon these the users used user-centred empirical methodology came up with their own communications that were interactive which were an anti ragging video based stop-motion animation communication, a violence campaign, an anti litter campaign, an event promotional, a discrimination awareness using an eraser as a central character, a self realisation animation, a corruption awareness, a superstition awareness and a green campaign using clay animation.

Design of Response Sheet

The response was designed such as to be able to allow the ease of use for the user, with a familiar response format such as to do away with any ambiguity a few test were conducted to reassure the need to evaluate the interaction in the test for awareness of interactive mediums of communication, test for Visualisation, imagination, creativity and aesthetic sensitivity, and to analyse the fundamental awareness a simple test of analyzing an empirical test of key-frames to be able to give the user the advanced empirical interactive multimedia communication, a favorable 98 % were successful to go it right, effectiveness of visual culture is important in design, due to its multidisciplinary approach of combining the culture studies, aesthetic background, the science of humans and how they work and the rational empiricism that rely on the visual imagery an empiricism of having to test the foundations from the introduction of the basic's of the Animation design, production, and scriptwriting along with the importance of the effectiveness of visual culture in the study of animation, from any one of the case studies showed a drop from the initial task of having to choose the right sequence from 5 scattered images, the empiricism resulted in 61 % interacting well, 28 % interacting in a average way while there was a increase in the decline in the interactivity 11 %, this could be due to the reason of having to perform more number of empirical than just choosing visually, the user had to interact based upon the understanding of production, scriptwriting of animation. The empiricism helped in understanding the user's capability in multimedia communication of design, of first interact with the medium as in learning by doing rather than studying the medium. To understand the user's empiricism of interaction of mediums to a given context to evaluate its effectiveness was evaluated by giving an empirical task of having to design for the kind of ranges or the types or styles of Animation techniques that you can use with each different mediums to be able to effectively design traditional animation practices where each frame is drawn by hand and their importance and relation to contemporary animation techniques such as Stop motion animation technique, flip book, cutout animation, 3D animation, from the examples analysed to effectively communicate the essence through a campaign for a green campus, was effective in 81 % of the users using their creativity of thinking out of the box and bringing out concepts due to the motivation of having to interact for a cause 'campaign for a green campus', the empiricism also brought forward the users who could not come up with creative concepts i.e., 19 % who needed more interaction for them to be able to design better. An empiric method to evaluate the inferences of the introduction to specialized areas in Multimedia such as light effects, cel-animation, character animation, clay animation and puppet animation how can it be effective in spreading an awareness of harms of the use of carcinogens such as tobacco, how can the design effectively communicate it to the viewer, showed 78 % of the users who could bring out concepts that were effective while there was still 22 % who could not do so, due to the lack of visualization, this can be corrected with interactive multimedia to increase its effectiveness. An empiricism to evaluate the collective inventiveness of the users in applying technology to the imaginative capability of being able to design a character as had been done earlier with Fido Dodo or Baiti, the users were given a task to design for the software such as if the user can use their own character such or their own cartoon by simplifying the outlines of the features to depict a character doing a trick as Courtney does by waving her hand riding on a flying car and communicating with a text box 'I've got the top down, let's go', Earl who riding on a surf board communicating with a text box 'Surf's up, lets ride', Rover moves his head allowing the long ears to wave and wags his tail and communicating though a bark and a text box 'I love play to fetch, throw me a question', Merlin bows and communicates 'while a little old fashioned, a little digital magic is always helpful', this assisted in encouraging the user to think imaginatively as well as different 81 % of the users could interact well with the ideas of effectiveness in design with the use of cartoon characters, while 19 % failed to do so. The response was effective in being able to assimilate the users understanding of the mediums of communication such as to design an effective medium for a viewer.

Participant Observation

The survey was to understand the users perception of the knowledge domains of the software, principles to empirically develop the final effective design.

Software Knowledge

From the survey it was observed that a majority of 55 % of the users has a good knowledge, 18 % had very good knowledge of the task, 9 % were manageable while a 18 % had a poor knowledge of the interaction with the software to complete the task. Which needs the user to interact more with the mediums such as to be able to be able to evaluate the empiricism for its effectiveness.

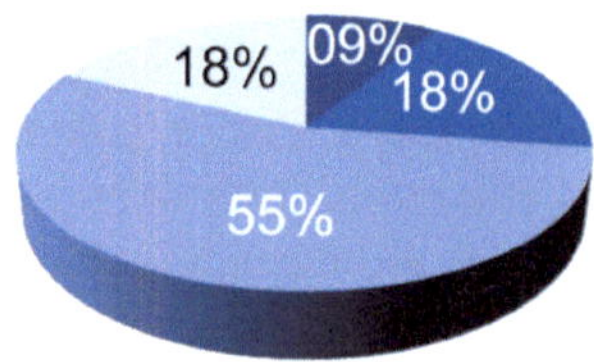

Knowledge of the Design Principles

Knowledge of the final design of having to use the mediums of Video + SFX, Puppetry/2D–animation, 3D/character/physical FX, CGI animation, Cel-animation Stop motion effectively to be able to make the interaction effective, From the survey it was observed that a majority of 70 % of the users has a good knowledge of the final design, 10 % had an average knowledge of the final design, 10 % had a poor knowledge while a 10 % had a very poor knowledge of the final design of the interaction with the software to develop the final design.

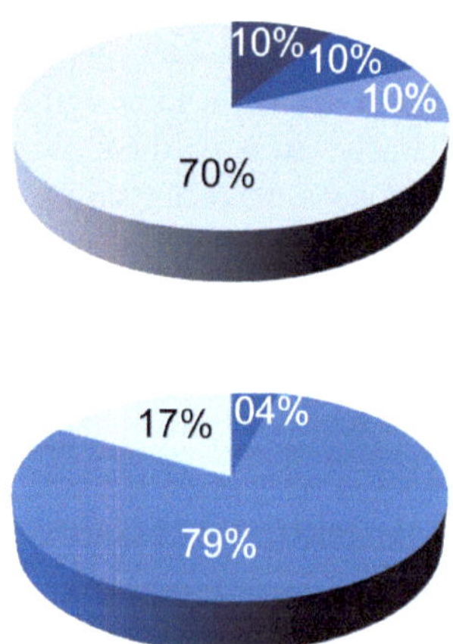

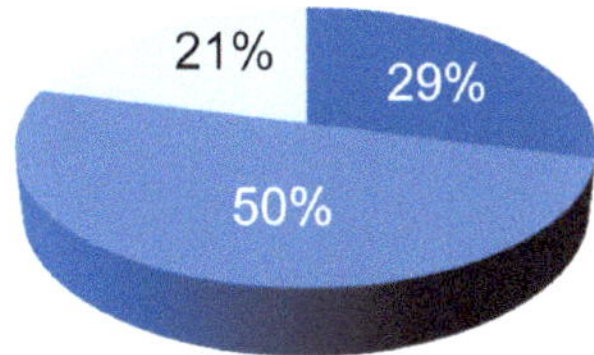

Which needs the user to interact more with the mediums such as to be able to be able to evaluate the empiricism for its effectiveness. The task to test out the effectiveness of the users comprehension of the mediums of interaction to analyse the acceptance of the different mediums of Interaction design task for their effectiveness. From the survey it was observed that a majority of 04 % of the users has an average knowledge clarity of the design task, 17 % had an good knowledge clarity of the design task, 79 % had a good knowledge clarity of the design task of the interaction with the software to develop the final design. Which needs the user to interact more with the mediums such asto be able to be able to evaluate the empiricism for its effectiveness. From the survey it was observed that a majority of 29 % of the users has an average knowledge of the Necessary literature survey conducted 21 % had an good knowledge of the necessary literature survey conducted, 50 % had a good knowledge of the Necessary literature survey conducted of the interaction with the task to develop the final design.

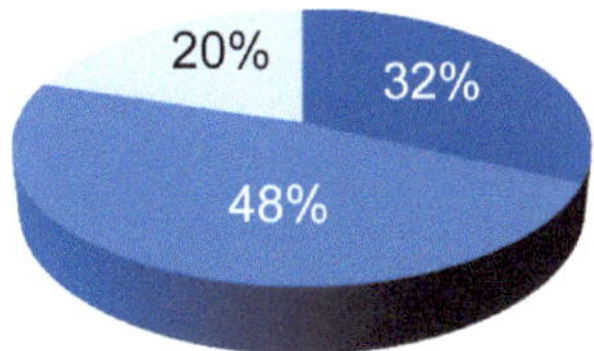

Which helps one understand the effectiveness based upon the need for the literature survey has an average knowledge clarity of the up to date literature analysis, 32 % had a very good knowledge up to date literature analysis, 48 % had a good knowledge clarity of the up to date literature analysis conducted of the interaction with the task to develop the final design.

Non-participant Observation

Non-participant observation, on the other hand, is when the researcher do not get involved in the activities of the group but remains as a passive observer, watching and listening to its activities and drawing conclusions from this. For example, you might want to study the moral imparted to the user in a situation. As an observer, one could watch, follow, and record the activities as they are performed. After making a number of observations, conclusions could be drawn about the functions the user carry out in the situation. The user group in any setting can be observed in the same manner. From the survey it was observed that again a majority of 55 % of the users has a good knowledge, 18 % had very good knowledge of the task, 09 % were manageable while a 18 % had a poor knowledge of the interaction with the software to complete the task. Which needs the user to interact more with the mediums such as to be able to be able to evaluate the empiricism for its effectiveness.

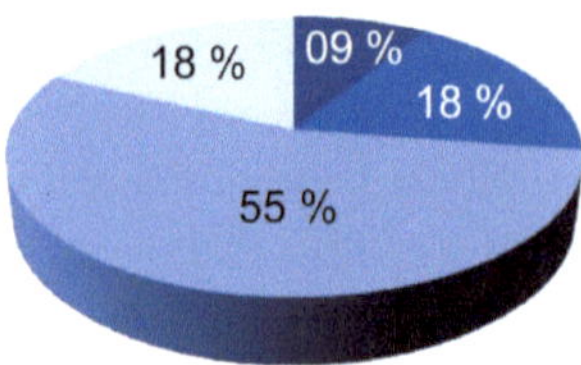

The Design Task

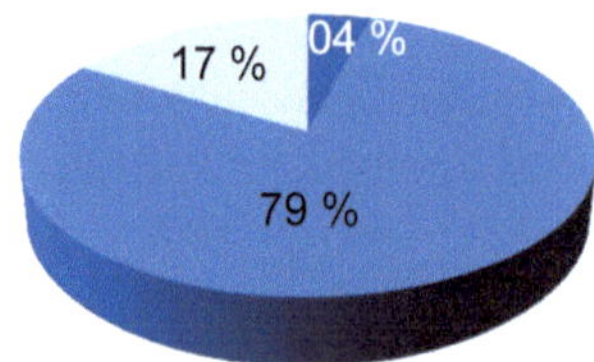

The task to test out the effectiveness of the users comprehension of the mediums of interaction to analyse the acceptance of the different mediums of Interaction design task for their effectiveness. From the interaction it was observed that a majority of 04 % of the users has an average knowledge clarity of the design task, 17 % had an good knowledge clarity of the design task, 79 % had a good knowledge clarity of the design task of the interaction with the software to develop the final design. Which needs the user to interact more with the mediums such as to be able to be able to evaluate the empiricism for its effectiveness.

The Medium

The task to test out the effectiveness of the users comprehension of the mediums of interaction to analyse the acceptance of the different mediums of Interaction design task for their effectiveness.

About the Medium

A medium can be a Communication which addresses the storage and transfer of data used to store and deliver information or data, in the field of Natural science a medium is a Processing in which industrial engineering use materials which play a part in the process of manufacturing, medium is excitable which can communicate a data of a description that can assist the interaction of another for a period, a medium of entertainment has data from the creative field of art, opera, television, movies etc. and medium of spirituality addresses the interaction between the physical world and the spiritual world, a medium is essential to interact design. To study the empiricism of a group of users of their assumption of the effectiveness of many mediums which have been used by them to bring effectiveness to their designs was taken up. From the survey it was observed that a majority of 04 % of the users has an average knowledge clarity of the design task, 17 % had an good knowledge clarity of the design task, 79 % had a good knowledge clarity of the design task of the interaction with the software to the mediums such as to be able to be able to evaluate the empiricism develop the final design. Which needs the user to interact more with for its effectiveness. From the survey it was observed that a majority of 29 % of the users has an average knowledge clarity of the Necessary literature survey conducted, 21 % had an good knowledge clarity of the Necessary literature survey conducted, 50 % had a good knowledge clarity of the Necessary literature survey conducted of the interaction with the task to develop the final design. Which helps one understand the effectiveness based upon the need for the literature survey. From the survey it was observed that a majority of 73 % of the users has an average knowledge acceptability of the medium of the user, 27 % had an good knowledge acceptability of the medium of the user. From the survey it was observed that a majority of 36 % of the users has an average knowledge time taken for the output for the final result medium of the user, 36 % had an good knowledge time taken for the output for the final result of the medium of the user, 10 % had a very good knowledge time taken for the output for the final result of the medium, 09 % had a very poor knowledge of the time taken for the output for the final result of the medium, 09 % had a very poor knowledge of the time taken for the output for the final result of the medium of the user analysis conducted of the interaction with the task to develop the final design. Which helps one understand the effectiveness based upon the need for the literature survey.

Acknowledgments I would like to acknowledge all the design aspirants who, think differently along with the young designers inspiring one and all with their queries.

References

1. http://www.waset.org/author/manoj-majhi
2. Niparko, J.K.: KISS {Keep It Simple Stupid} Cochlear Implants: Principles & Practices, p. 111 (2009)
3. Willmore, J.: "TIMTOWTDI There is more than one way to do it" No Magic Bullet: Seven Steps to Better Performance, p. 118 (2009)
4. Bowater, L., Yeoman, K.: "recipient" Science Communication: A Practical Guide for Scientists, p. 75 (2012)
5. Siddiqui, M.H.: Multimedia is the exciting combination. Educ. Technol., 66 (2009)

Handheld Isobaric Aesthesiometer for Measuring Two-Point Discrimination

M. Manivannan, R. Periyasamy and Devasahayam Suresh

Abstract Two Point Discrimination test is notorious for their unexplained variations within subjects, between subjects, and between studies. This paper presents a systematic study of these variations and proposes a simple design to control the variations, thereby improving the TPD measurement accuracies. In the proposed Isobaric Aesthesiometer, monofilaments are used replacing the rigid prongs in a handheld aesthesiometer. Compared to the force variation (0–500*g*) in the normal aesthesiometer with rigid prongs which approximately follows the second-order system, the force variations in the isobaric aesthesiometer is found to be nearly constant within a tight tolerance. Although the proposed handheld instrument could be used in controlling forces applied in a TPD measurement, unless a standard protocol is developed for measuring the TPD, the variation cannot be controlled even with this device.

Keywords Monofilaments · Handheld clinical devices · Tactile sensitivity

1 Introduction

The two-point discrimination (TPD) test originated from Weber [1] is one of the most frequently used tests for the clinical assessment of the somatosensory system. It is defined as the minimum separation between two stimuli placed on the skin that can be perceived as two separate points. Its wide spread use is due to the simplicity of the test; it does not require any fancy equipment, just a paper clip may be enough, and it is easily portable and provides immediate information about the

M. Manivannan (✉)
Touchlab, Indian Institute of Technology Madras, Chennai, India
e-mail: mani@iitm.ac.in

R. Periyasamy
Department of Biomedical Engineering, NIT Raipur, Chattisgarh, India

D. Suresh
Department of Bioengineering, Christian Medical College, Vellore, India

A. Chakrabarti (ed.), *ICoRD'15 – Research into Design Across Boundaries Volume 2*, Smart Innovation, Systems and Technologies 35, DOI 10.1007/978-81-322-2229-3_58

sensory system. However, there is enormous and implausible variability in the TPD measurements. Long clinical experience of TPD testing, especially in patients with major nerve injuries, verifies the persistence of the same problems that plagued the original investigators of the TPD test [2, 3], i.e. unexplained variations within subjects, between subjects, and between studies [4]. No systematic study of these variations has been reported in literature so far. In this paper we attempt to reason out for the variation and a simple method to control the variation, thereby improving the TPD measurement accuracies.

There could be two reasons for the variations. First is the variation of the stimulus itself at the stimulated site of the skin due to the manual test. It might be difficult for the conventional manual method of TPD to provide two-point stimulation identically at each time, since the receptive fields of skin receptors would easily vary with how a stimulus probe contacts the skin [5]. Second, the preceding two-point stimulus, as the conditioning stimulus, could affect the cognitive response to the target stimulus. In addition to the effects of those factors, the response to the stimulus around true TPD threshold might result alternate judgments. Our focus in this paper is to address the first reason: controlling the variation of the stimulus itself.

TPD is usually measured using a paper clip or handheld aesthesiometers, modified vernier calipers. The problems with the repeatability of force in all handheld instruments for TPD testing have been extensively investigated and described [6, 7]. Spectral analysis of the force produced by hand held sensibility measurement instruments shows they all produce both high and low frequency signals sufficient in strength to stimulate both slowly adapting and quickly adapting end organs, and are not capable of stimulating one particular group. These dynamic properties of testing stimuli explain why our tests are not as repeatable and sensitive as desired. The understanding of these dynamic properties in sensibility measurement is a key for improved instruments and more repeatable findings in clinical testing [7]. There are at least two suggested procedures regarding the pressure applied during TPD testing; The first is that the application force should be very light, 10–15*g*, which corresponds with the force producing the very first small "blanching" around the prongs [8, 9]. Another described method is; "Just sufficient pressure is utilized for the subject to appreciate the stimulus" [10].

Apparently there is a lack of standardization of the technique of TPD assessment, and the test is probably performed in different ways by different clinicians. This is a striking and very serious problem since the TPD test is so frequently used to compare different repair techniques. Probably the most common mistake with TPD testing is to apply too much pressure, which can completely change the result: more pressure will bring more receptors into the field of stimulation and cause more deformation on the skin [11, 12]. Movement of skin by the patient can influence the result, and it may be tempting to apply sufficient pressure to provoke a result, with the risk of including non-denervated receptors in adjacent skin territories.

Dellon has presented an elegant solution to the pressure problem associated with TPD measurement by introducing computerized equipment which applies a standardized pressure [13–17]. Such sophisticated equipment may be very useful in a laboratory setting for diagnostic purposes in various types of neuropathies, but is difficult to use in routine clinical practice. In this paper we present a very simple technique for applying a standardized pressure without the use of sophisticated computer, and still be able to control the variability associated with handheld instruments, to some extent. We have published elsewhere our work in understanding the variability of TPD over different body sites, and diseased conditions, specifically with Diabetic Mellitus [18, 19]. The scope of this paper is to analyze and control the variability of forces in a TPD measurement at the same anatomical site, across patients, and across observers.

2 Analysis of Force Variations in TPD Measurement

Various extrinsic and intrinsic factors influence the variability of TPD measurements. Extrinsic factors include the testing procedure adopted (e.g., the number and location of testing sites, on which little evidence and no consensus exist) [20]. Clearly, some extrinsic factors remain difficult or impossible to measure, such as the variability of the subject's response. However, the intrinsic factors can and should be addressed. In this paper we mainly focus on the intrinsic parameters for controlling the variability.

Unequivocally any hand-held instruments produce variations in application force from one stimulation to another, one instrument to another, and from one examiner to another. These variations of application force cannot be compensated for by care in technique and need to be controlled for measurement reliability.

The lack of a standardized procedure to perform TPD test is also problematic; for example should the test start with smallest or the widest distance, and how many applications should be used? Omer and Bell-Krotoski [21] suggest 3 times, and Moberg [9] suggests 10 times. The importance of blunted ends on the testing instrument to ensure that touch and not pain is assessed has been emphasized [10]. The examination is usually performed using an unfolded paper clip or a Disk-CriminatorTM (Post Office box 16392, Baltimore, MD 21210).

The variation in the terminal probe shape and area has been investigated by author [22], accounting for the instrument to instrument variation. They conclude that spherical probes give the largest variation in two-point discrimination while pointed end probes provide the least variation. Different terminal probes will produce different recruitment of the sensory end organs in the hand, and strongly suggests the need for standardization of TPD measurement itself.

Measurement of TPD involves inherent hand movements of observers mostly in XY plane, as shown in Fig. 1. This planar movement includes both transverse movements and rotation about Z axis. It is important that the two pressure points are applied at exactly the same time. Even a very small time difference in application may introduce a critical error. However, due to rotational movements one point might apply more pressure than the other. The patient may easily discriminate in time between two non-synchronous pressure applications, thus detecting more than

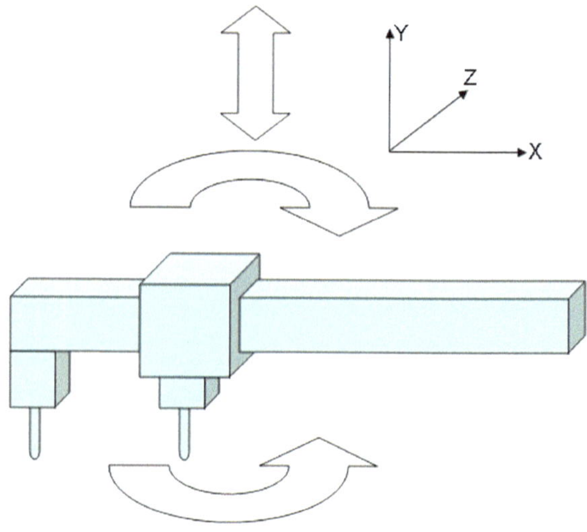

Fig. 1 Measurement of TPD involves inherent hand movements of observers mostly in XY plane

one pressure point with factors related to time-dynamics in application rather than spatial orientation. The human upper extreme's dominant mode of movement during TPD measurement is extension and flexion of the wrist joint, which creates motion along the axis of the aesthesiometer. In a typical aesthesiometer grasp, the hand can be approximated as a rigid body that pivots around a two-degree-of-freedom revolute joint in the wrist, ignoring up-down movements, as shown in Fig. 2. For a normal range of frequencies up to 30 Hz spectral analysis indicates that the passive human wrist holding a stylus behaves like a second-order LTI system, as described by the following equation for a translational motion:

$$m\ddot{x} + b\dot{x} + kx = F \tag{1}$$

where F is the applied aesthesiometer force on an anatomical location, m is the effective end-point mass, b damping, k stiffness of the observers hand, and the corresponding equation for a rotational motion is:

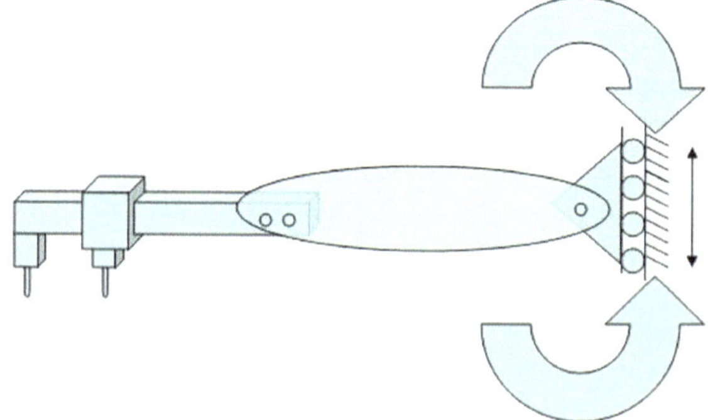

Fig. 2 Typical aesthesiometer grasp can be approximated as a rigid body that pivots around a two-degree-of-freedom revolute joint in the wrist

$$J\ddot{\theta} + \beta\dot{\theta} + \kappa\theta = \tau \quad J = mR^2 \quad \beta = bR^2 \quad \kappa = kR^2 \tag{2}$$

where J is the wrist's rotational inertia, B, the damping co-efficient, k the stiffness, all are related to the translational parameters by R^2, the square of the distance from the wrist joint to each prong in the aesthesiometer.

Time response of the wrist for an input pulse of short duration and high magnitude has been found to be well predicted by the second-order model identified for the subject using frequency domain analysis. This is true for first 50 ms after the pulse input; however, the muscle reflex come into play after 50 ms and small amplitude motion dissipates quickly.

The value of each of these parameters varies depending on the muscle force. Increased muscle activation changes the observer's effective stiffness and damping. The parameter values found to be linearly varying with the muscle activation force. The slope of this linearity and the intercept varies for each observer, accounting for the inter-observer variability.

3 Human Factors in TPD Measurement

Two humans are involved in the handheld TPD measurements. While the subject is passively feeling forces on the skin, the observer is actively manipulating forces applied on the subject. Therefore, both the human factors related to passive and active forces are relevant for our analysis.

The subject in a TPD test experiences a large variety of tactile sensations when aesthesiometer is pressed on the skin; these sensations are really combinations of a few building blocks or primitives. For simplicity, normal indentation, lateral skin stretch, relative tangential motion, and vibration are the primitives for conditions of contact with an object. The subject perceives many of these primitives though tactile information conveyed by mechanoreceptors in the skin.

When an observer contacts or presses aesthesiometer through active motion of the hand on the subject's skin, the contact forces are sensed by both the tactile and kinesthetic sensory systems. Overall contact force is probably the single most important variable that determines both the neural signals in the sensory system, as well as the control of contact conditions through motor action. It appears that the JND for contact force is 5–15 % of the reference force value over a wide range of conditions involving substantial variation in force magnitude, muscle system, and experimental method, provided that the kinesthetic sense is involved in the discrimination task [23–25].

4 Control of Variability in TPD Measurements

It is to be noted from the earlier sections that variation of stimulus and the cognitive state are the main factors resulting in the variation of TPD measurements. In this paper we focus only on controlling the varying stimulus in the handheld instruments, specifically forces under the two prongs. Controlled delivery of forces, under a single prong, on the skin has been addressed well in the literature for measuring the sensibility, not TPD, using Monofilaments. When tip of the monofilament of given length and diameter is pressed against the skin at right angles, the force of application increases as long as the operator continues to advance the probe, until the filament buckles. After the fiber bends, continued advance creates more bend, but not more force of application. This principle makes it possible for an operator using a hand held probe to apply a reproducible force, within a tight tolerance, to the skin surface. A monofilament is mathematically represented by a circular column. The radius, the length, and the modulus of elasticity of a material affect the critical force at which it buckles [26, 27].

The Semmes Weinstein set of monofilaments (SWMF) provide an approximately logarithmic scale of actual force, and a linear scale of perceived intensity. They have a long history of effective use in clinical settings, and can be used to diagnose pathologies of hyper or hypo aesthesia. They are originally intended to detect areas of skin that have abnormal sensation and to record the degree of severity of the sensory disturbance occurring as a result of brain injury. Since their development, the SWMF monofilaments have demonstrated their ability to detect abnormal function in peripheral nerves. Bell-Krotoski and Tomancik [28] validated the forces applied by the SWMF monofilaments. They concluded that if the monofilaments as supplied by the manufacturer were of correct length and diameter, the tests are repeatable within a small, predictable range.

The ability of monofilaments to deliver controlled forces has been used in our Isobaric aesthesiometer. The two rigid prongs in the conventional aesthesiometer have been replaced with two monofilaments in the isobaric aesthesiometer. The new aesthesiometer has provision to easily replace monofilaments of equal length. As in the monofilament test, the new aesthesiometer has to be pressed against skin surfaces until both the monofilaments buckle. When buckled the new aesthesiometer is delivering the required force, corresponding to the length of the monofilaments, within the same tolerance of the individual monofilaments.

5 Measurement of Forces in Isobaric Aesthesiometer

Isobaric Aesthesiometer delivers controlled forces, within a tight tolerance, on the skin surfaces. In order to validate the claims, we have designed a simple low-force measurement cantilever with strain gages. The arrangement is shown in Fig. 3. The force transducer consists of a steel cantilever (x mm wide, y mm thick, and z mm

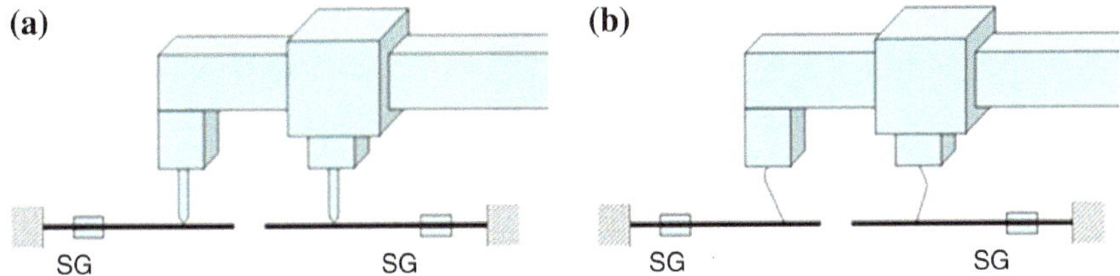

Fig. 3 Schematic diagram of the cantilever arrangement for measuring the forces under the two prongs of **a** a conventional aesthesiometer, **b** our isobaric aesthesiometer. *SG* Strain Gages

long) with two strain gauges (type XY) glued to it on either side of each cantilever. A XX adhesive (type Z make Y) is used for this purpose. The strain gauges are glued to either side of the cantilever and are connected to form a full/half bridge. If a force of 1 N is applied, the deflection of the cantilever is approx. 80 μm. To obtain a 0–10 V DC signal which is proportional to the force, our strain gauge amplifier is connected to the transducer (type of Instrumentation amplifier). Zero level and gain of this amplifier can be adjusted within a wide range. The output signal of the amplifier is connected to the analog input of the computer and displayed to the user. Figure 4a, b shown the force variation when a typical subject is asked to maintain a force at 20 g, using both conventional aesthesiometer and our isobaric aesthesiometer.

It could be noticed from Fig. 4a that a rotation about the wrist causes opposite forces, one increasing and the other decreasing, in the two prongs and that a translational movement of the wrist, either up or down, causes both increasing or both decreasing forces in the two prongs. The rotational moment could even result in the loss of contact at one of the prongs, shown in Fig. 4a with force nearly zero. The force variation in the normal aesthesiometer with rigid prongs shows that it approximately follows the second-order system, as discussed earlier, and the forces could be anywhere between the minimum (0 g) to maximum threshold of the cantilever arrangements (500 g) as shown in Fig. 4b. However, the force is nearly constant in the isobaric aesthesiometer as shown in Fig. 4b. The maximum variation of the force, after buckling of the filament, is the variation of the filament used. It is to be noted that this variation is not same for both left and right filaments; this difference could be due to many reasons. The filaments are undergoing different rate

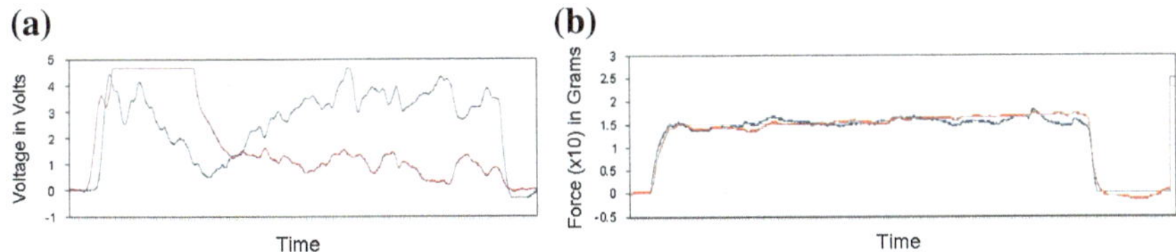

Fig. 4 Force delivered at both the prongs of aesthesiometer **a** conventional, with gain XX **b** isobaric, with gain YY. *Red colored data* is for the left prong and the *black colored data* is for the right prong. The data is for the same subject performing the test

of compression due to the moment about the wrist. However this difference is very small (5 %) and negligible. We have tested the isobaric aesthesiometer in seven subjects and found that the force is constant as shown in the Fig. 4b in all the subjects.

6 Discussion

The mechanical properties of monofilaments depend on humidity, temperature and age [29]. Variation of forces produced by single mono-filaments leads to errors in precision and accuracy. The nylon polymers (such as those used to make mono-filaments) have a polarized chemical structure, is water absorbent, and are not recommended for use in humid conditions if stability of the material is required [30]. Repeated loading lowers the bending forces of the filaments as plasticity of filaments is increased. Commercially available monofilaments have significant variability within and between devices from different manufacturers. Their actual bending force varies widely from their designated buckling threshold value. How much variation in the forces is acceptable for the TPD is unknown. This will lead to the standardization of the TPD measurement procedure [31]. Apart from the variation of the forces due to the monofilaments, the forces in the isobaric aesthesiometer however will also depend on the rotational movement of the observers' wrist. This causes different forces in the two monofilaments installed in the aesthesiometer. There could be three possible scenarios depending on which part of loading-curve each monofilament is currently positioned. Ideal loading-curve of monofilaments is divided into three regions: Region 1, before Point A—the point at which bending commences; Region 2, in-between Point A and Point B—the point at which the buckling commences; Region 3, beyond Point B which maintains the force at constant as shown in Fig. 5.

While measuring TPD the two monofilaments in the isobaric aesthesiometer could be in different operating range, and this leads to six possible combinations, as shown in Table 1.

The condition 6 in which both the monofilaments in region 3 (>B) is ideal for measurement of TPD. Although the hands of the observers rotate, if they could maintain the filaments in this region, the measurements will be repeatable and reliable. If the rotation results in the two monofilaments operating in other regions, TPD measurement could be erroneous. The Absolute Pressure sensitivity of skin under the monofilaments and the just noticeable difference (JND) of Pressure Sensitivity matters when two monofilaments are used for TPD measurements. This is especially valid for the monofilaments maintaining two forces, one before buckling and the other after buckling. When the filaments are in different force levels, then the JND pressure at the site of measurement will dictate if the measurement is valid. If the difference in force before and after buckling is greater than the JND, the subject would erroneously perceive the two filaments as single.

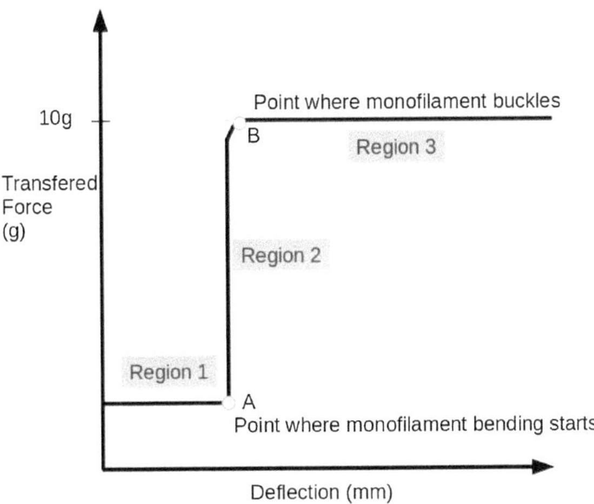

Fig. 5 Ideal loading curve of monofilaments

Table 1 Possible operating regions for two monofilaments while measuring TPD

Condition	Monofilament 1	Monofilament 2
1	<A	<A
2	<A	<B
3	<A	>B
4	<B	<B
5	<B	>B
6	>B	>B

7 Summary

Two Point Discrimination test is notorious for their unexplained variations within subjects, between subjects, and between studies. This paper presents a systematic study of these variations and proposed a simple design to control the variation, thereby improving the TPD measurement accuracies. Monofilaments are used replacing the rigid prongs in a handheld aesthesiometer. Although the proposed handheld instrument could be used to avoid variations, unless a standard protocol is developed for measuring the TPD, the variation cannot be controlled even with this device.

References

1. Weber, E.H.: De pulsu, resorptione, auditu et tactu. In: Ross, H.E., Murray, D.J., Hove (eds.) On the tactile senses. Taylor & Francis, UK (1834)
2. Craig, J.C., Johnson, K.O.: The two-point threshold: not a measure of tactile spatial resolution. Curr. Dir. Psychol. Sci. **9**, 29–32 (2000)

3. Lundborg, G., Rosén, B.: The two-point discrimination test—time for a re-appraisal? J Hand Surg. Br. Eur. **29**(5), 418–422 (2004)
4. Johnson, K., Van Bowen, R., Hsiao, S.: The perception of two points is not the spatial resolution threshold. In: Boivie, J., Hansson, P., Lindblom, U. (eds.) Touch, temperature and pain in health and disease: mechanisms and assessments, progress in pain research and management, vol. 2, pp. 389–404. IASP Press, Seattle (1994)
5. Tamura, Y., Hoshiyama, M., Inui, K., Kakigi, R.: Central mechanisms for two-point discrimination in humans. Neurosci. Lett. **342**(3), 187–190 (2003)
6. Krotoski, J.B.: Tendon and nerve surgery in the hand, In: Hunter, J.M., Schneider, L.H., Mackin, E.J. (eds.) Correlating sensory morphology and tests of sensibility with function, pp. 49–62. (Mosby, St Louis) (1997)
7. Bell-Krotoski, J.A., Buford Jr, W.L.: The force/time relationship of clinically used sensory testing instruments: a revision and update. J. Hand Ther. **10**(4), 297–309 (1997)
8. American Society of Hand Therapists: Clinical assessment recommendations, ASHT (1992)
9. Moberg, E.: Two-point discrimination test a valuable part of hand surgical rehabilitation in tetreplegia. Scand. J. Rehabil. Med. **22**, 127–134 (1990)
10. Dellon, A.L.: Sensibility, re-education of sensation in the hand. Williams & Wilkins, Baltimore (1981)
11. Moberg, E.: Reconstructive hand surgery in tetraplegia, stroke and cerebral palsy: some basic concepts in physiology and neurology. J. Hand Surg. **1**, 29–34 (1976)
12. Moberg, E.: The upper limb in tetraplegia: a new approach to surgical rehabilitation. Thieme, Stuttgart (1978)
13. Dellon, A.L.: A numerical grading scale for peripheral nerve function. J. Hand Ther. **6**, 152–160 (1993)
14. Tassler, P.L., Dellon, A.L.: Correlation of measurements of pressure perception using the pressure specified sensory device with electro diagnostic testing. J. Occup. Environ. Med. **37** (7), 862–866 (1995)
15. Dellon, E.S., Keller, K.M., Moratz, V., Dellon, V.L.: Validation of cutaneous pressure threshold measurements for the evaluation of hand function. Ann. Plast. Surg. **38**(5), 485–492 (1997)
16. Dellon, A.L.: Management of peripheral nerve problems in the upper and lower extremity using quantitative sensory testing. Hand Clin. **15**, 697–715 (1999)
17. Barber, M.A., Conolley, J., Spaulding, C.M., Dellon, A.L.: Evaluation of pressure threshold prior to foot ulceration: one-versus two-point static touch. J. Am. Podiatr. Med. Assoc. **91**(10), 508–514 (2001)
18. Periyasamy, R., Manivannan, M., Narayanamurthy, V.B.: Changes in two point discrimination and the law of mobility in diabetes mellitus patients. J. Brachial Plexus Peripheral Nerve Inj. **3**, 3 (2008)
19. Manivannan, M., Periyasamy, R., Narayanamurthy, V.B.: Vibration perception threshold and the law of mobility in diabetic mellitus patients. Prim Care Diab. **3**(1), 17–21 (2009)
20. Young, M., Matthews, C.: Neuropathy screening: can we achieve our ideals? Diab. Foot **1**, 22–25 (1998)
21. Omer, G.E., Bell-Krotoski, J.: Management of peripheral nerve problems. In: Omer, G.E., Spinner, M., Van Beek, M.A. (eds.) Sensibility testing, pp 11–28. W.B. Saunders Company, London (1998)
22. Levin, L.S., Regan, N., Pearsall, G., Nunley, J.A.: Variations in two-point discrimination as a function of terminal probes. Microsurgery **10**(3), 236–241 (1989)
23. Jones, L.A.: Matching forces: constant errors and differential thresholds. Perception **18**, 681–687 (1989)
24. Pang, X.D., Tan, H.Z., Durlach, N.I.: Manual discrimination of force using active finger motion. Percept. Psychophys **49**, 531–540 (1991)
25. Tan, H.Z., Pang, X.D., Durlach, N.I.: Manual resolution of length, force, and compliance. In: Proceedings of ASME WAM′ 1992, DSC-vol. 42, pp. 13–18

26. Timonshenko, S.P., Young, D.H.: Elements of strength of materials. Princeton, NJ, Van Nostrand (1962)
27. McGill, M., Molyneaux, L., Yue, D.K.: Use of the Semmes-Weinstein 5.07/10 gram monofilament: the long and the short of it. Diabet. Med. **15**, 615–617 (1998)
28. Bell-Krotoski, J., Tomanick, E.: The repeatability of testing with Semmes-Weinstein monofilaments. J. Hand. Surg. Am. **12A**, 155–161 (1987)
29. Haloua, M.H., Sierevelt, I., Theuvenet, W.J.: Semmes-Weinstein monofilaments: influence of temperature, humidity, and age JHS, vol. 36a, pp. 1191–1196. July 2011
30. Brydson, J.A.: Plastics materials, pp. 399–401. Butterworths, London (1975)
31. Lavery, L.A., Lavery, D.A., Lavery, D.C., LaFontaine, J., Bharara, M., Najafi, B.: Accuracy and durability of Semmes–Weinstein monofilaments: what is the useful service life, Diabetes Research and Clinical Practice vol. 97(3), pp. 399–404, Sep 2012

An Application of Particle Swarm Optimization Technique for Optimization of Surface Roughness in Centerless Grinding Operation

Subhas Chandra Mondal and Prosun Mandal

Abstract There is growing need among the manufacturers to optimize performance of a centerless grinding operation to get high level of accuracy for micro-finishing of product. High quality surface finish is desired for manufacturing crane-hook-pin used in the bottom block of the crane subjected to bending and torsional load. This paper presents an application of particle swarm optimization techniques to identify optimal parameter settings to minimize surface roughness of C40 steel crane-hook-pin in centreless grinding operation. Experiments have been carried out and the effect of various process parameters (depth of cut, regulating wheel speed and coolant flow) has been studied in a full factorial design (three factors at three level) of 27 experiments for all possible combination of these factors. Particle Swarm Optimization (PSO) technique combining with response surface modelling (RSM) are used to find optimal parameter settings. The optimal parameter corresponds to 12 rpm regulating wheel speed, 1/3rd opening of coolant valve opening and 20 μm depth of cut.

Keywords Centreless grinding · Surface roughness · RSM · Particle swarm optimization

1 Introduction

Today's manufacturing industries are very much concerned about the quality of the products. Manufacturing industries are focused on producing better quality products in time at a minimum cost. Surface finish is one of the crucial performance parameters that have to be controlled within suitable limits for manufacturing crane-hook-pin used in the bottom block of the crane subjected to bending and torsional load. Optimal level of surface finish is desired in the crane-hook-pin for better

S.C. Mondal (✉) · P. Mandal
Department of Mechanical Engineering, Indian Institute of Engineering Science and Technology, Shibpur, Howrah, India
e-mail: scmondall@gmail.com

A. Chakrabarti (ed.), *ICoRD'15 – Research into Design Across Boundaries Volume 2*, Smart Innovation, Systems and Technologies 35, DOI 10.1007/978-81-322-2229-3_59

performance, to increase quality with an enhanced service life and reliability where frequent failure caused breakdown of the equipment and chances of accident. Therefore, prediction or monitoring of the surface roughness at optimal level of machined components (pin) has been an important area of research.

The enhancement of such complex process performance requires model based process optimization technique, which is a powerful tool for evaluating the performance of complex systems. Empirical models, such as the regression analysis model, the fuzzy logic model, and the neural network model, have, generally, shown satisfactory prediction accuracy, particularly useful for the on-line response evaluation and control. In many cases, data from design of experiment (DOE) were used to establish the regression models or to develop the fuzzy rule sets or to train the neural networks. For better performance researcher have done lot of work on surface roughness modeling in grinding operation. Aguiar et al. [1] predicted surface roughness in grinding using artificial neural network. Shrivastava et al. [2] developed an intelligent modeling of surface roughness during diamond grinding of advanced ceramics by two different approaches multiple regression analysis (MRA) and artificial neural network (ANN) and compared the same [3–5]. Practical optimization is the art and science of allocating scarce resources to the best possible effect.

Optimization techniques play a major role in every day for industrial planning, resource allocation; scheduling, and decision-making. New optimization techniques are developed and often stimulated by fascinating insights from other fields. Particle Swarm Optimization was first introduced by Kennedy and Eberhart [6] and used for optimization of continuous nonlinear functions. The swarm is composed of volumeless particles with stochastic velocities, each of which represents a feasible solution. The algorithm finds the optimal solution through moving the particles in the solution space.

This paper proposed an application of particle swarm optimization technique and response surface methodology to identify optimal parameter settings in centerless grinding operation to minimize surface roughness of C40 steel-pin.

2 Literature Review

The domain of applicability of Particle Swarm Optimization (PSO) has steadily increased since the algorithm was first introduced in 1995 by Kennedy and Eberhart [6]. While initial experiments were restricted to optimizing simple functions, further development has enabled PSO to be used for real life applications. This type of optimization process is used to search the entire problem space and thus helpful in solving complex problems. This optimization method has its own advantages, i.e. (a) It is insensitive to scaling of the design variables. (b) Its implementation is simple. The particle with the best coordinate values after all the particles have been compared is termed as the global best value (gbest). This global best value gives the best solution of the optimization problem. Every particle has a velocity component and a position component. Say, for *i*th particle the position component is represented by x_i and the velocity component is represented by v_i. Now, this particle will keep on changing its velocity, along with which the position of the particle also changes. The velocity component is represented by $v_i = v_{i1}, v_{i2}, v_{i3}, \ldots, v_{in}$ and the position is represented by $x_i = x_{i1}, x_{i2}, \ldots, x_{in}$. The particles change their velocity and position.

Malviya and Pratihar [7] applied PSO for tuning of the neural networks that were utilized for carrying out both forward and reverse mapping of different metal inert gas welding processes. Four different approaches were developed and their performances were compared while solving the problems. The first and second approaches dealt with PSO-tuned multilayer fed forward neural networks (MLFFNN) and radial basis function neural networks (RBFNN) respectively, whereas, some hybrid schemes were adopted in the last two approaches. Both the third and fourth approaches showed better results as compared to first and second approaches. Singh and Wang [8] have proposed a fuzzified multi-objective PSO algorithm and applied that algorithm to dispatch the electric power, considering both economic as well as environmental issues. Tripathi et al. [9] described a novel PSO approach for multi-objective optimization called time-variant multi-objective particle swarm optimization (TV-MOPSO) for solving multi-objective optimization problems. The performance measure of the new approach was compared with other multi-objective PSO techniques and it was found that the new approach gave better optimization results. Li et al. [10] used particle swarm optimization (PSO) for cutting parameters optimization. They first introduced the fundamental principle of PSO; then, the algorithm for PSO application in cutting parameters optimization was developed; thirdly, cutting experiments with and without optimized cutting parameters were conducted to demonstrate the effectiveness of optimization, respectively. The results show that the machining process was improved obviously [11].

Several researchers and practitioners applied RSM for optimization of grinding operation [12–15]. Mamun and Dhar [12] developed numerical model of surface roughness in grinding under Minimum Quantity Lubricants (MQL) using response surface method (RSM). Kumar et al. [15] also developed model to predict wear and surface roughness in electro-discharge diamond grinding using two techniques, namely design of experiments and neural network [14, 15]. Based on the literature review it is obvious several researchers and practitioners are applied PSO algorithm for optimization of process parameters in a variety of manufacturing process but application of PSO in centerless grinding process while matching C40 steel crane-hook pin is rare.

3 Experimentation

The component material was C40 steel rod of 17 mm diameter and 100 mm length, used as crane-hook-pin as shown in Fig. 1. In this experiment C40 steel used because this grade of steel offers better forming and bending quality. It was used for

Fig. 1 Finished product of C40 steel pin

applications, where critical bending operations are required. Due to carbon range of 0.4 % and manganese 0.9 %, it is very suitable for components where, critical bending has to be achieved and high tensile and toughness can be obtained by means of quenching and tempering. The component used in the experimentation, was supported by specially made work rest blade with a 30° angle.

3.1 Experimental Design

Grinding experiments were conducted on centreless grinding machine-tool as shown in Fig. 2. A vitrified grinding wheel A463V5L10, with an abrasive of aluminum oxide capacity to ground of 2–100 mm diameter was used. The machine is equipped with dynamic grinding wheel balancing. The measurements were carried out with a Mitutoyo Surftest SJ-301 with cutoff length 2.5 mm and number of sampling length 5, and a stylus type surface texture-measuring instrument.

Fig. 2 Experimental set-up

Table 1 Input process parameters and their levels

Factors	Symbols	Low level	Medium level	High level
Wheel speed	d_c	12	14	16
Depth of cut	v_w	20	25	30
Coolant flow	c_f	1/3	2/3	1

The measurement results are displayed digitally or graphically on the touch panel, and output to the built in printer. It has a maximum measuring range of 350 μm (−200 to +150 μm). The input process parameters and there level is shown in Table 1. A series of experiments have been conducted to evaluate which grinding parameters affect the surface roughness. In Table 2 there are 27 experimentations are performed for different machining conditions.

3.2 Optimization

Particle swarm optimization is an innovative distributed intelligent concept based on algorithm for solving optimization problems that originally took its motivation from the biological examples of swarming of bees, flocking of birds and herding in animals. Centerless grinding experiments are carried out varying three input parameters: depth of cut, regulating wheel speed and coolant valve opening (Table 3).

An important area in the statistical design of experiment is the response surface methodology (RSM) which is a collection of mathematical and statistical techniques useful for the modeling and analysis of problems in which a response of interest is influenced by several variables and the objective is to optimize this response. Three controlling input process parameters namely depth of cut, regulating wheel speed and coolant valve opening and one quality characteristics, surface roughness are considered in the model. Total 27 no. experiments were carried out according to the experimental design plan. The second order mathematical models in terms of machining parameters were developed for surface roughness prediction using RSM on the basis of experimental results. The RSM based equations are shown in Eq. 1. Regression analysis is performed to model the surface roughness. The results of the regression analysis (Table 3) and the subsequent ANOVA analysis (Table 4) show that the proposed response surface model is adequate (R-Sq = 87.41 % and R-Sq (adj) = 80.75 %) to express the real centerless grinding process.

$$\begin{aligned} Y = & -1.6759 + 0.0256 d_c + 0.1846 v_w + 0.6355 c_f - 0.0005 d_c^2 - 0.0068 v_w^2 \\ & - 0.0507 c_f^2 + 0.0006 d_c v_c - 0.0112 v_w c_f - 0.0075 c_f d_c \end{aligned} \quad (1)$$

Table 2 Experimental design matrix

Experimental no.	d_c (μm)	V_w (rpm)	C_f (% of valve opening)
1	20	12	1/3
2	20	12	2/3
3	20	12	1
4	20	14	1
5	20	14	2/3
6	20	14	1/3
7	20	16	1/3
8	20	16	2/3
9	20	16	1
10	25	12	1
11	25	12	2/3
12	25	12	1/3
13	25	14	1/3
14	25	14	2/3
15	25	14	1
16	25	16	1
17	25	16	2/3
18	25	16	1/3
19	30	12	1/3
20	30	12	2/3
21	30	12	1
22	30	14	1
23	30	14	2/3
24	30	14	1/3
25	30	16	1/3
26	30	16	2/3
27	30	16	1

Table 3 Regression coefficients in centerless grinding process

Predictor	Coefficient	SE coefficient	T	P
Constant	−1.6759	0.5898	−2.841	0.011
d_c	0.0256	0.02176	1.176	0.256
v_w	0.1846	0.0702	2.631	0.018
c_f	0.6356	0.2116	3.004	0.008
d_c^2	−0.0005	0.0004	−1.265	0.223
v_w^2	−0.0068	0.0024	−2.817	0.012
c_f^2	−0.0507	0.0869	−0.583	0.567
$dc * v_w$	0.0006	0.0007	0.854	0.405
$v_w c_f$	−0.012	0.0041	−2.926	0.009
$d_c c_f$	−0.0075	0.0102	−0.731	0.475

Table 4 Analysis of variance

Source	DOF	SS	MS	F	P
Regression	9	0.0425	0.00472	4.00	0.007
Residual error	17	0.02006	0.00118		
Total	26	0.06256			

The error in the predication model proposed in Eq. (1) is follows normal distribution. The assumption of normality is checked by using normal probability plot of the residuals and percent shown in Fig. 3. From Fig. 3 it is clear that the proposed centerless grinding process follows normal distribution. From the response surface plots (Fig. 4) it is observed that the minimum surface roughness (0.12 μm) is achieved at the low level of the depth of cut (depth of cut: 20 μm) and combined with the low levels of the regulating wheel speed (wheel speed: 12 rpm) and coolant valve opening (valve opening: 1/3rd).

Particle Swarm Optimization is now applied to optimize surface roughness using RSM-based model (Eq. 1). The code for the computer is developed in MATLAB 7.1. The control parameters of PSO algorithm for solving the parametric optimization problems are set as below: swarm size: 24; no of repetitions: 100 and number of runs: 1. The PSO algorithm is now used to optimize the above-mentioned RSM-based equation (Eq. 1). The objective is to minimize surface roughness (Ra). It is observed that for minimum surface roughness optimum parameters are depth of cut 20.0017 μm, regulating wheel speed is 12 rpm and coolant valve opening 0.333 means 1/3 opening valve. These values are similar to the optimum parameters

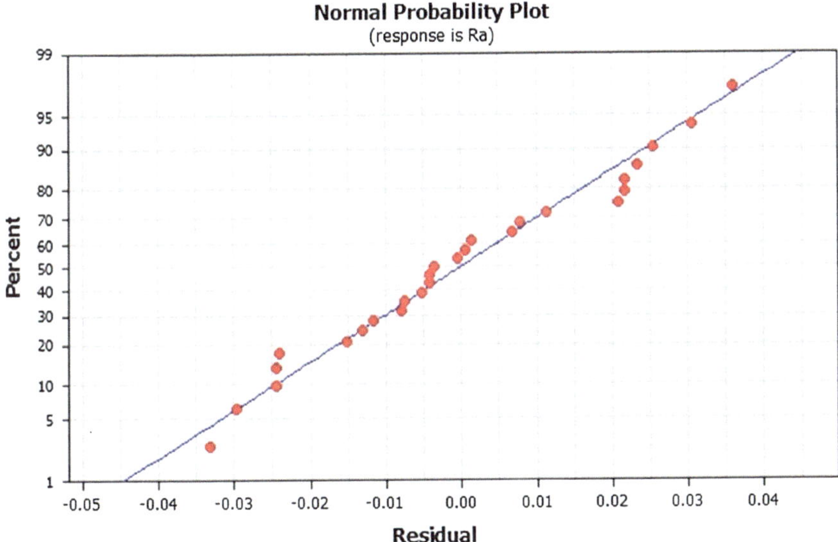

Fig. 3 Normal probability plot of residual

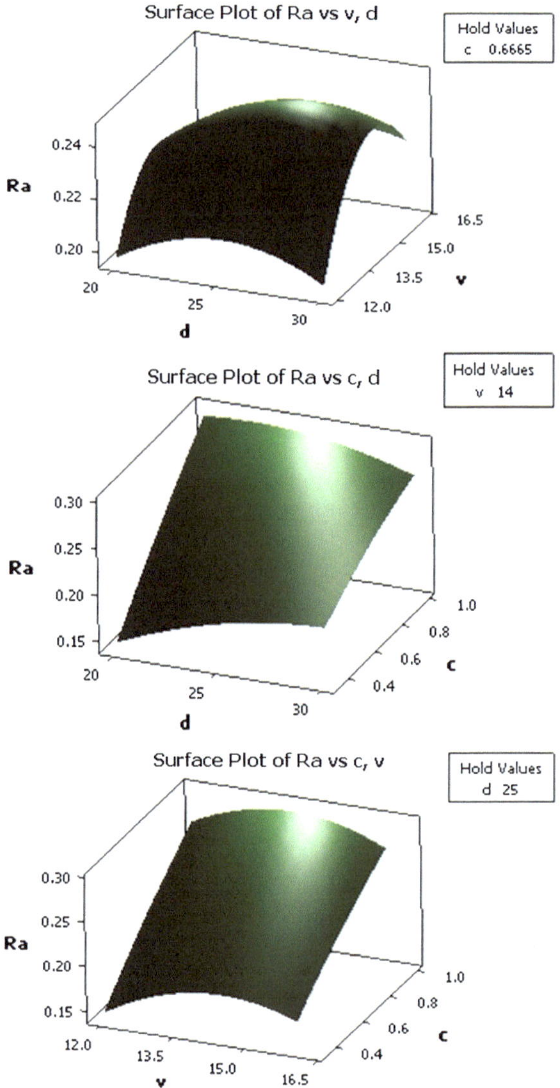

Fig. 4 Response surface plots of surface roughness

obtained from response surface methodology (RSM).That mean Particle Swarm Optimization method verified the Response Surface Methodology.

Figure 5 shows the convergence of the developed PSO algorithm for surface roughness and it is observed that minimum surface roughness is 0.12116 μm. Figure 6 shows that the predicted surface roughness curve is almost similar and it lies between the ranges of 0.1 and 0.3 μm for maximum no. of runs, whereas for experiment run no. 3 and 4 surface roughness is maximum. That are when depth of cut is low and full open coolant valve surface roughness is maximum.

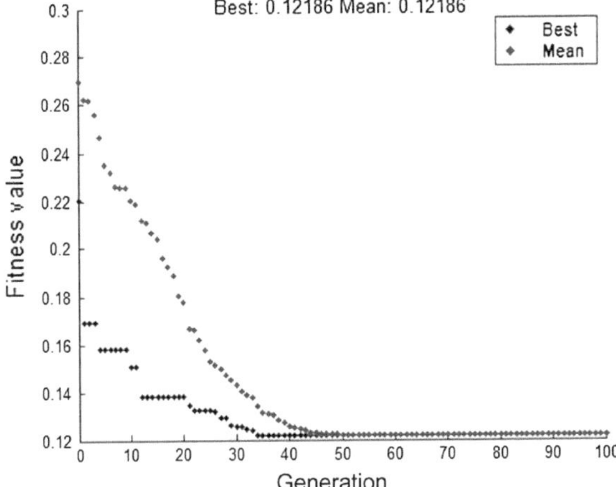

Fig. 5 Convergence of PSO algorithm for surface roughness (μm)

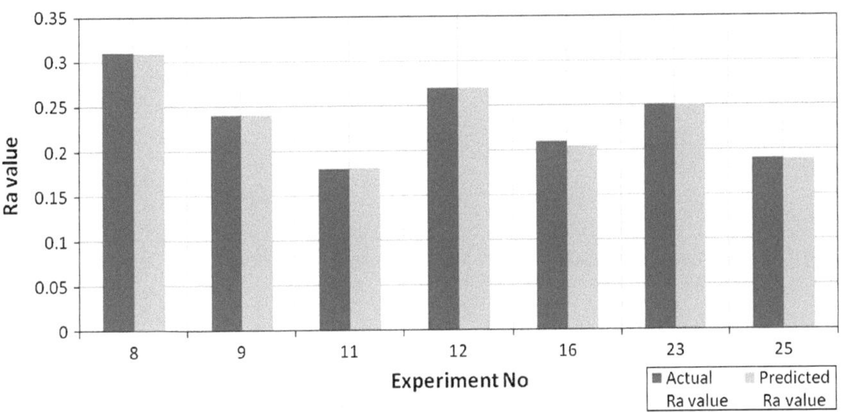

Fig. 6 Bar chart for predicted and experimental *Ra value*

4 Conclusions

This paper presents a non-traditional optimization method (PSO) for the optimization of surface roughness in centerless grinding operation. High quality surface finish is desired for manufacturing crane-hook-pin used in the bottom block of the crane subjected to bending and torsional load. Optimal level of surface finish (0.12116 μm) is obtained for the optimal settings (depth of cut 20.0017 μm, Regulating wheel speed 12 rpm and 1/3 coolant valve opening) in the crane-hook-pin for better performance with an enhanced service life and reliability. From the Analysis of variance and the response surface plots, it is observed that the optimal parameter settings are depth of cut (20 μm), regulating wheel speed (12 rpm), and coolant valve opening (1/3) for the surface roughness of 0.12186 μm. Hence, it may be concluded that RSM method is verified by particle swarm optimization and PSO gives better results in term of performance than RSM. The results of the regression analysis and the subsequent ANOVA analysis show that the proposed response surface model is adequate (R-Sq = 87.41 % and R-Sq (adj) = 80.75 %) to express centerless grinding process.

The proposed PSO and RSM based results can be further verified by extending this research work and compared with other non conventional optimization techniques (meta heuristic techniques) such as genetic algorithm, simulated annealing, Tabu search. Combined Taguchi and Response surface methodology may be applied for optimization of the surface roughness of the crane-pin and the results may be compared with PSO. In the proposed research three process parameters, depth of cut (d_c), regulating wheel speed (v_w), and coolant valve opening (c_f) are considered as input parameters and surface roughness as output characteristics. Due to machine constraints some other important process parameters like the work-piece centre height, the longitudinal dressing feed-rate and in-feed speed are not considered in this study. These may be considered in future research work.

Acknowledgments The author gratefully acknowledges the kind support and cooperation provided by the technical staffs at the Workshop of IIEST, Shibpur, Howrah.

References

1. Augier, P.R., Cruz, E.D.C., Paula, W.C.F., Bianchi, C.E.: Predicting Surface Roughness in Grinding using Neural Networks Advances in Robotics, Automation and Control. CC BY-NC-SA 3.0 (2008)
2. Shrivastava, P.K., Dubey, A.K.: Intelligent modeling of surface roughness during diamond grinding of advanced ceramics. In: Proceedings of the World Congress on Engineering, vol. 1 (2011)
3. Fredj, N.B., Amamou, R.: Ground surface roughness prediction based upon experimental design and neural network models. Int. J. Adv. Manuf. Technol. **31**, 24–36 (2006)
4. Kwak, J.S., Ha, M.K.: Neural network approach for diagnosis of grinding operation by acoustic emission and power signals. J. Mater. Process. Technol. **147**(1), 65–71 (2004)
5. Ali, Y., Zhang, L.C.: Estimation of residual stress induced by grinding using a fuzzy logic approach. J. Mater. Process. Technol. **63**, 875–880 (1997)
6. Kennedy, J., Eberhart, R.: Particle swarm optimization. Inst. Electr. Electron. Eng. **3**, 1942–1948 (1995)
7. Malviya, R., Pratihar, D.K.: Tuning of neural networks using particle swarm optimization to model MIG welding process. Swarm Evol. Comput. **1**, 223–235 (2011)
8. Singh, C., Wang, L.: Environmental/economic power dispatch using a fuzzified multi-objective particle swarm optimization algorithm. Electr. Power Syst. Res. **77**, 1654–1664 (2007)
9. Tripathi, P.K., Bandyopadhyay, S., Pal, S.K.: Multi-objective particle swarm optimization with time variant inertia and acceleration coefficients. Inf. Sci. **177**, 5033–5049 (2007)
10. Li, J.G., Yao, Y.X., Gao, D., Liu, C.Q., Yuan, Z.J.: Cutting parameters optimization by using particle swarm optimization (PSO). Appl. Mech. Mater. **10–12**, 879–883 (2008)

11. Deepa, S.N., Sugumaran, G.: Model order formulation of a multivariable discrete system using a modified particle swarm optimization approach. Swarm Evol. Comput. **1**, 204–212 (2011)
12. Mamun, A.A., Dhar, N.R.: Numerical modeling of surface roughness in grinding under minimum quantity lubricants (MQL) using response surface method (RSM). J. Mater. Prod. Technol. **12**(5), 1–21 (2012)
13. Lio, Y.S., Huang, J.T., Su, H.C.: A study on the machining parameters optimization of wire electrical discharge machining. J. Mater. Prod. Technol. **71**, 487–493 (1997)
14. Thomas, M., Beauchamp, Y., Youssef, A.: An experimental design for surface roughness and built-up edge formation in lathe dry turning. Int. J. Qual. Sci. **2**(3), 167–180 (1997)
15. Kumar, S., Choudhury, S.K.: Prediction of wear and surface roughness in electro-discharge diamond grinding. J. Mater. Process. Technol. **191**, 206–209 (2007)

Author Index

A
Abeele, Vero Vanden, 409
Abel, Marie-Hélène, 507
Agrawal, Mehul, 409
Albarran, Diana, 445
Amrit, Mannu, 375
Andriankaja, Hery, 227
Aoussat, Améziane, 27
Aravind, M.A., 579
Arduin, Pierre-Emmanuel, 507
Arumugam, Gomathinayagam, 569
Ashraf, Abdul Sameer, 421

B
Babu, S., 569
Balachandran, Swaminathan, 303
Bandi, Kirthi, 159
Bansal, Himanshu, 375
Bauer, Wolfgang, 483
Bavan, D. Saravana, 533
Bersano, Giacomo, 27
Bhattacharya, Soma, 653
Bockelmann, Paul, 495
Bordegoni, Monica, 397
Boruah, Dipanka, 237

C
Carulli, Marina, 397
Cassim, Fatima, 385
Chakrabarti, Amaresh, 251, 327, 433, 459, 495
Chakrabarti, Debkumar, 621, 663
Chandra, Satish, 3
Chandran, Kumari Moothedath, 459
Charan, P. Ram, 579
Charkrabarti, Debkumar, 631
Chatterjee, Jayanta, 181
Chilbule, Sanjay, 559
Chowdhury, Anirban, 631
Chucholowski, Nepomuk, 483
Cluzel, François, 49
Cugini, Umberto, 397

D
Das, Amarendra Kumar, 75, 237
Deshmukh, Bhagyesh, 521
Dinesh, N. S., 559, 579
Divya, N., 1
Drechsler, Klaus, 495
Duhovnik, Jože, 171

E
Ebrahim, S. Mohamed, 569
Elara, Mohan Rajesh, 365
Elezi, Fatos, 483
Eynard, Benoît, 227, 507

F
Faludi, Jeremy, 201
Farel, Romain, 49
Fayemi, Pierre-Emmanuel, 27

G
Gürtler, Matthias, 291, 339
Gangadharan, Kiran, 601
Geurts, Luc, 409
Gurumoorthy, B., 327, 433

A. Chakrabarti (ed.), *ICoRD'15 – Research into Design Across Boundaries Volume 2*,
Smart Innovation, Systems and Technologies 35, DOI 10.1007/978-81-322-2229-3

H
Honiball, Nina, 385
Hwee, Teo Kiah, 365

J
Jagtap, Sachin, 191
Jagtap, Santosh, 191
Jha, Raushan Kumar, 545
Joshi, Shraddha, 471

K
Kamalbabu, P., 533
Kammardi, Pradyumna K., 559
Kammerl, Daniel, 39
Kapoor, Ajay, 89
Karmakar, Sougata, 631
Katoppo, Martin L., 137
Kota, Srinivas, 99, 159
Kumar, Bijendra, 99
Kumar, Prakash, 621
Kundal, Amit, 181
Kundu, Amar, 601
Kuys, Blair, 89

L
Lakshmi Narayana, B., 15
Larsson, Andreas, 191
Le Duigou, Julien, 227, 507
Lee, Jun Bum, 445
Lindemann, Udo, 291, 339

M
Mörtl, Markus, 39
Madhuri, Vrisha, 559
Madhusudanan, N., 433
Mafalda, R., 113
Mahalingam, Aswhin, 89
Majhi, Manoj, 663
Malaschewski, Oliver, 39
Mandal, Prosun, 687
Mani, Monto, 459
Manivannan, M., 675
Mantovani, G. L., 113
Maranzana, Nicolas, 27
Marconcini, J. M., 113
Maurer, Maik, 483
McAloone, Tim C., 125
McDade, Justin R., 591
Melles, Gavin, 89
Mistry, Roohshad, 521
Mocko, Gregory, 275
Mohan Kumar, G. C., 533
Mondal, Subhas Chandra, 687
Mujumdar, Purva, 653
Murthy, Lakshmi, 213
Muthu, Selvaraji, 545

N
Nagesh, G., 15
Narayana, G.S., 569
Nori, Chalapathi Rao, 559

O
Olander, Elin, 613
Otto, Kevin, 445

P
Patil, Aniket, 521
Patil, Vikas, 521
Pereira, L., 113
Periyasamy, R., 675

R
Rabaa, Rashed S., 591
Rajanayagam, James, 89
Rajili, Noor Adila Mohd, 613
Ranganath, R., 15
Rao, Nori Chalapathi, 579
Ray, Gaur G., 601
Reddy, Chiru Venkat, 591
Reddy, Swathi Matta, 631
Renu, Rahul Sharan, 275
Roth, Michael, 291
Roy, Satyaki, 151
Roy, Shatarupa Thakurta, 181

S
Saavedra, Cristina Carro, 339
Santhanakrishnan, Deepak, 191
Santhi, B., 327
Sarkar, Prabir, 99
Schenkl, Sebastian A., 39
Schrieverhoff, Phillip, 339
Sen, Dibakar, 327
Sharma, Gaurav, 15
Sharma, Nishant, 601
Shende, Avinash, 75
Shukla, Parth, 151
Singh, Vikrant, 591
Sivakumar, S., 545
Sivaloganathan, Sangarappillai, 313
Sondur, Kaushik D., 559
Sorathia, Keyur, 421
Sosa, Ricardo, 365, 445
Sridhara, C. D., 15
Srinivasan, V., 339

Srivastava, Anmol, 353
Summers, Joshua D., 471
Suresh, D., 675

T
Tan, Ning, 365
Thomas, Joseph, 89
Thomsen, Jakob, 125
Tschaut, Rita, 483

U
Uchil, Praveen, 251
Uma Maheswari, J., 653

V
Valencia, Phebe, 137
Vanderloock, Karen, 409
Varma, Surbhit, 421
von Kessel, Sabina, 641
Vukašinović, Nikola, 171
Vyas, Parag K., 263

W
Warell, Anders, 191, 613
Weibin, Ding, 397

Y
Yammiyavar, Pradeep, 353, 375
Yanis, Rana, 313
Yannou, Bernard, 49

Z
Zadnik, Žiga, 171